ELECTRICITY INFRASTRUCTURES IN THE GLOBAL MARKETPLACE

Authored by **T. J. Hammons**

Electricity Infrastructures in the Global Marketplace
Authored by T. J. Hammons

Published by InTech
Janeza Trdine 9, 51000 Rijeka, Croatia

Publishing Process Manager Jelena Marusic

Technical Editor Teodora Smiljanic

Cover Designer InTech Design Team

Additional hard copies can be obtained from orders@intechopen.com

Electricity Infrastructures in the Global Marketplace,
Authored by T. J. Hammons p. cm.
ISBN 978-953-307-155-8

For EU product safety concerns:
IN TECH d.o.o., Prolaz Marije Krucifikse Kozulić 3, 51000 Rijeka, Croatia,
info@intechopen.com or visit our website at intechopen.com.

Meet the author

Thomas James Hammons (Fellow IEEE 1996) received the B.Sc. degree in Engineering (1st Class Honors), and the DIC, and Ph.D. degrees from Imperial College, London, UK He was Professor of Electrical and Computer Engineering at McMaster University, Hamilton, Ontario, Canada in 1978-1979. He is the author/co-author of over 440 scientific articles and papers on electrical power engineering and is Editor of a book on Renewable Energy that was published by INTECH in December 2009. He has lectured extensively in North America, Africa, Asia, and both in Eastern and Western Europe.

Dr Hammons is Past Chair of the United Kingdom and Republic of Ireland (UKRI) Section IEEE and Past Chair of International Practices for Energy Development and Power Generation of IEEE. He is also a Past Chair of the IEEE PES Task Force on harmonizing power-engineering standards worldwide and Past Permanent Secretary of the International Universities Power Engineering Conference. He is a Chartered Engineer (CEng) and a registered European Engineer in the Federation of National Engineering Associations in Europe.

Contents

1

The Role of Nuclear in the Future Global Energy Scene

1.1 Introduction

Energy and human life are closely liked. Civilization, present and future, depends on energy to provide the facilities the human race needs.

The world being created today will determine the outcome of a number of issues and conflicting demands, which we only now beginning to identify. Whilst their resolution will fashion the future world, the immediate challenge is to provide enough energy, water and food, to raise the standard of living of the ever-increasing world population without "imperiling our irreplaceable environment".

1.1.1 The Greenhouse Effect

In the last few years global warming, caused by the build up of greenhouse gases, has been the issue on everyone's agenda. As stated by Sir David King, the former UK Government's Chief Scientific Advisor, it is a bigger global threat than terrorism.

Carbon dioxide accounts for half of the human race's contribution to global warming. Carbon emissions have been rapidly increasing since the industrial revolution. In 2002 carbon equivalent emissions from human activity were about 6,500 million tonnes per year with the prediction this would double by 2050. Carbon dioxide emissions come from various sources, such as humans breathing, the natural world and the burning of fossil fuels, either in the generation of electricity or directly in transport. Today, the supply of electricity is responsible for 16% of worldwide carbon dioxide emissions. For the developed world this proportion is greater. For example, in the UK, electricity generation from fossil fuels is responsible for 33% of emissions.

1.1.2 The Global Scene

Powered by countries where rapid economic change is already underway, population, on a global basis, will increase from its present level of 6 billion to over 9 billion by 2025 and in addition the life span of the global population is increasing.

This brings with it a greater degree of urbanization and increased demand for energy. In 1950 only New York had a population of over 10 million. By 2015 there will be 21 cities of more than 10 million, whilst the number with populations between 5 and 10 million will go from 7 to 73. Asia and Africa, currently two-thirds rural, will be half urban by 2025.

Of the 6 billion people in the world only 2 billion have reliable access to electricity; 2 billion have unreliable access, leaving 2 billion with little or no access, of whom one billion live in slums. 2.4 billion people are dependent on wood, crop residues and dung to cook their food.

From 1980 to 2004 total world primary energy demand grew by 54% and to 2030 it is projected to grow at much the same rate (average 1.6% per year, from 469EJ to 716EJ). Electricity growth is even stronger and is projected to almost double from 2004 to 2030 (growing at average 2.6% per year from 17,408 TWh to 33,750 TWh). Due to population growth the increased demand is most dramatic in developing countries.

So what are the sources of fuel that will be used to produce this electricity? The current world demand for electricity is heavily dependent on fossil fuels with coal at 40%, gas 15% and oil 10% whilst hydro is 17% and nuclear is 16% and renewables are minimal. This dependence on fossil fuels will intensify. (Figure 1.1) What is significant is coal produces twice the quantity of CO_2 than does oil or gas whilst hydro and renewables produce far less CO_2 with nuclear producing only 0.4% of that produced by coal.

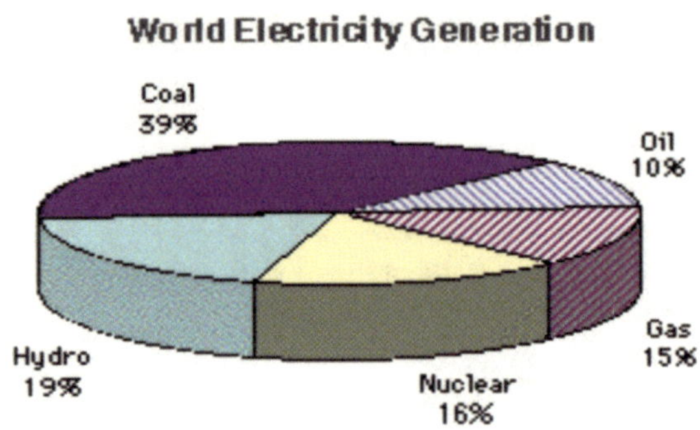

Figure 1.1

Using the UK as an example about 20% of its electricity is currently produced by nuclear power, including that imported from France via a cross channel cable. Had all that power been produced by fossil fuel, a staggering additional 50 million tonnes of carbon dioxide would have been pumped into the atmosphere per year. The saving of CO_2 emissions by the UK's nuclear power stations is equivalent to allowing 120 million people to breathe continuously. It would mean taking 50% of British cars off the road to make equivalent savings.

The Asian nations account for about 30% of the world's coal reserves, China alone has 11% of the total and India 6%. So their energy needs, now and in the long-term future, will come mainly from coal. For example, in the case of China 68.3% of their energy came from indigenous coal in 2005.

No one would deny the developing nations their chance to improve their standard of living. But if they increase energy consumption at the rate suggested, using their indigenous reserves of fossil fuels, emissions of carbon dioxide will rise well above sustainable levels. If this happens, without any reductions elsewhere, the world will destroy its own environment.

Many people, politicians as well as engineers, face the challenge now as to how to convert natural resources into a form of energy which will least affect the environment. Power engineers can produce electricity from a number of sources, none of which are totally environmentally benign. In reality, none of the so-called "benign" energy sources can provide a significant contribution to the global demand for energy. In fact a recent study predicts that renewables, other than hydro, will only contribute 4.4% of the world's electricity by 2030. For example in the UK to meet a Government target of 60% reduction in greenhouse gases, by 2050, assuming no change in technology, the UK should use only 30% of fossil fuel for power generation. So even if nuclear stayed at the present figure of less than 20%, renewables would have to be almost 50%. Clearly an impossible target.

So for the next few decades, there are only a few realistic options for reducing carbon dioxide emissions from electricity generation:

- Expand the use of renewable energy sources such as wind, solar, biomass and geothermal;
- Increase the efficiency of electricity generation and usage;
- Capture carbon dioxide emissions at fossil-fuelled stations and permanently store the carbon dioxide;
- Use of carbon offset permits;
- Increased use of nuclear power;
- Or the rationing of electricity, which would not be very popular.

Against the time scale faced, only nuclear power offers a proven long-term environmentally acceptable solution, able to produce the quantity of energy that the world will need in future.

1.1.3 The Role of Nuclear Today

At this time nuclear provides over 16% of the world's electricity, almost 24% in OECD countries and 35% in the EU.

Many countries have a significant nuclear component in their energy mix, (Figure 1.2) giving them a hedge against imported fuel price increases and also a degree of security of supply. For example France has developed a strong nuclear program, which it will maintain. France currently generates almost 80% of its electricity from nuclear and it has the cheapest electricity in Europe enabling it to export up to 15% of its electricity to its neighbors. The US, with 20% nuclear is, because of its high-energy demand, a major consumer of nuclear electricity.

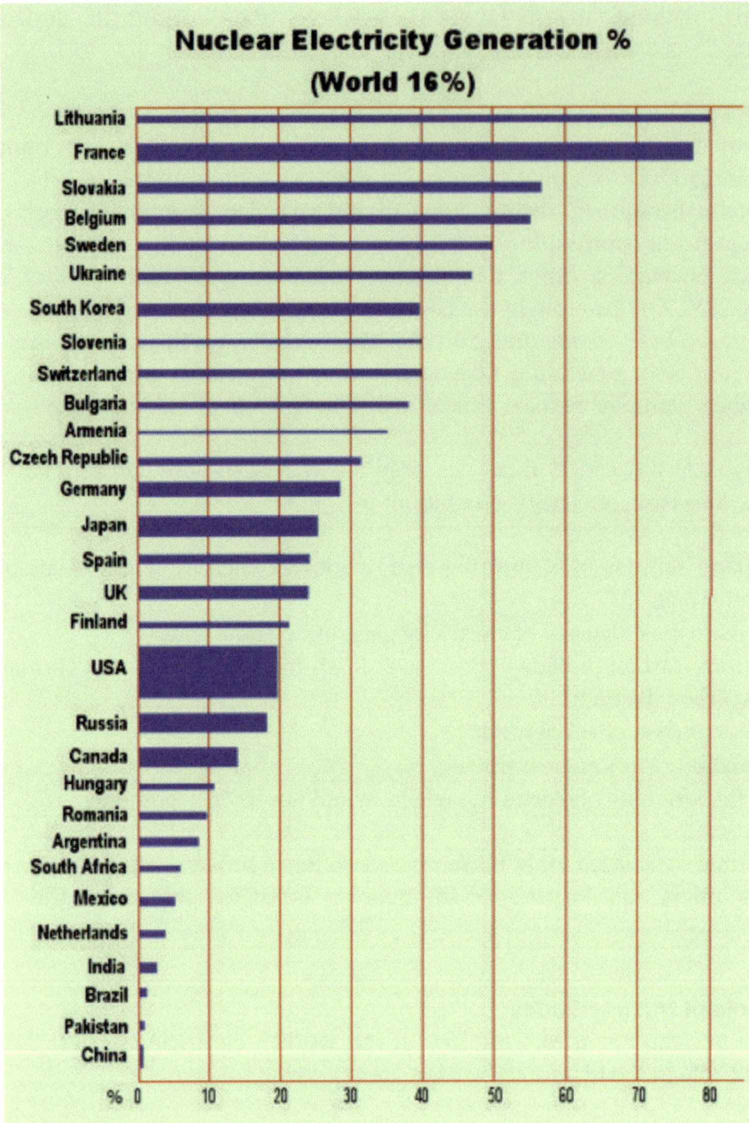

NOTE: Taiwan has 22% nuclear generation

Figure 1.2 Percentage of Nuclear Generation

Today there are:

- 439 Power Reactors in 30 countries and Taiwan, China, with a total capacity of 371,936MWe
- Supplying 16% of the world's electricity
- 12,500 reactor years of experience
- Three new reactors on line in 2006
- 35 under construction, with a capacity of over 25,000 MWe

- 94 planned with a capacity of 102,000 MWe
- 222 proposed with a capacity of 193,000 MWe
- The world produces as much electricity today from nuclear energy as it did from all the other sources combined in 1960

In the US, incentives for nuclear power have led to statements of interest for 33 new reactors, on 22 sites, with the first reactors planned to be in operation by 2015. In Canada there are plans for 6,000 MWe of new plant.

Argentina and Brazil have declared their intentions to restart their programs and to co-operate in the development of new reactors and fuel cycle capabilities.

South Africa has authorized the construction of 20,000 MWe of plant.

From the current nuclear capacity of 9,000 MWe China is expected to reach 40,000 MWe by 2020 and between 120,000 – 160,000 MWe by 2030. Incredible as it may seem this will only provide 5% of China's needs, fossil fuel will supply 75%. To put it in context, China commissioned 105,000 MWe of new plant (90% fossil); almost double the UK capacity in 2006 alone.

India plans to move from 3,500 MWe today to 21,000MWe by 2020.

Many European countries are building new nuclear, for example Russia, France, Romania, Bulgaria and Finland. The nuclear option is under discussion in at least 30 countries, which currently have no nuclear reactors. Some of this interest is linked to seawater desalination.

1.2 Public Perception

Whilst nuclear plays a significant role in the energy mix today in many countries the public perception of its advantage and disadvantages, and hence future national policy, vary widely across the world. For example, Finland, conscious of the need to be less dependent on imported energy from Russia, recently voted in favor of building a new nuclear station, which is now under construction, with another under consideration.

Developments in the US are also particularly significant. There public opinion is now in favor of nuclear power. On the 8th August 2005 President Bush signed a bill promoting the use of nuclear energy to allow America to produce cleaner energy, to be less reliant on foreign suppliers of fuel and to move closer to building more nuclear power stations by the end of this decade.

A survey of 1,100 people living within 16 km of a nuclear plan in the USA showed that 83% are in favor of nuclear energy, 76% are happy to see a further reactor building on their local site, and 88% are confident of that plant's safety. Employees of electric companies were excluded from the survey. Overall 81% said they felt well informed about their local plant, correlating with an absence of NIMBYism.

The arguments, as seen by the public, against new nuclear build fall into the following heading:

- Economics;
- Disposal of nuclear waste;
- Safety;
- Proliferation; and
- Decommissioning of Nuclear Facilities

1.2.1 Economics of Nuclear Power

For nuclear power plants any cost figures normally include spent fuel management, plant decommissioning and final waste disposal. In contrast coal and gas fired economics take no account of the effects of acid rain or global warming. Only nuclear can claim cost benefits if carbon credits are taken into account as it does not produce any CO_2.

Nuclear decommissioning costs are about 9-15% of the initial capital cost of a nuclear power plant. But when discounted, they contribute only a few percent to the investment costs and even less to the generation cost. In the USA they account for 0.1-0.2 cent/kWh, which is no more than 5% of the cost of the electricity produced.

The back-end of the fuel cycle, including spent fuel storage or disposal in a waste repository, contributes up to another 10% to the overall costs per kWh, less if there is a direct disposal of spent fuel rather than reprocessing. The $26 billion US spent fuel program is funded by a 0.1 cent/kWh levy.

French generation costs, published in 2002 show (EUR cents/kWh): nuclear 3.20, gas 3.05-4.26, coal 3.81-4.57. Nuclear costs benefit from the use of large standardized plants in France.

In addition the cost of nuclear power generation has been dropping over the last decade. This is because of declining fuel (including enrichment), operating and maintenance costs. In general the construction costs of nuclear power plants are significantly higher than for coal- or gas-fired plants because of the need to use special materials, and to incorporate sophisticated safety features and back up control equipment and the longer time scale of construction. These contribute much of the nuclear generation cost but once the plant is built the cost variables are minor, and due to the long life of nuclear plants once the plant is totally depreciated the overall operating costs drop sharply.

In the past long construction periods pushed up financing costs. However today, for example in Asia, construction times have been shorter, for instance the new-generation 1300 MWe Japanese reactors which began operating in 1996 and 1997 were built in a little over four years, and 48 to 54 months is typical projection for plants today. (Figure 1.3)

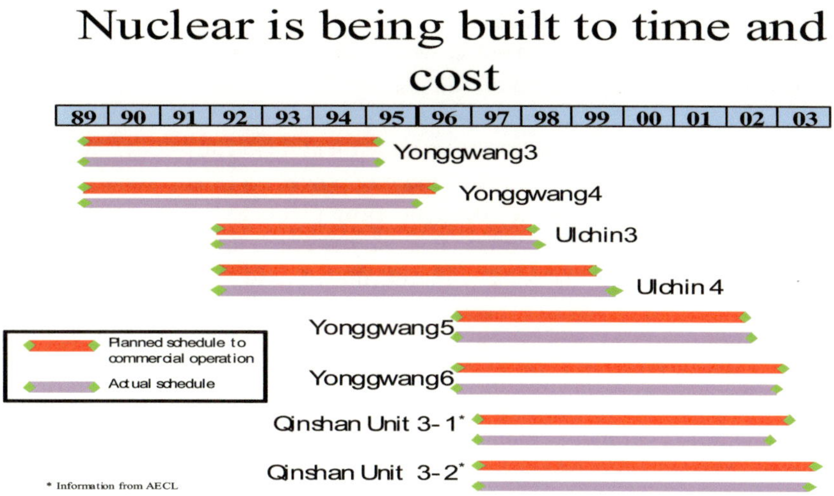

Figure 1.3

US figures for 2007, published by the Energy Utility Cost Group, showed nuclear utility generating costs averaging 2.866 c/kWh, comprising 1.832 c/kWh operation and maintenance, 0.449 c/kWh fuel and 0.585 c/kWh capital expenditure. US figures from a different source for 2007, published by NEI, gave 1.68 c/kWh for fuel plus O&M.

These figures are for fuel plus operation and maintenance costs only, they exclude capital costs, since these vary greatly among utilities and states, as well as with the age of the plant. (Figure 1.4)

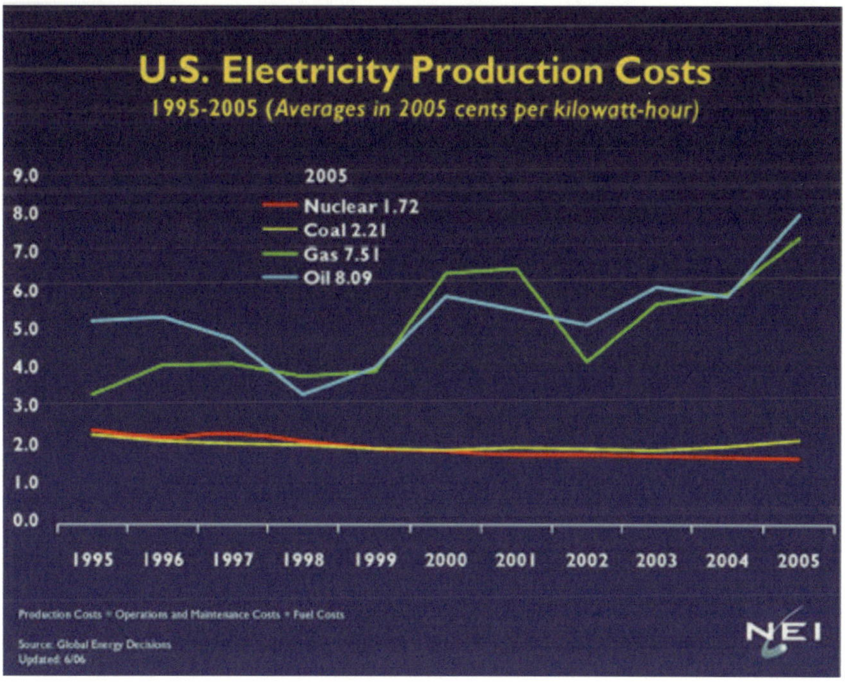

Figure 1.4 US Electricity Production Costs

1.2.1.1 Future Cost Competitiveness

The OECD does not expect investment costs in new nuclear generating plants to rise, as advanced reactor designs become standardized.

Assuming equipment and construction costs increase at the same rate for nuclear and fossil fired plants, the future competitiveness of nuclear power will depend substantially on the additional costs which may accrue to coal generating plants to ensure clean coal generation and the cost of gas for gas-fired plants. It is at present uncertain how the real costs of meeting targets for reducing emissions will be attributed to fossil fuel plants.

Overall, and under current regulatory measures, the OECD expects nuclear to remain economically competitive with fossil fuel generation, except in regions where there is direct access to low cost fossil fuels. In Australia, for example, coal-fired generating plants are close to both the mines supplying them and the main population centers, and large volumes of gas are available on low cost, long-term contracts.

A 2005 OECD comparative study showed that nuclear power had increased its competitiveness over the previous seven years. The principal changes since 1998 were increased nuclear plant capacity factors and rising gas prices. The study did not factor in any costs for carbon emissions from fossil fuel generators, and focused on over one hundred plants able to come on line 2010-15, including 13 nuclear plants. Nuclear overnight construction costs ranged from US$ 1000/kW in Czech Republic to $2500/kW in Japan, and averaged $1500/kW. Coal plants were costed at $1000-1500/kW, gas plants $500-1000/kW and wind capacity $1000-1500/kW.

OECD electricity generating cost projections for year 2010 on - 5% discount rate are shown in Table 20.1

	nuclear	coal	gas
Finland	2.76	3.64	-
France	2.54	3.33	3.92
Germany	2.86	3.52	4.90
Switzerland	2.88	-	4.36
Netherlands	3.58	-	6.04
Czech Rep	2.30	2.94	4.97
Slovakia	3.13	4.78	5.59
Romania	3.06	4.55	-
Japan	4.80	4.95	5.21
Korea	2.34	2.16	4.65
USA	3.01	2.71	4.67
Canada	2.60	3.11	4.00

Source: OECD/IEA NEA 2005.

Table 1.1 US 2003 cents/kWh, Discount rate 5%, 40 year lifetime, 85% load factor.

Nuclear costs were highest by far in Japan. Nuclear is comfortably cheaper than coal in seven of ten countries, and cheaper than gas in all but one. At 10% discount rate (Table 1.2) nuclear ranged 3-5 cents/kWh (except Japan: near 7 cents, and Netherlands), and capital becomes 70% of power cost, instead of the 50% with 5% discount rate. Here, nuclear is again cheaper than coal in eight of twelve countries and cheaper than gas in all but two. Among the technologies analyzed for the report, the new EPR if built in Germany would deliver power at about 2.38 c/kWh - the lowest cost of any plant in the study.

	nuclear	coal	gas
Finland	4.22	4.45	-
France	3.93	4.42	4.30
Germany	4.21	4.09	5.00
Switzerland	4.38	-	4.65
Netherlands	5.32	-	6.26
Czech Rep	3.17	3.71	5.46
Slovakia	4.55	5.52	5.83
Romania	4.93	5.15	-
Japan	6.86	6.91	6.38
Korea	3.38	2.71	4.94
USA	4.65	3.65	4.90
Canada	3.71	4.12	4.36

Source: OECD/IEA NEA 2005.

Table 1.2 US 2003 cents/kWh, Discount rate 10%, 40 year lifetime, 85% load factor.

Based partly on these figures the European Commission in January 2007 published comparative cost estimates for different fuels (Table 1.3):

	2005	Projected 2030 with EUR 20-30/t CO2 cost
Gas CCGT	3.4-4.5	4.0-5.5
Coal - pulverised	3.0-4.0	4.5-6.0
Coal - fluidised bed	3.5-4.5	5.0-6.5
Coal IGCC	4.0-5.0	5.5-7.0
Nuclear	4.0-5.5	4.0-5.5
Wind onshore	3.5-11.0	2.8-8.0
Wind offshore	6.0-15.0	4.0-12.0

Table 1.3 Comparative generating cost in EU – 10% discount rate (EUR)

A 1997 European electricity industry study compared electricity costs from nuclear, coal and gas for base-load plant commissioned in 2005. At a 5% discount rate nuclear (in France and

Spain) at 3.46 cents/kWh (US), was cheaper than all but the lowest-priced gas scenario. However at a 10% discount rate nuclear, at 5.07 c/kWh, was more expensive than all but the high-priced gas scenario. (ECU to US$ @ June '97 rates)

In 1999 Siemens (now Framatome ANP) published an economic analysis comparing combined-cycle gas plants with new designs, including the European Pressurized Water Reactor (EPR) and the SWR-1000 boiling water reactor. Both the 1550 MWe EPR, if built as a series in France/Germany, and the SWR-1000 (with an 8% discount rate) would be competitive with gas-combined cycle, at EUR 2.6 cents/kWh. The current-generation Konvoi plants operating in Germany produce power at 3.0 cents/kWh including full capital costs, falling to 1.5 c/kWh after complete depreciation.

A detailed study of energy economics in Finland published in mid 2000 showed that nuclear energy would be the least-cost option for new generating capacity. The study compared nuclear, coal, gas turbine combined cycle and peat. Nuclear has very much higher capital costs than the others --EUR 1749/kW including initial fuel load, which is about three times the cost of the gas plant. But its fuel costs are much lower, and so at capacity factors above 64% it was the cheapest option.

An August 2003 study (Figure 1.5) put nuclear costs at EUR 2.37 c/kWh, coal 2.81 c/kWh and natural gas at 3.23 c/kWh (on the basis of 91% capacity factor, 5% interest rate, 40 year plant life). With emission trading @ EUR 20/t CO2, the electricity prices for coal and gas increase to 4.43 and 3.92 c/kWh respectively:

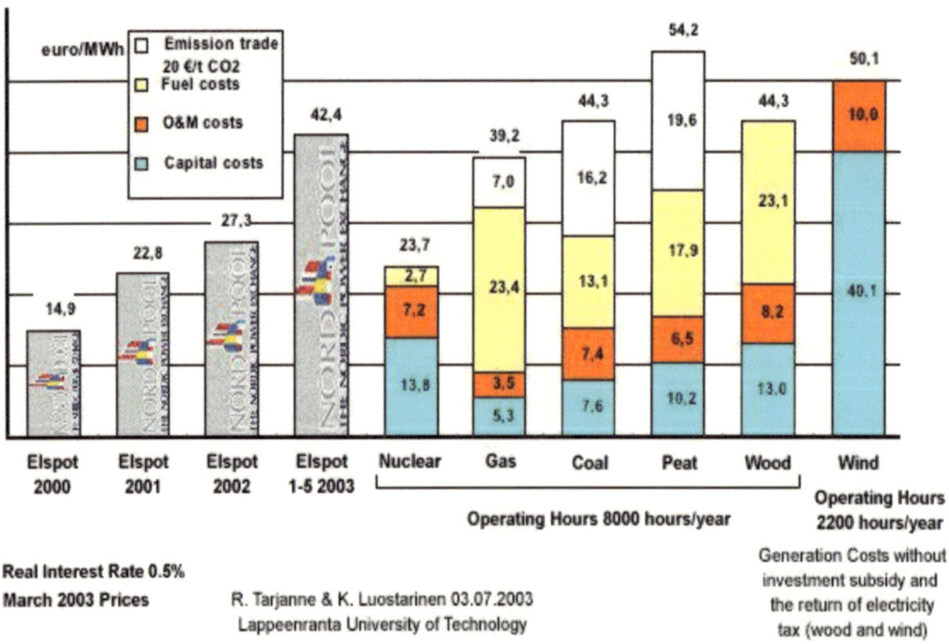

Figure 1.5 Tarjamme and Luostarmen Study 2003

In the middle three bars of Figure 1.5 the relative effects of capital and fuel costs can be clearly seen. The relatively high capital cost of nuclear power means that financing cost and time taken in construction are critical, relative to gas and even coal. But the fuel cost is very much lower, and so once a plant is built its cost of production is very much more predictable than for gas or even coal. The impact of adding a cost for carbon emissions can also be seen.

The UK Royal Academy of Engineering carried out an authoritative study in March 2004 on the costs of generating electricity in the UK which took into account capital costs, running costs, fuel, and maintenance costs. Decommissioning costs were assumed to be neutral except in the case of nuclear where these costs were allowed for. In the case of wind, the cost of standby generation was included. For base-load plant, the costs of nuclear were marginally greater than those of combined cycle turbine plant. However, taking into account possible future carbon emission taxes, based on £30 per tonne, nuclear generation became the clear winner.

Also, current designs of nuclear reactors, are being considerably simplified, thereby reducing the capital build times resulting in less financing costs. In addition these new designs produce considerably less waste that in turn reduces back end costs.

1.2.1.2 Nuclear Fuel Costs

From the outset the basic attraction of nuclear energy has been its low fuel costs compared with coal, oil and gas fired plants. Uranium, however, has to be processed, enriched and fabricated into fuel elements, and about half of the cost is due to enrichment and fabrication. Allowances must also be made for the management of radioactive spent fuel and the ultimate disposal of this spent fuel or the wastes separated from it.

In January 2007, the approx. US $ cost to get 1 kg of uranium as UO_2 reactor fuel at likely contract prices (about one third of current spot price) are shown in Table 1.4.

Uranium:	8.9 kg U_3O_8 x $53	472
Conversion:	7.5 kg U x $12	90
Enrichment:	7.3 SWU x $135	985
Fuel fabrication:	per kg	240
Total, approx:		US$ 1787

(If assuming a higher uranium price, say two thirds of current spot price: 8.9 kg x 108 = 961, this gives a total of $2286 or 0.635 c/kWh.)

Table 1.4 At 45,000 MWd/t burn-up this gives 360,000 kWh electrical per kg, hence fuel cost: 0.50 c/kWh.

A Finnish study in 2000 quantified fuel price sensitivity to electricity costs (Figure 1.6)

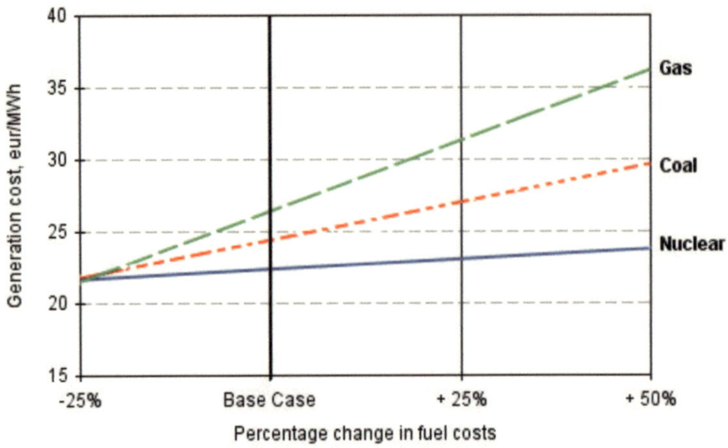

Figure 1.6 The impact of fuel costs on electricity generation costs.
Finland, early 2000

These show that a doubling of fuel prices would result in the electricity cost for nuclear rising about 9%, for coal rising 31% and for gas 66%. These are similar figures to those from the 1992 OECD report. Oil and hence gas prices have already risen significantly since the study, partly reflected in the 2003 figures above.

1.2.2 Disposal of Nuclear Waste

A major concern in the minds of the public is the long-term disposal of nuclear waste. Radioactive wastes arise from many sources - such as:

- Materials and equipment which have become contaminated during the operation of nuclear power stations and the manufacture of nuclear fuel and nuclear weapons;
- Waste arising from reprocessing nuclear fuel after it has been used in a reactor;
- Decommissioning nuclear reactors and other nuclear facilities;
- Use of radioactive materials in university research and medicine;
- Industrial manufacture and use of isotopes for tracing;
- It also arises from coal fired electricity generation and oil exploration.

Every part of the nuclear fuel cycle produces some radioactive waste and the cost of managing and disposing of this 'radwaste' is built into the process. Uranium mining for example generates fine sandy tailings, which contain virtually all the naturally occurring radioactive elements found in the uranium ore.

A large portion of radioactive waste produced from the nuclear fuel cycle has radiation levels similar to, or not much higher than, the natural background level. This waste is relatively easy to deal with. Only a small proportion is highly radioactive and requires isolation from people. The general considerations for classifying radioactive wastes are; a) how long the waste will remain at a hazardous level, b) what the concentration of the radioactive material in the waste and c) whether the waste is heat generating.

The persistence of the radioactivity determines how long the waste requires management. The concentration and heat generation dictate how the waste should be handled. These considerations also result in the disposal methods.

1.2.2.1 Classification of Nuclear Waste

There are several systems of nomenclature in use, but the following is generally accepted:

- Exempt waste, excluded from regulatory control because radiological hazards are negligible.
- Low-level Waste (LLW) contains enough radioactive material to require action for the protection of people, but not so much that it requires shielding in handling or storage.
- Intermediate-level waste (ILW) requires shielding. If it has more than 4000 Bq/g of long-lived (over 30 year half-life) alpha emitters it is categorized as "long-lived" and requires more sophisticated handling and disposal.
- High-level waste (HLW) sufficiently radioactive to require both shielding and cooling, generates >2 kW/m 3 of heat and has a high level of long-lived alpha-emitting isotopes.

Very low level waste or exempt waste. These categories contain negligible amounts of radioactivity and may be disposed of with domestic refuse.

Low-level Waste comprises the bulk of waste from the nuclear fuel cycle. It comprises paper, rags, tools, clothing, and filters etc that contain small amounts of mostly short-lived radioactivity. It does not require shielding during handling and transport and is suitable for shallow land burial. To reduce its volume, these wastes are often compacted or incinerated before disposal. Disposal sites for low-level waste are in operation in many countries. Worldwide they make up 90% of the volume but have only 1% of the total radioactivity of all radioactive wastes.

Intermediate-level Waste contains higher amounts of radioactivity and normally requires shielding. Shielding can be barriers of lead, concrete or water to give protection from penetrating radiation such as gamma rays. Intermediate-level wastes typically comprise resins, chemical sludges and metal fuel cladding, as well as contaminated materials from reactor decommissioning. It may be solidified in concrete or bitumen for disposal. Generally short-lived waste (mainly from reactors) is buried, but long-lived waste (from fuel reprocessing) will be disposed of underground.

High-level Waste (HLW) contains the fission products and transuranic elements generated in the reactor core that are highly radioactive and hot. (Figure 1.7) High-level waste accounts for over 95% of the total radioactivity produced though the actual amount of material is low, 25-30 tonnes of spent fuel, or 3 cubic meters per year of vitrified waste for a typical large nuclear reactor (1000 MWe, light water type), i.e., 2.8% of the total volume of radioactive waste. All the high level waste produced to date in the UK from the military and the civil programs would only fill 4 double decker buses.

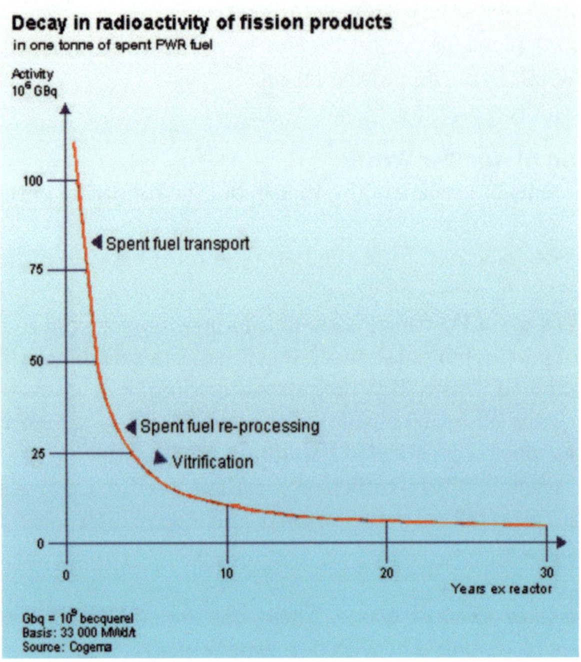

Figure 1.7 Decay in Radioactivity of Fission Products

It is important to realize that the future reactor designs use less fuel hence produce less waste than previous reactors. Ten APR1000 PWR reactors could replace all the UK current reactors. Assuming these reactors operated for 60 years the new waste arising would only be a very small amount to what is already in existence and already safely stored. (Figure 1.8)

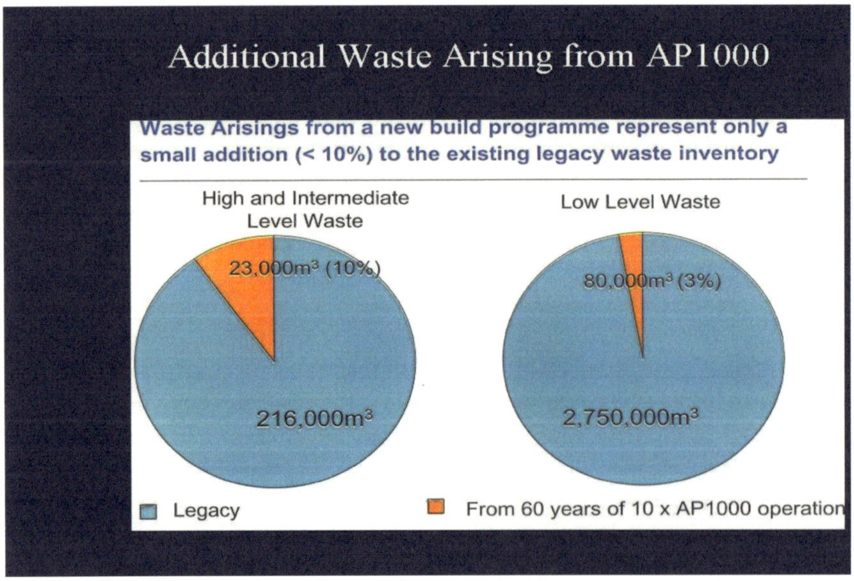

Figure 1.8 UK Waste Arising

1.2.2.2 Management of High Level Waste

There are two types of high level waste, fission products and transuranics separated from the spent fuel and the spent fuel elements themselves from the reactor core when they are not reprocessed. Both types of HLW must be treated prior to disposal. HLW from reprocessing is incorporated into solid blocks of borosilicate glass. This process is known as vitrification. For direct disposal, spent fuel requires encapsulation in containers made, for example, of stainless steel or copper.

For reprocessing when the fission products are first extracted from the spent fuel they are in liquid form, having been dissolved in acid (usually nitric acid). This liquid can be safely retained in stainless steel tanks that are equipped with cooling systems until it is converted into a solid, which is a more convenient material for management, storage, transport and disposal. After drying it is incorporated into molten borosilicate glass that is allowed to solidify inside corrosion resistant canister. Vitrification produces a stable solid that has the high-level waste incorporated its structure.

In either case however there is a cooling period of 20 to 50 years between removal from the reactor and disposal, with the conditioned spent fuel or conditioned HLW being retained in interim storage. This is because the level of radioactivity and heat from the used fuel fall rapidly in these years down to about one thousandth of the level at discharge in 40 years. Such long-term storage facilities may be at one central place as in Sweden or at the reactor site, as in the US. They may again be underwater or dry storage, where circulating air removes the heat generated by the spent fuel. The structure and design of both the building and containers protects the outside world from radiation exposure and the fuel from potential outside hazards.

1.2.2.3 Disposing of High Level Wastes

Final disposal of high level wastes is required in due course but there is no technical or logistical reason why this is urgent. Rather the contrary, the longer HLW waste is in storage, the easier it is to handle safely. HLW is accumulating at about 12,000 tonnes a year worldwide. High-level wastes are highly radioactive for a long time so must be isolated from people for thousands of years while their radiation levels drop.

Geological repositories are planned in stable rock formations in the main countries utilizing nuclear energy. It is the responsibility of each country to dispose of its wastes. Typically a repository will be 500 meters down in rock, clay or salt. The idea is a multiple barrier concept:

- The waste, either as a ceramic oxide (e.g. the spent fuel itself) or through vitrification (separated HLW from reprocessing) is immobilized.
- It is then sealed in a corrosion resistant canister such as stainless steel or copper.
- Finally it is buried in a sold rock formation.

Other means of stabilizing high-level waste are at the research stage. One of the more advanced is a substance called Synroc. This is an advanced ceramic principally comprising three natural titanate minerals which are geo-chemically stable and which together have the capacity to incorporate into their crystal structures nearly all of the elements present in high-level radioactive waste, thereby immobilizing them.

There is an interesting example from nature of long term geologically storage over millions of years. Several nuclear reactors were discovered in 1972 at the Oklo uranium mine in the West African republic of Gabon. The deposit of ore, which contained about 3% U-235, began a self-sustaining chain reaction millions of years ago. Like all reactors, this one created its own high-level waste, up to 5,000 kg of fission products and transuranic elements, which today are found only in, used fuel. The Oklo chain reaction occurred intermittently for more than 500,000 years. Despite its location in a wet, tropical climate, Oklo's uranium deposit and high-level waste has remained securely locked in this natural repository for the past 2000 million years. Many of the waste products stayed where they were created or moved only a few centimeters before decaying into harmless products.

1.2.2.4 Management of Low and Intermediate Waste

The intermediate-level waste (ILW) along with the low-level waste represent some 90% of the total volume of radioactive waste generated during the lifetime of a nuclear power plant. This relatively large volume of long-lived and short-lived ILW contains only about 1% of the total radioactivity. Only a small proportion of the intermediate-level waste remains significantly radioactive for years but all ILW requires shielding when it is handled. Low-level waste (LLW) and short-lived intermediate-level waste is of three kinds:

Process wastes result from the treatment, purification and filtration systems of fluids in direct contact with the parts of the reactor that may be contaminated by radioactivity. These wastes include:

- Filters in the cooling water circuits of the nuclear power plant;
- Resins that trap radioactive materials in the water circuits;
- Radioactive particulates that are retained by air filters installed in the ventilation stacks of nuclear facilities.

Technological wastes arise from the necessary maintenance carried out on a nuclear power plant. Technological waste represents half the volume of LLW and short-lived ILW, but contains little radioactivity.

Solid technological wastes might contain rags, cardboard, plastic sheets, bags, tools and protective clothing. Liquid technological wastes comprise mainly oils, small amounts of lubricants and organic solvents used for decontamination.

Decommissioning wastes occur at the end of a nuclear reactor's life. After the spent fuel is removed the plant is decommissioned and eventually demolished. During this process, large amounts of wastes are generated, though most is not radioactive. About a tenth of it contains some radioactivity up to the intermediate level.

Plant operators make constant efforts to reduce the quantities of waste that are generated. Waste is collected, sorted and then conditioned. The management strategy chosen depends upon the origin and radioactivity level of the waste. LLW, with the lowest concentrations of radioactivity, is usually retained in metal drums, which are often compacted after filling to reduce the volume. Other techniques may also be used to effect volume reduction. These include: melting of metallic waste, incinerating of the combustible parts of waste (whilst retaining the radioactive ash) and super-compacting waste to reduce the total volume further.

Low-level wastes that contain slightly higher radioactivity levels are stabilized by cement or an organic solid (bitumen or resin) and then placed in concrete containers for shielding. Disposal sites for such wastes are in operation in many countries. Typically, these are shallow earth burial sites, which provide a suitable facility to contain the wastes safely. A 1000 MWe nuclear power reactor can be expected to produce around 100m^3 of low-level waste every year.

1.2.2.5 Long-Lived Intermediate Level Waste

Typically, these wastes arise from dismantled internal structures of the reactor core, which become radioactive after prolonged operation. They also include: the control rods, which regulate the nuclear reaction, the source assemblies, which are used to initiate a nuclear reaction, after new fuel has been loaded, and other rods that limit the reactivity of fresh fuel. ILW is treated and conditioned by incorporating it into cement and then placing it in concrete containers. In some instances, the conditioned waste might subsequently be placed into an additional container, made of metal. Special packages are used for transporting long-lived intermediate level waste. These packages meet internationally approved standards that ensure that the waste is safely contained.

Ultimately long-lived ILW will go to deep geological disposal as with high-level waste.

Sweden has already done this but in most countries, long-lived waste is being safely stored and contained at interim storage facilities. The maintenance of a 1000 MWe nuclear power reactor produces less than 0.5 cubic meters of long-lived ILW each year. If the spent fuel goes for reprocessing, then the cladding from the spent fuel adds an additional 3 cubic meters of ILW.

1.2.2.6 Spent Fuel: Reprocessing and Recycling

Fresh Uranium oxide fuel contains up to 5% U-235. When the fuel reaches the end of its useful life, it is removed from the reactor. At this point it typically contains about 95% U-238, 3% fission products (the residues of the fission reactions) and transuranic isotopes, 1% plutonium and 1% U-235. The plutonium is produced by the neutron irradiation of U-238.

Spent fuel still contains about a quarter of the original fissile U-235 as well as much of the plutonium that has been formed in the reactor. Reprocessing separates out this uranium and plutonium. Several reprocessing facilities, Sellafield in the UK, La Hague in France, and Chelybinsk in Russia are in operation. The wastes left after reprocessing can then be disposed of, while the uranium and plutonium may be recycled for use in a nuclear reactor as mixed oxide (MOX) fuel. This is called the 'closed fuel cycle' because the useful ingredients of spent fuel are recycled.

With the recycling option the energy potential can be realized in new nuclear fuel since Pu-239 and U-235 contained in the spent fuel are fissile.

1.2.2.7 Waste from Reprocessing

The reprocessing of spent fuel gives rise to low, intermediate and high level wastes:

High-level waste comprises the non-reusable part of the spent nuclear fuel itself both fission products and transuranic elements other than plutonium. The fission product leftovers are vitrified, i.e. incorporated into glass. Hulls and end fittings from the fuel assemblies are compacted, to reduce the total volume of the waste, and are frequently incorporated into cement before being placed into containers for disposal as ILW.

The major commercial reprocessing plants operating in France and UK also undertake reprocessing for utilities in other countries, notably Japan. Most Japanese spent fuel is reprocessed in Europe, with the vitrified waste and the recovered uranium and plutonium (as MOX) being returned to Japan to be recycled.

1.2.2.8 Recycling

Among the benefits of recycling identified by those countries that are utilizing MOX fuel are conservation of uranium, minimizing the amount of high-level radioactive, reducing reliance on new uranium supply, reducing the fissile plutonium inventory and reduction of spent fuel storage requirements.

1.2.2.9 Plutonium Recycling

Plutonium is recycled through a special fuel fabrication plant to produce mixed oxide (MOX) fuel. MOX fuel is a mixture of plutonium and uranium oxides (formed from natural, depleted or reprocessed uranium). MOX fuel containing 5 to 7% plutonium has characteristics that are similar to uranium oxide based fuel and used as part of a reactor's fuel loading. There are 34 reactors licensed to use MOX fuel across Europe with seventy-five others in the licensing process. Japan for example planned to introduce MOX fuel into twenty of its reactors by the year 2010. It should be noted that plutonium arising from the civil nuclear fuel cycle is not suitable for bombs because it contains far too much of the Pu-240 isotope, due to the length of time the fuel has been in the reactor.

1.2.2.10 Uranium Recycling

Uranium from reprocessing, sometimes referred to as Rep-U, must usually be enriched, and to facilitate this it must first be converted to UF_6.

1.2.3 Safety

Although Chernobyl blemished the image of nuclear energy, the accident's positive legacy is an even stronger system of nuclear safety worldwide. In 1989, the nuclear industry established the World Association of Nuclear Operations (WANO) to foster a global nuclear safety culture. Through private-sector diplomacy, WANO has built a transnational network of technical exchange that includes all countries with nuclear power. Today every nuclear power reactor in the world is part of the WANO system of operational peer review. The aim of WANO's peer-review system standards is set by the UN's International Atomic Energy Agency (IAEA).

Advances in safety practice are unmistakable. At most plants worldwide, reportable safety-related 'events' are near zero. National and international insurance laws assign responsibility to nuclear plant operators. In the US for example, reactor operators share in a 'pooled' private insurance system that has never cost taxpayers a penny.

Today, nuclear power plants have a superb safety record - both for plant workers and the public. In the transport of nuclear material, highly engineered containers - capable of withstanding enormous impact - are the industrial norm. More than 20,000 containers of spent fuel and high-level waste have been shipped safety over a total distance exceeding 30 million kilometers. During the transport of these and other radioactive substances - whether for research, medicine or nuclear - there had never been a harmful radioactive release.

Compare this safety record to other industries such as coal mining, the chemical or transport industries or the risks of smoking or drinking.

1.2.4 Proliferation

Proliferation is a major consideration. Nuclear power entails potential security risks, notably the possible misuse of nuclear facilities to acquire technology or materials as a precursor to the acquisition of a nuclear weapons capability. This is a subject of current major international concern. Fuel cycles that involve the chemical reprocessing of spent fuel to separate weapons-useable plutonium and uranium enrichment technologies are of special significance. An international response is required to reduce the proliferation risk. The response should:

- Re-appraise and strengthen the institutional underpinnings of the International Atomic Energy Agency safe-guards regime, including sanctions;
- Guide nuclear fuel cycle development in ways that reinforce shared non-proliferation objectives.

Civil nuclear power has a role to play in these objectives. The estimated 1500 tonnes of highly enriched uranium from Russia's nuclear weapons could be diluted to supply sufficient PWR fuel for all the world's PWR reactors for 8-9 years whilst plutonium, which represents 95% of energy left in non-reprocessed fuel, can be burned by turning it into mixed oxide fuel again to supply PWR reactors. This is already happening in the US with 174 tonnes of high-enriched uranium and 225 tonnes of Russian material being converted to civil use.

Terrorism cannot be ignored. But nuclear power is not an easy target for terrorists. Reactor core are massively shielded by concrete and computer tests have shown them resistant to 500 mph impacts from aircraft. The only reason for terrorists attacking nuclear power stations would be to prey on fears generated by militant greens rather than produce a lot of dead bodies. Gas and Oil terminals are much more likely targets.

1.2.5 Decommissioning of Nuclear Facilities

To date, 100 mines, 90 commercial power reactors, over 250 research reactors and a number of fuel cycle facilities, have been retired from operation.

At the end of 2005, IAEA reported that eight power plants had been completely decommissioned and dismantled, with the sites released for unconditional use. A further 17 had been partly dismantled and safely enclosed, 31 were being dismantled prior to eventual site release and 30 were undergoing minimum dismantling prior to long-term enclosure.

The International Atomic Energy Agency has defined three options for decommissioning, the definitions of which have been internationally adopted:

- Immediate Dismantling (or Early Site release/Decon in the US): This option allows for the facility to be removed from regulatory control relatively soon after shutdown or termination of regulated activities. Usually, the final dismantling or decontamination activities begin within a few months or years, depending on the facility. Following removal from regulatory control, the site is then available for re-use.
- Safe Enclosure (or Safestor): This option postpones the final removal of controls for a longer period, usually in the order of 40 to 60 years. The facility is placed into a safe storage configuration until the eventual dismantling and decontamination activities occur.
- Entombment: This option entails placing the facility into a condition that will allow the remaining on-site radioactive material to remain on-site without the requirement of ever removing it totally. This option usually involves reducing the size of the area where the radioactive material is located and then encasing the facility in a long-lived structure such as concrete, that will last for a period of time to ensure the remaining radioactivity is no longer of concern.

There is no right or wrong approach, each having its benefits and disadvantages. National policy determines which approach is adopted. In the case of immediate dismantling (or early site release), responsibility for the decommissioning is not transferred to future generations. The experience and skills of operating staff can also be utilized during the decommissioning program. Alternatively, Safe Enclosure (or Safestor) allows significant reduction in residual radioactivity, thus reducing radiation hazard during the eventual dismantling. The expected improvements in mechanical technique should also lead to a reduction in the hazard and also costs.

In the case of nuclear reactors, about 99% of the radioactivity is associated with the fuel which is removed following a permanent shutdown. Apart from any surface contamination of plant, the remaining radioactivity comes from "activation products" such as steel components that have long been exposed to neutron irradiation. Their atoms are changed into different isotopes such as iron-55, cobalt-60, nickel-63 and carbon-14. The first two are highly radioactive, emitting gamma rays. However, their half-life is such that after 50 years from closedown their radioactivity is much diminished and the risk to workers largely gone.

EDF in France, in particular have a great deal of experience in decommissioning their early nuclear stations.

There are three stages in the Safestor process for decommissioning nuclear power stations:

- **Stage 1** comprises monitored shut down of the installation. Before this level is reached, the power plant is shut down during an initial two to three year period. Non-nuclear equipment and buildings are dismantled. The fuel is unloaded from the reactor and transferred to the reprocessing plant. Finally, all the plant systems are drained down, leaving the power plant "inert". Any residual radioactive material area is contained. By this stage, 99% of the radioactivity has been removed. Although access to the plant is restricted, the equipment is necessary for monitoring of radioactivity is maintained.
- **Stage 2** comprises partial and conditional clearance of the site. This takes around four to five years. The auxiliary systems and fuel handling equipment, which can only be contained for a few years, can be decontaminated before dismantling. The radioactive waste is packaged before dispatch to the storage facility. The part of the plant around the reactor is isolated, contained and placed under surveillance.
- **Stage 3** comprises total and unconditional clearance of the plant site after the third stage of dismantling, which lasts four to five years, and takes place after a forty-year break. The rest of the plant is completely dismantled, and all remaining radioactive materials and equipment are removed. The buildings themselves are dismantled, and the nuclear equipment cut up (using eclectic arc or thermal lance equipment, or by remote control in the case of highly radioactive materials).

Dismantling a reactor produces a considerable amount of materials requiring processing (steel, concrete, pipes, electric cables, etc), in addition to a large quantity of very low active waste, mainly from the final stage of dismantling. Once this phase is completed, the site no longer requires monitoring, and can be returned to use.

1.3 Advantages Of Nuclear Power

So against these concerns what are the advantages of nuclear power, apart from helping to reduce global warming effects?

The UK situation is again an interesting case study as the Government has come to realize the need for security of supply. Currently the generation mix in the UK is 32% coal, 22% nuclear, 38% gas, 4% oil and 4% others and renewables. In other words, a diversified supply.

However, there was a lack of coherent strategy for UK future energy demands and that this is now a major concern not only in the UK but globally. In the UK, demand is increasing by 1 to 1½% per year, coal and nuclear plants are closing down, and the market does not see the certain economic returns required to build new power stations. Yet windmills are being subsidized at £50/60 per MWh at total extra costs to electricity consumers of £30 billion by 2020, more than twice the cost of a 10GW nuclear power program.

Without new power plant, by 2010, standby surplus plant margin will have fallen from a secure position of 25% to a mere 6%. But worse still, by 2020, the UK will be almost totally dependent on imported gas supplies, mainly from Russia, as there are only small amounts of strategic gas and oil reserve within the UK. And these imports will be at the end of a very long supply chain traversing areas of potential political instability giving rise to risks of serious supply shortages and price instability, particularly when Russia is rapidly becoming the major supplier of oil and gas to China, Korea and Japan.

Currently the UK is the highest amongst G8 countries for security of supply because it is largely independent of imported fuels. By 2024 this situation would be completely reversed, the UK would be dependent on imported gas, and so would be the least secure of the G8 countries. The imported gas supply costs are linked to oil prices that are rapidly increasing. On 11th August 2004 UK oil imports exceeded exports for the first time in 11 years. Oil reserves world wide will soon peak, as was so clearly demonstrated by Shell in 2004, and as of June 2008 oil prices had reached $139 a barrel up from $65 in May 2007.

It is difficult to see how a nation such as the UK's, that was totally energy self sufficient, with the exception of uranium ore which is in plentiful supply from stable countries such as Canada and Australia, a nation that was blessed with coal, oil, gas and nuclear, that enabled it to ride through a succession of energy crises, including the oil price increases in 1973, and coal strikes in the early 1980s, allowed itself to be at risk not only on the price of imported energy, that will affect its industrial base, but also has the potential for major blackouts. Also with an average trade deficit of roughly £4 billion a month how would the UK pay for all the gas it would need to import? It is against this background that the Government in the UK decided in 2007/2008 to give a green light for new nuclear construction in the UK. Many other nations also have ongoing nuclear programs to combat such risks and many are now considering the need for a nuclear component in their energy mix.

1.4 Nuclear Power Reactors

1.4.1 Components

The principles for using nuclear power to produce electricity are the same for most types of reactor. The energy released from continuous fission of the atoms of the fuel is harnessed as heat in either a gas or water, and is used to produce steam. The steam is used to drive the turbines that produce electricity.

There are several components common to most types of reactors:

Fuel; usually pellets of uranium oxide (UO_2) arranged in tubes to form fuel rods. The rods are arranged into fuel assemblies in the reactor core. In the case of the Pebble Bed Reactor the fuel is in the form of 60 mm diameter spheres.

Moderator; this is material which slows down the neutrons released from fission so that they cause more fission. It is usually water, but may be heavy water or graphite.

Control rods; these are made with neutron-absorbing material such as cadmium, hafnium or boron, and are inserted or withdrawn from the core to control the rate of reaction, or to halt it. (Secondary shutdown systems involve adding other neutron absorbers, usually as a fluid, to the system.)

Coolant; a liquid or gas circulating through the core so as to transfer the heat from it. . In light water reactors the water moderator functions also as primary coolant. Except in BWRs, there is secondary coolant circuit producing the scheme.

Pressure vessel or pressure tubes; usually a robust steel vessel containing the reactor core and moderator/coolant, but it may be a series of tubes holding the fuel and conveying the coolant through the moderator.

Steam generator; part of the cooling system where the heat from the reactor is used to make steam for the turbine.

Containment; the structure around the reactor core which is designed to protect it from outside intrusion and to protect those outside from the effects of radiation in case of any malfunction inside. It is typically a meter-thick concrete and steel structure.

1.5 The Development History Of Current Nuclear Reactors

Man's understanding of the science of atomic radiation, atomic structure and nuclear fission has developed since 1895 with much of it in the early 1940s. Between 1939 and 1945, development was focused on the atomic bomb. It was Enrico Fermi, at the University of Chicago, took the first major step in the building of the atomic bomb when he supervised the design and assembly of an "atomic pile", a code word for an assembly that in peacetime would become known as a "nuclear reactor".

However, in the course of the developing nuclear weapons, the West and the Soviet Union acquired a range of new technologies and engineers soon realized that the tremendous heat produced by the nuclear fission process could be tapped either for direct use or for generating electricity.

It was also clear that such thermal reactors would allow development of compact long-lasting power sources that could have various applications, especially in powering submarines.

Another type of reactor is the fast breeder reactor that produces more fuel than it uses. It was this type of experimental reactor that first produced a small amount of electricity in December 1951, almost 60 years ago, in the USA.

At that time work in the Soviet Union refined existing thermal reactor designs and developed new ones for commercial energy production.

Their existing graphite-moderated channel-type reactor, for producing plutonium, was modified for heat and electricity generation and in 1954 the world's first nuclear power station began operation, with a design capacity of 5MW. This served as a prototype for other graphite channel reactor designs, including the Chernobyl-type reactor known as an RBMK. (Figure 1.9)

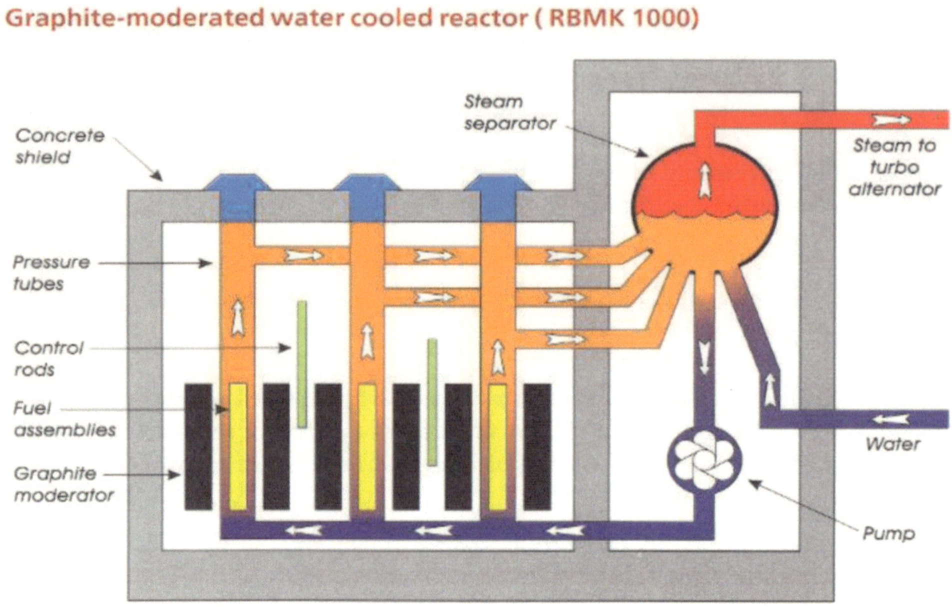

Figure 1.9 RBMK Reactors

In the 1950s the Russians were also developing fast breeder reactors.

In 1964 the first two Soviet commercial nuclear power plants were commissioned, a 100 MW boiling water reactor and a small 210 MW pressurized water reactor, known in Russia as a VVER. The first large RBMK started up in 1973 and the same year saw the commissioning of the first of four small 12 MW boiling water channel-type units for the production of both power and heat.

In the northwest Arctic a slightly bigger VVER, with a rate capacity of 440 MW began operating and this became a standard design. The world's first commercial prototype fast breeder reactor started up in 1972 producing 120 MW electricity and heat to desalinate seawater. A prototype fast neutron reactor started generating 12 MW in 1959. So a vast amount of effort that developed many different designs, took place in Russia.

In 1953 President Eisenhower proposed his "Atoms for Peace" program, which set the course for civil nuclear energy development in the USA.

The main US effort up to that time, under Admiral Rickover, was to develop the Pressurized Water Reactor (PWR) for submarine use. The PWR uses enriched uranium oxide fuel and is moderated and cooled by ordinary light water. (Figure 1.10)

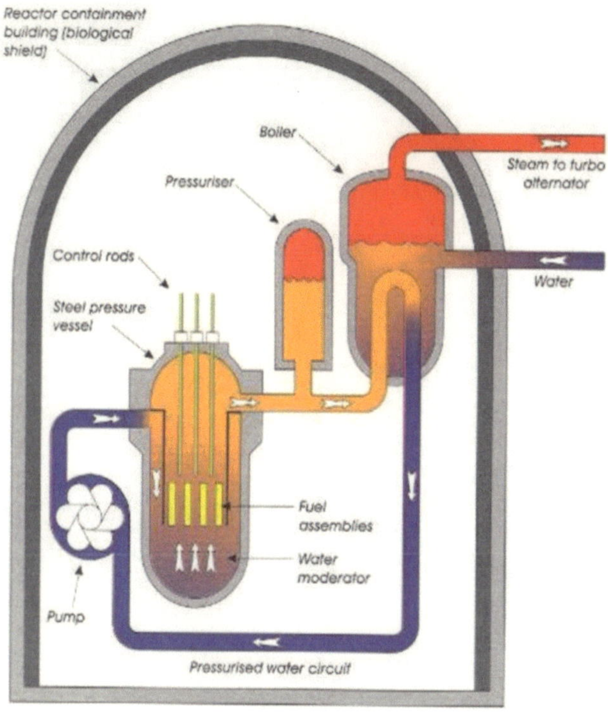

Figure 1.10 Pressurized Water Reactor (CPWR)

The Mark 1 prototype naval reactor started up in March 1953 and the first nuclear-powered submarine, USS Nautilus, was launched in 1954. In 1959 both the USA and the USSR launched their first nuclear-powered surface vessels, ranging from icebreakers to aircraft carriers. The Mark 1 naval reactor led to the building of the 90 MW Shipping Port demonstration PWR reactor, for electricity generation, which started up in 1957 and operated until 1982.

Westinghouse designed the first fully commercial PWR of 250 MW, which started up in 1960 and operated to 1992. Meanwhile the Argonne National Laboratory developed a Boiling Water Reactor (BWR) (Figure 1.11). The first commercial unit, designed by General Electric, was started up in 1960.

By the end of the 1960s international orders were being placed for PWR and BWR reactor units of outputs up to 1,000 MW.

Because, at that time, the USA had a virtual monopoly on uranium enrichment, UK development took a different approach, which resulted in a series of reactors, the Magnox Reactors, fuelled by natural uranium, moderated by graphite and cooled by carbon dioxide. (Figure 1.12)

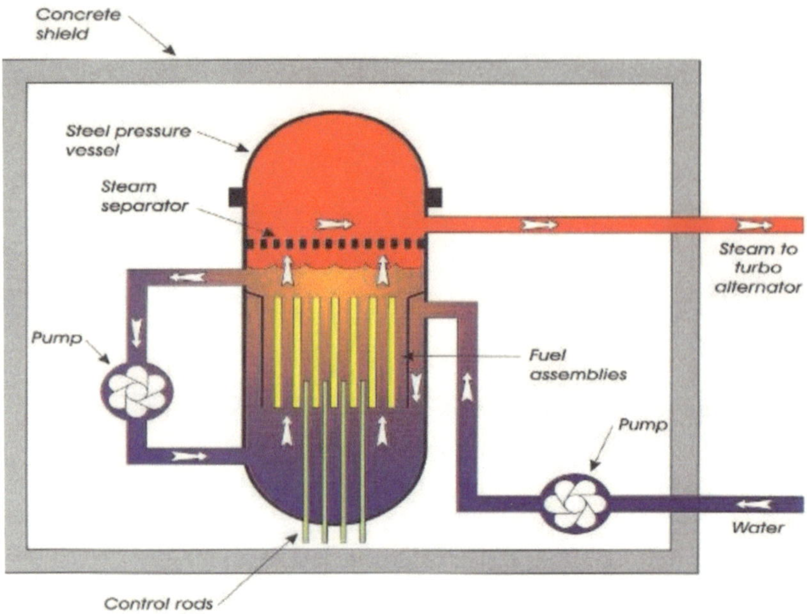

Figure 1.11 Boiling Water reactor (BWR)

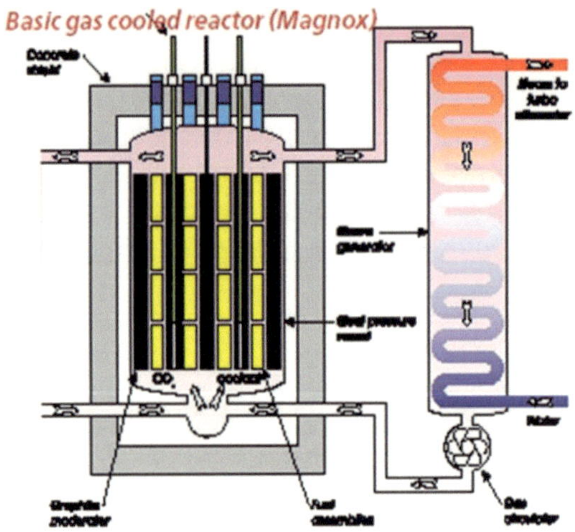

Figure 1.12 Magnox Reactor

The first of these 50 MW Magnox reactors, Calder Hall-1, started up in 1956 and was closed in 2002. A total of 26 Magnox units were built between the 1950s and the 1970s. Eighteen were closed and the remaining 8 are scheduled to be closed by 2011.

However, after 1963, based on the Magnox designs, the UK developed the Advanced Gas Cooled Reactors (AGR). (Figure 1.13) These were to become the backbone of the UK nuclear generation program with 14 AGR reactors providing 8,380 MW.

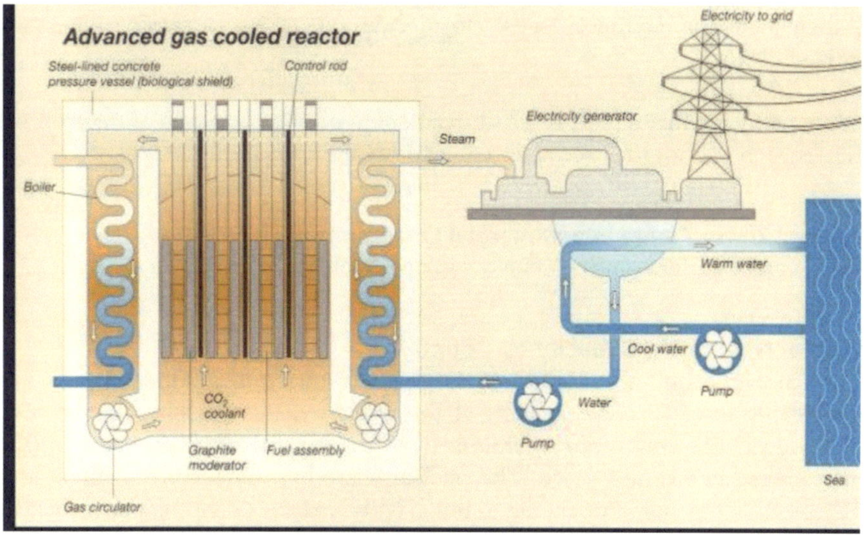

Figure 1.13 Advanced Gas Cooled Reactor (AGR)

Canadian reactor development headed down a different track, using natural uranium fuel and heavy water, both as a moderator and as a coolant. The first CANDU unit started up in 1962 and was followed by 32 more worldwide. (Figure 1.14)

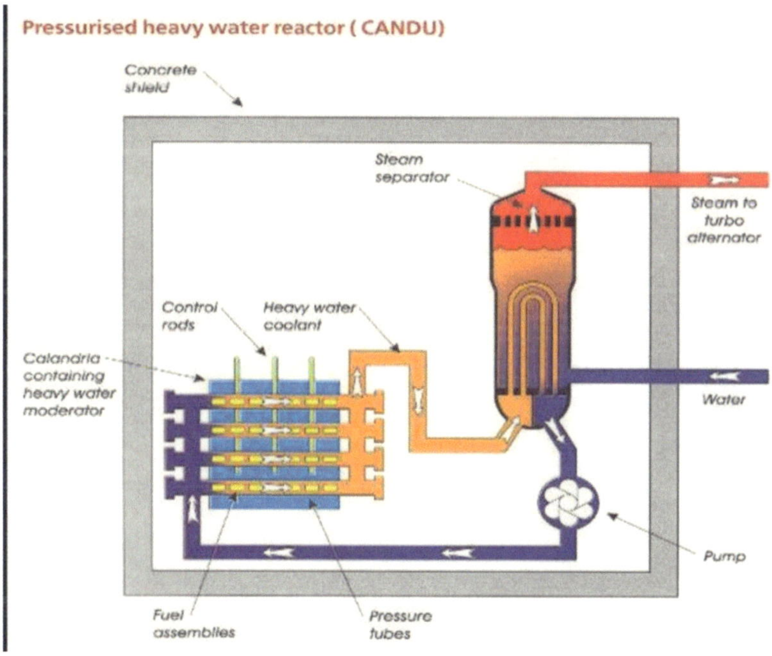

Figure 1.14 CANDU Reactor

France started with a gas-graphite design similar to Magnox, using a different fuel cladding and her first reactor commenced operation in 1956, with commercial models operating from 1959.

France then had the common sense to decide on three successive generations of standardized PWRs.

In addition, many countries built research reactors to provide a source of neutron beans for scientific research and for the production of medical and industrial isotopes.

1.5.1 Nuclear Power Plants in commercial Operation

There are several different types of reactors in operation today as shown in Table 1.5

1.5.2 Nuclear Generating Capacity by Country

As shown in Figure 1.2 the United States has 103 reactors in operation and nuclear generating capacity of 97 GWe, making it the world's leading nuclear nation. Only one reactor, however, has come into operation over the past decade and some smaller, less efficient reactors have closed down. The nuclear share has, however, remained at around 20% of US electricity generation, owing to much better reactor operating performance.

In the remainder of the Americas, Canada stands out with 17 reactors currently in operation and nuclear capacity of 12 GWe. 13% of Canada's electricity generation is nuclear. Elsewhere, Mexico, Brazil and Argentina all have small nuclear programs. South Africa is the only African nation with a small nuclear component in its energy mix. However, it now plans to considerably increase its nuclear generating capacity by the installation of further PWRs or Pebble Bed Reactors.

Reactor type	Main Countries	Number	GWe	Fuel	Coolant	Moderator
Pressurized Water Reactor (PWR)	US, France, Japan, Russia	264	250.5	Enriched UO_2	water	water
Boiling Water Reactor (BWR)	US, Japan, Sweden	94	86.4	Enriched UO_2	water	water
Pressurized Heavy Water Reactor 'CANDU' (PHWR)	Canada	43	23.6	Natural UO_2	heavy water	heavy water
Gas-cooled Reactor (AGR & Magnox)	UK	18	10.8	Natural U (metal),enriched UO_2	CO_2	graphite
Light Water Graphite Reactor (RBMK)	Russia	12	12.3	enriched UO_2	water	graphite
Fast Neutron Reactor (FBR)	Japan, France, Russia	4	1.0	PuO_2 and UO_2	liquid sodium	none
Other	Russia	4	0.05	Enriched UO_2	water	graphite
	TOTAL	439	384.6			

Table 1.5 Nuclear Power Plants in Commercial Operation

At approaching 80%, France has the highest nuclear share in its electricity generation of any country, with 59 reactors in operation and generating capacity of 63 GWe. Three successive generations of PWRs have been built and the first of a new generation of European Pressurized Water Reactors (EPR) will come into operation around 2012.

Many other European countries have substantial nuclear generating capacity, notably Germany, United Kingdom, Spain, Sweden and Belgium. Within the European Union (EU) as a whole, the nuclear share exceeds 30% of total electricity generation and five of the ten 2004 EU accession states (Czech and Slovak Republics, Hungary, Slovenia and Lithuania) have nuclear power. Finland is building the only new reactor under construction in the EU apart from France.

Japan has 54 nuclear reactors in operation with capacity of 45 GWe providing a nuclear share of around 25%. Nuclear power has become a key element in Japan's energy security and environmental policy, as it has no access to substantial indigenous energy resources. Plans exist for substantial numbers of new reactors in the future.

In Asia, Korea also has a maturing nuclear power sector, but the main growth areas for nuclear are undoubtedly China and India, the biggest developing countries in the world. In both cases, the programs are starting at low bases in terms of shares of total electricity generating capacity but they are targeting nuclear capacities of 40 GWe and 20 GWe by 2020 respectively.

Russia has an important nuclear sector and exports its technology and nuclear materials to many other countries. Its reactor program, however, became stalled at the fall of the Soviet Union and is only now getting back on track. There are currently 31 reactors in operation with generating capacity of 22 GWe, giving a nuclear share of about 17% in total electricity.

Ukraine has substantial nuclear generating capacity and remains close to the Russian industry. The East European countries remain dependent on Soviet-era technology but are gradually breaking away as they enter the EU. Bulgaria and Romania entered the EU in January 2007 and both are interested in adding to their existing stock of reactors.

1.5.3 Nuclear Growth Since 1970

The biggest factor in the continued rise in the quantity of nuclear electricity has, however, been the improved operating performance of nuclear reactors. The United States demonstrates this most strongly, as reactor load factors (showing plant utilization level compared with the theoretical maximum) typically languished in the 60-70% range in the 1980s. The onset of power market liberalization forced reactor operators to improve or go out of business and average load factors in Union States are now around 90%. Other countries had long demonstrated that this is possible and good practice continues to spread, such that world load factors have risen by ten percentage points since 1990.

Over the past five years, world nuclear electricity production has risen by 300 TWh, similar to the output from 40 new nuclear reactors, yet the net increase in the number of reactors has been only 5.

1.6 CURRENT REACTOR TYPES

1.6.1 Light Water Reactors

1.6.1.1 The Pressurized Water Reactor (PWR) (Figure 1.10)

This is the most common reactor type, with over 230 in use for power generation and a further several hundred in naval propulsion. The design originated as a submarine power plant. It uses ordinary water as both coolant and moderator. The design is distinguished by having a primary cooling circuit which flows through the core of the reactor under very high pressure, and a secondary circuit in which steam is generated to drive the turbine.

A PWR has fuel assemblies of 200-300 rods each, arranged vertically in the core, and a large reactor would have about 150-250 fuel assemblies with 80-100 tonnes of uranium.

Water in the reactor core reaches about 325°C; hence it must be kept under about 150 times atmospheric pressure to prevent it boiling. Pressure is maintained by steam in a pressuriser (see diagram). In the primary cooling circuit the water is also the moderator, and if any of it turned to steam the fission reaction would slow down. This negative feedback effect is one of the safety features of the type. The secondary shutdown system involves adding boron to the primary circuit.

The secondary circuit is under less pressure and the water here boils in the heat exchangers that are thus steam generators. The steam drives the turbine to produce electricity, and is then condensed and returned to the heat exchangers in contact with the primary circuit.

1.6.1.2 Boiling Water Reactor (BWR) (Figure 1.11)

This design has many similarities to the PWR, except that there is only a single circuit in which the water is at lower pressure (about 75 times atmospheric pressure) so that it boils in the core at about 285°C. The reactor is designed to operate with 12-15% of the water in the top part of the core as steam, and hence with less moderating effect and thus efficiency there.

The steam passes through drier plates (steam separators) above the core and then directly to the turbines, which are part of the reactor circuit. Since the water around the core of a reactor is always contaminated with traces of radionuclides, it means that the turbine must be shielded and radiological protection provided during maintenance. The cost of this tends to balance the savings due to the simpler design. Most of the radioactivity in the water is very short-lived, so the turbine hall can be entered soon after the reactor is shut down.

A BWR fuel assembly comprises 90-100 fuel rods, and there are up to 750 assemblies in a reactor core, holding up to 140 tonnes of uranium. The secondary control system involves restricting water flow through the core so that steam in the top part means moderation is reduced.

1.6.2 Pressurized Heavy Water Reactor (PHWR or CANDU) (Figure 1.14)

The CANDU reactor design has been developed since the 1950s in Canada. It uses natural uranium (0.7% U-235) oxide as fuel, hence needs a more efficient moderator, in this case heavy water (D_2O).

The moderator is in a large tank called a calandria, penetrated by several hundred horizontal pressure tubes that form channels for the fuel, cooled by a flow of heavy water under high pressure in the primary cooling circuit, reaching 290°C. As in the PWR, the primary coolant generates steam in a secondary circuit to drive the turbines. The pressure tube design means that the reactor can be refueled progressively without shutting down, by isolating individual pressure tubes from the cooling circuit. This ability to refuel on load, as opposed to other reactor types that have to shut down to reload, is a big operating advantage.

A CANDU fuel assembly consists of a bundle of 37 half-meter long fuel rods (ceramic fuel pellets in zircaloy tubes) plus a support structure, with 12 bundles lying end to end in a fuel channel. Control rods penetrate the calandria vertically, and a secondary shutdown system involves adding gadolinium to the moderator. The heavy water moderator circulating through the body of the calandria vessel also yields some heat (though this circuit is not shown on the diagram above).

1.6.3 Advanced Gas Cooled Reactor (AGR) (Figure 1.13)

These are the second generation of British gas-cooled reactors, using graphite moderator and carbon dioxide as coolant. The fuel is a uranium oxide pellet, enriched to 2.5-3.5%, in stainless steel tubes. The carbon dioxide circulates through the core, reaching 650°C and then past steam generator tubes outside it, but still inside the concrete and steel pressure vessel. Control rods penetrate the moderator and a secondary shutdown system involves injecting nitrogen to the coolant.

The AGR was developed from the Magnox reactor (Figure 1.12) also graphite moderated and CO_2 cooled, and a number of these are still operating in UK, albeit they are now planned to progressively close. They use natural uranium fuel in metal form.

1.6.4 Light Water Graphite-Moderated Reactor (RBMK) (Figure 1.9)

This is a Soviet design, developed from plutonium production reactors. It employs long (7 meter) vertical pressure tubes running through graphite moderator, and is cooled by water, which is allowed to boil in the core at 290°C, much as in a BWR. Fuel is low-enriched uranium oxide made up into fuel assemblies 3.5 meters long. With moderation largely due to the fixed graphite, excess boiling simply reduces the cooling and neutron absorption without inhibiting the fission reaction, and a positive feedback problem can arise.

1.6.5 Fast Neutron Reactors

Some reactors (only one in commercial service) do not have a moderator and utilize fast neutrons, generating power from plutonium while making more of it from the U-238 isotope in or around the fuel. While they get more than 60 times as much energy from the original uranium compared with the normal reactors, they are expensive to build and await resource scarcity to come into their own.

1.7 Small Nuclear Power Reactors

As nuclear power generation has become established since the 1950s, the size of reactor units has grown from 60 MWe to more than 1300 MWe, with corresponding economies of scale in operation. At the same time there have been many hundreds of smaller reactors built both for naval use (up to 190 MW thermal) and as neutron sources, yielding enormous expertise in the engineering of small units.

Today, due partly to the high capital cost of large power reactors generating electricity via the steam cycle and the need for nuclear in developing countries where the demand is not high and whose transmission systems are not capable of handling large centralized units of power, there is a move to develop smaller units. These may be built independently or as modules in a larger complex, with capacity added incrementally as required. The IAEA defines "small" as under 300 MWe.

The most prominent modular project is the South African-led consortium developing the Pebble Bed Modular Reactor of 170 MWe. Chinergy is preparing to build a similar unit, the 195 MWe HTR-PM in China. A US-led group is developing another design with 285 MWe modules. Both drive gas turbines directly, using helium as a coolant and operating at very high temperatures. They build on the experience of several innovative reactors in the 1960s and 1970s.

Generally, modern small reactors for power generation are expected to have greater simplicity of design, economy of mass production, and reduced siting costs. Many are also designed for a high level of passive or inherent safety in the event of malfunction. Traditional reactor safety systems are 'active' in the sense that they involve electrical or mechanical operation on command. Some engineered systems operate passively, e.g. pressure relief valves. Both require parallel redundant systems. Inherent or full passive safety depends only on physical phenomena such as convection, gravity or resistance to high temperatures, not on functioning of engineered components.

Some are conceived for areas away from transmission grids and with small loads, others are designed to operate in clusters in competition with large units. The cost of electricity from a 50 MWe unit is estimated by DOE as 5.4 to 10.7 c/kWh (compared with charges in Alaska and Hawaii from 5.9 to 36.0 c/kWh).

Already operating in a remote corner of Siberia are four small units at the Bilibino co-generation plant. These four 62 MWt (thermal) units are an unusual graphite-moderated boiling water design with water/steam channels through the moderator. They produce steam for district heating and 11 MWe (net) electricity each. They have performed well since 1976, much more cheaply than fossil fuel alternatives in the Arctic region.

The US Congress is funding research on both small modular nuclear power plants (assembled on site from factory-produced modules) and advanced gas-cooled designs (which are modular in the sense that up to ten or more units are progressively built to comprise a major power station).

1.7.1 Light Water Reactors

US experience has been of very small military power plants, such as the 11 MWt, 1.5 MWe (net) PM-3A reactor that operated at McMurdo Sound in Antarctica 1962-72, generating a total of 78 million kWh. There was also an Army program for small reactor development and some successful small reactors from the main national program commenced in the 1950s. One was the Big Rock Point BWR of 67 MWe that operated for 35 years to 1997.

Of the following, the first three designs have conventional pressure vessel plus external steam generators (PV/loop design). The others mostly have the steam supply system inside the reactor pressure vessel ('integral' PWR design). All have enhanced safety features relative to current PWRs.

The Russian KLT-40S is a reactor well proven in icebreakers and now proposed for wider use in desalination and, on barges, for remote area power supply. Here a 150 MWt unit produces 35 MWe (gross) as well as up to 35 MW of heat for desalination or district heating (or 38.5 MWe gross if power only). These are designed to run 3-4 years between refueling and it is envisaged that they will be operated in pairs to allow for outages (70% capacity factor), with on-board refueling capability and spent fuel storage. At the end of a 12-year operating cycle the whole plant is taken to a central facility for overhaul and storage of spent fuel. Two units will be mounted on a 20,000 tonne barge.

Although the reactor core is normally cooled by forced circulation, the OKBM design relies on convection for emergency cooling. Fuel is uranium aluminum silicide with enrichment levels of up to 20%, giving up to 4-year refueling intervals.

A larger Russian factory-built and barge-mounted unit (requiring a 12,000 tonne vessel) is the VBER-150, of 350 MW thermal, 110 MWe. It has modular construction and is derived by OKBM from naval designs, with two steam generators. Uranium oxide fuel enriched to 4.7% has burnable poison; it has low burnup (31 GWd/t average, 41.6 GWd/t max) and 8 year refueling interval.

OKBM's larger VBER-300 PWR is a 295 MWe unit, the first of which will be built in Kazakhstan. It was originally envisaged in pairs as a floating nuclear power plant, displacing 49,000 tonnes. As a cogeneration plant it is rated at 200 MWe and 1900 GJ/hr. The reactor is designed for 60-year life and 90% capacity factor. It has four steam generators and a cassette core with 85 fuel assemblies enriched to 5% and 48 GWd/tU burn-up. Versions with three and two steam generators are also envisaged, of 230 and 150 MWe respectively. Also with more sophisticated and higher-enriched (18%) fuel in the core, the refueling interval can be pushed from 2 years out to 15 years with burn-up to 125 GWd/tU. A 2006 joint venture between Atomstroyexport and Kazatomprom sets this up for development as a basic power source in Kazakhstan, then for export.

Another larger Russian reactor is the VK-300 boiling water reactor being developed specifically for cogeneration of both power and district heating or heat for desalination (150 MWe plus 1675 GJ/hr) by the Research & Development Institute of Power Engineering (NIKIET). It has evolved from the VK-50 BWR at Dimitrovgrad, but uses standard components wherever possible, and fuel elements similar to VVER. Cooling is passive, by convection, and all safety systems are passive. Fuel burn-up is 41 GWday/tU. It is capable of producing 250 MWe if solely electrical. In September 2007 it was announced that six would be built at Kola and at Primorskaya in the Far East, to start operating 2017-20.

A smaller OKBM PWR unit under development is the ABV, with 45 MW thermal, 10-12 MWe output. The ABV-6M is said to be 18 MWe. The units are compact, with integral steam generator and enhanced safety. The whole unit of some 600 tonnes will be factory-produced for ground or barge mounting - it would require a 2500 tonne barge. The core is similar to that of the KLT-40 except that enrichment is 16.5% and average burn up 95 GWd/t. Refueling interval is about 8 years, and service life about 50 years.

The CAREM (advanced small nuclear power plant) being developed by CNEA and INVAP in Argentina is a modular 100 MWt /27 MWe pressurized water reactor with integral steam generators designed to be used for electricity generation (27 MWe or up to 100 MWe) or as a research reactor or for water desalination (with 8 MWe in cogeneration configuration). CAREM has its entire primary coolant system within the reactor pressure vessel, self-pressurized and relying entirely on convection. Fuel is standard 3.4% enriched PWR fuel, with burnable poison, and is refueled annually. It is a mature design that could be deployed within a decade.

1.7.2 High Temperature Gas-Cooled Reactors

Building on the experience of several innovative reactors built in the 1960s and 1970s, new high-temperature gas-cooled reactors (HTRs) are being developed which will be capable of delivering high-temperature (up to 950°C) helium either for industrial application via heat exchanger or directly to drive gas turbines for electricity (the Brayton cycle) with almost 50% thermal efficiency possible (efficiency increases 1.5% with each 50°C increment). Technology developed in the last decade makes HTRs more practical than in the past, though the direct cycle means that there must be high integrity of fuel and reactor components.

Fuel for these reactors is in the form of TRISO particles less than a millimeter in diameter. Each has a kernel (c0.5 mm) of uranium oxycarbide, with the uranium enriched up to 20% U-235, though normally less. This is surrounded by layers of carbon and silicon carbide, giving a containment for fission products that is stable to 1600°C or more. With negative temperature coefficient of reactivity (the fission reaction slows as temperature increases) and passive decay heat removal, this makes the reactors inherently safe. They do not require any containment building for safety.

The reactors are sufficiently small to allow factory fabrication, and will usually be installed below ground level.

There are two ways in which these particles are arranged: in blocks - hexagonal 'prisms' of graphite, or in billiard ball-sized pebbles of graphite encased in silicon carbide, each with about 15,000 fuel particles and 9g uranium. There is a greater amount of spent fuel than from the same capacity in a light water reactor. The moderator is graphite.

The Japan Atomic Energy Research Institute's (JAERI) High-Temperature Test Reactor (HTTR) of 30 MW thermal started up at the end of 1998 and has been run successfully at 850°C. In 2004 it achieved 950°C outlet temperature. Its fuel is in 'prisms' and its main purpose is to develop thermo chemical means of producing hydrogen from water.

Based on the HTTR, JAERI is developing the Gas Turbine High Temperature Reactor (GTHTR) of up to 600 MW thermal per module. It uses improved HTTR fuel elements with 14% enriched uranium achieving high burn-up (112 GWd/t). Helium at 850°C drives a horizontal turbine at 47% efficiency to produce up to 300 MWe. The core consists of 90 hexagonal fuel columns 8 meters high arranged in a ring, with reflectors. Each column consists of eight one-meter high elements 0.4 m across and holding 57 fuel pins made up of fuel particles with 0.55 mm diameter kernels and 0.14 mm buffer layer. In each 2-yearly refueling, alternate layers of elements are replaced so that each remains for 4 years.

On the basis of four modules per plant, capital cost is projected at US$ 1300-1700/kWe and power cost about US 3.4 c/kWh.

China's HTR-10, a small high-temperature pebble-bed gas-cooled experimental reactor at the Institute of Nuclear & New Energy Technology (INET) at Tsinghua University north of Beijing started up in 2000 and reached full power in 2003. It has its fuel as a 'pebble bed' (27,000 elements) of oxide fuel with average burn up of 80 GWday/t U. Each pebble fuel element has 5g of uranium enriched to 17% in around 8300 particles. The reactor operates at 700°C (potentially 900°C) and has broad research purposes. Eventually it will be coupled to a gas turbine, but meanwhile it has been driving a steam turbine.

Construction of a larger version, the 200 MWe (450 MWt) HTR-PM, was approved in principle in November 2005, with construction starting in 2009. This will have two reactors modules, each of 250 MWt, using 9% enriched fuel (520,000 elements) giving 80 GWd/t discharge burn up. With an outlet temperature of 750°C the pair will drive a single steam cycle turbine at about 40% thermal efficiency. The size was reduced to 250 MWt from earlier 458 MWt modules in order to retain the same core configuration as the prototype HTR-10 and avoid moving to an annular design like South Africa's PBMR. This Shidaowan demonstration reactor at Rongcheng in Shandong province is to pave the way for an 18-unit (3x6x200MWe) full-scale power plant on the same site at Weihei, also using the steam cycle. Plant life is envisaged as 60 years with 85% load factor.

China Huaneng Group, one of China's major generators, is the lead organization involved in the demonstration unit with 47.5% share; China Nuclear Engineering & Construction (CNEC) will have a 32.5% stake and Tsinghua University's INET 20% - it being the main R&D contributor. Projected cost is US$ 385 million (but later units falling to US$1500/kW with generating cost about 5c/kWh). Start-up is scheduled for 2013. The HTR-PM rationale is both eventually to replace conventional reactor technology for power, and also to provide for future hydrogen production. INET is in charge of R&D, and is aiming to increase the size of the 250 MWt module and also utilize thorium in the fuel. Eventually a series of HTRs, possibly with Brayton cycle directly driving the gas turbines, will be factory-built and widely installed throughout China.

In 2004 the small HTR-10 reactor was subject to an extreme test of its safety when the helium circulator was deliberately shut off without the reactor being shut down. The temperature increased steadily, but the physics of the fuel meant that the reaction progressively diminished and eventually died away over three hours. At this stage a balance between decay heat in the core and heat dissipation through the steel reactor wall was achieved and the temperature never exceeded a safe 1600°C. This was one of six safety demonstration tests conducted then. The high surface area relative to volume, and the low power density in the core, will also be features of the full-scale units (which are nevertheless much smaller than most light-water types).

Between 1966 and 1988, the AVR experimental pebble bed reactor at Juelich, Germany, operated for over 750 weeks at 15 MWe, most of the time with thorium-based fuel. The fuel consisted of about 100,000 billiard ball-sized fuel elements. The thorium was mixed with high-enriched uranium (HEU). Maximum burnups of 150 GWd/t were achieved. It was used to demonstrate the inherent safety of the design due to negative temperature coefficient: the helium coolant flow was cut off and the reactor power fell rapidly.

The 300 MWe THTR reactor in Germany was developed from the AVR and operated between 1983 and 1989 with 674,000 pebbles, over half containing Th/HEU fuel (the rest graphite moderator and some neutron absorbers). These were continuously recycled and on average the fuel passed six times through the core. Fuel fabrication was on an industrial scale. Several design features made the AVR unsuccessful, though the basic concept was again proven. It drove a steam turbine.

An 80 MWe HTR-module was then designed by Siemens as a modular unit to be constructed in pairs. It was licensed in 1989, but was not constructed. This design was part of the technology bought by Eskom in 1996 and is a direct antecedent of PBMR.

South Africa's Pebble Bed Modular Reactor (PBMR) is being developed by a consortium led by the utility Eskom, and drawing on German and previous UK expertise. (Figure 1.15)

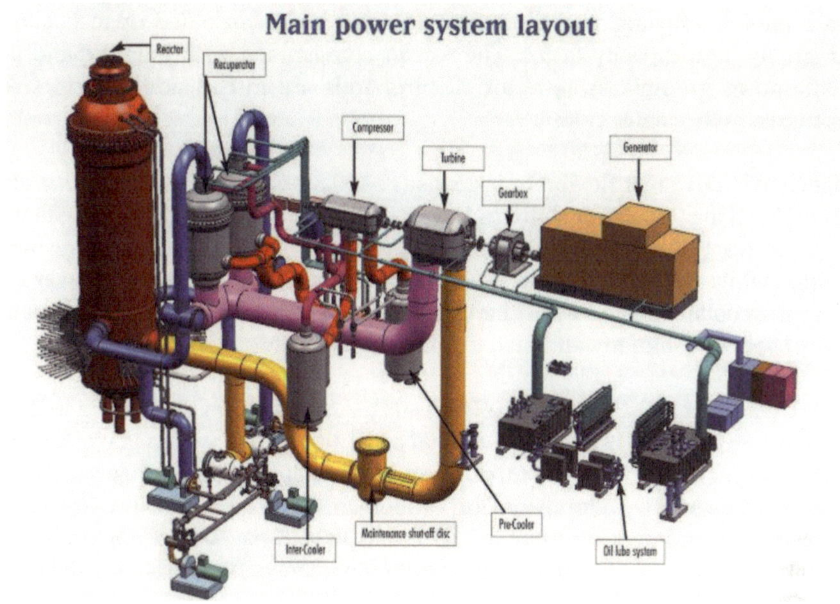

Figure 1.15 Pebble Bed Modular Reactor (PBMR)

It aims for a step change in safety, economics and proliferation resistance. Production units will be 165 MWe. The PBMR will have a direct-cycle gas turbine generator and thermal efficiency about 41%, the helium coolant leaving the bottom of the core at about 900°C. Up to 450,000 fuel pebbles 60 mm diameter, 210 g mass and containing 9g uranium enriched to 10% U-235 recycle through the reactor continuously (about six times each, taking six months) until they are expended, giving an average enrichment in the fuel load of 5% and average burn-up of 80 GWday/t U (eventual target burn-ups are 200 GWd/t). (Figure 1.16)

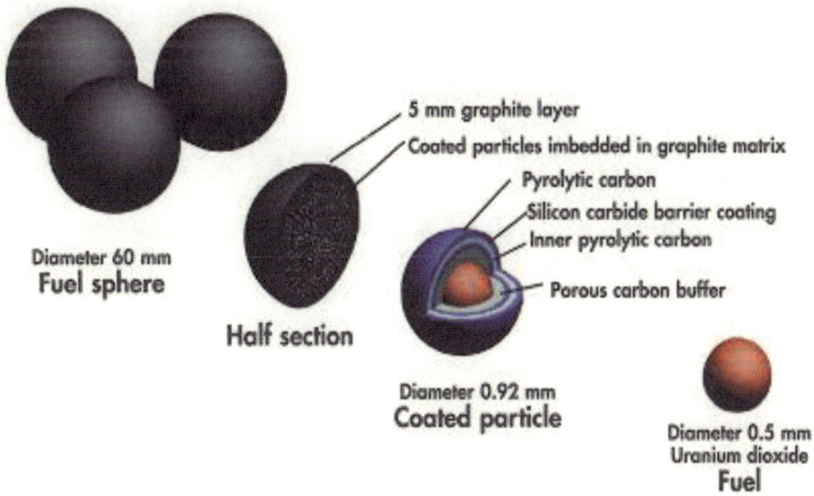

Figure 1.16 Fuel Element Design for PBMR

This means on-line refueling as expended pebbles (which have yielded up to 91 GWd/t) are replaced, giving high capacity factor. The reactor core is lined with graphite and there is a central column of graphite as reflector. Control rods are in the side reflectors and cold shutdown units in the center column.

Performance includes great flexibility in loads (40-100%) without loss of thermal efficiency, and with rapid change in power settings. Power density in the core is about one tenth of that in light water reactor, and if coolant circulation ceases the fuel will survive initial high temperatures while the reactor shuts itself down - giving inherent safety. Power control is by varying the coolant pressure and hence flow. Each unit will finally discharge about 35 tonnes/yr of spent pebbles to ventilated on-site storage bins.

The PBMR Demonstration Power Plant (DPP) started construction at Koeberg in 2009 and is expected to achieve criticality in 2013. Eventual construction cost (when in clusters of four or eight units) is expected to be very competitive. Investors in the PBMR project are Eskom, the South African Industrial Development Corporation and Westinghouse. The first commercial units are expected on line soon after the DPP and Eskom has said it expects to order 24, which justify fully commercial fuel supply and maintenance. A contract for the pebble fuel plant at Pelindaba has been let.

Each 210g-fuel pebble contains about 9g U and the total uranium in one fuel load is 4.1 t. MOX and thorium fuels are envisaged. With used fuel, the pebbles can be crushed and the 4% of their volume which is micro spheres removed, allowing the graphite to be recycled. The company says microbial removal of C-14 is possible (also in the graphite reflectors when decommissioning).

In 2006 the PBMR Board formalized the concept of a higher-temperature PBMR Process Heat Plant (PHP) with reactor output temperature of 950°C. The first plants are envisaged for 2016 and the applications will be oil sands production, petrochemical industry (process steam), steam methane reforming for hydrogen and eventually thermo chemical hydrogen production. This design will be submitted to US Department of Energy as a candidate Next-Generation Nuclear Plant.

A design certification application to the US Nuclear Regulatory Commission was considered in 2008, with approval expected in 2012, opening up world markets.

A larger US design, the Modular Helium Reactor (MHR, formerly the GT-MHR), will be built as modules of up to 600 MWt. In its electrical application each would directly drive a gas turbine at 47% thermal efficiency, giving 280 MWe. It can also be used for hydrogen production (100,000 t/yr claimed) and other high temperature process heat applications. The annular core consists of 102 hexagonal fuel element columns of graphite blocks with channels for helium coolant and control rods. Graphite reflector blocks are both inside and around the core. Half the core is replaced every 18 months. Burn-up is up to 220 GWd/t, and coolant outlet temperature is 850°C with a target of 1000°C.

The MHR is being developed by General Atomics in partnership with Russia's OKBM, supported by Fuji (Japan) and Areva NP. Initially it will be used to burn pure ex-weapons plutonium at Seversk (Tomsk) in Russia. A burnable poison such as Er-167 is needed for this fuel. The preliminary design stage was completed in 2001, but the program to construct a prototype in Russia seems to have languished since. Areva is working separately on a version of this called Antares.

The development timeline was for a prototype to be constructed in Russia 2006-09 following regulatory review there.

A smaller version of this, the Remote-Site Modular Helium Reactor (RS-MHR) of 10-25 MWe has been proposed by General Atomics. The fuel would be 20% enriched and refueling interval would be 6-8 years.

A third full-size HTR design is Areva's Very High Temperature Reactor (VHTR) being put forward by Areva NP. It is based on the MHR and has also involved Fuji. Reference design is 600 MW (thermal) with prismatic block fuel like the MHR. Target core outlet temperature is 1000°C and it uses and indirect cycle, possibly with a helium-nitrogen mixes in the secondary system. This removes the possibility of contaminating the generation or hydrogen production plant with radionuclides from the reactor core.

HTRs can potentially use thorium-based fuels, such as HEU or LEU with Th, U-233 with Th, and Pu with Th. Most of the experience with thorium fuels has been in HTRs. General Atomics say that the MHR has a neutron spectrum is such and the TRISO fuel so stable that the reactor can be powered fully with separated transuranic wastes (neptunium, plutonium, americium and curium) from light water reactor used fuel. The fertile actinides enable reactivity control and very high burn-up can be achieved with it - over 500 GWd/t - the Deep Burn concept and hence DB-MHR design. Over 95% of the Pu-239 and 60% of other actinides are destroyed in a single pass.

The three larger HTR designs, with the AHTR described below, are contenders for the US Next-Generation Nuclear Plant.

A small US HTR concept is the Adams Atomic Engines 10 MWe direct simple Brayton cycle plant with low-pressure nitrogen as the reactor coolant and working fluid, and graphite moderation. The reactor core will be a fixed, annular bed with about 80,000 fuel elements each 6 cm diameter and containing approximately 9 grams of heavy metal as TRISO particles, with expected average burn-up of 80 GWd/t. The initial units will provide a reactor core outlet temperature of 800°C and a thermal efficiency near 25%. Limiting coolant flow controls power output. A demonstration plant is proposed for completion by 2011 with series production by 2014.

1.7.3 Liquid Metal Cooled Fast Reactors

Fast neutron reactors have no moderator, a higher neutron flux and are normally cooled by liquid metal such as sodium, lead, or lead-bismuth, with high conductivity and boiling point. They operate at or near atmospheric pressure and have passive safety features (most have convection circulating the primary coolant). Automatic load following is achieved due to the reactivity feedback - constrained coolant flow leads to higher core temperature that slows the reaction. Primary coolant flow is by convection. They typically use boron carbide control rods.

The Encapsulated Nuclear Heat Source (ENHS) is a liquid metal-cooled reactor concept of 50 MWe being developed by the University of California. The core is at the bottom of a metal-filled module sitting in a large pool of secondary molten metal coolant that also accommodates the 8 separate and unconnected steam generators. There is convection circulation of primary coolant within the module and of secondary coolant outside it. Outside the secondary pool the plant is air-cooled. Control rods would need to be adjusted every year or so and load-following would be autonomous. The whole reactor sits in a 17-meter deep silo. Fuel is a uranium-zirconium alloy with 13% U enrichment (or U-Pu-Zr with 11% Pu) with a 15-20 year life. After this the module is removed, stored on site until the primary lead (or Pb-Bi) coolant solidifies, and it would then be shipped as a self-contained and shielded item. A new-fuelled module would be supplied complete with primary coolant. The ENHS is designed for developing countries and is highly proliferation-resistant but is not yet close to commercialization.

A related project is the Secure Transportable Autonomous Reactor – STAR being developed by Argonne under the leadership of Lawrence Livermore Laboratory (DOE). It a lead-cooled fast neutron modular reactor with passive safety features. Its 400 MWt. size means it can be shipped by rail and cooled by natural circulation. It uses U-transuranic nitride fuel in a cassette that is replaced every 15-20 years. The STAR-LM was conceived for power generation, running at 578°C and producing 180 MWe.

STAR-H2 is an adaptation for hydrogen production, with reactor heat at up to 800°C being conveyed by a helium circuit to drive a separate thermo chemical hydrogen production plant, while lower grade heat is harnessed for desalination (multi-stage flash process). Any commercial electricity generation then would be by fuel cells, from the hydrogen. Its development is further off.

A smaller STAR variant is the Small Sealed Transportable Autonomous Reactor - SSTAR, being developed in collaboration with Toshiba and others in Japan (see 4S four paragraphs below). It has lead or Pb-Bi cooling, runs at 566°C and has integral steam generator inside the sealed unit, which would be installed below ground level. Conceived in sizes 10-100 MWe, main development is now focused on a 45 MWt/ 20 MWe version as part of the US Generation IV effort. After a 20-year life without refueling, the whole reactor unit is then returned for recycling the fuel. The core is one-meter diameter and 0.8m high. SSTAR will eventually be coupled to a Brayton cycle turbine using supercritical carbon dioxide. Prototype envisaged 2015.

For all STAR concepts, regional fuel cycle support centers would handle fuel supply and reprocessing, and fresh fuel would be spiked with fission products to deter misuse. Complete burn up of uranium and transuranics is envisaged in STAR-H2, with only fission products being waste.

Japan's LSPR is a lead-bismuth cooled reactor of 150 MWt / 53 MWe. Fuelled units would be supplied from a factory and operate for 30 years, then be returned. Concept intended for developing countries.

A small-scale design developed by Toshiba Corporation in cooperation with Japan's Central Research Institute of Electric Power Industry (CRIEPI) and funded by the Japan Atomic Energy Research Institute (JAERI) is the 5 MWt, 200 kWe Rapid-L, using lithium-6 (a liquid neutron poison) as control medium. It would have 2700 fuel pins of 40-50% enriched uranium nitride with 2600°C melting point integrated into a disposable cartridge. The reactivity control system is passive, using lithium expansion modules (LEM), which give, burn up compensation, partial load operation as well as negative reactivity feedback. As the reactor temperature rises, the lithium expands into the core, displacing an inert gas. Other kinds of lithium modules, also integrated into the fuel cartridge, shut down and start up the reactor. Cooling is by molten sodium, and with the LEM control system, reactor power is proportional to primary coolant flow rate. Refueling would be every 10 years in an inert gas environment. Operation would require no skill, due to the inherent safety design features. The whole plant would be about 6.5 meters high and 2 meters diameter.

The Super-Safe, Small & Simple - 4S 'nuclear battery' system is being developed by Toshiba and CRIEPI in Japan in collaboration with STAR work and Westinghouse in USA. It uses sodium as coolant (with electromagnetic pumps) and has passive safety features, notably negative temperature and void reactivity. The whole unit would be factory-built, transported to site, installed below ground level, and would drive a steam cycle. It is capable of three decades of continuous operation without refueling. Metallic fuel (169 pins 10mm diameter) is uranium-zirconium enriched to less than 20% or U-Pu-Zr alloy with 24% Pu for the 10 MWe version or 11.5% Pu for the 50 MWe version. Steady power output over the core lifetime is achieved by progressively moving upwards an annular reflector around the slender core (0.68m diameter, 2m high in the 10 MWe version, 1.2m diameter and 2.5m high in the 50 MWe version) at about one millimeter per week. Burn up will be 34,000 MWday/t. After 14 years a neutron absorber at the center of the core is removed and the reflector repeats its slow movement up the core for 16 more years. Burn up will be 34,000 MWday/t. In the event of power loss the reflector falls to the bottom of the reactor vessel, slowing the reaction, and external air circulation gives decay heat removal. A further safety device is a neutron absorber rod that can drop into the core. After 30 years the fuel would be allowed to cool for a year, then it would be removed and shipped for storage or disposal.

Both 10 MWe and 50 MWe versions of 4S are designed to automatically maintain an outlet coolant temperature of 550°C - suitable for power generation with high temperature electrolytic hydrogen production. Plant cost is projected at US$ 2500/kW and power cost 5-7 cents/kWh for the small unit - very competitive with diesel in many locations. The design has gained considerable support in Alaska and toward the end of 2004 the town of Galena granted initial approval for Toshiba to build a 4S reactor in that remote location. A pre-application NRC has been underway with a view to application for design certification in 2009 and construction and operating license (COL) application by 2012. Its design is sufficiently similar to PRISM - GE's modular 150 MWe liquid metal-cooled inherently-safe reactor which went part-way through US NRC approval process for it to have good prospects of licensing. Toshiba plans a worldwide marketing program to sell the units for power generation at remote mines, desalination plants and for making hydrogen. Eventually it expects sales for hydrogen production to outnumber those for power supply. The L-4S is Pb-Bi cooled version of 4S.

The Hyperion reactor is a small self-regulating hydrogen-moderated and potassium-cooled reactor fuelled by powdered uranium hydride. A US design certification application is possible in 2012.

A significant fast reactor prototype was the EBR-II, a fuel recycle reactor of 62 MWt at Argonne which used the pyrometallurgically refined spent fuel from light water reactors as fuel, including a wide range of actinides. The objective of the program is to use the full energy potential of uranium rather than only about one percent of it. It is shut down and being decommissioned. An EBR-III of 200-300 MWe was proposed but not developed.

Russia has experimented with several lead-cooled reactor designs, and has used lead-bismuth cooling for 40 years in its submarine reactors. Pb-208 (54% of naturally-occurring lead) is transparent to neutrons. A significant Russian design is the BREST fast neutron reactor, of 300 MWe or more with lead as the primary coolant, at 540°C, and supercritical steam generators. The core sits in a pool of lead at near atmospheric pressure. It is inherently safe and uses a U+Pu nitride fuel. No weapons-grade Pu can be produced (since there is no uranium blanket), and spent fuel can be recycled indefinitely, with on-site facilities. A pilot unit is being built at Beloyarsk and 1200 MWe units are planned.

A smaller and newer Russian design is the Lead-Bismuth Fast Reactor (SVBR) of 75-100 MWe. This is an integral design, with the steam generators sitting in the same Pb-Bi pool at 400-480°C as the reactor core, which could use a wide variety of fuels. The unit would be factory-made and shipped as a 4.5m diameter, 7.5m high module, then installed in a tank of water that gives passive heat removal and shielding. A power station with 16 such modules is expected to supply electricity at lower cost than any other new Russian technology as well as achieving inherent safety and high proliferation resistance. (Russia built 7 Alfa-class submarines, each powered by a compact 155 MWt Pb-Bi cooled reactor, and 70 reactor-years operational experience was acquired with these.)

1.7.4 Molten Salt Reactors

During the 1960s the USA developed the molten salt breeder reactor concept as the primary back-up option for the fast breeder reactor (cooled by liquid metal) and a small prototype MSR Experiment (8 MW) operated at Oak Ridge over four years. There is now renewed interest in the concept in Japan, Russia, France and the USA, and one of the six generation IV designs selected for further development is the MSR.

In the Molten Salt Reactor (MSR) the fuel is a molten mixture of lithium and beryllium fluoride salts with dissolved enriched uranium, thorium or U-233 fluorides. The core consists of unclad graphite moderator arranged to allow the flow of salt at some 700°C and at low pressure. Heat is transferred to a secondary salt circuit and thence to steam. It is not a fast reactor, but with some moderation by the graphite is epithermal (intermediate neutron speed). The fission products dissolve in the salt and are removed continuously in an on-line reprocessing loop and replaced with Th-232 or U-238. Actinides remain in the reactor until they fission or are converted to higher actinides which do so. A full-size 1000 MWe MSR breeder reactor was designed but not built. In 2002 a Thorium MSR was designed in France with a fissile zone where most power would be produced and a surrounding fertile zone where most conversion of Th-232 to U-233 would occur.

The FUJI MSR is a 100 MWe design operating as a near-breeder and being developed internationally by a Japanese, Russian and US consortium.

The attractive features of this MSR fuel cycle include: the high-level waste comprising fission products only, hence shorter-lived radioactivity; small inventory of weapons-fissile material (Pu-242 being the dominant Pu isotope); low fuel use (the French self-breeding variant claims 50kg of thorium and 50kg U-238 per billion kWh); and safety due to passive cooling up to any size.

The Advanced High-temperature Reactor (AHTR) is a larger reactor using a coated-particle graphite-matrix fuel like that in the GTMHR (see above section) and with molten fluoride salt as primary coolant. While similar to the gas-cooled HTR it operates at low pressure (less than 1 atmosphere) and higher temperature, and gives better heat transfer than helium. The salt is used solely as coolant, and achieves temperatures of 750-1000°C while at low pressure. This could be used in thermo chemical hydrogen manufacture. Reactor sizes of 1000 MWe/2400 MWt are envisaged, with capital costs estimated at less than $1000/kW.

Molten fluoride salts are a preferred interface fluid between the nuclear heat source and any chemical plant. The aluminum smelting industry provides substantial experience in managing them safely. The hot molten salt can also be used with secondary helium coolant generating power via the Brayton cycle.

1.7.5 Modular Construction

The IRIS developers have outlined the economic case for modular construction of their design (about 330 MWe), and the argument applies similarly to other smaller units. They point out that IRIS with its size and simple design is ideally suited for modular construction. The economy of scale is replaced here with the economy of serial production of many small and simple components and prefabricated sections. They expect that construction of the first IRIS unit will be completed in three years, with subsequent reduction to only two years.

Site layouts have been developed with multiple single units or multiple twin units. In each case, units will be constructed so that there is physical separation sufficient to allow construction of the next unit while the previous one is operating and generating revenue. In spite of this separation, the plant footprint can be very compact so that a site with three IRIS single modules providing 1000 MWe is similar or smaller in size than one with a comparable total power single unit.

Eventually IRIS is expected to have a capital cost and production cost comparable with larger plants. But any small unit such as this will potentially have a funding profile and flexibility otherwise impossible with larger plants. As one module is finished and starts producing electricity, it will generate positive cash flow for the next module to be built. Westinghouse estimates that 1000 MWe delivered by three IRIS units built at three year intervals financed at 10% for ten years require a maximum negative cash flow less than $700 million (compared with about three times that for a single 1000 MWe unit). For developed countries small modular units offer the opportunity of building as necessary, for developing countries it may be the only option, because their electric grids cannot take 1000+ MWe single units.

1.7.6 Floating Nuclear Power Plants

Apart from over 200 nuclear reactors powering various kinds of ships, Rosatom in Russia has set up a subsidiary to supply floating nuclear power plants ranging in size from 70 to 600 MWe. These will be mounted in pairs on a large barge, which will be permanently moored where it is needed to supply power and possibly some desalination to a shore settlement or industrial complex. The first will have two 40 MWe reactors based on those in icebreakers and will operate at Severodvinsk, in the Archangel region. Gazprom will use five of the next seven for offshore oil and gas field development and for operations on the Kola and Yamal peninsulas. One is for Pevek on the Chukotka peninsula, another for Kamchatka region, both in the far east of the country. Further Far East sites being considered are Yakutia and Taimyr. Electricity cost is expected to be much lower than from present alternatives.

The Russian KLT-40S is a reactor well proven in icebreakers and now proposed for wider use in desalination and, on barges, for remote area power supply. Here a 150 MWt unit produces 35 MWe (gross) as well as up to 35 MW of heat for desalination or district heating. These are designed to run 3-4 years between refueling and it is envisaged that they will be operated in pairs to allow for outages, with on-board refueling capability and used fuel storage. At the end of a 12-year operating cycle the whole plant is taken to a central facility for overhaul and removal of used fuel. Two units will be mounted on a 20,000 tonne barge. A larger Russian factory-built and barge-mounted reactor is the VBER-150, of 350 MW thermal, 110 MWe. The larger VBER-300 PWR is a 325 MWe unit, originally envisaged in pairs as a floating nuclear power plant, displacing 49,000 tonnes. As a cogeneration plant it is rated at 200 MWe and 1900 GJ/hr.

1.8 Advanced Nuclear Power Reactors

The nuclear power industry has been developing and improving reactor technology for more than five decades and is starting to build the next generations of reactors to fill orders now materializing.

Several generations of reactors are commonly distinguished. Generation I reactors were developed in 1950-60s, and outside the UK none are still running today. Generation II reactors are typified by the present US fleet and most in operation elsewhere. Generation III (and 3+) are the Advanced Reactors discussed in this section. The first are in operation in Japan and others are under construction or ready to be ordered. Generation IV designs are still on the drawing board and will not be operational before 2020 at the earliest.

About 85% of the world's nuclear electricity is generated by reactors derived from designs originally developed for naval use. These and other second-generation nuclear power units have been found to be safe and reliable, but they are being superseded by better designs.

Reactor suppliers in North America, Japan, Europe, Russia and South Africa have a dozen new nuclear reactor designs at advanced stages of planning, while others are at a research and development stage. Fourth-generation reactors are at concept stage. Third-generation reactors have:

- a standardized design for each type to expedite licensing, reduce capital cost and reduce construction time,
- a simpler and more rugged design, making them easier to operate and less vulnerable to operational upsets,
- higher availability and longer operating life - typically 60 years,
- reduced possibility of core melt accidents,
- minimal effect on the environment,
- higher burn-up to reduce fuel use and the amount of waste,
- burnable absorbers ("poisons") to extend fuel life.

The greatest departure from second-generation designs is that many incorporate passive or inherent safety features which require no active controls or operational intervention to avoid accidents in the event of malfunction, and may rely on gravity, natural convection or resistance to high temperatures.

Advanced Thermal Reactors being marketed in 2008 are given in Table 1.6.

Traditional reactor safety systems are 'active' in the sense that they involve electrical or mechanical operation on command. Some engineered systems operate passively, e.g. pressure relief valves. They function without operator control and despite any loss of auxiliary power. Both require parallel redundant systems. Inherent or full passive safety depends only on physical phenomena such as convection, gravity or resistance to high temperatures, not on functioning of engineered components.

Country and Developer	Reactor	Size MWe	Design Progress	Main Features (improved safety in all)
US-Japan (GE-Hitachi, Toshiba)	ABWR	1300	Commercial operation in Japan since 1996-7. In US: NRC certified 1997, FOAKE.	Evolutionary design. More efficient, less waste. Simplified construction (48 months) and operation.
USA (Westinghouse)	AP-600 AP-1000 (PWR)	600 1100	AP-600: NRC certified 1999, FOAKE. AP-1000 NRC certification 2005.	Simplified construction and operation. 3 years to build. 60-year plant life.
France-Germany (Areva NP)	EPR US-EPR (PWR)	1600	Future French standard. French design approval. Being built in Finland. US version developed.	Evolutionary design. High fuel efficiency. Low cost electricity.
USA (GE)	ESBWR	1550	Developed from ABWR, under certification in USA	Evolutionary design. Short construction time.
Japan (Utilities, Mitsubishi)	APWR US-APWR EU-APWR	1530 1700 1700	Basic design in progress, planned for Tsuruga US design certification application 2008.	Hybrid safety features. Simplified Construction and operation.
South Korea (KHNP, derived from Westinghouse)	APR-1400 (PWR)	1450	Design certification 2003, First units expected to be operating c 2012.	Evolutionary design. Increased reliability. Simplified construction and operation.
Germany (Areva NP)	SWR-1000 BWR)	1200	Under development, pre-certification in USA	Innovative design. High fuel efficiency.
Russia (Gidropress)	VVER-1200 (PWR)	1200	Replacement for Leningrad and Novovoronezh plants	High fuel efficiency.
Russia (Gidropress)	V-392 (PWR)	950-1000	Two being built in India, Bid for China in 2005.	Evolutionary design. 60-year plant life.
Canada (AECL)	CANDU-6 CANDU-9	750 925+	Enhanced model Licensing approval 1997	Evolutionary design. Flexible fuel requirements. C-9: Single stand-alone unit.
Canada (AECL)	ACR	700 1080	undergoing certification in Canada	Evolutionary design. Light water cooling. Low-enriched fuel.
South Africa (Eskom, Westinghouse)	PBMR	170 (module)	prototype due to start building (Chinese 200 MWe counterpart under const.)	Modular plant, low cost. High fuel efficiency. Direct cycle gas turbine.
USA-Russia et al (General Atomics - OKBM)	GT-MHR	285 (module)	Under development in Russia by multinational joint venture	Modular plant, low cost. High fuel efficiency. Direct cycle gas turbine.

Table 1.6 Advanced Thermal Reactors Being Marketed in 2008

1.8.1 Licensing

Many of the future designs are larger than predecessors. Increasingly they involve international collaboration. Certification of designs is on a national basis, and is safety-based. In Europe there are moves towards harmonized requirements for licensing.

However, in Europe reactors may also be certified according to compliance with European Utilities Requirements (EUR). These are basically a utilities' wish list of some 5000 items needed for new nuclear plants. Plants certified as complying with EUR include Westinghouse AP1000, Gidropress' AES-92, Areva's EPR, GE's ABWR, Areva's SWR-1000, and Westinghouse BWR 90.

In the USA a number of reactor types have received Design Certification (see below) and others are in process: ESBWR from GE-Hitachi, US EPR from Areva and US-APWR from Mitsubishi. Early in 2008 the NRC said that beyond these three, six pre-application reviews would get underway until about 2010. These include IRIS from Westinghouse, PBMR from Eskom and 4S from Toshiba as well as Canada's ACR and General Atomics' GT-MHR apparently

Longer term, NRC expected to focus on the Next-Generation Nuclear Plant (NGNP) for the USA - essentially the Very High Temperature Reactor (VHTR) among the Generation IV designs.

Two major international initiatives have been launched to define future reactor and fuel cycle technology, mostly looking further ahead than the main subjects of this chapter.

Generation IV International Forum (GIF) is a US-led grouping set up in 2001 that has identified six reactor concepts for further investigation with a view to commercial deployment by 2030.

The IAEA's International Project on Innovative Nuclear Reactors and Fuel Cycles (INPRO) is focused more on developing country needs, and initially involved Russia rather than the USA, though the USA has now joined it. It is now funded through the IAEA budget.

At the commercial level, by the end of 2006 three major Western-Japanese alliances had formed to dominate much of the world reactor supply market:

- Areva with Mitsubishi Heavy Industries(MHI) in a major project and subsequently in fuel fabrication,
- General Electric with Hitachi,
- Westinghouse had become a 77% owned subsidiary of Toshiba (with Shaw group 20%).

Then in March 2008 Toshiba signed a technical cooperation agreement on civil nuclear power with Russia's Atomenergoprom- the single vertically-integrated state holding company for Russia's nuclear power sector created in 2007. This could lead to a "strategic partnership" and include designing and engineering of commercial nuclear power plants, as well as manufacturing and maintenance of large equipment.

1.8.2 Light Water Reactors

In the USA, the federal Department of Energy (DOE) and the commercial nuclear industry in the 1990s developed four advanced reactor types. Two of them fall into the category of large "evolutionary" designs that build directly on the experience of operating light water reactors in the USA, Japan and Western Europe. These reactors are in the 1300-megawatt range.

One is an advanced boiling water reactor (ABWR) derived from a General Electric design. Two examples built by Hitachi and two by Toshiba are in commercial operation in Japan, with another under construction there and two in Taiwan. Four more are planned in Japan and another two in the USA. Though GE and Hitachi have subsequently joined up, Toshiba retains some rights over the design. Both GE-Hitachi and Toshiba (with NRG Energy in USA) are marketing the design.

The other type, System 80+, is an advanced pressurized water reactor (PWR), which was ready for commercialization but is not now being promoted for sale. Eight System 80 reactors in South Korea incorporate many design features of the System 80+, which is the basis of the Korean Next Generation Reactor program, specifically the APR-1400 that is expected to be in operation soon after 2010 and marketed worldwide.

The US Nuclear Regulatory Commission (NRC) gave final design certification for both in May 1997, noting that they exceeded NRC "safety goals by several orders of magnitude". The ABWR has also been certified as meeting European requirements for advanced reactors.

Another, more innovative US advanced reactor is smaller - 600 MWe - and has passive safety features (its projected core damage frequency is nearly 1000 times less than today's NRC requirements). The Westinghouse Advanced Passive 600, AP-600, gained NRC final design certification in 1999.

These NRC approvals were the first such generic certifications to be issued and are valid for 15 years. As a result of an exhaustive public process, safety issues within the scope of the certified designs have been fully resolved and hence will not be open to legal challenge during licensing for particular plants. US utilities will be able to obtain a single NRC license to both construct and operate a reactor before construction begins.

Separate from the NRC process and beyond its immediate requirements, the US nuclear industry selected one standardized design in each category - the large ABWR and the medium-sized AP-600, for detailed first-of-a-kind engineering (FOAKE) work. The US$ 200 million program was half funded by DOE. It means that prospective buyers now have firm information on construction costs and schedules.

The Westinghouse AP-1000, scaled-up from the AP-600, received final design certification from the NRC in December 2005 - the first generation 3+ type to do so. It represents the culmination of a 1300 man-year and $440 million design and testing program. In May 2007

Westinghouse applied for UK generic design assessment (pre-licensing approval) based on the NRC design certification, and expressing its policy of global standardization. The application was supported by utilities including E.ON.

Overnight capital costs are projected at $1200 per kilowatt and modular design will reduce construction time to 36 months. The 1100 MWe AP-1000 generating costs are expected to be below US$ 3.5 cents/kWh and it has a 60-year operating life. It has been selected for building in China (4 units) and is under active consideration for building in Europe and USA, and is capable of running on a full MOX core if required.

General Electric has developed the ESBWR of 1390 MWe with passive safety systems, from its ABWR design. Originally the European Simplified Boiling Water Reactor, this is now known as the Economic & Simplified BWR (ESBWR) and a 1560 MWe version is at preliminary stage of NRC design certification in the USA, so that design approval was expected at the end of 2008, with formal certification 12 months later. It is favored for early US construction and could be operational in 2014. It uses 4.2% enriched fuel and has a design life of 60 years.

Another US-origin but international project which is a few years behind the AP-1000 is the International Reactor Innovative & Secure (IRIS). Westinghouse is leading a wide consortium developing it as an advanced 3rd Generation project. IRIS is a modular 335 MWe pressurized water reactor with integral steam generators and primary coolant system all within the pressure vessel. It is nominally 335 MWe but can be less, e.g. 100 MWe. Fuel is initially similar to present LWRs with 5% enrichment and burn-up of 60,000 MWd/t with fuelling interval of 3 to 3.5 years, but is designed ultimately for 10% enrichment and 80 GWd/t burn-up with an 8-year cycle, or equivalent MOX core. The core has low power density. IRIS could be deployed in the next decade, and US design certification is at pre-application stage. Multiple modules are expected to cost US$ 1000-1200 per kW for power generation, though some consortium partners are interested in desalination, one in district heating.

In Japan, the first two ABWRs, Kashiwazaki Kariwa-6 & 7, have been operating since 1996 and are expected to have a 60-year life. These GE-Hitachi-Toshiba units cost about US$ 2000/kW to build, and produce power at about US 7c/kWh. Two more started up in 2004 & 2005. Future ABWR units are expected to cost US$ 1700/kW. Several of the 1350 MWe units are under construction and planned in Japan and Taiwan.

To complement this ABWR Hitachi-GE has completed systems design for three more of the same type - 600, 900 and 1700 MWe versions of the 1350 MWe design. The smaller versions will have standardized features that reduce costs. Construction of the ABWR-600 is expected to take 34 months - significantly less than the 1350 MWe units.

Mitsubishi's large APWR (1538 MWe) - advanced PWR - was developed in collaboration with four utilities (Westinghouse was earlier involved). The first two are planned for Tsuruga. It is simpler, combines active and passive cooling systems to greater effect, and has over 55 GWd/t fuel burn-up. It will be the basis for the next generation of Japanese PWRs.

The US-APWR will be 1700 MWe, due to higher thermal efficiency (39%) and has 24 month refueling cycle and target cost of $1500/kW. US design certification application was in January 2008 with approval expected in 2011. The first units may be built for TXU at Comanche Peak near Dallas, Texas. In March 2008 MHI submitted the same design for EUR certification, as EU-APWR.

The Atmea joint venture has been established by Areva NP and Mitsubishi Heavy Industries to develop an 1100 MWe (net) three-loop PWR with extended fuel cycles, 37% thermal efficiency and the capacity to use mixed-oxide fuel only. Fuel cycle is 12-24 months and the reactor has load-following capability. They expect to have this ready for license application by 2010. The reactor is regarded as mid-sized relative to other generation III units and will be marketed primarily to countries embarking upon nuclear power programs.

In South Korea, the APR-1400 Advanced PWR design has evolved from the US System 80+ with enhanced safety and seismic robustness and was earlier known as the Korean Next-Generation Reactor. Design certification by the Korean Institute of Nuclear Safety was awarded in May 2003. The first of these 1450 MWe reactors will be Shin-Kori-3 & 4, expected to be operating about 2012. Fuel has burnable poison and will have up to 60 GWd/t burn-up. Projected cost is US$ 1400 per kilowatt, falling to $1200/kW in later units with 48-month construction time. Plant life is 60 years.

In Europe, several designs are being developed to meet the European Utility Requirements (EUR) of French and German utilities, which have stringent safety criteria. Areva NP (formerly Framatome ANP) has developed a large (1600 and up to 1750 MWe) European pressurized water reactor (EPR), which was confirmed in mid 1995 as the new standard design for France and received French design approval in 2004. It is derived from the French N4 and German Konvoi types and is expected to provide power about 10% cheaper than the N4. It will operate flexibly to follow loads, have fuel burn-up of 65 GWd/t and the highest thermal efficiency of any light water reactor, at 36%. It is capable of using a full core load of MOX. Availability is expected to be 92% over a 60-year service life. It has four separate, redundant safety systems rather than passive safety.

The first EPR unit is being built at Olkiluoto in Finland, the second at Flamanville in France. A US version, the US-EPR, is undergoing review in USA with intention of a design certification application in 2007. It is now known as the Evolutionary PWR (EPR). Overnight capital cost is quoted as $2400 per kilowatt, levelised over the first four units. Albeit due to delays in the construction program the costs of Olkiluoto have risen sharply.

Together with German utilities and safety authorities, Areva NP (Framatome ANP) is also developing another evolutionary design, the SWR 1000, a 1200-1290 MWe BWR with 60-year design life. The design was completed in 1999 and US certification was sought, but then deferred. As well as many passive safety features, the reactor is simpler overall and uses high-burn up fuels enriched to 3.54%, giving it refueling intervals of up to 24 months. It is ready for commercial deployment and the prospects of that will be helped by a 2008 agreement with Siemens and the major German utility E.On (Siemens built the Gundremmingen plant on which the design is based, for E.On).

Toshiba has been developing its evolutionary advanced BWR (1500 MWe) design, originally BWR 90+ from ABB then Westinghouse, working with Scandinavian utilities to meet EUR requirements.

In Russia, several-advanced reactor designs have been developed - advanced PWR with passive safety features.

Gidropress late-model VVER-1000 units with enhanced safety (AES 92 & 91 power plants) are being built in India and China. Two more are planned for Belene in Bulgaria. The AES-92 is certified as meeting EUR.

A third-generation standardized VVER-1200 reactor of 1150-1200 MWe is an evolutionary development of the well-proven VVER-1000 in the AES-92 plant, with longer life, greater power and efficiency. The lead units will be built at Novovoronezh II, to start operation in 2012-13 followed by Leningrad II for 2013-14. An AES-2006 plant will consist of two of these OKB Gidropress reactor units expected to run for 50 years with capacity factor of 90%. Capital cost is said to be US$ 1200/kW and construction time 54 months. They have enhanced safety including that related to earthquakes and aircraft impact with some passive safety features, double containment and core damage frequency of $1x10^{-7}$.

Atomenergoproekt say that the AES-2006 conforms to both Russian standards and European Utilities Requirements (EUR).

The VVER-1500 model was being developed by Gidropress. It will have 50-60 MWd/t burn-up and enhanced safety. Design was expected to be complete in 2007 but this schedule has slipped in favor of the evolutionary VVER-1200.

OKBM's VBER-300 PWR is a 295-325 MWe unit developed from naval power plants and was originally envisaged in pairs as a floating nuclear power plant. It is designed for 60-year life and 90% capacity factor. It now planned to develop it as a land-based unit with Kazatomprom, with a view to exports, and the first unit will be built in Kazakhstan.

Canada has had two designs under development that are based on its reliable CANDU-6 reactors, the most recent of which are operating in China.

The CANDU-9 (925-1300 MWe) was developed from this also as a single-unit plant. It has flexible fuel requirements ranging from natural uranium through slightly-enriched uranium, recovered uranium from reprocessing spent PWR fuel, mixed oxide (U & Pu) fuel, direct use of spent PWR fuel, to thorium. It may be able to burn military plutonium or actinides separated from reprocessed PWR/BWR waste. A two-year licensing review of the CANDU-9 design was successfully completed early in 1997, but the design has been shelved.

Some of the innovation of this, along with experience in building recent Korean and Chinese units, was then put back into the Enhanced CANDU-6 - built as twin units - with power increase to 750 MWe and flexible fuel options, plus 4.5 year construction and 60-year plant life (with mid-life pressure tube replacement). This is under consideration for new build in Ontario.

The Advanced Candu Reactor (ACR), a 3rd generation reactor, is a more innovative concept. While retaining the low-pressure heavy water moderator, it incorporates some features of the pressurized water reactor. Adopting light water cooling and a more compact core reduces capital cost, and because the reactor is run at higher temperature and coolant pressure, it has higher thermal efficiency.

The ACR-700 design was 700 MWe but is physically much smaller, simpler and more efficient as well as 40% cheaper than the CANDU-6. But the ACR-1000 of 1080-1200 MWe is now the focus of attention by AECL. It has more fuel channels (each of which can be regarded as a module of about 2.5 MWe). The ACR will run on low-enriched uranium (about 1.5-2.0% U-235) with high burn-up, extending the fuel life by about three times and reducing high-level waste volumes accordingly. It will also efficiently burn MOX fuel, thorium and actinides.

Regulatory confidence in safety is enhanced by a small negative void reactivity for the first time in CANDU, and utilizing other passive safety features as well as two independent and fast shutdown systems. Units will be assembled from prefabricated modules, cutting construction time to 3.5 years. ACR units can be built singly but are optimal in pairs. They will have 60 year design life overall but require mid-life pressure tube replacement.

ACR is moving towards design certification in Canada, with a view to following in China, USA and UK. In 2007 AECL applied for UK generic design assessment (pre-licensing approval). The first ACR-1000 unit is expected to be operating in 2016 in Ontario.

The CANDU X or SCWR is a variant of the ACR, but with supercritical light water coolant (e.g. 25 MPa and 625°C) to provide 40% thermal efficiency. The size range envisaged is 350 to 1150 MWe, depending on the number of fuel channels used. Commercialization envisaged after 2020.

India is developing the Advanced Heavy Water reactor (AHWR) as the third stage in its plan to utilize thorium to fuel its overall nuclear power program. The AHWR is a 300 MWe reactor moderated by heavy water at low pressure. The calandria has 500 vertical pressure tubes and the coolant is boiling light water circulated by convection. Each fuel assembly has 30 Th-U-233 oxide pins and 24 Pu-Th oxide pins around a central rod with burnable absorber. Burn-up of 24 GWd/t is envisaged. It is designed to be self-sustaining in relation to U-233 bred from Th-232 and have a low Pu inventory and consumption, with slightly negative void coefficient of reactivity. It is designed for 100-year plant life and is expected to utilize 65% of the energy of the fuel.

Once it is fully operational, each AHWR fuel assembly will have the fuel pins arranged in three concentric rings arranged:

Inner: 12 pins Th-U-233 with 3.0% U-233,
Intermediate: 18 pins Th-U-233 with 3.75% U-233,
Outer: 24 pins Th-Pu-239 with 3.25% Pu.

The fissile plutonium content will decrease from an initial 75% to 25% at equilibrium discharge burn-up level.

1.8.3 High Temperature Gas-Cooled Reactors

These reactors use helium as a coolant that at up to 950ºC drives a gas turbine for electricity and a compressor to return the gas to the reactor core. Fuel is in the form of TRISO particles less than a millimeter in diameter. Each has a kernel of uranium oxycarbide, with the uranium enriched up to 17% U-235. This is surrounded by layers of carbon and silicon carbide, giving a containment for fission products that is stable to 1600°C or more. These particles may be arranged: in blocks - hexagonal 'prisms' of graphite or in billiard ball-sized pebbles of graphite encased in silicon carbide.

As described earlier South Africa's Pebble Bed Modular Reactor (PBMR) is being developed by a consortium led by the utility Eskom, and drawing on German expertise. It aims for a step change in safety, economics and proliferation resistance. Production units will be 165 MWe. They will have a direct-cycle gas turbine generator and thermal efficiency about 42%. Up to 450,000 fuel pebbles recycle through the reactor continuously (about six times each) until they are expended, giving an average enrichment in the fuel load of 4-5% and average burn-up of 90 GWday/t U (eventual target burn-ups are 200 GWd/t). This means on-line refueling as expended pebbles are replaced, giving high capacity factor. The pressure vessel is lined with graphite and there is a central column of graphite as reflector. Control rods are in the side reflectors and cold shutdown units in the central column.

Performance includes great flexibility in loads (40-100%), with rapid change in power settings. Power density in the core is about one tenth of that in light water reactor, and if coolant circulation ceases the fuel will survive initial high temperatures while the reactor shuts itself down - giving inherent safety. Each unit will finally discharge about 19 tonnes/yr of spent pebbles to ventilated on-site storage bins.

Overnight construction cost (when in clusters of eight units) is expected to be US$ 1000/kW and generating cost below 3 US cents/kWh. Investors in the PBMR project are Eskom, the South African Industrial Development Corporation and Westinghouse. A demonstration plant is due to be built in 2007 for commercial operation in 2010.

A larger US design, the Gas Turbine - Modular Helium Reactor (GT-MHR), will be built as modules of 285 MWe each directly driving a gas turbine at 48% thermal efficiency. The cylindrical core consists of 102 hexagonal fuel element columns of graphite blocks with channels for helium and control rods. Graphite reflector blocks are both inside and around the core. Half the core is replaced every 18 months. Burn-up is about 100,000 MWd/t. It is being developed by General Atomics in partnership with Russia's Minatom, supported by Fuji (Japan). Initially it will be used to burn pure ex-weapons plutonium at Tomsk in Russia. The preliminary design stage was completed in 2001.

1.8.4 Fast Neutron Reactors

Several countries have research and development programs for improved Fast Breeder Reactors (FBR), which are a type of Fast Neutron Reactor. These use the uranium-238 in reactor fuel as well as the fissile U-235 isotope used in most reactors.

About 20 liquid metal-cooled FBRs have already been operating, some since the 1950s, and some supply electricity commercially. About 290 reactor-years of operating experience have been accumulated.

Natural uranium contains about 0.7 % U-235 and 99.3 % U-238. In any reactor the U-238 component is turned into several isotopes of plutonium during its operation. Two of these, Pu 239 and Pu 241, then undergo fission in the same way as U 235 to produce heat. In a fast neutron reactor this process is optimized so that it can 'breed' fuel, often using a depleted uranium blanket around the core. FBRs can utilize uranium at least 60 times more efficiently than a normal reactor.

They are however expensive to build and could only be justified economically if uranium prices were to rise to pre-1980 values, well above the current market price. For this reason research work on the 1450 MWe European FBR has almost ceased. Closure of the 1250 MWe French Superphenix FBR after very little operation over 13 years also set back developments.

In the UK, the Dounreay Fast Reactor started operation in 1959 using sodium-potassium coolant. The much larger Prototype Fast reactor that operated for 20 years until the UK Government withdrew funding followed this.

Research continues in India. At the Indira Gandhi Center for Atomic Research a 40 MWt fast breeder test reactor has been operating since 1985. In addition, the tiny Kamini there is employed to explore the use of thorium as nuclear fuel, by breeding fissile U-233. In 2004 construction of a 500 MWe prototype fast breeder reactor started at Kalpakkam. The unit is expected to be operating in 2010, fuelled with uranium-plutonium carbide (the reactor-grade Pu being from its existing PHWRs) and with a thorium blanket to breed fissile U-233. This will take India's ambitious thorium program to stage 2, and set the scene for eventual full utilization of the country's abundant thorium to fuel reactors.

Japan plans to develop FBRs, and its Joyo experimental reactor that has been operating since 1977 is now being boosted to 140 MWt. The 280 MWe Monju prototype commercial FBR was connected to the grid in 1995, but was then shut down due to a sodium leak.

The Russian BN-600 fast breeder reactor has been supplying electricity to the grid since 1981 and has the best operating and production record of all Russia's nuclear power units. It uses uranium oxide fuel and the sodium coolant delivers 550°C at little more than atmospheric pressure. The BN 350 FBR operated in Kazakhstan for 27 years and about half of its output was used for water desalination. Russia plans to reconfigure the BN-600 to burn the plutonium from its military stockpiles.

Construction has started at Beloyarsk on the first BN-800, a new larger (880 MWe) FBR from OKBM with improved features including fuel flexibility - U+Pu nitride, MOX, or metal, and with breeding ratio up to 1.3. It has much enhanced safety and improved economy - operating cost is expected to be only 15% more than VVER. It is capable of burning 2 tonnes of plutonium per year from dismantled weapons and will test the recycling of minor actinides in the fuel.

Russia has experimented with several lead-cooled reactor designs, and has used lead-bismuth cooling for 40 years in reactors for its 7 Alfa class submarines. Pb-208 (54% of naturally-occurring lead) is transparent to neutrons. A significant new Russian design is the BREST fast neutron reactor, of 300 MWe or more with C, and supercritical steam generators. It lead as the primary coolant, at 540 is inherently safe and uses a high-density U+Pu nitride fuel with no requirement for high enrichment levels. No weapons-grade plutonium can be produced (since there is no uranium blanket - all the breeding occurs in the core). The initial cores can comprise Pu and spent fuel - hence loaded with fission products, and radio logically 'hot'. Subsequently, any surplus plutonium, which is not in pure form, can be used as the cores of new reactors. Used fuel can be recycled indefinitely, with on-site reprocessing and associated facilities. A pilot unit is planned for Beloyarsk and 1200 MWe units are proposed.

In the USA, GE was involved in designing a modular 150 MWe liquid metal-cooled inherently safe reactor - PRISM. GE and Argonne have also been developing an advanced liquid-metal fast breeder reactor (ALMR) of over 1400 MWe, but both designs at an early stage were withdrawn from NRC review. No US fast neutron reactor has so far been larger than 66 MWe and none has supplied electricity commercially.

The Super-PRISM is a GE advanced reactor design for compact modular pool-type reactors with passive cooling and decay C, heat removal. Modules are 1000 MWt and operate at higher temperature - 510 than the original PRISM. The pool-type modules contain the complete primary system with sodium coolant. The Pu & DU fuel can be oxide or metal, but minor actinides are not removed in reprocessing so that even fresh fuel is intensely radioactive and hence resistant to misappropriation. The fission products are removed in reprocessing and resultant wastes are shorter-lived than usual. Fuel stays in the reactor six years, with one third removed every two years. The commercial plant concept uses six reactor modules to provide 2280 MWe, and the design meets Generation IV criteria including generation cost of less than 3 cents/kWh.

Korea's KALIMER (Korea Advanced Liquid Metal Reactor) is a 600 MWe pool type sodium-cooled fast reactor designed to operate at over 500°C. It has evolved from a 150 MWe version. It has a transmuter core, and no breeding blanket is involved. Future development of KALIMER as a Generation IV type is envisaged.

In the USA Mitsubishi Heavy Industries (MHI) is involved with a consortium to develop the Advanced Recycling Reactor, a fast reactor that will burn actinides with uranium and plutonium. This will be based on MHI's Japan Standard Fast reactor concept, though with breeding ration less than 1:1. In this connection MHI has also set up Mitsubishi FBR Systems (MFBR).

1.8.5 Accelerator-Driven Systems

A recent development has been the merging of accelerator and fission reactor technologies to generate electricity and transmute long-lived radioactive wastes.

A high-energy proton beam hitting a heavy metal target produces neutrons by spallation. The neutrons cause fission in the fuel, but unlike a conventional reactor, the fuel is sub-critical, and fission ceases when the accelerator is turned off. The fuel may be uranium, plutonium or thorium, possibly mixed with long-lived wastes from conventional reactors.

Many technical and engineering questions remain to be explored before the potential of this concept can be demonstrated.

1.9 Generation IV Nuclear Reactors

The Generation IV International Forum (GIF) was initiated in 2000 and formally chartered in mid 2001. It is an international collective representing governments of countries where nuclear energy is significant now and also seen as vital for the future. They are committed to joint development of the next generation of nuclear technology. Led by the USA, Argentina, Brazil, Canada, France, Japan, South Korea, South Africa, Switzerland, and the UK are members of the GIF, along with the EU. Russia and China were admitted in 2006.

After some two years' deliberation, GIF (then representing ten countries) late in 2002 announced the selection of six reactor technologies that they believe represent the future shape of nuclear energy. These are selected on the basis of being clean, safe and cost-effective means of meeting increased energy demands on a sustainable basis, while being resistant to diversion of materials for weapons proliferation and secure from terrorist attacks. They will be the subject of further development internationally.

In addition to selecting these six concepts for deployment between 2010 and 2030, the GIF recognized a number of International Near-Term Deployment advanced reactors available before 2015.

Most of the six systems employ a closed fuel cycle to maximize the resource base and minimize high-level wastes to be sent to a repository. Three of the six are fast reactors and one can be built as a fast reactor, one is described as epithermal, and only two operate with slow neutrons like today's plants.

Only one is cooled by light water, two are helium-cooled and the others have lead-bismuth, sodium or fluoride salt coolant. The latter three operate at low pressure, with significant safety advantage. The last has the uranium fuel dissolved in the circulating coolant. Temperatures range from 510°C to 1000°C, compared with less than 330°C for today's light water reactors, and this means that four of them can be used for thermo chemical hydrogen production.

The sizes range from 150 to 1500 MWe (or equivalent thermal), with the lead-cooled one optionally available as a 50-150 MWe "battery" with long core life (15-20 years without refueling) as replaceable cassette or entire reactor module. This is designed for distributed generation or desalination.

At least four of the systems have significant operating experience already in most respects of their design, which may mean that they can be in commercial operation well before 2030.

In February 2005 five of the participants signed an agreement to take forward the R&D on the six technologies. The USA, Canada, France, Japan and UK agreed to undertake joint research and exchange technical information.

While Russia was not initially part of GIF, one design corresponds with the BREST reactor being developed there, and Russia is now the main operator of the sodium-cooled fast reactor for electricity - another of the technologies put forward by the GIF.

India is also not involved with the GIF but is developing its own advanced technology to utilize thorium as a nuclear fuel. A three-stage program has the first stage well established, with Pressurized Heavy Water Reactors (PHWRs, elsewhere known as CANDUs) fuelled by natural uranium to generate plutonium. Then Fast Breeder Reactors (FBRs) use this plutonium-based fuel to breed U-233 from thorium, and finally advanced nuclear power systems will use the U-233. The spent fuel will be reprocessed to recover fissile materials for recycling. The two major options for the third stage, while continuing with the PHWR and FBR programs, are an Advanced Heavy Water Reactor and sub critical Accelerator-Driven Systems.

Closely related to GIF is the Multinational Design Evaluation Program (MDEP) set up in 2005, led by the OECD Nuclear Energy Agency and involving the IAEA. It aims to develop multinational regulatory standards for design of Gen IV reactors. The US Nuclear Regulatory Commission (NRC) has proposed a three-stage process culminating in international design certification for these. Ten countries are involved so far: Canada, China, Finland, France, Japan, Korea, Russia, South Africa, UK, USA, but others which have or are likely to have firm commitments to building new nuclear plants may be admitted. In September 2007 the NRC called for countries involved in development of Gen IV reactors to move to stage 3 of design evaluation, which means developing common design requirements so that regulatory standards can be harmonized. NRC has published its draft design requirements.

1.9.1 Generation IV International Forum Reactor Technologies

Gas-cooled fast reactors; like other helium-cooled reactors which have operated or are under development, these will be high-temperature units - 850°C, suitable for power generation, thermo chemical hydrogen production or other process heat. For electricity, the gas will directly drive a gas turbine (Brayton cycle). Fuels would include depleted uranium and any other fissile or fertile materials. Spent fuel would be reprocessed on site and all the actinides recycled to minimize production of long-lived radioactive wastes.

Lead-cooled fast reactors; liquid metal (Pb or Pb-Bi) cooling is by natural convection. Fuel is depleted uranium metal or nitride, with full actinide recycle from regional or central reprocessing plants. A wide range of unit sizes is envisaged, from factory-built "battery" with 15-20 year life for small grids or developing countries, to modular 300-400 MWe units and large single plants of 1400 MWe. Operating temperature of 550°C is readily achievable but 800°C is envisaged with advanced materials and this would enable thermo chemical hydrogen production.

This corresponds with Russia's BREST fast reactor technology that is lead-cooled and builds on 40 years experience of lead-bismuth cooling in submarine reactors. Its fuel is U+Pu nitride. More immediately the GIF proposal appears to arise from two experimental designs: the US STAR and Japan's LSPR, these being lead and lead-bismuth cooled respectively.

Molten salt reactors; the uranium fuel is dissolved in the sodium fluoride salt coolant which circulates through graphite core channels to achieve some moderation and an epithermal neutron spectrum. Fission products are removed continuously and the actinides are fully recycled, while plutonium and other actinides can be added along with U-238. Coolant temperature is 700°C at very low pressure, with 800°C envisaged. A secondary coolant system is used for electricity generation, and thermo chemical hydrogen production is also feasible.

During the 1960s the USA developed the molten salt breeder reactor as the primary back-up option for the conventional fast breeder reactor and a small prototype was operated. Recent work has focused on lithium and beryllium fluoride coolant with dissolved thorium and U-233 fuel. The attractive features of the MSR fuel cycle include: the high-level waste comprising fission products only, hence shorter-lived radioactivity; small inventory of weapons-fissile material (Pu-242 being the dominant Pu isotope); low fuel use (the French self-breeding variant claims 50kg of thorium and 50kg U-238 per billion kWh); and safety due to passive cooling up to any size.

Sodium-cooled fast reactors; this builds on more than 300 reactor-years experienced with fast neutron reactors over five decades and in eight countries. It utilizes depleted uranium in the fuel and has a coolant temperature of 550°C enabling electricity generation via a secondary sodium circuit, the primary one being at near atmospheric pressure. Two variants are proposed: a 150-500 MWe type with actinides incorporated into a metal fuel requiring pyrometallurgical processing on site, and a 500-1500 MWe type with conventional MOX fuel reprocessed in conventional facilities elsewhere.

Early in 2008, the USA, France and Japan signed an agreement to expand their cooperation on the development of sodium-cooled fast reactor technology. The agreement relates to their collaboration in the Global Nuclear Energy Partnership, aimed at closing the nuclear fuel cycle through the use of advanced reprocessing and fast reactor technologies, and seeks to avoid duplication of effort.

Supercritical water-cooled reactors. This is a very high-pressure water-cooled reactor that operates above the thermodynamic critical point of water to give a thermal efficiency about one third higher than today's light water reactors from which the design evolves. The supercritical water (25 MPa and 510-550°C) directly drives the turbine, without any secondary steam system. Passive safety features are similar to those of simplified boiling water reactors. Fuel is uranium oxide, enriched in the case of the open fuel cycle option. However, it can be built as a fast reactor with full actinide recycle based on conventional reprocessing. Most research on the design has been in Japan.

Very high-temperature gas reactors; these are graphite-moderated, helium-cooled reactors, based on substantial experience. The core can be built of prismatic blocks such as the Japanese HTTR and the GTMHR under development by General Atomics and others in Russia, or it may be pebble bed such as the Chinese HTR-10 and the PBMR under development in South Africa, with international partners. Outlet temperature of 1000°C enables thermo chemical hydrogen production via an intermediate heat exchanger, with electricity cogeneration, or direct high-efficiency driving of a gas turbine (Brayton cycle). There is some flexibility in fuels, but no recycle. Modules of 600 MW thermal are envisaged.

1.9.2 INPRO

As well as the GIF, another program with similar aims is coordinated by the IAEA. This is the International Project on Innovative Nuclear Reactors and Fuel Cycles (INPRO). It was launched in 2001 and has 22 members including Russia, aiming "to support the safe, sustainable, economic and proliferation-resistant use of nuclear technology to meet the global energy needs of the 21st century." It does this by examining issues related to the development and deployment of Innovative Nuclear Energy Systems (INS) for sustainable energy supply.

One of the case studies in phase 1 of INPRO was undertaken by Russia on its BN-800 fast reactor, though the emphasis was on the methodology rather than the technology. Nevertheless, fast reactor systems will feature in further INPRO work.

1.9.3 Global Nuclear Energy Partnership (GNEP)

This concept, announced in 2006, builds on earlier US work with the Integral Fast Reactor (IFR) project and international work on fast reactors. Its main thrust in to counter proliferation concerns, but will have the effect of much greater resource utilization as well.

It envisages fabrication and leasing of fuel for conventional reactors, with the used fuel being returned to fuel supplier countries and pyro-processed to recover uranium and actinides, leaving only fission products as high-level waste. The actinide mix is then burned in on-site fast reactors.

1.10 The Hydrogen Economy

1.10.1 Nuclear Energy and Hydrogen Production

Hydrogen can be burned in a normal internal combustion engine, and some test cars are thus equipped. Trials in aircraft have also been carried out. However, its main use is likely to be in fuel cells.

A fuel cell is conceptually a refuelable battery, making electricity as a direct product of a chemical reaction. But where the normal battery has all the active ingredients built in at the factory, fuel cells are supplied with fuel from an external source. They catalyze the oxidation of hydrogen directly to electricity at relatively low temperatures and the claimed theoretical efficiency of converting chemical to electrical energy to drive the wheels is about 60% (or more). However, in practice about half that has been achieved, except for the higher-temperature solid oxide fuel cells – 46%.

Like electricity, hydrogen is an energy carrier (but not a primary energy source). As oil becomes more expensive, hydrogen may replace it as a transport fuel and in other application. This development becomes more likely as fuel cells are developed, with hydrogen as the preferred fuel. If gas also becomes expensive, or constraints are put on carbon dioxide emissions, non-fossil sources of hydrogen will become necessary.

Similar to electricity, hydrogen for transport use will tend to be produced near where it is to be used. This will have major geo-political implications as industrialized countries become less dependent on oil and gas from distant parts of the world.

While a growing hydrogen economy already exists, linked to the worldwide chemical and refining industry, a much greater one is in sight. With new uses for hydrogen as a fuel, the primary energy demand for its production may approach that for electricity production.

Certainly the most exciting future prospect is that of cogeneration of electricity and hydrogen, so called "hydricity". If we are to address the problem of global warming then we have to reduce carbon emissions from both the electricity generation and transport sectors. The use of hydrogen, via fuel cells for transportation, has great potential. Here co generating nuclear reactors can provide the answer to reducing CO_2 production from generation and transportation. It is then possible that nuclear's role in hydrogen production may evolve in several stages, such as the use of nuclear heat to assist steam reforming of natural gas, then the high-temperature electrolysis of steam, using heat and electricity from nuclear reactors, then finally, high-temperature thermo chemical production using nuclear heat from high temperature reactors. So because reactor types, such as the VHTR, can produce hydrogen as a by-product of electricity production these could be a potential source of reasonably priced hydrogen, created without emitting carbon dioxide, which can be used to fuel the hydrogen powered vehicles that are already being developed. This is a long-term solution.

The advent of the hydrogen economy also has another incidental benefit for the nuclear industry in that it can stabilize the economics by shielding them from the problems in the market place due to it being a base load generator. This is because, whilst electricity is sold in a simple commodity market, it cannot be stored. However, the hydrogen economy solves this problem, because whilst a nuclear generator operates continuously any excess power, at times of light load, can be diverted into electrolysis plants to generate hydrogen. Incidentally this is why pumped storage and nuclear are natural partners.

Thus whilst a VHTR plant will generate electricity and hydrogen 24 hours per day, hydrogen production can be increased at night by producing hydrogen rather than electricity. There will also be the advantage that electricity can be sold according to market price. This is a complete reversal of the situation today and has the potential to permanently enhance the economics for nuclear power generation.

1.11 The Nuclear Fuel Cycle

The various activities associated with the production of electricity from nuclear reactions are referred to collectively as the nuclear fuel cycle. The nuclear fuel cycle starts with the mining of uranium and ends with the disposal of nuclear waste. With the reprocessing of used fuel as an option for nuclear energy, the stages form a true cycle (Figure 1.17).

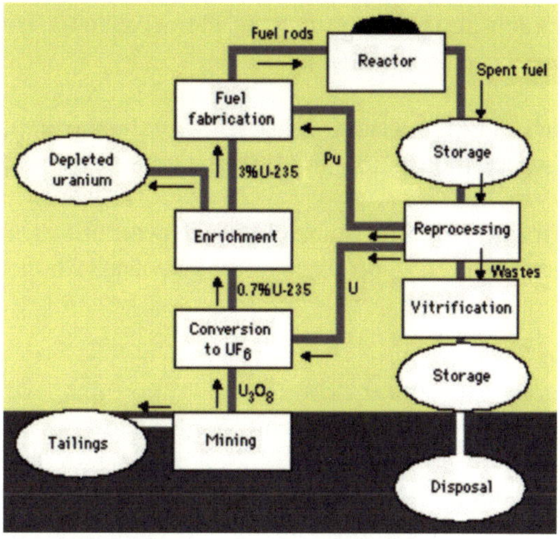

Figure 1.17 The Nuclear Fuel Cycle

1.11.1 Uranium

Uranium is a slightly radioactive metal that occurs throughout the earth's crust. It is about 500 times more abundant than gold and about as common as tin. It is present in most rocks and soils as well as in many rivers and in seawater. It is, for example, found in concentrations of about four parts per million (ppm) in granite, which makes up 60% of the earth's crust. In fertilizers, uranium concentration can be as high as 400 ppm (0.04%), and some coal deposits contain uranium at concentrations greater than 100 ppm (0.01%). Most of the radioactivity associated with uranium in nature is in fact due to other minerals derived from it by radioactive decay processes, and which are left behind in mining and milling.

There are a number of areas around the world where the concentration of uranium in the ground is sufficiently high that extraction of it for use as nuclear fuel is economically feasible. Such concentrations are called ore.

1.11.2 Uranium Mining

Both excavation and in situ techniques are used to recover uranium ore. Excavation may be underground and open pit mining.

In general, open pit mining is used where deposits are close to the surface and underground mining is used for deep deposits, typically greater than 120 m deep. Open pit mines require large holes on the surface, larger than the size of the ore deposit, since the walls of the pit must be sloped to prevent collapse. As a result, the quantity of material that must be removed in order to access the ore may be large. Underground mines have relatively small surface disturbance and the quantity of material that must be removed to access the ore is considerably less than in the case of an open pit mine.

An increasing proportion of the world's uranium now comes from in situ leaching (ISL), where oxygenated groundwater is circulated through a very porous orebody to dissolve the uranium and bring it to the surface. ISL may be with slightly acid or with alkaline solutions to keep the uranium in solution. The uranium is then recovered from the solution as in a conventional mill.

The decision as to which mining method to use for a particular deposit is governed by the nature of the orebody, safety and economic considerations.

In the case of underground uranium mines, special precautions, consisting primarily of increased ventilation, are required to protect against airborne radiation exposure.

1.11.3 Uranium Milling

Milling, which is generally carried out close to a uranium mine, extracts the uranium from the ore. Most mining facilities include a mill, although where mines are close together, one mill may process the ore from several mines. Milling produces a uranium oxide concentrate that is shipped from the mill. It is sometimes referred to as 'yellowcake' and generally contains more than 80% uranium. The original ore may contain as little as 0.1% uranium.

In a mill, uranium is extracted from the crushed and ground-up ore by leaching, in which either a strong acid or a strong alkaline solution is used to dissolve the uranium. The uranium is then removed from this solution and precipitated. After drying and usually heating it is packed in 200-litre drums as a concentrate.

The remainder of the ore, containing most of the radioactivity and nearly all the rock material, becomes tailings, which are emplaced in engineered facilities near the mine (often in mined out pit). Tailings contain long-lived radioactive materials in low concentrations and toxic materials such as heavy metals; however, the total quantity of radioactive elements is less than in the original ore, and their collective radioactivity will be much shorter-lived. These materials need to be isolated from the environment.

1.11.4 Conversion

The product of a uranium mill is not directly usable as a fuel for a nuclear reactor. Additional processing, generally referred to as enrichment, is required for most kinds of reactors. This process requires uranium to be in gaseous form and the way this is achieved is to convert it to uranium hex fluoride, which is a gas at relatively low temperatures.

At a conversion facility, uranium is first refined to uranium dioxide, which can be used as the fuel for those types of reactors that do not require enriched uranium. Most is then converted into uranium hex fluoride, ready for the enrichment plant. It is shipped in strong metal containers. The main hazard of this stage of the fuel cycle is the use of hydrogen fluoride.

1.11.5 Enrichment

Natural uranium consists, primarily, of a mixture of two isotopes (atomic forms) of uranium. Only 0.7% of natural uranium is "fissile", or capable of undergoing fission, the process by which energy is produced in a nuclear reactor. The fissile isotope of uranium is uranium 235 (U-235). The remainder is uranium 238 (U-238).

In the most common types of nuclear reactors, a higher than natural concentration of U-235 is required. The enrichment process produces this higher concentration, typically between 3.5% and 5% U-235, by removing over 85% of the U-238. Separating gaseous uranium hex fluoride into two streams, one being enriched to the required level and known as low-enriched uranium does this. The other stream is progressively depleted in U-235 and is called 'tails'.

There are two enrichment processes in large-scale commercial use, each of which uses uranium hex fluoride as feed: gaseous diffusion and gas centrifuge. They both use the physical properties of molecules, specifically the 1% mass difference, to separate the isotopes. The product of this stage of the nuclear fuel cycle is enriched uranium hex fluoride, which is reconverted to produce enriched uranium oxide.

1.11.6 Fuel Fabrication

Reactor fuel is generally in the form of ceramic pellets. These are formed from pressed uranium oxide that is sintered (baked) at a high temperature (over 1400°C). The pellets are then encased in metal tubes to form fuel rods, which are arranged into a fuel assembly ready for introduction into a reactor. The dimensions of the fuel pellets and other components of the fuel assembly are precisely controlled to ensure consistency in the characteristics of fuel bundles.

In a fuel fabrication plant great care is taken with the size and shape of processing vessels to avoid criticality (a limited chain reaction releasing radiation). With low-enriched fuel criticality is most unlikely, but in plants handling special fuels for research reactors this is a vital consideration.

1.11.7 Uranium Requirements

Table 1.7 is a review of current uranium requirements.

Concerns are being expressed about the uranium requirement for future nuclear generation.

The uranium resource is sustainable, with adequate known resources being continuously replenished at least as fast as they are being used. The essential dynamic is the strength of market forces when the market is constantly evolving through advances in human knowledge and the technologies of exploration, mining, and resource utilization. Depletion of today's known uranium resources will be more than counterbalanced by replenishment from new discoveries, technical progress and possible substitution.

Country and Developer	Reactor	Size MWe	Design Progress	Main Features (improved safety in all)
US-Japan (GE-Hitachi, Toshiba)	ABWR	1300	Commercial operation in Japan since 1996-7. In US: NRC certified 1997, FOAKE.	Evolutionary design. More efficient, less waste. Simplified construction (48 months) and operation.
USA (Westinghouse)	AP-600 AP-1000 (PWR)	600 1100	AP-600: NRC certified 1999, FOAKE. AP-1000 NRC certification 2005.	Simplified construction and operation. 3 years to build. 60-year plant life.
France-Germany (Areva NP)	EPR US-EPR (PWR)	1600	Future French standard. French design approval. Being built in Finland. US version developed.	Evolutionary design. High fuel efficiency. Low cost electricity.
USA (GE)	ESBWR	1550	Developed from ABWR, under certification in USA	Evolutionary design. Short construction time.
Japan (Utilities, Mitsubishi)	APWR US-APWR EU-APWR	1530 1700 1700	Basic design in progress, planned for Tsuruga US design certification application 2008.	Hybrid safety features. Simplified Construction and operation.
South Korea (KHNP, derived from Westinghouse)	APR-1400 (PWR)	1450	Design certification 2003, First units expected to be operating c 2012.	Evolutionary design. Increased reliability. Simplified construction and operation.
Germany (Areva NP)	SWR-1000 BWR)	1200	Under development, pre-certification in USA	Innovative design. High fuel efficiency.
Russia (Gidropress)	VVER-1200 (PWR)	1200	Replacement for Leningrad and Novovoronezh plants	High fuel efficiency.
Russia (Gidropress)	V-392 (PWR)	950-1000	Two being built in India, Bid for China in 2005.	Evolutionary design. 60-year plant life.
Canada (AECL)	CANDU-6 CANDU-9	750 925+	Enhanced model Licensing approval 1997	Evolutionary design. Flexible fuel requirements. C-9: Single stand-alone unit.
Canada (AECL)	ACR	700 1080	undergoing certification in Canada	Evolutionary design. Light water cooling. Low-enriched fuel.
South Africa (Eskom, Westinghouse)	PBMR	170 (module)	prototype due to start building (Chinese 200 MWe counterpart under const.)	Modular plant, low cost. High fuel efficiency. Direct cycle gas turbine.
USA-Russia et al (General Atomics - OKBM)	GT-MHR	285 (module)	Under development in Russia by multinational joint venture	Modular plant, low cost. High fuel efficiency. Direct cycle gas turbine.

Sources: Reactor data: WNA to 23/05/08.
IAEA- for nuclear electricity production & percentage of electricity (% e) 5/07.
WNA: Global Nuclear Fuel Market (reference scenario) - for U.

Table 1.7 World Nuclear Power Reactors 2006-2008 and Uranium Requirements

In addition, a huge increase in efficiency is readily possible through the technological step to fast neutron reactors. This option - unique among mineral resources - offers the nuclear industry a special kind of insurance against future resource shortage.

It may therefore be fairly concluded that uranium supplies will be more than adequate to fuel foreseeable expansions of nuclear power.

1.12 Thorium as a Nuclear Fuel

Thorium, as well as uranium, can be used as a nuclear fuel. Although not fissile itself, thorium-232 (Th-232) will absorb slow neutrons to produce uranium-233 (U-233), which is fissile (and long-lived). Hence like uranium-238 (U-238) it is fertile.

In one significant respect U-233 is better than uranium-235 and plutonium-239, because of its higher neutron yield per neutron absorbed. Given a start with some other fissile material (U-235 or Pu-239), a breeding cycle similar to but more efficient than that with U-238 and plutonium (in normal, slow-neutron reactors) can be set up. However, there are also features of the neutron economy that counter this advantage. In particular Pa-233 is a neutron absorber that diminishes U-233 yield. The Th-232 absorbs a neutron to become Th-233 that quickly beta decays to protactinium-233 and then more slowly to U-233. The irradiated fuel can then be unloaded from the reactor, the U-233 separated from the thorium, and fed back into another reactor as part of a closed fuel cycle.

Over the last 30 years there has been interest in utilizing thorium as a nuclear fuel since it is more abundant in the Earth's crust than uranium. Also, all of the mined thorium is potentially useable in a reactor, compared with the 0.7% of natural uranium, so some 40 times the amount of energy per unit mass might theoretically be available (without recourse to fast breeder reactors).

A major potential application for conventional Pars involves fuel assemblies arranged so that a blanket of mainly thorium fuel rods surrounds a more-enriched seed element containing U-235 that supplies neutrons to the sub critical blanket. As U-233 is produced in the blanket it is burned there. This is the Light Water Breeder Reactor concept that was successfully demonstrated in the USA in the 1970s.

It is currently being developed in a more deliberately proliferation-resistant way. The central seed region of each fuel assembly will have uranium enriched to 20% U-235. The blanket will be thorium with some U-238, which means that any uranium chemically separated from it (for the U-233) is not useable for weapons. Spent blanket fuel also contains U-232, which decays rapidly and has very gamma-active daughters creating significant problems in handling the bred U-233 and hence conferring proliferation resistance. Plutonium produced in the seed will have a high proportion of Pu-238, generating a lot of heat and making it even more unsuitable for weapons than normal reactor-grade Pu.

A variation of this is the use of whole homogeneous assembles arranged so that a set of them makes up a seed and blanket arrangement. If the seed fuel is metal uranium alloy instead of oxide, there is better heat conduction to cope with its higher temperatures. Seed fuel remains three years in the reactor, blanket fuel for up to 14 years.

Since the early 1990s Russia has had a program to develop a thorium-uranium fuel, which more recently has moved to have a particular emphasis on utilization of weapons-grade plutonium in a thorium-plutonium fuel.

The thorium-plutonium fuel claims four advantages over MOX: proliferation resistance, compatibility with existing reactors - which will need minimal modification to be able to burn it and the fuel can be made in existing plants in Russia. In addition, a lot more plutonium can be put into a single fuel assembly than with MOX, so that three times as much can be disposed of as when using MOX. The spent fuel amounts to about half the volume of MOX and is even less likely to allow recovery of weapons-useable material than spent MOX fuel, since less fissile plutonium remains in it. With an estimated 150 tonnes of weapons plutonium in Russia, the thorium-plutonium project would not necessarily cut across existing plans to make MOX fuel.

1.12.1 Thorium R&D History

The use of thorium-based fuel cycles has been studied for about 30 years, but on a much smaller scale than uranium or uranium/plutonium cycles. Basic research and development has been conducted in Germany, India, Japan, Russia, the UK and the USA. Test reactor irradiation of thorium fuel to high burnups has also been conducted and several test reactors have either been partially or completely loaded with thorium-based fuel.

Noteworthy experiments involving thorium fuel include the following, the first three being high-temperature gas-cooled reactors:

- Between 1967 and 1988, the AVR (Atom Versuchs Reaktor) experimental pebble bed reactor at Julich, Germany, operated for over 750 weeks at 15 MWe, about 95% of the time with thorium-based fuel. The fuel used consisted of about 100 000 billiard ball-sized fuel elements. Overall a total of 1360 kg of thorium was used, mixed with high-enriched uranium (HEU). Maximum burnups of 150,000 MWd/t were achieved.
- Thorium fuel elements with a 10:1 Th/U (HEU) ratio were irradiated in the 20 MWth Dragon reactor at Winfrith, UK, for 741 full power days. Dragon was run as an OECD/Euratom cooperation project, involving Austria, Denmark, Sweden, Norway and Switzerland in addition to the UK, from 1964 to 1973. The Th/U fuel was used to 'breed and feed', so that the U-233 formed replaced the U-235 at about the same rate, and fuel could be left in the reactor for about six years.
- General Atomics' Peach Bottom high-temperature, graphite-moderated, helium-cooled reactor (HTGR) in the USA operated between 1967 and 1974 at 110 MWth, using high-enriched uranium with thorium.
- In India, the Kamini 30 kWth experimental neutron-source research reactor using U-233, recovered from ThO_2 fuel irradiated in another reactor, started up in 1996 near Kalpakkam. The reactor was built adjacent to the 40 MWt Fast Breeder Test Reactor, in which the ThO_2 is irradiated.

- In the Netherlands, an aqueous homogenous suspension reactor has operated at 1MWth for three years. The HEU/Th fuel is circulated in solution and reprocessing occurs continuously to remove fission products, resulting in a high conversion rate to U-233.
- There have been several experiments with fast neutron reactors.

1.12.2 Thorium Power Reactors

Much experience has been gained in thorium-based fuel in power reactors around the world, some using high-enriched uranium (HEU) as the main fuel:

- The 300 MWe THTR (Thorium High-Temperature Reactor) reactor in Germany was developed from the AVR and operated between 1983 and 1989 with 674,000 pebbles, over half containing Th/HEU fuel (the rest graphite moderator and some neutron absorbers). These were continuously recycled on load and on average the fuel passed six times through the core. Fuel fabrication was on an industrial scale.
- The Fort St Vrain reactor was the only commercial thorium-fuelled nuclear plant in the USA, also developed from the AVR in Germany, and operated 1976 - 1989. It was a high-temperature (700°C), graphite-moderated, helium-cooled reactor with a Th/HEU fuel designed to operate at 842 MWth (330 MWe). The fuel was in micro spheres of thorium carbide and Th/U-235 carbide coated with silicon oxide and pyrolytic carbon to retain fission products. It was arranged in hexagonal columns ('prisms') rather than as pebbles. Almost 25 tonnes of thorium was used in fuel for the reactor, and this achieved 170,000 MWd/t burn-up.
- Thorium-based fuel for Pressurized Water Reactors (PWRs) was investigated at the Shippingport reactor in the USA using both U-235 and plutonium as the initial fissile material. It was concluded that thorium would not significantly affect operating strategies or core margins. The light water breeder reactor (LWBR) concept was also successfully tested here from 1977 to 1982 with thorium and U-233 fuel clad with Zircaloy using the 'seed/blanket' concept.
- The 60 MWe Lingen Boiling Water Reactor (BWR) in Germany utilized Th/Pu-based fuel test elements.

1.12.3 Emerging Advanced Thorium Reactor Concepts

Concepts for advanced reactors based on thorium-fuel cycles include:

- Light Water Reactors; with fuel based on plutonium oxide (PuO_2), thorium oxide (ThO_2) and/or uranium oxide (UO_2) particles arranged in fuel rods.
- High-Temperature Gas-cooled Reactors (HTGR) of two kinds: pebble bed and with prismatic fuel elements.

  Gas Turbine-Modular Helium Reactor (GT-MHR); research on HTGRs in the USA led to a concept using a prismatic fuel. The use of helium as a coolant at high temperature, and the relatively small power output per module (600 MWth), permit direct coupling of the MHR to a gas turbine (a Brayton cycle), resulting in generation at almost 50% thermal efficiency. The GT-MHR core can accommodate a wide range of fuel options, including HEU/Th, U-233/Th and Pu/Th. The use of HEU/Th fuel was demonstrated in the Fort St Vrain reactor (see above).

Pebble-Bed Modular reactor (PBMR) - Arising from German work the PBMR was conceived in South Africa and is now being developed by a multinational consortium. It can potentially use thorium in its fuel pebbles.

- Molten salt reactors (MSR) - This is an advanced breeder concept, in which the fuel is a molten mixture of lithium and beryllium fluoride salts with dissolved enriched uranium, thorium or U-233 fluorides. The core consists of unclad graphite moderator arranged to allow the flow of salt at some 700°C and at low pressure. Heat is transferred to a secondary salt circuit and thence to steam. It is not a fast reactor, but with some moderation by the graphite is epithermal (intermediate neutron speed). The fission products dissolve in the salt and are removed continuously in an on-line reprocessing loop and replaced with Th-232 or U-238. Actinides remain in the reactor until they fission or are converted to higher actinides which do so. The MSR was studied in depth in the 1960s, but is now being revived because of the availability of advanced technology for the materials and components.
- There is now renewed interest in the MSR concept in Japan, Russia, France and the USA, and one of the six generation IV designs selected for further development is the MSR. In 2002 a Thorium MSR was designed in France with a fissile zone where most power would be produced and a surrounding fertile zone where most conversion of Th-232 to U-233 would occur.
- Advanced Heavy Water Reactor (AHWR); India is working on this, and like the Canadian ACR the 300 MWe design is light water cooled. The main part of the core is sub critical with Th/U-233 oxide and Th/Pu-239 oxide, mixed so that the system is self-sustaining in U-233. The initial core will be entirely Th-Pu-239 oxide fuel assemblies, but as U-233 is available, 30 of the fuel pins in each assembly will be Th-U-233 oxide, arranged in concentric rings. It is designed for 100-year plant life and is expected to utilize 65% of the energy of the fuel. About 75% of the power will come from the thorium.
- CANDU-type reactors; AECL is researching the thorium fuel cycle application to enhanced CANDU-6 and ACR-1000 reactors. With 5% plutonium (reactor grade) plus thorium high burn-up and low power costs are indicated.
- Plutonium disposition; today MOX (U,Pu) fuels are used in some conventional reactors, with Pu-239 providing the main fissile ingredient. An alternative is to use Th/Pu fuel, with plutonium being consumed and fissile U-233 bred. The remaining U-233 after separation could be used in a Th/U fuel cycle.

Much development work is still required before the thorium fuel cycle can be commercialized, and the effort required seems unlikely while (or where) abundant uranium is available.

1.13 Nuclear Fusion Power

Fusion powers the sun and stars as hydrogen atoms fuse together to form helium, and matter is converted into energy. Hydrogen, heated to very high temperatures changes from a gas to a plasma in which the negatively charged electrons are separated from the positively charged atomic nuclei (ions). Normally, fusion is not possible because the positively charged nuclei naturally repel each other. But as the temperature increases the ions move faster, and they collide at speeds high enough to overcome the normal repulsion. The nuclei can then fuse, causing a release of energy.

In the sun, massive gravitational forces create the right conditions for this, but on Earth they are much harder to achieve. Fusion fuel - different isotopes of hydrogen - must be heated to extreme temperatures of over ten million degrees Celsius, and must be kept dense enough, and confined for long enough (at least one second) to trigger the energy release. The aim of the controlled fusion research program is to achieve "ignition" which occurs when enough fusion reactions take place for the process to become self-sustaining, with fresh fuel then being added to continue it.

1.13.1 Basic Fusion Technology

With current technology, the reaction most readily feasible is between the nuclei of the two heavy forms (isotopes) of hydrogen - deuterium (D) and tritium (T). Each D-T fusion event releases 17.6 MeV (2.8 x 10^{-12} joule, compared with 200 MeV for a U-235 fission). Deuterium occurs naturally in sea water (30 grams per cubic meter), which makes it very abundant relative to other energy resources. Tritium does not occur naturally and is radioactive, with a half-life of around 12 years. It can be made in a conventional nuclear reactor, or in the present context, bred in a fusion system from lithium. Lithium is found in large quantities (30 parts per million) in the Earth's crust and in weaker concentrations in the sea. While the D-T reaction is the main focus of attention, long-term hopes are for a D-D reaction, but this requires much higher temperatures.

In a fusion reactor, the concept is that neutrons will be absorbed in a blanket containing lithium that surrounds the core. The lithium is then transformed into tritium and helium. The blanket must be thick enough (about 1 meter) to slow down the neutrons. This heats the blanket and a coolant flowing through it then transfers the heat away to produce steam that can be used to generate electricity by conventional methods. The difficulty has been to develop a device that can heat the D-T fuel to a high enough temperature and confine it long enough so that more energy is released through fusion reactions than is used to get the reaction going.

At present, two different experimental approaches are being studied: fusion energy by magnetic confinement (MFE) and fusion by inertial confinement (ICF). The first method uses strong magnetic fields to trap the hot plasma. The second involves compressing a hydrogen pellet by smashing it with strong lasers or particle beams.

1.13.2 Magnetic Confinement (MFE)

In MFE, hundreds of cubic meters of D-T plasma at a density of less than a milligram per cubic meter are confined by a magnetic field at a few atmospheres pressure and heated to fusion temperature.

Magnetic fields are ideal for confining plasma because the electrical charges on the separated ions and electrons mean that they follow the magnetic field lines. The aim is to prevent the particles from coming into contact with the reactor walls as this will dissipate their heat and slow them down. The most effective magnetic configuration is toroidal, shaped like a thin doughnut, in which the magnetic field is curved around to form a closed loop. For proper confinement, this toroidal field must have superimposed upon it a perpendicular field component (a poloidal field). The result is a magnetic field with force lines following spiral (helical) paths, along and around which the plasma particles are guided. There are several types of toroidal confinement system, the most important being tokamaks, stellarators and reversed field pinch (RFP) devices.

In a tokamak, the toroidal field is created by a series of coils evenly spaced around the torus-shaped reactor, and the poloidal field is created by a strong electric current flowing through the plasma. In a stellarator the helical lines of force are produced by a series of coils which may themselves be helical in shape. But no current is induced in the plasma. RFP devices have the same toroidal and poloidal components as a tokamak, but the current flowing through the plasma is much stronger and the direction of the toroidal field within the plasma is reversed.

In tokamaks and RFP devices, the current flowing through the plasma also serves to heat it to a temperature of about 10 million degrees Celsius. Beyond that, additional heating systems are needed to achieve the temperatures necessary for fusion. In stellarators, these heating systems have to supply all the energy needed.

The tokamak (*toroidalnya kamera ee magnetnaya katushka* - torus-shaped magnetic chamber) was designed in 1951 by Soviet physicists Andrei Sakharov and Igor Tamm. Tokamaks operate within limited parameters outside which sudden losses of energy confinement (disruptions) can occur, causing major thermal and mechanical stresses to the structure and walls. Nevertheless, it is considered the most promising design, and research is continuing on various tokamaks around the world, the two largest being the Joint European Torus (JET) in the UK and the tokamak fusion test reactor (TFTR) at Princeton in the USA.

Research is also being carried out on several types of stellarator. The biggest of these, the Large Helical Device at Japan's National Institute of Fusion Research, began operating in 1998. It is being used to study of the best magnetic configuration for plasma confinement. At Garching in Germany, plasma is created and heated by electromagnetic waves, and this work will be progressed in the W7-X stellerator, to be built at the new German research center in Greifswald. Another stellarator, TJ-II, is under construction in Madrid, Spain. Because stellarators have no toroidal current there are no disruptions and they can be operated continuously. The disadvantage is that, despite the stability, they do not confine the plasma so well.

RFP devices differ from tokamaks mainly in the spatial distribution of the toroidal magnetic field, which changes sign at the edge of the plasma. The RFX machine in Padua is used to study the physical problems arising from the spontaneous reorganization of the magnetic field, which is an intrinsic feature of this configuration.

1.13.3 Inertial Confinement (ICF)

In ICF, which is a newer line of research, laser or ion beams are focused very precisely onto the surface of a target, which is a sphere of D-T ice, a few millimeters in diameter. This evaporates or ionizes the outer layer of the material to form a plasma crown that expands generating an inward-moving compression front or implosion that heats up the inner layers of material. The core or central hot spot of the fuel may be compressed to one thousand times its liquid density, and ignition occurs when the core temperature reaches about 100 million degrees Celsius. Thermonuclear combustion then spreads rapidly through the compressed fuel, producing several times more energy than was originally used to bombard the capsule. The time required for these reactions to occur is limited by the inertia of the fuel (hence the name), but is less than a microsecond. The aim is to produce repeated micro explosions.

Recent work at Osaka in Japan suggests that 'fast ignition' may be achieved at lower temperature with a second very intense laser pulse through a millimetre-high gold cone inside the compressed fuel, and timed to coincide with the peak compression. This technique means that fuel compression is separated from hot spot generation with ignition, making the process more practical.

So far most inertial confinement work has involved lasers, although their low energy makes it unlikely that they would be used in an actual fusion reactor. The world's most powerful laser fusion facility is the NOVA at Lawrence Livermore Laboratory in the US, and declassified results show compressions to densities of up to 600 times that of the D-T liquid. Various light and heavy ion accelerator systems are also being studied, with a view to obtaining high particle densities.

1.13.4 Cold Fusion

In 1989, spectacular claims were made for another approach, when two researchers, in USA and UK, claimed to have achieved fusion in a simple tabletop apparatus working at room temperature. Other experimenters failed to replicate this "cold fusion", however, and most of the scientific community no longer considers it a real phenomenon. Nevertheless, research continues. Cold fusion involves the electrolysis of heavy water using palladium electrodes on which deuterium nuclei are said to concentrate at very high densities.

1.13.5 Fusion History

Today, many countries take part in fusion research to some extent, led by the European Union, the USA, Russia and Japan, with vigorous programs also under way in China, Brazil, Canada, and Korea. Initially, fusion research in the USA and USSR was linked to atomic weapons development, and it remained classified until the 1958 Atoms for Peace conference in Geneva. Following a breakthrough at the Soviet tokamak, fusion research became big science in the 1970s. But the cost and complexity of the devices involved increased to the point where international co-operation was the only way forward.

In 1978, the European Community (with Sweden and Switzerland) launched the JET project in the UK. JET produced its first plasma in 1983, and saw successful experiments using a D-T fuel mix in 1991. In the USA, the PLT tokamak at Princeton produced a plasma temperature of more than 60 million degrees in 1978 and D-T experiments began on the Tokamak Fusion Test Reactor (TFTR) there in 1993. In Japan, experiments have been carried out since 1988 on the JT-60 Tokamak.

1.13.6 ITER

In 1985, the Soviet Union suggested building a next generation tokamak with Europe, Japan and the USA. Collaboration was established under the auspices of the International Atomic Energy Agency (IAEA). Between 1988 and 1990, the initial designs were drawn up for an International Thermonuclear Experimental Reactor (ITER) with the aim of proving that fusion could produce useful energy. The four parties agreed in 1992 to collaborate further on Engineering Design Activities for ITER (ITER is both an acronym, and means 'a path' or 'journey' in Latin). Canada and Kazakhstan are also involved through Euratom and Russia respectively.

Six years later, the ITER Council approved the first comprehensive design of a fusion reactor based on well-established physics and technology with a price tag of US$ 6 billion. Then the USA decided pull out of the project, forcing a 50% reduction in costs and a redesign. The result was the ITER - Fusion Energy Advanced Tokomak (ITER- FEAT) - expected to cost $3 billion but still achieve the targets of a self-sustaining reaction and a net energy gain. The energy gain is unlikely to be enough for a power plant, but it will demonstrate feasibility (Figure 1.18).

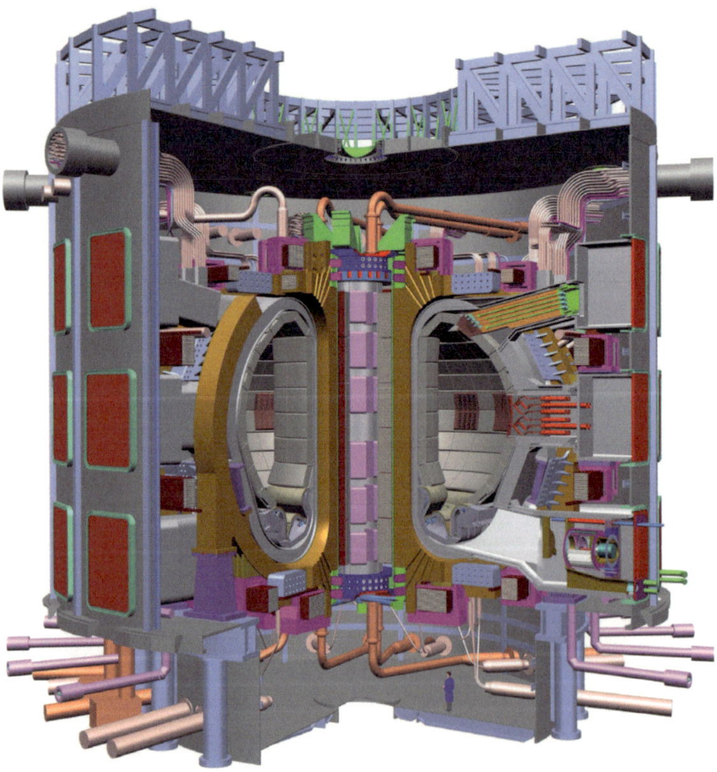

Figure 1.18 International Tokamak Experimental Reactor (ITER)

In 2003 the USA rejoined the project and China also announced it would do so. After deadlocked discussion, the six partners agreed in mid 2005 to site ITER at Cadarache, in southern France. The deal involved major concessions to Japan, which had put forward Rokkasho as a preferred site. The EU and France will contribute half of the EUR 12.8 billion total cost, with the other partners - Japan, China, South Korea, USA and Russia - putting in 10% each. Japan will provide a lot of the high-tech components, will host a EUR 1 billion materials testing facility and will have the right to host a subsequent demonstration fusion reactor. The total cost of the 500 MWt ITER comprises about half for the ten-year construction and half for 20 years of operation.

In November 2006 China, India, Japan, Russia, South Korea, the USA and the European Union - signed the ITER implementing agreement. The French President praised the attempt to "tame solar fire to meet the challenge of ecological energy".

1.13.7 Assessing Fusion Power

The use of fusion power plants could substantially reduce the environmental impacts of increasing world electricity demands since, like nuclear fission power, they would not contribute to acid rain or the greenhouse effect. Fusion power could easily satisfy the energy needs associated with continued economic growth, given the ready availability of fuels. There would be no danger of a runaway fusion reaction as this is intrinsically impossible and any malfunction would result in a rapid shutdown of the plant.

However, although fusion generates no radioactive fission products or transuranic elements and the unburned gases can be treated on site, there would a short-term radioactive waste problem due to activation products. Some component materials will become radioactive during the lifetime of a reactor, due to bombardment with high-energy neutrons, and will eventually become radioactive waste. The volume of such waste would be similar to that due to activation products from a fission reactor. The radiotoxicity of these wastes would be relatively short-lived compared with the actinides (long-lived alpha-emitting transuranic isotopes) from a fission reactor.

There are also other concerns, principally regarding the possible release of tritium into the environment. It is radioactive and very difficult to contain since it can penetrate concrete, rubber and some grades of steel. As an isotope of hydrogen, it is easily incorporated into water, making the water itself weakly radioactive. With a half-life of 12.4 years, tritium remains a threat to health for about 125 years after it is created, as a gas or in water. It can be inhaled, absorbed through the skin or ingested. Inhaled tritium spreads throughout the soft tissues and tritiated water mixes quickly with all the water in the body. Each fusion reactor could release significant quantities of tritium during operation through routine leaks, assuming the best containment systems. An accident could release even more. This is one reason why long-term hopes are for the deuterium-deuterium fusion process, dispensing with tritium.

While fusion power clearly has much to offer when the technology is eventually developed, the problems associated with it also need to be addressed if is to become a widely used future energy source. Much will change before fusion power is commercialized, including the development of new materials.

1.14 Nuclear Energy And Seawater Desalination

It is estimated that one fifth of the world's population does not have access to safe drinking water, and that this proportion will increase due to population growth relative to water resources. The worst affected areas are the arid and semiarid regions of Asia and North Africa. Wars over access to water, not simply energy and mineral resources, are conceivable. Fresh water is a major priority in sustainable development. Where it cannot be obtained from streams and aquifers, desalination of seawater or mineralized groundwater is required.

Most desalination today uses fossil fuels, and thus contributes to increased levels of greenhouse gases. Total world capacity is approaching 30 million m^3/day of potable water, in some 12,500 plants. Half of these are in the Middle East. The largest produces 454,000 m^3/day.

Desalination is energy-intensive. Reverse osmosis needs about 6 kWh of electricity per cubic meter of water (depending on its salt content), while other techniques require heat at 70-130°C and use 25-200 kWh/m^3. A variety of low-temperature heat sources may be used, including solar energy. The choice of process generally depends on the relative economic values of fresh water and particular fuels.

Small and medium sized nuclear reactors are suitable for desalination, often with cogeneration of electricity using low-pressure steam from the turbine and hot seawater feed from the final cooling system. The main opportunities for nuclear plants have been identified as the 80-100,000 m^3/day and 200-500,000 m^3/day ranges.

The feasibility of integrated nuclear desalination plants has been proven with over 150 reactor-years of experience, chiefly in Kazakhstan, India and Japan.

The BN-350 fast reactor at Aktau, in Kazakhstan, successfully produced up to 135 MWe of electricity and 80,000 m^3/day of potable water over some 27 years, about 60% of its power being used for heat and desalination. The plant was designed as 1000 MWt but never operated at more than 750 MWt, but it established the feasibility and reliability of such cogeneration plants. (In fact, oil/gas boilers were used in conjunction with it, and total desalination capacity through ten MED units was 120,000 m^3/day.)

In Japan, some ten desalination facilities linked to pressurized water reactors operating for electricity production has yielded 1000-3000 m^3/day each of potable water, and over 100 reactor-years of experience have accrued. MSF was initially employed, but MED and RO have been found more efficient there. The water is used for the reactors' own cooling systems.

India has been engaged in desalination research since the 1970s and in 2002 set up a demonstration plant coupled to twin 170 MWe nuclear power reactors (PHWR) at the

Madras Atomic Power Station, Kalpakkam, in southeast India. This Nuclear Desalination Demonstration Project is a hybrid reverse osmosis / multi-stage flash plant, the RO with 1800 m^3/day capacity and the higher-quality MSF 4500 m^3/day. They incur a 4 MWe loss in power from the plant.

Much relevant experience comes from nuclear plants in Russia, Eastern Europe and Canada where district heating is a by-product.

Large-scale deployment of nuclear desalination on a commercial basis will depend primarily on economic factors. The UN's International Atomic Energy Agency (IAEA) is fostering research and collaboration on the issue, and more than 20 countries are involved.

One obvious strategy is to use power reactors which run at full capacity, but with all the electricity applied to meeting grid load when that is high and part of it to drive pumps for RO desalination when the grid demand is low.

South Korea has developed a small nuclear reactor design for cogeneration of 90 MWe of electricity and potable water at 40,000 m^3/day. The 330 MWt SMART (System integrated Modular Advanced Reactor) reactor (an integral PWR) has a long design life and needs refueling only every 3 years. The feasibility of building a cogeneration unit employing MSF desalination technology for Madura Island in Indonesia is being studied. Another concept has the SMART reactor coupled to four MED units, each with thermal-vapor compressor (MED-TVC) and producing total 40,000 m^3/day.

Spain is building 20 RO plants in the southeast to supply over 1% of the country's water.

In the UK, a 150,000-m3/day RO plant is proposed for the lower Thames estuary, utilizing brackish water.

In India plants delivering 45,000 m^3/day are envisaged, using both MSF and RO desalination technology.

China is looking at the feasibility of a nuclear seawater desalination plant in the Yantai area producing 160,000 m^3/day by MED process, using a 200 MWt reactor.

Russia has embarked on a nuclear desalination project using dual barge-mounted KLT-40 marine reactors (each 150 MWt) and Canadian RO technology to produce potable water.

Pakistan is continuing efforts to set up a demonstration desalination plant coupled to its KANUPP reactor (125 MWe PHWR) near Karachi and producing 4500 m^3/day.

Tunisia is looking at the feasibility of a cogeneration (electricity-desalination) plant in the southeast of the country, treating slightly saline groundwater.

Morocco has completed a pre-project study with China, at Tan-Tan on the Atlantic coast, using a 10 MWt heating reactor which produces 8000 m^3/day of potable water by distillation (MED).

Egypt has launched a feasibility study of a cogeneration plant for electricity and potable water at El-Dabaa, on the Mediterranean coast.

Algeria is considering a 150,000-m^3/day MSF desalination plant for its second-largest town, Oran (though nuclear power is not a prime contender for this).

A 200,000 m^3/day MSF desalination plant was designed for operation with the Bushehr nuclear power plant in Iran in 1977, but appears to have lapsed due to prolonged construction delays.

Argentina has also developed a small nuclear reactor design for cogeneration or desalination alone - the 100 MWt CAREM (an integral PWR).

Large-scale deployment of nuclear desalination on a commercial basis will depend primarily on economic factors. One obvious strategy is to use power reactors which run at full capacity, but with all the electricity applied to meeting grid load when that is high and part of it to drive pumps for reverse osmosis desalination when the grid demand is low.

There are now a large number of prospective projects, most of which have requested technical assistance from IAEA under its technical cooperation project on nuclear power and desalination. This was initiated in 1998 with a review of reactor designs intended for coupling with desalination systems as well as advanced desalination technologies. This program is expected to enable further cost reductions of nuclear desalination.

1.15 Acknowledgements

This Chapter has been prepared by Dr. Hawley, Vice-Chancellor of The World Nuclear University and Chairman of Berkeley Resources Ltd, Welsh Power Group Ltd and Lister Petter Investment Holdings Ltd and a Non-Executive Director of Colt Telecom Group SA. He has been an Advisory Director to HSBC Bank plc, Managing Director of CA Parsons and NEI plc, CEO of Nuclear Electric and British Energy, a Board Member of Rolls-Royce plc and Chairman of several companies, including Taylor Woodrow plc and until 2004 an Advisory Director to HSBC Bank plc. He is an acknowledged international expert on power generation, nuclear energy and the environment. He is the author of many books and papers on aspects of power generation and dielectrics.
Much of the information contained in this chapter has been extracted from the World Nuclear Association website Information Papers section http://www.world-nuclear.org and ably edited by Ian Hore-Lacy to whom I owe grateful thanks.
Deep thanks are given to Michelle Brider who has so ably turned all Dr Hawley's scribbling into this chapter.

1.16 References

[1] Barre Bertrand and Bauguis Pierre-Rene, "Understanding the Future - Nuclear Power", Editions Hirlé, ISBN 978-2-914729-53-6
[2] "Energy for the Future", Philosophical Transactions of the Royal Society, Vol 365, 2007
[3] Barre Bertrand, "All about Nuclear Energy from Atom to Zirconium" Areva, 2003
[4] Beck, P.,"Prospects and Strategies for Nuclear Power", The Royal Institute of International Affairs, 1994, ISBN 1-85383-217-0
[5] Ellioh D, "Nuclear or Not?" Palgrave Macmillan, 2007, ISBN - 13: 978-0-230-50764-7
[6] Grimston M C and Beck P, "Double or Quits - The Global Future of Civil Nuclear Energy", The Royal Institute of International Affairs, 2002, ISBN 1-85383-908-6
[7] Hawley R, "Nuclear Power in the UK - Past, Present and Future", World Nuclear Association Annual Symposium, 2006
[8] Hawley R, "Nuclear Power - What has Changed", FST Journal, Vol 5, P7, 2006
[9] Hawley R, "The Future of Nuclear Power", Nuclear Future, Vol 01, pp 235-240, 2005
[10] Hawley R, "The UK Nuclear Option", Int. J Global Energy Issue, Vol 25, pp. 4-13, 2006
[11] Hewitt G F and Collier J G, "Introduction to Nuclear Power", Taylor and Francis, 2000, ISBN 1-56032-454-6
[12] Hore-Lacy I, "Nuclear Energy in the 21st Century", World Nuclear University Press, 2006, ISBN 0-12-373622-6
[13] IAEA - TECDOC - 1536, January 2007, "Status of Small Reactor Designs Without On-Site Refuelling", ISBN 92-0-115606-5, ISSN 1011-4289
[14] Kidd S, "Core Issues - Dissecting Nuclear Power Today", Nuclear Engineering International Special Publications 2008, ISBN 978-1-903077-56-6
[15] Lillington J, "The Future of Nuclear Power", Elsevier, 2004,ISBN 0-7506-7744-9
[16] Lovelock J, "The Revenge of Gaia" Penguin Books, 2007
[17] Massachusetts Institute of Technology (MIT), "The Future of Nuclear Power", An Interdisciplinary MIT Study, 2003
[18] Nuttall W J, "Nuclear Renaissance - Technologies and Policies for the Future of Nuclear Power", Institute of Physics Publishing, 2005, ISBN 0-7503-0936-9
[19] Patterson W C, "Nuclear Power", Penguin Books Ltd, 1986
[20] Price T, "Political Electricity - What Future for Nuclear Energy?", Oxford University Press, 1990, ISBN 0-19-217780-X
[21] "Projected Costs of Generating Electricity" 2005 Update, Nuclear Energy Agency / IEA / OECD
[22] Robinson A B, Robinson N and Soon W, "Environmental Effects of Increased Atmospheric Carbon Dioxide", J. Amer Physicians and Surgeons, Vol 12, pp. 79-90, 2007
[23] Socolow Robert, "Solving the Climate Problem", Environment, Vol 46, no. 10, 2004
[24] Taylor S, "Privatisation and Financial Collapse in the Nuclear Industry", Routledge, 2007, ISBN 10: 0-415-43175-1
[25] "The New Economics of Nuclear Power", WNA Report www.world-nuclear.org,
[26] University of Chicago, "The Economic Future of Nuclear Power", August 2004
[27] Wilson R, "Sustainable Nuclear Energy: Some Reasons for Optimism", Int J Global Energy Issues, Vol 28, Nos 2/3, pp. 138-160, 2007
[28] World Nuclear Association, http://www.world-nuclear.org

2

Harnessing Untapped Hydropower

This chapter has its foundation in papers given at the IEEE PES Summer Meeting, Vancouver, British Columbia in 2001. Since that conference, the world has embraced more strongly a new awareness of the challenge of global warming. In reviewing the opportunities to address the problem of climate change, hydroelectric power generation may be seen in a more positive light compared to other alternative power generation technologies, not withstanding the responsible attention that the hydro industry must continue pay to the environmental and social consequences of the construction and operation of a major hydroelectric project.

In the light of the ongoing review of economic development and the dominant role that electricity generation takes in such progress, this chapter seeks to review the opportunities for hydropower development in selected countries/locations worldwide. The authors are well aware that no single volume or chapter can fully catalogue the potential, but this is an attempt to record some of the aspects of the technology that influence the opportunities for development and some examples from the more promising markets.

There are a number of unique benefits from hydropower, rarely found in other sources of energy. Debates continue on the emissions from reservoirs, but on balance, few would now argue that the environmental benefits of hydropower are less than fossil-fuel power generation. The calculated savings of Greenhouse Gases (GHG) in 1997 by the operation of hydropower were equivalent to all the automobiles on the planet (in terms of avoided fossil fuel generation).

Development of all the remaining hydroelectric potential could not hope to satisfy total future world demand for electricity, but implementation of even half of this potential could have enormous environmental benefits in terms of avoided generation of GHG by fossil fuels.

With the mature technologies available to mankind, a combination of Nuclear power, hydropower and pumped storage could be the basis for addressing the issues of global warming.

2.1 General

Carefully planned hydropower development can make a vast contribution to improving living stands in the developing world (Asia, Africa, Latin America), where the greatest potential still exists. Approximately 2 billion people in rural areas of developing countries are still without an electricity supply.

In the context of this chapter, large hydro is defined as a plant capacity between 10 megawatts (MW) and 18,000 MW, while small hydro is defined as a plant capacity between 1 and 10 MW. While focusing on large hydro, it is acknowledged that, in the recent past, much emphasis has been put on the environmental integration of small hydro plants into river systems in order to minimize environmental impacts, incorporating new technology and operating methods.

As the most mature and important of the clean, renewable energy options, hydropower is often one of a number of benefits of a multipurpose water resources development project. As hydro schemes are generally integrated within multipurpose development schemes, they can often help to subsidize other vital functions of a project. Typically, a dam and its associated reservoir results in a number of secondary benefits associated with human well-being, such as secure water supply, irrigation for food production and flood control, and societal benefits such as increased recreational opportunities, improved navigation, the development of fisheries, cottage industries, etc. This is not the case for any other source of energy.

As indicated, existing hydropower represents a significant potential contributor to world energy resources. In 2007 renewables (excluding large hydro) accounted for 3.4% of world energy generation, while large hydro accounted for 15% [1]. This reference indicates that this was

> *"... down from 19 percent a decade ago. Large hydro grew during the five-year period 2002–2006 at a global average of 3 percent per year (less than 1 percent in developed countries). China has seen the highest growth, at over 8 percent per year during the period."*

At the commencement of the period reviewed (2002) 44% of the world's hydropower was generated in four countries. The largest generators were Canada with 315 GWh and China with 309 GWh, followed closely by Brazil with 282 GWh[2] and the United States with 255 GWh. By continents, Asia accounted for 24% of the world's hydro generation, with 618 GWh, followed by North America with 23% or 595 GWh and Europe with 20% or 537 GWh.

At the end of the period studied;

> *"The top five hydropower producers [in 2006] were China (14 percent of world production), Canada and Brazil (12 percent each), the United States (10 percent), and Russia (6 percent). China's hydro growth has kept pace with its rapidly growing power sector, with about 6 Gigawatt (GW) of large hydro and 6 GW of small hydro added in 2006. Many other developing countries continue to actively develop hydro".*

Asia is constructing more and more hydropower capacity as its economies mushroom and although Canadian hydro generation is also growing, China has already overtaken Canada, to become the largest hydro generator in the world [3]. Russia lies in fifth place with 180 GWh and Norway is sixth with 125 GWh. Norway is regarded by many as having the best managed hydro system in the world, which accounts for 99.3% of the total power generated in that country.

An additional US $15–20 billion continues to be invested annually in large hydropower. 12 to 14 GW was added in 2006 for a world total of 770 GW added in 2006.

The contribution of hydro reflected by the above numbers are however difficult to monitor because of the lack of a central register.

Notwithstanding the reliance on hydropower by the northern European countries, overall, on a regional basis, hydroelectricity contributes the highest proportion to total electricity supply in Central and South America where it accounts for 68% of electricity produced.

Hydropower provided at the beginning of the new Millennium 20% (2600 TWh/year) of the electricity world consumption (12900 TWh/year). It plays a major role in many countries. Of 175 countries, which have available data, more than 150 have hydropower resources. For 65 of them, hydro produces more than 50% of electricity; for 24, more than 90% and for 10, practically the total.

According to the Hydropower & Dams Atlas [4], untapped world hydro potential is as follow:

- Gross hydro potential: 40,500 TWh/year
- Technically feasible: 14,300 TWh/year
- Economically feasible: 8,100 TWh/year

The remaining exploitable capacity represents 1500 GW (producing 5500 TWh/year). It is estimated that by the middle of this century, the consumption of electricity in the world will be multiplied by a factor of 2.5 to 3.0. For the power generation sector, large hydropower remains an available energy technology and will contribute largely to this development, although environmental constraints, resettlement impacts, and the availability of sites have limited further growth in many countries. Particularly attractive hydropower sites i.e. high capacity factor sites are often used to directly supply high demand heavy industries such as smelters.

It is possible to harmonize the implementation of hydro plants with conservation of the environment and to ensure that the plants represent a net benefit to those affected by their construction.

The main advantages of hydroelectricity can be summarized briefly as follows:

- It is a reliable and mature technology, proven by one century of construction and operation.
- It is easily accessible, particularly for developing countries.
- It plays a major role in reducing greenhouse gas emissions in terms of avoided generation by fossil fuels.
- The fact that unit speeds are slow, and other design factors, contribute to a very low operation and maintenance costs.

- It is competitive and the kWh cost (once completed) does not depend on variations in combustible costs and on international economic politics.
- It is an important factor in supporting energy and economic independence for a country, because of the effective zero fuel costs and absence of reliability on foreign fuel sources.

Finally, hydropower stations are very often integrated within multipurpose schemes, which satisfy other fundamental human needs (irrigation, domestic and industrial water supply, flood protection) and hydropower can help finance these other functions.

A drawbacks of large hydro development is the generally longer construction lead time, higher construction cost compared to combustion turbines and some other fossil fuel plants, although in some countries coal and oil plants (and of course nuclear plants) also have long lead times. The long lead times with the correspondingly long "payback" is a severe constraint to utilization of non-government funding.

The impact of higher construction cost is more than offset by zero fuel cost and hence, hydro power plants generally yield higher operating margins. Overall returns should better match those of thermal plants.

2.2 System Benefits

The overriding challenge of global warming will likely factor heavily into system planning throughout the developed and developing world, and indications are that hydro power will have a strong part to play in the various system portfolios.

Almost all power generation technologies already developed, and those under development suffer from one or both of the following constraints that do not affect hydropower - lack of flexible response without significant effects on efficiency, or lack of sustainability because of the energy source. The efficiency penalties associated with ramping up output apply to the (generally) polluting technologies that need to be maintained at the highest efficiencies to lower their carbon imprint (such as coal fired generation) while the relative unpredictability of wind power and solar power and the intermittent (while predictable) nature of tidal power - three of the promising new technologies - necessitates some form of energy storage.

For all the above technologies, as well as nuclear generation - which may turn out to be the only way to maintain the required level of energy generation to maintain current standards of living in the developed world and to increase standards to an acceptable level in the developing world - the benefits of the flexibility of hydropower and the storage potential are, at present irreplaceable.

The degree to which hydropower contributes to the system stability and response will depend on the economic growth - and the change in standard of living - in each country. As countries develop, three drivers (among many) affect their requirements for system flexibility; the mix of heavy industry that the economy embraces; the speed at which constant (residential) heavy power users such as refrigerators and air conditioning are installed; and the adoption at different times of technology with intermittent heavy power demands such as recharging of battery operated cars. The relative lack of flexibility of some of the other power sources ensures that hydropower stations, either river based or off-stream pumped storage seems certain to continue to be vital in system operation. The only counter trend to this driver would seem to be the interconnection of systems - either through land connections or undersea HVDC transmission, which also must continue and accelerate.

Hydropower provides vital benefits to an electrical system. When water is stored in large quantities in the reservoir behind a dam, or has been pumped up to the upper reservoir of a pumped storage scheme, it is immediately available for use when required. Pumped storage technology has been developed sufficiently to minimize the losses in storage, which principally arise because of the hydraulic losses in the conduit system. The difference between the "value" of the energy during the pumping cycle and the "value" during the generating cycle are usually large enough to overcome the efficiency penalties, making the technology ideal for smoothing the output of a large base load plant such as a nuclear station. Thus the benefits of hydro extend beyond simple generation.

The first and most obvious benefit is the flexibility of a large hydro power station to provide large amounts of power within a minute or so of demand (sometimes seconds depending on how the plant is configured and operated). This flexibility and the flexibility of ramping, allow the hydro plant to respond to the demand curve, and to efficiently "lop the peaks" of demand allowing other fossil plants to operate at their constant efficient output. Pumped storage can of course act on the obverse of this problem by converting excess power from the fossil plants to stored energy.

The second benefit is truly a family of benefits, known as secondary benefits. These secondary benefits include:

- Spinning reserve - the hydropower plant can operate at a zero load (and of course zero fuel consumption) while synchronized to the electric system. When loads increase, additional power can be loaded rapidly into the system to meet demand.
- Non-spinning reserve - Hydropower has a "quick start" time measured in seconds or minutes, compared with as much as 30 minutes for other turbines and hours for steam generation.
- Voltage support – Hydro is very useful in providing reactive power, thereby assuring that power will flow from generation to load.
- Regulation and frequency response - the ability to support the system during moment-to-moment fluctuations in system power requirements is met by a hydropower' plant's fast response characteristic making the technology especially desired to provide regulation and frequency response in a system with sensitive equipments.
- Black start capability – most hydro power plants have the ability to start generation without an outside source of power enabling the system to be brought up during catastrophes and providing auxiliary power to more complex generation sources that could take hours or even days to start.

2.3 Situation at Present

Fig. 2.1 Earth at Night

In reviewing the world energy demand, it is useful to examine Figure 2.1 showing the world from space at night. It can immediately be seen that almost the whole of Africa, most of the South American continent and large parts of China are without lights. This immediately underscores the main long-term markets for the utilizing of untapped hydro.

Hydropower is not the largest available renewable primary source. This mantle is held by biomass, but hydropower is the largest renewable source of electricity generation. Hydropower accounts for 6% of primary energy supply and 17% of electricity generation.

Although there are hydroelectric projects under construction in many countries, most of the remaining hydro potential may be found in the developing countries particularly in South and Central Asia, Latin America, and Africa. Other countries with significant remaining hydropower potential are Canada, Turkey and Russia.

Hydropower constitutes about 20% of the world's electricity generating capacity. The theoretical potential of worldwide hydropower is 2,800 GW, about four times greater than that which has been exploited.

However, the actual amount of electricity that will ever be generated by hydropower will be much lower than the theoretical potential, because of the environmental concerns and economic constraints.

A study by the Utility Data Institute, USA, predicts that a world total of 695 GW of new electricity capacity will come on line in the next ten years from all sources, 22 per cent of which will be hydro, 26 per cent gas, and 27 per cent coal, with the remainder coming from a variety of sources.

The world's total technical feasible hydro potential is estimated at about 14 300 TWh/year, of which about 8080 TWh/year is currently considered economically feasible for development. About 723 GW (or about 2600 TWh/year) is already in operation, with a further 108 GW under construction (Table 2.1) [4].

Area	Technically feasible	In operation & under construction
Asia	4,225,479	699,636
China	1,923,304	198,700
Lao	210,000	3,037
Myanmar	160,000	1,450
Japan	129,840	91,654
Cambodia	83,000	0
CIS & Russia	2,105,600	323,760
North America	1,007,713	601,791
South & Central America	3,933,770	550,658
Peru	1,091,540	12,615
Europe	1,158,029	486,819
Africa	1,590,828	64,043
Oceania	206,366	42,637
World total	14,227,785	2,769,344

Table 2.1 Hydropower Potential (GWh/year)[5]

The majority of the remaining hydro potential is found in developing countries in the regions mentioned, South and Central Asia, Latin America and Africa. In most of the European countries the economically feasible hydropower potential has mostly been harnessed.

A number of countries, such as China India, Iran and Turkey, are currently undertaking large-scale hydro development programs, and there are projects under construction in about 80 countries. According to the recent world surveys, a number of countries see hydropower as the key to their future economic development. Examples include Sudan, Rwanda, Mali, Benin, Ghana, Liberia, Guinea, Myanmar, Bhutan, Cambodia, Armenia, Kyrgyzstan, Cuba, Costa Rica, and Guyana.

In North America, hydropower is the most widely used form of renewable energy. The installed hydropower capacity amounts to 175 GW (67 GW in Canada, 99 GW in the US, and 10 GW in Mexico).

Hydropower accounts for 57% of the electricity generated in Canada, 7% in the US (the US uses hydropower for peaking not base load) and 12% in Mexico. Canada's economical hydropower potential is second only to that of Brazil in the Western Hemisphere and still has several projects under either construction or planning, amounting to 6.6 GW.
Latin America has a very large hydropower potential. Many countries rely heavily on hydropower for their electricity supply. For instance, hydropower makes up 80% of Brazil's electricity generation.

Brazil has plentiful hydropower resources. Its installed hydropower capacity is already 64 GW. The capacity under construction or planning is more than 25 GW. One of the hydropower plants under construction is the giant 11.18 GW Belo Monte power plant. Capacity under construction or planning in other South American countries, particularly Argentina, Bolivia, Chile, Colombia, Guyana, Peru, and Venezuela, amounts to 9.7 GW together with 4.4 GW of hydropower capacity under construction or planning in Central American countries.

China has the largest hydropower resources in the world, with a host of rivers. Its installed hydropower capacity rested at 83 GW at the end of 2002. A large number of hydropower plants are under construction or planning, amounting to 77.7 GW. The giant 18.2 GW Three Gorges Dam with a dam height of 181 m on the Yangtze River (the country's longest river) is the world's largest hydropower project so far.

Russia holds fifth place with 180 GWh and Norway in sixth with 125 GWh. Norway is regarded by many as having the best managed hydro system in the world, which accounts for 99.3% of the total power generated in that country.

2.4 Prior Development Methods

The approach for development of hydropower generation worldwide has progressed through three significant phases since the beginning of the 19th century. These phases have been observed in all regions of the world and correlate directly with the historical developmental phases (colonial, independent and multilateral etc), the type of projects selected for development, and the resources available for implementation.

In general, the phases can be described as follows:

Phase I can be thought of as the birth of modern power systems and comprises the time from the first development of the electric generation industry through to the major expansions required after the Second World War. This period was characterized by project development by largely private sector utilities, colonial development in discrete parts of territories and industrial companies seeking to meet immediate demands. Development was often managed or controlled by engineers due to the unique nature of the projects and the state of knowledge in the industry. Financing was limited and projects were developed as needed, often for specific industrial projects in the developing world or for the limited demand from colonial outposts. The configuration and capacity of the projects considered was driven by economic and technical factors usually leading to modest scale projects that could be financed from the resources of the relatively small utilities in existence at that time. The developed world was creating interconnecting grids during this process but typically in the developing world, the grids remained isolated because of the limited colonial development objectives.

Phase II was ushered in by the rapid economic growth and industrialization following the Second World War as well as decolonization and independence. Energy use expanded significantly in this period as development accelerated in the already industrialized societies and spread internationally. The rate of growth envisaged and promoted by economic planners exceeded the capability of the nascent private utilities to finance the required generation expansion. Accordingly, many governments in the developed world started to take a direct role in the power sector through the formation and/or expansion of publicly owned utilities. Regulation of the power sector accompanied this transition as governments sought to control the price of retail energy. In the developing world, the major financing needs in this period were supported by the multilateral financing agencies such as the World Bank. During the 1960s and 1970s utilities embarked on a program of building much larger projects supported by government financing resources, in an effort to keep pace with development and in many cases to foster development in emerging economies. Many projects were configured to be very large. They were often multi purpose designed to meet several needs including water supply, flood control and irrigation as well as power generation, and were intended to be national development "engines" as well as for the simple purpose of generation. Unfortunately, many of the difficulties that have been assessed by the World Commission on Dams in the selected review projects relate to these mega projects.

In the most recent Phase III that has evolved in the last ten to fifteen years, the world has in many ways returned to the development model used during the emerging years of the power industry. This can be characterized the world over by the privatization of power generation and distribution systems and the implementation of "private" projects driven by investors as well as a retreat by multinationals from direct funding of large hydropower. One of the most important elements driving this transition is the realization that foreign direct investment under appropriate conditions can be an important source of financing the large capital requirements of power sector expansion. In many ways, this reflects on the success of the efforts described here as "Phase II". Multi-lateral financing has ensured that many nations have matured politically and commercially so that large-scale foreign private investment has becoming more viable. This current phase has several variants and the extent to which each country has moved down the road of market driven investment governs investment strategy adopted by private power developers. Today energy sales from independent power projects use various vehicles ranging from direct power purchase agreements with a utility at the outset of privatization to a sophisticated power pool or merchant market in the more developed markets. Though investment in power generation and in distribution has been significantly influenced by privatization, the creation of large-scale transmission interconnection has not attracted the market funds necessary for realization.

It is also conceivable that the private capital market may not be able to supply capital at the rate needed to address the issues of global warming.

The three phases of development orientation and philosophy presented above and reflected in the activity in the regions is of course a generalization and, perhaps, an oversimplification of the complex circumstances of national and international economic development. However, there is no question that during this period there has been a relatively linear chronological progress. This movement from initial limited and focused utility investment - through focus on major "economy-stimulating" projects - and, currently, back to a more market-driven investment profile is evident throughout most of the developing nations in the world and most certainly in Asia. An indication of the extent of dam building (which approximates to hydropower development) during these phases can be gained from Figure 2.2.

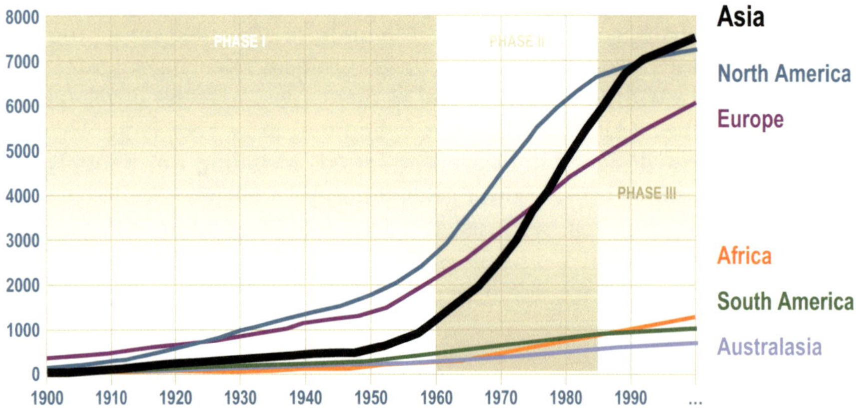

(Source: ICOLD, 1998. Excludes the time-trend of dams in China)

Fig. 2.2. Extent of Dam Building Worldwide (1900~2000)[6]

The urgent challenges of global climate change, and the general acceptance that energy generation must relay less on fossil fuels in the future, together with the realization that economic development is a forceful factor in the stabilization of the political world now necessitates a reconsideration of hydro development and consideration of the benefits of the sort of larger projects that cannot be funded by private financing, including large scale transmission interconnection.

2.5 Review of Selected Regional Prospects

Hydro is a mature technology, and has been developed all around the world. Resources are being developed at a rate of approximately 2.5% per annum, and the USA has achieved the greatest development of its resources with a total installed hydro capacity of 73,500 MW.

The World Energy Council estimated that in 1990 world energy demand was approximately 12,000 TWh, and postulates that in 2020 it will be nearly double at 23,000 TWh.

In 1990, hydro contributed 2,240 TWh of energy, representing 18.5% of the total, and if 50% of the total economically feasible resources were developed, in 2020, hydro would contribute approximately 28% of energy generation worldwide.

Table 2.2 indicates world hydro potential (in TWh) of each region:

	Gross	Economic	Feasible
Europe	5,584	2,070	1,655
Asia	13,399	3,830	3,065
Africa	3,634	2,500	2,000
America	11,022	4,500	3,600
Oceania	592	200	160
Total	**34,231**	**13,100**	**10,480**

Table 2.2. World Hydro Potential by Region

Hydropower development in the regions reflects, as expected, economic development achieved by each region. The total hydro developed as of 1997 as a proportion of the economic hydro potential is shown in Table 2.3.

Africa	6 %
South and Central America	18 %
Asia	18 %
Oceania	22 %
North America	55 %
Europe	65 %

Table 2.3. Proportion of Hydro Developed, by Region

There are commercially available sector reports noting in detail the various projects that have been identified region-by-region and nation-by-nation. These commercial reports are recommended for those developers seeking the viable opportunities particularly in the prospective markets in Africa, Asia and Central and South America. It is not the scope of this chapter to repeat the sector reports but merely to address some of the interesting aspects of development in particular regions.

The following sections provide an overview of recent and upcoming possibilities in a few markets.

2.6 Canada

About half of the Canadian provinces are responsible for the majority of hydroelectric energy production in Canada, with Quebec being a market leader. 60% of the countries electricity is supplied by hydropower and the total installed capacity is approximately 72,000 MW. The largest producers are provincially-owned electric utilities such as:

- Hydro-Quebec
- BC Hydro
- Manitoba Hydro
- Ontario Power Generation
- Newfoundland and Labrador Hydro.

These utilities have already developed a series of large-scale hydro sites across the country. One of the most significant hydroelectric developments in the world is La Grande complex on the Quebec side of James Bay. It has a capacity of over 15,000 MW. Some of the other large-scale hydro sites in Canada include:

- Churchill Falls station in Labrador
- Manicouagan-Outardes complex on the Quebec North Shore
- Sir Adam Beck station on the Niagara River in Ontario
- Nelson River development in Manitoba
- Gordon Shrum station in Northern British Columbia
- Columbia River complex in the southern part of British Columbia.

Electric utilities are the main generators of hydroelectric energy, however there are other electrical producers of hydroelectric energy. Several industrial companies own and operate hydroelectric facilities for their own use.

Canada's remaining (economic and environmentally acceptable) potential is spread throughout the country but is predominantly in the provinces of Alberta (12,000 MW), and Quebec and the Maritimes (25,000 MW). The others have approximate potential for about 4 to 5000 MW each.

Major projects are under construction in Quebec:

- Eastmain 1 - 480 MW
- Chûte-Allard/RDC - 138 MW
- Péribonka - 385 MW

and Ontario:

- Niagara Tunnel - C$ 600 M

Near term projects in Québec & Manitoba include:

- Eastmain 1A - 888 MW
- Wuskwatim – 200 MW

Planned Projects include:

- Québec
  - La Romaine - 1550 MW
- Manitoba
  - Keeyask – 620 MW
  - Conawapa – 1250 MW
- Newfoundland & Labrador
  - Lower Churchill (Gull Island) -- 2000 MW
  - Lower Churchill (Muskrat Falls) -- 824 MW.

For the purposes of this chapter, we examine three projects in planning by Manitoba Hydro.

Potential projects considered include three smaller or mid range hydro plants at Gull Rapids on the Nelson River, Notigi on the Rat River and Wuskwatim on the Burntwood River.

The three hydro options considered are all low impact projects with potential of 600 MW at Gull and 1300 MW at Conawapa (and possibly 1900 MW more) on Nelson River. Notigi potential is 100 MW and 200 MW at Wuskwatim. There is possibly 650 MW more on Burntwood River. Other potential from partially developable hydro sites in the province amounts to about 4000 MW.

Wuskwatim and Gull can be constructed to minimize their environmental impacts. Flooding would be less than 1-km square at Wuskwatim, Gull would be in the order of 48-km square, and Notigi Plant would have no additional flooding as the water control structure already exits.

Methyl mercury production due to flooding would be minimal and Green House Gas (GHG) emissions would essentially be zero.

Construction cost including the transmission cost would be in the order of $ 500 million for Notigi, $1,000 million for Wuskwatim, and $3,000 million for Gull. By comparison additional combustion turbines with a generation capability of 140-280 MW would cost about $100 -$200 million. The larger hydroelectric plant at Conawapa would cost about $5,000 million including transmission to the load centers in the south.

Historic variability of water supply in the Nelson and Churchill Rivers Drainage basin over the last 80 to 100 years shows a range of percent average flow from a low of 55% to a high of 160%. Manitoba Hydro's long term planning criteria must consider the lowest historical level for power resource planning, but plant structures including the spillways must be designed for maximum flood conditions.

Figure 2.3 shows the Manitoba Hydro hydraulic system with respect to the elevation levels along with the existing plant locations indicated. Figure 2.4 shows an illustrative view of Burntwood and Nelson Rivers areas in reference to potential new generation.

Transmission is a major consideration. Drivers for Manitoba Hydro include strong growth in its export sales, while provincial requirements are growing steadily. Export revenues grew almost 50 percent in the four years from 1996 to 2000 from $250 million to $376 million. Regulatory changes in the U.S. market, coupled with a burgeoning economy and a lack of new generation and transmission in the U.S. have meant steadily increasing prices for wholesale electricity exported to the U.S. from Canada.

Manitoba Hydro was been able to benefit from these changes, and now has the ability to market to approximately 35 export customers, compared to seven export customers in 1996. With domestic electricity consumption in Manitoba growing slowly but steadily, this means that over time the utility will have less and less available energy to sell on the export market. If it wants to maintain its current level of export sales, it will have to build new sources of generation.

Strong export sales have been a key reason that electricity rates in Manitoba have not risen as strongly as they might (for industrial customers, the rates remained stable for nearly ten years).

Manitoba Hydro has the capability of a number of east - west interconnection schemes which have been studied in the past such as the Manitoba – Ontario connection that included new generation in northern Manitoba with a north – south HVDC line. This project had an estimated cost of $2 B.

Another east - west scheme that is viable is a 500 kV HVDC multi terminal concept that covered Manitoba, Saskatchewan and Alberta Provinces, also known as western grid.

There has also been a number of north – south interconnection studies in the past to the States of Minnesota, North Dakota, Nebraska and Wisconsin. One current project which has recently been completed is a 230 kV connection between western Manitoba and central

North Dakota with a transmission line length of 200 km (and a cost of $55 million). This project will re-establish Manitoba Hydro's import capability to 500 MW; it will also increase the import and export capabilities by 200 MW.

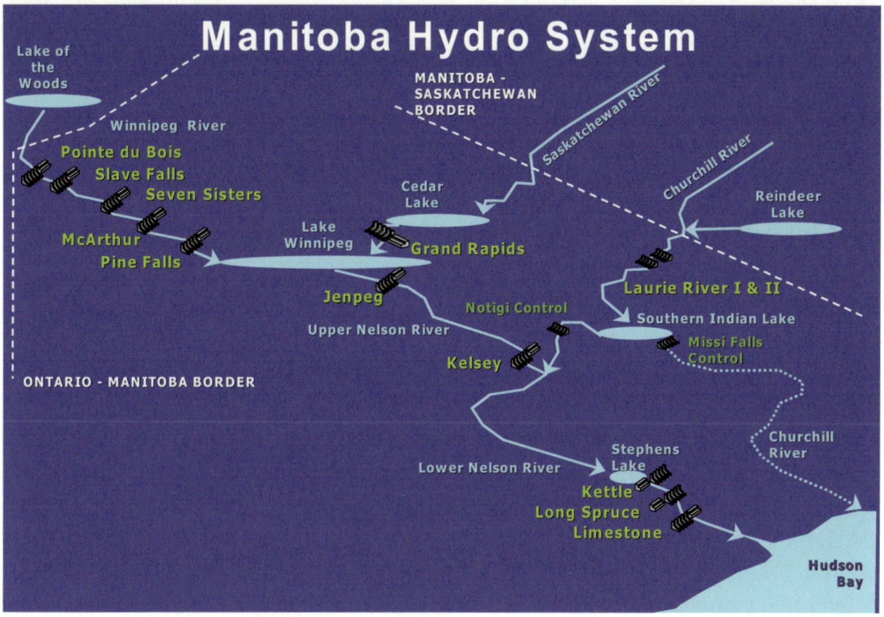

Fig. 2.3. Manitoba Hydro System

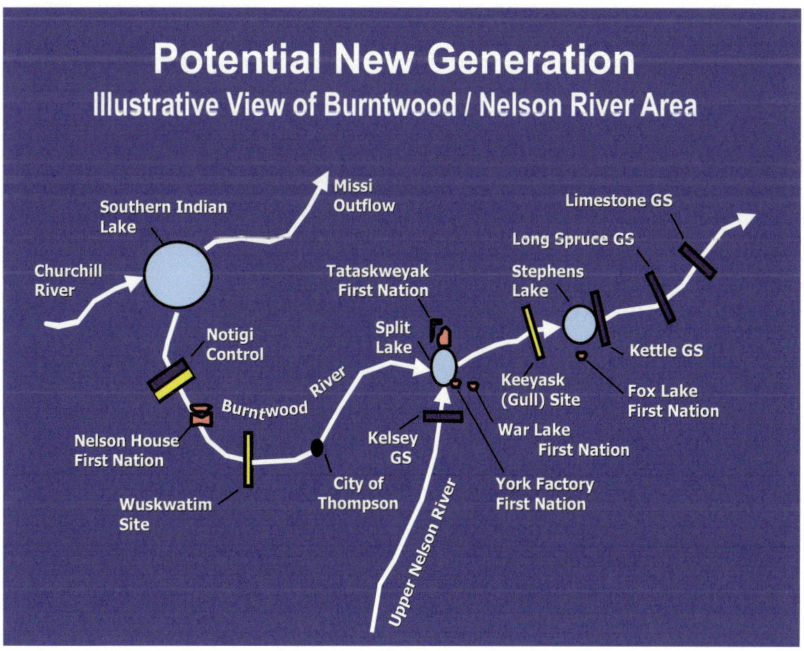

Fig. 2.4. Potential New Generation- Burntwood/Nelson River Area

2.7 South and South East Asia

South and South East Asia, encompasses the SE Asian Tigers - with strong development and industrialization that will continue to ensure a strong power market – together with the Indian subcontinent and its peripheral countries, which are also poised to contribute to India's power. Notable potential exists (2005/2006 figures) as follows:

2.7.1 Bhutan

Development in Bhutan has resulted in an installed capacity of 445 MW, with a further 1,020 MW under construction. The capacity in planning – which varies from report to report is between 1660 MW and 7,805 MW. Notable projects in planning include Bunakha (180 MW) & Wang Chu (900 MW), both in the Wang Chu River valley. A further potential project Sankosh (of installed capacity 4,060 MW) has been identified.

2.7.2 India

India's economy is developing fast - with the consequent necessity for generation – and the country is endowed with economically exploitable and viable hydro potential assessed to be about 148,700 MW installed capacity. The major enterprise charged with developing it is National Hydropower Corporation but the National Thermal Power Corporation is developing Kol Dam and power plant and there are private power companies in the sector. As of 2005, the installed capacity in India is 31,982 MW with an additional 13,245 MW under construction. The largest projects under construction are Subansari lower (2,000 MW) Parbati (800 MW), and Omkareshwar (520 MW). 8,860 MW are planned approximating to 17 schemes.

2.7.3 Laos

Laos is developing hydro in order to provide power to the Thai and perhaps Vietnam grids. There are various private power companies developing sites and there are some substantial projects in construction and in the later stage of development. In 2005 the installed capacity was 673 MW with a further 2,011 MW under construction. Between 3,000 and 5,000 MW are planned.

2.7.4 Malaysia

The installed capacity in 2005 was 2,078 MW. Most of the hydro potential of peninsular Malaysia has been developed and the potential now remains in Sarawak and Sabah. Although Ulu Terengganu is proceeding on the peninsular, the massive Bakun project – which has an installed capacity of 2,400 MW – is the prime project under construction. Bakun has been under development for many years, and at various times an HVDC cable to peninsular Malaysia has been proposed. It appears that this cable may be implemented in the near future, most notably because existing gas supplies supporting combustion turbines are being depleted, rendering hydropower from Sarawak a desirable energy option. A new Sarawak Corridor of Renewable Energy ("SCORE") has been announced including Murum (900 MW), Limbang (150 MW), Barang (1,000 MW), and Baleh (1,000 MW) all planned for development in the near to medium future.

2.7.5 Myanmar

The installed capacity in Myanmar is 745 MW, with an additional 1780 MW under construction. The Myanmar Electric Power Enterprise has identified the potential for nearly 40,000 MW but capacity additions that are in planning, though unlikely to be realized in the foreseeable future are limited to approximately 8,000 MW.

2.7.6 Nepal

Nepal has very limited installed capacity, some 560 MW in 2005, with 69 MW under construction. However, as a result of the Himalayan topography, the hydro potential is enormous - some 43,000 MW is postulated. Although the 2007 World Energy Survey records no planned hydro, a cursory examination of the project list from the Nepal Electricity Authority indicates many projects, including West Seti, Arun, Kali Gandaki Pansheswar and many other projects of 200 MW plus. The limited financial capacity of Nepal is a brake on development.

2.7.7 Pakistan

Pakistan had in 2005 an installed capacity of some 6,499 MW, dominated by the installations at Tarbela and Mangla. The potential economical installed capacity has been identified as 41,722 of which 734 MW is under construction and some 8,900 to 27,000 MW are in various stages of planning. To date, large scale hydro has been dominated by the Indus basin treaty signed between India and Pakistan in 1960, but projects such as Neelum Jhelum (969 MW) which will commence design this year, Dasu (2,712 MW) Bunji (1,500 MW), Pattan (1,172), Thakot (1,043 MW) and Kohala (740 MW) are included in the 8,900 MW planning figure. Kalabagh dam was conceived in 1953 and remains feasible for 2,400 MW installed capacity - increasing to 3,600 MW. However there remains significant social difficulty in implantation. Another large project, Bashar could have an installed capacity of 3,600 MW,

2.7.8 Vietnam

The potential in Vietnam is put at 300,000 GWh/yr with an economically feasible potential of about 80,000 GWh/yr. Total installed capacity in Vietnam is approximately 4,198 MW, and some 20 plants totaling 7,768 MW are currently reported to be under construction. In planning are another 4,600 MW.

2.8 Africa

As noted only 4 % of Africa's technically feasible hydroelectric power has been developed. This is symptomatic of the continents lack of economic growth. At present only 10% of Africans have access to electricity - with the lowest per capita consumption in the world (450 kWh, compared to the world average of 2,326 kWh) - and there is an uneven distribution of power supply.

The existing infrastructure is under strain. The bulk of the region's generating capacity and transmission network was built in the 1950s and 1960s, and continued under investment in maintenance and renewal has left, in many countries, a system creaking. Technical and non-technical losses typically exceed 20%. Power outages, brown outs, and power surges are commonplace, whilst load shedding in times of drought are routine. A prolonged drought in 1999/2000 led to severe power outages in Kenya and Ethiopia, for example, whilst silting up of dams has led to a steady decline in capacity and concerns about dam failure. It is estimated that some of Kenya's dams are up to 70% silted and Malawi's, up to 80%.

Africa's power sector is dominated by South Africa in Southern Africa, Egypt and Morocco in North Africa, and Nigeria in West Africa. 82 % of Africa's power comes from the northern and southern regions alone, with three-quarters coming from five countries - Egypt, South Africa, Libya, Morocco and Algeria. Current production in Africa is of the order of 76 TWh/yr, of which the leading producers are Egypt and Mozambique with 11 TWH/yr.

The gross theoretical potential is 4,000 TWh/yr, while the technically feasible potential is 1,750 TWh/yr. The economically feasible potential is 1,000 TWh/yr. Current installed capacity is 20,300 MW and Hydro under construction is greater than 2,400 MW.

The region possesses some of the largest watercourses in the world (the Nile, the Congo, the Niger, the Volta, the Zambezi), and whilst hydroelectricity is by far the single largest source of electricity in a number of countries (see below), the resources remain largely untapped. An example is that of Cameroon which has hydro potential of up to 115 TWh/yr and yet installed hydro capacity currently stands at less than 800 MW.

The feasible hydroelectric potential in the Democratic Republic of Congo alone is reported to be sufficient to provide three times as much power as Africa presently consumes.

Approximately 50 % of power in Africa is generated from coal-based facilities, although this figure is projected to decrease to 36 % by 2020. Hydro is the significant contributor in many countries and can be a major contributor to the future growth in generation. Current contribution is seen in Figure 2.5.

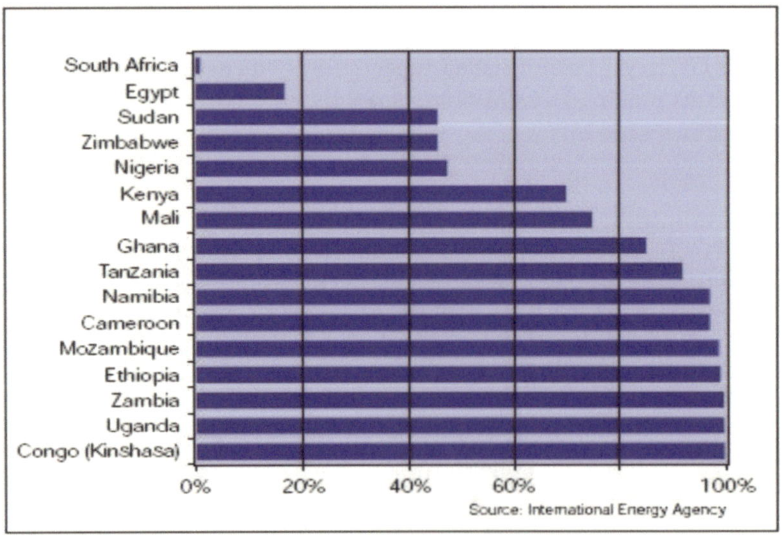

Fig. 2.5. Contribution of Hydropower to net Electricity Generation 2002[7]

There are significant barriers to development, among them the lack of transmission facilities, the somewhat unpredictable weather patterns in parts of the Continent, and the low level of development at present.

Power sharing, through transmission interconnection will provide a substantial boost to the continent's development. It is becoming more prevalent in recent years. Development in regional economies, such as the formation of the Southern African Democratic Community (SADC) and the Economic Community of West African States (ECOWAS) has also encouraged the export and import of power. South Africa, who has surplus power, is able to operate more efficiently.

Regional sharing has occurred on an informal basis, and on a more formalized basis, such as with the Southern African Power Pool (SAPP), where a number of countries have been linked and purchase power at regulated prices. Countries linked through the SAPP network are: South Africa, Mozambique, Zimbabwe, Zambia, Namibia, Botswana, the Democratic Republic of Congo, Swaziland, Tanzania, Lesotho and Malawi. Kenya and Tanzania may connect to the Zambian power grid, which would bring Kenya into the SAPP. Ongoing connection is a goal of projects initiated and implemented by the New Partnership for Africa's Development (NEPAD), an initiative that sees investment in Africa's infrastructure as a high priority. All NEPAD power projects are aimed at boosting electrical power generation, distribution and transmission in Africa.

South Africa, Ghana and Zambia are the biggest net exporters of power.
Further discussion on transmission can be found later.

The extremes of precipitation also constitute a problem for the development of hydroelectric in parts of Africa. Droughts in East Africa in 1999-2000 had a serious impact on the hydroelectric facilities in the region, with Kenya in particular suffering from severe power shortages. Ghana's existing large hydroelectric facilities (Akosombo and Kpong) were also adversely affected by droughts during the late 1990's. However, droughts tend to affect regions, and countries, at different times to others. Improved transmission would allow countries experiencing power shortages to import electricity from other regions recouping the cost of these imports when hydroelectric power production attains its design level by exporting.

The low level of economic development also presents a challenge to power developers - a chicken and egg situation. Africa's electricity consumption is expected to grow at a rate of 4.4 % per year over the next 25 years, which is low compared to that experienced by the South East Asian (Tiger) economies, but will nevertheless require capacity expansions of 270 GW by 2030. The expansion is of course dependent on the robustness of the regional economies, which - in the case of sub Saharan Africa - have had recent growth rates in excess of 4%. Renewed interests in the region's oil sector, with the United States and China in particular vying to take significant portion of their oil from Africa, is expected to reinforce this growth.

The potential for hydro is mainly focused in central and eastern Africa. Potential for hydro capacity growth is everywhere, but completely dominated by the Democratic Republic of the Congo, Ethiopia, Madagascar, and the Cameroons as seen in Figure. 2.6.

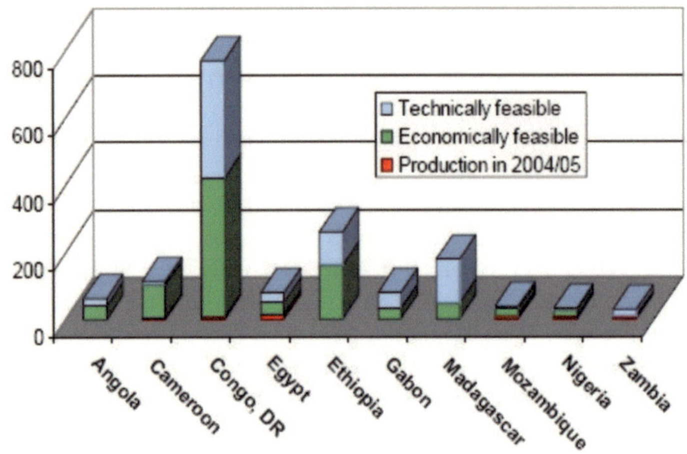

Fig. 2.6. Estimated Potential for Hydropower in Selected African Countries (TWh/yr)

The overwhelming potential (of 774 TWh/yr) of the Democratic Republic of the Congo deserves a special mention.

The enormous potential of the Congo River could play an important role in providing power to Central and Southern Africa, as well as exporting electricity to North Africa and possibly Southern Europe. The Congo River accounts for nearly 30% of Africa's surface water reserves

The "lions share" of the available potential of the Congo River is at the series of rapids at Inga. Development at Inga is preferred because the specific topography leads to relatively benign environmental impacts, the ultimate capacity of 39,000 MW is attractive, and the huge area of the drainage basin is a guard against overwhelming drought.

The Democratic Republic of Congo currently has 1,775 MW of capacity at its two Inga hydroelectric facilities completed in 1972, and in 1982 respectively. Inga provides power to the DRC, the Republic of Congo's (Congo) and Southern Africa - including Zambia, Zimbabwe and South Africa.

The Energy Ministry, DRC estimates that the country earned $500,000 monthly in 2001 from its energy exports.

Increasing the generating capability of the Inga facility is high on the priority list. Completion of the upgrading of Inga 1 and 2, together with the construction of Inga 3 slated to have a capacity of 3,500 MW would provide enough excess generating capacity for the creation of a new regional electricity export scheme.

At the heart of the Inga proposals however is the proposed Grand Inga scheme that would be the largest generating facility in Africa. The construction of Three Gorges Dam, in China, has reawakened hope that Grand Inga can be constructed. The project with a capacity of 39,000 MW, in 53 units, would be nearly as large as South Africa's existing generating capacity, and provide the possibility for a pan-African electricity-exporting project.
The World Energy Council has articulated the goals as

> *"The ultimate objective of Grand Inga is to bring affordable and clean energy to the African continent to facilitate economic development to improve the standards of living. It offers a unique opportunity to move the African continent closer to achieving its sustainable development goals."*[1]

There are other proposed regional hydro projects but the potential is focused on Grand Inga

2.8.1 Ethiopia

Ethiopia has a potential for approximately 161 TWh/yr and has embarked on a series of projects to try to develop its enormous potential. The resources are second only to the Congo in Africa.

Substantial hydro capacity is set to be developed. Hydro schemes already contribute 670 MW out of the 713 MW of capacity on the national grid but the government has finally succeeded in attracting investment in new hydro projects. Currently the 435 MW Beles, 420 MW Gilgel Gibe II and 300 MW Tekeze hydro projects are all being developed by private sector companies.

Ethiopia possesses a host of other potential large-scale hydro sites and so could become a massive electricity provider to large swathes of Africa that are currently power poor. The Ministry of water resources estimates the country has more than 30,000 MW of untapped economically feasible hydroelectric potential. Two that have recently been funded for study are the 896 MW Baro scheme and the 1,000-1,600 MW Karadobi project. The Chemoga Yeda I and II (440 MW), Halele Werabesa I and II (374 MW) and Aleltu East (189 MW) are also currently in the planning stage.

2.8.2 Uganda

Uganda already has two hydroelectric schemes on the stretch of Nile within the country between Lake Victoria and the Sudanese border: the 180 MW Nalubaale project, previously called Owen Falls, and the 120 MW Kiira dam scheme, which was brought on stream in 1999. The Bujagali project is set to be completed by 2011 at an estimated cost of US$ 800 M

2.8.3 Zambia

Zambia continues to seek ways to increase generating capacity through public/private partnerships. Construction of the Kafue Gorge Lower Hydroelectricity facility (installed capacity 750-900 MW) is estimated to cost more than US$1 B and development partners have been called upon to support the Government in implementation.

The existing 900 MW Kariba North Bank power station is to be expanded by an additional 360 MW. In addition the 120 MW Itezhi-Tezhi will be initiated shortly.

2.8.4 Mozambique

Cahora Bassa northern expansion (550 MW). Mozambique is seeking to embark on the construction of a 1,400 MW hydropower plant on the country's Zambezi River at Mphanda Nkuwa which is estimated to cost US$ 1.1 B to add to the Cahora Bassa station, which lies upstream of the new site. The Chinese have pledged to provide funding.

2.8.5 Ghana

Ghana has identified 17 Hydropower sites, but only two have so far been developed, principally for Aluminum production (Akisombo - 1,038 MW and Kpong - 160 MW). The government of Ghana has created a development company to implement 400 MW of the hydro project at an estimated cost of US$ 620 Million. The 400 MW But hydro project at an estimated cost of US$ 620 Million. The rest of the potential remains to be implemented.

2.9 Latin America

Latin America is the third of the major regions that provides significant and valuable development opportunities for hydropower implementation.

Of all regions, perhaps the developing economies of Latin America and particularly South America are most reliant on Hydropower, but there remain many exploitable resources untapped.

As mentioned Brazil is one of the countries with the most reserves, but there remains many projects in Brazil and in these other principal markets:

2.9.1 Argentina

Argentina's gross theoretical hydropower potential is reported to be 354,000 GWh/yr, and so far about 26% of the potential has been developed. At present there are 9,921 MW of installed capacity, with a significant part of the capacity in the projects it shares with Uruguay (Salto Grande) and with Paraguay (Yacyreta). Although it is reported that there are no projects presently under construction, some 2,400 MW of capacity is in planning - again focused on bi-national schemes with Brazil and Paraguay. Argentina provides an excellent example of developments that can be achieved across borders.

2.9.2 Brazil

Brazil gross theoretical capacity exceeds 3,000 TWh/yr of which it is postulated that 800 TWh/yr can be developed. Brazil has an installed capacity of approximately 71,000 MW, including the massive Itaipu project (shared with Paraguay), and some other very large installations. Approximately 5,000 MW of hydro capacity is under construction, and approximately 36,600 MW is in planning stages.

2.9.3 Chile

The installed hydro capacity in Chile is approximately 4,700 MW with only 300 MW under construction. 3,000 MW more capacity is in planning, including Baker (1,000 MW) Pascua (1,200 MW), Neltume (400 MW), Choshuenco (150 MW) and Punilla (100 MW).

2.9.4 Colombia

Colombia is relatively well developed with 9,000 MW of installed capacity and another 10,000 MW in planning. The country enjoys a theoretical potential of 1,000 TWh/yr, but the exploitable is only 20% of that. The current construction will only add 660 MW however. The capacity under planning includes Sogamoso (840 MW), Nechi (750 MW), Miel II (410 MW) and Andaqui (705 MW).

2.9.5 Venezuela

Venezuela gross theoretical capacity of 320 TWh/yr has only 130 TWH/yr that is economically exploitable. At present approximately 74% of the load is supplied by hydro. Venezuela has invested heavily, most notably in Guri, which includes 10,000 MW of installed capacity. The country's total installed capacity is 14,413 MW with some 2,250 MW under construction. Planned projects total 12,100 MW including Tocoma on the lower Caroni with a projected installed capacity of 2,160 MW.

2.10 China

The potential for hydropower in China is huge. The possible capacity is estimated as 402GW of which only 28% is currently (2005) utilized. The Government plans to triple hydro capacity from the 2004 level (105GW) to approximately 300GW by 2025, and some 50,000 + MW is under construction. The 2004 and 2005 installed capacity includes approximately 34.5 GW and 38.5 GW respectively indicating an addition of some 7,000 MW of large hydro. The projected 300 GW of hydro in 2025 will represent 30% of total capacity.

The potential for development in China is so large that it is difficult for this chapter to do justice to the country's resources and plans, so it will concentrate on Southwest China, including the projected AC transmission at 750kV and UHV and HVDC at +/-600kV and higher.

The abundant hydropower resources in Southwest China will be exploited in large-scale in the next three decades. Among a number of large hydropower projects in this region to be completed during the period is a project with two large hydropower stations approximately 90 km apart on the located on the Jinsha River, the upper reach of Yangtze, named Xi Luodu and Xiang Jiaba. The design of this project has been commenced recently with one hydro power station having the same total capacity as the Itaipu Power Station in Brazil and the other with close to half the capacity of the first. More than half of their power will be transmitted to East China, a distance of close to two thousand kilometers, and the remainder will be transmitted to Central China, a distance about one thousand kilometers. Moreover, further hydropower projects in the region will be developed with the requirement for transmission of large blocks of power too. The merits and drawbacks of different schemes, such as 750kV or 1150kV AC power transmission as well as the +/-600kV or +/-750kV DC power transmission, have been discussed over many years.

2.10.1 Precipitation and Topographical Conditions in Southwest China

Exploitable hydropower resources in Southwest China account for 53% of China's total. The precipitation and topographical conditions that contribute to hydropower resources in Southwest China are, in summer, the southeast monsoon originating from the Pacific Ocean and the southwest monsoon originating from the Indian Ocean heavy laden with moisture. Annual precipitation is 1500-2000 mm in this region. The topography of China is high in the West and low in the East. The Qinghai-Tibet Plateau has an average elevation of 4000 m above sea level with the world's highest mountains, the Himalayas, on its southwestern border. Many rivers originating from the Qinghai-Tibet Plateau or other high mountain ranges flow down to the low plains in the East. Thus, the rivers passing through Southwest China present great opportunities for hydropower development.

In the region, many rivers possess both abundant runoffs and very steep slopes; especially the Jinsha river system in this region, including its main tributaries such as the Yalong and Dadu rivers, which are very rich in hydropower resources and possesses about 40% of the country's total exploitable hydropower resources.

2.10.2 Prospective Large Projects in Southwest China

The exploitable hydropower resources of the Jinsha River System are shown in Table 2.4. Among them the Xi Luodu Hydropower Station will contain 18 generating units, 700 MW each, and the Xiang Jiaba Hydropower Station will contain 8 generating units, 750 MW each. Some others in Southwest China are shown in Table 2.5.

River	Length km	Total Head m	Economically feasible capacity (GW)	Annual Energy TWhr
Jinsha main stream	3481	5142	58.91	323.4
Lower Jinsha	1326	1680	50.33	274.7
Yalong main stream	1571	3870	22.16	135.5
Dadu main stream	1062	4177	20.46	108.8

Table 2.4. Exploitable Hydropower Resources of the Jinsha River System

River	Length km	Total Head m	Economically feasible capacity (GW)	Annual Energy TWhr
Wujiang	1310	3418	8.56	42.1
Yuanshui	1022	1462	3.49	15.6
lancang	2153	4583	20.88	113.3

Table 2.5 Exploitable Hydropower Resources of other River Systems in SW China

On the Yalong main steam, near the confluence of the Yalong and the Jinsha, there exists already a large project Ertan, containing six units, 550 MW each. Hydropower developments along the Jinsha River System over the next three decades are depicted in Table 2.6.

Project	Capacity	Commissioning
Xi Luodu	12.6	2020
Xiang Jiaba	6.0	2025
Hutiaoxia	6.0	undecided
Baihetan	12.5	undecided
Wudongde	7.4	undecided
Jinping	3.8	Around 2020
Pubugou	3.3	Around 2020

Table 2.6. Major Developments in the Jinsha River System

2.10.3 Associated Transmission

Different schemes for the transmission of hydropower from the two power stations have been discussed in recent years including an all HVDC, and all HVAC and a hybrid with HVDC being used for transmission to East China, while Central China receives power via HVAC. The selection of which system to use is critical, particularly in the light of the very large blocks of power to be moved. Some important correlated issues, such as the rational overall arrangement for nationwide power grid interconnections, determination of the highest voltage level going to be introduced in the coming three decades, etc., will be investigated and studied more carefully, and more widely debated before selecting the alternatives for the power transmission.

2.11 Transmission

Although all large-scale generation have constraints on location, such as ease of delivery of high volumes of fuel, and the desirability to be remote from centers of population, hydro is significantly constrained by the topographical requirements of the development. Thus the transmission of power from hydro projects often assumes greater importance (than for a fossil fuel project) in the assessment of feasibility and in the final cost of a project.

It is a challenge for any single hydro plant – other than those massive developments - to support the transmission infrastructure required for the overall utilization of hydro in a region or country, so mechanisms must be found to enable this infrastructure to be put in place independent of the individual generation investments.

In addition regional initiatives and joint investments by governments and international agencies can result in more attractive investments for those considering generation investment.

In additions interconnection of national and regional systems allows generation to iron out different demand patterns and different generation mixes to the communal benefit of all. A prime example of this cooperation is the interconnected transmission on the west coast of the USA enabling the USA and Canada to jointly benefit from the development of the upper reaches of the Columbia River, and the firming of the power in the lower reaches.

The countries of the EU are also significant exporters and importers of power even though each of them has a different architecture of their power generation.

Table 2.7 indicates the percentage of each country's production that is imported and exported:

Nation	**Export/ Production**	**Import/ Consumption**
Germany	6.5%	7.4%
France	14.7	0.7
U K	0.0	4.7
Italy	0.5	13.9
Spain	1.9	4.5
Sweden	6.4	5.3
Netherlands	0.7	13.0
EU Total	6.3	6.9

Table 2.7. Import and Export of Power in Europe [8]

In considering the harnessing of untapped hydro throughout the world, the requirement for and the potential for long distance transmission of electrical energy is paramount.

As noted elsewhere, energy transmission may be achieved through HV AC or DC lines. For more than half a century, after the struggle between AC and DC for universal production, transmission and utilization of electrical energy had been won by AC; this position remained unchallenged until the 1950s. At that time the availability of large, high powered, mercury arc rectifiers gave birth to a new interest in the application of DC for transmission purposes. One of the primary drivers for this reawakening of interest was cost - the historical justification for many of the early HVDC schemes.

The explanation is illustrated in Figure 2.7, which plots the cost of AC and DC transmission against distance. Because DC requires only two conductors (or only one if ground return is used) compared to three for normal three-phase AC transmission, the incremental cost of conductors is reduced in the same ratio. The cost of the supporting towers and insulators is correspondingly also reduced. Finally the transmission losses are also reduced in a similar fashion. These reduced incremental cost advantages are offset, however, by the greatly increased cost requirements of DC converter stations in comparison to AC substations.

Nevertheless, if the transmission distance is sufficiently long a "breakeven distance" is reached beyond which HVDC becomes more economical than AC.

Depending on the prices of copper and aluminum as well as other materials, the threshold distance changes year by year, but the approximate breakeven point is 800 km for EHV overhead lines. However, for underground or submarine cables, which cost an order of magnitude more than overhead lines, the breakeven distance is only of the order of 50km

Hence the new technology was born and was given greater impetus, both technically and economically, by the introduction of thyristor technology and other subsequent developments.

However, it is not only for cost reasons that HVDC proves beneficial to the transmission process. Sometimes it presents the only possible option:

- Long sea crossing (Cable charging current prevents the use of AC transmission over distances > 75-100km, depending on voltage)
- Frequency conversion (i.e., 50Hz to 60Hz)
- Where synchronization is impossible (i.e., due to stability reasons or frequency control regimes).

AC power transmission sometimes exhibits stability problems in long overhead transmission lines.

Examples from past, present and future of how HVDC has and might benefit long distance transmission - and the inclusion of substantial hydro in the generation mix - are presented below.

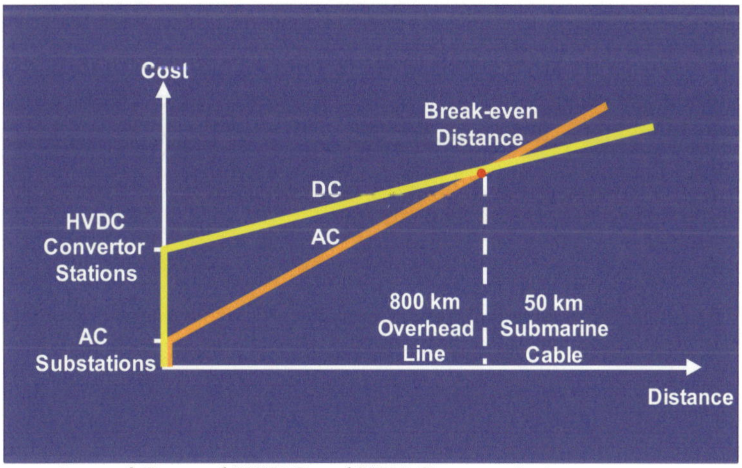

Fig. 2.7. Comparison of Costs of HVDC and HVAC transmission

Planners and interested parties have developed many proposals throughout the world for the creation of HV grids, and the interconnection of these grids, to facilitate export and import of electricity. Many such exports and imports would be provided by Independent power producers.

A legacy of the colonial period in the developing world has been that the transmission networks of each state are often isolated from those of neighboring countries. Since independence, many national power companies have struggled to maintain power supplies to the main towns and progress on extending power grids has suffered, with the development of cross-border transmission interconnectors well down the priority list. Malaysia and Singapore are an example however of countries that interchanges electricity at peak times. Cross border transmission could have great benefit - Japan's annual load factor is for example approximately 55% leading to great inefficiencies in supply capability for the peak demand. Typical advantages articulated for the extension of the Indian grid described below also apply to inter country's connection;

Economic

- Shared peak generation and exchange of energy means reduced generation plant margins in each region
- Peak load diversity permits export of surplus energy and operation of thermal plant at base load with higher efficiency
- Investment in new generation is avoided (or postponed) at a cost of only one fifth
- Improved reliability of the AC system through availability of emergency power
- Reduced industrial disruption due to load shedding.

Environmental

- Fuller use is made of seasonal high rainfall run-off, transferring extra hydro capacity to the other regions/nations
- By importing thermal energy in time of drought precious hydro resources are conserved and more efficiently managed
- Less coal burning is achieved in non hydro countries/regions if secondary energy from the hydro resources can be imported
- Construction of fewer generating stations is possible by sharing generation reserves for outages and peaks, reducing environmental impact in all regions/nations.

Some countries are focusing on the transmission aspect of development, such as Laos because "*Electric power production contributes both foreign currency accumulation and domestic social progress.*"

Within countries some major transfers of electricity are in various stages of development, including the proposed cable between Sarawak (east Malaysia) and peninsular Malaysia.

2.11.1 North America

The largest capacity HVDC scheme in North America, and one of the two earliest, is the Nelson River Scheme in Canada (Figure 2.8). The two bi-poles of this scheme carry 3,800MW over a distance of 900km from the hydro stations on the Nelson River in northern Manitoba to Dorsey Converter Station near Winnipeg. Some of this energy is dispatched further south into the industrial heartland of the US Midwest. The original bi-pole 1 employed mercury arc valves, but Pole 1 was refurbished and upgraded in 1990-1992 with modern thyristor technology. Not only does this HVDC scheme provide long distance bulk power transfer but also Power Oscillation Damping to assist with frequency control and stability of the receiving AC system.

Contemporary with the Nelson River Scheme is the Pacific Intertie that transports 3,100 MW of Columbia River hydro energy from Celilo in Oregon to Sylmar, near Los Angeles, 1,400km to the south. This HVDC tie runs in parallel with EHV AC transmission and can also be used to provide additional control and stabilization to the Pacific Coast AC system. Similar capabilities are provided by the 1,920 MW Inter Mountain HVDC Project, which brings hydro energy from the Boulder Dam on the Colorado River across 800 km into Los Angeles.

The final North American example is the 2,250 MW Quebec - New England HVDC Project that taps the rich hydro resources of the James Bay region of northern Quebec. This scheme is unusual for being multi-terminal: feeding its power not only 1500km south to Sandy Pond in New England, but also providing vital supplies to the metropolis of Montreal via an intermediate converter station at Nicolet.

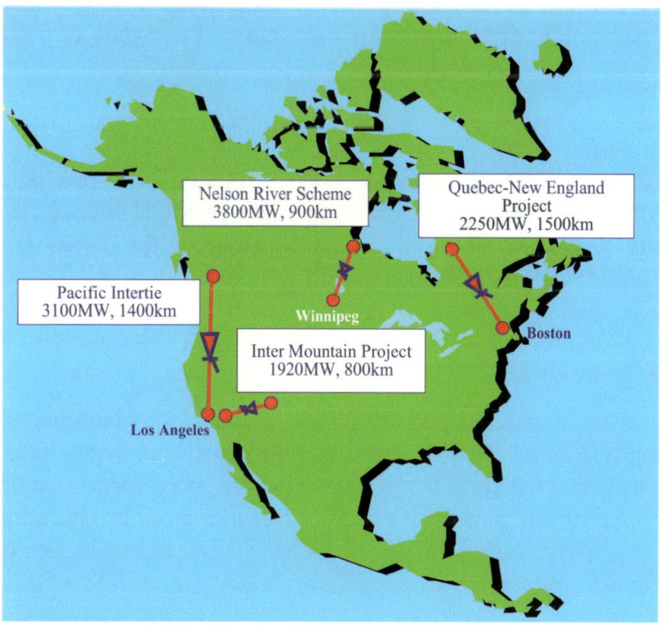

Fig. 2.8 Canadian HVDC Transmission

2.11.2 South America

South America is the location of the world's largest power station, the hydroelectric installation at Itaipu on the border of Brazil and Paraguay. Originally 12,000 MW, but recently upgraded to 13,400 MW by the addition of two 700 MW generating units, the bulk of this capacity is fed to the industrial center of Brazil near Sao Paolo. The distance is 800km and around half of the energy is transmitted by two 750kV AC circuits. However, for stability reasons it is not possible to transmit the total output by AC and 6,300 MW is supplied over two bi-poles of ±600kV, which is the world's highest DC voltage to date. The stabilizing effect of HVDC's fast acting controls ensures the stability of the overall transmission. Note also the ability of DC to transmit a similar capacity at a lower voltage than AC – and in a smaller transmission corridor.

Brazil is presently experiencing a crisis of energy supply and is in urgent need of reinforcement from both internal and external resources. Some of its neighbors (Argentina, Uruguay and Venezuela) are rich in hydro capacity, but suffer from problems of extreme remoteness and/or a different operating frequency. Brazil, a minority among South American nations, employs 60Hz whereas its neighbors operate at 50Hz. Thus, the many recent and proposed in feeds to Brazil, whether from near or far, are by means of HVDC – both long distance and back-to-back. These include existing and future interconnections with Argentina and Uruguay, together with proposed links with Venezuela. There has also existed for many years a Brazilian dream to tap the vast untapped hydro resources of the Amazon region, for which HVDC would be the prime choice on account of the extreme distance. Brazil, however, is not alone in South America in having its hydro resources located remote from the load centers. Chile is rich in hydro in the Andes Mountains of Patagonia in the south of that exceptionally long country. The Chileans are therefore studying the economics of transmitting this energy by means of HVDC to the capital Santiago.

2.11.3 Scandinavia

Scandinavia, the mountainous north of Europe, is rich in hydro potential tapped and untapped. Continental European industry, far to the south, is hungry for cheap energy and has looked to these resources for many years. In this case the distances are not extreme in overhead line terms but the difficulties remain two-fold:

- In some cases lengthy sea crossings
- In most cases unsynchronized AC systems.

All of the examples considered necessitate sea crossing by submarine cable. Although the shorter crossings are of insufficient length to either justify or necessitate the use of DC the complication exists that the large Scandinavian power pool (Nordel) cannot be synchronized to the Western European power system due to differences in operating regimes and, in particular, in frequency control strategies. Hence HVDC is in all cases the only option for tapping these hydro resources.

The existing schemes in the category of hydro deliverers are five. There are also other HVDC schemes that are not discussed, as their functions are simply interconnectors. The five are (Figure 2.9):

Skagerrak,	1,190 MW,	240 km
Konti-Skan,	550 MW,	180 km
Fenno-Skan,	500 MW,	200 km
Baltic Cable,	600 MW,	260 km
Sweden-Poland,	600 MW,	230 km.

Four of these have their primary purpose as the delivery of cheap hydro energy to continental Europe. Their secondary purpose, for which they are employed on a seasonal basis when water levels are low in the summer months and on a daily basis in winter when the domestic electric storage heating loads are high in Scandinavia, is to utilize available European thermal resources to conserve hydraulic stocks in a cost-effective manner. The exception to this pattern is the Fenno-Skan link whose purpose is to deliver hydro energy from Finland to Stockholm in Sweden.

The recent deregulation and privatization of European utilities has encouraged the increase in cross-border energy trading. Thus, many new projects have been initiated to deliver the abundant hydro resources of Scandinavia to industrial Europe. Moreover, advances in cable technology now permit much greater lengths of submarine cables, permitting delivery to even further destinations. Of the following three projects the first two have been ordered but not yet commenced and the third is currently at the bidding stage:

NorNed,	600 MW,	650km
Viking Cable,	800 MW,	600km
North Sea Interconnector,	1,200 MW or 800 MW,	700km.

The final example is uncertain until economic circumstances change, on account of the distance involved and the difficulty of the cable route due to extreme depth and currents in the Faroes Trench. This is the long discussed Iceland to Scotland link (950km) for which a single cable would permit the transport of 800 MW and two cables 1600 MW. Iceland has vast untapped hydro resources together with substantial geothermal reserves, but access thereto must continue to remain some way in the future.

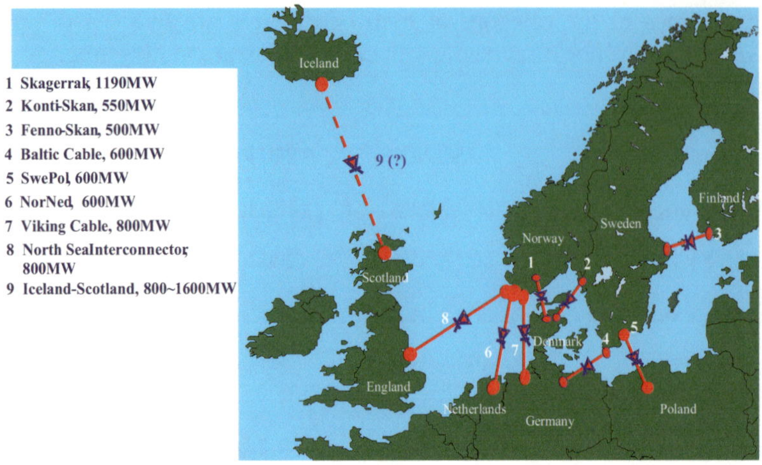

Fig. 2.9. Scandinavian Connections

2.11.4 India

In the case of India the examples of HVDC facilitating the release of hydro resources are of a completely different nature to those that have been considered so far. India comprises four distinct regions each with its own independent electric power system operating nominally at 50Hz but, because of an endemic national energy shortage, these systems are subjected to sustained variations in frequency between 47Hz and 52 Hz on a daily basis. Historically these regional systems were not interconnected because, although regional interconnections are vitally necessary to share scarce national resources, it was not possible to connect them synchronously because of their disparate frequency control regimes and for reasons of stability. The PowerGrid Corporation of India was charged by the Indian government with establishing a National Grid to facilitate the sharing of resources in order to avoid or minimize the problems of energy shortage. The only solution was to base the grid on asynchronous, HVDC interconnections. In this case the interconnections are of zero distance, with the AC-DC-AC converters located in the same station in a Back-to-Back configuration. To date the following ties link the regions (Figure 2.10):

- Vindyachal, 500 MW
- Chandrapur, 2 x 500 MW
- Visakhapatnam, 500 MW
- Sasaram, 500 MW (entered service 2002).

Repeat installations are planned for Visakhapatnam and Sasaram, with the bidding process already initiated. The South region is relatively rich in hydro resources while the generation in the other regions is essentially based on thermal with limited capacity. The benefits introduced by these links are as both economic and environmental:

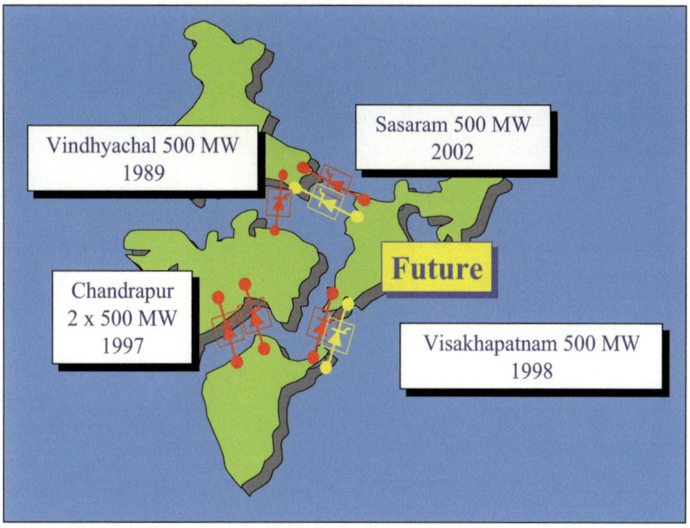

Fig. 2.10. Indian HVDC Connections

2.11.5 China

China is a country of almost continental dimensions. Its energy resources, both hydro and thermal, are located in the center or Far West whereas the centers of population and industry are located essentially on the coast. The distances across which bulk energy must be transported are enormous, ranging from hundreds of km to more than 2,000 km. The first applications of HVDC link to provide access to remote hydro resources were (Figure 2.11):

- Gezhouba - Shanghai, 1,200 MW 700 km
- Tian - Guang, 1,800 MW 970 km.

However the biggest driver for expansion of this process has been the initiation of the Three Gorges Dam project on the Yangtze River, exceeding Itaipu to become the world's largest power station, with a capacity of 18,000MW. In order to evacuate this vast capacity the following schemes have been ordered or initiated:

- Three Gorges - Shanghai (I), 3,000 MW 800 km
- Three Gorges - Guangzhou, 3,000 MW 1,000 km
- Three Gorges - Shanghai (II), 3,000 MW 800 km.

Export of China's resources to neighboring countries is also under consideration, utilizing the ample hydro potential of Yunnan province to supply the needs of Bangkok in Thailand:

- Yunnan - Bangkok, 2,000 MW 1,100 km.

Many more projects have already been identified and wait in their turn for the allocation of the necessary priority and resources for development.

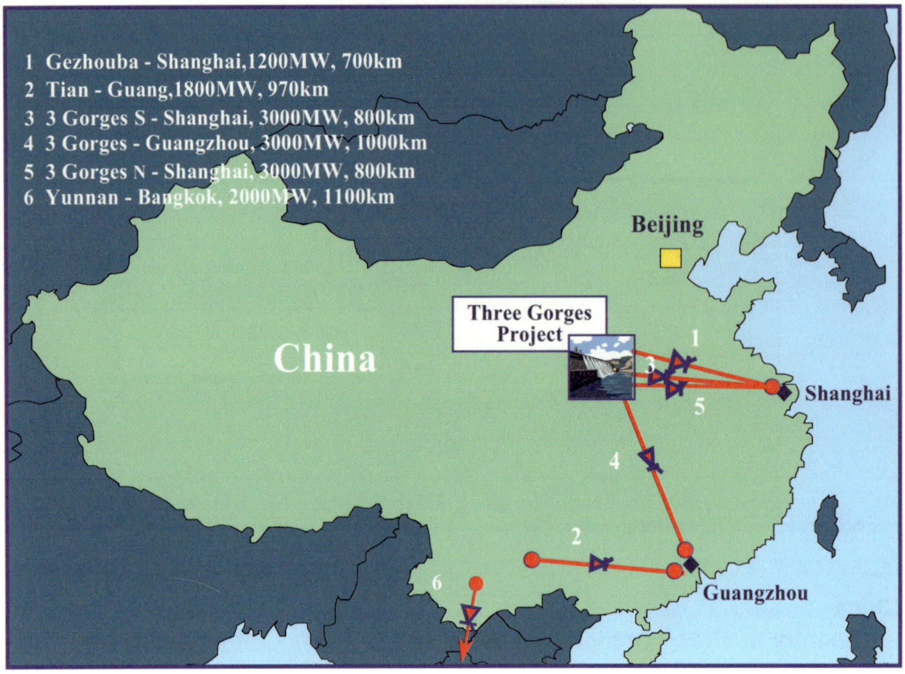

Fig. 2.11. Chinese HVDC Connections

2.11.6 Africa

Regional transmission projects in Africa are currently being supported by NEPAD.

The NEPAD strategic framework document arises from a mandate given to the five initiating Heads of State (Algeria, Egypt, Nigeria, Senegal, and South Africa) by the Organization of African Unity (OAU) to develop an integrated socio-economic development framework for Africa. The 37th Summit of the OAU in July 2001 formally adopted the strategic framework document).

Proposed Regional Transmission Projects many of which will be HVDC because of the distances and the significant energy flow are shown on Figure 2.12 and include:

- Upgrade of Zambia -DR Congo -S. Africa Interconnection
- Zambia-Tanzania Interconnection
- Namibia-Botswana Interconnection
- W. Africa Grid Network and Power Pool.

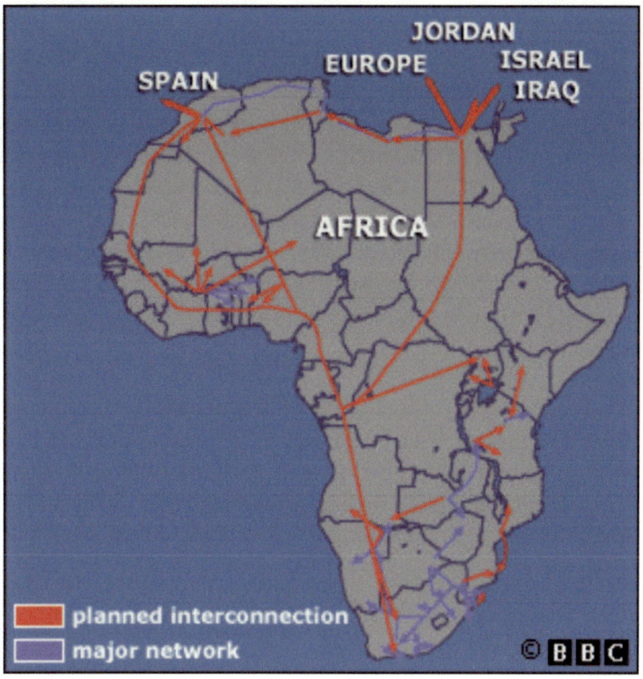

Fig. 2.12. African Transmission

2.11.7 South East Asia

A prime example of a regional project that is gradually moving forward is one sponsored by ASEAN within which a Power Forum has developed a Memorandum to increase the electricity trade. Within this forum, study groups are investigating the future of an ASEAN power system (Figure 2.13).

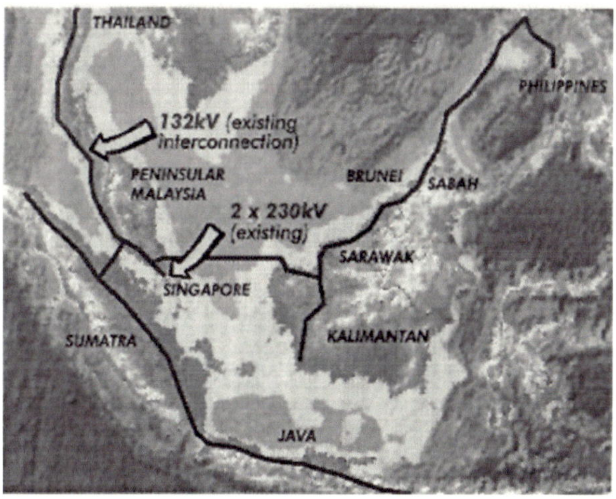

Fig. 2.13. ASEAN Power System Interconnection

2.12 Environmental

The development of water resources by humanity - including the development of hydro resources - has a range of environmental effects beyond those intended. These effects can be beneficial, or adverse, and are not simply limited to the aquatic environment. The environmental lobby has driven the debate on these matters and there has been an almost inexhaustible litany over the last few decades on the denigration that hydroelectric power plants (and in particular the dams that are an integral part of many storage schemes) may have reeked on the environment, fauna and flora and indigenous peoples. Meanwhile other expanding forms of human activity - not least the generation of electricity using fossil fuels - have inexorably continued to move mankind to the crisis that now overwhelms, in the form of irreversible climate change.

The Energy Information Agency at the United States Department of Energy and the World Energy Council monitor global energy consumption on a regular basis. EIA's report, "International Energy Outlook 2000" included a forecast that total energy consumption, worldwide, from all sources, will grow by 60 percent between 1997 and 2020. Consumption was expected to increase from 111,000 TWh/year to 178,000 TWh/year.

When the electricity share of total energy consumption is considered, the increase becomes even more dramatic. The International Energy Outlook 2000 forecast that global consumption of electricity would be 76 percent higher in 2020 than in 1997. Consumption will increase from 12,000 TWh (1997) to 22,000 TWh (2020).

By the year 2050, the world population is expected to increase by 50 per cent, from 6 to 9 billion. Energy consumption per inhabitant per year is generally in correlation with the standard of living of the population. Today the less developed countries in the world, with 2.2 billion inhabitants, have an annual per capita consumption of primary energy 20 times less than those of the industrialized countries (with 1.3 billion inhabitants), and per capita electricity consumption that is 35 times less.

World energy consumption, and especially electricity consumption, will increase considerably during this century, not only because of the demographic pressure, but also because of the development in living standards, The raising of overall global economic prosperity continues to drive the consumption of energy (including generated electricity) to record levels, with electricity consumption anticipated to increase at rates faster than overall energy supply. The majority (80 per cent) of energy available today is provided from thermal sources, i.e. coal, gas and oil; but there are growing global concerns regarding the lack of sustainability of these forms of energy that casts doubts their use in a long-term energy strategy.

Developed nations will probably seek more energy – not least for "plug in hybrids" and zero emission electric automobiles – while the rest of the world (which will represent 7 billion inhabitants by 2050 and 78 per cent of world population) will seek the basic energy consuming necessities of life that have been enjoyed for years in the West, such as lighting, refrigerators, treated water, etc.

In this scenario it is not viable to simply exclude hydroelectric power, and methodologies must be put in place to defuse and mitigate the environmental aspects of hydropower development. Essentially – just as with all human activity - the concept of sustainability must be significant part of the discussion of power generation as a whole. Under these conditions the incorporation of the most mature renewable, hydro, into the system mix is invaluable.

In comparison with hydropower, thermal plants take less time to design, obtain approval, build and recover investment. However, they have higher operating costs, typically shorter operating lives (about 25 years), are major sources of air, water and soil pollution and greenhouse gases, while providing fewer opportunities for project economic spin-offs. A comparison supplied by Manitoba Hydro is shown in Figure 2.14.

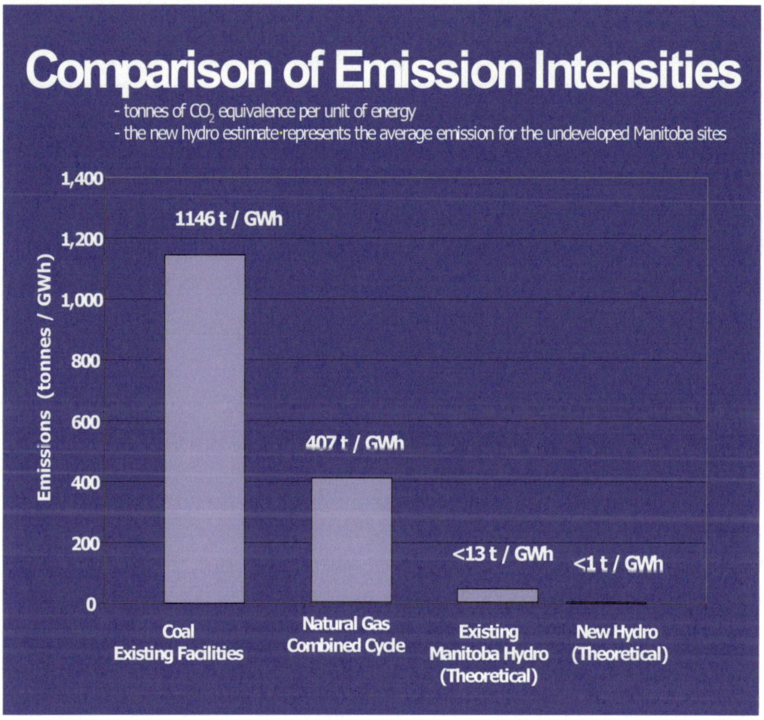

Fig. 2.14 Comparison of Emissions (Manitoba)

The average hydro power plant obviously has a place in a sustainable energy mix, but the perceived problems must be addressed.

In the late 90s an attempt to air all the environmental issues associated with the construction of dams was made by the assembly of the World Commission on Dams (WCD). This body produced a report "Dams and Development - A New Framework for Decision Making" that crafted some guidelines for development of dams. Essentially the guidelines are a restatement of the importance of dialogue with all affected parties, which of course is a central element of a mature democracy.

A recent perspective is that of sustainability, and embracing this approach may hold the key to the implementation of hydropower.

Key elements of sustainability seem to be hard to define. To date much of the debate around what constitutes "environmental friendly" generation devolves into a debate as to the definition of "green energy", "renewables" and "alternative energy", each with characteristics that make them somehow preferable to "conventional" forms of generation, such as coal-fired, nuclear, or large hydroelectric facilities. Much has been written on this subject without ever finally resolving the question of what types of electric generation technologies are "best". Best, one might ask from whose point of view? Best, according to what criteria? Such debate is endless and of little value. There is no answer to the question "how green is green?" Most jurisdictions include wind power and solar power to be green, but one state in the US (for example) considers combined-cycle gas turbines to be green.

And why - it is reasonable to ask - is large hydro not considered to be "green" by some. It is overwhelmingly renewable. Large hydro has been attacked by various people for a host of sins, including destruction of habitat, species extirpation, social upheaval, and being a source of greenhouse gas (GHG) emissions. It is simple to show that the benefits of large oligotrophic reservoirs capable of multi-year storage of energy, created in steep-sided valleys, vastly outweigh the damage done. Impacts can be amortized over many decades, equal to the life of the facility. Eventually, the impounded riverine system that has become a reservoir adapts to a new kind of ecological equilibrium as a quasi-liminetic system. In contrast, thermal generation produces impacts through gaseous pollutants every hour that the plant runs. Coal plants add ash disposal and mining-related impacts to the list. Emissions can be scrubbed, but they cannot be amortized. Any hydro-based system, whether large or small, has of course associated operating impacts due to flow restrictions, fluctuating reservoir levels, blockage of fish passage and sometime-elevated levels of mercury in reservoirs. These are difficulties that must be dealt with.

Other impacts of large hydro projects, such as habitat destruction and methane production from the decay of submerged plant matter, are harder to deal with. Habitat destruction can never really be compensated for, and any projects that contribute to the extirpation or worse, extinction of species need to be very carefully considered. To put the problem in perspective though, the amount of land areas sacrificed to reservoirs is minuscule compared to that devoted to agriculture and habitation. Is it "worse" to cover existing habitat with cornfields, parking lots, housing and factories than with water?

In the production of methane etc, all large hydro projects must not be treated equally. Methane production is greatest in reservoirs in shallow, tropical regions, and least in steep-sided valleys in cold climates. In fact the production of GHG may be less than some indicate. Although there is decomposition of biomass in the waters of a reservoir, recent research in North America [Gagnon, 1999] indicates that the GHG emission factor for hydro plants is typically 30 - 60 times less than factors for fossil fuel generation. Development of half the world's economically feasible hydropower could reduce GHG emissions by about 13 per cent, together with beneficial avoidance of sulphur dioxide ($SO2$) emissions (the main cause of acid rain) and nitrous oxide emissions. Even including the fuel required to build hydropower stations, a coal-fired plant can emit 1000 times more $SO2$ than a hydropower system. An analysis has been completed for the Tucurui plant in Brazil, including 'worst case' assumptions concerning the decomposition of flooded biomass and even under these worst case assumptions, the emission factor for Tucurui would be five times lower than that for coal plant.

Assuming that a sustainable generation policy includes significant hydro - as argued herein - those environmental effects that need to be addressed when planning and implementing hydro are:

2.12.1 River Barriers

An obvious effect of a dam is to form a barrier to movement of fish and other aquatic organisms up and downstream. This problem has been recognized for years, in terms of both the barrier itself, the reduction in water velocity upstream of the dam and loss in spawning habitat. These aspects affect different species and the different life cycle stages of species in different ways. Systems to assist upstream movement, through fish ladders, lifts, locks, trap and haul, etc. have had varying success around the world, and where species numbers have dropped additional methods such as artificial spawning channels and hatcheries have a place etc.

The attempts to assist various salmon species are most well known, but other species with the reverse life cycle, known as catadromous species, including shrimps and eels need to be considered so that the effects of dams minimized.

Much research has been done on the specific risks to different sizes and species of fish. Measures commonly used include fish screens at turbine inlets, and many countries require this by law. Finer meshed screens can be placed at times of year when fish are actively migrating. Various types of self-cleaning screen have been developed to cope with the build-up of debris. Behavioral methods have also been developed to defer fish from the intake, and guide them to the safety of a bypass channel. These include: louvre screens (which generate turbulence), bubble curtains acoustic barriers, electrical fields, and underwater lights [Turnpenny, 1999]. Well-designed behavioral systems (e.g., louvre screens or the latest acoustic screening techniques), can achieve better than 90 per cent exclusion for certain species. However, certain dams have proven to have significant impacts to native fish.

Knowledge from experimental studies about the mechanisms of fish damage in turbines has in recent years led the development of 'fish-friendly' turbines. (Pressure and velocity characteristics within a rotating turbine can be modeled and the probability of different risk conditions estimated).

2.12.2 Alteration of Flow Regimes and Temperature

Alteration of the flow regime is particularly acute in large storage reservoirs in climactic situations with accentuated wet and dry seasons, and somewhat less in those used for peaking or daily storage operation. Of course run of river facilities have the least impact.

Alteration is further accentuated if as is often the case there is a diversion component of the project such as for irrigation or water transfer.

Methodologies and policies exist to try to mitigate these effects as far as possible and should be considered. These methodologies include incremental flow in-stream methodology to determine a reasonable in-stream flow to restrict the effects to downstream fisheries.

Water emerging from a dam tends to be colder, and often has altered levels of dissolved gases, minerals and chemical content, different from those present prior to the dam. The result, in some cases, is the native fish cannot tolerate the new conditions and are forced to relocate, or suffer mortality losses. Temperature variations or excesses can sometimes be mitigated by the drawing off of water from particular levels in the reservoir that avoids the worst stress on indigenous species.

Consideration should be given to ramping rates, particularly for daily cycling and or peaking plants. A downstream reregulating dam can mitigate this, but topography may not allow this solution. It is reported that some success has been achieved by simply "stepping" of ramping.

2.12.3 Flow Diversion

When a project includes a significant diversion such as a long canal, or as a secondary factor water is drawn for irrigation or transfer, the problem of mortality of younger, weaker or larval (or egg) states can sometimes be a danger. Proper siting of the diversion, and careful screening will lessen the problem.

2.12.4 Sedimentation

Sedimentation from weathered rock, organic and chemical materials being transported in a river can become trapped in a reservoir. Over time these sediments may build up and begin to occupy a significant volume of the original storage capacity. In addition, since they are trapped, the soils cannot continue to refresh the river system downstream of the dam. The lack of the transported sediments may have adverse impacts to sustainable riparian vegetation, and to the continued use of lands for agriculture. It is considered imperative to assess as accurately as possible at the conceptual stage of a project the average annual sediment load entering a reservoir, or passing through a run-of-river project, so that appropriate measures can be taken. A number of measures can be taken such as periodic flushing or dredging from reservoirs (successful flushing has been reported in many countries, and especially in China).

2.12.5 Nutrients

The long-term operation of storage facilities can also influence the recruitment of not only sediments but also nutrients and gravel into rivers downstream of reservoirs. The loss affects river productivity; but can be offset by restoration programs.

2.12.6 Water Quality

Changes in water quality are potential outcomes from locating a dam in a river. Effects are often experienced both upstream and downstream of a dam. Some of the effects can be increased or decreased dissolved oxygen, increases in total dissolved gases, modified nutrient levels, thermal modification and heavy metal levels. Relatively few reservoirs have acute problems, and mitigation measures can be adopted if necessary.

Again the effects are highly dependent on size, shape, depth and operation rules. Narrow reservoirs with high inflows relative to outflows will tend to have minimal effects on water quality. In contrast large reservoirs with greater storage capacity and large surface area subject to seasonal solar gain allow development of seasonal stratification resulting in significant changes in water quality at various depths. At depth - particularly if biomass is present where light does not penetrate sufficiently for photosynthesis, oxygen levels can become depleted.

Solutions to these complications include the removal of biomass by careful clearing before impounding, the use of multi level intakes, and discharge through oxygenating facilities such as Howell Bunger Valves.

Unfortunately an opposite problem may occur from that of lack of oxygen, that is an excess of nitrogen. Deep spillway plunge pools can allow air-entrained water to plunge to a depth at which the pressure is sufficient to supersaturate the water with nitrogen. Simplistically, fish in the area can suffer similar afflictions to that sometimes-affecting deep-sea diver, which is the bends, (known as gas bubble disease in fish).

The solution to this difficulty is to use turbines to discharge and to try to use energy dissipation devices that avoid excessive plunging.

Despite the various attributes of reservoirs that must be addressed, many reservoirs provide an excellent environment for fish that develop in the new, expanded aquatic ecosystems. In several situations game management agencies have stocked fish in and below the reservoir, with high economic or recreational value.

2.12.7 Social Aspects

As with other forms of economic activity, hydro projects can have both positive and negative social aspects. Social costs are mainly associated with transformation of land use in the project area, and displacement of people living in the reservoir area.

Relocating people from a new reservoir area is, undoubtedly, the most challenging social aspect of hydropower, leading to significant concerns regarding local culture, reasonable spreading of economic benefit and pain, religious beliefs, and effects associated with inundating burial sites.

While there can never be a 100 percent satisfactory solution to involuntary and resettlement, enormous progress has been made in the way the problem is handled. Developed nations tend to ignore the fact that many of them addressed similar problems of involuntary resettlement (or at least resettlement driven by unstoppable economic forces). Human history has been punctuated by resettlement. The key to this problem is sensitivity and fairness, accompanied by timely and continuous communications between developers and those affected; adequate compensation, support and long term contact. It is vitally important to ensure that the disruption of relocation is balanced by some direct benefits from the project.

The countries in Asia and Latin America, where resettlement is a major issue, have developed strategies for compensation and support for people who are impacted, and an increasing number of examples are demonstrating that current strategies may be proving successful.

Although displacement by hydropower can be significant and must of course be well handled, the reader must keep in mind that other electrical generating options can also cause significant resettlement: coal mining and processing and coal ash disposal, also displace communities. GHG-induced climate change may eventually cause massive population migrations, if sea levels rise substantially.

As with the other environmental effects, social effects of hydro schemes are variable and project specific. A private developer must closely work with national and regional governments to provide for this aspect early in the planning stage of a project mobilizing sufficient resources and ensuring that the plan aligns directly with established national political and social policy. It is appropriate for the national and/or regional host government to lead and direct the required relocations. Whenever adverse impacts cannot be avoided or mitigated, compensation measures can be implemented.

A developer can often ensure that benefits are distributed, at least in the short term by utilizing local labor for the construction phase of a hydro scheme (which often lasts several years). Required access roads lead to easy influx of outside labor and the development of new economic activities, with resulting tensions if local and potentially resettled populations in the area are unprepared.

2.12.8 A Sustainable Portfolio

In conclusion, the environmental disbenefits, and benefits of hydro and the development of hydro around the world must be considered in the light of the sustainability of any given energy generation portfolio, whether the sample is restricted to an individual nation or is regional.

Some authorities have described four system conditions that allow us to test whether a generation portfolio meets the conditions for sustainability, at least with respect to its environmental dimension. The four system conditions are:

Substances from the earth's crust must not systematically increase in nature

Does a generating system including hydro meet this test? Yes. The greenhouse gas intensity of our system is substantially driven by fossil fuel generation. As an example BC Hydro, which is substantially hydro based, contributes only 42 tonnes CO2e/GWh (carbon dioxide equivalent per gigawatt hour) compared to the Canadian average of 230 and the US average of 610. As an example outside of the North America, it is reported that fossil-fuel generation, in China, contributed 23 million tons of SO2 in

1995, causing 40 per cent of the total land area to be seriously affected by acid rain. The resulting damage to crops, forests, materials and human health was calculated, in 1995, to be more than US$ 13 billion.

In North America the consumption of coal is at approximately the same level, though with somewhat more advanced emission "scrubbing".

Substances produced by society must not systematically increase in nature

Does a hydro's generating system meet this test? Yes -again using the example of BC Hydro, the only significant pollutants other than CO2 from the BC Hydro generation system is nitrogen oxides (NOx). Efforts are ongoing to substantially reduce NOx emissions using with selective catalytic reduction technology.

The physical basis for the productivity and diversity of nature must not systematically be diminished

Does Hydro generation meet this test? Yes- although undoubtedly, reservoirs have diminished productivity and diversity to some extent. Properly organized mitigation programs that enhance habitat productivity and diversity using techniques like spawning channels and minimum flows go a long way to keeping impacts within tolerable bounds.

Fair and efficient in meeting basic human rights

Does hydro generation system meet this test? It is difficult to say in general. To pass this test, the principles discussed above with respect to relocation, etc. must be addressed. Hydro generation clearly provides long term affordable energy to meet economic and lifestyle objectives, and with appropriate attention to the societal effects by responsible governments can be minimized.

2.13 Project Development

Although hydropower perfectly fulfils the requirements of sustainable development and is a major tool to reduce global warming, the technically feasible global potential is very little used at present (see Section. 2.2). Hydropower development is mainly hindered by the high and long-term investments required and by the fact that potential hydropower sites are often located at great distances from the dense consumer areas. Furthermore, large projects, especially these with large reservoirs, invoke severe discussions concerning their environmental impacts.

The strategies to overcome these disadvantages in the competition market in energy sectors are as follows:

- Privatization of the energy market and innovative financing of hydropower projects for example on the basis of BOO (Build-Operate-Own) and BOT (Build-Operate-Transfer) models.

- Developing hydraulic schemes as multipurpose projects and splitting the costs.
- Development of revolutionary technologies based on superconductors for the transportation of electricity over long distances with insignificant loss.
- Taking into consideration of environmental and socio-economical issues from the very beginning of prefeasibility studies and involvement of ecologists as well as of all persons concerned by the project at its early stage of design.

2.14 The Future

This chapter has highlighted the three phases of the development of hydropower and has examined some of the opportunities to harness the untapped potential of the world.

Two facts are well understood by economists; first that of the world infrastructure stocks, the electricity sector needs to form a greater percentage (compared with for example roads and railways) and secondly that as a percentage of those infrastructure stocks, higher and middle income countries demonstrate nearly twice the value in the electricity sector than low income countries. A third aspect, highlighted in this chapter is the relative abundance of hydro potential in those countries in most need of power, and the final part of the equation is the fact that hydro is relatively benign to the climate compared to other generation.

The world has become increasingly aware of the overall damage being inflicted on the environment from a plethora of activities of mankind. Although hydro has drawbacks in terms of inundation, interruption of sedimentation, water quality etc., mankind has begun to understand that climate change and environmental degradation is a complex topic, perhaps - at present - too complex for any of us to fully understand, and perhaps hydro's advantages outweigh its disadvantages.

In the context of the scientific community's recognition that perhaps the main threat to biodiversity and food production is global climate change, the main issue appears to be to what degree will society accept some local impacts of hydropower, in order to mitigate the global impacts of climate change and other environmental threats from thermal pollution. In short we cannot afford to dismiss any form of renewable energy from the supply, and power generation solutions must be found that have the minimal impact on the climate.

Unfortunately in this period when there should be a beneficial acceleration of hydro development, the retreat of the major international agencies - such as the World Bank – from participation in major hydro development, in no small part because of the eloquence of the environmental community, has created a hiatus in the flow of funding of development, at least that funding based in the West.

Meanwhile the demand for increased power generation continues to climb, particularly in those regions of the world striving to "catch up" with the standard of living of the West.

There are only four main forms of finance available for the construction of hydropower:

- Reinvested capital from existing utilities (both private and public)
- Host nation government capital
- Multilateral agency capital
- Private finance both from within the host country and from without.

There are challenges in attracting capital from these four sources to hydro that affect all of them to one extent or another:

1 Significant investment is required for rehabilitation of existing facilities and for "catch up" maintenance
2 The necessity of investing almost 100% of the capital before any return (compared to "pay as you go" for fossil fuel)
3 Uneconomic and unbalanced tariff structures, rendering the whole power sector financially unstable
4 Lack of creditworthiness in customers whether they are government institutions, industry or private purchasers
5 Significant associated infrastructure development needs such as access roads and transmission.
6 Small markets
7 The necessary addressing of environmental issues, often aggravated by external groups.

In the developing nations that have the greatest hydro potential, reinvested capital from the existing utilities and capital from the host nation government are often not available. Governments encouraging development have huge demands for capital from all sectors of infrastructure both for new works and for rehabilitation. The lack of such capital has been a key problem for development and often responsible for the challenges facing the power generation industry, and many governments have realized that public sector funds are simply inadequate concluding that the payment burden needs to be shifted as far as possible from taxpayer to user. But constraints to private sector investment are many and progress on regulatory, restructuring and privatization reform has yet to bring the dividends that are needed

As discussed the private sector has been invited to invest in hydro in the developing world but there are significant difficulties for private financing. It is well known that hydro engineering has reached a level of sophistication and maturity such that, given previous experience in the development of hydro, most technical difficulties of hydro implementation are well understood and can be solved (at a price). The main difficulties pertain to accurately forecasting and quantifying the risks and associated costs of each individual project. Numerous different factors control whether and to what extent private funding is available for a development in this "Phase III' of hydropower project development throughout the world.

One of the difficulties with attracting private investment and finance to hydropower projects is the need for a higher return on equity than was traditionally sought by utilities and the multi lateral agencies. This has led to a system where heavy debt leveraging is essential. The large size of power sector investments and the shorter-term outlook of private investors also affect the nature of the projects that can be undertaken in the private sector. With the necessity of attracting private finance, controlling factors in development of power generation, and particularly of hydro are: (i) the scale of the capital investment, (ii) possibility for an attractive return on equity and minimum feasible debt service characteristics, (iii) security of project revenue during debt service, and (iv) management of the major project risk factors.

Table 2.8 indicates the principal risks associated with a hydro development.

Political/Economic	Government Rules and Regulations
	Inflation
	Tax rate Variations
	Economic Force Majeure
Commercial	Demand
	Power Purchaser Credit
	Power Purchaser Longevity
	Interest
	Refinancing
	Capital and Credit Availability
Currency	Exchange Rates
	Repatriation
Technical (Geology and Hydrology)	
Environmental	Inundation and Loss of Land Base
	Impacts on terrestrial and aquatic Species
	Approvals procedures
Social	Resettlement
	Public Attitudes to development
	Project Area impacts and compensation
Return on Investment	
Construction Time	Schedule delays and associated costs

Table 2.8. Hydro Development Risks

All the difficulties must be addressed in order for private capital to be mobilized more fully, and to more efficiently use the available government and multi lateral finance. Assistance is needed from the international funding community if progress is to be made.

At the most basic level, hydropower participates in a worldwide intense competition for capital. The capital market does not give "preference" to infrastructure and power development as the World Bank and other multilateral and bilateral agencies have been doing.

In fact power development, and particularly hydro is at a significant disadvantage compared to many other investments. Hydro projects of necessity often require a relatively long period of negative cash flow before any return can be realized, and investors must somehow be tempted to invest preferentially in hydro instead of (for example) factories producing domestic and export goods readily marketable and profitable in western countries.

Accordingly, the nature of the hydro projects to be undertaken in the private sector will be different from the mega projects previously considered by the major national utility companies. A review of the risks inherent in development can lead to an understanding of the projects more likely to be attractive to investors

The multilateral agencies have in the last ten years been less enthusiastic in funding hydro power, often as a result of the organized onslaught of criticism from opposition groups, which have at times protested directly to potential contractors and suppliers associated in providing implementation expertise.

The following characteristics are apparent in projects that have been demonstrated to be "bankable", or considered desirable by private investment:

- High Head - so that minimal amounts of water are needed, and Pelton wheels (i.e. simple and easily maintained equipment) can be used. High head also tends to require less reservoir area, which can reduce environmental impacts and approvals procedures
- Run of River - so that diversion structures are small and storage is minimized, again keeping costs low and reducing the environmental impacts associated with large reservoirs
- Surface Based Configuration - to minimize the construction and geological risks attendant to tunnels and underground powerhouse caverns
- Compact - so that the smallest stretch of river is affected
- Appropriate Size - to minimize exposure to potential future slowdown in the regional electricity demand
- Short development cycle and debt repayment.

Developers are no longer exclusively engineers and thus have had less exposure to the technical aspects of development. In the contemporary scenario developers are often financial experts with a focus on minimizing or avoiding risk that will look to a power project merely as a business investment that can be evaluated on the same basis as any other competing investment in other sectors of the economy. Such investors do not have an inherent technical connection with the industry other than its opportunity to meet attractive investment conditions. Therefore the typical developer will be seeking to offset risk, and place it with appropriate parties (who can manage it) along with meeting investment objectives.

A developer will be fully prepared to pay for offsetting risk, on condition of course that those placement costs can ultimately be recouped. As a result, development philosophy and practice are currently directed toward the Engineer/Procure/Construct (EPC) form of contracting in which much of the construction and design risk is placed on the contractor who is assumed to be more capable of managing this risk. It is also notable that the contractor would be much more familiar with these risks than would be the investors who often do not have long connections to the power or construction industry.

In general the ideal placement of commercial risk would be with the power purchaser (or the market) while the political risk is managed by selecting investment locations meeting minimum acceptable conditions. The remaining political risk may be mitigated by purchasing some cover through insurers or from multilateral agencies such as World Bank, Asian Development Bank, and other institutions.

Political and other market risks do typically decline as a host economy maintains its development. It is, therefore, not surprising that the power generation sector is moving forward more vigorously (in general) in those countries that have the potential to raise significant or all the required debt in their own financial markets. In other cases, as noted, the multilateral agencies have an important function that they are increasingly exercising in accepting the political risks attendant to a particular development proposal.

The scale of projects that may be expected to be developed by private financing in a particular locale or country is a subject of some interest. Given the list of desirables characteristics described earlier, and developer's orientation toward limiting their risk exposure, it is not surprising that in general hydropower project developments in Asia have been and can be expected to continue to be of limited size. Apart from one or two notable exceptions, privately funded development to date tends to be less than about 180 MW. Few privately funded projects larger than 250 MW are anticipated in the foreseeable future other than under very special conditions where the national government may take a direct role in risk management in partnership with the developer.

As economies become more developed, as power prices more fully reflect real investment costs, and as the equity and reinsurance markets develop further, gradually larger projects may be expected. However, it is worth noting that some of the geo-technical and cash flow difficulties and risks that are attendant on hydro projects are less important for thermal projects. Unless there are other constraints on thermal development, such as those related to international agreements on global warming, thermal project proposals will continue to be regarded by private developers as more viable than hydro and will take precedence.

What have been termed "mega Projects" (an arbitrary definition might be those above 1000 MW) clearly are not favored under the present scenario for private development, and will for the moment remain outside of the pattern. Projects of this scope and size encompass extraordinary market risk, often have significant geotechnical and construction risk, and of course may become a lightning rod for enhanced political risk. However, as shown by the example at Bakun, in many cases a mega project private development proposal is unlikely to succeed in the absence of extraordinary support from the government and special power purchase and contractual terms.

In the meantime, in the absence of funding from the international agencies and the difficulties of attracting private finance, a powerful force has appeared that may facilitate rescue of major hydro development. The Chinese government through numerous agencies such as the China Exim Bank and quasi government organizations such as Sinohydro, and the Three Gorges Corporation, are supporting many projects particularly in Asia and Africa, by financing, and constructing the projects.

As the other countries and international organizations shy away from hydropower development assistance Chinese companies and banks are now involved in billions of dollars worth of contracts and memos of understanding to construct nearly 50 major projects in 27 countries. It has been reported that officially China does not attach "strings" to its loans and grants.

In Southeast Asia alone, some 21 Chinese companies are involved in 52 hydropower projects of various sizes, according to research issued this year at the China-ASEAN Power Cooperation & Development Forum.

There will eventually be an end to China's largesse, and in order to mobilize finance from the greater international community, it is imperative to make the environmental process more predictable. Not only that, but the market must give clear price signals to the financial community that the development of resources that have low emissions present less risk and greater reward. Renewable Energy credits and carbon offsets can also help. In the various markets in which Hydro plays a part some or all of the following challenges need to be addressed:

- Clear Energy Policy (National, regional and global)
- Simplifying and streamlining regulatory requirements and approvals (the Decision Making Process)
- Furthering Public-Private Partnerships
- Transparent and equitable regulation
- Fully, but efficiently, engage stakeholders (including benefit sharing)
- Provide fiscal incentives (tax holidays, tax credits, green credit (carbon offset) programs)
- Market signals favoring low emissions (consistent signals for sustainable development)
- Strengthening of local financial markets to allow for minimization of exchange rate risks
- Transmission infrastructure investment
- Significant investment for rehabilitation and catch up maintenance
- Reform of uneconomic and unbalanced tariff structures, which render electricity markets financially unstable.

Attending to these aspects, cumulatively and with the global pricing signals, could form the basis of guidelines for the development and management of hydropower projects and constitute a sustainable approach to renewable hydropower resource development.

A significant number of developed countries now have legislation, regulations and incentive packages to encourage the development of various renewable generation within their own countries - perhaps now is the time to enhance the conditions for overseas development assistance for renewables and medium to large scale hydro by similar practices encouraging cross border hydro investment in developing nations.

As part of the restructuring of the energy markets, the creation of a spot market sometimes occurs, but spot markets in energy are too volatile to signal investment in hydro with perhaps the special case of pumped hydro which can take advantage of high differential prices during the day.

Hydroelectric power has an important role to play in the future, and provides considerable benefits to an integrated electric system. The worlds remaining hydroelectric potential needs to be considered in the new energy mix, with planned projects taking into consideration social and environmental impacts, so that necessary mitigation and compensation measures can be taken. Clearly, the population affected by a project should enjoy a better quality of life as a result of the project.

Any development involves change and some degree of compromise, and it is a question of assessing benefits and impacts at an early enough stage, and in adequate detail, with the full involvement of those people affected, so that the right balance can be achieved.

Two billion people in developing countries have no reliable electricity supply, and especially in these countries for the foreseeable future, hydropower offers a renewable energy source on a realistic scale.

2.15 Acknowledgement

This Chapter has been prepared by Brian Sadden, Consulting Civil Engineer, Montgomery Watson, Harza, USA.with contributions by David. A. Balser (Manager Environmental Group, BC Hydro, Canada), Olcay Unver (Regional Development, Southern Anatolia Project, Turkey), the late Jan Veltrop (Commissioner, World Commission on Dams, USA), Yang Haitao and Yao Guocan (EPRI, Beijing, China), Brian Gemmell (Marketing Manager (North America), ALSTOM Power Electronic Systems, New York, USA), John Loughran (GEC, Stafford, UK), and Hilmi Turanil (Manitoba Hydro, Canada).

2.16 References

[1] Renewables 2007 – a global status report by Renewable Energy Policy Network for the 21st Century.
[2] Boletim Energia No 206, published by ANEEL, February 2006
[3] Powering China's Development: The Role of Renewable Energy, Eric Martinot, Li Junfeng, November 2007[4.World Atlas and Industry Guide published annually by the International Journal on Hydropower and Dams
[5] International Water Power & Dam Construction Yearbook (1997)
[6] ICOLD 1998
[7] International Energy Authority
[8] United Nations "Energy Statistics Yearbook, United Nations, 1995

3

Harnessing Untapped Biomass Potential Worldwide

3.1 Introduction

Biomass includes all kinds of non-fossil organic matter that is available on a renewable basis for conversion to energy and products. It is an abundant, geographically widespread, low sulfur, and carbon neutral fuel resource. It includes crops and agricultural residues, commercial wood and logging residues, animal wastes, and organic portion of municipal sold waste, and methane gas from landfills. According to the United Nations, biomass accounts for about 14% of world energy use and over one third of energy use in developing nations. It is estimated that the renewable, above ground biomass that could be harvested for power production is many times the world's total annual consumption.

Biomass-to-electricity power generation is a proven electricity generation option. Today in North America, biomass has 11 GW of installed capacity and along with wind power is a significant source of non-hydro renewable electricity. More than 500 facilities around the U.S. are currently using wood or wood waste to produce combined heat and power. This installed capacity consists of about 7.0 GW from forest products and agricultural wastes, about 2.5 GW of municipal solid wastes (MSW) and 0.5 GW of landfill gas.

The majority of biomass used today is a residue produced either in the primary or secondary processing industries, or as post consumer residues. Many of the industries that process wood or sugar cane are themselves significant consumers of energy in the form of process heat and electricity so that this is a sector with a considerable amount of Rankine cycle combined heat and power (CHP) installations. However, many of them underutilize their residues. Post consumer residues, as urban wood and landfill gas, already make a significant power contribution in the United States, Europe and Japan. Large-scale expansion will require increased harvest residue collection and use in the form of forest thinnings, wood slash, straws and stalks from cereal crops, as well as the development of energy crops.

A U.S. supply curve for 2020 is discussed with its approximately 450 million tonne (Mt) potential, as well as a USA stretch potential for the middle of the century of a Gigatonne (Gt).

Energy generation through the combustion of municipal waste is gaining in use. Recovering energy from garbage has evolved over the years from the simple incineration of waste in an uncontrolled, environmentally unfriendly way to the controlled combustion of waste with energy recovery, materials recovery and sophisticated air pollution control equipment insuring that emissions are within US and EU limits. The waste-to-energy industry has proven itself to be an environmentally friendly solution to the disposal of municipal solid waste and the production of energy. Recovering energy from the waste is an excellent idea and waste-to-energy is a clean, renewable, sustainable source of energy, and a common sense alternative to land filling.

Biomass is proven in many power-producing applications for base and intermediate load. Relative to conventional fossil fuels, however, biomass has relatively low energy density, requires significant processing, is an unfamiliar fuel among potential customers and is relatively expensive at the burner tip. In a world driven by calculations of rates of return to capital, biomass fuels are relegated to the position of an opportunity fuel with a large untapped potential in mainstream energy markets. Motivating the power industry to use more biomass fuels - to tap into the biomass energy potential - will require policy interventions from R&D investments to tax and other policy incentives. Many policy interventions existing in the United States are compared to a few examples of the European approach.

Recent US experience on actual biomass demonstration projects illustrates the difference properly targeted policy incentive can have on biomass' ability to meet its untapped potential. As an example, the Antares Group Inc. is participating in several biomass power demonstration projects. These include switch grass co-firing in Iowa, willow and residue co-firing in New York State, and gasification for combined heat and power in Connecticut. It is policy incentives that make all these projects financially viable. An overview of these projects with and without the policy incentives makes that point clear.

The electricity production from biomass is and will continue to be used as base-load power in the existing electrical distribution system. A series of case studies are discussed for the three conversion routes for Combined Heat and Power applications of biomass—direct combustion, gasification, and co-firing. The cost of electricity and cost of steam as a function of variables such as plant size and feed cost are estimated using a discounted cash flow analysis described here.

Environmental considerations are also addressed. Two primary issues that could create a tremendous opportunity for biomass are global warming and the implementation of Phase II of Title IV of the Clean Air Act Amendment of 1990 (CAAA). The environmental benefits of biomass technologies are among its greatest assets. Global warming is gaining greater salience in the scientific community and among the general population. Co-firing biomass and fossil fuels and the use of integrated biomass gasification combined cycle systems can be an effective strategy for electric utilities to reduce their emissions of greenhouse gases.

As an example of a new bio-power option for distributed generation and CHP for rural enterprises, homes and small communities, the BioMax from Community Power Corporation (CPC) which uses a variety of biomass residues to provide power and heat is described, discussed, and evaluated. CPC's BioMax systems are skid-mounted, fully automated, environmentally friendly bio-power systems configured for combined heat and power applications that consist of an advanced and controllable downdraft gasifier integrated with an engine/generator that produces 5, 20 and 50kW from producer gas.

Included is an assessment of applicable technologies for rural development with Senegal Bio-Mass exploitation. This evaluates the latest technology options for utilizing feedstock from Senegal's groundnut industry in a mix with other government initiatives such as waste-to-energy programs. It assesses some of these technologies from the green power sector against local Senegal conditions. The implications for other Economic Community of West African States (ECOWAS) countries with similar rural supply challenges and other fuel source types are evaluated with recommendations.

3.2 An Overview of Biomass Combined Heat and Power Technologies

Bio-power is a commercially proven electricity generating option in the United States, and with about 11 GW of installed capacity is a significant source of non-hydro renewable electricity. The capacity encompasses about 7.5 GW of capacity using forest product and agricultural industry residues, about 3.0 GW of MSW-based generating capacity and 0.5 GW of other capacity such as landfill gas based production. Bio-power experienced a dramatic factor-of-three increase in grid-connected capacity after the Public Utilities Regulatory Policy Act (PURPA) of 1978 guaranteed small electricity producers (less than 80 MW) that utilities would purchase their surplus electricity at a price equal to the utilities' avoided cost of producing electricity. In the period 1980-1990, growth resulted in industry investment of $15 billion dollars and the creation of 66,000 jobs

Today's capacity is based on mature, direct combustion boiler/steam turbine technology. The average size of bio-power plants is 20 MW (the largest approaches 75 MW) and the average efficiency is 20%. The small plant sizes (which leads to higher capital cost per kilowatt-hour of power produced) and low efficiencies (which increase sensitivity to fluctuation in feedstock price) has led to electricity costs in the 8-12 ¢/kWh range.

The next generation of stand-alone bio-power production will substantially mitigate the high costs and efficiency disadvantages of today's industry. The industry is expected to dramatically improve process efficiency through biomass co-firing in coal-fired power stations, through the introduction of high-efficiency gasification combined cycle systems, and through efficiency improvements in direct combustion systems made possible by the addition of dryers and more rigorous steam cycles at larger scale of operation. Technologies presently at the research and development stage, such integrated gasification fuel cell systems, and modular systems are expected to be competitive in the future.

A series of case studies [1] have been undertaken on the three conversion routes for CHP applications of biomass—direct combustion, gasification, and co-firing. The studies are based on technology characterizations developed by NREL and EPRI [2], and much of the technology descriptions given are excerpted from that report. Variables investigated include plant size and feed cost; and both cost of electricity and cost of steam are estimated using a discounted cash flow analysis.

The nearest term and lowest-cost option for the use of biomass is co-firing with coal in existing boilers. Co-firing refers to the practice of introducing biomass as a supplementary energy source in high efficiency boilers. Boiler technologies where co-firing has been practiced, tested, or evaluated, include wall- and tangentially-fired pulverized coal (PC) boilers, cyclone boilers, fluidized-bed boilers, and spreader stokers. Extensive demonstrations and trials have shown that effective substitutions of biomass energy can be made up to about 15% of the total energy input with little more than burner and feed intake system modifications to existing stations. After tuning the boiler's combustion output, there is little or no loss in total efficiency, implying that the biomass combustion efficiency to electricity would be about 33-37%. Since biomass in general has significantly less sulfur than coal, there is a SO_2 benefit; and early test results suggest that there is also a NO_x reduction potential of up to 20% with woody biomass. Investment levels are very site specific and are affected by the available space for yarding and storing biomass, installation of size reduction and drying facilities, and the nature of the boiler burner modifications. Investments are expected to be in $100-700/kW of biomass capacity, with a median in the $180-200/kW range.

Another potentially attractive bio-power option is based on gasification. Gasification for power production involves the devolatilization and conversion of biomass in an atmosphere of steam or air to produce a medium- or low- calorific gas. This biogas is used as fuel in a combined cycle power generation cycle involving a gas turbine topping cycle and a steam turbine bottoming cycle. A large number of variables influence gasifier design, including gasification medium (oxygen or no oxygen), gasifier operating pressure, and gasifier type. The first generation of biomass GCC systems would realize efficiencies nearly double that of the existing industry. Costs of a first-of-a-kind biomass GCC plant are estimated to be in the $1800-2000/kW range with the cost dropping rapidly to the $1400/kW range for a mature plant in the 2010 time frame.

Direct-fired combustion technologies are another option, especially with retrofits of existing facilities to improve process efficiency. Direct combustion involves the oxidation of biomass with excess air, giving hot flue gases that produce steam in the heat exchange sections of boilers. The steam is used to produce electricity in a Rankine cycle. In an electricity-only process, all of the steam is condensed in the turbine cycle, while in CHP a portion of the steam is extracted to provide process heat. The two common boiler designs used for steam generation with biomass are stationary- and traveling-grate combustors (stokers) and atmospheric fluid-bed combustors. The addition of dryers and incorporation of more-rigorous steam cycles is expected to raise the efficiency of direct combustion systems by about 10% over today's efficiency, and to lower the capital investment from the present $2,000/kW to about $1275/kW.

Bio-power is unique among renewable energy sources because it involves combustion that releases air pollutants. Major emissions of concern from bio-power plants are particulate matter (PM), carbon monoxide (CO), volatile organic compounds (VOC), and nitrogen oxides (NO_x). Biopower sulfur dioxide emissions are typically low because of the low amount of sulfur usually found in biomass. Actual amounts and the type of air emissions depend on several factors, including the type of biomass combusted, the furnace design, and operating conditions.

Life cycle assessment studies [3] have been conducted on various power generating options in order to better understand the environmental benefits and drawbacks of each technology.

Material and energy balances were used to quantify the emissions, energy use, and resource consumption of each process required for the power plant to operate. These include feedstock procurement (mining coal, extracting natural gas, growing dedicated biomass, collecting residue biomass), transportation, manufacture of equipment and intermediate materials (e.g., fertilizers, limestone), construction of the power plant, decommissioning, and any necessary waste disposal.

The life cycle assessment studies have permitted the determination of where biomass power systems reduce the environmental burden associated with power generation. The key comparative results can be summarized as follows:

- The GWP of generating electricity using a dedicated energy crop in an IGCC system is 4.7% of that of an average U.S. coal system.
- Cofiring residue biomass at 15% by heat input reduces the greenhouse gas emissions and net energy consumption of the average coal system by 18% and 12%, respectively.
- The life cycle energy balances of the coal and natural gas systems are significantly lower than those of the biomass systems because of the consumption of non-renewable resources.
- Biomass systems produce very low levels of particulates, NO_x, and SO_x compared to the fossil systems.
- System methane emissions are negative when residue biomass is used because of avoided decomposition emissions.
- Biomass systems consume very small quantities of natural resources compared to the fossil systems.

3.3 Biomass Availability for BioPower Applications

The estimation of biomass supplies is confounded by the many ways in which biomass is generated and used, especially as today the biomass for energy stream is composed of residues from primarily industrial and societal activities. Thus, the production of biomass feedstocks and bio-energy use is very dependent on the functioning of some other component of the economy, the three major areas being: forestry, agriculture, and the urban environment. While this includes a wide range of resources, ranging from primary residues through to post consumer residues, energy crops also have a significant potential.

To simplify the discussion of biomass it is necessary to provide some definitions and characterization of where in the economy biomass is generated or utilized as bio-energy. One methodology is to identify the stage of processing/utilization since the creation of the biomass by photosynthesis.

It is also necessary to note that there is no biomass currency such as the tonne of oil equivalent (toe). However, the majority of biomass is composed of lignin, cellulose, and hemicellulose polymers in proportions such that most lignocellulosics have a calorific value in the range of 17.5-18.6 GJ t^{-1} when measured on a totally dry basis. Each tonne of biomass has 5 MWh_{th} energy content. A gigatonne has a 5 PWh equivalent of primary energy. The world Total Primary Energy Supply (TPES) in 2001 was about 120 PWh. Current global estimates of future biomass potential are of the same order, though today the world biomass consumption is estimated at about 13 PWH (TPES).

3.3.1 Energy Crops

Energy crops are a primary supply and involve the production and growth of biomass specifically for biomass to energy and fuels applications. This is widespread in developing countries for fuel wood, as well as examples of Eucalypt forestry for charcoal production in iron production in Brazil [4]. Also, in Brazil a significant fraction of the sugar cane crop is dedicated to ethanol production [5], while 9% of the U.S. corn harvest is used in the production of ethanol from starch [6]. Research and development in Europe and the United States is developing the use of woody or straw materials (lignocellulosics) as high yielding non-food energy crops. The impact of energy crops in moving the biomass supply away from what is available as a residue can be seen from the following example. Assuming a 38% efficiency, a 1 Mt annual supply base can support a generating capacity of 225 - 240 MW operating at a 90% capacity. Using an energy crop yielding 15 t ha^{-1} y^{-1} the area planted to the energy crop would need to be about 70 kha, representing less than 4% of the land area inside a circle of 80 km centered on the power plant. Typical ratios of energy out: fossil energy in, for such a plant, would be about 1:12 while the carbon dioxide emissions would be < 50 g kWh^{-1}, or even zero if the energy crop accumulates soil carbon at current anticipated rates.

3.3.2 Primary Residues

Primary residues are produced as a by-product of a primary harvest for another material or food use of grown biomass. A representative of this is the use of tops and limbs as well as salvage wood from forestry operations cutting saw-logs or pulpwood. This material along with forest thinning is a developing biomass supply system in Finland, for example [7]. Much of the research in the United States in recent years has focused on corn stover *(Zea mays)* as a large scale opportunity primary residue associated with the harvest of the principal grain crop [8].

3.3.3 Secondary Residues

The majority of biomass used today in the energy system is generated as secondary and tertiary residues. Secondary residues arise during the primary processing of biomass into other material and food products. Sugarcane bagasse is widely used to fuel CHP providing the heat and electricity needs of sugar processing as well as export of electricity to the grid. In the forest industries, black liquor from kraft pulping is a major fuel for CHP and the recovery of process chemicals. The meat, dairy, and egg production in concentrated animal feed operations (CAFO) is a rapidly growing area in which bio-energy production is part of the solution to environmental issues created by this landless food production system.

3.3.4 Tertiary Residues

Urban or post consumer residues are a major component of today's bio-energy system. In fact the official statistics of the IEA, for example, describe biomass as combustible renewables and waste, and in many countries the tertiary sector is captured under the title of municipal solid waste or MSW. The tertiary sector generates energy in combustion facilities as well as from the generation of methane as land fill gas (LFG) from properly managed burial of mixed wastes from cities. Methane is also produced in sewage treatment facilities. Individual rates of residue generation are currently about 22 MJ person^{-1} d^{-1} in the United States; this combined with the high population densities of metropolitan areas, results in very high bio-energy potentials in this sector [9].

3.3.5 Biomass Potential for 2020

There is a consensus biomass resource potential estimate for 2020 in the United States, which captures most of the sources described above, other than the CAFO potential [10]. This is described in the form of a supply curve and indicates that there are about 7-8 EJ of primary energy at # 4.0 $ GJ^{-1}. This represents about 450 Mt of dry lignocellulosic biomass potential, which can be compared with today's utilization of about 190 Mt. The ultimate technical potential for biomass in the United States is not yet established; however, work is underway on what is called the Gigatonne scenario, which would investigate the effect of seeking double the 2020 projection for say the 2040-2050 period.

3.4 Thermo-chemical Technologies for Biomass Energy

Biomass is a renewable resource that can be used for the production of a variety of products currently produced from fossil fuel resources [11]. Among these products are electric power, transportation fuels, and commodity chemicals. This diversity of products has encouraged development of "biorefineries" to replace traditional plants dedicated to the production of either electric power or manufactured products. Thermo chemical technologies, including combustion, gasification, and pyrolysis, will play important roles in the development of biorefineries.

3.4.1 Combustion

Combustion for the generation of electric power is familiar to the utility industry, although fossil resources, especially coal, have been more commonly employed than biomass. As illustrated in Figure 3.1, solid-fuel combustion consists of four steps: heating and drying, pyrolysis, flaming combustion, and char combustion [12]. No chemical reaction occurs during heating and drying. Water is driven off the fuel particle as the thermal front advances into the particle. Once water is driven off, particle temperature increases enough to initiate pyrolysis, a complicated series of thermally driven reactions that decompose organic compounds in the fuel. Pyrolysis proceeds at relatively low temperatures in the range of 225°–500° C to release volatile gases and form char. Oxidation of the volatile gases results in flaming combustion. The ultimate products of volatile combustion are carbon dioxide (CO_2) and water (H_2O) although intermediate products can include carbon monoxide (CO), condensable organic compounds, and soot.

Combustion of biomass in place of coal has several advantages including reduced emissions of sulfur and mercury [13]. Combustion of biomass has almost no net emission of greenhouse gases since the carbon dioxide emitted is recycled to growing biomass. Combustion of biomass, however, can still produce emissions of nitrogen oxides and particulate matter. Some biomass has high concentrations of chlorine, which is a precursor to dioxin emissions under poor combustion conditions. Although co-firing of biomass with coal offers some near-term opportunities for the utility industry, the need for higher efficiencies at smaller scales and the compelling opportunities for biorefineries suggest that gasification or pyrolysis will be better future options for using biomass.

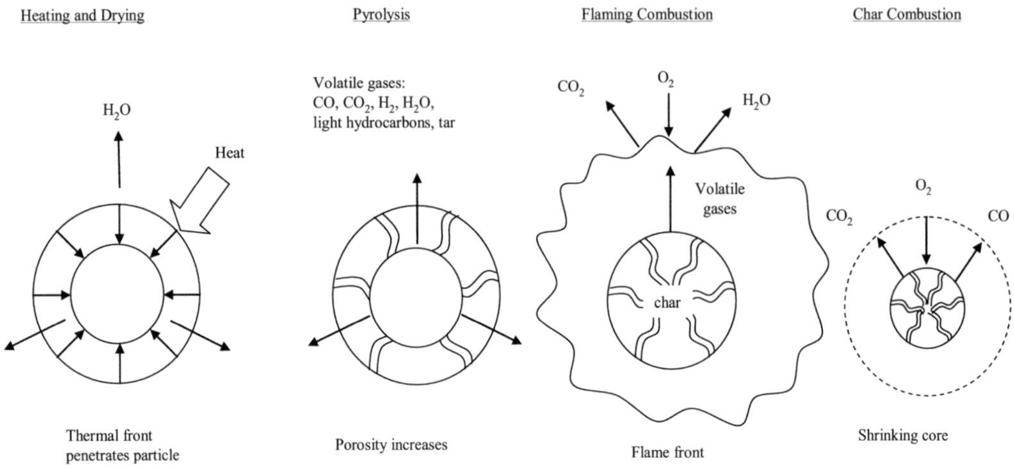

Fig. 3.1. Mechanism of Combustion

3.4.2 Gasification

Gasification is the partial oxidation of solid fuel at elevated temperatures to produce a flammable mixture of hydrogen (H_2), CO, methane (CH_4), and CO_2 known as producer gas.

Figure 3.2 illustrates the four steps of gasification: heating and drying, pyrolysis, solid-gas reactions that consume char and gas-phase reactions that adjust the final chemical composition of the producer gas [14]. Drying and pyrolysis are similar to those processes during direct combustion. Pyrolysis produces char, gases (mainly CO, CO_2, H_2, and light hydrocarbons) and condensable vapor. The amount of these products depends on the chemical composition of the fuel and the heating rate and temperature achieved in the reactor. Gas-solid reactions convert solid carbon into gaseous CO, H_2, and CH_4. Gas phase reactions adjust the final composition of the product gas. Chemical equilibrium is attained for sufficiently high temperatures and long reaction times. Under these circumstances, products are mostly CO, CO_2, H_2, and CH_4. Analysis of the chemical thermodynamics of gasification reveals that low temperatures and high pressures favor the formation of CH_4 whereas high temperatures and low pressures favor the formation of H_2 and CO.

Often gasifier temperatures and reaction times are not sufficient to attain chemical equilibrium and the producer gas contains various amounts of light hydrocarbons such as acetylene (C_2H_2) and ethylene (C_2H_4) as well as up to 10 wt-% heavy hydrocarbons that condense to tar [15].

Heating and drying, pyrolysis, and some of the solid-gas and gas-phase reactions are endothermic processes, requiring a source of heat to drive them. This heat is usually supplied by admitting a small amount of air or oxygen into the reactor, which burns part of the fuel, releasing sufficient heat to support the endothermic reactions.

Producer gas can be used to fuel high efficiency power cycles like combustion turbines, fuel cells, and various kinds of combined cycles. Producer gas can also be used in chemical synthesis of transportation fuels, commodity chemicals, and even hydrogen fuel [11]. In spite of these advantages; gasification has technical hurdles to overcome before widespread commercialization. Challenges include increasing carbon conversion; eliminating particulate matter, tar, and trace contaminants in the producer gas; and increasing plant availability by developing more reliable fuel feed systems and refractory materials. If producer gas is to be used as fuel in high-pressure combustion turbines, efficient and economical methods for compressing the gas during or after gasification must be developed.

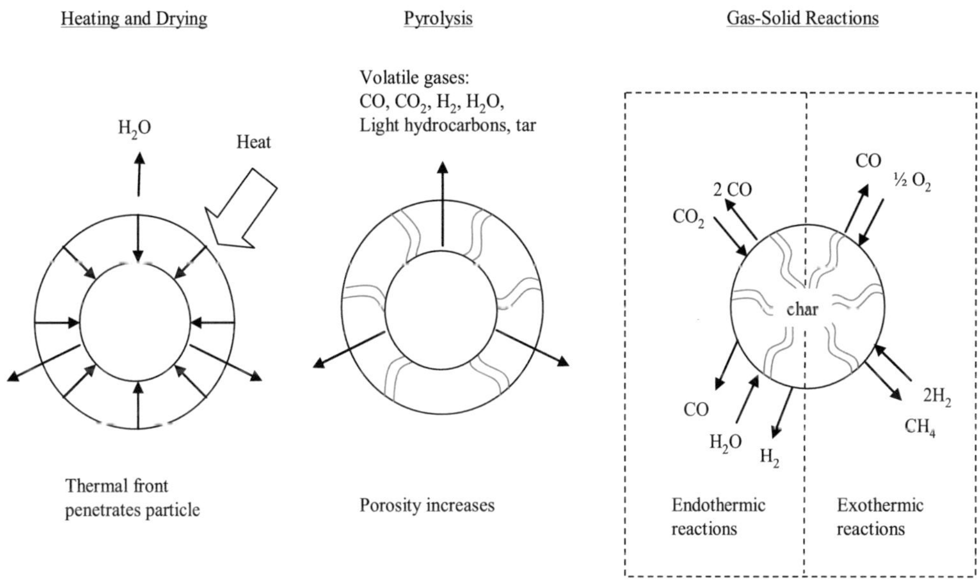

Fig. 3.2. Mechanism of Gasification

3.4.3 Pyrolysis

Pyrolysis is the heating of solid fuel in the complete absence of oxygen to produce a mixture of char, liquid, and gas. Although practiced for centuries in the production of charcoal, pyrolysis in recent years has been optimized for the production of liquids. In a process known as fast pyrolysis, chemical reaction and quenching proceed so rapidly that thermodynamic equilibrium is not attained, resulting in enhanced liquid yields on the order of 70 wt-% of the original biomass [16]. This mixture of organic compounds and water is known as bio-oil.

Bio-oil is a low viscosity, dark-brown fluid with up to 15 to 30% water, which contrasts with the black, tarry liquid resulting from slow pyrolysis or gasification. Fast pyrolysis liquid is a mixture of many compounds although most can be classified as acids, aldehydes, sugars, and furans, derived from the carbohydrate fraction, and phenolic compounds, aromatic acids, and aldehydes, derived from the lignin fraction. The liquid is highly oxygenated, approximating the elemental composition of the feedstock, which makes it highly unstable.

Figure. 3.3 illustrates the production of bio-oil, which begins with milling of biomass to fine particles of less than 1 mm diameter to promote rapid reaction. The particles are injected into a reactor, such as a fluidized bed, that has high heat transfer rates. The particles are rapidly heated and converted into condensable vapors, non-condensable gases, and solid char. These products are transported out of the reactor into a cyclone operating above the condensation point of pyrolysis vapors where the char is removed. Vapors and gases are transported to a quench vessel or condenser where vapors are cooled to liquid. The non-condensable gases are burned in air to provide heat for the pyrolysis reactor. A number of schemes have been developed for indirectly heating the reactor, including transport of solids into fluidized beds or cyclonic configurations to bring the particles into contact with hot surfaces.

Bio-oil can be used as a substitute for heating oil although its heating value is only about half that of its petroleum-based counterpart. Its handling and storage characteristics are inferior, as well. Nevertheless, the ability to produce liquid fuel from biomass offers opportunities for distributed production of a high-density fuel that can be easily pressurized for injection into combustion turbines. In addition, bio-oil contains a variety of organic compounds that, if they could be economically recovered, offer opportunities for pyrolysis-based bio-refineries.

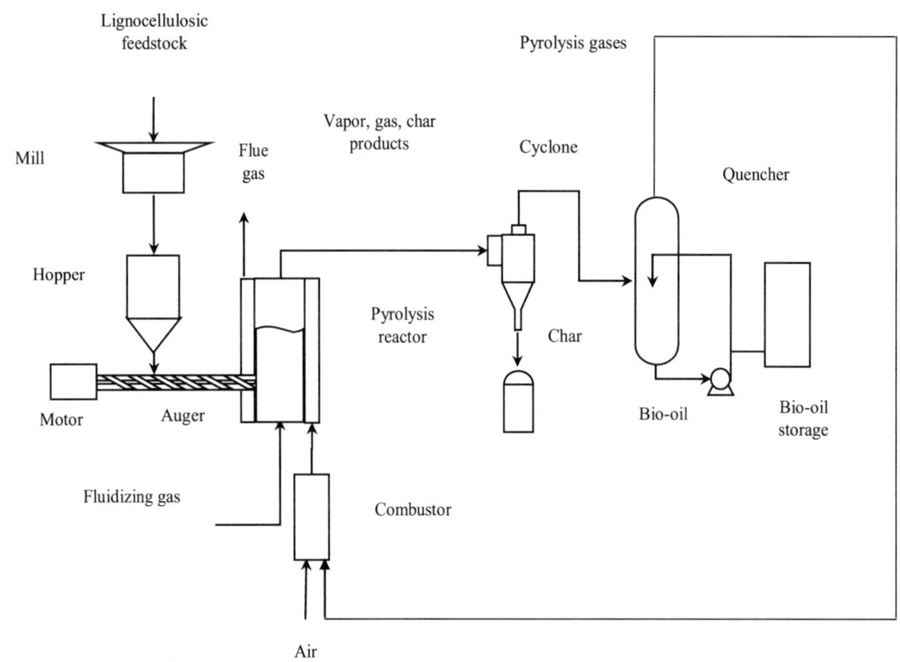

Fig 3.3 Schematic Illustration of Bio-Oil Production Facility

In summary, a number of thermo chemical conversion processes are available to meet the growing demand for biomass energy. Biorefineries offer an intriguing future opportunity for the electric utility industry to meet this demand.

3.5 The BioMaxTM---A New Biopower Option for Distributed Generation and CHP

Access to reliable, utility-grade electricity is key to improving the quality and economy of life of many rural communities throughout the world. Conventional approaches to rural electrification such as grid extension or small diesel generators are increasingly prohibitive in cost and often environmentally harmful. The Community Power Corporation's (CPC) new BioMax small modular biopower systems offer an affordable and environmentally friendly means of using a variety of local forest and agricultural biomass residues to generate on-site the right amount of electricity and thermal energy needed by most rural enterprises, homes, hospitals, clinics, government offices, water pumps and community micro-grids.

3.5.1 Technology

Beginning in 1999, CPC joined with the US National Renewable Energy Laboratory (NREL) followed by Shell Renewables, the California Energy Commission and the US Forest Service to develop and bring to market a new generation of environmentally friendly small modular bio-power systems. The first BioMax prototypes ranging from 5 kW to 20 kW are now deployed in the Philippines and six locations in the USA. In January 2004, CPC signed follow-on contracts with the California Energy Commission and the US Forest Service to develop an advanced 50kW BioMax system for prime-power, distributed generation applications.

CPC's fully automated BioMax systems use a variety of biomass fuels to generate electricity and thermal energy. CPC's BioMax system (Figure 3.4) is designed as a "green" alternative to conventional fossil fuel generators and to free the community/user from dependence on the supply and high cost of imported fossil fuels such as gasoline or diesel fuel. By eliminating the need for importing diesel fuel, the community's financial resources are retained in the community and there is no environmental damage from spillage of diesel fuel or exhaust emissions. BioMax users with on-site woody residues avoid the high cost of waste disposal by generating power and heat from that waste.

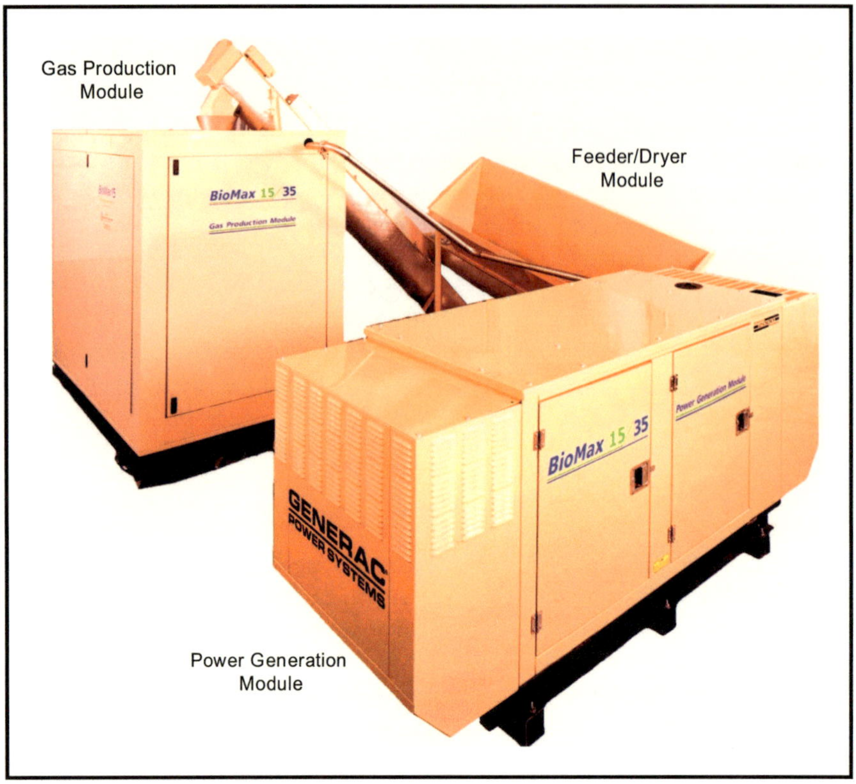

Fig. 3.4. BioMax 15/35

CPC's new bio-power technology incorporates the latest computer-based control technology and gasifier design to achieve unparalleled levels of clean-gas performance, turndown flexibility, and environ-mental friendliness. The "wood gas" is conditioned and fed into a standard internal combustion engine genset for conversion to mechanical, electrical, and thermal power. BioMax systems have also been used to operate a solid oxide fuel cell, a Stirling engine and a microturbine.

CPC's advanced design gasifier with fully integrated controls produces an extremely clean combustible gas from a variety of woody fuels including any kind of wood chips or densified biomass made from switch grass, sawdust, spent hops, grape skins, etc. Most nutshells including coconut, walnut, and pecan have proven to be an excellent fuel for the BioMax.

The small amount of byproduct char is entrained out of the gasifier and is removed from the producer gas stream by inertial separation and filtering. Very low tar levels in the producer gas are a result of automatic control of proper reactor temperatures over the full power range of the generator. The system does not produce condensed water nor does it use any form of liquid scrubbers. The only byproduct of the system is char and fine ash, the amount depending on the original ash content of the biomass feedstock.

Waste heat from the hot producer gas is recovered and used for drying the wood-chip feedstock or for space heating. The moisture content of the feedstock is reduced about 15 percentage points during delivery from the feed hopper to the gasifier. The BioMax gasifiers have been successfully operated with woodchips having between about 5% and 25% moisture. Additional thermal energy is available from the engine coolant and exhaust.

The computer-based control system adjusts the fuel/air ratio in the engine and makes necessary adjustments to the process variables of the gasifier to maintain the desired temperature profile and gasifier bed porosity. The controller remotely alerts the operator if it cannot operate the system within specifications and gives the operator ample time to make corrections. If the operator is not available to refill the feed hopper or if the gasifier or engine/generator system continues to operate improperly, the "expert" controller will automatically (and independently) shut down the gasifier and engine system in a safe manner.

The BioMax line is undergoing a field-based beta testing program with a wide variety of users including a high school, furniture factory, wood shavings company, forest service facility, and a rural enterprise in the Philippines. There are also two BioMax systems at research institutions in the USA.

In summary, the BioMax line represents a new level of fully automated and environmental friendly bio-power systems designed for the 21st century. On-going R&D at Community Power Corporation's product development facility in Denver, Colorado will continue to achieve upgrades and performance enhancements in the areas of hot-gas filtration, feedstock variety, control systems, and cost reductions to increase the commercial viability of the systems.

3.5.2 Summary of BioMax Features

- Electrical output in blocks from 5kWe to 50kWe; 120 and 240 VAC; 50 and 60 Hz
- Combined heat and power operation for rural electrification and distributed generation applications
- Environmentally friendly, non-condensing system without water scrubbers or liquid effluents
- Fully automatic, closed-loop control of all components including gasifier, gas conditioning and genset
- Dispatch able power within 30 seconds of auto-startup - uses no diesel fuel or gasoline
- Fuel flexible: wood chips, wood pellets, coconut shells, corn, corncobs, nutshells, etc.
- Optional automatic dryer/feeder for wood chips
- Modular, transportable, no need for on-site buildings or waste water disposal, 1 day installation.

3.5.3 Comparison of BioMax Bio-Power System with other Power Generation Technologies

BioMax Bio-Power is compared with other Power Generation Technologies in Figure 3.5 and in Table 3.1.

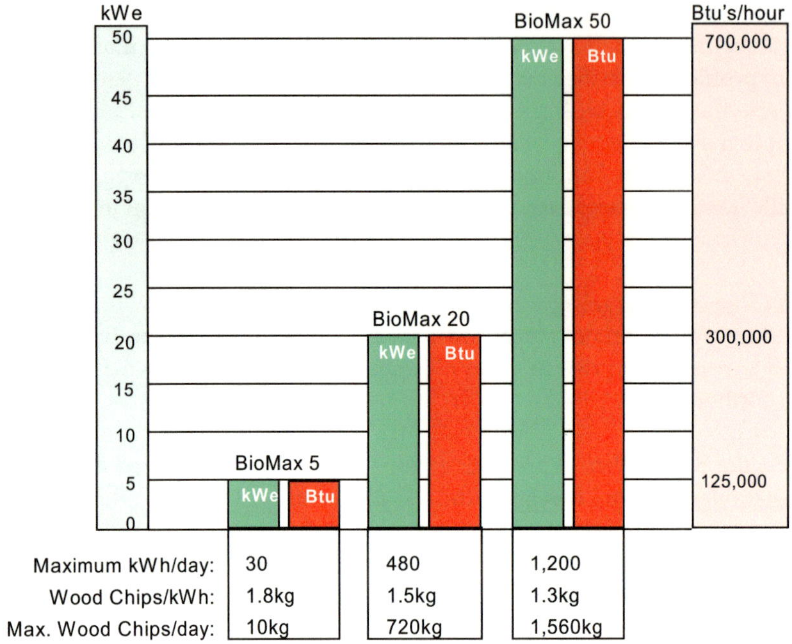

Source: Community Power Corporation
Fig. 3.5 BioMax Biopower CHP Systems

COMPARISON MATRIX	**BIOMAX Community Power Corp.**	PV SYSTEM	DIESEL GENERATOR	FUEL CELL	MICRO-TURBINE	SMALLWIND TURBINE
kW Range	**5-100**	2.5 - 15	5 – 6,000	5 – 3,000	30 - 400	3 – 200
Capacity compared	**20 kW**	15 kW	15 kW	15 kW	30 kW	10 kW
Stand-alone system	**yes**	yes	yes	yes	yes	yes
Dispatchable Power	**yes**	no	yes	yes	yes	no
Installed Capital Cost, $/kW	**$1,200 - $4,000**	$10,000 – $15,000	$200 - $650	$3,000 – $4,000	$1,200 - $1,700	$2,000 -$3,000
Combined Heat and Power?	**Yes**	No	Yes	Yes	Yes	No
Electrical system efficiency	**20 -22%**	6 - 12%	35%	36 - 50%	14 - 30%	25%
Overall Efficiency	**80-85%**	6 - 12%	80-85%	80-85%	80-85%	25%
Fuels	**Fuel flexible: straight biomass or dual fuel with a fossil generator: diesel or LPG**	None	Diesel fuel	Hydrogen, natural gas or propane	Natural gas or propane	None
Fuel cost	**Biomass: $ 0 -0.04/kWh at $0.02/kg Diesel: $ 0.10/kWh at $ 1.35/gal**	$0	$0.10/kWh @ $1.35/gal	$0.08/kWh @ $1.35/gal equivalent	$ 0.15/kWh at $1.35/gal equivalent	$0
Variable O&M ($/kWh)	**0.005 - 0.015**	$0.001 – 0.004	$0.005 – 0.015	$0.0019-0.0153	$0.003 – 0.008	$0.01
Energy density (kW/M2)	**30**	0.02	50	1 – 3	59	.01
Needs battery storage	**No**	Yes	No	No	No	Yes
Needs power conditioning	**No**	Yes	No	Yes	Yes	Yes

Table 3.1Equipment by Comparison

3.6 Motivating the Power Industry with Biomass Policy and Tax Incentives

Biomass is an abundant, geographically widespread, low sulfur, carbon neutral fuel resource. It is proven in many power-producing applications for base load and intermediate load. However, relative to conventional fossil fuels, biomass has relatively low energy density, requires significant processing, is an unfamiliar fuel among potential customers and is relatively expensive at the burner tip. In a world driven by calculations of rates of return to capital, biomass fuels are relegated to the position as an opportunity fuel with a large untapped potential in mainstream energy markets. Motivating the power industry to use more biomass fuels – to tap into the biomass energy potential – will require policy interventions from R&D investments to tax and other policy incentives. This discussion will focus on many of the policy interventions existing in the United States and those proposed in the Energy Bill before the U.S. Congress. By way of comparison, a few examples of the European approach will be discussed. Recent U.S experience on actual biomass demonstration projects will illustrate the difference properly targeted policy incentive can have on biomass' ability to meet its untapped potential.

The Antares Group Inc. is participating in several biomass power demonstration projects. Among them are switchgrass co-firing in Iowa, willow and residue co-firing in New York State, and gasification for combined heat and power in Connecticut. A recent "Healthy Forests" initiative proposed by the Bush Administration, led to a comprehensive review of commercially viable biomass systems for use with forest thinning in rural areas. It is policy incentives that make all these projects financially viable. An overview of these projects with and without the policy incentives will make that point clear.

3.7 Energy Generation through the Combustion of Municipal Solid Waste

3.7.1 The Concept

Recovering energy from waste, or waste-to-energy (WTE) as it's called in the U.S., isn't a new idea but it has evolved over the years from the simple incineration of waste in an uncontrolled, environmentally unfriendly way to the controlled combustion of waste with energy recovery, materials recovery and sophisticated air pollution control equipment insuring that emissions are within U.S. and EU limits [17]. This process took over 50 years of development and many improvements in design and technology, but the waste-to-energy industry has proven itself to be an environmentally friendly solution to the disposal of municipal solid waste and the production of energy. Modern WTE facilities reduce the volume of incoming municipal solid waste (MSW) by 90%-95% creating energy and jobs in the process and extending the life of landfills by generations. Many major metropolitan areas worldwide have facilities capable of processing 1,000, 2,000 and 3,000 tons of MSW a day with energy production in the form of steam and/or electricity.

Barlow Projects has come up with an innovative combustion system and facility design tailor made for those smaller communities with only 100 to 500 tons a day of waste. In some cases this means a collection of communities or districts pooling their waste and bringing it to a central location for processing. This may solve the landfill capacity problem for an entire region while simultaneously providing a predictable waste disposal fee, energy and the creation of new jobs.

3.7.2 Technical Challenges

Municipal solid waste is a difficult fuel to burn. Its non-homogenous nature complicates fuel handling and fuel feeding as well as ash handling. Controlling furnace temperature is critical to managing this process. Additionally, the fuel stream requires sophisticated air pollution control and emissions monitoring equipment to deal with acid gases and metals emissions that result from MSW combustion. Steam generation is accomplished with water wall or waste heat boilers. In both cases, tube corrosion is an issue due to high combustion temperatures and elevated levels of sulfur oxides and hydrogen chloride. Limiting tube metal temperatures is key to avoiding high temperature corrosion, particularly in super heater tubes. Air pollution control technologies for MSW combustion have advanced significantly in the last two decades. Stringent emissions standards for a variety of constituents mandate effective control and the industry has risen to these challenges. Most facilities are now equipped with scrubbers, activated carbon injection systems, bag houses and in some cases SNCR systems, but CO must be controlled at the front end by carefully monitoring and controlling the combustion process. Additionally, capital and operating costs must be minimized especially for smaller scale projects. Employing an "all-dry" reagent system provides excellent removal efficiencies without a sizeable increase in costs.

3.7.3 Biomass and Renewable Status

As far as the technology has come, one of the great challenges to WTE today is its status as a renewable fuel. Like any power plant, WTE facilities don't get built unless the economics work. WTE facilities must balance the revenue from accepting the waste, (referred to as the tipping fee) and the price of the energy it is able to sell in the form of steam or electricity, with the debt service and operational expenses of the facility. Because electricity rates are so low in many parts of the country, getting a 1.5 cent per KWh credit for electricity sold can make the difference between the project getting built or not. The DOE has classified MSW as Biomass for years [18] but, although qualifying as Biomass is generally the standard for gaining acceptance as Renewable, many environmental activists object to this notion and are having some success preventing these projects from benefiting from state or federal tax credits. The Energy Bill in Conference Committee as of the writing of this paper currently has sections specifically dealing with energy generated from Biomass, Municipal Solid Waste and a national Renewable Portfolio Standard (RPS) [19]. An RPS would require federal agencies to buy a certain percentage of their power from renewable technologies. A tax credit for power generated from new WTE facilities would, perhaps, provide the incentive needed to re-ignite the WTE industry and get more of these facilities built. An RPS would insure there was a client to purchase the power generated by these facilities. Additionally, non-governmental organizations like "Green-E" have set up programs to certify that certain types of energy are "green". Some utilities use this unofficial certification as their standard for marketing that energy as part of their green portfolio. Unfortunately, "Green-E" does not use the same guidelines as the US DOE for deciding what is renewable or "green" and what isn't and WTE is not currently eligible for their certification.

3.7.4 Public Acceptance

Despite the fact that the EPA recently released a report praising the WTE industry as being a "clean, reliable, renewable source of energy" [20], many still have a negative perception of waste-to-energy. This may be due, in part, to the fact that at one time there were over 700 incinerators operating in the United States burning trash without energy recovery or air pollution control equipment [21]. The majority of those chose to shut down when faced with installing the expensive air pollution control equipment mandated by the Clean Air Act, but not enough time has passed to erase these perceptions. Fortunately, the new air pollution control equipment being used is up to the task and modern facilities are meeting new stringent EPA emissions requirements. Additionally, EPA has done exhaustive studies to determine the safe exposure levels of the constituents that could potentially be emitted from a WTE facility and independent studies have determined that the actual amounts emitted do not present a significant threat to human health [22]. There are WTE facilities located in the middle of small communities, in large cities, on college campuses and near hospitals. Those who tour WTE facilities are often amazed that odors are minimal outside the facility and that there is no smoke coming out of the stack. It is those environmentalists that cling to old data and unsound concepts about what could be done with the waste rather tan presenting the challenge to public acceptance. Concepts like "zero waste" which advocate unrealistic recycling levels.

And what about recycling? Can't one just recycle everything so that there isn't anything left for landfilling or incineration? Don't WTE facilities compete with recycling and burn up valuable resources? With all the talk about recycling and the progress that's been made toward source reduction, composting and other forms of diversion, American's still send about 130 million tons of garbage to landfills every year [23]. Even the most efficient recycling programs are only diverting 50% of the waste stream leaving the other 50% to be managed in some other way. The bottom line is that some forms of waste are just not suitable for recycling because it isn't economical to do so. By removing those items from the waste stream that can and should be recycled one can improve the quality of the fuel and improve the efficiency of the combustion system. This does not mean that one shouldn't keep trying to manufacture goods in such a way that makes them more amenable to recycling and that one shouldn't continue to work towards higher diversion rates, but at best, only a portion of the problem is addressed. As a species, we are currently entombing millions of tons of fuel in the earth in the form of refuse that could be used to generate heat or electricity through the WTE process. This fuel has a heating value approaching ½ that of coal [24]. Why wouldn't we take advantage of that?

3.7.5 Potential

A Modern 500 ton/day Resource Recovery facility will generate approximately 10 MWs of energy. Currently only about 15% of America's municipal solid waste is combusted in 98 WTE facilities [25]. If additional WTE facilities were built to combust the waste we are currently landfilling, the potential for energy generation in America alone is approximately 8700 MW of what can legitimately be called renewable, sustainable energy. The demand overseas is even greater as many European countries are running out of landfill space and outlawing landfilling altogether. Outside the U.S. there have been over 60 new WTE plants built since 1996 [26]. Small island nations are also ripe for WTE projects because once their landfill is full, they have no other option than to ship it off island at great cost. Even in the United States, the landfills in rural areas are filling up, leaving many communities with no other option than to long-haul their waste to one of the many mega-landfills built to serve the big cities. This is not only expensive but contributes to the ever-growing number of semi-trailers on our roadways. Semi-trailers that are usually traveling empty one way.

The bottom line is that recovering energy from waste that we throw away is still a good idea and, despite some misconceptions, Waste-to-Energy is now a clean, renewable, sustainable source of energy, and a common sense alternative to landfilling. U.S. counties, municipalities, solid waste authorities and energy companies should re-consider this alternative to burying this fuel in a hole.

3.8 Senegal Bio Mass Exploitation: An assessment of applicable technologies for rural development

The UNDP World energy assessment report [27] is comprehensive in addressing the scope of energy options and their implications worldwide but its importance is the focus it places on the 2 billion people on the planet with no access to electricity.

In fact, 2 billion people—one third of the world's population—rely almost completely on traditional energy sources and so are not able to take advantage of the opportunities made possible by modern forms of energy [27-29]. Moreover, most current energy generation and use are accompanied by environmental impacts at local, regional, and global levels that threaten human well-being now and well into the future. In Agenda 21 the United Nations and its member states have strongly endorsed the goal of sustainable development, which implies meeting the needs of the present without compromising the ability of future generations to meet their needs. The importance of energy as a tool for meeting this goal was acknowledged at every major United Nations conference in the 1990s, starting with the Rio Earth Summit (UN Conference on Environment and Development) in 1992 [30].

Section 3.8 therefore attempts to illustrate partial solutions to produce sustainable electricity from Senegal's groundnut industry and the benefits to rural communities in the immediate vicinity of processing facilities.

3.8.1 Innovative Renewable Energy Technology for Rural Enterprise

The following excerpt from a recent study illustrates the reality of energy deficiencies and available opportunities:

With a groundnut production of about 1 million t per year, Senegal (Figure 3.6) is one of the most important countries of the world. With 80 % of this production for export, Senegal is the world's largest groundnut exporting country. Its major export products are groundnuts for eating, groundnut oil and groundnut pellets as a fodder, made from groundnut cake. The town of Kaolack being the provincial capital of the Kaolack province lies in the centre of the groundnut-growing region and has two groundnut processing plants with a combined capacity of 250,000 t per year. One of these plants is NOVASEN. In 1999, it started production in newly installed plant in the industrial zone at Kaolack harbour. NOVASEN uses a modern press process, which yields 92 % of oil in one go from the groundnut kernel (conventional procedures have a maximum yield of 80 %; therefore, in a second stage another 10 to 15 % has to extracted from the groundnut cake, by chemical means). The new process has increased energy efficiency, thus NOVASEN does not need to burn the groundnut shells to generate the energy needed for the plant, as in conventional groundnut processing. On account of this, the NOVASEN plant in Kaolack produces about 10,000t to

15,000t of groundnut shells per year, which are, at present, is not used in other ways. NOVASEN intends to carbonize these groundnut shells in a modern retort and to use all of its by-products (flue gases, pyrolysis oil) as a source of energy in its groundnut processing plant. An industrial briquetting plant is expected to produce about 3,000 to 4,000 t high-quality biocoal per year for the local charcoal market at Kaolack or other urban centres in Senegal. At present, about 360,000t charcoal are consumed each year in Senegal, which are produced locally in traditional earth kilns from Senegal's forest resource. According to official estimates, between 50,000 and 80,000 ha of forest cover are lost annually on account of charcoal production with goes in line with claiming agricultural land. Under this situation, the planned NOVASEN project does not only constitute a very effective rational use of energy measure in industry, but also an important contribution to the protection of natural resources in Senegal. In traditional charcoal kilns with only 17 % efficiency, some 18,000 t of wood would be needed to make 3,000 t of charcoal - this amount of wood could be saved every year through the planned NOVASEN carbonisation and briquetting plant [31, 32].

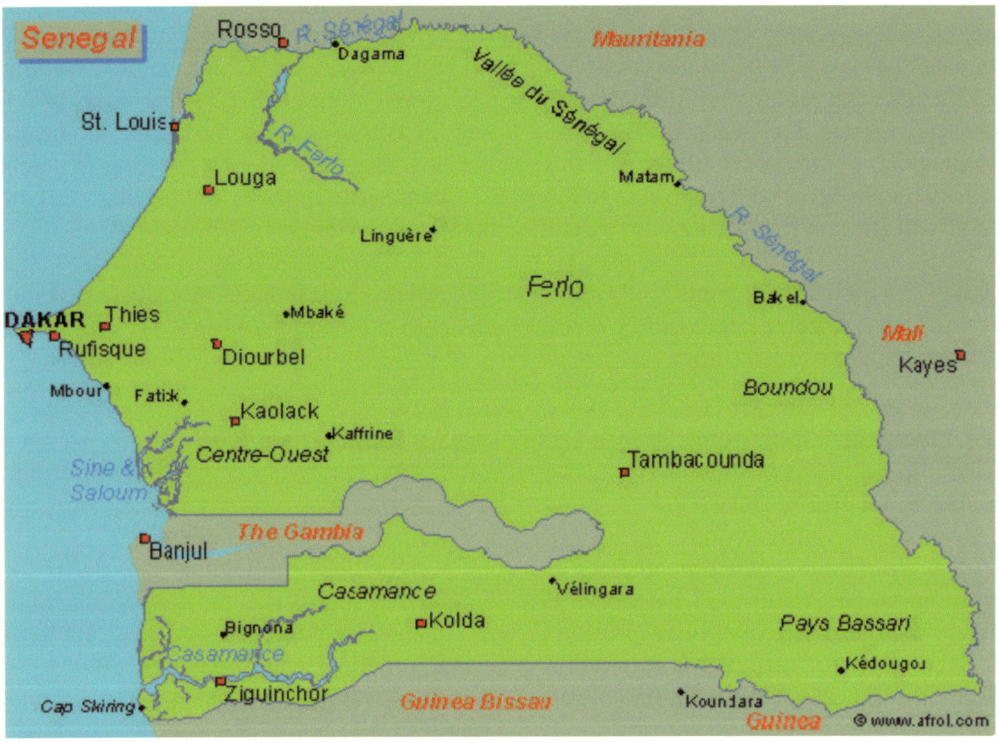

Fig. 3.6. Senegal

3.8.2 The BioMax System

An advanced small modular bio-power (SMB) system that may be very appropriate for Senegal and other ECOWAS countries has been developed by US-based Community Power corporation (CPC)1 for markets worldwide. The system is known as the *Community Productive Use Platform* (C-PUP), and produces thermal energy, shaft power, and electricity. The C-PUP incorporates a CPC BioMax bio-energy system. *BioMax* is the trade name used by Community Power Corp. for its small modular bio-power (SMB) systems that convert woody biomass residues to electricity, shaft power, and thermal energy. The C-PUP converts locally available biomass into useful mechanical, electrical, and thermal power that can be applied to a myriad of productive use applications. The heart of the CPUP is CPC's Gas Production Module (GPM) that converts coconut shells (and other dry woody biomass) to a product gas for delivery to a spark-ignited engine mounted on a power distribution platform. The gasifier converts biomass by low-oxygen thermal decomposition into a gas mixture that is primarily composed of hydrogen, carbon monoxide, methane, carbon dioxide, and water vapor. The platform can allocate shaft power as needed to various mechanical and electrical loads including motors and compressors.

The peak electrical output of the BioMax 15 unit in the CPUP is 15 kWe from the conversion of about 23 kg of coconut shells per hour. In addition, about 20 kW of thermal energy is available in the form of clean, hot air for drying crops and fish [33].

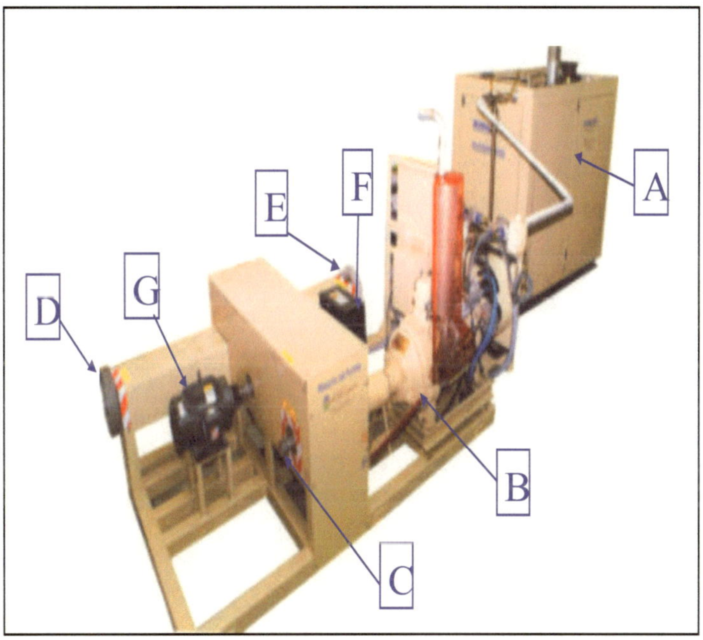

Fig. 3.7. The SMB Community Productive Use Platform

The Gas Production Module (A) (Figure 3.7) converts woody biomass to a fuel gas that is ignited in an engine (B) to turn a shaft. The shaft power is distributed to output (C) to run a small biomass fuel grinder (not shown); to outputs (D) and (E) for powering a variety of larger implements such as flour mills, rice mills, decorticators, composters, water pumps, etc. (not shown). Any combination of these mechanical outputs can be engaged, or disengaged simultaneously. If electrical power is needed, a 15 kW generator (F) can be engaged. If the Gas Production Module or engine are not available, motor (G) can be connected to a backup electrical power source to drive any combination of mechanical outputs (C, D and E). All rotating shafts, belts, pulleys, and heated surfaces are covered for worker safety.

	BioMax 5	BioMax 20	BioMax 50
Max. kWh/day:	30-50	480	1,200
Shells/kWh (est.):	3.5kg	3kg	2.5kg
Max.Shells/day:	120-175kg	1,440kg	3,000kg

Table 3.2 Groundnut Shell Feedstock

Senegal's yield in excess of 10,000T (9,071,847kg) groundnut shell feedstock can produce very significant amount of on and off grid electricity per the kWh production yields shown in Table 3.2.

The secondary use of waste being looked at needs to be evaluated not just as a source of available electricity supply but also as prime movers that can drive 'strategic loads'. Strategic loads in this context imply clusters of strategic entities. In turn those strategic entities as *collective units* themselves become 'Prime Movers' in that they have direct evolutionary impacts on society.

3.9 Acknowledgement

This Chapter has been prepared by Peter Meisen, President, Global Energy Network Institute (GENI), San Diego, CA, USA. Contributors include Richard L. Bain (Group Manager and Principal Researcher, Thermo Chemical Conversion Group, National Renewable Energy Laboratory (NREL), Golden, CO, USA). Ralph P. Overend (NREL, Golden, CO, USA), Robert C. Brown (Bergles Professor of Thermal Science, Department of Mechanical Engineering, Iowa State University (ISU), USA), Robb R. Walt (President and CEO, Community Power Corporation, Littleton, CO, USA), Christian Demeter (Vice President American Bio-energy Association, USA), Gregg Tomberlin (Barlow Projects, Inc., Fort Collins, CO, USA)., and Soungalo Sanoko (Technical Director, SONACOS, Dakar, Senegal). The Chapter is primarily based on an up-date of the invited panel session summary papers presented at the Panel Session on Harnessing Untapped Biomass Potential Worldwide at the IEEE-PES 2004 General Meeting (GM2004)

3.10 References

[1] Bain, R. L.; Amos, W. P.; Downing, M.; Perlack, R. L. (2003). Biopower Technical Assessment: State of the Industry and the Technology. 277 pp; NREL Report No. TP-510-33123.

[2] "Renewable Energy Technology Characterizations," (1997). EPRI-TR-109469, Electric Power Research Institute, Palo Alto, California, December.

[3] Mann, M.K.; Spath, P.L. (2000). A Comparison of the Environmental Consequences of Power from Biomass, Coal, and Natural Gas. First World Conference and Exhibition on Biomass for Energy and Industry. June 5-9, Seville, Spain.

[4] Rossillo-Calle, F. and G. Bezzon, *Production and use of industrial charcoal*, in *Industrial Uses of Biomass Energy: The example of Brazil*, F. Rossillo-Calle, S.V. Bajay, and H. Rothman, Editors. 2000, Taylor and Francis: London. p. 183 -199.

[5] Goldemberg, J., et al., *Ethanol learning curve--the Brazilian experience.* Biomass and Bioenergy, 2004. **26**(3): p. 301-304.
[6] NCGA, *World of Corn Consumption 2002*. 2003, National Corn Growers Association http://www.ncga.com/03world/main/consumption.hm
[7] Eriksson, H.M.H, Hall J.P., Helynen, Satu, *Rationale for forest energy production*, in *Forestry Sciences*, J.B. Richardson, Björheden R.; Hakkila, P.; Lowe, A.T. and Smith, C.T., Editors. 2002, KluverAcademic Publishers: Dordrecht, NL. p. 1 - 17.
[8] Sokhansanj, S., Turhollow, A., Cushman, J., and Cundiff, J. *Engineering aspects of collecting corn stover for bioenergy*. Biomass and Bioenergy, 2002. **23**(5): p. 347-355.
[9] Chum, H.L. and R.P. Overend, *Biomass and Bioenergy in the United States*, in *Advances in Solar Energy: an Annual Review*, Y. Goswami, Editor. 2003, American Solar Energy Society: Boulder, CO. USA. p. 83-148.
[10] Haq, Z., *Biomass for Electricity Generation*. 2002, Energy Information Administration http://www.eia.doe.gov/analysis/2001anal08.html
[11] Brown, R. C. (2003) Biorenewable Resources: Engineering New Products from Agriculture, Iowa State Press, Ames, IA.
[12] Borman, G. L. and Ragland, K. W. (1998) Combustion Engineering, McGraw-Hill.
[13] Jenkins, B. M., Baxter, L. L., Miles, Jr., T. R., and Miles, T. R. (1998) "Combustion properties of biomass, Fuel Process. Tech. 54, 17 - 46.
[14] Bridgwater, A. V. (1995) "The technical and economic feasibility of biomass gasification for power generation," Fuel 74, 631-653.
[15] Reed, T., editor (1981) Biomass Gasification: Principles and Technology, Noyes Data Corp., Park Ridge, N.J.
[16] Bridgwater, A., Czernik, S., Diebold, J., Meier, D., Oasmaa, A., Peacocke, C., Piskorz, J., and Radlein, D. (1999) Fast Pyrolysis of Biomass: A Handbook, CPL Press, Newbury, United Kingdom.
[17] Hickman, H. Lanier, "A Brief History of Solid Waste Management During the Last 50 Years", MSW Management, September/October 2001
[18] United States Department of Energy (DOE) Energy Efficiency and Renewable Energy (EERE) website at: http://www.eere.energy.gov/RE/bioenergy.html
[19] H.R. 6, the Energy Policy Act of 2003 which can be found at: http://thomas.loc.gov/cgi-bin/query/z?c108:H.R.6:
[20] From a letter Written to Maria Zannes at IWSA by Marianne Lamont Horinko with the Office of Solid Waste and Emergency Response and Jeffrey R. Holmstead of the Office of Air and Radiation on Feb. 14th, 2003.
[21] Hickman, H. Lanier, "A Brief History of Solid Waste Management During the Last 50 Years", MSW Management magazine, September/October 2001
[22] Rao, Chaudhuri, Garcia, Stormwind and Ruffle, "Multiple Pathway Health Risk Assessment of a Municipal Waste Resource Recovery Facility in Maryland", EM, August 2003
[23] U.S. E.P.A. "Municipal Solid Waste in The United States: 2001 Facts and Figures Executive Summary"
[24] ASSURE, "Energy from Waste Fact Sheets", www.assure.org
[25] U.S. E.P.A. "Municipal Solid Waste in The United States: 2001 Facts and Figures Executive Summary"

[26] Statistic from the WTE Research and Technology Council website at: http://www.seas.columbia.edu/earth/wtert/wtertfaq.html
[27] World Energy Assessment, Energy and the Challenge of Sustainability, UNDP, 2000
[28] WEC-FAO (World Energy Council and Food and Agriculture Organization of the United Nations). 1999, *The Challenge of Rural Energy Poverty in Developing Countries.* London.
[29] World Bank, 1996, Rural Energy and Development: Improving Energy Supplies for Two Billion People. Washington, D.C. 1997. World Development Indicators, 1997. Washington, D.C.
[30] WRI (World Resources Institute), 1998. *A Guide to the Global Environment.* Oxford: Oxford University Press.
[31] Biocoal from Groundnut Shells - An Industrial Approach to Household Energy in Senegal, *by Dipl.-Ing. Benjamin Jargstorf, FACTOR 4 ENEREGY PROJECTS, Wismar, G ermany*
[32] Coura Badiane, Senegal's Trade in Groundnuts: Economic, Social and Environmental Implications, TED Case Studies Number 646, December 2001
[33] *Renewable Energy Technology Annex.* Small Modular Biopower (SMB) for a Community Productive Use. Platform (C-PUP), ANNEX 2004, Ashden Trust Awards for Sustainable Energy Bioenergy Conversion for Sustainable Rural Enterprise in the Philippines A-1

4

Energy Potential of the Oceans in Europe and North America: Tidal, Wave, Currents, OTEC and Offshore Wind

This Chapter examines energy potential of the oceans in Europe and North America: tidal, wave, underwater currents, and ocean thermal energy conversion. It considers ocean wave and tidal power projects in San Francisco, wave power technologies (oscillating water column, overtopping devices, float systems, and hinged contour devices), and cost. Feasibility assessments of offshore wave and tidal current production are described, and wave project results are given. U.S wave energy resources, feasibility definition study sites, feasibility study for wave energy conversion (WEC) devices, demonstration scale plant design, commercial-scale plant design, learning curves, economics, and recommendations are discussed. Recent progress in offshore energy technology development including wind-power is evaluated. Also discussed is the role of tidal power in the United Kingdom to reduce greenhouse gas emissions. Here, the proposed Severn Barrage is focused on considering potential benefits, conditions for sustainable development, energy policy context and compliance with environmental legislation. UK tidal resource is reviewed: stream resource (that is KE contained in fast-flowing tidal currents) and tidal range resources (that refers to gravitation potential energy). A feasibility study for tidal range in the Mersey Estuary and other schemes in the UK is summarized. Also given is a strategic overview of the Severn Estuary resource, electric output and characteristics, carbon emissions (carbon payback and carbon reduction) and physical implications of a barrage.

4.1 Introduction

Renewable energy sources from the oceans include offshore wind, wave energy, tidal energy, Ocean Thermal Energy Conversion (OTEC) and underwater currents. Harvesting ocean energy is not a new concept, yet it has remained a marginal resource. Today there is serious interest in offshore technology in Europe and Asia but funding for projects in power from the oceans at this time in USA is lacking. Wind farm technology has moved offshore where the prevailing winds can be more consistent and out of sight. Offshore wind energy is the fastest growing sector in renewable energy.

Areas of great tidal differences can produce regular and predictable tidal currents of 5 knots or more, creating large energy potentials. However, there are many areas with great tidal differences and only very slow currents. Tidal currents of 5 knots or more typically require a somewhat narrow passageway between a bay or estuary and the ocean and a somewhat shallow depth of the passageway. Tidal range, therefore, is not the only factor in speed of the tidal current. The Physics of Tidal Power is reviewed in Reference [1].

France has had a 240MW tidal power generating facility for 40 years. It is a tidal head plant and the technology is quite different [1].

Projects harnessing tidal currents have shifted toward capturing tidal-driven coastal currents. A study of 106 possible locations in the EU countries for tidal turbines showed that such sites could generate power of the order of 50 TWh per year. The power density of a marine current is approximately four times that for a wind generator, so this marine current resource is potentially large. However, there are fewer places in the world where marine currents are usable [1] compared with access to winds.

Wave energy can be considered as a concentrated form of solar energy. Winds are generated by the differential heating of the earth's surface, and, because they blow over large spans of water, part of their energy is converted into waves. The first commercial-scale wave power facility turning wave energy into compressed air was established in Scotland. Some proposed schemes involve hinged pontoons with hydraulics, while others appear like floating pistons that rise and fall with the wave action. Several prototype demonstrations are planned in the next few years. Growth in this sector is anticipated to reach at least $100 million per annum by 2011.

The difference in temperature between the surface waters and the deeper ocean waters can produce significant thermal energy. Ocean-based renewable energy development lag land-based systems because of significant capital requirements and difficulty in obtaining the necessary financing due to risk and market barriers. The technical capabilities, both in engineering and management, exist in the offshore sector appear ready to undertake the size and scope of projects envisioned.

4.2 Ocean Wave and Tidal Power Projects in San Francisco

Ocean waves and Bay tides interacting with the Sacramento River flowing from the Sierra Nevada Mountains combine to create two excellent renewable energy resources, Long rolling ocean waves are a condensed form of wind energy. Tidal currents are driven primarily by the gravitational pull of the moon and are independent of local weather conditions [2].

San Francisco is undertaking two renewable energy projects: a pilot demonstration for tidal power that was commissioned in 2005 and a first U.S. commercial installation for wave energy that is proposed to produce up to 750kW. Both projects could be expanded in prudent phases to provide an important part of San Francisco's current 840MW peak demand. San Francisco is modeling these technologies for environmentally safe implementation in coastal and reverie communities round the world.

Wave power can be harvested by a variety of devices, with several unique approaches nearing commercialization. Most wave energy devices use air or seawater, and one devise produces electricity directly from motion, in addition to taking advantage of hydraulic pumps to generate power [3-9].

The Pelamis device for San Francisco [7] consists of a total of four cylindrical sections, which are connected together by three hydraulic power conversion modules. The total length of the device is 120m and device diameter is 4.6m. The power conversion module comprises four hydraulic rams (two heave, two sway), high-pressure accumulators for power smoothing storage, two variable displacement motors for power conversion, and two 125kW generators, with integrated transformer cabling AC power to shore. The device is secured by a compliant slack moored anchoring system. At a later date, a series of these devices can be installed to comprise a wave farm. By employing the Pelamis device, a generation capacity of 22 MW/km is anticipated. However, additional device optimization is anticipated.

The HydroVenturi tidal power generation device is comprised of a cube of venturi tubes attached to the marine bottom on a rack sited safely some 18m below the surface and outside the navigation channel in the Golden Gate Passage. The tidal current flowing through the device is accelerated through the Venturi tubes to create a 2.5Kg pressure drop, thus creating suction enough to pull air down to an air storage tank integrated into the cube below the Venturi tubes. The compressed air is then pushed through a pipe to an on-shore air turbine to produce electricity. A 150kW device has been demonstrated in England.

A 1MW Venturi commercial installation was expected for grid intertie by 2008, with expansion phased in 5-20MW increments. The air turbines in the commercial installation may be housed in a secure area on land under the Golden Gate Bridge. Several other tidal power generation locations have been identified in order to serve neighboring communities such as Marin County and Oakland.

San Francisco's interest in Hydro Venturi technology is because it has no moving parts underwater. A technology with no moving parts underwater makes tidal power attractive

4.3 Wave Power Technologies

Wave power technologies are now reviewed. The oceans contain a vast amount of mechanical energy in form of ocean waves and tides. The high density of oscillating water results in high energy densities, making it a favorable form of renewable energy. The total U.S. available incident wave energy flux is about 2,300 TWh/yr. The US Department of Energy (DOE) Energy Information Authority (EIA) estimated that 2003 hydroelectric generation in USA was about 270 TWh, which is a little more than a tenth of the offshore wave energy flux into the U.S. The fact that good wave and tidal energy resources can be found in close proximity to population centers and technologies being developed to harness the resource have a low visual profile makes this an attractive source of energy. Recent advances in offshore oil exploration technology and remote management of power generation systems have enabled significant progress in advancing technology development by simple technology transfer. However, despite enormous progress over recent years, underwater current and wave power conversion technologies are at an immature stage of development. This is because of a lack of accepted standards, a wide range of technical approaches, and large uncertainties on performance and cost of these systems. Further Research, Development and Demonstration (RD&D) and the creation of early adopter markets through government subsidies is required to move these technologies into a competitive market place.

Feasibility assessments of wave and tidal current power are made in Section 4.4.

4.3.1 Wave Power Conversion Devices and Technologies

Wave power conversion devices are installed either **on-shore** and embedded in a cliff or an existing harbor wall, **near-shore** in close proximity to shore standing on the seabed or **off-shore** in deep waters. Similar to offshore wind, a wider applicability and more consistent and concentrated resource of energy can be found offshore and is more suitable for large-scale deployments. Installing such devices away from the coastline solves many issues such as visual impact, permitting and environmental impact.

The device must be able to handle a wide range of incident wave power levels, from near-flat seas to the most extreme storm conditions (which produce waves power levels more than an order of magnitude above the average). Waves typically have a low frequency of the order of 0.1 Hz, while power generation equipment runs at hundreds of rpm. The device must change the slow-acting, multi-directional wave force into a high-speed, unidirectional force capable of powering a generator. Short-term storage becomes an important consideration to maintain consistent power output.

Technologies to convert ocean wave power into electricity are many. The most promising ones are summarized below:

- **Oscillating Water Column** - (OWC) systems consist of a partially submerged structure, which forms an air chamber, with an underwater opening that allows the seawater to flow into the chamber. The volume of air inside the chamber is compressed as the water rises inside the chamber, driving air through a turbine. As the water level in the chamber subsides, the air is drawn back through the turbine. Both directional and self-rectifying air turbines have been developed. The axial-flow Wells turbine is the best-known turbine for this kind of application and has the advantage of not requiring rectifying air valves.

- **Overtopping Devices** - guides incoming waves up a ramp and up into a reservoir raised slightly above sea level. The water trapped in the reservoir flows back to the sea through a conventional low-head hydroelectric generator.
- **Float Systems** - Their common feature is a buoy that sits on the ocean's surface. The motion of this buoy is converted into electricity typically by a hydraulic power take off such as a hydraulic ram. These float systems come in different shapes and forms.
- **Hinged Contour Devices** - contains different floating sections, which are hinged together. As the wave passes, the sections move relative to each other and the hinges produce power. The power conversion uses hydraulic elements.

As part of a nationwide collaborative program to demonstrate offshore wave power technologies, the Electric Power Research Institute (EPRI) reviewed available technology options in 2004. Some of the results are outlined in Table 4.1. The wide range of different specifications is a clear indicator of the immaturity of this emerging market. Average power output was assessed for a typical Oregon wave climate with an incident wave power level of 21kW/m. This is a typical US west coast wave power level.

The most important criteria assessing these devices are maturity of the development stage. This is indicated in Table 4.1 as Maturity Rating.

Maturity Rating	Company	Device Width (m)	Device Weight (Tons)	Average Power (kW)	Power Train
1	Ocean Power Delivery	4.6	380	153	Hydraulic
2	Energ-etech	35	450	259	Air Turbine
2	Wave Dragon	260	22,000	1369	Low Head Hydro
2	Wave Swing	9.5	NA	351	Linear Generator
3	Wave Bob	15	440	131	Hydraulic
3	Aqua Energy	6	22	17	Water Pump
3	OreCON	32	1250	532	Air/Hydraulic
3	INRI	5.4	112	16	Water Pump

Source: M. Previsic [10]

Table 4.1.Technology Comparison

4.3.2 Electrical Interconnection

Most wave power conversion devices under development incorporate frequency converters and step-up transformers to synchronize onto the grid. As a result, power quality tends to be good and power factors high. Short-term storage is incorporated to account for wave-to-wave variations. Storage options depend on the power take off train, and can incorporate hydraulic accumulators, storage through flywheel effects and capacitor banks. It remains to be seen how well these short-term storage options deal with the large variability of power levels in ocean waves.

Wave farm interconnection voltage levels depend on many variables, but are typically in the range of 12kV to 33kV. Recent offshore wind projects in Europe showed that environmental risks prohibit use of oil insulated cables in the sensitive coastal environment. XLPE insulations have proven to be an excellent alternative, having no such potential hazards associated with devise operation.

4.3.3 Cost

An Ocean Power Delivery Pelamis wave energy conversion device was used to establish costing models for a commercial scale (300,000 MWh/year) wave farm. Levelized cost components are shown in Figure 4.1. Cost breakdown shows that impact on cost of electricity of Operation and Maintenance (O&M) is significant and is the one component that has most uncertainty associated to it. The only way such O&M costs can be driven down and confidence established is by building demonstration projects.

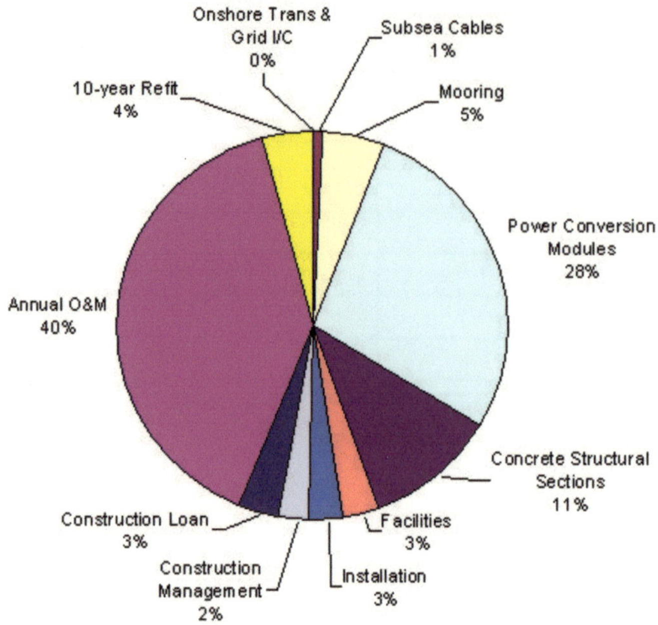

Source: M. Previsic [10]
Fig. 4.1. Breakdown of cost: Pelamis Wave Energy Conversion Device

An assessment of offshore wave energy conversion devices is made in References [10] and [11]. The methodology, guidelines and assumptions for conceptual design of offshore wave energy power plants is given in Reference [12]. System level design, preliminary performance and cost estimates for Hawaii, Oregon, Main, Massachusetts, and San Francisco Pelamis offshore wave power plants are given in References [13-17], respectively, and system level design, preliminary performance and cost estimate for the San Francisco Energetech offshore wave power plant is given in References [18]. Further, the state of the art for wave energy conversion is reviewed in Reference [19], and a technical assessment guide for ocean wave power is made in Reference [20]. A wave energy resource assessment for California is given in Reference [21].

Most of the EPRI Wave Power (WP) Reports [11, 13-18] are available on their website (www.epri.com).

4.4 Feasibility Assessment of Offshore Wave and TIDAL CURRENT Power Production: A Collaborative Public/ Private Partnership

Collaborative power production feasibility definition studies on offshore wave energy and tidal current energy on behalf of a number of public and private entities have been undertaken from 2004. The outcome of the offshore wave study under the Electric Power Research Institute (EPRI) is a compelling techno-economic case for investing in the research, development and demonstration (RD&D) of technology to convert the kinetic energy of ocean waves into electricity.

EPRI Wave Power Reports [11, 13-18] and References [22-29] summarize the activities in this area.

4.4.1 Feasibility of Wave and Tidal Current Energy

The elements of a wave and tidal current energy feasibility study are: a) Identify and characterize potential sites for assembling and deploying a power plant and for connecting the plant to the electric grid; b) Identify and assess wave energy conversion (WEC) devices; c) Conduct a conceptual design of a demonstration- and commercial-scale offshore wave power plant and, based on performance and cost estimates, assess the techno-economic viability of the wave energy source and the energy conversion technology; and d) Identify and assess the environmental and regulatory issues associated with implementing the technology.

Two characteristics of waves and tides important to the generation and dispatch of electricity from wave energy conversion devices are its variability and predictability. While the ocean is never totally calm, wave power is more continuous than the winds that generate it. The average power during the winter may be six times that obtained during the summer; however, power values may vary by a factor of a hundred with the random occurrences of storms. Therefore, the power of waves is highly variable. The predictability of wave energy is of the order of a few days. The waves resulting, for example, from storms that occur off the coast of Japan, will take that long to reach the northwest cost of the United States. The power from tidal currents, on the other hand, typically varies according to a diurnal cycle. The major benefit of tidal power is its high predictability for a given site years in advance, provided there is a thorough knowledge of the site. A drawback of tidal power is its low capacity factor, and that its peak availability misses peak demand times because of the 12.5-hour cycle of the tides.

Ocean waves are generated by the winds that result from uneven heating around the globe. Waves are formed by winds blowing over the water surface, which make the water particles adopt circular motions as depicted in Figure 4.2. this motion carries kinetic energy, the amount of which is determined by the speed and duration of the wind, the length of sea it blows over, the water depth, sea bed conditions and also interactions with the tides. Waves occur only in the volume of water closest to the water surface, whereas in tides, the entire water body moves, from the surface to the seabed.

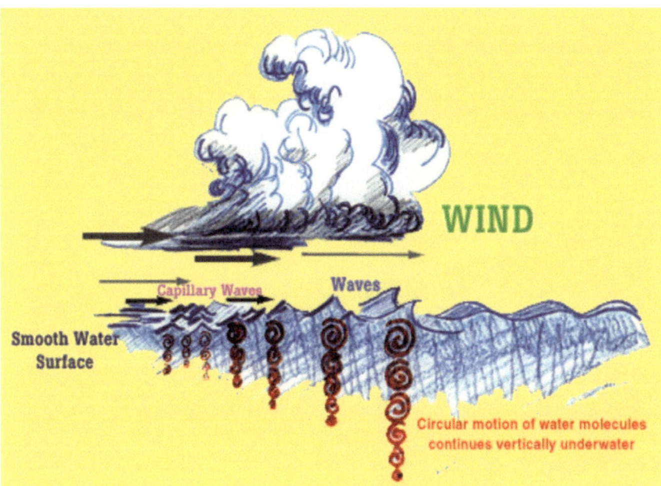

Source: M. Previsic [10]

Fig. 4.2. Wave generating forces based on wind-water interaction

The tides are generated by rotation of the earth within the gravitational fields of the moon and sun [1]. The relative motion of these bodies causes the surface on the oceans to be raised and lowered periodically, as illustrated in Figure 4.3. The physics of tidal power is explained in Reference [1].

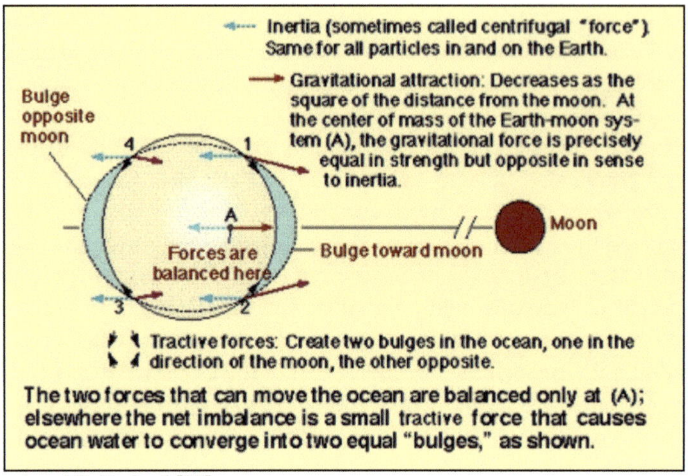

Source: O. Siddiqui and R. Bedard [30]

Fig. 4.3. Tide-generating forces based on earth-moon interactions.

In deep water, the wave power spatial flux (in kW/m of wave front crest) is given by significant wave height (H_s in m) and the peak wave period (T_p in sec). Based on these two parameters, the incident wave power (J in kW/m of wave crest length) associated with each sea state record is estimated by the following equation:

$$J = 0.42 \times (H_s)^2 \times T_p \qquad \text{(kW)}$$

It is significant to note that wave power varies with the square of wave height - that is, a wave whose height is doubled generates four times as much power.

The power of a tidal current is given by the following equation:

$$P_{water} = \tfrac{1}{2}\, r\, A\, V^3 \ \ \text{(W)}$$

where A is the cross- sectional area of flow intercepted by the turbine device (m^2), r is the water density (kg/m^3) and V is current velocity speed (m/s). The current velocity V varies in a precisely predictable manner as an additive function of period of the different sinusoidal tidal components.

4.4.2 Wave Project Results

4.4.2.1 U.S. Wave Energy Resources

An ideal site to deploy, operate and maintain an offshore wave energy power plant must have many attributes. First and foremost is a sufficient native energy and energy spectra potential.[1] The U.S. regional wave regimes and the total annual incident wave energy for each of these regimes are shown in Figure 4.4 (the total U.S. available incident wave energy flux is about 2,300 TWh/yr).

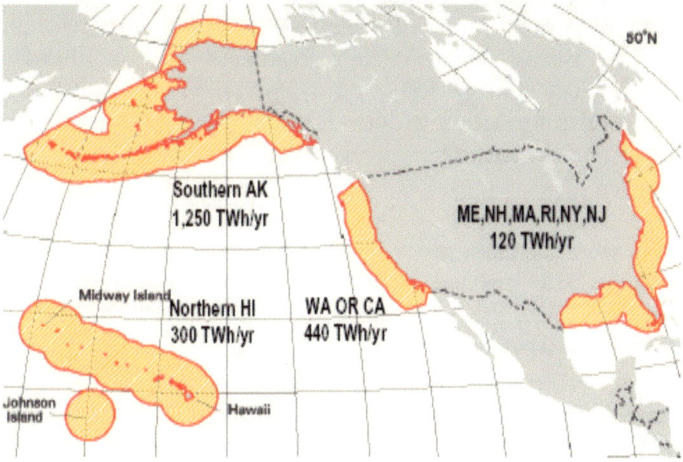

Source: O. Siddiqui and R. Bedard [30]
Fig. 4.4. U.S. energy resources

4.4.2.2 Feasibility Definition Study Sites

Site attributes characterized by the Project Team included offshore bathymetry[2] and seafloor surface geology, robustness of the coastal utility grid, regional maritime infrastructure for both fabrication and maintenance, conflicts with competing uses of sea space and existence of other unique characteristics that might minimize project development costs (e.g., existing ocean outfall easements for routing power cable and shore crossing).

	HI	OR	CA	Mass	Maine
County	Oahu	Douglas	SF	Cape Cod	Cumberland
Grid I/C	Wai-manalo Beach	Gardner	Wastewater Plant	Well Fleet	Old Orchard Beach S/S
Average Annual J (kW/m)	15.2	21.2	11.2[1]	13.8	4.9
Depth (m)	60	60	30	60	60
Distance from Shore	2	3.5	13	9	9
Cable Landing	Makai Pier	IPP out flow pipe	Water out flow	Dir Drill	Dir Drill

Source: O. Siddiqui and R. Bedard [30]
[1]Sited within the marine sanctuary exclusionary zone

Table 4.2 Estimated Performances of Pilot Demonstration Plants

---

[1] Energy as function of wave height and wave period or frequency
[2] Bathymetry is the depth of the seafloor below mean water height (i.e., the inverse of a topographic map)

Table 4.2 identifies the site selected in each of the five states that participated in the study, and also provides a few key characteristics of each selected site.

4.4.2.3 Feasibility Study - WEC Devices:

Twelve companies responded to EPRI's request for information. An initial screening considered two key issues: 1) technology readiness (i.e. readiness of device for demonstration in the 2006 time period); and 2) survivability in adverse conditions (i.e., sufficiency of technical information provided by the device manufacturer to prove the survivability in storm conditions). The eight devices that passed the initial screening criteria are shown in Table 4.3.

These eight devices were then assessed with the objective of determining any critical issues and recommending RD&D needed to achieve technological readiness for an at sea demonstration. As a result of this assessment, the eight devices were grouped into one of three levels of development categories:

- Level 1 – Development complete and full-scale testing in the ocean underway
- Level 2 – Development near complete. Only deployment, recovery and mooring issues are yet to be validated. There are funded plans for full-scale at sea testing.
- Level 3 – Most critical RD&D issues are resolved. Additional laboratory and sub-scale testing, simulations and systems integration work is needed prior to finalization of the full-scale design. There are no funded plans for full-scale at sea testing.

	Length (m)	Width (m)	Power (kW) [1]	Type	Rating
Ocean Power Delivery	120	4.6	153	Floating Attenuator	1
Energetech	25	35	259	OWC - Bottom Terminator	2
Wave Dragon	150	260	1369	Floating Overtopping	2
Wave Swing	9.5	9.5	351	Bottom Point Absorber	2
Wave Bob	16	15	131	Floating Point Absorber	3
Aqua-Energy	6	6	17	Floating Point Absorber	3
OreCON	32	32	532	Floating OWC	3
Ind Natural Resources Inc	5.4	5.4	112	Bottom Point Absorber	3

Source: O. Siddiqui and R. Bedard [30]

[1]Based on Oregon average annual wave energy resource

Table 4.3. Estimated Performance of Pilot Demonstration Plants

At the time of EPRI's analysis (March 2004), only one WEC device manufacturer had attained a Level 1 technology readiness status - Ocean Power Delivery with its Pelamis device.

4.4.2.4 Demonstration-Scale Plant Design – Oregon Example

Demonstration-scale (as well as commercial-scale) designs were based on the Ocean Power Delivery (OPD) Pemamis WEC device for the five sites listed in Table 4.2. The Pelamis WEC device consists of four cylindrical steel sections, which are connected by three hydraulic power conversion modules (PCM). Total length of the device is 120m and device diameter is 4.6m. Figure 4.5 illustrates the device being tested off the Scottish coast.

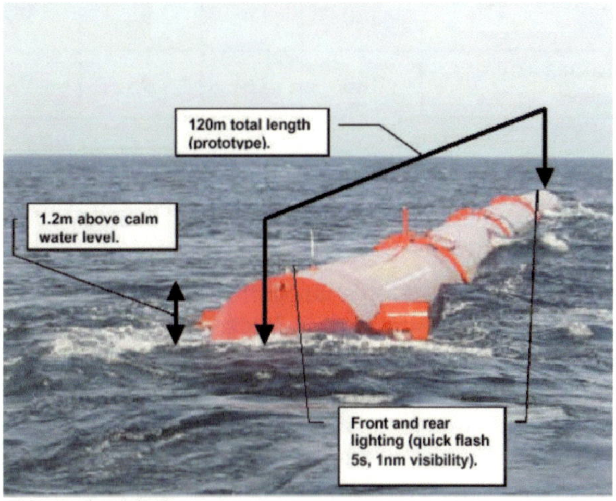

Source: O. Siddiqui and R. Bedard [30]
Fig. 4.5. OPD Pelamis WEC device.

A second San Francisco, CA design based on the Energetech OWC WEC device depicted in Fig. 4.6 has also been tested.

Source: O. Siddiqui and R. Bedard [30]
Fig. 4.6. Energetech WEC device.

The estimated performance of the single unit demonstration plant at each of the five sites is shown in Table 4.4.

	HI	OR	CA [1]	Mass	Maine
Device Rated Capacity (kW)	750	750	750	750	750
Annual Energy Absorbed (MWh/yr)	1,989	1,472	1,229	1,268	426
Annual Energy Produced (MWh/yr)	1,663	1,001	835	964	290
Average Electrical Power (kW)	180	114	95	98	33
Number of Homes Powered by Plant	180	114	95	98	33

Source: O. Siddiqui and R. Bedard [30]

[1]Energetech site numbers: 1000 kW, 1643 MWh/yr, 1264 MWh/yr, and 144 kW respectively

Table 4.4. Estimated Performance of Pelamis Pilot Demonstration Plants

4.4.2.5 Commercial-Scale Plant Design – Oregon Example:

The commercial system uses a total of 4 clusters, each one containing 45 Pelamis units (i.e., 180 total Pelamis WEC devices), connected to sub-sea cables. Each cluster consists of 3 rows with 15 devices per row. The other state designs are organized in a similar manner with 4 clusters. The number of devices per cluster varies such that each plant produces an annual energy output of 300,000 MWh/yr. The electrical interconnection of the devices is accomplished with flexible jumper cables, connecting the units in mid-water. The introduction of 4 independent sub-sea cables and the interconnection on the surface provides some redundancy in the wave farm arrangement.

The estimated performance of the commercial-scale plant at each of the five sites is shown in Table 4.5.

	HI	OR	CA	Mass	Maine
Device Rated Capacity (kW)	500	500	500	500	500
Annual Energy Absorbed (MWh/yr)	1,989	1,997	1,683	1,738	584
Annual Energy Produced (MWh/yr)	1,663	1,669	1,407	1,453	488
Average Electrical Power at Busbar (kW)	191	191	161	166	56
Number of OPD Pelamis Units Needed for 300,000 MWh/yr	180	180	213	206	615
Number of Homes Powered by Plant	34,000	34,000	34,000	34,000	34,000

Source: O. Siddiqui and R. Bedard [30]

Table 4.5. Estimated Performance of Pelamis Commercial Plants

The device rated capacity has been derated from 750 kW in the demonstration plant to 500 kW for the commercial plant. The performance assessment of the demonstration plants shows that the PCMs are overrated and reducing the rated power to 500 kW per device would yield a significant cost reduction and only a relatively small decrease in annual output (attributed to the fact that the U.S. sites have a lower energy level than UK sites for which the device was originally developed).

4.4.2.6 Learning Curves and Economics:

The costs and cost of electricity shown in the previous section are for the *first* commercial scale wave plant. Learning through production experience reduces costs - a phenomenon that follows a logarithmic relationship such that for every doubling of the cumulative production volume, there is a specific percentage drop in production costs. The specific percentage used in this study was 82%, which is consistent with documented experience in the wind energy, photovoltaic, shipbuilding, and offshore oil and gas industries.

The industry-documented historical wind energy learning curve is shown as the top line in Figure 4.7 [31]. The cost of electricity is about 4 cents/kWh in 2004 U.S. dollars based on 40,000 MW of worldwide installed capacity and a good wind site. The lower and higher bound cost estimates of wave energy are also shown in Figure 4.7. The 82% learning curve is applied to the wave power plant installed cost but not to the operation and maintenance part of cost of electricity (hence the reason that the three lines are not parallel).

Figure 4.7 shows the cost of wave-generated electricity: low band (bottom curve), upper band (middle curve); and wind generated electricity (top curve) at equal cumulative production volume under all cost estimating assumptions for the wave plant. It shows that the cost of wave-generated electricity is less than wind-generated electricity at any equal cumulative production volume under all cost estimating assumptions for the wave plant. The lower capital cost of a wave machine (compared to a wind machine) more than compensates for the higher O&M cost for the remotely located offshore wave machine. A challenge to the wave energy industry is to drive down O&M costs to offer even more economic favorability and to delay the crossover point shown at greater than 40,000 MW.

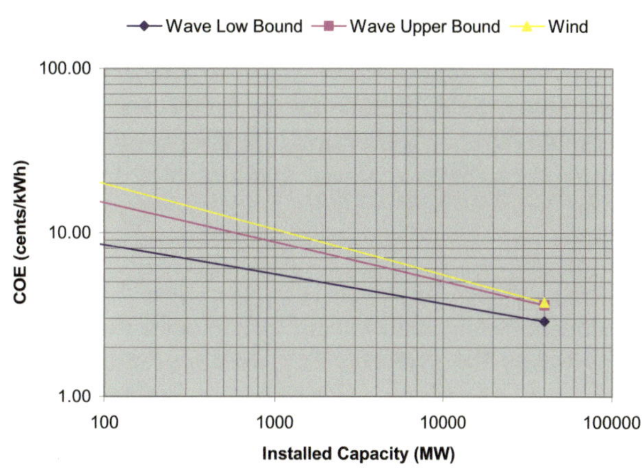

Source: O. Siddiqui and R. Bedard [30]

Fig. 4.7. Electrical interconnection of demo-plant - Oregon example.

The techno-economic forecast made by the Project Team is that wave energy will first become commercially competitive with the current 40,000 MW installed land-based wind technology at a cumulative production volume of 15,000 MW or less in Hawaii and northern California, about 20,000 MW in Oregon and about 40,000 MW in Massachusetts. This forecast was made on the basis of a 300,000 MWh/yr (nominal 90 MW at 38% capacity factor) Pelamis WEC commercial plant design and application of technology learning curves. Maine was the only state in the study whose wave climate was such that wave energy may never be able to economically compete with a good wind energy site.

In addition to economics, there are other compelling arguments for investing in offshore wave energy technology. First, with proper sitting, converting ocean wave energy to electricity is believed to be one of the most environmentally benign ways to generate electricity. Second, offshore wave energy offers a way to minimize the 'Not In My Backyard' (NIMBY) issues that plague many energy infrastructure projects, from nuclear to coal and to wind generation. Because these devices have a very low profile and are located at a distance from the shore, they are generally not visible. Third, because wave energy is more predictable than solar and wind energy, it offers a better possibility than either solar or wind of being dispatch able and earning a capacity payment.

A characteristic of wave energy that suggests that it may be one of the lowest cost renewable energy sources is its high power density. Processes in the ocean concentrate solar and wind energy into ocean waves making it easier and cheaper to harvest. Solar and wind energy sources are much more diffuse, by comparison.

Since a diversity of energy sources is the bedrock of a robust electricity system, to overlook wave energy is inconsistent with national needs and goals. Wave energy is an energy source that is too important to overlook.

4.5 Recent Progress in Offshore Renewable Energy Technology Development

Interest in marine renewable energy is at an all-time high, and prospects for ocean-based renewable energy development look brighter all the time. The recent progress in offshore renewable energy technology development is now examined and potential markets for tidal power, wave energy conversion, and offshore wind are considered. The analysis of market potentials for offshore renewable technology is based solely on identified projects. Therefore, the forecasts are relatively conservative, as the prospective markets could expand as technological advances are achieved and as regulatory environments improve.

4.5.1 Tidal Energy

Historically, tidal projects have been large-scale barrage systems that block estuaries. Within the last few decades, developers have shifted toward technologies that capture the tidally driven coastal currents or tidal stream. The challenge is, "*to develop technology and innovate in a way that will allow this form of low density renewable energy to become practical and economic*" [22].

Two groups of technologies are in operation or planning; these are tidal current turbines and tidal stream generators. Tidal current turbines are basically underwater windmills. The tidal currents are used to rotate an underwater turbine. First proposed during the 1970s' oil crisis, the technology has only recently become a reality. One company, Marine Current Turbines (U.K.) installed the first full-scale prototype turbine (300 kW) off Lynmouth in Devon, U.K. in 2003. Shortly thereafter, the Norwegian company Hammerfest Støm installed their first prototype device.

There are a great number of sites suitable for tidal current turbines. As tidal currents are predictable and reliable, tidal turbines have advantages over offshore wind counterparts. The ideal sites are generally within 1 km of the shore in water depths of 20-30 m.

Tidal stream generators use the tidal stream to generate power from, for example, the raising and lowering of a hydraulic arm. Several promising devices are at the advanced stage of development. For example, the UK firm, The Engineering Business Ltd. has developed and tested a simple concept of placing hydrofoils in tidal stream to produce an oscillatory motion in the vertical or horizontal plane. The device, known as the Stingray™ Tidal Current Generator, "*transforms the kinetic energy of the moving water into hydraulic power, which turns a generator by means of a hydraulic motor*" [22].

4.5.1.1 Tidal Forecasts

It is anticipated that multi-megawatt installations will emerge by the end of the decade (Figure 4.8). By 2008, a forecast of 14.8 MW installed capacity was expected with 65% of the capacity in the United Kingdom. Norway, which already has installed capacity, will be the second dominant player, but lacks defined projects over the next 5 years. Other countries (Canada, France and United States) have a minor role, but could expand prototype devices as the devices progress. Canada and the United States have potential locations, some of which are under negotiation for U.K. tidal generation technology [23].

Almost 70% of forecast capacity for 2008 was anticipated from tidal current turbines with approximately 30% from tidal stream generators. Tidal current turbines represent an extremely important sector for offshore renewables as there are several well-developed devices and such technology, once proven, could be installed in large numbers in the near future. However, a lack of identified projects distorts the forecast near the end of the 5-year period, precisely when significant projects could materialize.

It is conceivable that tidal current turbines such as those of Marine Current Turbines or Hammerfest Støm could eventually be installed in large projects comparable in size to offshore wind farms.

With fewer announced projects, tidal stream generators have a lower forecast. Many of the devices are at earlier stage of design. One system that is generating much interest and has potential is the '*Stingray*' device, designed and built by experienced offshore engineers [25]. Capital expenditures for tidal energy are forecast at $35 million over the next five-year period. The U.K. is the biggest market with $23 million of expenditures forecast to 2008. Forecasts for Norway at a level of $10 million of expenditures over the same period are anticipated. Successful projects could lead to further development later in the period and beyond [23]. Several projects await financial support and could significantly impact the installed capacity as the projects are in excess of 100 MW.

Revised global estimates for capital expenditure in tidal power technology is indicated in Figure 4.8.

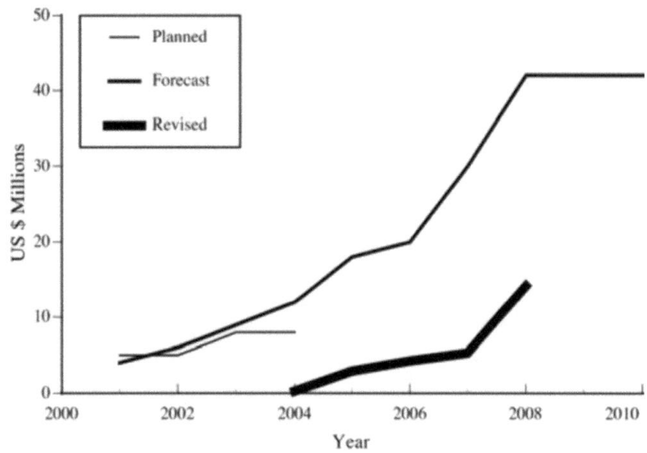

Source: A. T. Jones and A. Westwood [32]

Fig. 4.8. Revised global estimates of capital expenditure in tidal power technology (modified from [24]).

4.5.1.2 Projects

Shihwa Lake Tidal Power Plant, Korea: Korea has a plentiful tidal and tidal current energy resource. Under construction is a single stream style generator at Ansan City's Shiswa Lake, which will have a capacity of 252 MW, comprised of 12 units of 21MW generators. Annual power generation, when completed in 2008, was projected at 552 million kWh. If successful, this project will surpass La Rance (France) as the largest tidal power plant in the world. Korea is also planning a tidal current power plant in Uldol-muk Strait, a restriction in the strait where maximum water speed exceeds 6.5 m/s. The experimental plant will utilize helical or "*Gorlov*" turbines developed by GCK Technology [26].

Yalu River, China: By creating a tidal lagoon offshore, Tidal Electric has taken a novel approach to resolve environmental and economic concerns of tidal barrage technology [27]. Due to the highly predictive nature of the ocean tides, the company has developed simulation models with performance data from available generators to optimize design for particular locations. The recent announcement of a cooperative agreement with the Chinese government for ambitious 300 MW offshore tidal power generation facilities off Yalu River, Liaoning Province allows for an engineering feasibility study to be undertaken.

Tidal Electric also has plans under consideration for United Kingdom-based projects in Swansea Bay (30 MW), Fifoots Point 930 MW), and North Wales (432 MW).

4.5.2 Wave Energy

The true potential of wave energy will only be realized in the offshore environment where large developments are conceivable. Nearly 300 concepts for wave energy devices have been proposed.

Modular offshore wave energy devices that can be deployed quickly and cost effectively in a wide range of conditions will accelerate commercial wave energy. In the coming decade, wave energy will become commercially successful through multiple-unit projects.

Opportunities for expansion of offshore market are expected to increase. This is because the growth of shoreline wave energy devices will be increasingly limited by the low number of available sites and by high installation costs.

Deployment costs for shoreline wave energy devices are very high because they are individual projects and economies of scale are therefore not applicable. The site-specific demands of shoreline wave energy devices mean a further restriction of growth in this sub-sector. Whereas an offshore 50-MW wave farm is conceivable, and will in time be developed, no shoreline wave energy converter can offer such potential for deployment in this way. As such, individual coastal installations are expected to be few and far between [23].

Shoreline wave energy will, however, continue to be relevant, with approximately 25 percent of the forecast capacity over the next five years. The average unit capacity is generally higher than existing offshore technology. Individual devices can be very effective, especially for remote or island communities where, for example, an individual unit of 4MW could have a big impact [23].

Offshore locations offer greater power potential than shoreline locations. Shoreline technologies have the benefit of easy access for maintenance purposes, whereas offshore devices are in most cases more difficult to access. Improvements in reliability and accessibility will be critical to the commercial success of the many devices currently under development [23].

Most wave energy projects to date have been small, and few are connected to a power grid. Shoreline devices offer the advantage of easier access to a grid. For offshore devices, meeting this need will be challenging and costly, although not prohibitively so.

4.5.2.1 Wave Energy Forecast

Wave is a most promising sector over the 2004-2008 period and into the long-term future (Figurc 4. 9).

The development process for wave energy can be looked at in three phases. First, small-scale prototype devices, typically with low capacity, will be deployed. During the second stage, outside funding from government or private investors is possible for the most promising devices. The final stage is the production of full-scale, grid-connected devices that will in some cases be deployable in farm style configurations.

The United Kingdom is expected to be the dominant player over the next five years. In comparison with other countries, the UK has forecast capacity every year, whereas to 2008, installations elsewhere are more intermittent. Australia, Portugal, and Denmark are the next most significant markets and have several projected installations, but they lag far behind the UK. The United Kingdom government has shown reasonable levels of support, which have injected many technologies with valuable grants. The result is a number of advanced wave energy. Future prospects are for deployment of prototype devices. Coupled with a world-class natural resource, the United Kingdom could be the undisputed world leader in wave energy by 2008. Prospects after 2008 are even brighter [23].

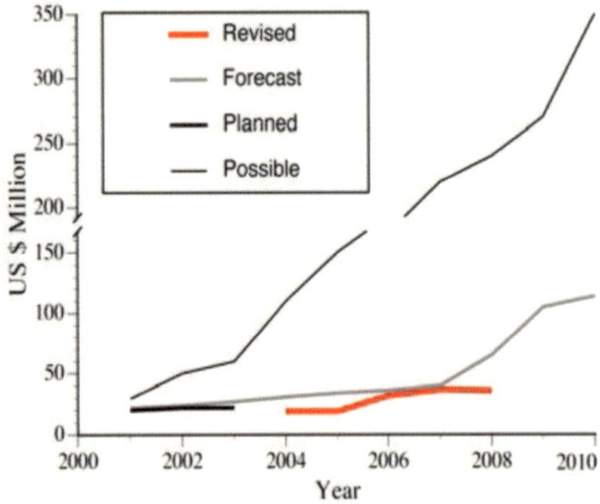

Source: A. T. Jones and A. Westwood [32]
Fig. 4.9. Revised estimates for capital expenditure in wave energy conversion technology. (modified from [24]).

The United States market shows encouraging levels of interest in wave technology; however, the market will be affected by the lack of positive government involvement [23].

4.5.3 Offshore Wind

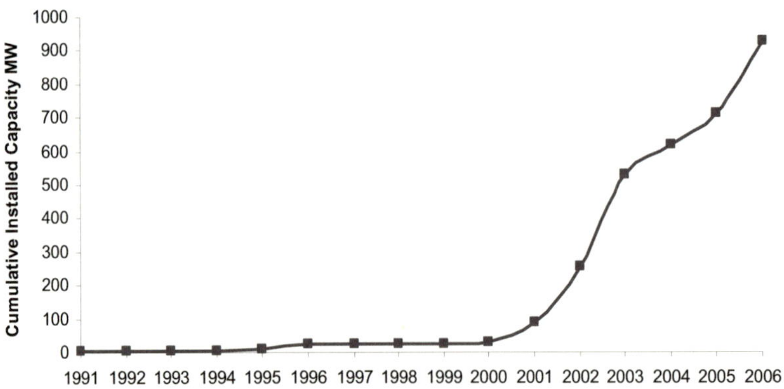

Source: Douglas-Westwood Ltd
Fig. 4.10. Cumulative worldwide offshore wind capacity

The total global offshore wind capacity forecast for installation between 2006 and 2010 stands at 7.4 GW (see Figures 4.10, 4.11). The UK is the world's largest market for the five-year period 2005-2010. The UK's prospects are expected to be twice those of Germany for this period, although the German market at 1.1 GW is still the second largest in the world.

Long-term prospects are excellent off Germany but in the short and mid-term future the industry has much to overcome. Denmark has only two main projects planned for completion by the end of the decade with 200 MW each at Horn Rev and Nysted that are now making progress. The Netherlands has just two projects that were commissioned in 2006 and 2007. No firm prospects have emerged from the last licensing round but long-term potential is there.

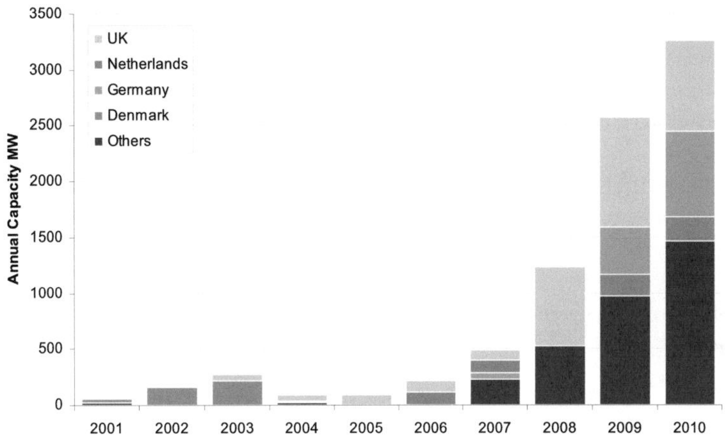

Source: Douglas-Westwood Ltd

Fig. 4.11. Forecast global offshore wind capacity

Technological progress is extremely important for the industry, and will drive developments. As better technology is implemented, large strides in capacity will be achieved using proportionally fewer turbines. For example, up to 1,225 turbines will be installed by 2010. Turbine capacity is increasing, from 2000-2003 the average turbine size was 2 MW, current projects are using 3 MW machines and the industry is pushing development of 5 MW turbines for installations from 2009. Prototype installations of these next-generation turbines have already taken place and the first two offshore units were commissioned off the UK at the Beatrice project in 2006.

Long-term signals are good for the UK market, whereas an air of uncertainty hangs over Germany despite its very promising future forecast. The United Kingdom's development is gradual, whereas Germany's depends on large, technologically challenging projects. Denmark's five-year forecast is disappointing, with only two projects scheduled for commissioning in the period, one in 2009 and one in 2010. Although the country showed initial promise for offshore development, a lack of government commitment has been harmful to the industry here. There are no firm plans for future projects after the coming two, so long term prospects are uncertain.

Offshore wind has a potentially large market in North America. Although the United States has considerable offshore wind potential, regulatory uncertainty is a source of concern. The United States has a significant number of projects in the planning stages [29]. These projects, many of which are very speculative, are not expected to arise until the end of the decade.

For the entire marine renewables sector, 7,500 MW of installed capacity is projected between 2006 and 2010. Some 98% of that capacity is in the form of offshore wind farms. Wind farms installed capacity was 213 MW in 2006. By 2010, this will grow to 3,200 MW - over a ten-fold growth within five-years. The value of the market over the next five-years is projected at $16 billion.

Wave and tidal power will only be a small percentage of the total expenditure on offshore renewables, of the order of $150 million in total expenditure between them. However, wave and tidal power currently attract higher expenditures per megawatt. This indicates higher costs of the immature developing industries. These costs will fall as time goes by and the industries progresses. The leading devices should be comparable with, and in some cases more competitive than offshore wind, by the end of the decade.

The dominance of offshore wind does not mean wave and tidal energy are not important, they are just less well developed, and the industry is much younger. If wave and tidal were compared to offshore wind market data from ten years ago, their market share would be much higher. Offshore wind is booming at present. From around 2010, wave and tidal could begin to see this rapid growth.

4.6 Role of TIDAL Power In The United Kingdom to reduce greenhouse gas emissions

Sections 4.6~4.13 discusses the role of Tidal Power in the UK in fulfilling the UK's requirements for reducing greenhouse gas emissions. Generating electricity from tidal range of the Severn Estuary has the potential to generate some 5% of UK electricity from a renewable indigenous resource.

These Sections focus primarily on the proposed Severn Barrage considering potential benefits, conditions for sustainable development, energy policy context and compliance with environment legislation. UK tidal resource is reviewed: stream resource and tidal range resource. The top tidal range and tidal stream sites in the UK with the resource (in TWh/year) are indicated.

A feasibility study for Tidal Range development in the Mersey Estuary is also summarized and other schemes including the Loughor Estuary (Wales), Duddon Estuary (located on the Cumbrian coast) and the Thames Estuary proposals are reported. Also given is a strategic overview of the Severn Estuary resource, electric output and characteristics, carbon emissions (carbon payback and carbon reduction potential) and physical implications of a barrage.

Approximately 40% of the UK's electricity will have to be generated from renewables (wind, tidal/wave, and plant energy) by 2020 as a result of a legally binding EU target under the Bali Protocol. It is likely to mean a six-fold increase in the amount of onshore wind turbines and a 50-fold increase in the number of offshore wind turbines. This is because the 20% target for all renewables by 2020 applies to energy across the board, including transport and heating, where the scope for renewables is less, implying the electric sector must do more. By 2050, the UK is planning to reduce its CO_2 emissions by at least 60% compared with its emissions in 1990.

A study is underway and is expected to last roughly two years (until January 2010). Under consideration is tidal range, including barrages, lagoons and other technologies, and includes a Strategic Environmental Assessment of plans for generating electricity from the Severn Estuary tidal range to ensure a detailed understanding of its environmental resource recognizing the nature conservation significance of the Estuary. The scheme would use proven technology of a hydroelectric dam but filled by the incoming tide rather than by water flowing downstream. The Severn Estuary has some of the best tidal potential in the world and could more than double the current UK supply of renewable electricity and contribute significantly to targets for renewable energy and CO_2 emissions reduction. The scheme would have a capacity of 8640 MW and produce roughly 17 TWh/year with a load factor of 0.22.

The physics of tidal power: types of tides, semidiurnal tides with monthly variation, diurnal tides with monthly variation, and mixed tides are examined. Variations in output from tidal power due to spring neap cycle is assessed, and technically available tidal energy resource in Europe is also estimated by parametric modeling. Existing tidal energy schemes and sites considered for development worldwide are reviewed. Then, harnessing tidal power (flow or basin, modes of operation and configuration, ebb generation, flood generation, two-way generation and pumping) is indicated. Tidal stream technology that is in the early stages of development but could harness half of the UK's tidal potential is reviewed. The proposed Severn barrage considering tidal resonance in the Severn Estuary, potential benefits, the conditions for sustainable development and energy policy context, compliance with environment legislation and UK tidal resource is also reviewed.

The electricity transmission system in the UK in the Severn area is evaluated where system constraints and upgrades and implications of tidal power are considered. The awareness of energy sources (wind, solar, coal, nuclear, gas, tidal/wave and bio-energy) that can generate electricity in the UK is outlined.

Concerns on Environment Impact considering the protected status of the Severn Estuary (Habitats Directive and Nature 2000), the Birds Directive defining biodiversity objectives, habitats and ecology are considered. Potential carbon savings for the two Severn proposals are then reviewed.

A consensus view is given on tidal power in the UK (tidal stream long-term potential {policy improvements, strategic planning and consenting}, tidal lagoons, and tidal barrages). Conditions for a sustainable Severn barrage (energy policy context, ensuring public interest, apportionment of risks and benefits, avoiding short-termism, regional impacts and priorities) complying with environmental legislation (applying environmental limits and providing compensatory habitats) is given. The final decision on whether this project that will contribute to the UK fulfilling its greenhouse gas emission targets will be given the go-ahead is reviewed.

4.6.1 Tidal Power

Tidal Power including the physics of tidal power (types of tide: diurnal tides with monthly variation, mixed tides, major periodic component, the resource), European energy potential, existing tidal energy schemes, world-wide energy potential, and harnessing tidal power (that includes flow or basin, existing tidal energy schemes, modes of operation and configuration, adaptation of tide-generated to grid network requirements, etc.) is considered first (see Reference [1]). A number of different barrage options worldwide are then summarized. These options include barrages in UK; La Ranch Tidal Barrage in France; and former Soviet Union, China, South Korea, India, Canada, and others. Development trends, economics, institutional constraints and development are discussed.

4.6.1.1 Physics of Tidal Power

Tidal energy is derived from the gravitational forces of attraction that operate between a molecule on the earth and moon, and between a molecule on the earth and sun. The force is $f = K M m / d^2$, where m is the mass of the molecule on the earth, M is the mass of the moon or sun, d is the distance between the bodies, and K is the universal constant of gravitation. The attractive force exerted by the sun is about 2.17 times less than that due to the moon due to the mass and much greater distance that separates the earth and sun. As the earth rotates, the distance between the molecule and the moon will vary. When the molecule is on the dayside of the earth relative to the moon or sun, the distance between the molecule and the attracting body is less than when the molecule is on the horizon, and the molecule will have a tendency to move away from the earth. Conversely, when the molecule is on the night side of the earth, the distance is greater and the molecule will again have a tendency to move away from the earth. The separating force thereby experiences two maxims each day due to the attracting body. It is also necessary to take into the account the beating effect caused firstly by difference in the fundamental periods of the moon- and sun-related gravitational effects, which creates the so-called spring and neap tides, and secondly the different types of oscillatory response affecting different seas. If the sea surface were in static equilibrium with no oscillatory effects, lunar forces, which are stronger than solar forces, would produce tidal range that would be approximately only 5.34 cm high.

4.6.1.2 Types of Tide

Tidal phenomena are periodic. The exact nature of periodic response varies according to the interaction between lunar and solar gravitation effects, respective movements of the moon and sun, and other geographical peculiarities. There are three main types of tide phenomena at different locations on the earth.

- ***Semidiurnal Tides with Monthly Variation:*** This type of tide has a period that matches the fundamental period of the moon (12 hr 25 min) and is dominated by lunar behavior. The amplitude of the tide varies through the lunar month, with tidal range being greatest at full moon or new moon (spring tides) when the moon, earth, and sun are aligned. At full moon, when moon and sun have diametrically opposite positions, the tides are highest, because the resultant center of gravity of moon and earth results in the earth being closer to the sun, giving a higher gravity effect due to the sun. At new moon, maximum tidal range is less. Minimum tides (neap tides) occur between the two maxims and correspond to the half-moon when the pull of the moon and sun is in quadrature, i.e., the resultant pull is the vector sum of the pull due to moon and sun, respectively. In this case, the resultant gravitation force is a minimum. A resonance phenomenon in relation to the 12 hr-25-min periods characterizes tidal range.

- ***Diurnal Tides with Monthly Variation.*** This type of tide is found in the China Sea and at Tahiti. The tidal period corresponds to a full revolution of the moon relative to the earth (24 hr- 50-min). The tides are subject to variations arising from the axis of rotation of the earth being inclined to the planes of orbit of the moon around the earth and the earth around the sun.

- ***Mixed Tides.*** Mixed tides combine the characteristics of semidiurnal and diurnal tides. They may also display monthly and bimonthly variation. Examples are of mixed tides are those observed in the Mediterranean and at Saigon.

4.6.1.3 Major Periodic Components

The following periodic components in tidal behavior can be identified: (i) a 14-day cycle, resulting from the gravitational field of the moon combining with that of the sun to give maxims and minima in the tides (called spring and neap tides, respectively); (ii) a ½ year cycle, due to the inclination of the moon's orbit to that of the earth, giving rise to a period of about 178 days between the highest spring tides, which occur in March and September, (iii) the Saros, a period of 18 2/3 years required for the earth, sun, and moon to return to the same relative positions, and (iv) other cycles, such as those over 1600 years which arise from further complex interactions between the gravitational fields.

Maximum height reached by high water varies in 14-day cycles with seven days between springs (large tide range) and neaps (small tide range). The spring range may be twice that of the neaps. Half-yearly variations are +/-11%, and over 18 2/3 years +/- 4%. In the open ocean, the maximum amplitude of the tides is less than 1 m. Tidal amplitudes are increased substantially particularly in estuaries by local effects such as shelving, funneling, reflection, and resonance. The driving tide at the mouth of the estuary can resonate with the natural frequency of tidal propagation up the estuary to give a mean tidal range of over 11 m in the Severn Estuary, UK and can vary substantially between different points on the coastline[3] The physics of tidal range is examined by Baker in more depth in [33].

4.6.2 European Energy Potential

The amount of energy available from a tide varies approximately with the square of tidal range. The energy available from a tidal power plant would therefore vary by a factor of four (eight for tidal stream) over a spring-neap tide cycle. Typical variation in output from tidal range and tidal stream power in the Severn Estuary due to the spring-neap cycle is indicated in Figures 4.12(a) and 4.12(b), respectively. Approximately 20 suitable regions for development of tidal power worldwide have been identified.

A parametric approach [34] has been used to estimate tidal energy potential for appropriate EU countries (Belgium, Denmark, France, Germany, Greece, Ireland, Portugal, Spain, The Netherlands, and UK). An assessment of all reasonably exploitable sites within the EU with a

---

[3] Tidal range is the tidal height between high-tide and low tide. Typical tidal ranges are Bay of Fundy (Canada) 19.6 m; Granville (France) 16.8 m; La Rance (France) 13.5 m.

mean range exceeding three meters yielded a total energy potential of about 105 TWh/year. This potential is mainly in the UK (50 TWh/year) and France (44 TWh/year), with smaller contributions in Ireland, The Netherlands, Germany and Spain. Technically available resource for tidal energy estimated using parametric modeling is given in Table 4.6.

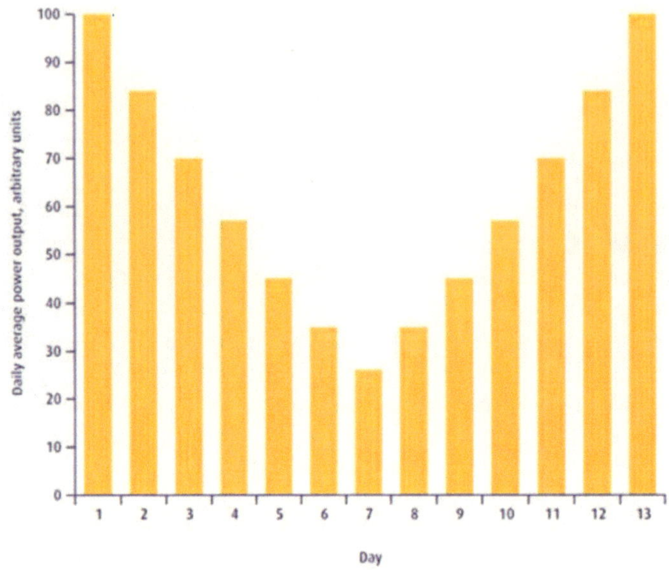

Fig. 4.12(a). Typical variation in output from tidal range power due to spring-neap cycle

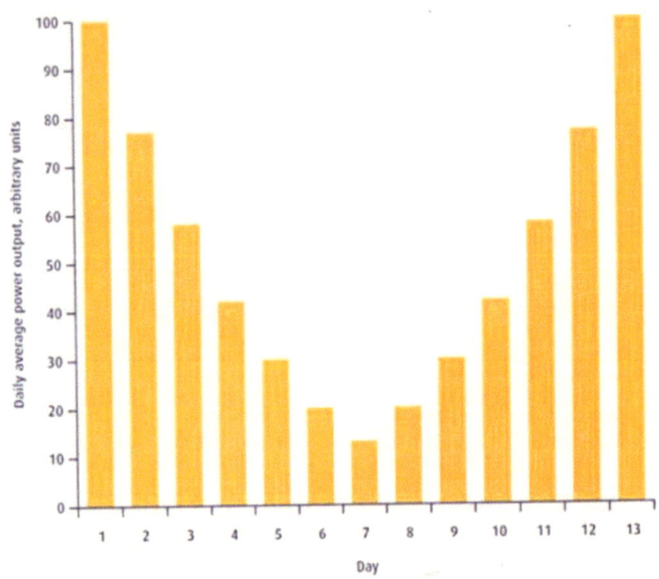

Fig. 4.12(b). Typical variation in output from tidal stream power due to spring-neap cycle.

Country	Technically Available Tidal Energy Resource		Percentage of European Tidal Resource
	GW	TWh/year	
United Kingdom	25.2	50.2	47.7
France	22.8	44.4	42.1
Ireland	4.3	8.0	7.6
Netherlands	1.0	1.8	1.8
Germany	0.4	0.8	0.7
Spain	0.07	0.13	0.1
Other W European	0	0	0
Total W European	63.8	105.4	100.0

Table 4.6. Technically Available Tidal Energy Resource in Europe Estimated by Parametric Modeling

4.6.3 Existing Tidal Energy Schemes

Relatively few tidal power plants have been constructed to date. The first and largest is the 240 MW barrage at La Ranch (France) [35], which was built for commercial production in the 1960s. Other tidal power plants include the 17.8 MW plant at Annapolis (Canada), the 400-kW experimental plant at Kislaya Guba (former Soviet Union), and the 3.2 MW Jiangxia station (China).

4.6.4 Sites Considered for Development Worldwide

Economic feasibility of tidal barrage schemes is dependent on the world market price of fossil fuels, interest rates over scheme expected life, and on level of fossil fuel levies based on the carbon content of fuel and electricity not produced by renewable energy sources, etc. Tidal power sites of capacity above 1GW considered for development with installed capacity and approximate annual output include: (i) Argentina San Jose, 6.8GW, 20.0 TWh; (ii) Canada Cobequid, 5.34 GW, 14.0 TWh; (iii) Canada Cumberland 1.4 GW, 3.4 TWh; (iv) Canada Shepody, 1.8GW, 4.8 TWh; (v) India Gulf of Cambay, 7.0 GW, 15 TWh; (vi) UK Severn, 8.6 GW, 17 TWh; (vii) USA Knit Arm, 2.9 GW, 7.4 TWh; (viii) USA Turnagain Arm, 6.5 GW, 16.6 TWh; (ix) Former Soviet Union Mezen, 15 GW, 50 TWh; (x) Former Soviet Union Tugur, 10 GW, 27 TWh; and (xi) Former Soviet Union Penzhinskaya, 50 GW, 200 TWh.

4.6.5 Harnessing Tidal Power (flow or basin, existing tidal energy schemes, modes of operation and configuration, adaptation of tide-generated to grid network requirements)

Devises include waterwheels; lift platforms, air compressors, water pressurization, etc. Energy can be extracted either directly by harnessing the kinetic energy of a tide flow, or by using a basin to capture potential energy of a rising and falling mass of water.

4.6.5.1 Tidal Flow

Tide flows have a poor energy density. Theoretical available power P is given by $P=D\ A\ V^3$, where D is the fluid density, A is the area swept out by the turbine rotor, and V is the undisturbed stream velocity [36]. The energy can be harnessed only with poor maximum efficiency, similar to a windmill, where an efficiency of 59.3% is possible. Directly harnessing power in this way, however, does not require expensive additional structures.

4.6.5.2 Basin

This method involves constructing a barrage and forming a basin from a natural bay or estuary. Considerable extra cost is incurred, but this is more than outweighed by the extra energy that is extractable. The energy available from a turbine in an effective barrage is one or two orders of magnitude greater than that from a similar size of turbine in a tide stream of, for example, 2 m/s. The extra cost of constructing the barrage may be only a third of scheme overall cost.

4.6.6 Modes of Operation and Configuration

The tide is the only factor that affects the generating activity of a tidal power plant that is programmed to produce maximum output. The output at any given time can be accurately calculated as far in advance as is necessary.

4.6.6.1 Single-Action Outflow (Ebb) Generation

Barrages can use either one basin or a combination of basins, and can operate by ebb, flood, or two-way generation, with or without pumping. The simplest method is ebb generation using a single basin. The basin is permitted to fill through sluices (gated openings). Generation takes place as the basin is emptied via turbines once the tide level has dropped sufficiently. There are two bursts of generation each day.

Typical day-to-day fluctuations are: (i) there are two bursts of generation activity each day, beginning approximately three hours after high tide and lasting 4-6 hours; (ii) for each cycle production levels rapidly increase with tidal range, the output characteristic therefore displaying a 14-day cycle; (iii) high-water times shift by about 1 hr per day; (iv) in each 14-day period, the generation will not be evenly distributed throughout the 24-hr of the day; (v) output levels will only show slight variation from one fortnightly period to the next; and (vi) annual production levels will show fluctuations of around +/-5% and will follow a cycle of 18 2/3 years.

4.6.6.2 Flood Generation

Here, power is provided as the basin fills. The basin empties through sluices as the tide falls. This method is not as efficient as ebb generation since it involves using the basin between existing low tide level and slightly above normal mid-tide level, thus producing less energy. An advantage of this mode is that it facilitates the production of energy out of phase with a neighboring ebb generation scheme, complementing its output and perhaps providing some firm capacity.

4.6.6.3 Two-way Generation:

This is a combination of ebb and flood generation, generating as the basin both fills and empties, but with a smaller power output for simple ebb generation (except at the highest tide ranges) due to reduced range within the basin. There is a resultant reduction in efficiency with two-way generation since turbines and water flow cannot be optimized. Two-way generation produces electricity in approximately 6-hr cycles, with smaller power output and a greater plant utilization factor.

4.6.7 Tidal Stream

Tidal current turbines are basically underwater windmills where tidal currents are used to rotate an underwater turbine. First proposed during the 1970s' oil crisis, the technology has only recently become a reality. Horizontal axis turbines are more commonly employed. Marine Current Turbines (MCT) {http//www.marineturbines.com/home.htm} installed the first full-scale prototype turbine (300kW) off Lynmouth in Devon, UK in 2003. Their second project, a 1 MW prototype, is expected soon. It will be followed by an array of similar systems (farm) to be installed in an open sea, where three turbines will be added to provide a total capacity of 5 MW. A similar project is the Hydro Helix project in France.

The Norwegian company Hammerfest Stom installed their first grid-connected 300kW devise that was tested and the concept proven {http//www.e-tidevannsenergi.com/} A tidal stream turbine has been designed for the Pentland Firth between the North of Scotland and the Orkney Islands [37] where the first design was for twin turbines with 20 m rotors and was rated at 1-2 MW depending on current speed. In today's design, the 60 m deep four 20 m rotors cover water flow rather than a pair to keep blade loads within practical limits and the whole power output is 4 MW. The SMD Hydrovision Tidal Project (UK) {http//www.smdhydrovision.com} consists of a pair of contra-rotating 500 kW turbines mounted together on a single crossbeam. The 1 MW units are designed to be mounted in an offshore tidal environment with a peak tidal velocity of 5 knots (2.5 m/s) or more and a water depth of greater than 30 m. The Lunar Energy Project (UK) and the HyroHelix Energies Project (France) {http//www.lunarenergy.co.uk http//www.hyrdohelix.fr/} feature a ducted turbine fixed to the seabed via gravity foundation. A 1/20th model was tested in 2004 and a 1 MW prototype is expected soon. The ideal sites are generally within several kilometers of the shore in water depths of 20-30 m.

There are also vertical axis turbines that are cross flow machines whose axis of rotation meets the flow of the working fluid at right angles. Cross flow turbines allow the use of a vertically oriented rotor that can transmit the torque directly to the water surface without need of complex transmission systems or an underwater nacelle. The vertical axis design permits the harnessing of tidal flow from any direction, facilitating the extraction of energy not only in two directions, the incoming and outgoing tide, but making use of the full tidal eclipse of the flow [38]. In these types of turbines, the rotational speed is very low, of the order of 15 rpm.

4.6.7.1 The Enermax Project (Italy) {http//www.pontediarchimede.com}:

This uses the Kobold turbine. Its main characteristic is its high starting torque that permits it to start even in loaded conditions. A pilot plant is located in the Straight of Messina, close to the Sicilian shore in Italy, in an average sea tidal current of 2m/sec.

4.6.7.2 The Blue Energy Project (Canada) {http//www.bluenergy.com}

Four fixed hydrofoil blades of the Blue Energy tidal turbine are connected to a rotor that drives an integrated gearbox and electrical generator assembly. The turbine is mounted in a durable concrete marine caisson that anchors the unit to the ocean floor, directs flow through the turbine further concentrating the resource supporting the coupler, gearbox, and generator above it. The hydrofoil blades employ a hydrodynamic lift principle that causes the turbine foils to move proportionately faster than the speed of the surrounding water. The rotation of the turbine is unidirectional on both the ebb and flow of the tide. A unit turbine is of the order of 200 kW output power. For large-scale power production, multiple turbines are linked in series to create a tidal fence across an ocean passage or inlet.

4.6.7.3 The Gorlov Helical Turbine (GHT) (USA.{http//www.gcktechnology.com/GCK/}

The turbine consists of one or more long helical blades that run along a cylindrical surface similar to a screw thread, having a airfoil or airplane wing profile. GHT blades provide a reaction thrust that can rotate the turbine faster than the water flow itself. The GHT is self-starting and can produce power from water current flow as low as 1.5 m/sec with power increasing in proportion to the water velocity cubed. Due to axial symmetry, the GHT always rotates in the same direction, even when tidal currents are reversed. The standard model (1 m in diameter, 2.5 m in length) can be installed either vertically or horizontally to the water current [39]. A single GHT has a rated power of 1.5 kW for 1.5 m/s water speed and 180 kW for 7.72 m/sec. A similar concept to the GHT is the Achard known as the Harvest project (France) {http//www.legi.hmg.inpg.fr/cavit/Deta/Harvest.html}.

4.6.8 Adaptation to Grid Network Requirements

The output from a tidal plant displays characteristics that are not compatible with those of conventional generation, transmission, and system load. A pumping system increases average output levels and enhances flexibility of the scheme. This in turn leads to improved economic efficiency as supply times can be varied to match energy cost levels.

Single-action outflow (ebb) generation barrages can use one basin or a combination of basins, and can operate by ebb, flood, or two-way generation, with or without pumping.

4.7 Proposed Severn Barrage

Few sites worldwide are as close to electricity users and the transmission grid as are the potential sites in the UK.

The Severn is probably the most well known of all potential tidal energy locations, and projects for damming the Severn estuary date back for over a century. The tide range is up to 11 m near the head, being amplified and funneled by the Bristol Channel. The channel and estuary form a resonator having an effective length equivalent to ¼ of that of the tidal wave. Most attention is focused on schemes further down the estuary where tide range is reduced and a longer barrage is needed, but where the energy extractable is many times greater. Tidal resonance in the Severn Estuary is illustrated in Figure 4.12.

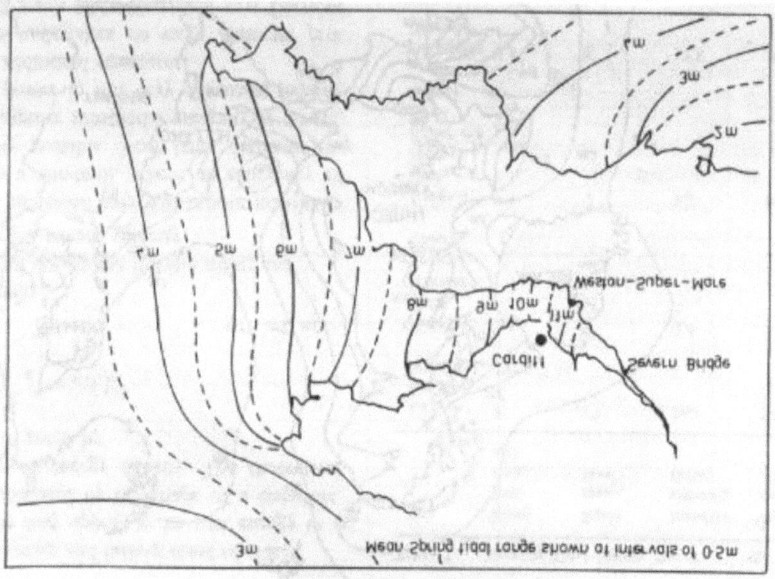

Fig. 4.13. Tidal resonance in the Severn Estuary.

A number of different barrage options have been proposed. The Cardiff-Weston scheme is one of the largest and would have a generating capacity of around 8.64GW. The Shoots scheme (which would run near to the two Severn road crossings) is 1.05GW with an annual output of around 2.75 TWh. Power output and cost summary for the two options are given in Table 4.7.

4.7.1 Potential Benefits

The assumption is that both barrages would be operated on the ebb tide, with the addition of flood pumping to increase *total energy* output. This means that they would be generating electricity for around 7~8 hours on each tide, and output would vary within this period. The annual output of each barrage is less than that implied by their size, around 4.4 % of UK electricity supply, about the same as would be produced by a 2 GW conventional fossil-fuel or nuclear power station.

The high capital cost of a barrage project leads to a very high sensitivity to the discount rate used (Table 4.7). At a discount rat of 2 % that could be justified for a climate change mitigation project, cost of electricity from both barrage proposals is highly competitive with other forms of generation. At a commercial discount rate of >8 %, these costs escalate significantly, making private investment unlikely without significant Government market intervention.

		Cardiff-Weston	Shoots
Length of Embankments		16.1 km	4.1 km
Generating Capacity		8.64 GW	1.05 GW
Annual Average Electricity Output		17 TWh	2.75 TWh
Contribution to UK Electricity Supply (2006 Data)		4.4 %	0.7 %
Estimated Cost of Construction		£15 bn	1.5 bn
Estimated cost of output at various discount rates (high case scenario)	2 %	2.31p/kWh	2.58 p/kWh
	3.5 %	3.68 p/kWh	3.62 p/kWh
	8 %	9.24 p/kWh	7.52 p/kWh
	10 %	12.37 p/kWh	9.54 p/kWh
	15 %	22.31 p/kWh	15.38 p/kWh

Table 4.7. Power Output and Cost Summary for the Two Main Severn Barrage Options

There would be substantial flood risk benefits. The timing of output is not optimal, but output is not a major problem for the electricity grid that can be managed at very low cost. The output would displace output from fossil-fuelled plants and would make a genuine and sizable contribution to meeting the UK's targets on renewable energy and on reducing carbon dioxide emissions.

4.7.2 Conditions for Sustainable Development

The issue has been approached from a general position that favors renewable energy under which its development might be sustainable. It has been done within a framework that places a high value on long-term public interest and on maintaining the overall integrity of internationally recognized habitats and species.

4.7.3 Energy Policy Contexts and Compliance with Environment Legislation

There is risk that development of a barrage might divert Government's attention away from other necessary solutions to the challenge of climate change, including development of a more decentralized energy system and the reduction of energy demand. A Severn barrage has a number of disadvantages that are similar to those of nuclear power, and developing such a large amount of electricity generating capacity in a single location would not itself move the UK any closer to a more decentralized energy system. The Government does not have policies in place at this time to deliver the carbon savings that will be required by 2050, and in particular the delivery of emissions reductions over the next 15 years. A Severn barrage could be pursued as part of a major drive to reduce emissions substantially over both the short- and long-term.

4.7.4 UK Tidal Resource

Available estimates of the UK's tidal range and tidal stream resource for potential electricity generation are given in Table 4.8. Estimating potential electricity output requires a number of assumptions on technical constraints of the devises installed, their efficiency, and effect of resource extraction on the remaining resource. This implies that there is a large degree of uncertainty in all resource estimates given in Table 4.8.

Tidal Range Sites		Tidal Stream Sites		
Site Name	*Resource (TWh/year)*	*Site name*	*Area*	*Resource (TWh/year)*
Severn	17	Pentland Skerries	Pentland Firth	3.9
Mersey	1.4	Stroma	Pentland Firth	2.8
Duddon	0.212	Duncans-by Head	Pentland Firth	2.0
Wyre	0.131	Casquets	Alderney	1.7
Conwy	0.06	South Ronaldsay	Pentland Firth	1.5
		Hoy	Pentland Firth	1.4
		Race of Alderney	Alderney	1.4
		South Ronaldsay	Pentland Firth	1.1
		Rathlin island	North Channel	0.9
		Mull of Galloway	North Channel	0.8

Table 4.8. Top Tidal Energy Sites in the UK with the Resource

4.8 Electricity Transmission System

The Electricity Transmission Network in Great Britain is illustrated in Reference [40]. Most generating plants in Great Britain are connected to the transmission system, with some 12 GW of generating capacity connected to the distribution networks. The capacity of the transmission system to connect generation and manage the flows of electricity depends on the capacity of the network. The process of connecting to the network is based around the principle of matching the Connection Entry Capacity (CEC) (the generating capacity of the power station) with the Transmission Entry Capacity (TEC) (the capacity of the network to accept a new generator). Connection offers are made on the basis of an *invest and connect* approach whereby CEC can never exceed TEC, so new lines must be built to connect new generation. At present, there are significant TEC constraints in the north of England and in Scotland, which are preventing the connection of new generation projects. Areas of the transmission network will need to be upgraded to higher voltage levels to increase the TEC.

These issues pose significant challenges for the connection of tidal stream projects. Existing capacity constraints and delays to network upgrades will further delay the date by which tidal stream projects might be connected. If the current approach to transmission connection is not modified, it is unlikely that the UK will see any significant level of tidal stream connection before 2020.

For tidal range, the situation is less significant. Firstly, tidal range resources are generally located in areas where grid constraints capacity transmission lines are less pronounced, and are closer to high capacity transmission lines and to centers of demand. Secondly, tidal barrages are likely to be larger one-off projects when compared to a tidal stream array, making the incorporation of grid connection costs a smaller part of the overall project cost and therefore more manageable.

The Severn Estuary area has significant network capacity for new generation, with negative transmission network use of system (TNUoS) charges currently in force. This applies for generators in the southwest of England due to a shortage of generation to meet local requirements.

The transmission network around the Severn Estuary is shown in Figure 4.14. It is quite well developed, with possible connections at both 400kV and 275kV on both sides of the estuary not far from the landing points for a barrage. For the larger Cardiff-Weston scheme, two connections into the 400kV network would be required at both north sides of the barrage (i.e. four connections in total) as the 275kV network on the north side is near to capacity and there is very limited capacity on the 132kV network on the south side.

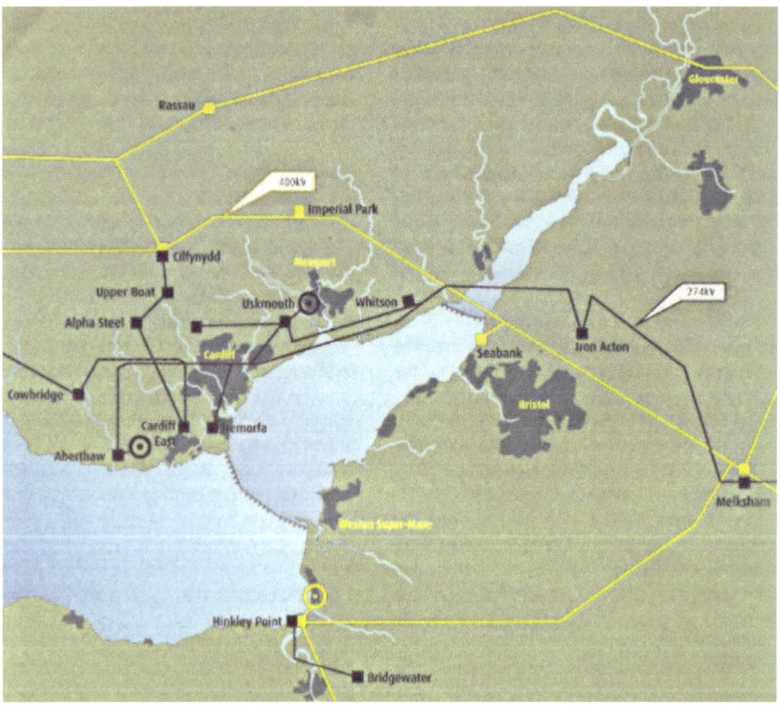

Fig. 4.14. Electricity transmission network (400 and 275kV) around the Severn Estuary.

For the much smaller-rated capacity of the Shoots scheme, all of the connection options seem to have sufficient capacity to accommodate this through just one connection. A connection to the Hinkley Point-Melksham 400kV double-circuit line would be appropriate due to high demand for new capacity. This could further increase with decommissioning of the nuclear power station at Hinkley Point (if it is not replaced by new nuclear generation). Both schemes would require some new transmission infrastructure to connect into the existing network. This requirement is far higher for the Cardiff-Weston scheme due to its higher rated capacity.

4.9 Feasibility Studies: Mersey Estuary, Loughor Estuary (Wales), Duddon Estuary (Cumbrian Coast), Wyre Barrage (Lancashire), and Thames Barrier

4.9.1 Mersey Barrage

The first stage of the feasibility study was completed in 1988 [41], which included hydraulic and energy modeling together with a preliminary examination of the geo-technical conditions, socio-industrial benefits and likely effects on shipping and the ecology of the estuary. No overriding impediments to the construction of a barrage were identified at that time. The Mersey Estuary has a mean spring tidal range of 8 m and a potential annual resource of about 1.4 TWh. The barrage would have 28 turbine-generators with 8-m turbines rated at 25 MW, giving an installed capacity of 700 MW [42]. The proposed barrage would be approximately 2 km long, with a design life of at least 120 years for the main structure, with two periods of turbine renewal at 40-year intervals.

Technology Option	Rated Capacity (MW)	Annual Electricity Output (GWh}
Tidal lagoon (Zone 1)	350	650
Tidal Barrage (Zone 2)	700	1200
Central reservation (Zone 2)	20	40
Constrained channel (Zone 2)	50	100
Tidal fence (Zone 2)	35	80
Tidal gate (Zone 3)	380	700
Water wheel (Zone 3)	200	500

Source: Peel Environmental Ltd

Table 4.9. Comparison of Main Tidal Power Options for the Mersey Estuary.

There is renewed interest as a result of a recent study commissioned by Peel Environmental Ltd in association with the North West Regional Development Agency (NWDA) and the Mersey Basin Campaign [43]. There are a number of potential options for harnessing energy from the Mersey. To assess the options, the estuary was divided into study zones. These are indicated in Figure 4.15. The only viable option for zone 1 was considered to be a tidal lagoon. This could be operated independently from the other options. For the remaining zones, the two most productive options were two tidal barrage options (one termed as a *tidal gate*), although several tidal stream options were also studied.

The capacity and estimated electricity output from each option is indicated in Table 4.9. The construction cost of a Mersey Barrage is estimated at £1.5bn (2006 prices). This results in a unit cost of output ranging from 12.27p/kWh to 15.79p/kWh when a commercial discount rate of 8-10% is assumed. This would reduce to about one third if a 2% discount rate were used. The costs using the higher discount rates would result in electricity that is not commercially competitive under current conditions.

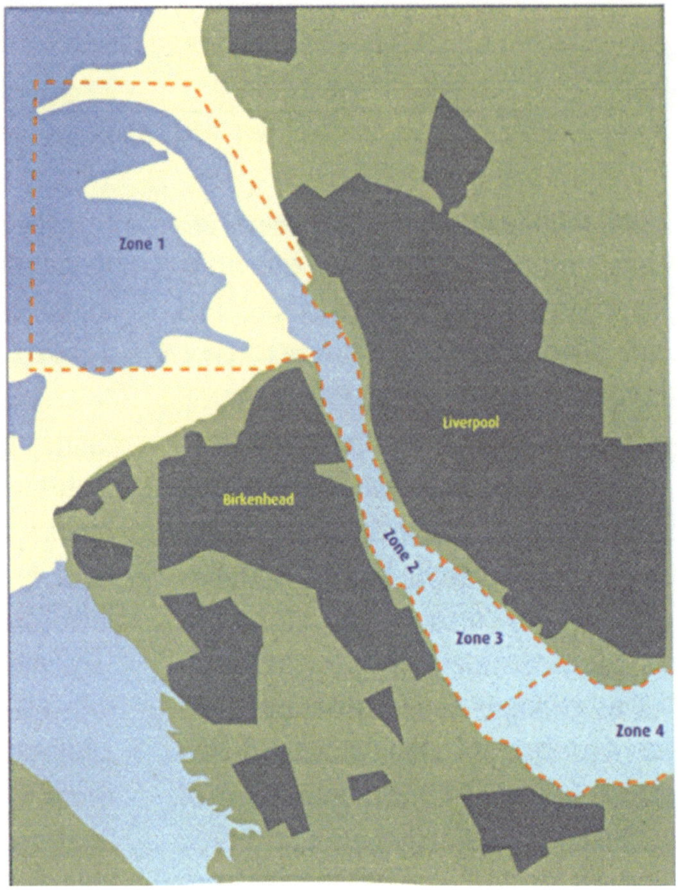

Source: Peel Environmental Ltd.
Fig. 4.15. Mersey showing study zones

4.9.2 Other UK Barrages

These include the Loughou Estuary in Wales which has an annual spring tide of 3.9 m and could generate 5 MW, the Duddon Estuary located on the Cumbrian coast that has a mean tidal range of 5.8 m and could generate 0.212 TWh/year from ten 10 MW turbines, the Wyre Barrage (Lancashire) with a mean tidal range of 6.6 m and installed capacity of around 60 MW that would generate about 0.131 TWh/year, the Thames Barrier that would form a new flood protection barrier that could generate possibly up to 800 MW, and the Conwy Barrage (North Wales) that would have six 5.5-MW generators giving an installed capacity of 33 MW. Here, the mean tidal range varies from 7.1 m (spring) to 3.5 m (neap) and average energy is 0.0568 TWh/year (0.0602 TWh/year with pumping)

4.10 Environment Impact

The Severn Estuary is a unique and dynamic environment. It has the second largest tidal range in the world, combined with a high-suspended sediment load, and has a number of special features, including extensive areas of salt marsh, and mobile sandbanks. It is an important site for migratory birds, and for fish movements in and out of the estuary's tributaries, such as the Wye and Usk. For these reasons the Severn Estuary has been designated a protected site under national and international legislation.

The most important pieces of conservation legislation for a prospective Severn barrage are the EU Directives on Birds and Habitats that protect sites designated as Special Protection Areas (SPAs) and Special Areas of Conservation (SACs). Identification of sites is a science-led process that is based on protecting important ecosystem types and threatened bird species. The Severn Estuary is a SPA and a candidate SAC. The aim of designation is to protect against biodiversity loss by conserving a series of important or at-risk habitats and species that make up the Europe-wide Natura 2000 network. The Natura 2000 network is based on the need to conserve biodiversity across Europe, and internationally.

A Severn barrage could lead to a loss of biodiversity, resulting in the need for a compensatory habitats package to maintain overall integrity of the Natura 2000 network. The EU Directives provide a clear and robust legal framework for achieving sustainable development and therefore compliance with the Directive is a central condition for a sustainable Severn barrage. Providing compensatory habitat would be a very significant undertaking on a scale hitherto unprecedented in the UK. It would have to be an integral part of any barrage proposal.

In summary, there is a strong case to be made for a sustainable Severn barrage. Much wider and stronger action on climate change is a pre-requisite for UK Sustainable Development Commission's (SDC) support. There may be an environmental opportunity available by linking a compensatory habitats package to climate change adaptation. A Severn barrage must be publicly led as a project and publicly owned as an asset to ensure long-term sustainability. The Government should consider a range of innovative financing mechanisms that would maintain overall public control and ownership of the project.

4.11 Carbon Emissions

One of the main arguments for building a Severn barrage is its potential contribution to reducing carbon dioxide emissions and therefore its ability to help the UK meet its national and international obligations on renewables and emissions of greenhouse gases[4]

The reduction in carbon dioxide emissions from a Severn barrage depends heavily on assumptions made on the carbon intensity of the displaced electricity. The output from a tidal barrage is intermittent, is highly predictable, and has very low operational cost. It would be treated as base-load generation, similar to that for nuclear power plants. Therefore, tidal power output is most likely to displace the output from large, centralized, fossil fuel plants.

---

[4] Under the recently-agreed EU target for 20% of all energy requirements to come from renewables by 2020 www.defra.gov.uk/news/latest/2007/climate-0309.htm , the UK will need to commit to developing at least this amount. On greenhouse gases, it is assumed that the UK will need to make substantial progress in its goal for a 60% reduction in carbon dioxide emissions by 2050, and that the UK 's commitment will most likely need to rise to a 80-90% cut in line with scientific evidence.

The long lifecycle of a Severn barrage has a positive impact on the carbon *emissions factor* as the embedded emissions from construction are counter-balanced by 120 years of zero emissions electricity generation. The emissions factor for the Severn Cardiff-Weston barrage is estimated to be 2.42gCO_2/kWh and 1.58gCO_2/kWh for the Shoots scheme, which translates into a carbon payback of around 5-8 months for the two schemes. It is in the very lowest category for power generation and compares well against other low carbon technologies such as nuclear power (16gCO_2/kWh) [44].

The Severn barrage would displace the need for some other form of new capacity, such as CCGT, as this is currently the preferred choice for new-build base-load generation. New-build gas-fired plant has a carbon intensity of around 90tC/GWh.

Table 4.10 presents the likely annual carbon savings (as both carbon and carbon dioxide) from the two Severn barrage proposals. Although it is possible to calculate the lifetime carbon savings (over the 120 years expected life of a barrage), the figures are unlikely to be realistic because over this period the generating capacity being displaced will be progressively less carbon intensive.

	Cardiff-Weston		**Shoots**	
	MtC	MtCO_2	MtC	MtCO_2
Annual carbon savings (based on 90tC/GWh)	1.53	5.60	0.25	0.91
Percentage reduction in UK carbon emissions (1990 baseline)	0.92%		0.15%	
For Comparison				
Annual carbon savings based on *average gas* displacement (100tC/GWh)	1.7	6.22	2.75	1.00
Annual carbon savings based on *grid mix* displacement (131tC.GWh)	2.23	8.15	0.36	1.32

Table 4.10. Potential Carbon Savings from a Severn Barrage.

4.12 Physical Implications of a Barrage

The construction, presence and operation of a Severn Barrage would involve major physical changes to water levels, geomorphology, and sedimentary processes. These physical changes underlie and have significant implications for (i) the environment—the estuarine ecosystem, inter-tidal and wetland habitats, birds, fish; and (ii) the economy and society at a local and regional scale—ports and navigation, land drainage and flooding, water quality, infrastructure and transport, employment, industry and recreation.

The changes that a barrage would cause extend well beyond the direct physical footprint of the structure, and involve physical changes to the estuary as a result of reducing the tidal range and changing the water levels within the barrage basin (upstream) and outside the barrage (downstream). The physical barrier across the estuary (Cardiff-Weston barrage is about 16km long), together with the changes to water levels, the tidal currents and the wave regime of the estuary, also mean that the sedimentary and morphological characteristics and processes of the estuary would be significantly changed.

The hyper-tidal nature of the estuary is responsible for creating a series of unique conditions and habitats such as extensive mud flats and mobile sand banks and extracting energy from this dynamic regime in the form of a tidal barrage would fundamentally change the nature of the Severn Estuary. On the whole, a barrage would raise the average water level inside the basin by raising the low tide levels to around present mean sea level and by reducing high tide levels by up to 1m (up to about 0.5m for the Shoots scheme). The mean sea level in the estuary would be raised by some 2.5m to 3m for the Cardiff-Weston scheme. The overall effect is to reduce the tidal range by about 50%. For the Cardiff-Weston scheme, the range would decline from 11.5m to 4.5m on spring tides, and 5.5m to 2.5m on neaps. For the Shoots scheme, the result would be a similar reduction in tidal range, from 12.5m to 4.5m on spring tides, and 6.5m to 3.5m on neaps. Down stream of a barrage, model predictions for the Cardiff-Weston alignment are that low water level would be raised somewhat and high water levels would be reduced, the effects declining with distance up to 75km seawards.

Morphology refers to the form and development of the landscape. Morphology and the sediment regime have implications for the environment, the engineering of a barrage, and in relation to ports and navigation.

4.13 Consensus View on Tidal Power in the UK

4.13.1 Tidal Power

The UK has considerable tidal power resource that could be exploited to produce renewable electricity. Current estimates suggest that the UK total resource is divided roughly equally between tidal range and tidal stream potential, with a combined output equal to around 10% of UK electricity supply. All options for exploiting this resource should be considered as a narrow focus on just one project (a Severn barrage) could be detrimental to the development of a whole class of emerging tidal stream technologies, some of which could be sizeable generators of renewable electricity. There is no conflict between the tidal range and tidal stream technologies that could be deployed in the Severn. Tidal stream devises are unlikely to be viable in the Severn Estuary, but there are more appropriate conditions further out in the Bristol Channel where this might be viable. Small-scale tidal lagoon development could take place alongside a tidal barrage. Large-scale tidal lagoon development in the Severn Estuary would not offer any economic or environmental advantage over a barrage.

4.13.2 Tidal Stream

The long-term potential of tidal stream technologies, subject to constraints that might be imposed due to location-specific impacts upon the environment, natural marine processes, and long-term costs being acceptable, should be exploited [45]. The UK is in a unique position with superior tidal stream resource combined with devices being developed or tested. Tidal stream technologies could make a substantial contribution to the sustainable energies of the UK. Considering the progress that has been made on tidal stream, the objective now should be to *stay the course*. In many ways, the tidal stream industry is the same as wind power was 20 years ago, and the timescale for bringing prototype technologies to large-scale deployment needs to be as fast-tracked as possible.

4.13.3 Policy Improvements

There are a number of areas where Government policy could be improved. The support and funding structures need to be reviewed and improved in line with circumstances as they develop and change. A flexible approach could be taken on the future of the Department for Business, Enterprise and Regulatory Reform's (BERR) Marine Renewables Deployment Fund (MRDF), which has so far not had any applicants due to delays in getting demonstration projects off the ground. Lessons could be learnt from the success of the Scottish Government's £8 m support package for marine energy technologies, which has had strong interest from both tidal and wave developers. Increased support for marine renewables under a branded Renewables Obligation is also very welcome. This may provide an opportunity to revise the support available under the MRDF so that it focuses on providing grant funding for project development and testing, with the aim of stimulating progress towards initial tidal arrays and pre-commercial schemes.

The European Marine Energy Center in the Orkney Island is an excellent example of public sector funding being used to stimulate public sector investment and innovation in a strategic and efficient manner. Looking to the future, it is thought that there is potential to exploit the activity entered around the European Marine Energy Center (EMEC) to develop a regional hub around the Orkney Island and parts of the Caithness coastline, away from the Pentland Firth for commercial testing of devices beyond the prototype stage. In the long-term, lack of transmission capacity would appear to be a serious constraint on development of the UK's tidal stream resource in the north of Scotland. This also impacts on the offshore and onshore wind industry and on wave power devises. There are a number of problems with the current regime for connecting renewable generation and a real absence of long-term thinking on solutions to overcome them. This has serious consequences for the UK's ability to meet its targets for renewable electricity and the more ambitious EU targets that will eventually be implemented.

4.13.4 Strategic Planning and Consenting

Lack of good baseline information on the marine environment and on effects of large-scale deployment of different devices is a real issue. The gaps have to be filled over time through research of a strategic and generic nature as well as by developers. The Scottish Government is in the process of completing a strategic environmental assessment for marine renewables around the west and north coasts of Scotland, and the Welsh Assembly Government is developing a marine renewable strategy.

4.13.5 Tidal Lagoons

It is difficult to come to a clear view on the long-term potential of tidal lagoons due mainly to the lack of authoritative evidence and that the concept remains unproven. Government should investigate options to encourage one or more tidal lagoon demonstration projects.

4.13.6 Tidal Barrages

Extensive information is already available on the Severn Barrage that contains the majority of the UK's tidal range resource, and also for the Mersey. There does not seem to be an extensive overlap between tidal barrages and tidal stream devices, leading to the conclusion that they can, on the whole, be considered separately. The UK's potential for developing different tidal barrage options other than the Severn is extensive, but the reason why this potential has not been developed in the past is that they have not appeared to be economically viable. Further investigation into UK tidal barrage options outside the Severn Estuary should be considered on a case-by-case basis, as potential benefits will differ considerably.

4.14 Acknowledgement

This Chapter has been prepared and coordinated by T. J. Hammons, Chair of International Practices for Energy Development and Power Generation IEEE; UK Sustainable and Renewable Energy Group; University of Glasgow, UK. The UK Sustainable Development Commission is thanked for information that has been cited in this Chapter. Contributors include: Peter Meisen (President GENI, San Diego, CA, USA), Peter O'Donnell (Senior Energy Specialist, Manager Generation Solar and Renewables Programs, San Francisco Environment Organization, CA, USA); Omar Siddiqui (Senior Associate, Global Energy Partners LLC, Lafayette, CA, USA); Roger Bedard (Offshore Wave Energy Project Manager, EPRI, CA, USA), Andrew Mill (Managing Director European Marine Energy Center, UK); Mirko Previsic (Consultant—Offshore Renewables, Sacramento, CA, USA); Anthony T Jones (Senior Oceanographer, oceanUS consulting, Palm Springs, CA, USA); and Adam Westwood (Renewable Energy Manager, Douglas Westwood Ltd, Canterbury, UK). The Chapter is primarily based on an up-date of the invited panel session summary papers presented at the Panel Session on Harnessing the Untapped Energy Potential of the Oceans: Tidal, Wave, Currents and OTEC at the IEEE-PES 2005 General Meeting (GM2005) and in the book on *Renewable Energy* edited by T. J. Hammons published by Intech in December 2009 [46] and in Reference [47].

4.15 References

[1] T. J. Hammons. "Tidal Power", *Proceedings IEEE*, vol. 81, (3), pp. 419-433, 1993.
[2] Barry V. Davis, "A Major Source of Energy from the World's Oceans," IECEC-97 Intersociety Energy Conversion Engineering Conference, 1997,
[3] N.H. Halvorson, "Evaluation of Nova Energy Ltd.'s Hydro Turbine for (Canadian) Ministry of Employment and Investment", N.H. Halvorson Consultants Ltd.
[4] "Renewable Energy: Power for a Sustainable Future; Technology Update: (2001) Tidal Current Power Update & Wave Power Update", Oxford University Press.
[5] Peter Fraenkel, "Renewables; Is the Tide Turning for Marine Current Turbines?", Modern Power System, Marine Current Turbines Ltd, London, UK, June 30, 2001.

[6[R. Bedard, "Final Summary Report: Offshore Wave Power Feasibility Demonstration Project", (E2I EPRI Global WP009 - US), 2005..
[7] Mirko Previsic, "System Level Design, Performance and Costs for San Francisco Pelamis Offshore Wave Power Plant", (E2I EPRI Global - 006A - SF), 2004.
[8] George Hagerman , "Offshore Wave Power in the US: Environmental Issues", (E2I Global EPRI - 007 - US), 2004.
[9] Bonnie Ram , "Wave Power in the US: Permitting and Jurisdictional Issues", (E2I Global EPRI DOE NREL - 008 - US)
[10] M. Previsic, "Wave Power Technologies", Proceedings of the IEEE PES 05 GM, San Francisco, paper 05GM0542, June 2005, pp. 1-6.
[11] E2i/EPRI WP-004-US Rev 1 Assessment of Offshore Wave Energy Conversion Devices
[12] E2i/EPRI WP-005-US Methodology, Guidelines and Assumptions for the Conceptual Design of Offshore Wave Energy Power Plants (Farms)
[13] E2i/EPRI WP-006-HI System Level Design, Preliminary Performance and Cost Estimate - Hawaii -
[14] E2i EPRI WP-006-OR System Level Design, Preliminary Performance and Cost Estimate - Oregon
[15] E2i/EPRI WP-006-ME System Level Design, Preliminary Performance and Cost Estimate - Maine
[16] E2i/EPRI WP-006-MA System Level Design, Preliminary Performance and Cost Estimate - Massachusetts
[17] E2i/EPRI WP-006-SFa System Level Design, Preliminary Performance and Cost Estimate - San Francisco, California Pelamis Offshore Wave Power Plant
[18] E2i/EPRI WP-006-SFb System Level Design, Preliminary Performance and Cost Estimate - San Francisco Energetech Offshore Wave Power Plant
[19] Sea Technology Magazine August 2003, Wave Energy Conversion - The State of the Art
[20] EPRI - RE Technical Assessment Guide Ocean Wave Power Section for the years 2001, 2002 and 2003
[21] California Energy Commission - Wave Energy Resource Assessment for the State of California
[22] A.D. Trapp and M. Watchorn, "EB development of tidal stream energy," in *Proceedings MAREC 2001*, p. 169-173, 2001.
[23] A.T. Jones and A. Westwood, "Economic Forecast for Renewable Ocean Energy Technologies," presented at EnergyOcean 2004, Palm Beach, Florida, 2004.
[24] A.T. Jones and W. Rowley, "Global Perspective: Economic Forecast for Renewable Ocean Energy technologies," MTS Journal, vol. 36. No. 4, pp. 85-90, Winter 2002.
[25] N.J. Baker, M.A. Mueller, M. Watchorn, D. Slee, L. Haydock and N. Brown, "Direct drive power take off for the Stingray tidal current generator," in *Proceedings MAREC 2002*, p. 1-10, 2002.
[26] A. M. Gorlov, "The Helical Turbine and its Applications for Tidal and Wave Power," in *Proc. OCEANS 2003*, p. 1996, 2003.
[27] P.W. Ullman, "Offshore Tidal Power Generation - A new approach to power conversion of the oceans' tides," *MTS Journal*, vol. 36. no. 4, pp. 16-24, Winter 2002.
[28] P. Breeze, "The Future of Global Offshore Wind Power," *Reuter Business Insight* 2004.
[29] L. Coakley, "Long Island Offshore Wind Park - 140 Megawatts of Offshore Wind Energy," presented at EnergyOcean 2004, Palm Beach, Florida, 2004.

[30] O. Siddiqui and R. Bedard, "Feasibility Assessment of Offshore Wave and Tidal Current Power Production", Proceedings of the IEEE PES 05 GM, San Francisco, paper 05GM0538, June 2005, pp. 1-7.

[31] E2i/EPRI WP-08-US- Identification of Permitting Issues

[32] A. T. Jones and A. Westwood, "Recent Progress in Offshore Renewable Energy Development", Proceedings of the IEEE PES 05 GM, San Francisco, paper 05GM0543, June 2005, pp. 1-6.

[33] A. C. Baker. Tidal Power. United Kingdom: Peter Peregrinus Ltd. 1991.

[34] A. C. Baker, The development of functions relating cost and performance of tidal power schemes and their application to small-scale sites. Tidal Power. London: Thomas Telford, 1986.

[35] M. Banai and A. Bichon. Tidal Power in France: The Ranch tidal power station: Some results after 15 years of operation, Energy Digest, pp. 39-45, October 1982.

[36] A. D.Grant. Power Generation from tidal flows for navigation buoys, 2nd Int. Symp, Wave and Tidal Energy. BHRA, Cambridge, September 1981, pp. 117-128.

[37] I. G. Bryden et.al. Assessing the Potential of a Simple Tidal Channel to Deliver Useful Energy, Applied Ocean Research, Vol. 26, 2003, pp. 198-204.

[38] S. Kiho, et. al. The Power Generation from Tidal Currents by Darrieus Turbine, Renewable Energy, Vol. 9, (1-4), 1996, pp. 1242-1245.

[39] A. M. Gorlov. The Helical Turbine and its Application for Tidal and Wave Power. Proceedings IEEE OCEANCS'03, Vol. 4, San Diego (USA), September 2003, pp. 1996.

[40] NGET (2007). GB 7 year Statement. www.nationalgrid.com/uk/electricity/sys/current/

[41] Tidal Power from the River Mersey: A feasibility Study—Stage 1 Report. Mersey Barrage Co., 1988

[42] N. W. Hammond and P. Wood. Tidal Power from the Mersey: History and Prospects, in Developments in Tidal Energy, Proc. 3rd Conf. Tidal Power, London, UK, November 1989.

[43] Peel Environmental Ltd (2007). Mersey Tidal Power Study. www.merseytidalpower.co.uk

[44] Sustainable Development Commission (SDC). Paper 2, Reducing CO_2 Emissions--Nuclear and the Alternatives. From the SDC Project 'The Role of Nuclear Power in a Low Carbon Economy' {www.sd-commission.org.uk/pages/060306.html}.

[45] T. J. Hammons. Energy Potential of the Oceans in Europe and North America: Tidal, Wave, Currents, OTEC, and Offshore Wind. International Journal of Power and Energy Systems, Paper 203-4142, Vol. 28, (4), 2008, pp. 416-428.

[46] T. J. Hammons. Book: *Renewable Energy (T J Hammons (Ed.))*, InTech, ISBN: 978-953-7619-52-7, 580 p.,2009. Available on Line.

[47] T. J. Hammons. Tidal Power in the UK and Worldwide to Reduce Greenhouse Gas Emissions, International Journal of Engineering Business Management (IJEBM), Vol.3, (2), May 2011, 11 p.

5

Geothermal Power Generation: Global Perspectives, Technology, Direct Uses, Plants, Drilling and Sustainability Worldwide

This Chapter discusses the state of art in harnessing geothermal power for medium- and large-scale generation of electricity and for space heating worldwide. It reviews current, probable, possible and potential developments both in developed and developing countries in near future and long term.

First, world energy consumption is reviewed. Then relative contribution of energy sources in the world in OECD countries with respect to renewables is discussed. Installed geothermal capacities for electricity generation worldwide are summarized. Direct use of geothermal energy worldwide is examined together with the technical potential of renewable energy sources.

Geothermal energy in USA is evaluated. Highlighted is heat flow, tectonic controls, types of geothermal systems, US geothermal potential, geothermal energy, operating conditions for electricity generation, and environmental constraints. The Chapter then examines the Geothermal Technologies Program in USA, reviewing the goals: geo-science and supporting technologies, exploration and drilling research, and energy systems research and testing. Discussed also is direct use of geothermal energy where district heating, agriculture and aquaculture applications, and future developments are considered.

Also examined is improving geothermal power plants where typical R&D projects are reviewed. Geothermal drilling R&D aimed at reducing drilling costs by 25%~50%, which often account for more than half of the total capital required for a project, is then discussed.

Enhanced power cycles for enhancing geothermal sustainability are reviewed. Deep drilling projects in Iceland--exploration of deep unconventional geothermal resources that requires drilling wells to depths of 3.5 to 5 km and at temperatures of 450-600°C and which will exceed by an order of magnitude the power typically obtained from conventional wells are examined. The phases of development of a 400 MW_{el} scheme in the Hengill area of Iceland is discussed. Geothermal schemes in NE Iceland are also discussed as are geothermal development in reducing CO_2 emissions. Perspectives on the future of geothermal energy in the United States is indicated as is technology of harnessing geothermal power now and future. Economics, availability, and reliability of geothermal plants are reviewed.

5.1 Introduction

Energy consumption in the world is now a little over 400 EJ per year. Available energy resources in the world are large, and energy shortage is not expected in the foreseen future. On the other hand, most of the energy used in the world at present is coming from finite energy resources, whereas renewable energy sources are more suitable for sustainable development. The highest share of the use of renewable energy resources is in Iceland, where renewable energy comprises approximately 70% of the primary energy resources and approximately 30% is derived from fossil fuels. This unique position has been achieved by an extensive and advanced use of geothermal energy.

The state of art in harnessing geothermal power for medium- and large-scale generation of electricity (and for space heating) worldwide is examined. It gives a global perspective on geothermal power where world energy consumption, consumption of renewable energy, consumption of geothermal energy, direct use of geothermal energy, geothermal resources, and cost is indicated. An in-depth review of geothermal energy in USA: the technology for harnessing geothermal energy, direct use of geothermal energy, improving geothermal plants, and drilling is made. On the worldwide basis, geothermal energy is considered to have the largest technical potential of the renewable energy sources. Furthermore, the production price of geothermal energy is favorable in comparison to all other energy sources.

5.2 Global Perspective on Geothermal Energy

Most of the renewable energy sources presently used and under development in the world are in one way or another connected to the energy that the Earth is receiving from the Sun (hydro, biomass, solar- and wind energy). Most of the energy resources used in the world at present are coming from finite energy sources embedded in the crust of the Earth (oil, gas, coal, and uranium). Only one energy resource of the crust is renewable, namely geothermal energy. The source of geothermal energy is the continuous energy flux flowing from the interior of the Earth towards its surface.

The use of finite energy sources is not in good harmony with the concept of sustainable development and most countries are aiming at increasing the use of renewable energy sources at the expense of the finite energy resources (the UK is aiming for 80% of its energy to come from renewables by 2050). Geothermal energy has many desirable properties that make it suitable as a replacement for fossil fuels. The technical potential of geothermal energy is very large and the production price of geothermal energy is very favorable in comparison with other energy sources.

5.2.1 World Energy Consumption

Consumption of energy is one of the characteristics of the present society. Table 5.1 shows the worldwide consumption of primary energy in the year 1999.

	EJ	Gtoe	%
Fossil Fuels	322	7.68	79.2
Nuclear	28	0.66	6.8
Renewables	57	1.36	14.0
Total	**407**	**9.70**	**100**

Source: IEA 2001[1]

Table 5.1. World Energy Consumption in 1999

For the world in 1999, the use of renewable energy sources was only 14% of the primary energy sources, whereas finite energy sources (fossil fuels and nuclear) comprised 86%. The consumption of renewable energy sources is even lower in the developed countries than in the world as a whole. This is because the use of traditional fuel wood is more common in the developing countries than in the more affluent OECD countries. Most of the worldwide use of nuclear energy takes place in the OECD countries.

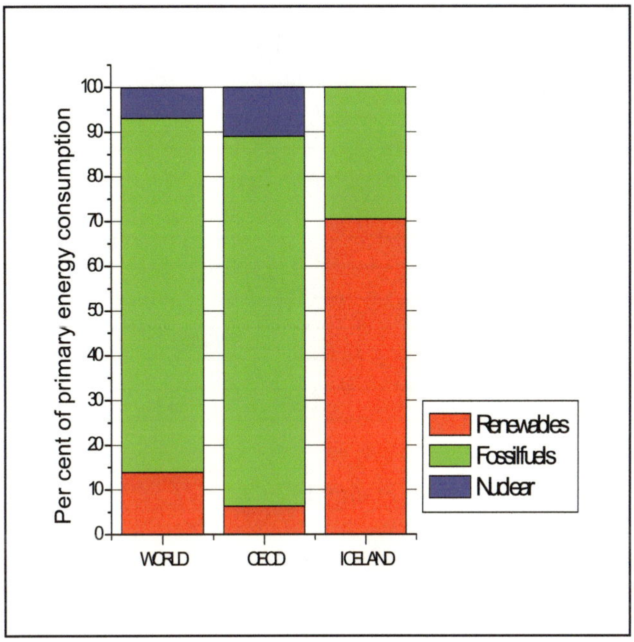

Fig. 5.1. Relative Contribution of Energy Sources in the World, OECD Countries, and Iceland.

Figure 5.1 shows relative contribution of energy sources in the world, in the OECD countries, and in Iceland. The highest share of the use of renewable energy resources is in Iceland. Table 5.2 illustrates primary energy supply in the OECD countries in 1999.

It is desirable to increase use of renewable energy sources at the expense of use of finite energy resources. Such development promotes the idea of sustainable development [2] and at the same time is expected to reduce the man-made emission of greenhouse gases.

	EJ	**Gtoe**	%
Fossil fuels	181	4.32	82.7
Nuclear	24	0.58	11.0
Renewables	14	0.33	6.3
Total	**219**	**5.23**	**100**

Source: IEA 2001[1]

Table 5.2. Primary Energy Supply in OECD Countries in 1999

5.2.2 Consumption of Renewable Energy Sources

Traditional biomass (fuel-wood) and hydro contribute the largest share to the use of "renewables" in the world (Table 5.1). Table 5.3 gives a further breakdown of the use of renewables.

	Electricity TWh	**Heat TWh**	**EJ**
Traditional biomass			38
Biomass-electricity	160		0.576
Biomass-heat		>700	>2.52
Biomass-ethanol			0.42
Wind-electricity	18		0.065
Solar-PHV-electricity	0.5		0.002
Solar-thermal-electricity	1		0.004
Solar-heat		14	0.05
Hydro	2,600		9.36
Geothermal-electricity	46		0.166
Geothermal-heat		40	0.144
Tidal	0.6		0.002
TOTAL	**2,826.1**	**>754**	**>51.3**

Source: WEA 2000 [3]

Table 5.3. Consumption of Renewable Energy in 1998

For the world, the share of renewable energy sources is about 14%. For the OECD countries, this ratio is only about 6% (see Tables 5.1 and 5.2). This is because about 80% of consumption of renewable energy sources in the world is the use of biomass (Table 5.3) and that use of traditional biomass is more common in the developing countries than in the OECD countries. Aside from hydro and biomass, the contribution of renewable energy sources at present is a very small fraction of the energy consumption in the world.

5.2.3 Consumption of Geothermal Energy

The use of geothermal energy is usually divided into the part used for electricity generation and the part used directly for heating purposes (direct use). Huttrer [4] has made a review of the electricity generation from geothermal energy, and Lund and Freestone [5] have reviewed the direct use of geothermal energy.

There are 21 countries in the world that use geothermal steam to generate electricity. Installed geothermal capacities for electricity generation worldwide is illustrated in Table 5.4. The largest installed capacities are in USA (2850 MWe in 2005: source US Information Agency) and the Philippines (1909 MWe in 2001: source Huttrer [4]) with lower values in other countries. The importance of this kind of electricity generation is different for these two countries. In the Philippines, electricity generated from geothermal is about 22% of the electricity generated in the country, whereas in USA it is only 0.36% of US annual electrical generation (16,010 GWh, see section 5.11). Table 5.5 lists the countries with the highest ratio of electricity generation from geothermal energy.

Country	Installed MWe	Generated GWh/a	Capacity Factor
Australia	0.17	0.9	0.60
China	29	100	0.39
Costa Rica	142	592	0.48
El Salvador	161	800	0.57
Ethiopia	8.5	30	0.40
France	4.2	25	0.68
Guatemala	33	216	0.75
Iceland	170	1,138	0.76
Indonesia	589	4,575	0.89
Italy	785	4,403	0.64
Japan	547	3,532	0.74
Kenya	45	366	0.93
Mexico	755	5681	0.86
New Zealand	437	2,268	0.59
Nicaragua	70	583	0.95
Philippines	1,909	9,181	0.55
Portugal	16	94	0.67
Russia	23	85	0,42
Thailand	0.3	1.8	0.68
Turkey	20	120	0.68
USA	2,228	1,5470	0.79
Totals	**7,972**	**49,262**	**0.71**

from Huttrer, 2001 [4]

Table 5.4. Installed Geothermal Capacities for Electricity Generation

Capacity factors for electricity generation for different renewable energy sources worldwide are: Geothermal 0.71, Hydro 0.42, Solar-thermal 0.30, Solar PV 0.15, and Wind 0.19 [6].

The main types of direct use of geothermal energy are bathing, space heating, greenhouses, fish farming, and in industry. Direct application can use both high- and low-temperature geothermal resources and is therefore much more widespread in the world than for electricity production. Direct application is, however, more site specific for the market, as steam and hot water is rarely transported over long distances. The longest geothermal hot water pipeline in the world is in Iceland (63 km). The production cost for direct utilization is highly variable, but is commonly lower than 2 US cents/kWh.

Table 5.6 shows direct use of geothermal energy worldwide in the in the year 1999 [5].

The large variation in the capacity factors in Table 5.6 is due to the different utilization mode of the direct use of geothermal energy. In USA and Switzerland, ground coupled heat pumps are the main sources of geothermal energy and the capacity factors for those countries are relatively low. Where geothermal energy is used for heating purposes in a moderate climate, the capacity factors are frequently in the range 0.4-0.7. The high capacity factors reported for India and Serbia might not be realistic.

Country	**Ratio of Electricity Generated from Geothermal Resources, %**
Philippines	22
El Salvador	20
Nicaragua	17
Iceland	15
Costa Rica	10
Kenya	8
New Zealand	6
Indonesia	5

Table 5.5. Countries with the Highest Share of Electricity Generated from Geothermal.

Country	**Installed MWt**	**Production, GWh/a**	Capacity **Factor**
China	2,282	10,531	0.53
Japan	1,167	7,482	0.73
USA	3,766	5,640	0.17
Iceland	1,469	603	0.44
Turkey	820	4,377	0.61
New Zealand	308	1,967	0.73
Georgia	250	1,752	0.80
Russia	308	1,707	0.63
France	326	1,360	0.48
Sweden	377	1,147	0.35
Hungary	473	1,135	0.27
Mexico	164	1,089	0.76
Italy	326	1,048	0.37
Romania	152	797	0.60
India	80	699	1.00
Switzerland	547	663	0.14
Serbia	80	660	0.94
Slovak Republic	132	588	0.51
other countries	2,118	4,731	0.25
Total	**15,145**	**52,976**	**0.40**

Source: Lund and Freeston, 2001 [5]

Table 5.6. Direct use of Geothermal Energy

5.2.4 World Energy Resources

In dealing with availability of energy resources, distinction has to be made between renewable energy sources and finite energy sources. The finite sources, fossil fuels and nuclear, are fixed amounts of energy stored in the Earth's crust; whereas the renewables are more or less a continuous current of energy. The finite sources can only be used once, but the exploitation of renewables will not affect the size of the energy current and this kind of energy can be utilized continuously without changing the amount of the available energy.

The term "reserves" means identified and economic resources and the term "resources" cover sub-economic and undiscovered resources. The sum of reserves and resources is denoted as a "resource base". For these resources, the reserves are best known, but the size of the resource base is more uncertain.

Due to the dynamic nature of the renewables, it is not possible to use the same classification for them as for the finite sources. For the renewable energy resources, names like "theoretical potential", "technical potential", and "economical potential" are frequently used. For these resources, it is usually easy to determine the size of the theoretical potential, but it is usually difficult to estimate how much of this energy is economical.

It can be convenient to compare the size of the resource base of the finite energy sources to the technical potential of the renewables. However, the resource base is a finite number whereas the technical potential is the yearly availability of the renewable energy source.

The resource base of the finite energy sources is shown in Table 5.7. Table 5.8 shows the technical potential of the renewables.

	EJ
Oil	32,422
Gas	49,805
Coal	199,666
Uranium	325,000
TOTAL	**606,893**

Source: WEA 2000 [3]

Table 5.7. Resource Base of Finite Energy Resources

It should be noted that during 100 years, the technical potential of renewables would produce the same amount of energy as stored in the resource base of the finite energy sources.

If it is assumed that 10% of the resource base is economical, Table 5.7 indicates that the present world energy consumption (407 EJ, see Table 5.1) could be maintained for some 150 years by the finite energy resources. Furthermore, if it is also assumed that 10% of the technical potential is economic; the renewable energy sources could maintain the present world energy consumption for a very long time.

	EJ per year
Hydropower	50
Biomass	276
Solar energy	1,575
Wind energy	640
Geothermal energy	5,000
TOTAL	**7,600**

Source: WEA 2000 [3]

Table 5.8. Technical Potential of Renewable Energy Sources

Geothermal energy is giving the largest share to the technical potential of renewables (Table 5.8). The technical potential is the yearly availability of the renewable resources. These estimates suggest that the technical potential of renewables is sufficiently large to meet future world energy requirements. The present annual consumption of primary energy in the world is about 407 EJ (Table 5.1). It is therefore expected that geothermal energy will be of large importance for the development of energy utilization in the future.

5.2.5 Cost of Renewable Energy

The range of energy cost is reported in WEA 2000 [3]. Table 5.9 shows the cost of electricity generation from renewable energy sources.

	Installed Cost USD/kW	**Energy Cost US cent/kWh**
Biomass	900 – 3,000	5 – 15
Solar photovoltaic	5,000 – 10,000	25 - 125
Solar thermal	3,000 – 4,000	12 – 18
Hydro	1,000 – 3,500	2 – 10
Geothermal	800 – 3,000	2 – 10
Wind	1,100 – 1,700	5 – 13
Tidal	1,700 – 2,500	8 – 15

Source: WEA 2000 [3]

Table 5.9. Current Installed Cost and the Cost of Electricity Generation

The installed cost and the energy cost for the generation of heat are shown in Table 5.10.

Tables 5.9 and 5.10 show clearly that the price of geothermal energy is favorable as compared to other energy sources, both renewable and finite energy sources.

	Installed Cost USD/kW	**Energy Cost US cent/kWh**
Biomass	250 – 750	1.0 – 5.0
Solar heat	500 – 1,700	3.0 – 20.0
Geothermal	200 – 2,000	0.5 – 5.0

Source: WEA 2000 [3]

Table 10. Current Installed Cost and Energy Cost for Generation of Heat

5.3 Geothermal Energy in USA

Temperature increases with depth in the earth at an average of 25°C/km. If the average surface temperature is 20°C, the temperature at 3 km is of the order of 95°C. Although direct-use applications of geothermal energy can utilize temperatures as low as about 35°C, the minimum temperature suitable for electrical generation is about 120°C.

Spatial variations of the thermal energy within the deep crust and mantle of the earth give rise to concentrations of thermal energy near the surface of the earth that can be used as an energy resource. Heat is transferred from the deeper portions of the earth by conduction through rocks, by movement of hot deep rock toward the surface, and by deep circulation of water.

In older areas of continents, such as much of North America east of the Rocky Mountains, heat flow is generally 40 to 60 mWm^{-2} (milliwatts per square meter). This heat flow coupled with the thermal conductivity of rock in the upper 4 km of the crust yields subsurface temperatures of 90 to 110°C at 4-km depth in the Eastern United States. Heat flow within the Basin and Range (west of the Rockies) is generally 70-90 mWm^{-2}, and temperatures are generally greater than 110°C at 4 km. There are large variations in the western United States, with areas of heat flow greater than 100 mWm^{-2} and mountain areas such as the Cascades and Sierra Nevada of generally lower heat flow. The large rainfall on the Cascades may suppress flow of heat to the surface in this relatively young volcanic area.

5.3.1 Tectonic Controls

The unifying geologic concept of plate tectonics provides a generalized view of geologic processes that move concentrations of heat from deep within the earth to drillable depths. The heat can be related to movement of magma within the crust, particularly when associated with recent volcanism, or deep circulation of water in active zones of faulting. Figure 5.2 shows the major plate boundaries, where much of the geothermal exploration occurring worldwide is focused, since most of the current volcanic activity of the earth is located near plate boundaries associated with spreading centers and subduction zones.

Solid, bold lines are extensional boundaries, hachured lines are zones of convergence with the hachures on the overriding plate, and dotted lines indicate translational or diffuse plate boundaries.

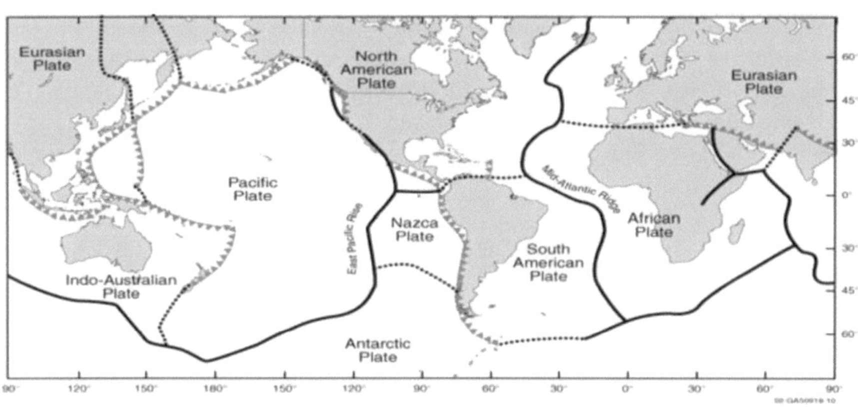

Fig. 5.2. Major Tectonic Plates of the World

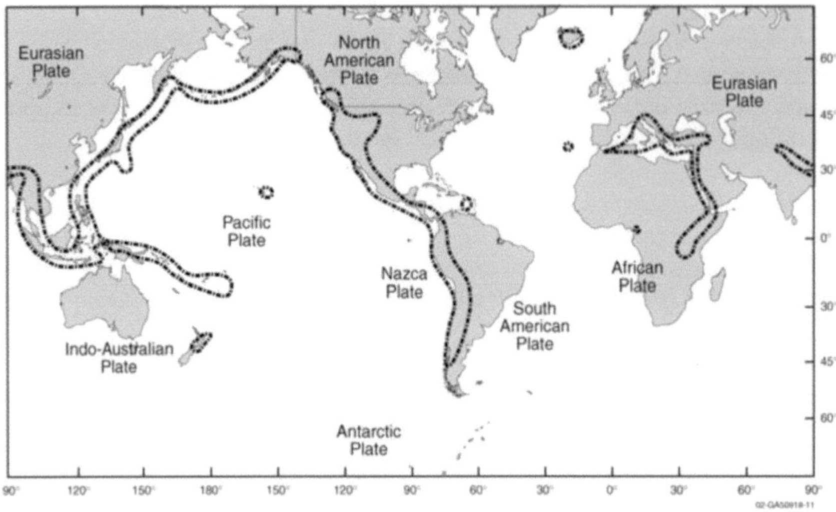

Fig. 5.3. Areas of the World with Potential for Producing Electricity using Geothermal Energy.

The brittle and moving plates of the lithosphere (crust and upper mantle) are driven by convection of plastic rocks beneath the lithosphere. Convection causes the crustal plates to break and move away in opposite directions from zones of up welling hot material. Magma moving upward into a zone of separation brings with it substantial amounts of thermal energy.

Rifting of the earth's crust can also occur in continental blocks. Two of the better-known examples are the East African Rift and the Rio Grand Rift in New Mexico. These rifts contain young volcanism and host several large geothermal systems.

Where continental plates converge, they crumple against each other. An example is the Himalayas, formed by the collision of the Indian and Asian plates.

Translational plate boundaries, which are locations where plates slide parallel to each other, may develop extensional troughs known as pull-apart basins, e.g., the Salton Trough of Southern California [7, p. 131]. Volcanism associated with the Salton Trough generated the heat in the Salton Sea, Cerro Prieto, and Imperial Valley geothermal fields. Tensional features further north on the San Andreas and related faults may be the cause of the volcanism thought to be the heat source for The Geysers geothermal field about 90 miles north of San Francisco.

A third source of elevated heat flow and volcanism are "hot spots." Several important geothermal systems are associated with recent volcanism caused by hot spots: Yellowstone, USA, the geothermal fields in Iceland, and those of the Azores.

Geothermal resources also have been developed in areas of anomalously high temperatures with no apparent active volcanism, such as the Basin and Range physiographic province in the western United States.

Areas of the world with geothermal potential are shown in Figure 5.3. Much of the world's potential for geothermal energy is associated with areas of volcanism caused by subduction and crustal spreading.

5.3.2 Types of Geothermal Systems

All commercial geothermal production is currently restricted to hydrothermal systems. Most hydrothermal resources contain water as liquid, but higher temperatures or lower pressures can create conditions where steam and water or only steam is the continuous phase in the reservoir. Successful, sustainable, geothermal energy usage depends on injection back into the reservoir of the maximum quantity of produced fluid to augment natural recharge of hydrothermal systems.

Other types of geothermal systems have been investigated for energy production:

- Geopressured-geothermal systems contain water with somewhat elevated temperatures (above normal gradient) and with pressures well above hydrostatic for their depth.
- Magmatic systems, with temperatures from 600 to 1400°C are associated with magmatic bodies beneath the surface of the earth.
- Hot dry rock geothermal systems, with temperatures from 200 to 350°C, are subsurface zones with low initial permeability and little water. These types of geothermal systems cannot be used economically for the production of energy.

5.3.3 US. Geothermal Energy Potential

A US. Geological Survey (USGS) assesses geothermal potential of the United States and provides an explanation of the terminology used to define various categories of the resources [8]. Resource base is all of the thermal energy contained in the earth. Accessible resource base is that part of the resource base shallow enough to be reached by production drilling. Resources are those portions of the accessible base that can be used at some reasonable future time. Reserves are that portion of the resource that has been identified and that can be used under current economic conditions. Reserves and resources are divided into categories of identified and undiscovered, based on knowledge of certainty of their existence.

The USGS published assessments of the moderate (90–150°C) and high-temperature (>150°C) geothermal resources of USA in 1975 [9] and 1979 [8] and published an estimate of low-temperature (<90°C) resources in 1983 [11].

The USGS assessment of low-temperature resources [10] estimated the beneficial heat in discovered and undiscovered hydrothermal systems less than 90°C to be about 41 and 30 GW_t for 3 years. The USGS [8] estimated that the identified high-temperature hydrothermal resource would operate power plants with an aggregate capacity of 23,000 MWe (megawatt electrical) for 30 years. The total U.S. hydrothermal resource, inferred from knowledge of earth science, was estimated to be 95,000 to 150,000 MWe for 30 years. Recent advances in the technology for converting geothermal energy into electricity have lowered the temperature needed for economic electrical production. As a result, lower-temperature resources will be included in the next USGS estimate of geothermal energy suitable for electrical production. The USGS initiated a new assessment of U.S. geothermal resources in 2003, beginning with those in the Great Basin.

5.3.4 Geothermal Energy Use in USA

Worldwide capacity for electrical generation using geothermal energy is 9,000 MWe (2008) of generating capacity on line in 21 countries [4,11]. Current net capacity in USA is 2,850 MWe. Electricity is produced in California (7 fields), Hawaii (1 field), Nevada (9 fields), New Mexico (1 field), and Utah (2 fields).

5.3.5 Operating Conditions for Electrical Generation

Most geothermal fields are liquid-dominated, i.e., water at high temperature and under high pressure but still in liquid form, is the pressure-controlling medium filling the fractured and porous rocks of the reservoir. In liquid-dominated geothermal systems used for electrical power production, water comes into the wells from the reservoir, and the pressure decreases as the water moves toward the surface, allowing part of the water to boil. As the wells produce a mixture of steam and water, a separator is installed between the wells and the power plant to separate the steam and water. The flashed steam goes into the turbine to drive the generator, and the water is injected back into the reservoir.

In several geothermal fields, the wells produce only steam. The separators and the system for handling the separated water are not needed. These systems are more economical, but are rare. Only two of the currently operating fields in the world, Larderello, Italy, and The Geysers, USA, are vapor-dominated.

Many water-dominated reservoirs below 175°C used for electricity generation are pumped to prevent the water from boiling as it is circulated through heat exchangers to heat a secondary liquid that then drives a turbine to produce electricity. Binary geothermal plants have no emissions because all of the produced geothermal water is injected back into the underground reservoir. The number of identified lower-temperature geothermal systems is many times greater than the reserves of high-temperature fluids. This provides an economic incentive to develop more efficient binary power plants.

5.3.6 Direct Use

Geothermal resources provide energy for agricultural uses, heating, industrial uses, and bathing. Fifty-five countries had 16,209 MWt (megawatt thermal) of total capacity for direct use in 1999 [5,12]. The total energy used is estimated at 162,000 TJ/y (terajoules per year). The US. capacity for direct use is about 3,766 MWt, and approximately 5,640 GWh per year are used [5,12].

The use of geothermal energy for direct uses is dominantly in the western states of Alaska, Arizona, California, Colorado, Hawaii, Idaho, Montana, Nevada, New Mexico, Oregon, Texas, Utah, Washington, and Wyoming. However, warm waters are also used for pools and spas and some space heating in Alabama, Arkansas, Georgia, Louisiana, Mississippi, New York, South Dakota, Texas, Virginia, and West Virginia.

Warm water, at temperatures above 20°C can be used directly for a host of processes requiring thermal energy. Thermal energy for swimming pools, space heating, and domestic hot water are the most widespread uses, but industrial processes and agricultural drying are growing applications of geothermal use (see Table 5.11). The cities of Boise, Idaho; Elko,

Nevada; Klamath Falls, Oregon; and San Bernardino and Susanville, California; have geothermal district-heating systems where a number of commercial and residential buildings are connected to distribution pipelines circulating water at 54 to 93°C from the production wells [13].

The use of geothermal energy through ground-coupled heat-pump technology has almost no impact on the environment and has a beneficial effect in reducing the demand for electricity. Geothermal heat pumps use the reservoir of constant temperature, shallow groundwater, and moist soil as the heat source during winter heating and as the heat sink during summer cooling. The energy efficiency of geothermal heat pumps is about 30 percent better than that of air-coupled heat pumps and 50 percent better than electric-resistance heating. Depending on climate, advanced geothermal heat pump use in USA reduces energy consumption, and correspondingly, power-plant emissions by 23 to 44 percent compared to advanced air-coupled heat pumps, and by 63 to 72 percent compared to electric-resistance heating and standard air conditioners [15].

	Capacity (MW [t])	Use (TJ/y)
District heating	268	4,232
Space heating	107	2,497
Agriculture	99	624
Balneology	92	947
Heat pumps	4,800	12,000
Total U.S.	**5,366**	**20,302**

Table 5.11. Direct use USA (Modified from [14])

5.3.7 Environmental Constraints

All known geothermal systems contain aqueous carbon dioxide species in solution. When a steam phase separates from boiling water, CO_2 is the dominant (over 90 percent by weight) non-condensable gas. In most geothermal systems, non-condensable gases make up less than five percent by weight of the steam phase. For each megawatt-hour of geothermal electricity produced in USA, the average emission of CO_2 is about 18% of that emitted when natural gas is burned to produce electricity.

A comparison of fossil and geothermal CO_2 emissions in kg CO_2 per kWh is: geothermal 0.082, coal 0.968, petroleum 0.709, natural gas 0.468 (data from [16]). Binary plants have no emissions, since all of the produced fluid is injected back into the reservoir.

Hydrogen sulfide can reach moderate concentrations of up to two percent by weight in the separated steam phase from some geothermal fields. This gas presents a pollution problem because it is easily detected by humans at concentrations of less than 1 part per million in air. H_2S concentrations are only high enough to require control at The Geysers, California, Coso, California, and Steamboat Springs, Nevada geothermal plants.

The incineration process burns the gas removed from the steam to convert H_2S to SO_2, the gases are absorbed in water to form SO_3 and SO_4 in solution, and iron chelate is used to form S_2O_3 [17]. The major product from the incineration process is a soluble thiosulfate, which is injected into the reservoir with the condensed water used for the reservoir pressure-maintenance program. Sulphur emissions for each megawatt-hour of electricity produced in 1991, as SO_2 by plant type in USA was 9.23 kg from coal, 4.95 kg from petroleum, and 0.03 kg from geothermal flashed-steam (from data of [18]). For each megawatt-hour of electricity produced in 1991, the average emission of nitrogen oxides by plant type in USA was 3.66 kg from coal, 1.75 kg from petroleum, 1.93 kg from natural gas, and zero from geothermal (from data of [18]).

In summary, geothermal energy provides a major economic source of base-load electrical energy for western USA as well as a clean source of energy for direct use over a broad area of USA. Geothermal energy produces about 2% of the electricity in Utah, 6% of the electricity in California, and 10% of the electricity in Northern Nevada. Further information concerning geothermal energy is available on many Web sites. Among the more informative are [19-23].

5.4 Geothermal Technologies Program: US. Department of Energy (DOE)

Geothermal energy is being used to generate electrical power; for direct use applications such as district heating, greenhouse heating, and aquaculture; among others.

The long-term sustainability of geothermal production has been demonstrated by continuous electrical power generation at the Lardarello field in Italy since 1913, at the Wairakei field in New Zealand since 1958, and at The Geysers field in USA since 1960. No geothermal field has been abandoned because of resource decline.

Today, geothermal energy amounts to more than 9,000 MW of installed electric power capacity worldwide. However, the ultimate promise of the geothermal energy is many times larger. With enhanced geothermal systems (EGS), using advanced techniques to engineer improved geothermal reservoirs; there is the potential to meet energy needs of approximately 17 percent of the world's population (see Sections 5.9~5.11).

P. Michael Wright of the Idaho National Engineering and Environmental Laboratory, Marshall Reed of DOE, and Karl Gawell of the Geothermal Energy Association have determined that with the advancement in technologies, geothermal energy has the capacity to produce 65GW to 138GW of electricity. World geothermal resources are estimated to be 15,000 times the world's oil reserves.

5.4.1 Comprehensive Research Program

The "next generation" technology currently under development will allow a greater portion of the geothermal resource base to be developed economically.
The Program adopted the following goals:

(i) Double the number of states with geothermal power facilities to eight
(ii) Reduce the levelized cost of generating geothermal power to 3~5 cents/kWh by 2010
(iii) Supply the electrical power or heat energy needs of 7 million homes and businesses in USA by 2015

Three business lines were formed to pursue those goals:

Geoscience and Supporting Technologies (Geoscience) - Geoscience research and development addresses characterization and management of the geothermal resource via improved understanding and enhancement of underground fracture systems, understanding the flow of hot fluids through reservoirs, and resource management through re-injection of spent geothermal fluid.

Exploration and Drilling Research - Exploration research seeks to improve the various geologic, geophysical, and geochemical methods used to find and define geothermal resources.

Energy Systems Research and Testing (ESR&T) - This business line concentrates on the means of converting geothermal heat into useful energy. Advanced cycles are developed to increase conversion efficiency (see Section 5.8). Improvements in equipment, such as condensers and heat exchangers, are made to reduce costs and improve performance. Operating problems are addressed to increase plant reliability.

5.5 Direct Use Geothermal Energy

Historically, direct use of geothermal resources has been on a small scale and on an individual basis, recent projects have focused more on the developments of major district heating systems, greenhouses or aquaculture complexes, or major industrial uses. Heat pumps utilizing very low-temperature geothermal fluids (120°) have extended geothermal developments into traditionally non-geothermal countries such as Denmark, Sweden, Switzerland, and large areas of the mid-western and eastern US. (Lund, 2002 [24]).

Worldwide (Lund and Freeston, 2000 [12]) the installed capacity of direct geothermal utilization in 2000 was 16,200 MWe and the energy use was approximately 162,000 TJ/year distributed among 60 countries. This amounted to a savings of an equivalent of 11.4 million metric tons of fuel oil per year. The worldwide distribution of direct use of geothermal energy is 14.33% Heat Pumps, 36.85% Space Heating, 11.75% Greenhouses, 6.50% Industrial, 6.64% Aquaculture, 22.15% Bathing, and 1.78% others. (Chandrasekharam and Bundschuh (Eds), 2002 [25]).

Internationally the largest uses of geothermal energy are for space heating (37%), 75 percent of which is in district heating systems and for swimming, bathing and balneology (22%) [26].

5.5.1 Direct Uses

The Lindal diagram (Figure 5.5) indicates the temperature ranges suitable for various direct uses of geothermal energy. Typically, agriculture and aquaculture use the lowest temperature resources. Space heating generally requires temperatures above 50°C, although temperatures as low as 40°C may be adequate in certain cases. Geothermal heat pumps can allow the use of temperatures as low as 4-6°C to provide space heating. Cooling, industrial processes and dehydration normally require temperatures above 100°C. Refrigeration based on ammonia absorption is possible at approximately 180°C. At temperatures over 110-120°C binary fluid electrical generation also becomes economically viable, and above 140°C conventional electric generation is viable. There is increased interest in coupling geothermal electrical generation and direct uses. Electrical generation may be either topping or bottoming cycle depending upon the requirements of the direct use application.

5.5.2 District Heating

Geothermal district heating systems are in operation in at least 12 countries including Iceland, France, Poland, Hungary, Turkey, Japan, China, Romania, Italy, the United States, Sweden and Denmark.

The first known geothermal district heating system was built in Chaudes-Aigues Cantal, France in the 14th century (Bloomquist, 1988 [27]) and is still in operation today. The first system in USA was the Artisan Hot and Cold Water Company built in Boise, Idaho in 1892.

The most famous geothermal district heating system in the world is the system supplying nearly 98% of the residents of Reykjavik, Iceland. The installed capacity is 830 MWe and is designed to meet the heating load down to -10°C. Increased load during colder periods is met by large storage tanks and an oil-fired peaking plant (Ragnarsson, 2000 [28]).

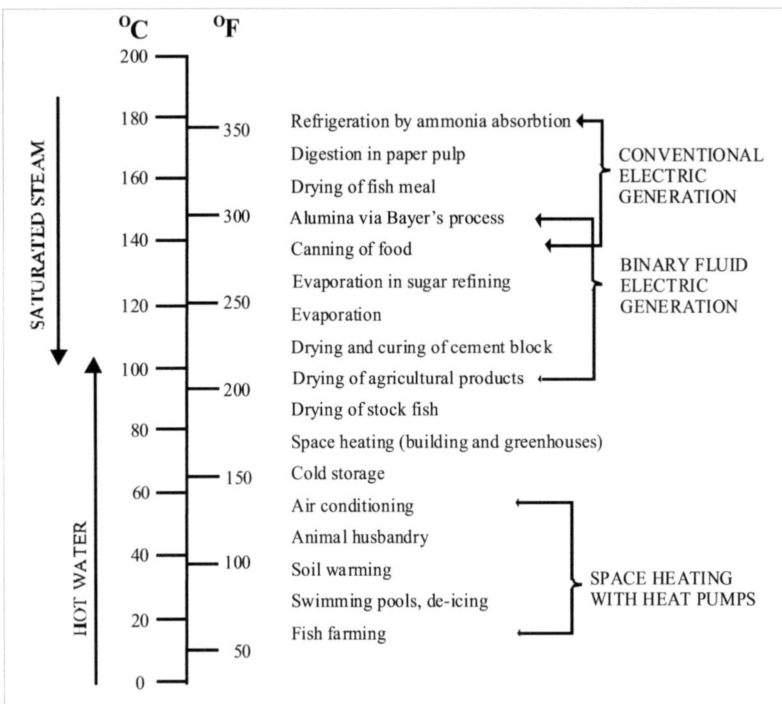

Source: Lund 2002 [24]

Fig. 5.5. Lindal Diagram Indicating Temperature Range for Direct Use of Geothermal Energy

5.5.3 Agriculture and Aquaculture Applications

Agriculture, including greenhouses and soil warming, and aquaculture uses of geothermal energy are increasing rapidly. They are particularly widespread since they require heating at the lower end of the temperature range where there is an abundance of geothermal resources (Chandrasekharam and Bundschuh (eds), 2002 [25]).

5.5.4 Future Developments

The potential for substantially increased use seems very promising. Future developments will depend upon:

- Increased resource information
- Increased knowledge of potential uses
- Prices of competing fuels, e.g. oil and gas
- The establishment of clear legal, institutional and regulatory framework conducive to geothermal development on a country by country basis
- Availability of capital, especially in the developing countries.

5.6 Improving Geothermal Power Plant

Large plants produce electricity at about 5 cents/kWh and small plants at about 7 cents/kWh. Geothermal appears poised for expansion. Some of the aspects of needs and research to significantly improve geothermal power plant for the generation of electricity are now presented.

Geothermal power plants are suitable for deployment in all types of terrain and environment. If the resource is typically above about 150 °C, a direct flash steam plant is used. For lower resource temperatures, a binary plant is used. They are typical Rankine plants, with either steam (flash plant) or hydrocarbon (binary plant) working fluids.

5.6.1 Goal and Objectives

The goal is to work with industry to make geothermal energy fully cost-competitive with fossil-fuelled alternatives. The objectives are to reduce investment, enhance operability, and to promote geothermal energy. Reducing plant investment requires better components, such as air-cooled condensers; lower cost materials and inexpensive coatings; and better cycles. Reducing operations and maintenance (O&M) costs can be achieved through actions to control brine chemistry; better instruments for tighter controls; and improved emissions control.

Efficiency can be enhanced by improved off-design operation; reduced parasitic losses; and improved cycles.

Plant revenue can be enhanced by co-production of a valuable by-product; and direct use of cascaded brine, first to produce electricity then as a thermal source.

The potential impacts of power plant Research and Development are shown in Table 5.12. Aggregated, these improvements have a potential for 18 to 25% reduction in the cost of electricity, or about 1¢/kWh reduction.

. Technical Area	Percent Reduction in Cost of Electricity
Heat Exchangers	8 to 10%
Cycle Efficiency	5 to 7%
Enhancement of Off-design Operation	3 to 5%
Reduced O&M	2 to 3%

Table 5.12. Potential Impacts of Power Plant Research and Development

5.7 Geothermal Drilling R&D Overview

Drilling is a critical element of the entire life cycle of geothermal development: exploration, production, injection, and well maintenance. The cost of drilling, logging, and completing geothermal wells is high compared to that of oil and gas because the rock is typically very hard, formations are highly fractured, and the temperatures encountered are very high. As these costs often account for more than half of the total capital required for a geothermal power project, reductions in the cost of drilling and completing wells have a very large impact on a project's overall commercial viability. Some of the techniques for reducing these costs include drilling faster, experiencing less idle time, increasing bit or tool life, achieving higher overall success rates, and producing more per well via the use of multi-laterals. The mission of geothermal drilling programs is to develop cost-cutting technologies for accessing geothermal resources.

The development of high-temperature instrumentation, lost circulation technology, hard-rock drill bit technology, and advanced drilling systems are all directed toward achieving this goal. Technologies developed include improved PDC bits, "Dewarless" high temperature instrumentation, slim hole drilling, polyurethane grout for lost circulation, acid-resistant cement, acoustic telemetry, and diagnosis-while-drilling (DWD) [29].

Geothermal drilling uses the same basic elements as land-based oil and gas drilling - rotary drill rigs, blow-out preventers, drill strings and drill bits - but geothermal resources are found in formations that are much more difficult to drill than those typical of the hydrocarbon industry. Geothermal rocks are typically very hard, fractured, and abrasive; formations are under-pressured and often contain corrosive fluids; hole diameters are large compared to oil and gas; and the drilling environment is often extremely hot and corrosive.

Drilling actually tracks the price of crude oil very closely because as energy prices rise, it becomes viable to drill deeper, more expensive wells. A way of normalizing this cost while accounting for inflation is to compare the cost of drilling a geothermal well with a similar (depth and location) oil well. Over the twenty-year period 1980-2000, technology improvements have lowered the cost of geothermal wells from approximately 1.75 times to 1.4 times that of an oil well. Some of the science and technology related breakthroughs associated with this program include PDC bits, rolling float meter (RFM), polyurethane foam, acoustic telemetry, HT electronics, slim-hole drilling demonstrations, insulated drill pipe, and improved cement.

The single most important barrier to commercialization and deployment is the small size of the geothermal industry. The number of geothermal wells drilled each year is less than 0.1% of the number of oil and gas wells, so it is clear that manufacturers and service companies can quickly identify their market. There are three factors that may mitigate this problem: (1) because geothermal problems are often more difficult than oil and gas problems, hardware developed and tested for geothermal use is sometimes considered "premium" grade, and therefore more reliable, for other drilling, (2) rising energy prices may expand the geothermal market - there is already some evidence of this in California and Nevada, and (3) deep gas drilling now regularly encounters hard rock and temperatures above 225^{0}C, so this may be a new market for what were previously considered "geothermal" tools.

5.8 Advanced Power Cycles for Enhancing Geothermal Sustainability

Until the early 1980s geothermal plants used steam turbines exclusively, operating on dry steam or separated steam. In the mid 1980s, advanced power cycles were introduced, initially to enable exploitation of lower enthalpy resources, then to recover heat from the separated water, and thereafter to handle high gas content resources as well as high enthalpy resources using combined steam/organic cycles [30]. Most of these plants are air-cooled, assuring 100% re-injection of geothermal fluids and thus enhancing sustainability as well as reducing environmental impact.

Of the 9,000 MW of geothermal plants installed worldwide, most use steam turbines operating on dry steam or steam produced by single or double flash. About 1,000 MW use Organic Rankine Cycle (ORC) or steam/ORC-combined cycles [31]. Examples of commercial plants range in capacity from 200 kW to 130 MW.

Operational experience has confirmed the advantages of the ORC plants, not only for the low enthalpy water-dominated resources, but also at high enthalpy for aggressive brine or brine with high non-condensable gas content. Air-cooled ORC plants are particularly well adapted to Engineered Geothermal Systems (EGS). The somewhat higher installed cost of these systems is often justified by environmental and long-term resource management considerations [32,33].

5.8.1 Optimization in Design of the Power Cycle

Optimization of the whole geothermal power plant system is accomplished by matching the working cycle and fluid properties with the characteristics of the resource, considering not only the resulting efficiency and cost, but also impact on the environment, long-term pressure support, requirements for make-up wells, and costs of operation and maintenance (O & M).

5.8.1.1 Heat Cycle Considerations

When the source is a liquid phase only (sensible heat) the ideal cycle would have a varying source temperature, being a succession of infinitesimal Carnot cycles. In a sub-critical Rankine cycle, the constant temperature of evaporation leads to a loss of energy. However, because of the lower latent heat of vaporization, this drawback is lower than in a steam cycle.

The super-critical binary cycle, the different total flow regenerative cycle, the cascaded binary cycle and the Kalina cycle are aimed at achieving the objective of getting closer to the ideal cycle. When dry steam is available, the most effective way is to use the conventional condensing steam cycle.

When the source is a mixture of steam and brine and/or has a high content of non-condensable gases, the most effective utilization of the resource is achieved through a combined cycle by expanding the steam first in a back pressure steam turbine and then the heat of condensation together with heat of the separated brine is used to drive a bottoming ORC.

It is necessary to consider the output net of the parasites, such as cycle pumps, production pumps, injection pumps, cooling systems and non-condensable gas extraction power consumption [34].

5.8.1.2 Resource Considerations

Sustainability is defined as ability to economically maintain the installed capacity over the life of a plant [35]. In case of geothermal power plants this is controlled by two factors: heat recharge and water recharge.

The decline of production in the Larderello, Geysers and Wairakei fields has focused attention on the necessity for long-term pressure support by re-injection as much of the geothermal fluid as possible.

Use of secondary loops and of down-hole and booster pumps, as employed in air cooled ORC plants, assures complete water recharge and reduces both fouling of the heat exchangers and scaling of the injection wells.

5.8.1.3 Enhanced Geothermal Systems

The value of the air-cooled ORC is particularly important in the case of Engineered Geothermal Systems, which are very much dependent on the water recovery ratio.

5.8.1.4 Environmental Considerations

Use of air-cooled ORC reduces impact on the environment by re-injection of:

- Non condensable gases (mainly H_2S released by the steam)
- Discharged fluids such as the separated brine (carrying off heavy metals) and blow-down from the cooling towers (chemicals).

5.8.2 Conventional Steam Turbine Geothermal Power Plants

Geothermal power plants operating from dry steam or from steam flashed from high temperature water employ either:

- Back pressure turbines which exhaust the spent steam to the atmosphere, or
- Condensing steam turbines that condense the steam in water-cooled condensers under vacuum, with the condensate used as make-up water, in the cooling tower (Figure 5.6).

Backpressure turbines have the lowest capital cost at the expense of lowest efficiency; condensing steam turbines are more expensive but operate at a higher efficiency than backpressure turbines.

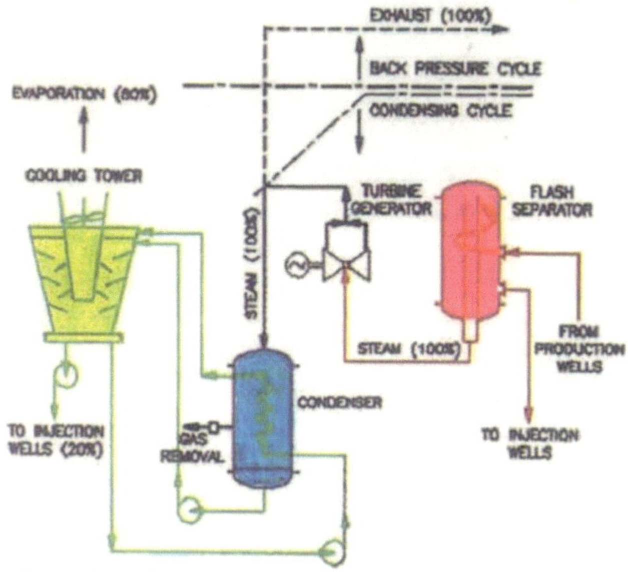

Source: Ormat Technologies, Inc., USA

Fig. 5.6. Conventional geothermal power plant - Backpressure or condensing

5.8.3 Geothermal Power Plants using Organic Rankine Cycle

The basic Organic Rankine Cycle (ORC) as used in an air-cooled binary geothermal plant is characterized by:

- 100 percent re-injection of the geothermal fluid
- Air-cooling for nearly zero environmental impact; and
- No surface discharge of fluids.

Different plant configurations have been developed to optimize use of the geothermal resource. A number of examples are given in 5.6.3.1 – 5.6.3.6.

5.8.3.1 Single Phase (Hot Water) Geothermal Power Plants

The irreversibility of a binary process on the hot side, namely the temperature difference between the heating fluid and the working fluid, is shown on the temperature (T) vs. heat withdrawn (Q) (from the liquid) diagram (Figure 5.7).

The marked parts between the two curves represent the irreversibility (losses) of the conversion process. Figure 5.7 shows that the similarity in shape of the two curves and the proximity between them are good indications of the process efficiency [36].

This loss can be reduced, as shown in Figure 5.8, by using a supercritical cycle as indicated earlier, by using a cascading approach [36] and/or by recovering some of the heat of the superheated exhaust vapor to preheat the motive fluid.

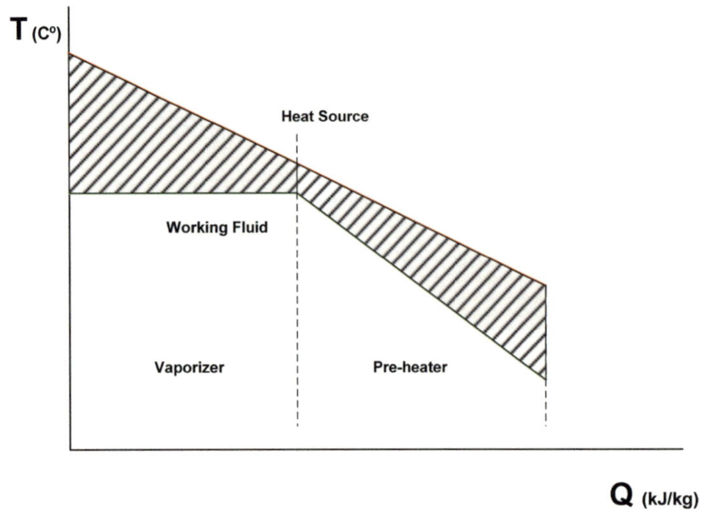

Source: Ormat Technologies, Inc., USA

Fig. 5.7. Typical T/Q diagram

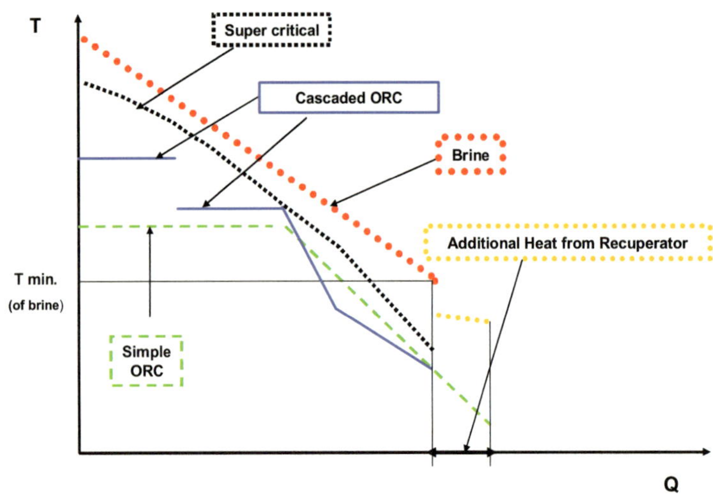

Source: Ormat Technologies, Inc., USA

Fig. 5.8. T/Q diagram: Reducing the irreversibility loss

5.8.3.2 Two-Phase Geothermal Power Plant

In the majority of geothermal fields worldwide, the geothermal fluid is separated in an aboveground separator into a stream of steam and a stream of brine.

In a low to moderate enthalpy resource the steam quality is 10-30 percent a function of the fluid enthalpy and the separation pressure. The two streams can very efficiently be utilized in a two-phase geothermal plant (Figure 5.9). Separated steam (usually with some percentage of Non-Condensable Gases (NCG)) is introduced in the vaporizer to vaporize the organic fluid.

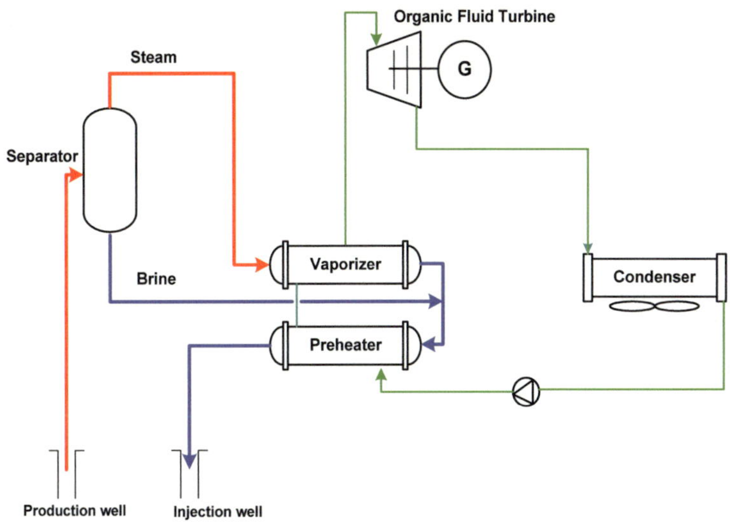

Source: Ormat Technologies, Inc., USA

Fig. 5.9. Two-phase power plant

The geothermal condensate is mixed with the separated brine to provide the preheating medium of the organic fluid. In the ideal case, the latent heat would be equal to the heat of vaporisation of the organic fluid, and the sensible heat of the brine plus condensate would be equal to the heat required to preheat the organic fluid.

5.8.3.3 Recuperated Organic Rankine Cycle

In most actual cases, the perfect match as above is not feasible, mainly because of limitation in the cooling temperature of the brine and condensate mixture. The limiting factor in most cases is silica-scaling risk, which is increased as the brine temperature drops. A method to partially overcome the cooling temperature limit is to add a recuperator that provides some of the preheating heat from the vapour exiting the turbine.

The recuperator is applicable when the organic fluid is of the "*dry expansion*" type, namely a fluid where the expansion in the turbine is done in the dry superheated zone and the expanded vapour contains heat that has to be extracted prior to the condensing stage. The recuperated Organic Rankine Cycle is typically 10-15 percent more efficient than the simple Organic Rankine Cycle [37]. This applies also to two-phase geothermal power plant.

5.8.3.4 Higher Enthalpy Two-Phase Geothermal Power Plant

When the resource enthalpy is higher and as a result the proportion of steam in the total fluid increases, the "*perfect match*" between the heat source and working fluid is not maintained. Thus, some of the available heat or the available energy is lost for power generation.

To utilize the two-phase heat source in a more efficient manner, a secondary organic loop, which uses the extra available steam, can be used. The cycle is shown in Figure 5.10. It is feasible when vapour extraction is possible within the expansion phase of the organic cycle. The simplest way to perform the extraction is with two turbines in series. In this case, some vapour is extracted between the high pressure and the low-pressure turbines and is condensed at an intermediate pressure (and temperature).

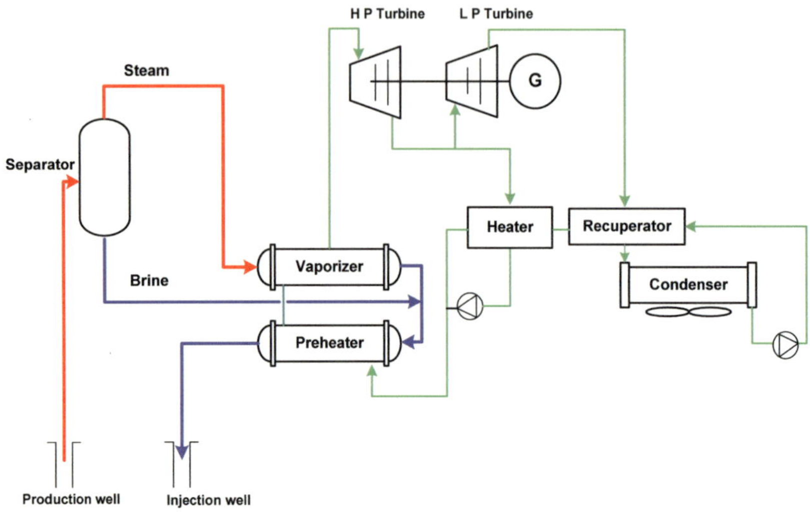

Source: Ormat Technologies, Inc., USA

Fig. 5.10. Secondary organic loop cycle

The condensed vapour preheats the main organic fluid stream as it exits the recuperator. The extracted organic fluid forms a secondary cycle that generates an additional 5 to 8 percent electrical power. When there is extra steam compared to brine (higher enthalpy) the above cycle is effective and the cooling temperature of the brine plus condensate is limited.

Figure 5.11 is a flow temperature diagram of the higher enthalpy cases. Line A is the simple two-phase cycle preheating phase. The significant irreversibility is represented by the large space between the steam and brine lines and line A. Line B shows the preheating phase in a recuperated two-phase cycle; the irreversibility is reduced and the cycle efficiency is increased accordingly.

The third line C demonstrates the additional gain in efficiency by using the two-phase/extraction cycle. The line moves further to the right, thus decreasing the gap between the heating line and the working fluid line. Another indication of the increase in efficiency from cycle A to B and to C, is the increasing heat quantity for heating the working fluid, as presented by points QA, QB, and QC.

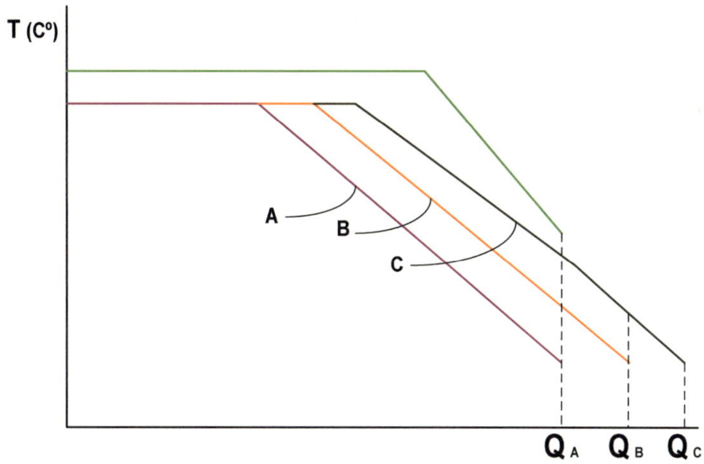

Source: Ormat Technologies, Inc., USA

Fig. 5.11. Higher enthalpy

5.8.3.5 Use of a Back Pressure Steam Turbine

Another approach for the higher enthalpy two-phase heat source is use of a back pressure steam turbine which generates extra power from excess steam not required for the vaporizer of the ORC.

Part of the preheating of the organic fluid is now done with low-pressure steam exiting the backpressure steam turbine (Figure. 5.12).

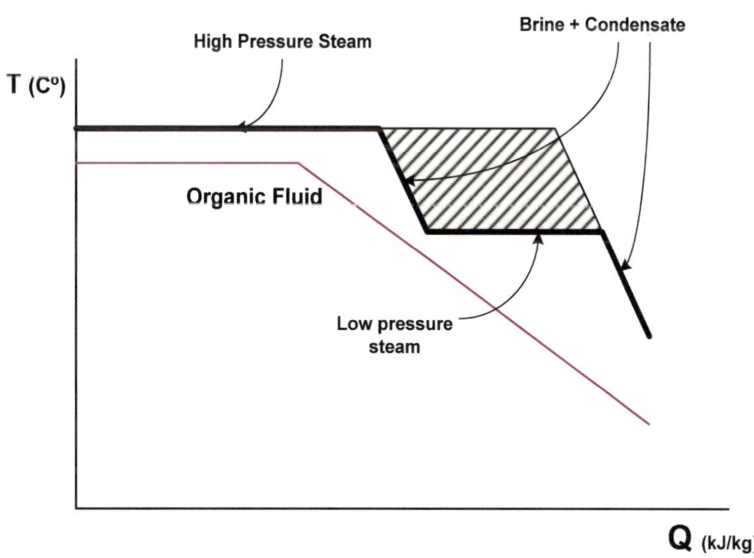

Source: Ormat Technologies, Inc., USA

Fig. 5.12. Pre-heating using exhaust in a backpressure steam turbine

The gap between the steam and the preheating line of the organic fluid could be filled even more efficiently by a multi-stage (two or more) backpressure steam turbine, with extraction of steam between the stages. But the decision on the number of stages is based on consideration of trade-off in process optimization between higher efficiency and complication (and cost) of the system.

5.8.3.6 Geothermal Combined Cycle [38]

For high enthalpy fluids with very high steam content a solution is the geothermal combined cycle configuration where the steam flows through the back pressure turbine to the vaporizer, while the separated brine is used for preheating or in a separated ORC (Figure 5.13) [38].

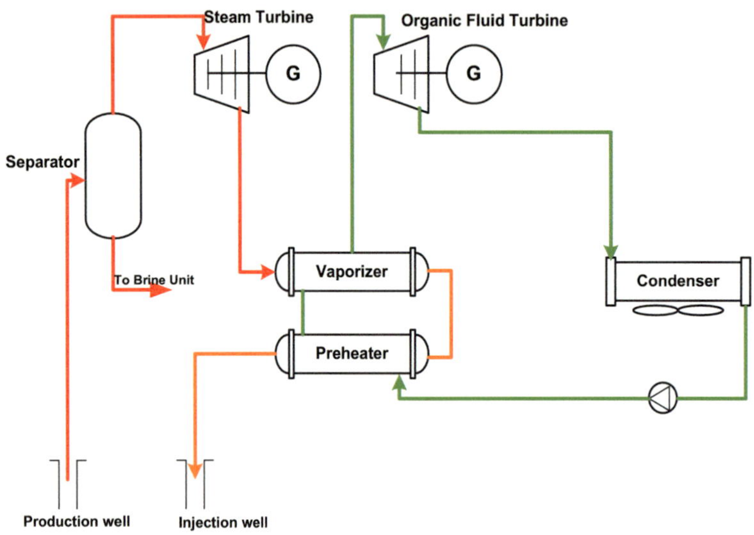

Source: Ormat Technologies, Inc., USA

Fig. 5.13. Geothermal combined cycle

5.8.4 Deployment

As of 2007, the capacity of geothermal plants using advanced power cycles worldwide is close to 1,000 MW, approximately 10% of the total geothermal capacity installed in the last 50 years.

A breakdown of the 1,000 MW of plants in commercial operation is as follows: 60 MW of ORC plants designed or built by Ben Holt, Turboden and Barber-Nichols; one 2 MW of Kalina cycle plant and more than 900 MW of ORC and combined cycle plants.

5.8.5 Enhancing Sustainability and Cost Effectiveness

Geothermal resources are complex geological structures that provide conduits for natural heat of the earth to heat underground waters that may then be utilised to convey heat to the surface. Technology to assess the heat content of geothermal resources is available, along with drilling technologies to access this heat and mature proven power technologies to convert this heat to commercial electricity.

The key to sustainability of this power generation lies in not depleting the waters that convey this energy to the surface.

The use of the field-proven air-cooled Organic Rankine Cycle based geothermal power plant enables these objectives to be achieved by extending the lifespan of the wells and reducing emissions.

Hence cost-effective power is generated with enhanced sustainability, mitigating depletion of geothermal resources. This element is particularly important in proposed Engineering Geothermal Systems.

5.9 Iceland Deep Drilling Project, Exploration of Deep Unconventional Geothermal Resources

The Iceland Deep Drilling Project (IDDP) is a long-term research and development program aimed to improve the efficiency and economics of geothermal power generation by harnessing deep natural supercritical hydrous fluids obtained at drillable depths. Producing supercritical fluids will require drilling wells and sampling fluids and rocks to depths of 3.5 to 5 km, and at temperatures of 450-600°C. The current plan is to drill and test a series of such deep boreholes in Iceland at the Krafla, the Hengill, and the Reykjanes high temperature geothermal fields. Investigations have indicated that the hydrothermal system extends beyond the three already developed target zones, to depths where temperatures should exceed 550-650°C. A deep well producing $0.67m^3$/sec steam (~$2400m^3$/h) from a reservoir with a temperature significantly above 450°C could yield enough high-enthalpy steam to generate 40-50 MW_{el} of electric power. This exceeds by an order of magnitude the power typically obtained from conventional geothermal wells.

The Project was initiated in 2000 by an Icelandic energy consortium, consisting of Hitaveita Sudurnesja Ltd. (HS), Landsvirkjun (LV), Orkuveita Reykjavikur (OR) and the Icelandic National Energy Authority Orkustofnun (OS). In 2007, Alcoa Inc. joined the IDDP consortium. The principal aim of the IDDP is to enhance the economics of high temperature geothermal resources by producing from deep reservoirs at supercritical conditions.

5.9.1 Supercritical Geothermal Fluids

Large changes in physical properties of fluids occur near the critical point in dilute systems. Orders of magnitude increases in the ratio of buoyancy forces to viscous forces occur that can lead to extremely high rates of mass and energy transport. Because of major changes in the solubility of minerals above and below the critical state, supercritical phenomena can play a major role in high temperature water/rock reaction and the transport of dissolved metals.

At temperatures and pressures above the critical point, which for pure water is at 221 bars and 374°C, only a single-phase supercritical fluid exists. Figure 5.14 shows the pressure-enthalpy diagram for pure water, showing selected isotherms. Steam turbines in geothermal plants generate electricity by condensing the steam separated from the two phase field (liquid and steam field in Figure 5.14) which, depending upon the enthalpy and pressure at which steam separation occurs, is often only 20-30% of the total mass flow. The concept behind the Deep Drilling program is to bring supercritical fluid to the surface in such a way that it transitions directly to superheated steam along a path like F-G in Figure 5.14, resulting in a much greater power output than from a typical geothermal well.

The conditions under which steam and water coexist is shown by the shaded area, bounded by the boiling point curve to the left and the dew point curve to the right. The arrows show different possible cooling paths (from Fournier[1], 1999).

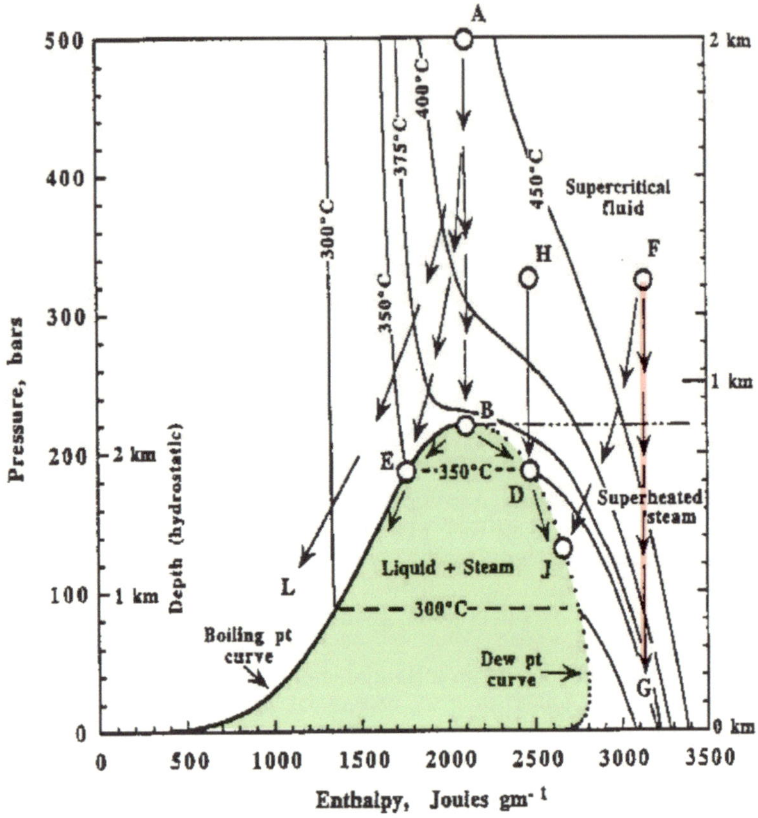

Fig. 5.14. Pressure enthalpy diagram for pure H_2O with selected isotherms

Supercritical conditions have been encountered during drilling in a small number of geothermal fields, like in Larderello in Italy, Kakkonda in Japan, and at Nesjavellir in Iceland, where they have presented problems for commercial exploitation and were sealed off from the conventional part of the systems. Apart from the high P-T conditions where underground blowout was involved, like at Nesjavellir [39] (Steingrimsson et al., 1990), the problems include low permeability, hole instability due to thermal creep, and the presence of acid volcanic gases. However, the drilling technology used in these cases was not designed to handle the conditions encountered when supercritical hydrous fluids were unexpectedly penetrated.

---

[1] Fournier, R. *Hydrothermal Processes Related to Moment of Fluid Flow from Plastic into Brittle Rock in the Magmatic-Epithermal Environment*, Economic Geology, Vol. 94, (8), 1999, pp. 1193-1211.

The IDDP intends to meet the hostile conditions expected in supercritical geothermal reservoirs by a conservative well design and by adopting the necessary safety measures. The safety casing will be cemented down to 2.4 km before drilling down to 3.5 km depth or deeper to reach the critical point. Once beyond that, the production casing will be cemented in order to produce only the supercritical fluid. By releasing the pressure, the supercritical fluid will expand and move upwards to the surface through the well bore as a superheated dry steam, following a path like F-G in Figure 5.14. The deep casings will prevent the fluid from mixing with the two-phase zone and as the pressure decreases, condensation is less likely to occur. A pilot study for harnessing the fluid will need to be undertaken, especially with respect to the fluid chemistry that will only be known after drilling.

5.9.2 Drilling in IDDP Wells

5.9.2.1 Design

Conventional geothermal drilling techniques will be used in drilling the IDDP wells. The first well was designed as a dual-purpose hole. To meet the engineering goals of the power companies, it is designed as an exploration/production well, and to meet the scientific goals of understanding the supercritical environment, some spot cores will be taken in the lowest part of the drill hole, which hopefully will be the supercritical zone.

5.9.2.2 Potential Drill Sites

Geothermal reservoirs at supercritical conditions are potentially to be found worldwide in any active volcanic complex. However, the depth to such reservoirs may vary greatly from shallow to deep, and the simplest approach would be to seek supercritical reservoirs in active high-temperature geothermal fields, closest to the earth's surface, in both sub aerial and submarine settings. Each high temperature hydrothermal system requires site-specific attention to target drill sites for reaching deep unconventional geothermal resource (DUGR) reservoirs with supercritical conditions and permeable rocks at drillable depths.

All active volcanic complexes are potential targets for finding deep geothermal systems at supercritical conditions. These volcanic complexes are of different ages and at different stages in their evolution; some are at infancy, others are mature and some are close to extinction.

The three Icelandic fields deemed to be prime targets for DUGR exploration, the Reykjanes, Hengill and Krafla geothermal systems, demonstrate different stages in the evolution of their magma-hydrothermal evolution, the first being at infancy, the second being "middle aged" and the third being mature. Deep drilling at all three will permit studying different stages in the development of supercritical conditions at depth. Additionally, they exhibit different fluid compositions, the first involving modified seawater, but the other two dilute fluids of meteoric origin. Extensive production in all three-drill fields has led to the hottest parts of the hydrothermal up-flow zones. However, the nature of their heat sources is poorly known except in the mature case of the Krafla system where a magna chamber has been identified at only 3-4 km depth [40].

5.9.3 Potential Benefits

5.9.3.1 Power Generation

The high-temperature fluids expected from the IDDP wells offer two advantages over fluids from conventional wells for generation of electric power, (i) higher enthalpy, which promises high power output per unit mass, and (ii) higher pressure which keeps the fluid density high and thus contributes to a high mass-flow rate.

The electric power output that can be expected from an IDDP well compared with that from a conventional has been estimated by Albertsson et. Al. [41,42].

The choice of technology to be applied for the power generation cannot be decided until the physical and chemical properties of the fluid are determined. Nonetheless, it appears likely that the fluid will be used indirectly, in a heat exchange circuit of some kind. In such a process the fluid from the well would be cooled and condensed in a heat exchanger and then injected back into the field. This heat exchanger would act as an evaporator in a conventional closed power-generating cycle.

5.9.3.2 Scientific Studies

In addition to investigations and sampling of fluids at supercritical conditions the IDDP will permit scientific studies of a broad range of important geological issues, such as investigation of the development of a large igneous province, and the nature of magma-hydrothermal fluid circulation on the landward extension of the Mid-Atlantic Ridge in Iceland. In addition, the IDDP will require use of techniques for high-temperature drilling, well completion, logging, and sampling, techniques that will have a potential for widespread applications in drilling into oceanic and continental high-temperature hydrothermal systems.

5.9.3.3 Economic Benefits

The potential economic benefits of the IDDP project may be listed as follows:

1) Increased power output per well, perhaps by an order of magnitude, and production of higher-value, high-pressure, high-temperature steam.
2) Development of an environmentally benign, high-enthalpy energy source beneath currently producing geothermal fields.
3) Extended lifetime of the exploited geothermal reservoirs and power generation facilities.
4) Re-evaluation of the geothermal resource base.
5) Industrial, educational, and economic spin-off.
6) Knowledge of permeability within drill fields deeper than 2-3 km depth.
7) Knowledge of heat transfer from magma to water.
8) Heat sweeping by injection of water into hot, deep wells.
9) Possible extraction of valuable chemical products.
10) Advances in research on ocean floor hydrothermal systems (the Reykjanes field).

Amongst approaches to improve the economics of the geothermal industry, three of the most significant are: (i) to reduce the cost of drilling and completing geothermal production wells as far as possible, (ii) to cascade the usage of thermal energy by using the effluent water for domestic heating and for industrial processes, and (iii) to reduce the number of wells needed by increasing the power output of each well, by producing supercritical fluids. Accordingly, the completion of the IDDP project is of considerable importance for the geothermal industry at large.

5.9.3.4 Environmental Issues

Developing environmentally benign high-enthalpy energy sources below the depth of currently producing geothermal fields is not only of economic value in relation to the already installed infrastructures, but it is also of environmental value by diminishing environmental impact of geothermal utilization. Producing more power without increasing the *footprint* of the exploited drill field is a significant benefit.

5.9.4 Potential Impacts

5.9.4.1 Global Impacts

Potential impact of utilizing geothermal resources at supercritical conditions could become quite significant. Not only would this call for re-evaluation of the geothermal energy resource base on a local scale, but also on a global scale. If producing supercritical fluids became widespread it would lead to a major enlargement of the accessible geothermal resource base.

It is conceivable that, in the more distant future, utilization of ocean floor geothermal systems might become viable. Submarine geothermal systems are abundant along the world's mid-ocean ridge systems and some of them (the black smokers) expel ~400°C hot seawater direct into the deep oceans, and precipitate chimneys of sulphide-ore deposits. The pressure of 2.5-3 km deep seawater results in supercritical hydrostatic pressures, and allows almost supercritical fluids to be expelled directly into the oceans. Tapping energy through shallow drill holes on the mid-ocean ridges using techniques initially developed by the international IDDP program is an exciting prospect.

5.9.4.2 Potential Impact on Greenhouse Gases

In the Stern Review to the British Government 2006 [43] (www.sternreview.org.uk) it is reported that since industrialization, greenhouse gas (GHG) levels have risen from 280 ppm CO_2 equivalent (CO_2e) to 430 ppm CO_2e today, and they increase by 2 ppm each year. The risks of the worst impacts of climate change can be substantially reduced, according to the review, if the GHG levels can be stabilized between 450 and 550 ppm CO_2e. Stabilization in this range would require emissions to be at least 25% below current levels by 2050, and perhaps much more. According to the Review, three measures need be taken, (1) taxation on GHG emission, (2) new techniques, and (3) removal of hindrances against economic energy usage. According to the Stern Report the main sources of the polluting greenhouse gases are 24% in the Power Sector, 14% in the Industry sector, another 14% in the Transport sector, and 5% in other energy related activities, altogether some 57%. Attempting to decrease CO_2e emission in any of these sectors would be a logical step to respond to the Stern Review.

The World Energy Council (WEC) has presented several scenarios for meeting future energy requirements with varying emphasis on economic growth rates, technological progress, environmental protection and international equity [44] (Nakicenovic et al., 1998). In all WEC´s scenarios, the peak of the fossil fuel era has already passed (Nakicenovic et al., 1998). Oil and gas are expected to continue to be important sources of energy in all cases, but the role of renewable energy sources and nuclear energy vary highly in the scenarios and the level to which these energy sources replace coal. In all the scenarios, the renewables are expected to become very significant contributors to the world primary energy consumption, providing 20-40% of the primary energy in 2050 (UK 80%) and 30-80% in 2100. They are expected to cover a large part of the increase in energy consumption and to replace coal.

Evidently, a large opportunity to cut GHG emission exists with the geothermal energy sector. However this estimate did not include innovations such as IDDP.

In summary, the long-term program to improve efficiency and economics of geothermal energy by harnessing deep unconventional geothermal resources is an ambitious project to produce electricity from natural supercritical hydrous fluids from drillable depths. Producing higher-temperature fluids for generation of electric power offers two advantages over using the fluids from conventional wells: (i) higher enthalpy, which promises high power output and higher efficiency per unit mass, and (ii) higher pressure, which keeps the fluid density high and thus contributes to higher mass-flow rates. The choice of technology to be applied for power generation from these high-temperature fluids will be decided after determining the physical and chemical properties of the fluids that are produced.

There are three approaches to improve the economics of the geothermal industry worldwide: (i) cascading the usage of geothermal energy by using the effluent water from electricity production for industrial processes and for domestic heating, (ii) reducing the cost of drilling and completing geothermal production wells, and (iii) reducing the number of wells needed by increasing the power output of each. The best way to achieve the latter is to produce supercritical fluids. Successful completion of the IDDP project is of considerable importance for the geothermal industry at large. A successful outcome would be a major step forward for the geothermal industry on a global scale, which in turn, could help counterbalance the threat of global warming by increased use of sustainable, non-polluting energy resources.

5.10 Geothermal Power Plants in Iceland in the Hengill Area

Geothermal plants in Iceland are now discussed. The Hengill area in SW-Iceland is one of the most extensive geothermal areas in Iceland. It is located 25 km east of Reykjavik. It has an area of approximately 110 km^2 and is estimated to sustain 700 MW_{el} power production in several power plants [45]. Two power plants operate in the area. Environmental impact assessment for two new power plants is being worked on. Power plants in the Hengill area will produce at least 600 MW_{el} and 433 MW_{th} by end of 2011. Research projects connected with the power plant project: (i) the Carb-Fix project, and (ii) the IDDP project, is being worked on.

Research drilling started at Nesjavellir in the north of the Hengill area in 1965. Hot water production for district heating in Reykjavík started at the Nesajvellir plant in 1990. Power production started there in 1998. Today, the Nesjavellir power plant produces 120 MW_{el} and 300 MW_{th}. The Nesjavelir plant was built in several stages.

To meet increasing demand for electricity and hot water for space heating in the industrial and domestic sectors, Orkuveita Reykjavíkur (OR) is currently building a CHP geothermal power plant at Hellisheiði. The approach for the Hellisheiði plant is the same as for Nesjavellir, i.e., it will be built in several stages. The first stage, which came on line in 2006, consist of two 45 MW_{el} units. The second stage of the Hellisheiði power plant, which consists of a 33 MW_{el} Low Pressure unit, started operating in November 2007. Construction of the third stage of the plant is in progress, that is the erection of a two additional high-pressure units, 45 MW_{el} each. Erection of the thermal plant, the fourth stage, started at the beginning of 2008.

At least two new geothermal power plants are planned for the Hengill area, at Hverahlíð and Bitra.

An environmental impact assessment (EIA) for the power plants at Hverahlíð and Bitra was published towards end 2007.

The capacity of the Hellisheiði power plant will be 300 MW_{el} electric and 400 MW_{th} thermal. Estimated capacity of the power plants in Hverahlíð and Bitra will be 90 MW_{el} and 135 MW_{el}, respectively.

With more knowledge of the Hengill geothermal area accumulated through running the Nesjavellir and Hellisheiði power plants and research drilling, new opportunities arise which can be utilized both in future power plants in the area and in other projects.

5.10.1 The Hengill Area

The Hengill area is a rural mountainous area in the middle of the western volcanic zone of Iceland that runs from Reykjanes in a northerly direction to Langjökull (Figure. 5.14). The Hengill region is one of the most extensive geothermal areas in Iceland. Surface measurements and the heat distribution estimate that the region will sustain 690 MW_{el} power production in several power plants [45]. The high temperature geothermal area at Hengill covers three central volcanoes and their surroundings. The youngest one is the most active, whereas the oldest one is eroded but still geothermal active.

5.10.2 Nesjavellir Power Plant

The first geothermal power plant in the Hengill area is the Nesjavellir power plant. Construction of the power plant began in early 1987, with the first stage being completed in May 1990. Four holes, generating about 100 MW_{th}, were then connected to the processing cycle, The next stage of power harnessing was brought online in 1995 when the fifth hole was connected; heat exchangers and a deaerator were added; and the production capacity was increased to 150 MW_{th} of geothermal power [46].

In fall 1998, the first steam turbine was commissioned and the second at the end of the year, producing total of 60 MW_{el}. Five additional holes were put online, increasing the total processing power of the power station to 200 MW_{th}, In June 2001 the third turbine was put into operation. The turbines are 30 MW_{el} each, making the total production of electricity 90 MW_{el} [19].

Early 2008, Nesjavellir power plant generates 300 MW_{th} and 120 MW_{el}. The Nesjavellir area is being researched to see if it is possible to add one more turbine to the power plant.

5.10.3 Hellisheiði Power Plant

The first research drilling for the Hellisheiði power plant was in 1985 and then again in 1994. These boreholes indicated that the geothermal fields could sustain power production but more drilling was needed before decisions could be made. In 2001 and 2002 five boreholes were drilled. Based on the results from these boreholes it was decided to start preparations for a power plant with total capacity of 120 MW_{el} and 400 MW_{th} with the objective to meet increasing demand for electricity and hot water for space heating in the industrial and the domestic sectors.

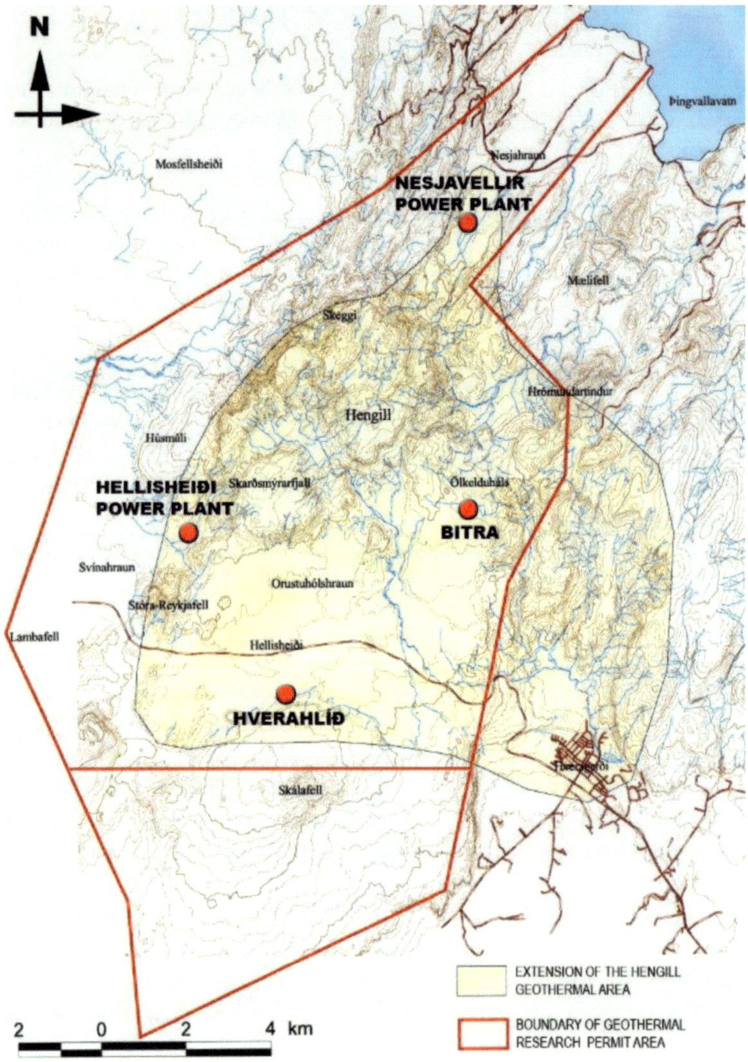

Fig. 5.14. Detailed map of the Hengill area.

Drilling continued and by end of 2005 18 new boreholes had been drilled. In light of the results of these drillings it was decided to enlarge the development area further north towards the main volcano. With this new area, estimated capacity of the geothermal area was increased by 120 MW_{el}. The first stage from this new area is 90MW_{el} to be ready in 2008. With this enlarged potential more geothermal water was available than initially estimated and more than is needed for the thermal plant. It was decided to add one low-pressure unit to increase utilization of the geothermal energy. Its size ended as 33 $MW_{el.}$

The first stage started operating in 2006 and comprises two 45 MW_{el} units. The second stage, a 33 MW_{el} Low Pressure unit, started operating in November 2007. The construction of the third stage, the erection of two additional high-pressure units rated at 45 MW_{el} each, is in progress. Erection of the thermal plant started at beginning 2008.

5.10.3.1 Construction Plan

The Hellisheiði power plant is being constructed similar to the Nesjavellir power plant. It is a cogeneration plant and will be comprised of modular units. The power plant capacity can expand as market demand increases, and can utilize greater knowledge of the geothermal capacity of the area that is being provided by drilling.

Commis-sioning	2006 MW_{el}	2007 MW_{el}	2008 MW_{el}	2009 MW_{th}	2910 MW_{el}	>2011 MW_{el}
Electricity						
High Pressure	1st. 90		3rd. 90		5th 90.	
Low Pressure		2nd. 33				
Thermal unit				4th 133		267

Table 5.13. Main Construction Stages for Hellisheiði Power Plant

The power production capacity of each electric unit will be 45 MW_{el} and 33 MW_{el} for the Low Pressure unit. For each thermal unit the capacity will be 133 MW_{th}. Table 5.13 shows the main construction stages for the Hellisheiði power plant and when each stage is scheduled to start operating.

5.10.3.2 Technical Description

The total development area of the Hellisheiði power plant is 820 ha. The development consists of geothermal utilization, access roads, service roads, production wells, the water supply system, steam transmission pipes, steam separator stations, power house, cooling towers, steam exhaust stacks, a fresh groundwater supply system, water tanks, hot-water transmission pipes, quarrying, discharge system, injection areas, and connection to the power grid.

5.10.3.3 Production Wells and Directional Drilling

Production wells are drilled both vertically and directionally, up to five wells per drilling site. With directional drilling it is possible to reach under valleys and in the direction of the mountain Hengill, without disrupting the valleys. Production wells can be up to 3.000 m and with directional drilling it is possible to drill 1,200 m from the center.

Production wells are grouped on drilling sites, up to five wells on predefined areas. The mean number of production wells per drill site is four with an area of about 12,000 m^2. The location of drill sites depends on geothermal and geophysics researches. Visual appearance of drill sites on the landscape has impact on where the drill sites are positioned. Minimum distance between production wells on a drill site is around 10m.

5.10.4 Hverahlíð and Bitra

Because of a growing demand for electricity in the industrial sectors, for example aluminum smelters, planning for two new power plants at the Hengill area has started, the Bitra power plant and the Hverahlíð power plant.

5.10.4.1 Environmental Policy

Construction of new geothermal power plants are subject to Environmental Impact Assessment (EIA) according to Article 5 and item 2 of Annex 1 of the Icelandic EIA Act No. 106/2000. Preliminary EIA proposals for the project at Bitra and Hverahlíð were presented in August 2006, and work on the EIA has been in progress since then. It was expected that the Planning Agency would issue their conclusion regarding the EIA early 2008.

5.10.4.2 Power Plant at Bitra

The development area is located about 8 km northeast of the Hellisheiði power plant. The development area of the plant was reduced from its original size on account of environmental reasons. Because of reduction of the development area and environmental policy, the effect of the Bitra power plant on its surroundings has been minimized [47].

Three research boreholes have been drilled in the area. The size of the power plant was estimated from information gathered from those boreholes and from results from a model of the geothermal area [48]. Estimated capacity of power plant at Bitra is 135 MW_{el}.

5.10.4.3 Power Plant at Hverahlíð

The development is located about 3 km southeast of the Hellisheiði power plant. The development area of the power plant was also reduced from its original size because of environmental reason. Like at Bitra, reduction of the development area and environmental policy will result in a minimal effect of the Hverahlíð plant on its surroundings [49].

Three research boreholes have been drilled in the area. The size of the power plant was estimated from information gathered from those boreholes and the results from a model of the geothermal area [48]. Estimated capacity of the power plant in that area is 90 MW_{el}.

5.10.5 Research Projects in the Hengill Area

5.10.5.1 New research Areas in the Hengill Area

Research boreholes have been drilled south of the Hellisheiði power plant. It was thought that these should be on the edge of the defined geothermal area. Results from the research drilling showed that the geothermal area extends further south. Because of this, it was decided to research the areas that are called Gráuhnúkar and Meitill. If research drilling gives positive results, it will be possible to extend the operation area of the Hellisheiði power plant to those sites or construct new smaller power plants in those areas. Ether way an EIA will be necessary.

5.10.5.2 Carb-Fix--Nature Imitated in Permanent CO_2 Storage Project

In fall 2007 a project was launched with the aim at storing CO_2 in Iceland's lavas by injecting greenhouse gases into basaltic bedrock where it literally turns to stone. CO_2 turning into calcite is a well-known natural process in volcanic areas. Now scientists of the University of Iceland, Columbia University N.Y. and the CNRS in Toulouse, France are developing methods to imitate and speed up this transformation of $CO_{2\ into}$ calcite that is a prevalent contributor to global warming.

Injecting CO_2 at carefully selected geological sites with large potential storage capacity can be a long lasting and environmentally benign storage solution. To date, CO_2 is stored as gas in association with major gas production facilities. The uniqueness of the Icelandic project is that whereas other projects store CO_2 mainly in gas form, where it could potentially leak back into the atmosphere, the current project seeks to store CO_2 by creating calcite in the subsurface. Calcite, a major component of limestone, is a common and stable mineral in the Earth that is known to persist for tens of millions of years.

In the project at the Hengill area, a mixture of water and steam is harnessed from 2000 m deep wells at the Hellisheidi power plant. The steam contains geothermal gases, i.e. CO_2. It is planned to dissolve the CO_2 from the plant in water at elevated pressure and then inject it through wells down to 400-800 m just outside the boundary of the geothermal system

It is estimated that the project will take three to five years. It was scheduled to start a full scale CO_2 injecting end 2008 or beginning 2009.

At least 600 MW_{el} and 433 MW_{th} are planned at power plants in the Hengill area by end of 2011.

5.11 A Perspective on the Future of Geothermal Energy in USA

Geothermal production began in USA at The Geysers field about 140 km north of San Francisco, California at what is now the world's largest geothermal field. USA continues to be the world leader in online capacity of geothermal energy and generation of electric power from geothermal energy. According to US Energy Information Agency, geothermal energy in 2005 generated approximately 16,010 GWh of electricity or about 0.36% of US annual electricity generation. Generation capacity is about 2850 MW.

Numerous exploration and development projects are underway which, if successful, would double this capacity. Beyond this growth there is still untapped potential for development of additional hydrothermal resources. The US Geological Survey (USGS) [50] estimated about 23,000 MWe capacity for 30 years of identified hydrothermal resources suitable for generation of electricity in USA and suggested that another 100,000 MWe of resources may be present but not yet identified. A more recent estimate prepared by a panel of experts hosted by the US. Department of Energy National Renewable Energy Laboratory [51] estimated that the identified accessible hydrothermal resource suitable for electrical generation is 30,000 MWe for 30 years with an additional 120,000 MWe unidentified. In addition, the US coastal region of Texas and Louisiana contains a significant amount of hot water nearly saturated with methane and with high wellhead pressures. A recent study by the Massachusetts Institute of Technology [52] reported that the thermal energy and energy in the methane may represent as much as 1,000 MWe capacity for 100 years.

5.11.1 Future Resources

Although these numbers are significant, they represent only a small fraction of the thermal energy underlying the USA. Current geothermal development is limited to geothermal systems driven by the convective flow of hot water associated with active volcanoes or with deep circulation of fluids. However, the majority of the earth's thermal energy is contained in areas where heat is transferred by conduction. It is this energy that is truly the future of geothermal energy in USA.

Researchers throughout the world have looked since the 1970s for ways to tap the conductive heat in the earth. The conceptual model, termed enhanced geothermal systems (EGS), is to drill wells and create or enhance subsurface fractures by use of reservoir stimulation practices pioneered by the petroleum industry. Such technology offers the promise of tapping the enormous amount of heat contained within the earth.

The US Department of Energy (USDOE) is working with private developers to investigate stimulation technology in poorly productive areas of commercial geothermal fields. They commissioned a study of the potential for enhanced geothermal systems in USA. The Massachusetts Institute of Technology lead team published their findings in December 2006 [52]. Report is available at http://geothermal.inl.gov or http://www1.eere.energy.gov/geothermal/future_geothermal.html .The study found that EGS represents a large, indigenous resource that could provide 100 GWe of electrical generation in the next 50 years with a reasonable investment in R&D. The report estimates that the EGS resource base is more than 13 million exajoules of which about 200,000 exajoules may be extractable. That represents 2,000 times the annual consumption of primary energy in the United States.

The USDOE is evaluating the findings of the report and comments from the geothermal and petroleum industries.

5.12 Acknowledgements

This Chapter has been prepared and coordinated by T. J. Hammons, Chair of International Practices for Energy Development and Power Generation IEEE; University of Glasgow, UK.

Contributors included Valgardur Stefansson (Orkustofnun, National Energy Authority of Iceland, Iceland); Arni Gunnarsson (Landsvirkjun, National Power Company, Iceland); Jay Nathwani (Department of Energy, Idaho Operations Office, USA); R. Gordon Bloomquist (Washington State University Cooperative Extension Energy Program, Washington, USA); Luciem Y. Bronicki and Daniel N. Schochet (ORMAT Technologies, Inc., USA); R. Gerald Nix (National Renewable Energy Laboratory, USA); Ed Hoover (Sandia National Laboratory, USA); Karl Urbank (Calpine Corporation, USA); Kenneth H. Williamson (UNOCAL Corporation, CA, USA); William E. Lewis (Power Engineers, USA), Björn Stefánsson and Bjarni Pálsson (Landsvirkjun, Iceland), Guðmundur Ómar Friðleifsson (Hitaveita Suournesja, Iceland); Ingólfur Hrólfsson and Sigurgeir Bjorn Geirsson (Orkuveita Reykjavikur, Iceland); and Allan Jelacic and Joel L. Renner (Idaho National Laboratory, US Department of Energy, USA). The Chapter is primarily based on an up-date of the invited panel session summary papers presented at the Panel Session on Harnessing Untapped Geothermal Power; and on Developments in Geothermal and Hydro Power in Iceland, Europe and Worldwide at the IEEE-PES 2008 General Meeting (GM2008) organized and chaired by T. J. Hammons and A. Gunnarsson [53].

5.13 References

[1] IEA 2001: Key World Energy Statistics from the IEA. 2001 Edition.

[2] Bruntland, Gro Harlem, Chairman of the World Commission on Environment and Development, 1987: *Our Common Future,* Oxford University Press, Oxford, 400 p.

[3] WEA 2000: World energy assessment: energy and the challenge of sustainability. Ed. by J. Goldemberg. United Nation Development Programme, United Nations Department of Economic and Social Affairs, World Energy Council, 2000, 508 pages.

[4] Huttrer, G.W., 2001: The status of world geothermal power generation 1995-2000. Geothermics, Vol. 30, no.1, pp. 1-27.

[5] Lund, J.W. and Freeston, D.H., 2001: Worldwide direct uses of geothermal energy 2000. Geothermics, Vol. 30, no.1, pp. 29-68.

[6] WEC 1998: Survey of Energy Resources 1998. 18th Edition, World Energy Council.

[7] P. Keary and F. J. Vine, *Global Tectonics,* 2nd Edition, Blackwell Science Ltd., London, 1996.

[8] L. J. P. Muffler, ed., "Assessment of Geothermal Resources of the United States—1978," U.S. Geological Survey Circular 790, 1979.

[9] D. E. White and D. L. Williams, "Assessment of Geothermal Resources of the United States—1975," U.S. Geological Survey Circular 726, 1975.

[10] M. J. Reed, "Assessment of Low-Temperature Geothermal Resources of the United Sates—1982," U.S. Geological Survey Circular 892, 1983.

[11] G. W. Huttrer, "The Status of World Geothermal Power Generation 1995-2000," *Proceedings of the World Geothermal Congress 2000,* Kyushu-Tohoku, Japan, International Geothermal Association, Auckland, N.Z., pp. 23-38, May 28-June 10, 2000.

[12] J. W. Lund and D. H. Freeston, "Worldwide Direct Uses of Geothermal Energy 2000," *Proceedings of the World Geothermal Congress 2000,* Kyushu-Tohoku, Japan, International Geothermal Association, Auckland, N.Z., pp. 1-22, May 28-June 10, 2000.

[13] K. Rafferty, K., "A Century of Service: The Boise Warm Springs Water District System," *Geothermal Resources Council Bulletin*, v. 21, no.10, pp.339–344, 1992.
[14] J. W. Lund and T. L. Boyd, "Geothermal Direct Use in the United States Update," *Proceedings of the World Geothermal Congress 2000*, Kyushu-Tohoku, Japan, International Geothermal Association, Auckland, N.Z., pp. 297–305, May 28–June 10, 2000.
[15] M. L'Ecuyer, C. Zoi, and J. S. Hoffman, "Space Conditioning: The Next Frontier," U.S. Environmental Protection Agency, EPA430-R-93-004, Washington, D.C., 1993.
[16] K. K. Bloomfield and J. N. Moore, "Production of Greenhouse Gases from Geothermal Power Plants," *Geothermal Resource Council Transactions*, V. 23, pp. 221–223, 1999.
[17] S. A. Bedell and C. A. Hammond, "Chelation Chemistry in Geothermal H_2S Abatement," *Geothermal Resources Council Bulletin*, v. 16, no. 8, pp. 3–6, 1987.
[18] J. G. Colligan, "U.S. Electric Utility Environmental Statistics, Electric Power Annual 1991," U.S. Department of Energy, Energy Information Administration, DOE/EIA-0348 (91), Washington, D.C., 1993.
[19] http://geothermal.marin.org/
[20] http://www.eren.doe.gov/geothermal
[21] http://www.geothermal.org
[22] http://www.geotherm.org
[23] http://iga.igg.cnr.it/index.php
[26] Mariusson, J.M. 2001, Iceland's Power Potential, Landsvirkjun.
[24] Lund, J.W., 2002. Direct Heat Utilization of Geothermal Resources. Geothermal Energy Resources for Developing Countries. D. Chandrasekharam & J. Bundschuh, 2002.
[25] Chandrasekharam and Bundschuh (eds), 2002. Direct Heat Utilization of Geothermal Resources, Geothermal Energy Resources for Developing Countries, Swets and Zeitlinger, Lisse, Netherlands, pp 129-147.
[26] Mariusson, J.M. 2001, Iceland's Power Potential, Landsvirkjun.
[27] Bloomquist, R. Gordon, 1988. District Heating Development Guide. Washington State Energy Office.
[28] Ragnarsson, A., 2000. Iceland Country Update. Proc. World Geothermal Congress, 2000, Japan.
[29] Glowka, D., The Role of R&D in Geothermal Drilling Cost Reduction, Geothermal Resources Council Transactions, Vol. 21, 1997, pp. 405-410.
[30] Bronicki, L.Y. *Geothermal Power Stations*, Encyclopedia of Physical Science and Technology, third edition, vol. 6, 2002.
[31] Bronicki, L. *Innovative Geothermal Power Plants, Fifteen Years of Experience*, World Geothermal Conf., Florence, 1995.
[32] Blaydes, Paula E. *Environmental Advantages of Binary Power Plants can Enhance Development Opportunities*, Trans. Geothermal Resources Council, vol. 18, October 1994.
[33] Flynn, Tom. *Geothermal Sustainability, Heat Utilization and Advanced Organic Rankine Cycle,* Trans.Geothermal Resources Council, vol. 26, Number 9, September/October 1997.
[34] Wilson, S.S. and Radwan, M.S. *Appropriate Thermodynamics for Heat Engine Analysis and Design*, Int. J. Mech. Eng. Educ., vol.5, 1977.
[35] Sanyal, Subir K. *Sustainability and Renewability of Geothermal Power Capacity,* Proc. World Geothermal Congress 2005, Antalya, Turkey, 24-29 April 2005.

[36] Krieger, Z and Moritz, A. *Cascaded Power Plant using Low and Medium Temperature Source Fluid*, U.S. Patent. 4 578 953, April 1, 1986.

[37] Tabor, H and Bronicki, L. Y. *Vapor Turbines*, U.S. Patent. 3 040 528, June 26, 1962.

[38] Krieger, Z and Kaplan, U. *Apparatus and Method for Producing Power using Geothermal Fluid*, U.S. Patent. 6 009 711, January 4, 2000.

[39] Steingrímsson, B., Guðmundsson, Á., Franzson, H. and Gunnlaugsson, E. *Evidence of a Supercritical Fluid at Depth in the Nesjavellir Field.* Proc. Fifteenth Workshop on Geothermal Reservoir Engineering, Stanford University, Stanford California, January 23-25 1990, SGP-TR-130, pp. 81-88.

[40] Einarsson, P. and Brandsdottor, B., *The Krafla Magma tic and Tectonic Episode of 1974-1989 at the Divergent Plate Boundary in North Iceland.* . Eos Trans.AGU 87 (52) Fall Meet. Suppl., 2006, Abstract T33E-07.

[41] Albertsson, A., Bjarnason, J.Ö., Gunnarsson, T., Ballzus C. and Ingason, K., *Fluid Handling and Evaluation, Part III,* 33 p. In: Iceland Deep Drilling Project, Feasibility Report, ed. G.O.Fridleifsson. *Orkustofnun Report OS-2003-007.*

[42] Albertson, A., Bjarnason, J.Ö, Gunnarsson, T., Ballzus, C. and Ingason, K., *The Iceland Deep Drilling Project: Fluid Handling, Evaluation, and Utilization.* Proceedings of the International Geothermal Conference IGC-2003 Reykjavik, September 2003, Session 6, pp. 23-30.

[43] Stern Review, *Stern Review on the Economics of Climate Change, Cambridge University Press,* October 2006. www.cambridge.org/9780521700801.

[44] Nakicenovic, N., Grübler, A., and McDonald, A. (editors), *Global Energy Perspectives.* Cambridge Univ. Press, 1998, 299 pp.

[45] Iðnaðarráðuneytið, *Innlendar Orkulindir til Vinnslu Raforku.* Iðnaðarráðuneytið,1994

[46] Orkuveita Reykjavíkur, *Nesjavellir Power Plant.* Orkuveita Reykjavíkur, 2003

[47] VSÓ, *Bitruvirkjun, allt að 135 MW_e jarðvarmavirkjun. Frummatsskýrsla.* VSÓ for Orkuveita Reykjavíkur, 2007.

[48] Grímur Björnsson, *Endurskoðað hugmyndalíkan af jarðhiakerfum í Hengli og einfalt mat á vinnslugetu nýrra borsvæða.* Orkuveita Reykjavíkur, 2007

[49] VSÓ, *Hverahlíðarvirkjun, allt að 90 MW_e jarðvarmavirkjun. Frummatsskýrsla.* VSÓ for Orkuveita Reykjavíkur, 2007.

[50] Muffler, L. J. P. ed., *Assessment of Geothermal Resources of the United States - 1978,* U. S. Geological Survey Circular 790, 163 p., 1979. Available: http://pubs.er.usgs.gov/usgspubs/cir/cir790

[51] Green, B. D. and Nix, R. G., *Geothermal - The Energy under our Feet,* National Renewable Energy Laboratory Technical Report NREL/TP-840-40665, Golden, Colorado, 16 p., 2006. Available: http://www1.eere.energy.gov/geothrmal/pdfs/40665.pdf .

[52] Tester, J. W., Anderson, B. J., Batchelor, A. S., Blackwell, D. B., Di Pippo, Ronald, Drake, E.L., Garnish, J., Livesay, B., Moore, M. C., Nichols, K., Petty, S., Toksöz, M. N., and. Veatch, R. W. Jr., *The Future of Geothermal Energy,* Idaho National Laboratory External Report INL/EXT-06-11746, Idaho Falls, ID, 396p., 2006. Available: http://geothermal.inl.gov or http://www1.eere.energy.gov/geothermal/future_geothermal.html.

[53] T. J. Hammons and A. Gunnarsson. *Geothermal Power Developments and Sustainability in Iceland and Worldwide,* International Journal of Power and Energy Systems, ACTA Press, Vol. 30, (2), 2010, pp. 94-107.

6

Reliability Modelling and Assessment of Power System Operation in the Competitive Electric Energy Market

This Chapter describes the main concepts and features of appropriate computational methodologies that have been developed for assessing the overall reliability and operational performance of electric power systems operating under the framework of the competitive electric energy market. These methodologies are based on the Monte - Carlo sequential simulation approach and are used for conducting the appropriate assessment studies on the composite generation and transmission power systems. A special attention is given to the impact of renewable energy sources while appropriate case studies are presented and examined thoroughly in order to describe more effectively the basic features of the developed methodologies.

6.1 Introduction

In recent years, electric power systems are adopting new technologies in their structure in order to achieve better performance and efficiency in the electricity production, transmission and distribution [1]. One of the most important aspects of the competitive electric energy market is the operation of independent power producers that can be connected at any system voltage level. This fact together with additional financial incentives being developed in many countries has increased considerably the number of power generating units using renewable energy sources [2].

Additionally, the role of transmission networks receives a great attention under the framework of this new environment being applied. Their role is to enable the competition by allowing an open access of all participants while keeping the system reliability and operational performance within appropriate standards [1,2].

An increased need for simulating the operation of power systems has emerged in order to take into consideration all the special operating features and procedures that are being incorporated as a result of this new environment. For this purpose, appropriate computational methods are being developed to be used for examining alternative operational schemes of power systems and investigating the impact of certain features (such as the penetration level of renewable energy sources) to their overall reliability and operational performance. These methods are mainly based on the Monte - Carlo sequential simulation approach and it was proved that they are very useful for comparing thoroughly different planning and operational schemes of power systems in order to deduce the optimal one.

6.2 Monte–Carlo Sequential Simulation Approach

The Monte - Carlo sequential simulation approach is a stochastic simulation procedure and can be used for calculating the operational and reliability indices of a power system by simulating its actual behavior [3-6]. The problem is treated as a series of real experiments conducted in simulated time steps of one hour. A series of system scenarios is obtained by hourly random drawings on the status of each system component and the determination of the hourly load demand. The operational and reliability indices are calculated for each hour with the process repeated for the remaining hours in the year (8760 hours). The annual reliability indices are calculated from the year's accumulation of data generated by the simulation process. The year continues to be simulated with new sets of random events until obtaining statistical convergence of the indices. The sequential simulation approach steps through time chronologically, by recognising that the status of a system component is not independent of its status in adjacent hours. Any event occurring within a particular time step is considered to occur at the end of the time step and the system state and statistical counters are updated accordingly. This approach can model any issues of concern that involve time correlations and can be used to calculate frequency and duration reliability indices. One very important advantage of the sequential simulation is the simplification of a particular system state simulation by considering information obtained from the analysis of the previous system states. This can only be applied when the system states change very little from one time step to the next. Such an assumption can be made for the transmission system that does not suffer large changes very often.

An efficient computational methodology has been developed at the National Technical University of Athens (NTUA) for the operational and reliability assessment of power systems applying the above principles of the Monte - Carlo sequential simulation approach. This methodology has the following main features [6]:

- The random events are determined by using appropriate pseudo-random numbers that are generated applying the mixed multiplicative congruent method. The antithetic sampling technique is also used for variance reduction.
- The classical two-state Markovian model is generally used to represent the operation of the system generating units. The generating units of certain thermal plants (for example, combined cycle plants) or the large thermal units may be represented by a multiple state model in order to recognise their derated states.
- The generating units may be taken out for scheduled maintenance during certain time periods of the year by using their appropriate data being specified.

The prime objective of the above computational methodology is to calculate appropriate indices that quantify the operational and reliability performance of a power system. It is generally considered that the following indices are the most important system and load-point indices while they have the corresponding units and acronyms in parentheses:

- Loss Of Load Expectation (LOLE) in hours/year.
- Loss Of Energy Expectation (LOEE) in MWh/year.
- Frequency of Loss Of Load (FLOL) in occurrences/year.
- Expected Demand Not Supplied (EDNS) in MW.
- Average Duration of Loss of Load (ADLL) in hours.

6.3 Reliability Modelling and Operational Performance of Isolated Power Systems with an Increased Penetration of Renewable Energy Sources

6.3.1 General

An important aspect of power generation systems operating under the framework of the competitive electric energy market is the increased use of renewable energy sources. However, their integration into the existing and future power systems represents an enormous technological challenge since there are serious limitations to the use of renewable energy sources due to the uncertainty of the weather conditions that constitute the main features of their operation. For this purpose, during the recent years, additional research effort has been devoted concerning the impact of the respective generating units on the operational performance of these systems. Wind has proven to be the most successful of all available renewable sources, since it offers relatively high capacities with generation costs that are becoming competitive with conventional energy sources. However, a major problem to its effective use as a power source is the fact that it is both intermittent and diffuse as wind speed is highly variable and site specific [7]. Additionally, an increased use of small hydroelectric plants that operate continuously throughout a year has also been identified.

The isolated power systems face specific problems that are related to their planning and operation when they are compared with the interconnected systems. In general, the customers of these systems face higher costs and poorer quality of supply than the customers of large interconnected systems. The main problems being identified concern the security and reliability of the systems while additional difficulties are expected due to high wind power penetration. Therefore, the introduction of high wind power penetration in the generation system of isolated power systems may reduce their operational reliability and dynamic security. A common aspect to all these problems is the requirement to ensure that sufficient reserve capacity exists within the system to compensate for sudden loss of generation. It is therefore evident that generation planning and operation is very critical on isolated systems.

6.3.2 General Features of Isolated Power System Operation

The reliability and operational assessment studies of isolated power systems require appropriate modeling of their features that affect their operation. These basic features are the following:

- The generation system mainly consists of thermal generating units of various types while hydroelectric power plants may exist that consist of appropriate generating units and reservoirs. Additionally, wind parks may be connected to the system busbars and they consist of appropriate units.
- The existing steam turbines and the internal combustion engines mainly supply the base-load demand. The combustion turbines have high production cost and they normally supply the daily peak load demand or the load demand that cannot be supplied by the other system units in outage conditions.
- The thermal generating units are called on to operate in order to supply the relevant load demand of the system according to a priority order that is determined by their production cost. Additionally, a level of spinning reserve capacity must be available

for use in emergency conditions that is usually determined by using an appropriate deterministic security criterion. It can be equal to either a certain percentage of the system load demand (e.g. 10%) or the capacity of the largest unit in operation or a constant value.

- A hydro chain may exist that consists of hydroelectric plants being located on the same river or water flow. The respective topographical sites or the construction facilities mainly determine the storage capability of reservoirs. Pump storage facilities may also exist.
- A variety of operating and water management policies can be implemented and they have a significant impact on the reliability and operational performance of the systems that have limitations in the energy being produced by the hydroelectric power plants.
- Several factors affecting the actual dispatch of generating units must be considered such as fuel costs, energy states of reservoirs, restrictions in use of water, energy used to pump water and irrigation requirements.
- Isolated power systems with a large wind penetration margin might face large voltage and frequency excursions and dynamically unstable situations when fast wind power changes and very high wind speeds result in sudden loss of wind power generation. This loss can be compensated with additional production from the system conventional generating units that are in a spinning reserve mode of operation. These system operational features determine the wind penetration margin that is expressed as a fraction of the wind power generation to the respective load demand. This penetration margin can be increased by considering the operation of hydroelectric plants that incorporate pumping facilities.

6.3.3 Computational Methodology

An efficient computational methodology has been developed at NTUA for the reliability and cost assessment of isolated power systems. This methodology integrates the operating aspects of wind parks and small hydroelectric power plants and evaluates their impact on the reliability and production cost of the system. These aspects are the following:

- The system conventional generating units are divided into two groups (fast, slow) according to their technical characteristics to change their power output. Additionally, these units are divided into two groups that can either supply the base-load demand or not supply it respectively.
- A number of wind parks can be installed at various geographical sites and each wind park consists of a certain number of groups of identical wind generating units. These wind parks are connected to the appropriate system busbars applying the existing connection rules of the system.
- The hourly wind speed of a geographical site is represented by an appropriate normal distribution which means that the values of the mean and standard deviation need to be given as input data for each hour of the year (8760 points). For simplicity reasons, the standard deviation may be assumed constant (e.g. 5%). The available power output of a wind-generating unit at any time point is calculated by using either a linear or nonlinear relationship between the power output and the wind speed of the respective geographical site.

- The classical two-state Markovian model simulates the operation of wind generating units.
- The total wind power generation of the system at any simulation time period of the year is not allowed to exceed a certain fraction of the respective system load demand. This fraction expresses the wind penetration constraint (margin) being assumed in order to retain acceptable service reliability, security and efficient operation of the conventional generating units. If the total wind power generation of the system is higher than the limiting value, it is necessary to reduce this generation level. For this purpose, the system control center will send appropriate orders to each wind park to reduce its total power output by a certain amount. This amount of power output is calculated so that the same percentage reduction will be applied for each wind park being in operation. As a result, it is assumed that a certain number of wind generating units in each wind park will be either disconnected from the system or decrease their power output by using appropriate procedures that take into account the technical characteristics of the respective units.
- A number of small hydroelectric power plants can be connected at appropriate system busbars and each plant consists of a certain number of identical generating units. It is considered that these units operate continuously throughout the year and their hourly power output is calculated by taking into account the respective monthly production and a typical curve for the hourly production of one day.
- The operational performance of wind and hydroelectric generating units is quantified by taking into account the events which occur when they fail to produce their available output capacity due to their existing limitations (failure, maintenance).
- The available spinning reserve of the system for each simulation time period is calculated by taking into account the operational features of system generation during the previous time period. For this purpose, two criteria are used. Criterion 1 assumes that the spinning reserve is equal to a certain percentage of the total wind power generation in order to compensate a sudden loss of this generation output in the cases of very fast wind speed changes. Criterion 2 assumes that the spinning reserve is equal to a certain percentage of total system loads in order to compensate for a sudden loss of the generation provided by the system conventional generating units (reliability criterion). The actual value for the spinning reserve is calculated as the greatest value being obtained by the two criteria. For this purpose, only the fast conventional generating units are taken into account.

An appropriate algorithm was developed and incorporated in the above methodology in order to simulate the dispatch procedures of system generating units for supplying the respective load demand in each simulation time period. This algorithm takes into account only the generating units of the system that are available (not being either in a repair or maintenance state) and has the following basic steps:

1. The small hydroelectric generating units are called on to operate and their power output is calculated according to their technical characteristics.
2. The conventional generating units, that are assigned to supply the base-load demand, are called on to operate at their minimum output capacity according to their priority order.

3. The power output of wind generating units is calculated by using the relevant wind speed data in each geographic site.
4. The wind penetration level is taken into account and, if it is necessary, appropriate reduction orders are applied to the power output of wind generating units.
5. The remaining load demand to be supplied is allocated to the conventional generating units. Firstly, the units supplying the base-load demand are called on to operate with the appropriate power output by using their priority order. If additional power generation is required, the other available conventional generating units are called on to operate by using their priority order.

It must be noted that the criteria for the system spinning reserve are also taken into account. These criteria determine the power output of the conventional generating units and the operation of additional units if it is required.
Using the developed computational methodology, the following additional system indices are calculated which have the corresponding units and acronyms in parentheses:

a) Four indices quantifying the system generation capability and production cost:
- Expected total energy supplied by conventional generating units (EGSM) in GWh/year.
- Expected energy supplied by wind generating units (EWSM) in GWh/year.
- Expected energy supplied by small hydroelectric power plants (EHSM) in GWh/year.
- Expected production cost of the generation system (PCSM) in Euro/MWh.

b) Four indices quantifying the operational performance of wind generating units by taking into account the events that may occur (failures, maintenance):
- Frequency of events being occurred (FNSWS) in occurrences/year.
- Expected annual duration of events being occurred (DNSWS) in hours/year.
- Expected load demand not supplied during the events being occurred (PNSWS) in MW.
- Expected energy not supplied during the events being occurred (ENSWS, ENSWM) in MWh/year.

c) Three indices quantifying the operational performance of the wind generation system by taking into account the events that occur when an order is issued for wind power reduction:
- Frequency of such events (AVPRF) in hours/year.
- Expected energy not supplied (AVPRE) in MWh/year.
- Expected wind power output not produced (AVPRL) in MW.

d) Four indices quantifying the operational performance of the small hydroelectric generating units by taking into account the events that may occur (failures, maintenance):
- Frequency of events being occurred (FNSHS) in occurrences/year.
- Expected annual duration of events being occurred (DNSHS) in hours/year.
- Expected load demand not supplied during the events being occurred (PNSHS) in MW.
- Expected energy not supplied during the events being occurred (ENSHS, ENSHM) in MWh/year.

e) Three indices quantifying the available spinning reserve by applying the respective criterion:
- Available spinning reserve (AVSPRES) as a percentage of the respective load demand.
- Percentages of applying Criteria 1 and 2 for evaluating spinning reserve (FWIND, FLOAD).

6.3.4 Assessment Studies

The developed computational methodology was applied for conducting reliability assessment studies on a typical isolated power system that is based on the power system of a Greek island. The main features of the system are the following:

- The installed generating capacity of the conventional units is equal to 522.6 MW. The system peak load demand is equal to 430 MW and occurs on a winter day while the overnight loads are approximately equal to 25% of the corresponding daily peak load demands.
- There are twenty conventional generating units of four different types that are installed in two power plants (I and II). These plants are located near to the major load points of the island for various reasons such as electrical design, environmental issues, etc. The complete characteristics of the generating units are shown in Table 6.1.
- There are six wind parks being installed at five different geographic sites of the island with 131 generating units having various power output capacities and their total generating capacity is equal to 66.85MW. The wind parks are usually installed in geographic sites with favorable wind conditions and their major characteristics are shown in Table 6.2.
- There are eight small hydroelectric power plants with eight generating units having various power output capacities and their total installed capacity is equal to 10.05 MW. The main characteristics of these units are shown in Table 6.3.
- The power generation system is only considered since it is assumed that the transmission network of the isolated system is fully reliable with an unlimited transmission capacity during any outages.
- The prices used for the calculation of the production cost for the generation system are only the supply prices of fuel.
- The available spinning reserve is calculated assuming that reserves are either equal to 100% of total wind power generation according to Criterion 1 or they are equal to 10% of the system load demand according to Criterion 2.
- The typical curve for the hourly production of hydroelectric plants for one day is a constant straight line.

This system provides a good example for illustrating the different operating features of isolated power systems. The full set of system indices was evaluated for the following eight alternative case studies:

Case 1: Base case study assuming a wind penetration margin of 20%.
Case 2: As in Case 1 but the wind penetration margin is decreased to 10%.

Case 3: As in Case 1 but the number of system wind generating units increases to 200 and their installed power output capacity increases by 35MW (52.3%). The respective data are shown in Table 6.4.

Case 4: As in Case 3 but the load demand increases by 8% (34.4 MW).

Case 5: As in Case 1 but the typical curve for the hourly production of hydroelectric plants for one day shows a greater power output during the daily peak load demand hours.

Case 6: As in Case 1 but the typical curve for the hourly production of hydroelectric plants for one day shows a greater power output during the daily base-load demand hours.

Case 7: As in Case 1 but the number of system small hydroelectric generating units increases to 16 and their installed power output capacity increases by 100% (10.05MW). The respective data are shown in Table 6.5.

Case 8: As in Case 1 but no small hydroelectric power plants exist.

Unit No.	**Plant No.**	**Type (*)**	**Load Pick-up**	**Priority Order**	**Output Capacity (MW)**		**Maintenance Duration (days)**
					Max	**Min**	
G1	II	T	Fast	12	13.0	0.0	28
G2	II	T	Fast	8	16.0	0.0	28
G3	II	T	Fast	10	19.0	0.0	28
G4	II	C	Fast	5	125.0	24.0	28
G5	I	S	Slow	1	23.5	14.0	35
G6	I	S	Slow	1	23.5	14.0	35
G7	I	S	Slow	1	23.5	13.0	35
G8	I	T	Fast	6	40.0	10.0	28
G9	I	D	Fast	4	11.5	0.0	42
G10	I	D	Fast	4	11.5	0.0	42
G11	I	D	Fast	4	11.5	0.0	42
G12	I	D	Fast	4	11.5	0.0	42
G13	I	T	Fast	9	15.0	0.0	28
G14	I	T	Fast	9	15.0	0.0	28
G15	I	T	Fast	11	15.0	0.0	28
G16	I	S	Slow	3	5.9	3.9	28
G17	I	S	Slow	2	14.1	7.0	28
G18	I	S	Slow	2	14.1	7.0	28
G19	II	T	Fast	7	57.0	0.0	28
G20	II	T	Fast	7	57.0	0.0	28
(*) T: Combustion Turbine. S: Steam Turbine C: Combined Cycle Engine D: Diesel Engine							

Table 6.1. Data for the Conventional Generating Units of the System

Wind Park No.	Group No.	Geographic Site	Number of Units	Output Capacity of Units (MW)	Installed Power Capacity (MW)
1	1	1	17	0.30	5.10
	2	1	3	0.50	1.50
2	3	3	17	0.60	10.20
3	4	3	17	0.60	10.20
4	5	2	50	0.50	25.00
5	6	4	18	0.55	9.90
6	7	5	9	0.55	4.95
Total	7		131		66.85

Table 6.2. Data of System Wind Generating Units

Power Plant No.	Numbe r of Units	Output Capacity of Units (MW)	Installed Power Capacity (MW)	Annual Energy Produced (GWh)
1	1	0.75	0.75	3.304
2	1	1.00	1.00	4.405
3	1	0.55	0.55	2.467
4	1	2.00	2.00	8.810
5	1	1.30	1.30	5.726
6	1	2.85	2.85	12.423
7	1	0.60	0.60	2.643
8	1	1.00	1.00	4.405
Total	8		10.05	44.183

Table 6.3. Data of System Small Hydroelectric Plants

Vind Park No.	Group No.	Geographic Site	Number of Units	Output Capacity of Units (MW)	Installed Power Capacity (MW)
1	1	1	27	0.30	8.10
	2	1	5	0.50	2.50
2	3	3	26	0.60	15.60
3	4	3	26	0.60	15.60
4	5	2	75	0.50	37.50
5	6	4	27	0.55	14.85
6	7	5	14	0.55	7.70
Total	7		200		101.85

Table 6.4. Data of System Wind Generating Units (Cases 3 and 4)

Power Plant No.	Number of Units	Output Capacity of Units (MW)	Installed Power Capacity (MW)	Annual Energy Produced (GWh)
1	2	0.75	1.50	6.608
2	2	1.00	2.00	8.810
3	2	0.55	1.10	4.934
4	2	2.00	4.00	17.620
5	2	1.30	2.60	11.452
6	2	2.85	5.70	24.846
7	2	0.60	1.20	5.286
8	2	1.00	2.00	8.810
Total	16		20.10	88.366

Table 6.5. Data of System Small Hydroelectric Plants (Case 7)

The results being obtained for the above eight Case studies are presented in Table 6.6. A considerable number of comments can be drawn from these results but the most important ones are the following:

- The decrease of wind penetration margin decreases the wind generation system indices and also decreases the system reliability performance. Comparing the results of the respective indices for Cases 1 and 2 demonstrates this.
- The increase of wind penetration margin and the addition of wind generating units always improve the system reliability indices as there is more available power to supply the load demand. Furthermore, the energy supplied by wind generating units is increased while the energy supplied by the conventional units of the system is decreased. However, the system production cost increases as indicated by the respective results for Cases 1, 3, 7 and 8.
- The addition of small hydroelectric units improves the system reliability indices, as there is more available power to supply the load demand, and increases the system hydroelectric generation indices as it can be seen by comparing the results of Cases 1 and 7. Furthermore, the total energy being supplied by both the wind generating units and conventional units decreases.
- When the power output of hydroelectric generating units is greater during the time periods of the daily peak load demands (Case 5), the LOEE index is smaller compared with that for Case 6, where their power output is greater during base load demand hours.
- The percentages of criteria 1 and 2 for the calculation of the available spinning reserve of the system (FWIND) vary according to the power output capacity of wind generating units and the wind penetration margin. When an increased level of system wind penetration margin is assumed (Case 1) or the installed power output capacity of system wind generating units is assumed to increase (Case 3), criterion 1 mainly determines the available spinning reserve. However, when the wind penetration margin is decreased (Case 2), criterion 2 mainly contributes to the evaluation of spinning reserve.

Case Study Index	1	2	3	4	5	6	7	8
LOLE	5.105	6.000	2.945	6.87	5.160	5.115	4.355	6.120
LOEE	142.24	164.05	72.738	196.48	141.32	143.57	116.79	172.94
EDNS	10.251	10.490	6.115	13.771	10.191	10.328	8.902	11.584
FLOL	2.200	2.365	1.430	2.955	2.195	2.205	1.920	2.480
EGSM	1766.04	1863.72	1683.40	1829.42	1765.57	1766.55	1726.44	1806.28
EWSM	300.535	202.834	383.247	401.918	301.007	300.022	296.909	303.423
EHSM	43.152	43.152	43.152	43.152	43.154	43.154	86.403	-
PCSM	101.73	99.265	103.43	100.17	101.71	101.73	102.72	100.77
FNSWS	1.975	1.975	1.215	1.215	1.975	1.975	2.635	1.975
DNSWS	8731.55	8731.55	8749.09	8749.09	8731.55	8731.55	8731.55	8731.55
PNSWS	1.5056	1.0183	2.0269	2.1252	1.5080	1.5030	1.5087	1.5206
ENSWS	15.84	10.773	20.353	21.325	15.866	15.815	15.657	15.988
ENSWM	9.795	7.832	13.574	13.939	9.805	9.785	9.736	9.839
AVPRF	2503.0	7965.9	6077.2	5328.06	2501.4	2504.0	2672.2	2339.5
AVPRE	24.267	121.967	113.291	94.620	23.795	24.780	27.893	21.380
AVPRL	9.6951	15.3111	18.6419	17.7588	9.5128	9.8960	10.4382	9.1385
AVSPRES	14.73	10.01	18.55	18.03	14.75	14.70	14.85	14.58
FWIND	90.9	46.9	99.34	98.87	90.9	90.88	91.4	90.2
FLOAD	9.1	53.1	0.66	1.13	9.1	9.12	8.6	9.8
FNSHS	31.365	31.365	31.365	31.365	31.365	31.365	58.230	-
DNSHS	663.65	663.65	663.65	663.65	663.65	663.65	1287.17	-
PNSHS	0.648	0.648	0.648	0.648	0.648	0.647	0.669	-
ENSHS	0.436	0.436	0.436	0.436	0.436	0.436	0.885	-
ENSHM	1.022	1.022	1.022	1.022	1.022	1.022	1.874	-

Table 6.6. System Reliability and Operational Indices

6.4 Reliability and Cost Assessment of Power Transmission Networks in the Competitive Electric Energy Market

6.4.1 General

Power transmission systems are increasingly involved in the competitive electric energy market and their role becomes more important since it must enable the open access of several players even though in a regulated way. For this purpose, certain strategies must be developed to insure that the system will be able to optionally utilize its total facilities since it will experience an increase in the number of energy sales and purchases. This increase is due to a number of reasons and includes a growing number of wheeling transactions between investor-owned utilities, cogeneration facilities and industrial consumers [8]. Present day practice on wheeling is that consumers and suppliers make direct commercial contracts across wholly third party owned transmission networks. In this way, pricing and investments of transmission systems are the most important problems [9, 10] together with the rules allowing their shared use. One of the most important areas that need an increased attention is the reliability assessment of transmission systems by taking into account both the respective generation and transmission facilities. The impact of transmission services can be quite significant on both the load-point and the system reliability performance. The reliability evaluation techniques at Hierarchical Level Two (HLII) [3] can be utilized to assess the impact of wheeling transactions on the system adequacy. Such a study has been published using a simplified analytical technique [11]. However, a more detailed system representation is always necessary considering the particular operating features of transmission services so that their impact on the reliability performance of an electric power system can be explored [12]. Pricing of transmission services plays a crucial rule in determining whether providing transmission services is economically beneficial to both the wheeling utility and customers. Some methods have been reported in the literature to allocate transmission fixed costs [9,10,13,14] and can be classified as embedded cost, incremental cost and composite embedded/ incremental cost methods. The cost to deliver electric energy varies as a function of time, system status and location within a power system.

6.4.2 Main Features of the Reliability and Cost Allocation Methods

In the competitive electric energy market, several players have access to the transmission networks even though in a regulated way. This means that the commercial and the energy/power wheeling transactions through the transmission networks become more and more frequent tasks, superimposing the original network task aimed at feeding the demand. The task of quickly and accurately evaluating the merits of wheeling transactions is becoming an important function of utility system planners and operators. This is because it affects vital system attributes (security, reliability, losses, etc.), puts a strain on the existing transmission network and may restrict the economic generation dispatching. Furthermore, the reliability performance of the transmission network together with the reliability indices of service supplied to system customers constitute one of the major aspects that are taken into account in the system planning and operating phases. It is therefore evident that there is a need for planning and operating considerations that must be followed by all the various transmission system customers in order to assure that all of them will be able to operate reliably and safely together.

A transmission transaction refers to the transmission component of the service being provided and it is associated with a power sale, a power purchase or a wheeling transaction. The following two categories of transmission transactions are mainly used which affect differently the reliability assessment of the transmission network being used:

- **Firm transmission transactions**: These are not subject to discretionary interruptions and they entail reservation of capacity on transmission facilities to meet transaction needs. They are the result of contractual agreements between the transmission system operator (or owner) and the wheeling customers.
- **Non - Firm transmission transactions**: They may be subject to load curtailment at the appropriate direction of the responsible body.

A wheeling customer can be connected anywhere in the transmission system which must transmit the power injected to the input busbars of wheeling transactions to appropriate output busbars. However, in outage conditions, the system may not be able to transmit the full amount of power required by each transaction due to limitations in the transmission circuits after the outage has occurred. The generation/load patterns of each wheeling customer have usually two elements of equal magnitude, the power injection and the contracted load. These patterns are always balanced which means that the total injection equals total load. The operating characteristics of each wheeling transaction can be modeled by assuming that its operation will follow a daily capacity curve, which may not exist in certain days in each week or certain weeks in the year. This curve indicates the capacity agreed to be transmitted by the utility and is determined by the respective number of capacity levels given for the twenty-four time intervals being considered. The transmission network must ensure open access in generation services and, in order to achieve this, it should provide adequate service charges, which remunerate investments and induce the best overall use of the system resources. Transmission of electricity must be offered and priced separately from the power itself and delivery of power must stand on its own as a business.

Several criteria have been used to justify transmission charge schemes such as simplicity, fairness of cost allocations, economic theory and others [13], [15]. The accurate knowledge of the actual costs of providing transmission transactions is vital to transmission network operators for a number of reasons such as the economic aspects of operating decisions and the business decisions with regard to operations, investment and commitments to customers [15].

The cost of power delivered at a transmission network busbar varies from place to place within the network and can be calculated using the following 'unbundled' services:

Total cost = energy cost + transmission cost + losses cost + ancillary services cost + dispatching cost + congestion cost
where:

- The energy cost is the system production cost (fixed and variable) that varies according to the time period of the day being concerned and the respective generating units in operation.
- The transmission cost includes the capital recovery cost of the system equipment and the maintenance cost.
- The losses cost is the cost of producing additional generation for supplying the transmission network losses that vary with loading, flow pattern, location and distance. All network users should pay this cost.
- The ancillary services cost is the cost for providing all services that are required for the safe and reliable operation of the system which include the generation reserve, the load frequency control, the voltage control, the system restoration, etc. The most important element of this cost is the generation reserve cost which has two parts covering the variable cost of the spinning generating units and the fixed cost of all the system stand-by generating units. This cost is allocated to all system busbars in proportion of their load demand.
- The dispatching cost includes the charges for the transmission system operator's services of scheduling, controlling and dispatching the transmission system.
- The congestion cost is associated with the transmission network equipment and security constraints and concerns the system operator actions to redispatch the generation in order to accommodate the additional transmission transactions. This cost changes the marginal cost of power at some system busbars and it is allocated to each new transaction.

The mesh structure of transmission networks provides a large number of possible routes by which electrical power can flow from the sources to the loads. Until recently, the question of tracing electricity was of a limited interest. However, the open access requirements have posed new aspects of system operation and the problem of tracing electricity gains additional importance. Computational techniques have been published [16, 17] that allow the tracing of electricity flow in meshed networks. Another more efficient technique was also developed at NTUA [18,19]. All these three techniques produce appropriate tables that show how much of the power output from a particular generating unit goes to a particular load and the contributions of individual generating units or load to individual line flows. Apart from giving additional insight into how power flows in the network, these techniques can be used as a tool for determining the charges for transmission losses and the actual usage of the system by a particular generating unit or load. Using these techniques for each generating unit bus, the set of buses is deduced that are reached by power produced by this generating unit. Power from a generating unit reaches a particular bus if it is possible to find a path through the network from the generating unit to the bus for which the direction of travel is always consistent with the direction of the flow as computed by a power flow computer program. The same techniques can be used to contribute the load of each busbar and the system losses to individual generating units in the network. Therefore, the geographically differentiated price charged to consumers could be computed on the basis of the relative contribution of each generating unit to their load and the price of each of these generating units.

The three tracing techniques are topological in nature that means that they deal with a general transportation problem of how the flows are distributed in a meshed network. The basic assumption used by tracing algorithms is the proportional sharing principle. A proportional sharing principle is assumed which states that each network node is a perfect 'mixer' of incoming flows so that the power flow leaving a node contains the same proportion of the inflows as the total nodal inflow. This is true since it is impossible to identify which particular inflowing electron goes into which particular outgoing branch and seems to agree with the generally accepted view that electricity is indistinguishable. Figure 6.1 presents an example for implementing the above-described principle. There are four branches that are connected to node i (nodes j, k, l, m are the other nodes). The total power flowing through node i is 100 MW while 40 MW and 60 MW are supplied from branches j-i and k-i respectively. According to the assumption being made, 70 MW flow through branch i-m and include 28 MW being supplied by branch j-i (70x40/100=28) and 42 MW being supplied by branch k-i (70x60/100=42) while 30 MW flow through branch i-l and include 12 MW being supplied by branch j-i (30x40/100=12) and 18 MW being supplied by branch k-i (30x60/100=18).

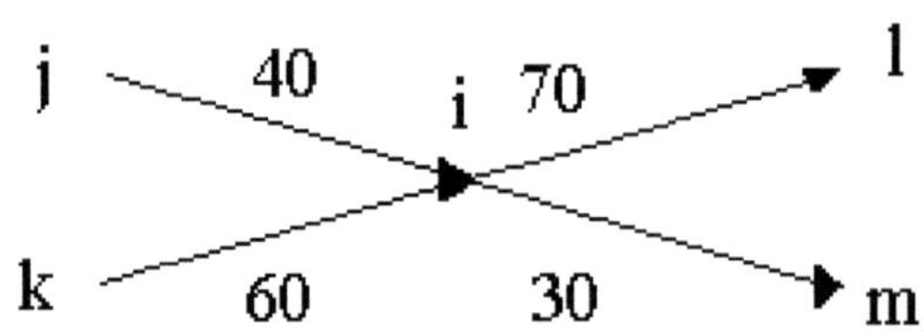

Fig. 6.1. Node with 100 MW power flow (power values in MW)

Technique No. 1 [16] calculates a table A from the equation AuP = PG where P is the vector of nodal through-flows, PG is the vector of nodal generations and the (i, j) element of the sparse and nonsymmetrical matrix A is given as:

$$\left[A_u\right]_{ij} = \begin{cases} 1 & for\ i=j \\ -c_{ji}=-\left|P_{j-i}\right|/P_j & for\ j\in d_i^{(u)} \\ 0 & otherwise \end{cases} \quad (1)$$

assuming that $d_i(u)$ is the set of nodes supplying directly node i. If A^{-1} exists, the ith element of vector is

$$P_i = \sum_{k=1}^{n} \left[A_u^{-1}\right]_{ik} P_{G_k} \qquad for\ i=1,2,\ldots,n \quad (2)$$

where n is the number of system nodes. A line outflow in line j from node i can be calculated using the proportional sharing principle, as:

$$\left|P_j\right| = \frac{\left|P_j\right|}{P_i} P_i = \frac{\left|P_j\right|}{P_i} \sum_{k=1}^{n} \left[A_u^{-1}\right]_{ik} P_{G_K} = \sum_{k=1}^{n} b_{kj} P_{G_k} \quad (3)$$

In this equation, the coefficient b_{kj} shows the respective contributions of each generating unit k to the power flow of each line j. The load demand P_{Li} can be calculated from P_i as:

$$P_{L_i} = \frac{P_{L_i}}{P_i} P_i = \frac{P_{L_i}}{P_i} \sum_{k=1}^{n} \left[A_u^{-1}\right]_{ik} P_{G_K} = \sum_{k=1}^{n} a_{ki}\ P_{Gk} \quad (4)$$

In this equation, the coefficient a_{ki} shows the respective contributions of each generating unit k to the i load demand and can be used to trace where the power of a particular load comes.

Technique No. 2 [17] is based on a set of definitions which are domains (set of buses getting power from a particular generating unit), commons (set of contiguous buses getting power from the same set of generating units), links (lines connecting commons), inflow (the sum of the generation by sources connected to busses located in this common and of the power imported in this common from others commons by links) and outflow (the sum of the loads connected to the busses located in this common and of the power exported through links from this commons to others commons). The basic assumption used in this method is the proportional sharing principle which assumes that for a given common, the proportion of the inflow which can be traced to generating unit i is equal to proportion of every bus load and to every line flow within this common and equal to the proportion of the outflow of this common. If C_{km} is the contribution of generating unit k to the common m, C_{kn} is the contribution of generating unit k to the common n, F_{mn} is the flow of the link between commons m and n, F_{kmn} is the flow on the link between commons m and n due to generating unit k and I_n is the inflow of common n, the following equations can be derived:

$$F_{kmn} = C_{km} * F_{mn} \tag{5}$$

$$I_n = \sum_m F_{mn} \tag{6}$$

$$C_{kn} = \frac{\sum_m F_{kmn}}{I_n} \tag{7}$$

Knowing the common to which a bus belongs and the contributions of each generating unit to each common, the power that each generating unit contributes to each load can be computed. It also makes it possible to compute what proportion of the use of each line can be apportioned to each generating unit. For lines linking busses in separate commons, the proportion of usage should be based on the contribution of the generating unit to the common, which is the starting end of this line. Therefore, the coefficient $a_{ki}=C_{kn}$ shows the respective contributions of each generating unit k to the i load demand which is located within common n. The coefficient $b_{kj}=C_{kn}$ shows the respective contributions of each generating unit k to the power flow of each line j which is located within common n. For line j linking commons, the coefficient $b_{kj}=C_{kn}$ shows the respective contributions of each generating unit k to the power flow of line j, which the sending end is the n common.

The developed technique No. 3 [18,19] is based on the computation of two characteristic tables A and B which express the respective contributions of each generating unit k to the load of each bus i and power flow of each line j. The coefficients of these tables (a_{ki} and b_{kj}) are computed as follows:

$$d_{ki} = \sum_j d_{km} \frac{P_j}{P_{om}} \tag{8}$$

where P_j is the power flow in line j supplying bus k, P_{om} is the total nodal flow of node m that is the sending end of line j

$$a_{ki} = d_{ki} \frac{L_i}{P_{oi}} \tag{9}$$

where L_i and P_{oi} is the load and the total inflow of bus i

and

$$b_{kj} = d_{ks} \frac{P_J}{P_{os}} \tag{10}$$

where P_j is the power flow of line j and P_{os} is the total inflow of bus s, which is the sending end of the line j. This iterative procedure starts from the corresponding generating unit busbar where the coefficient d_{kk} becomes unity and continues the computation of coefficients a_{ki} and b_{kj} following the direction of electricity flow through the network. With the assumption that generating units contribute to the losses in a line in proportion of their use of this line, this method computes the proportion of the generating unit output that is dissipated. The proportion of the total system load, the total transmission use and the total system losses contributed to each generating unit can also be computed using the coefficients a_{ki} and b_{kj}.

6.4.3 Computational Methodology

An improved computational method was developed at NTUA applying the features of the Monte - Carlo sequential approach [18,19]. Its objective is to integrate the operating aspects of wheeling transactions into the reliability and cost assessment of composite generation and transmission power systems. An important aspect of this method is the incorporation of independent computational techniques that incorporate the three alternative techniques mentioned before for tracing the power flows through the meshed transmission networks. This approach has the following main steps:

- For each generating unit bus, the set of buses is deduced which are reached by power produced by this generating unit. Power from a generating unit reaches a particular bus if it is possible to find a path through the network from the generating unit to the bus for which the direction of travel is always consistent with the direction of the flow as computed by a power flow computer program.
- The amount of power that each generating unit contributes to each load is computed.
- The proportion of the use of each branch that can be apportioned to each generating unit is computed.
- Since it is reasonable to assume that generating units contribute to the losses in a branch, the proportion of each generating unit output that can be dissipated in system losses is computed.
- The ancillary services cost is computed considering only the generation reserve cost. The ancillary services cost and the dispatching cost are distributed to each system bus, proportionately to its load demand.
- The total energy cost of each generating unit is allocated to all loads that are being supplied by it.
- The transmission network limitations (overloaded branches) are deduced applying a suitable DC load flow algorithm. This is mainly for limiting the computational time being required for the significant number of such calculations. The generating units that are "responsible" for these limitations are identified and appropriate operating procedures are simulated for alleviating the overloads in the most economical way. The first procedure to be considered is the generation rescheduling and, if it is still necessary, the procedure for load curtailment is applied at appropriate busbars. When all these limitations have been removed, the total system cost is calculated by applying the above described steps. The additional cost provides the congestion cost which is allocated appropriately to all "responsible" generating units while the cost of all the other generating units is also modified.
- A proportional sharing principle is assumed which states that each network node is a perfect 'mixer' of incoming flows so that the power flow leaving a node contains the same proportion of the inflows as the total nodal inflow. This is true since it is impossible to identify which particular inflowing electron goes into which particular outgoing branch and seems to agree with the generally accepted view that electricity is indistinguishable.

Furthermore, the basic features of the developed methodology are the following:

1. Each wheeling transaction is modeled as an equivalent generating unit connected to the input busbar of the transaction. This unit is assumed to be 100% reliable and its output is variable since, in each hour of the simulation period, its output capacity level is determined by the capacity curve being agreed. Furthermore, an additional load is assumed to be connected at the output busbar of the transaction and is equal to the capacity of the equivalent generating unit in each hour of the simulation period.
2. The existing generation rescheduling technique has been extended to include the particular operating characteristics of the wheeling transactions. The equivalent generating units of the wheeling transactions are assigned the lowest priority orders in the respective list used for generation redispatch. Furthermore, a certain amount of spinning reserve is always assigned applying the reliability criterion of satisfying the loss of the largest generating unit in operation.
3. The existing branch overloading technique has been extended so that it can take into account the wheeling transactions. In outage conditions, the generation/load patterns of the non-firm wheeling transactions may be decreased up to the firm capacity percentage being agreed in order to alleviate the possible overloads of the system branches.
4. In each hour of the simulation period, the network branch flows are calculated in two steps applying the corresponding computational technique (transmission system, wheeling transaction). The overall network branch flows are calculated appropriately. Additionally, the relative contributions of each generating unit to each system load and system branch are calculated together with the allocation of transmission losses to each generating unit. The technique being developed for tracing electricity is used for this purpose.
5. The operational performance of wheeling transactions is quantified by taking into account two types of events that cause a system failure to transmit the capacity agreed for each wheeling transaction due to the respective system limitations (split system, overloaded branches).

The prime objective of the developed computational methodology is to calculate appropriate indices that quantify the operational and reliability performance of a power transmission system accommodating wheeling transactions together with the respective indices for each transaction under consideration. These indices refer to system adequacy and can be described as follows while they have the corresponding units and acronyms given in parentheses:

- Four sets of reliability indices: The first set forms load-point and system indices which reflect their respective adequacy. The second set forms load-point and system interruption indices which reflect the characteristics of the interruptions occurred while the third set of indices is calculated for each hydroelectric plant. The system health indices are included in the fourth set.
- Four indices quantifying the transmission system capability and the operational performance of wheeling transactions:

- Expected annual energy of wheeling transactions being performed (EETRS) in GWh/year.
- Expected frequency of wheeling transactions not being performed (FNTRS) in occurrences/year.
- Expected annual duration of wheeling transactions not being performed (DNTRS) in hours/year.
- Expected annual not transmitted energy of wheeling transactions (ENTRS) in GWh/year.

This set of indices is calculated for each wheeling customer separately and for all categories of events considered in the above paragraph that cause a wheeling transaction not to be performed. Furthermore, additional information can be obtained concerning the transmission branches that were overloaded due to the wheeling transactions as well as the expected frequency of these events given in occurrences/year.

- Additional information are obtained concerning the system production cost (PC in k€/year) and the congestion cost (CNG in k€/year) of the generating units that were required to be redispatched or started-up.
- Three indices concerning the application of reliability criterion of the level of spinning reserve of the system generating units and the respective cost being required:
  - Expected frequency of spinning reserve criterion not being satisfied (FLNSP) in occurrences/year.
  - Expected annual duration of spinning reserve criterion not being satisfied (EDNSP) in hours/year.
  - Expected annual cost of spinning reserve criterion being satisfied (CSPIN) in k€/year.
- Two characteristic tables indicating the relative contributions (in percentages) of each system generating unit to the load of each busbar and the relative contributions (in percentages) of each busbar load to the power output of each generating unit. Using these two tables, three overall indices are calculated expressing the contributions of each system generating unit (in percentages) to the total system load (CLOAD), system losses (CLOSS) and transmission cost (CTRANS). These three indices are also calculated for each wheeling transaction separately.
- Two additional characteristic tables A and B are calculated which indicate the relative contributions (in percentages) of each system generating unit to the load of each busbar and the relative contributions (in percentages) of each busbar load to the generation of each generating unit. A more effective computational procedure is used applying equations 1-3 which calculates three additional overall indices expressing the contributions of each system generating unit (in percentages) to the total system load (CLOAD), system losses (CLOSS) and transmission cost (CTRANS). These three indices are also calculated for each wheeling transaction separately.
- An additional index concerning the expected annual cost of power delivered to each load point together with the five parts in to which it is divided. Additionally, an hourly curve can be obtained for an average day of the year showing the cost variation of the power delivered to each system load-point.

6.4.4 Assessment Studies

The developed computational method was applied for conducting reliability assessment studies on the IEEE-RTS [20] that was modified slightly in order to consider wheeling transactions and evaluate their impact on the system reliability and cost indices. The IEEE-RTS was assumed to be a wheeling transmission system and one non-firm wheeling transaction was considered having a firm capacity of 70% of the agreed capacity (450 MW) that remains constant throughout a year. This transaction assumes an injection at bus 10 and removal at bus 20. Tables 6.7 and 6.8 present the generating unit contributions to system loads, lines and transmission use for the following six case studies:

Case 1: Tacing technique No. 1 is used, no wheeling transactions.
Case 2: Tracing technique No. 2 is used, no wheeling transactions.
Case 3: Tracing technique No. 3 is used, no wheeling transactions.
Case 4: As in Case 1, one wheeling transaction exists.
Case 5: As in Case 2, one wheeling transaction exists.
Case 6: As in Case 3, one wheeling transaction exists.

Tables 6.7 and 6.8 show that the results of Case 3 are identical with those obtained for Case 1 and very close to the respective ones of Case 2. However, the application of the tracing techniques No. 1 and No. 2 (equations 1 - 7) require a considerable amount of computational time when they are applied in large practical systems because they need the inversion of a very large in size matrix. This problem has been overcome by applying the developed tracing technique No. 3 and the time being required is decreased significantly. This allows the efficient incorporation of this procedure into the computational method for the reliability and cost assessment of composite power systems (Case 3).

The full set of system indices described in the previous section was evaluated for the above six case studies and the obtained system results are presented in Table 6.9 while the reliability indices of certain load-points are presented in Table 6.10. These results clearly indicate that the wheeling transactions agreed and their specific design and operating features employed significantly affect the reliability and cost indices of the utility system. Each of these factors has a different effect on the system transmission capability. A considerable number of comments and conclusions can be drawn from these results but the most important ones are the following:

- The level of firm load to be satisfied in each wheeling transaction as well as the profile of its capacity curve affect the reliability indices but their impact is not significant. However, the indices of the wheeling transactions are affected in a greater extent.
- An important system characteristic is the selection of system busbars that will become the input and output busbars of the wheeling transactions agreed.
- The wheeling customer indices are also affected by the system topology and characteristics.
- The differences obtained in the index of the system congestion cost (CNG) may be positive or negative according to the characteristics of each wheeling transaction but

their actual numerical values are not significant enough. However, this situation may change in other system configurations.

- The reliability indices of transmission system load-points may be affected by the wheeling transaction being applied as is noticed by the results in Table 6.10 but the actual impact differs.
- The contributions of generating units to busbar loads and cost indices depend on their size and, additionally, on the priority level of dispatching them (Table 6.7).
- The cost of power delivered at system busbars depends mainly on the production cost of the generating units located near to the busbar (Table 6.9).

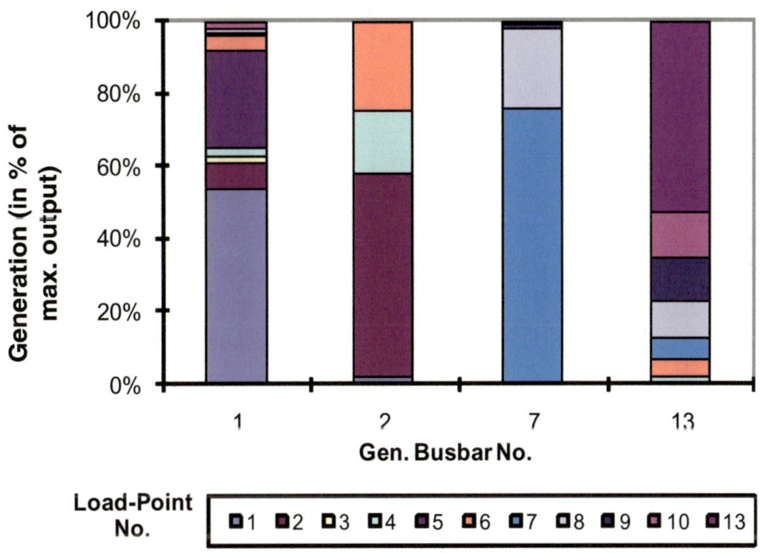

Fig. 6.2. Relative contributions of busbar loads to each system generating unit output (Case 3)

Bus	Gen.	Case 1		Case 2		Case 3	
No	MW	Load4	Line6	Load4	Line6	Load4	Line6
1	192	2.8	0.5	2.8	0.6	2.8	0.5
2	192	20.0	0.0	19.8	0.4	20.0	0.0
7	300	0.0	1.4	0.0	1.4	0.0	1.4
13	591	2.3	5.2	2.9	6.5	2.3	5.2
15	215	2.6	2.6	2.4	3.0	2.6	2.6
16	155	2.7	4.7	2.9	4.4	2.7	4.7
18	400	9.9	12.0	12.1	13.4	9.9	12.0
21	400	12.5	11.2	11.6	12.8	12.5	11.2
22	300	13.1	14.9	11.5	13.2	13.1	14.9
23	660	34.1	47.5	34.0	44.3	34.1	47.5
		100.0	100.0	100.0	100.00	100.0	100.00

Table 6.7. Generating Unit Contributions to Load L4 and Line 6 (in %)

A greater number of detailed sensitivity studies can be easily conducted for establishing the impact of each system parameter and obtaining a more realistic reliability assessment. Additionally, Figures 6.2 and 6.3 show the relative contributions of busbar loads to each system generating unit output and the relative contributions of each generating unit output to system load, transmission use and losses for Case 3. It can be noticed that these generating unit output contributions are larger in the load of busbars and loading of branches that are located near to the generating units busbars.

Bus	Gen.	Case 1		Case 2		Case 3	
No	MW	LOAD	TRAN.	LOAD	TRAN.	LOAD	TRAN.
1	192	5.19	2.59	5.19	2.58	5.19	2.59
2	192	4.54	3.11	4.54	3.03	4.54	3.11
7	300	4.29	0.05	4.29	0.05	4.29	0.05
13	591	8.92	1.96	8.92	1.93	8.92	1.96
15	215	5.84	3.10	5.84	3.22	5.84	3.10
16	155	5.81	2.29	5.81	2.29	5.81	2.29
18	400	16.46	9.55	16.46	10.66	16.46	9.55
21	400	16.46	16.68	16.46	17.62	16.46	16.68
22	300	11.70	26.55	11.70	25.98	11.70	26.55
23	660	20.79	34.12	20.79	32.69	20.79	34.12
10	450	0	0	0	0	0	0
		100.0	100.0	100.0	100.0	100.0	100.0
Bus	Gen.	Case 4		Case 5		Case 6	
No	MW	LOAD	TRAN.	LOAD	TRAN.	LOAD	TRAN.
1	192	4.38	3.42	4.38	3.39	4.38	3.42
2	192	3.83	3.40	3.84	3.30	3.83	3.40
7	300	3.62	0.05	3.63	0.05	3.62	0.05
13	591	7.52	2.43	7.53	2.29	7.52	2.43
15	215	4.93	2.65	4.93	2.65	4.93	2.65
16	155	4.90	2.08	4.93	2.07	4.90	2.08
18	400	13.89	7.70	13.88	8.32	13.89	7.70
21	400	13.88	15.42	13.87	15.35	13.88	15.42
22	300	9.87	24.77	9.86	23.94	9.87	24.77
23	660	17.54	18.46	17.53	18.11	17.54	18.46
10	450	15.64	19.62	15.62	20.53	15.64	19.62
		100.0	100.0	100.0	100.00	100.0	100.00

Table 6.8. Generating Unit Contributions to System Loads and Transmission Use (in %)

System and Wheeling Transactions

Case Study Index	1	2	3	4	5	6
Utility system						
LOLE	19.221	19.221	19.221	19.35	19.35	19.35
LOEE	2271.1	2271.1	2271.1	2335.6	2335.6	2335.6
EDNS	70.26	70.26	70.26	76.31	76.31	76.31
FLOL	4.266	4.266	4.266	4.297	4.297	4.297
EIR	.99985	.99985	.99985	.99985	.99985	.99985
PC-CNG	11973	11973	11973	-7.83	-7.83	-7.83
CLOAD	3	3	3	84.37	84.39	84.37
CTRANS	100.0	100.0	100.0	80.39	79.46	80.39
CLOSS	100.0	100.0	100.0	75.17	75.29	75.17
FLNSP	100.0	100.0	100.0	66.00	66.00	66.00
EDNSP	66.00	66.00	66.00	386.17	386.17	386.17
CSPIN	386.17	386.17	386.17	13760	13760	13760
	13760	13760	13760			
Wheeling Customer 1						
EETRS				3930.9	3930.9	3930.9
ENTRS				0.0857	0.0857	0.0857
DNTRS				1.786	1.786	1.786
FNTRS				0.234	0.234	0.234
CLOAD				15.63	15.61	15.63
CTRANS				19.61	20.54	19.61
CLOSS				24.83	24.71	24.83

Table 6.9. Reliability and Operational Indices for the Transmission

Bus No.	L4	L7	L8	L9	L15	L18
Case 1						
LOLE	0.2	19.0	14.1	13.3	14.9	14.1
LOEE	1.3	116.1	93.5	88.1	499.3	624.7
EDNS	0.2	4.4	4.5	4.4	18.4	21.7
FLOL	0.1	4.3	3.3	3.1	3.3	3.1
Case 4						
LOLE	0.2	19.7	14.7	13.9	15.3	14.7
LOEE	0.6	116.8	96.2	89.8	510.4	640.5
EDNS	0.2	4.4	4.6	4.5	19.1	21.9
FLOL	0.1	4.4	3.4	3.2	3.4	3.3

Table 6.10. Reliability Indices of Transmission System Load-Points

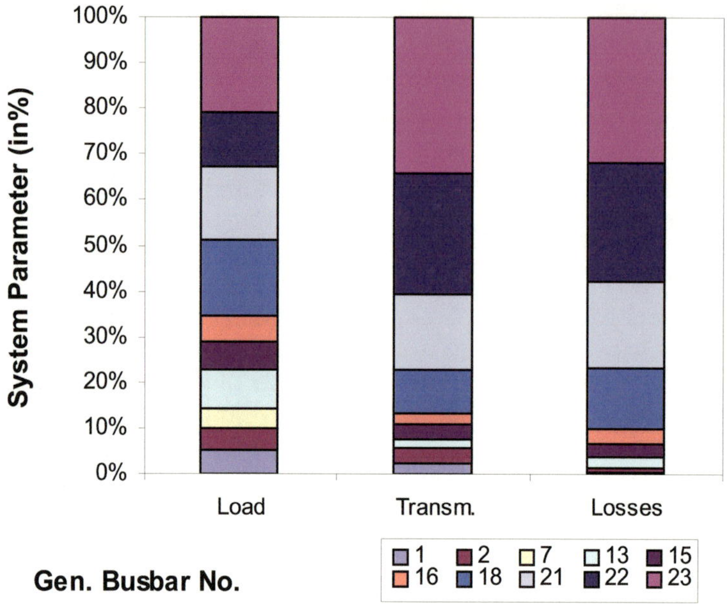

Fig. 6.3. Relative contributions of each generating unit output to system load, transmission use and losses (Case 3)

6.5 Conclusions

The application of renewable energy sources in isolated power systems may significantly affect their operational characteristics and inevitably their reliability performance. This Chapter describes the main concepts and features of an efficient computational methodology that is based on the sequential Monte - Carlo simulation approach. This methodology can be used as a useful tool for quantifying the reliability and operational performance of isolated power systems with high penetration of renewable sources. It also presents the results that were obtained from the reliability assessment studies conducted for an isolated power system which is based on the power system of a Greek island. It is shown that the system adequacy is critically dependent on the wind penetration margin that is assumed by the system operator. In addition, a sufficiently large percentage of spinning reserve being supplied by the system conventional generating units can improve the system reliability indices and prevent the occurrence of dynamically unstable situations in cases of a sudden loss of a large amount of power being produced by the wind generating units. Furthermore, it is clearly shown that the addition of small hydroelectric generating units improves the reliability performance of the system according to their characteristics.

Additionally, this Chapter describes the main concepts and features of an improved computational methodology that can be used for conducting reliability and cost assessment studies of transmission systems under the framework of the competitive energy market. This methodology is based on the sequential Monte - Carlo simulation approach and can analyse systems with any given configuration and operational features. An important emphasis is given in the realistic simulation of the operational behaviour of both the transmission system and the wheeling transactions so that a more accurate reliability assessment is achieved. For this purpose, three alternative techniques for power tracing are incorporated. An extended set of indices is calculated to quantify the system reliability performance and transmission capability and they can be used to measure the changes obtained in system reliability due to the wheeling transactions being accommodated. The analysis of various case studies is also presented for a modified IEEE-RTS that includes wheeling transactions. Finally, the obtained results demonstrate clearly the increased information being gained.

6.6 Acknowledgements

This Chapter has been compiled by Evangelos Dialynas who is a Professor in Electric Energy Systems at the School of Electrical and Computer Engineering of the National Technical University of Athens and the Director of the Electric Energy Systems Laboratory (EESL). Contributing authors are Dr. D. Papakammenos and Ph. D. candidate L. Daoutis who are research associates of the EESL and participate in its various research programs in the area of reliability assessment studies and applications of renewable energy sources.

6.7 References

[1] M. Shahidehpour, M. Alomoush, "Restructured Electrical Power Systems", Marcel Dekker, New York, 2001.

[2] M. Shahidehpour, H. Yamin, Z. Li, "Market Operations in Electric Power Systems", Wiley, New York, 2002.

[3] R. Billinton, R. N. Allan, "Reliability Assessment of Large Electric Power Systems", Kluwer Academic Publisher, Accord Station, 1988.

[4] G. J. Anders, J. Endrenyi, M. V. F. Pereira, L. Pinto, C. Oliveira, S. Cunha, "Fast Monte - Carlo Simulation Techniques for Power Systems Reliability Studies", 1990 CIGRE Session, Paris, Paper 38-205.

[5] J. R. Ubeda, R. N. Allan, "Sequential Simulation Applied to Composite System Reliability Evaluation", IEE Proceedings, Vol. 139, 1992, pp. 81 - 86.

[6] E. N. Dialynas, N. C. Koskolos, "Comparison of Contingency Enumeration and Monte - Carlo Simulation Approaches Applied to the Reliability Evaluation of Composite Power Systems", European Journal of Diagnosis and Safety in Automation, Hermes, Vol. 5, 1995, pp. 25 - 48.

[7] F. Castro Sayas, R. N. Allan, "Generation Availability Assessment of Wind Farms", IEE Proc. Gen. Trans. Dist., Vol. 143, No. 5, 1996, pp. 507-518.

[8] Glossary of Terms and Definitions Concerning Electric Power Transmission Access and Wheeling, IEEE PES Publication, No. 96TP110 - 0.

[9] H. H. Happ, "Cost of Wheeling Methodologies", IEEE Trans., Vol. PWRS - 9, 1994, pp. 147 - 156.

[10] D. Shirmohammadi, C. Rajagopalan, E. Alward, C. L. Thomas, "Cost of Transmission Transactions: An Introduction", IEEE Trans., Vol. PWRS - 6, 1991, pp. 1546 - 1560.

[11] N. A. Mijuskovic, "Reliability Indices for Electric - Power Wheeling", IEEE Trans., Vol. R - 43, 1994, pp. 207 - 209.

[12] CIGRE T.F. 38.03.11, "Adequacy Assessment of Interconnected Systems In Presence of Wheeling and Trading: Indices, Modelling and Techniques", CIGRE Symposium, Tours, France, 1997, paper 341 - 08.
[13] J. W. Marangon Lima, "Allocation of Transmission Fixed Charges: An Overview", IEEE Trans., Vol. PWRS -11, 1996, pp. 1409 - 1418.
[14] P. Jiuping, T. Yonael, R. Saifur, "Review of Usage-Based Transmission Cost Allocation Methods under Open Access", IEEE Trans., Vol. 15, 2000, pp. 1218 - 1224.
[15] L. Willis, J. Finney, G. Ramon, "Computing the Cost of Unbundled Services", IEEE Computer Applications in Power, 1996, pp. 16 - 21.
[16] J. Bialek, "Tracing the Flow of Electricity", IEE Proceedings, Part C, Vol. 143, No. 4, 1996, pp. 313 - 320.
[17] D. Kirschen, R. Allan, G. Strbac, "Contributions of Individual Generators to Loads and Flows", IEEE Trans., Vol. PWRS - 12, 1997, pp. 52 - 60.
[18] D. Papakammenos, E. N. Dialynas, "Reliability and Cost Allocation Methods of Transmission Systems Under Open Access", ISAP 2003, Lemnos, Greece
[19] D. Papakammenos, E. N. Dialynas, "Reliability and Cost Assessment of Power Transmission Networks in the Competitive Electrical Energy Market", IEEE Trans., Vol. PWRS - 19, 2004, pp. 390 - 398.
[20] IEEE APM Subcommittee Report: "The IEEE-Reliability Test System - 1996", IEEE Trans., Vol. PWRS - 14, 1999, pp. 1010 - 1020.

7

Europe: Status of Integrating Renewable Electricity Production into the Grid

The visionary targets of the European Community were to increase the share of renewable energy resources between 1997 and 2010 from 14 to 22 % as well as to double the contribution of cogeneration plants for heat and power (CHP) on total electricity production from 9 to 18 %. Consequently, the share of dispersed and renewable energy resources (DER) would cover 40 % of the whole electricity production in 2010. All countries have set their own targets to reach the common goal.

The DER in distribution systems will achieve an additional growth of more than 300 TWh/a to meet the challenging European targets. Additionally, wind power will grow primarily in the form of large wind farms that centrally feed into the transmission grids with 20-30 GW installed power by 2010. Large offshore wind farm sites with rated power up to 1,000 MW are currently under investigation to be installed in the North and in the Baltic Seas.

However, the output of most of the renewable energy sources depends on meteorological conditions and the CHP output is normally driven by the demand for heat that is higher in winter and lower in summer periods. The full load hours of the installed wind power capacity, for example, are approximately 1,400-1,600 h/a at onshore locations and between 800 and 1,000 h/a for photovoltaic plants.

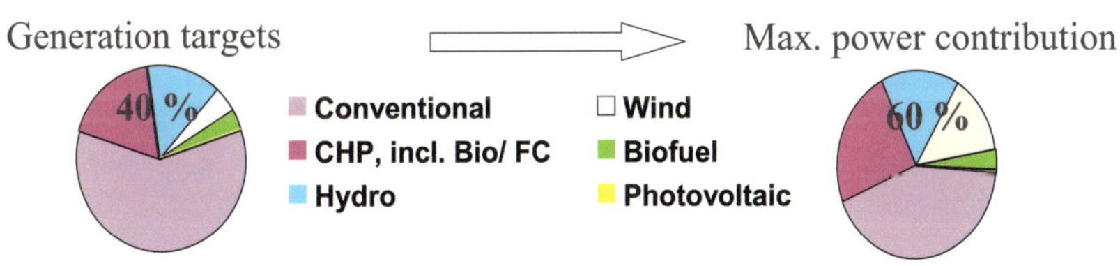

Conventional (bottom), CHP, incl. Bio/FC, Hydro, Wind, Biofuel, Photovoltaic (clockwise)

Figure 7.1. Generation Targets of the European Communities for 2010 and the Related Maximum Power Contribution of Renewable and CHP Generation

Thus, if the contribution of DER in electric energy generation shall achieve 40 %, their maximum possible contribution in the power balance must achieve 60 % of the European system peak load. A possible scenario is shown in Figure 7.1.

Such a large-scale penetration of DER in the power balance requires a sustainable restructuring of the present operation practice in power systems. A large number of different dispersed generation units in the range of some kW up to large centrally feeding wind farms of some 100 MW partially with an intermitting power output will be connected to all levels of the power system as shown in Figure 7.2. The question arises, how can the existing high level of power quality be maintained under these fundamental changing circumstances?

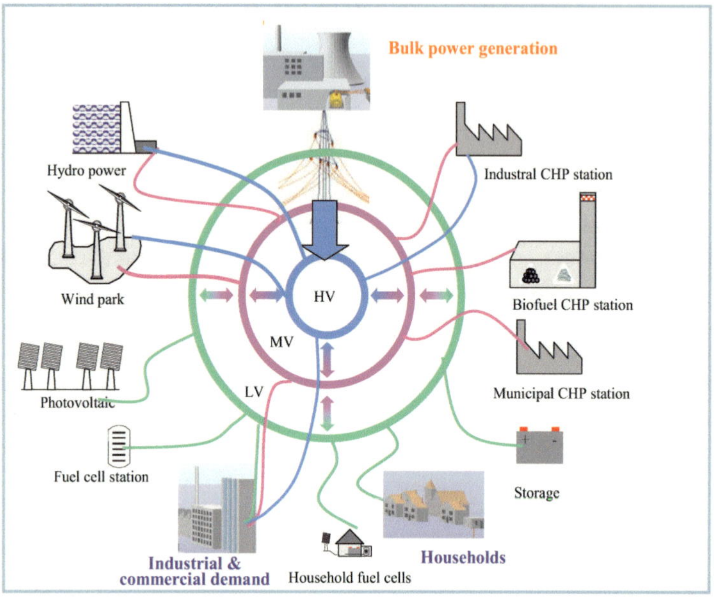

Figure 7.2. Power System Structure with Integration of Dispersed and Renewable Generation

To answer this question the experience of different countries with a large share of renewable and dispersed power generation is analyzed.

Based on the experience in these countries new recommendations and rules regarding grid conformity, reliability including the "fault ride through behavior" and dispatching of renewable and dispersed generation units are necessary. In some countries new guidelines are in development or already exist, for example in Germany.

7.1. The German Experience Of The Grid Integration Of Renewable Energy Sources

7.1.1 Prospective Development of Renewable Energy Generation in Germany

In Germany today, annual energy generation of the order of 520 TWh/a comes from approximately

- 56 % coal fired power plants,
- 28 % nuclear power plants,
- 8.5 % renewable energy sources,
- 7.5 % gas and oil fired power plants.

In the coming decades a fundamental change in generation structure will occur as the result of the political decision to shut down all nuclear power stations and, further, the need to replace most of the present power stations for reasons of aging. In Figure 7.3 the expected decommissioning of generation capability is presented.

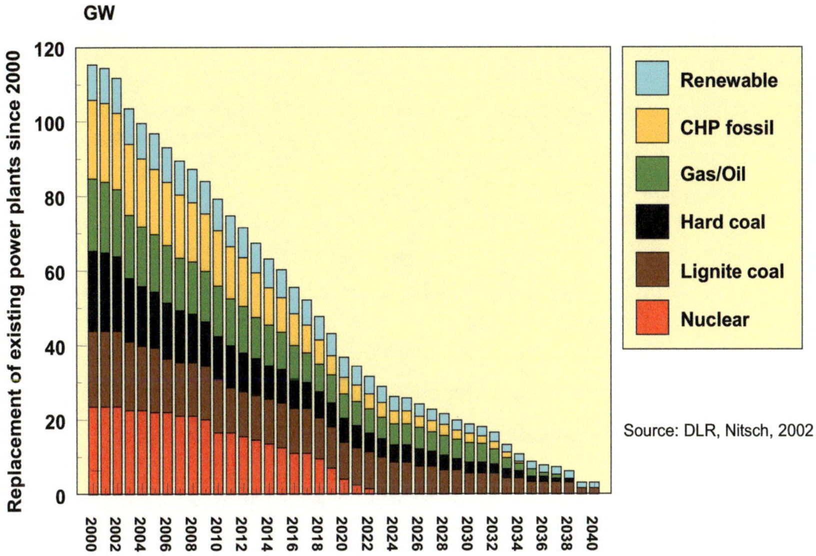

Top to Bottom: Renewable, CHP fossil, Gas/Oil, Hard coal, Lignite coal, Nuclear

Figure 7.3. Need for Replacement of Existing Power Plants

It is a political goal that only a portion of the traditional power stations will be replaced by fossil fired generation plants. In this situation the targeted growth of renewable and dispersed generation plays a significant role. In the field of renewable energy the official goals are to achieve shares of 12.5% in 2010 and 20% in 2020 of the overall electric energy generation.

However, this process seems to be ahead of schedule and the above targets were and will be gained in 2007 and 2015, respectively [1]. In Figure 7.4 the actual development scenario of the installed renewable power capability in Germany is shown.

As presented, the share of hydropower will be kept at the present level because of the lack of possible locations for new large hydropower stations. But all other renewable sources will grow significantly.

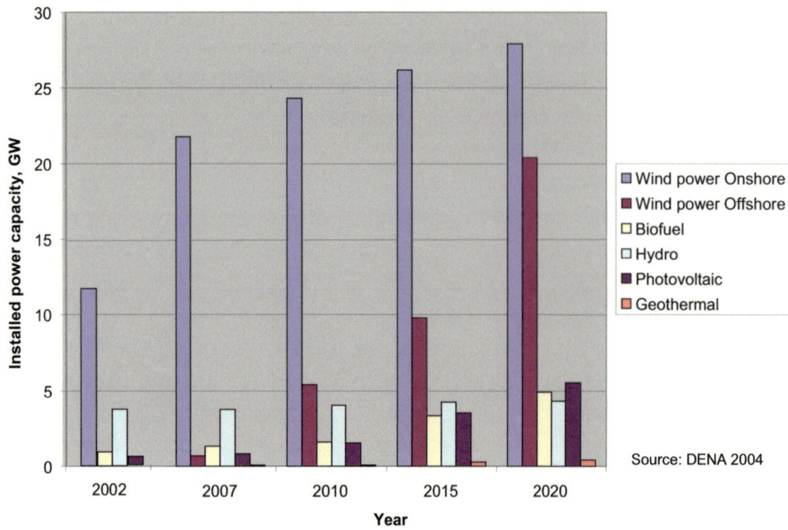

Left to right: Wind power Onshore, Wind power offshore, Biofuel, Hydro, Photovoltaic, Geothermal

Figure 7.4. Development of Renewable Generation Capability in Germany

A special high growth is expected in the wind power sector where Germany currently has a share of approximately 50 % of whole European capacity. Beginning in 2007, further growth was focused primarily on offshore locations where large wind farms with some hundred Megawatt installed power is being erected. Taking into account the achievable full load hours of the different generation technologies, the contribution of renewable energy sources in the whole energy balance of Germany will achieve about 30 % in 2020. However, their maximum contribution in strong wind situations may achieve:

- 70 % for peak load, and
- 100 % for weak load coverage,

where 58% and 83 % of the shares, respectively, come from fluctuating sources with an intermitting output depending on meteorological conditions (wind, solar). Such a large share of fluctuating power contribution requires advanced solutions to maintain power system security.

7.1.2 Economic Incentives

The generation of renewable energy is co-financed by fixed prices at high levels for the different renewable power sources and with subsidies for heat and power cogeneration.

The fixed prices are:

- wind energy - 8.7 €Ct/ kWh
- solar energy - 54 €Ct/ kWh
- biofuel energy - 6.6-10.2 €Ct/kWh (depending on the plant size)

independent of network level where the connection is provided. These much higher prices for renewable energy are paid by an additional charge of 0.54 €Ct/ kWh (level 2003/04) from all customers in accordance with Figure 7.5. Additionally, investors benefit from tax incentives for all capital expenses into renewable energy generation plants. Consequently, high profitability is the driver for the fast growth of renewable energy generation in Germany.

On the other hand, the network operators are obliged by law to ensure unlimited renewable power in-feed. This obligation incurs additional costs for:

- network enhancement from 2 Mil € in 2003 up to 40 Mil.€ in 2010
- spinning reserve to compensate power fluctuations from 130 Mil.€ in 2003 up to ~400 Mil.€ in 2010

in the case of one of the four transmission operators [2].

Figure 7.5 demonstrates the mean household price structure for electric energy in Germany.

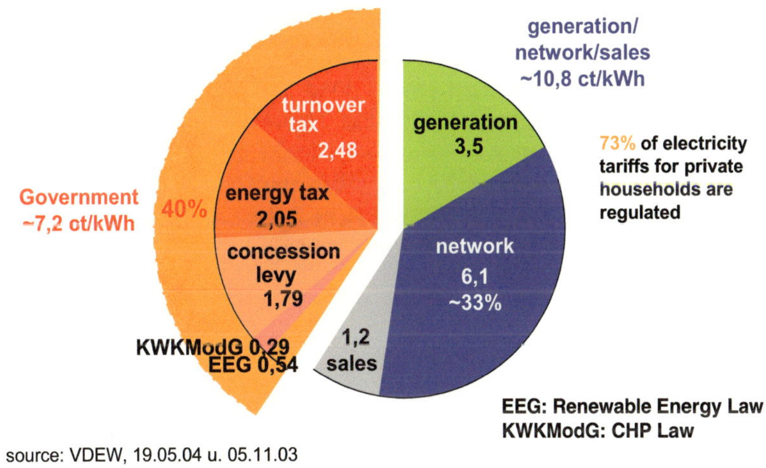

Figure 7.5. Shares of the Energy Price for Household Customers

7.1.3 Grid Integration of Large Scale Wind Power at the Transmission Level

In-feed of power by large wind farms is fundamentally subject to different patterns as is the case with conventional power sources such as thermal, gas turbine or hydroelectric generating plants. Three major problems need to be solved as the first priority:

Wind power output depends on meteorological conditions and may be intermittent. In addition to application of prediction tools for power schedule planning, a higher level of reserve power than before should be provided.

As wind power in-feed increases, the transmission capacity of the network becomes a further problem. Wind farms are mostly constructed in relatively underdeveloped regions in the north of Germany. The transmission networks in these regions have been expanded to only a limited extent.

Appropriate transmission capacities must be created in order for the power to reach the load centers.

The fault ride through of wind power plants should be adapted in such a way that wind generators contribute during short circuit currents and during network recovery after fault clearing.

In most cases, the wind velocities in northern Germany, as well as over the Baltic and North Seas, are mostly within the range of 3 to 12 m/s. Within this range, the power produced by a wind generator depends greatly on the wind velocity. Wind power producers basically feed in the maximum possible power obtainable from the wind and they receive a statutory payment.

Thus, planning the power balance of a transmission system depends substantially on the precision of weather forecasts, quite particularly if the share of wind power generation accounts for a significant portion of the network load. Special prediction tools for wind power generation have been developed and applied. However, their accuracy is limited as shown in Figure 7.6 and additional reserve power, significantly over the level, which is required for primary reserve to compensate outages in the UCTE grid (German share 750 MW), shall be provided for ensuring reliable system operation. Ongoing work towards improving the prediction accuracy is directed at minimizing the reserve power.

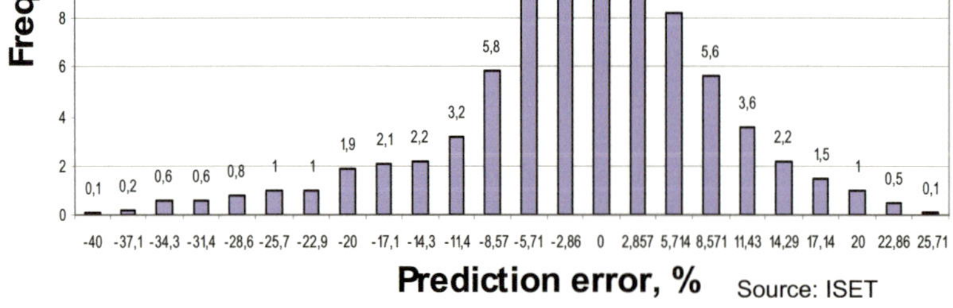

Figure 7.6. Distribution of Prediction Errors for Next Day Wind Power Forecasts

The solution to the second problem, grid enhancement, is restricted by legal difficulties and the long-term permission process for installation of new transmission lines. Moreover, existing conventional power stations will offer their generating capacity on the free market, selling it throughout Germany or Europe. Consequently, free energy trading is suffering increasing constriction owing to a lack of transmission capacity, and the installation of new transmission capacity will become mandatory. But, from an economic viewpoint, it is just the time to rethink the situation and to consider whether the network ought to be expanded for about 60 strong wind days per year, or whether generation management for wind power installations ought to be approved for this relatively short time.

Thirdly, apart from local impacts wind power also has a number of system-wide impacts because it affects

- power system dynamic and stability,
- reactive power control and voltage control,
- frequency control and load following/dispatch of conventional units.

The wind generators should fulfill three main aspects:

- no excitation of power oscillations after grid disturbances,
- in-feed of reactive power during and after system faults,
- maintaining system stability, minimize grid disruption.

Wind turbines installed in the German power system before 2003 had a single response to fault situations on the grid that resulted in instantaneous voltage drops: they were tripped off to protect their function until the grid recovered. The immediate loss of generation can impact on system stability and lead to cascaded tripping of some thousand MW of wind power. Reference [3] shows that faults in some grid locations can cause power tripping that is much higher than the whole spinning (primary) reserve of the European Network of Transmission System Operators for Electricity (ENTSO-E) grid of 3000 MW. For this reason, new rules for grid connection of wind power plants were established [4]. These rules describe the requirements regarding *Fault-Ride-Through* capability of wind turbines.

In Figure 7.7, rules regarding behavior of wind turbines connected to the German Power System during and after faults are illustrated.

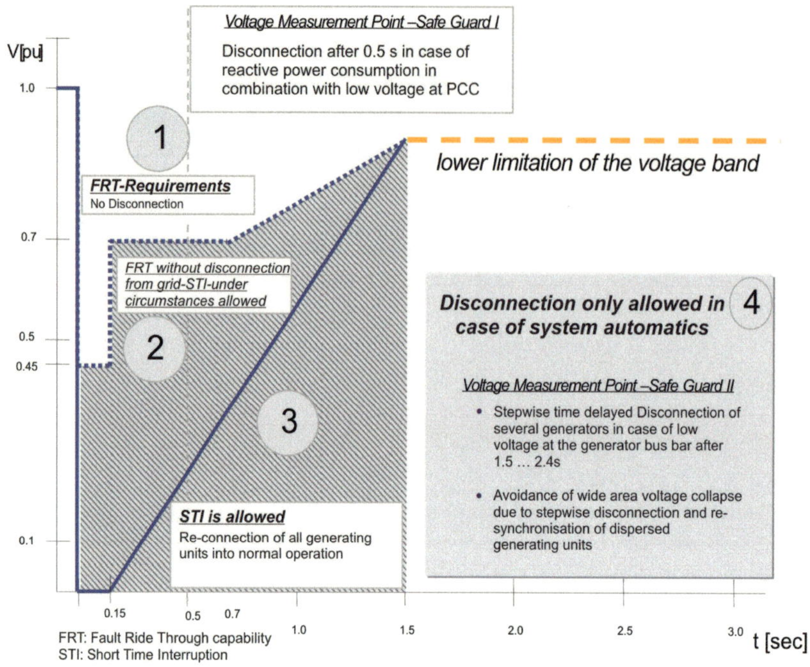

Figure 7.7. Requirements for the Fault Ride Through Capability of Wind Turbines Connected to the German Power System [5, 6]

According to the new German Grid Code [6], wind turbines must not disconnect from the grid and must not lead to instability in the event of three-phase faults even if the residual voltage is equal to zero at the grid connection point and for the time period of 150 ms, which corresponds to Region 1 in Figure 7.7.

In Region 2 of Figure 7.7, a short disconnection of the wind turbine (STI) is allowed if the generation unit becomes unstable during the fault. However, from the time point at which the disconnection occurred, the unit has to be resynchronized with the grid within 2 seconds maximum. Furthermore, the value of the generated active power by the wind turbine has to be restored to the pre-fault value with gradient of at least 10% of the generator rated power per second. In Region 3 the disconnection is allowed. However, in some special cases the grid operator can require the fulfilling of the resynchronization conditions for Region 3, which is similar to the case of Region 2.

Generally, all units that remain connected to the grid during a fault have to restore the value of the produced active power to their pre-fault value with a gradient of at least 20% of the rated power per second. Moreover, according to the new Grid Code, wind turbines have to support the voltage level of the grid during a voltage drop by in-feed of the reactive current.

7.1.4 Dispersed Generation in Distribution Systems

In addition to connection of large onshore and offshore wind farms to the transmission grid, a fast growth of dispersed energy resources (DER) in distribution systems is expected.

The problems to be solved at distribution level are:

- Ensuring network conformity in accordance with special rules of DER connection in medium and low voltage networks [7], e.g. regarding voltage quality, avoidance of equipment overloads, ability to withstand short circuits, influence on ripple control etc.
- Contribution for reliability of supply through provision of high availability and support of network recovery after faults.
- Compensation of power fluctuations and dispatch of a stable power balance in clusters of different DER, storage units and controllable loads.

These main requirements are presented in Figure 7.8.

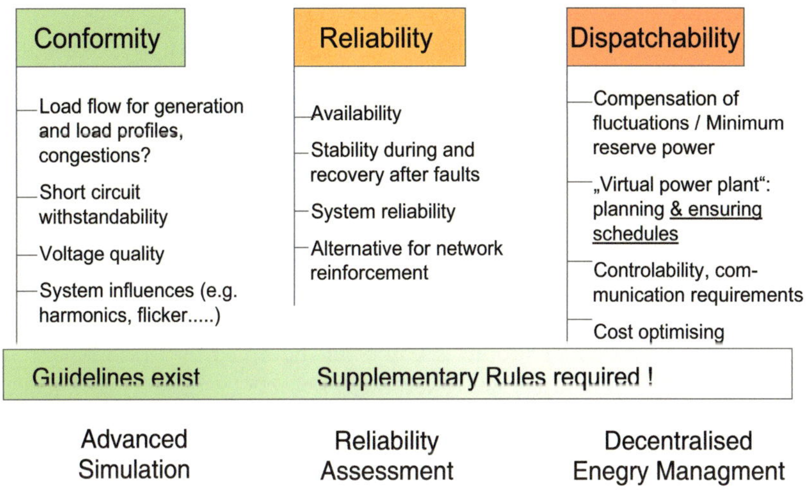

Figure 7.8. Requirements and Provision Means for a Large Scale Penetration of DER

The response regarding the first two requirements has to be analyzed by typical network planning methods. The simulation and assessment tools are available and have been approved in pilot projects [8].

The dispatch-ability requires more:

At present DER units are operated without higher-level control, feeding in maximum power as supported by current political and regulatory framework conditions. The transmission system operator is obliged to ensure power balance. This task will become more and more difficult under conditions of a growing contribution of uncertain and intermitting power output of DER. In the future, stable grid operation, economical considerations and environmental benefits will require intelligent energy management to be able to plan generation profiles at the distribution level as well. Those decentralized energy management systems have to balance required and available power in particular supply areas based on offline schedules for DER, storage units, demand side management capabilities and contractual power exchange.

The central dispatching of power balance will be supported by one of the decentralized dispatching systems as shown in Figure 7.9.

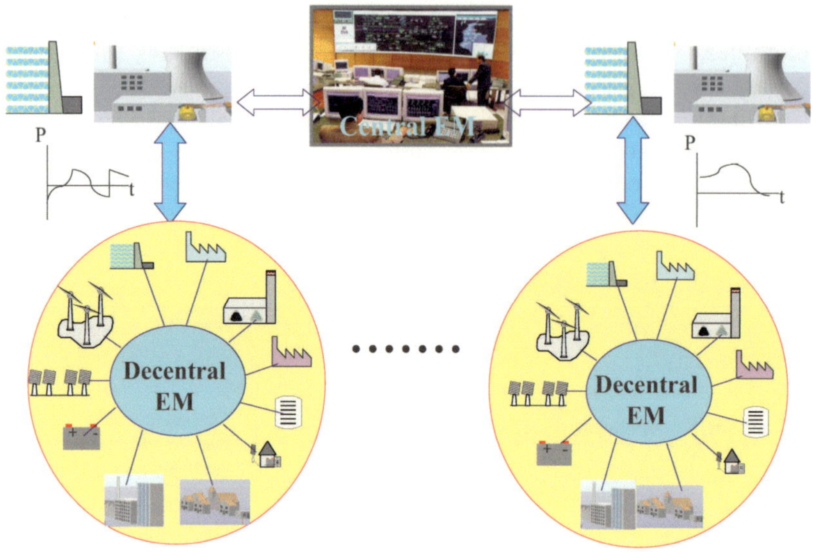

Figure 7.9. Future Task Splitting between Centralized and Decentralized Energy Management

Online monitoring and control of the units based on the schedules form balanced supply areas for different supply scenarios, i.e. different combinations of DER, storage, and load units. For higher-level management systems these balanced *"self sufficient cells"* appear as *"virtual power plants"* which show similar reliable, plan able, and controllable behavior like traditional power plants. There are various possibilities for vertical and horizontal integration of these locally optimized cells into central control centers.

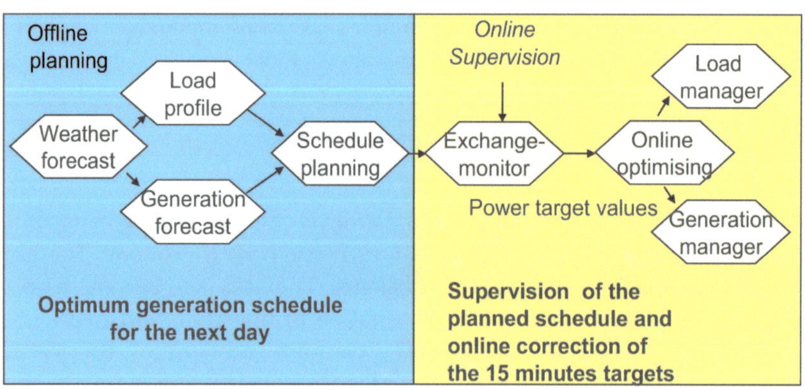

Figure 7.10. Principle of the Decentralized Power Management of DER

Adherence to the schedules has to be guaranteed online in operation to enable exactly defined contractual power exchange in the balanced supply areas. Unplanned power fluctuations and deviations from the schedules require fast adjustment of the real power flow within the individual period by dispatching controllable generation, storage units and demand in a one-minute time interval. The principle of the considered decentralized power management is presented in Figure 7.10.

To cope with unavoidable prediction errors for generation and demand, unit commitment accounts for the determined reserve power locally, while meeting all technical constraints. Thus, central power reserves can be reduced.

From the technical point of view, all of the means needed for operation with large-scale integration of DER are available and have been proven in practice [9].

However, the actual legal and incentive situation acts against an introduction of "virtual power plants". The legal and incentive frameworks have to be adapted so that the idea of the "virtual power plants" can become reality.

In summary, the increasing share of renewable and dispersed generation has no technical limits if *Conformity and Reliability* in context of the new guidelines [4], [7] is ensured and if their *Dispatch-ability* can be reached by technical means within an adapted legal and incentive framework.

7.2. Options for Large Scale Integration of Wind Power

The worldwide development of wind power installations now includes the planning of large-scale wind farms ranging in the magnitude of 100 MW, and is considered to constitute a significant part of the renewable power production planned in Europe and in the world. This is a challenging development that will have an impact on the power system stability and operation as outlined in section 7.2. The development is sound however; wind power is a cost-effective renewable source that can smoothly be integrated into the power system by applying adequate control technologies and market based solutions. Two cases are applied to demonstrate this. One considers the connection of a large wind farm to a fairly week regional grid (section 7.3), and the other considers the power system balancing of large magnitudes of wind power (section 7.4). It is demonstrated that local control actions enable quite large wind farms to be operated on fairly week grids, and that market based balancing tackles large magnitudes of wind power.

7.2.1 Impact of Wind Power on Power System Stability and Operation

Voltage control - reactive power compensation:

A main challenge related to voltage control is to maintain acceptable steady-state voltage levels and voltage profiles in all operating conditions, ranging from minimum load and maximum wind power production to maximum load and zero wind power. Capacitor banks and transformer tap changers represent the most common means to control voltage profiles. Another challenge in this context is related to the control (or limitation) of the exchange of reactive power between the main transmission grid and the regional distribution grid.

Voltage stability:

The output power from wind farms may vary significantly within a few seconds and, depending on the applied wind turbine technology, the reactive demand will also vary significantly. If the power system cannot supply this demand, a voltage instability or collapse may occur. Sufficient and fast control of reactive compensation is required to relax such possible voltage stability constraints related to wind farms, which can be provided through the use of wind turbines with active voltage control, or by using external compensators, such as Static Var Compensators (SVCs).

Transient stability:

Traditionally, the protection systems of wind turbines have been designed to disconnect and stop the units whenever a grid fault (temporary or permanent) is detected. With increasing integration of wind power there are and will be system requirements implying that wind turbines must be able to "ride through" temporary faults, and contribute to the provision of important system services, such as momentary reserves and short circuit capacity. This puts emphasis on transient stability performance, power oscillations and system damping. Control equipment within wind farms enabling both power and voltage control becomes increasingly important in this context.

Thermal transmission capacity constraints:

Thermal transmission capacity problems associated with wind power integration may typically be of concern in only a small fraction of the total operating time. Applying control systems to limit the wind power generation during critical hours may be a possible solution, or if other controllable power plants are available within the congested area, coordinated automatic generation control (AGC) may be applied. The latter alternative may be beneficial as energy dissipation may then be avoided.

Power fluctuations – frequency control:

Wind energy is by nature a fluctuating source of power. In a system where a significant part of the power generation comes from wind, system operational issues, such as frequency regulation and congestion management become a challenge due to the normal variations in the available wind power. Systems with substantial supply from wind farms thus call for flexible and improved solutions with respect to secondary generation control.

Adverse impact from interaction of power electronic converters:

Modern wind turbines utilizing power electronic converters provide enhanced performance and controllability compared to traditional fixed speed solutions. With increasing use of power electronics, however, there may be uncertainties with respect to possible adverse control interactions within the wind farm itself. Converter modulation principles and filter design are important issues that must be addressed and analyzed as part of the wind farm design and installation.

In summary, most of the challenges described above may result in operational conditions that adversely affect the quality of the voltage and power supplied to customers. Additionally, there may be system operational problems, such as congestion management and secondary control that not only affect the wind farm in question but the entire network. Thus, the problems suggest coordinated control solutions that maintain secure operation of the network, and at the same time allow for maximized and profitable integration of wind power. Indeed, large scale integration of wind power does not only set requirements on the power system, but also the wind power technology must be developed according to the system needs. The development of IEC 61400-21 [10] specifying procedures for characterizing the power quality of wind turbines and the various grid codes setting system requirements on wind farms, e.g. Eltra [11], are examples of such development.

7.2.2 Case – Local Control

The case study considers the connection in Norway of a large 200 MW wind farm to a typical regional distribution grid (see Figure 7.11. The study is based on an actual system, though slightly modified to serve the purpose of this Chapter. The regional distribution grid is connected to the main transmission grid via a long 132 kV line with a thermal power capacity limit of about 200 MW. Considering that the hydropower plant that is already connected is rated 150 MW and that the local load may be as small as 14 MW, a conservative approach would suggest that the wind farm capacity should not exceed 64 MW (i.e. 200 – 150 +14), or indeed 50 MW (i.e. 200-150) to ensure operation if the local load disconnects. However, contrary to such conservative planning, this case demonstrates that installation of a much larger wind farm is viable.

Due to environmental constraints, it is not an option in this instance to upgrade the 132 kV line for higher thermal power capacity. Hence, power electronics and control systems are applied to allow connection of the large wind farm.

Reference [12] shows that as long as the thermal capacity of the 132 kV line is respected, voltage control and stability is ensured by the application of a Static Var Compensator (SVC) and/or the utilization of the reactive control capabilities of modern wind turbines with frequency converters.

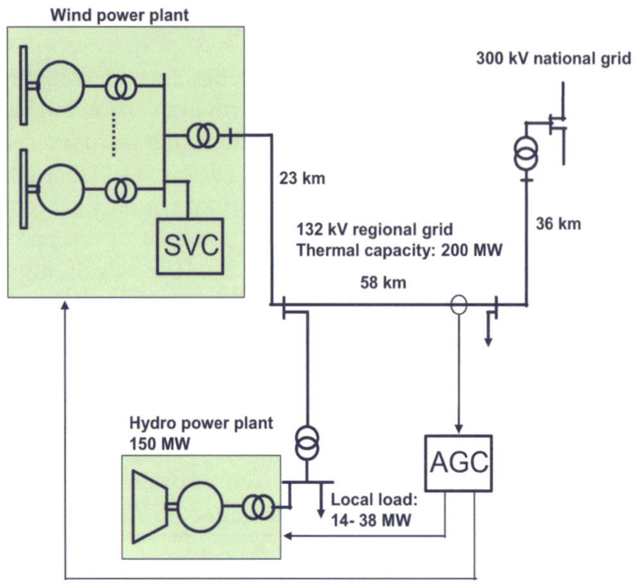

Operation of a 200 MW wind farm is viable using the Static Var Compensator or built in reactive control capabilities of modern wind turbines for securing voltage stability, and using Automatic Generation Control (AGC) for controlling that the thermal capacity of regional grid is respected.

Figure 7.11. Outline of Case Study Regional Grid.

Figure 7.12 illustrates that reactive support enables a stable voltage for feed-in of 0 to 200 MW of wind power, whereas without reactive support, the wind farm size would have to be restricted to about 50 MW.

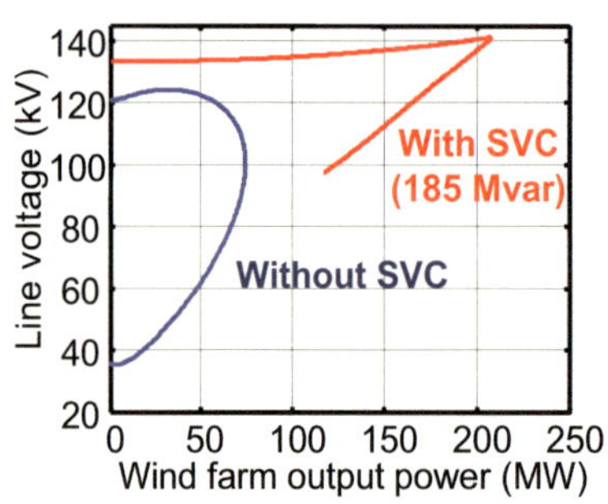

Figure 7.12. Result of Dynamic Simulations of Power System with 0-200 MW of Wind Power [12]

Ref [13] demonstrates that Automatic Generation Control (AGC) of the hydropower plant can be used to avoid overloading the 132 kV line. This is illustrated in Figure 7.13, showing a result of a dynamic simulation verifying the performance of the AGC.

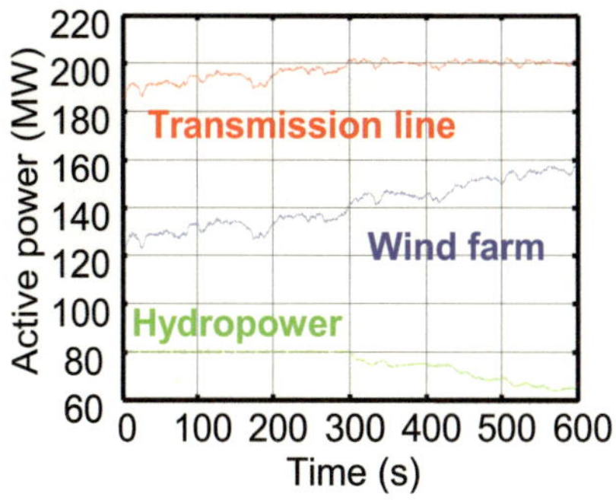

Figure 7.13. Result of Dynamic Simulation of Power System with 200 MW Wind Farm and AGC Control of Hydropower Plant [12]

The AGC operation influences the annual output and energy sales from the hydro and wind power plants. As found in [14] however, the impact on the energy sales is (surprisingly) moderate (see Table 7.1).

	Control hydro	**Control wind**	**Non-congested**
Wind power (GWh/y)	609	551	609
Hydropower (GWh/y)	616	657	657
Local load (GWh/y)	219	219	219
Line load (GWh/y)	1036	989	1047

Table 7.1. Case Study Results of 200 MW Wind Farm for Two Cases of AGC Control, i.e. Control Hydro (Reschedule Production) or Control Wind (Reduce Production), and for the Case of Unlimited Grid Capacity (Non-Congested Case) [14]

7.2.3 Case – Market Based Power Balancing

EU regulation requires that market based principles should be used for congestion management. In the Nordic power system the real time frequency control is also handled through a joint balancing market.

This case considers real operational data from the Nordic power system (see Figures 7.14 and 7.15). On January 8, 2005 there was a storm affecting southern Scandinavia initially causing high wind power production in Denmark. At a certain time however, the wind turbines started to cutout due to excessive wind speeds and the wind power production was reduced from 1800 MW to 100 MW during the afternoon hours. The loss of wind power production amounted to more than half of the consumer loads in western Denmark. Figure 7.15 shows how this situation was handled in operation. The loss of generation was compensated through the balancing power market (mostly activated in southern Norway) and by regulating the HVDC link between Norway and Denmark from full export to full import in the same hours. The example illustrates clearly that the Nordic power system can handle large amounts of wind power through the existing marked based mechanisms.

Secure operation requires that sufficient reserves and transmission capacity are available in such situations. In a future system with high penetration of wind power throughout Europe, the operational challenges with respect to operating reserves, frequency control and transmission capacity are expected to become increasingly important.

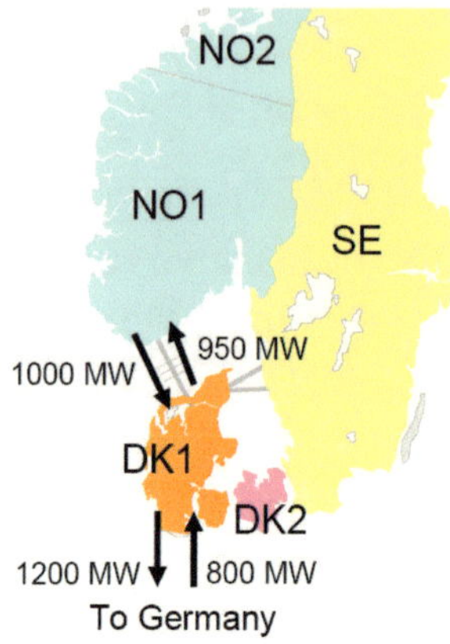

Figure 7.14. Map Showing Parts of Nordic Market (Elspot) Areas and Normal Transmission Capacities between western Denmark and Germany and between Denmark and Norway

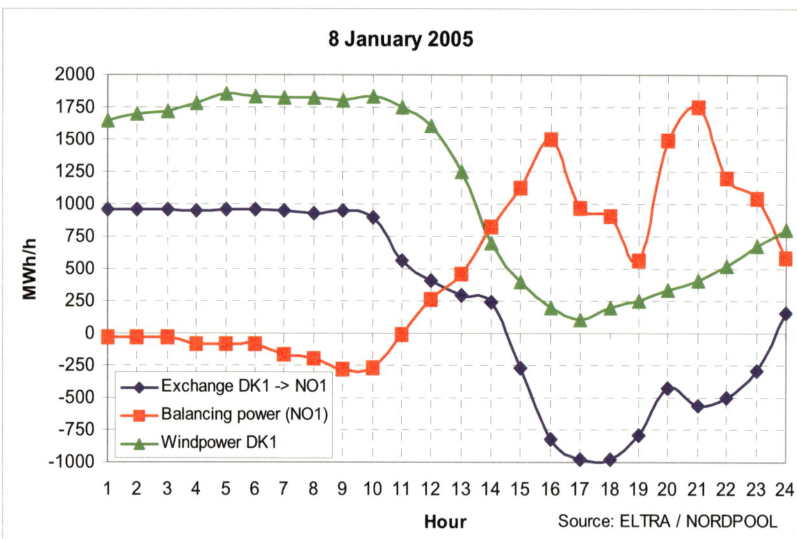

Figure 7.15. Actual Hour-by-Hour Data of Wind Power in western Denmark (DK1), Balancing Power in southern Norway (NO1) and Power Exchange over the HVDC Line between southern Norway and western Denmark

Section 7.2 has demonstrated options for large-scale integration of wind power. Local control enables the operation of a large wind farm on a fairly weak regional grid, and market based balancing tackles large magnitudes of wind power. Thus, a future with a high penetration of wind power throughout Europe seems viable, though the operational challenges with respect to operating reserves, frequency control and transmission capacity are expected to become increasingly important.

7.3. Spanish Experience of Grid Integration of Wind Energy Sources

Until recently, installed wind power was anecdotic, and its influence on the system insignificant. Over the last few years, however, the installation of wind power generation connected to the Spanish electric power system has expanded fast. This growth has proven more rapid than average growth within the European Community, as illustrated in Figure 7.16.

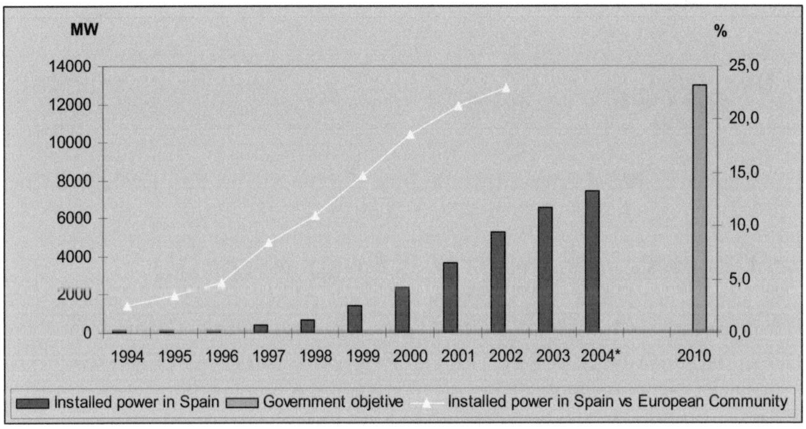

* up to 20th July

Figure 7.16. Evolution of Wind Power Generation Connected to the Spanish Electric Power System and Comparison with Growth in the European Community

By relating wind-installed power with other figures, we can demonstrate that the importance of wind generation in Spain is not less than in other countries like Germany or Denmark (Figure 7.17). When wind-installed power is compared with population (indirect way of comparing installed power with the size of the electric system), Spain appears to have a size comparable to Germany. If wind installed power is compared with import exchange capability, Spain fares well above other countries. This means that the transient support that Spain can receive from other countries, due to the *Principle of Joint Action*, is small compared with the wind-installed power.

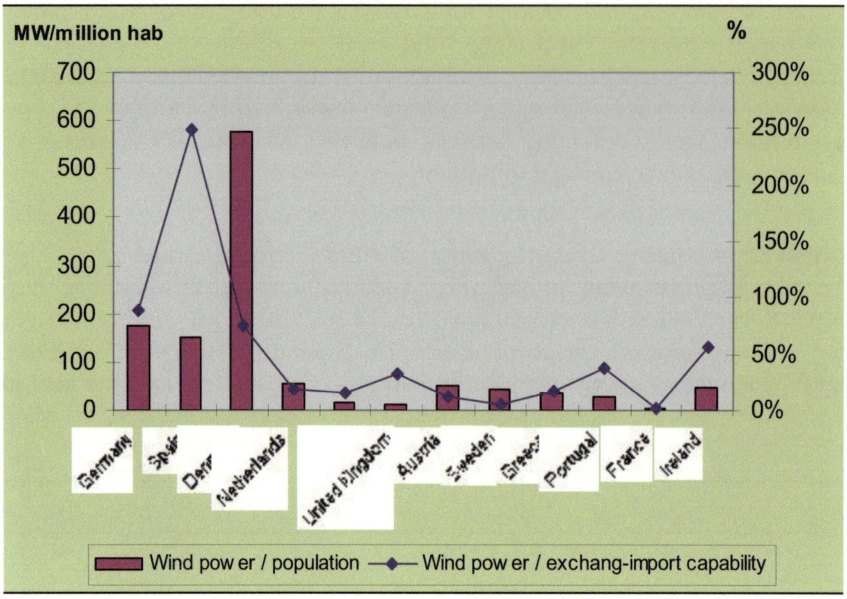

Figure 7.17. Relation of Wind Power Installed vs. Population and vs. Exchange Capability

7.3.1 Present Economic Incentives for Wind Energy in Spain [15]

Wind power producers are entitled to transfer their production to the system through the electricity distribution or transmission company whenever the absorption of the energy by the network is "technically possible". Wind power producers may chose from two different options in order to incorporate their production into the system. They can opt to participate directly in the wholesale electricity market or to sell the energy to distributors.

The first option of participating directly in the Spanish Wholesale Electricity Market involves either presenting bids or establishing bilateral contracts. In both cases wind power producers have the same treatment as the "ordinary regime" as far as ancillary services are concerned. If they opt to participate directly in the Spanish Wholesale Electricity Market presenting bids, their production has the following treatment concerning congestion management:

- Their production cannot be withdrawn on the grounds of network congestion problems (except for real time management) if they bid as price takers (bids at a price of 0 € / MWh).
- Their production shall be incorporated for solving technical constraints, provided their bid price is less than 70% of the reference tariff as defined in [15] article 2 (except for real time management). The producers shall be connected to a distribution company that in turn is connected to a point of the transmission network in which the System Operator (REE) has identified a constraint problem.

The second option is to sell the energy to the distributors. Wind power producers are entitled to sell their production to the distribution companies, which are obliged to buy this energy. The distribution companies deduct this production from the buying bids that they have to present to the Spanish Wholesale Electricity Market in order to supply their captive customers. The above is also what currently applies to production from all renewable and high efficiency plants, integrated in the so-called "*special regime*", as opposed to the "*ordinary regime*".

Depending on the option chosen, wind power producers are recompensed as follows (Table 7.3):

Participating directly in the Spanish Wholesale Electricity Market: Hourly marginal price of the wholesale market or price negotiated in bilateral contracts + subsidy + incentive + complement for reactive power + complement for fault ride through capability--deviation from production programs (see "Treatment of deviations from production programs"):

- Subsidy: percentage (40%) of the yearly electricity average tariff or reference tariff as defined in [15] article 2.
- Incentive: percentage (10%) of the yearly electricity average tariff or reference tariff as defined in [15] article 2.
- Complement for reactive power: percentage of the yearly electricity average tariff or reference tariff as defined in [15] article 2, (Table 7.2). Producers can also renounce to this complement and participate in the reactive power market (not in place yet).
- Complement for fault ride through capability (withstanding voltage sags): for 4 years, 5% of the yearly electricity average tariff or reference tariff as defined in [15] article 2.

Selling the energy to the distributors: Regulated tariff + complement for reactive power + complement for fault ride through capability - deviation from production programs (see "Treatment of deviations from production programs"):

- Regulated tariff: percentage of the yearly electricity average tariff or reference tariff as defined in [15] article 2. Irrespective of onshore or offshore installations, the above percentage is established as follows:

  - For installed capacities < 5 MW: 90% of the tariff during the first 15 years after commissioning, and 80% thereafter;
  - For installed capacities > 5 MW: 90 % of the tariff during the first 5 years, 85% during the following 10 years and 80% thereafter.

- Complement for reactive power: Percentage of the yearly electricity average tariff or reference tariff as defined in Table 7.2.
- Complement for withstanding voltage sags: Same as above.

Power factor	Active & reactive energy	%		
	Power factor	Peak	Plain	Off-peak
Inductive (lag)	< 0.95	-4	-4	8
	< 0.96 & ≥ 0.95	-3	0	6
	< 0.97 & ≥ 0.96	-2	0	4
	< 0.98 & ≥ 0.97	-1	0	2
	< 1 & ≥ 0.98	0	2	0
	1	0	4	0
Capacitive (lead)	< 1 & ≥ 0.98	0	2	0
	< 0.98 & ≥ 0.97	2	0	-1
	< 0.97 & ≥ 0.96	4	0	-2
	< 0.96 & ≥ 0.95	6	0	-3
	< 0.95	8	-4	-4

Table 7.2. Complement for Reactive Power

The reported tariffs, subsidies, incentives and complements were reviewed in 2006 and then again every 4 years. Irrespective of this, prices will also be reviewed when wind power generation reaches 13 000 MW of total installed capacity.

The treatment given to deviations differs, depending on the option chosen to incorporate the production in the system, as follows (Table 7.3):

Participating directly in the Spanish Wholesale Electricity Market: Same treatment as ordinary regime, which basically follows the principle that those installations that deviate from their programs pay the overall cost of solving the deviation of the whole system, in proportion to its own deviation. Selling the energy to the distributors: Wind power producers exceeding 10 MW of installed capacity are permitted a deviation of 20% from their forecast (they are obliged to give this forecast to the distribution company to which they are connected). Deviations exceeding that range are paid at a price consisting of a percentage (10%) of the yearly electricity average tariff or reference tariff as defined in [15] article 2.

	Participating directly in the Spanish Wholesale Electricity Market	Selling the energy to the distributors
Hourly marginal price or price negotiated bilaterally	Depends on the market	-------------------------------
Regulated tariff	-------------------------------	5.76576-6.48648 c€/kWh
Subsidy	2.88288 c€/kWh	-------------------------------
Incentive	0.72072 c€/kWh	-------------------------------
Complement for reactive power	Depends on power factor and time of the day	Depends on power factor and time of the day
Complement for withstanding voltage sags	0.36036 c€/kWh during first 4 years	0.36036 c€/kWh during first 4 years
Deviation from programs	Depending on deviations	Depending on deviations
TOTAL (not including complement for withstanding voltage sags)	3.60360 c€/kWh + market or negotiated price + complement for reactive power - cost of deviations	From 5.76576 to 6.48648 c€/kWh + complement for reactive power cost of deviations

Table 7.3. Summary of the Retribution Schemes for Wind Energy

7.3.2 The Spanish Experience

The minimum voltage protection systems in Spanish wind farms must comply with the specifications of the Ministerial Order of 5th September 1985 [16]. In accordance with this Order, it is mandatory to install three instantaneous minimum voltage relays between phases in the connection point of wind farms. The relays must provoke instantaneous disconnection of the wind farm when voltage drops below 85% of the average value between phases.

In order to integrate as much generation as possible, a delay in the disconnection of wind parks during disturbances has been considered. However, it has been confirmed that some technologies cannot stand such a delay.

Wind penetration levels are currently being reached in Spain, so that in the event of a short-circuit in the transmission network –even if it is correctly cleared, the minimum voltage protection system may cause instantaneous disconnection of a significant number of wind farms, with the consequent loss of power generation. Studies that have been carried out [17] show the importance of minimum voltage protection systems in wind farms and system stability. Figure 7.18 (real experience, not theoretical or simulation) presents the total wind production in the Spanish Peninsular Electrical System during the 18th of January of 2004.

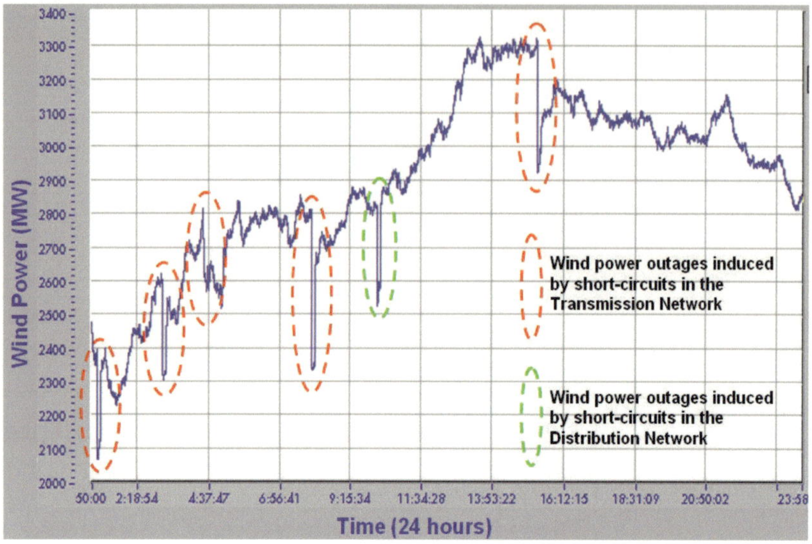

Figure 7.18. Wind Power Trips induced by Faults on the Network (MW)

The curve shows the wind production in peninsular Spain, with some sudden trips of production coincident with correctly cleared short-circuits in the transmission network. In this case, the interrupted production does not exceed 500 MW, but should the short-circuit occur on a day with more wind, or if the wind installed power increases, the amount of production disconnected will also increase.

In order to evaluate the influence of these trips in system security, the amount of connected ordinary regime plants is very important because they help contain the disturbance and recover the system parameters after the disturbance. For this reason, the same amount of wind production loss would be more severe in low demand conditions than in peak demand conditions.

Being aware of this drawback, REE has proposed new technical requirements to the regulator [18] in order to integrate a large amount of wind generation in the Spanish electric system, maintaining the actual security and quality standards. Of course, for this purpose it is necessary that wind generators meet some requirements for improving the fault ride through capability.

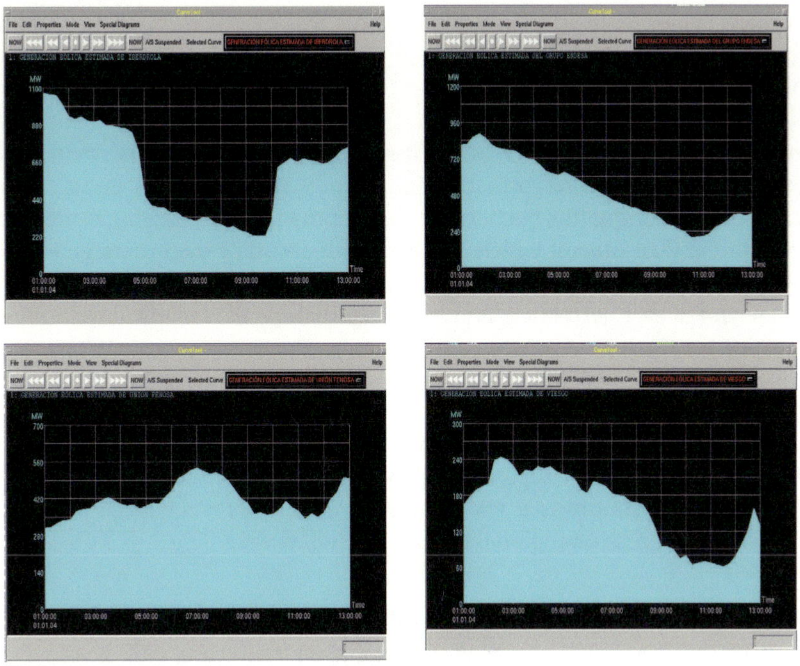

Figure 7.19. Wind Production in Four Different Distribution Areas

REE regularly evaluates the maximum wind power penetration that is compatible with system security according to transient stability analysis in different situations. According to this evaluation, it was sometimes required to reduce wind generation, for example on January 1 2004.

Figure 7.19 shows the different responses of wind generators in four distribution areas in Spain to a request for limiting the production.

These graphs display the aggregated production of the distribution zones. The first of the four graphs shows the production of a zone that has a control center, with a very good response. Conversely, the request to limit the production has not been correctly followed in the other zones.

These case studies confirm the importance of connecting all wind generation plants to a control center to effectively inter-act with the system operator.

In summary, wind power differs from conventional sources of energy in three main ways: the prime mover is wind, the location of resources, and the electrical machines. Controllability and availability of wind power significantly differs from thermal or hydro generation because the primary energy source cannot be stored and is uncontrollable. Wind power does not complicate very much short term balancing and all wind turbine types can be used for it, although variable speed wind turbines have better capabilities. Long term balancing is problematic. The power generated by wind turbines depends on actual value of the wind speed. When there is no wind, no power from wind turbines is available. Wind turbines complicate the long term balancing task, particularly at high wind power penetrations.

More than 7900 MW in wind mills were connected (2005) to the Spanish peninsular power system networks, this "*enormous*" amount requires advanced solutions in order to maintain the actual level of power quality, such as the development of dispatching centers (under the ownership of the TSO or others) which transmit with accuracy the orders given by the TSO to the wind farms. Integration of wind power is possible, but it requires the development of adequate procedures that harmonize and make compatible the technical requirements with the market rules.

Considering the reduced contribution of wind generators to short-circuit power and the high meshed level of the European networks, a short-circuit on the transmission network can lead to widespread voltage dips to neighboring TSOs. Therefore, the "*fault ride through capability*" of wind generators is a useful requirement to prevent large outages of wind power dependent on the given regional potential gradient area.

7.4. From the Kyoto Protocol to the Future Power Grid

The decision of the Russian Parliament (or Duma) to ratify the 1997 Kyoto Protocol on climate change has re-energized international cooperation on cutting greenhouse gas emissions.

Russian ratification ensured that the Protocol is legally binding on its 128 Parties on 16 February 2005 and launches an exciting new phase in the global campaign to reduce the risks of climate change. All must get down to the serious business of reducing emissions of carbon dioxide and other greenhouse gases, by giving industry, local authorities and consumers incentives to take action on climate change. Russia and the 29 other industrialized countries that have joined the Protocol will set themselves on a path to greater economic efficiency. Accelerating the development of the clean technologies that will dominate the global economy of the 21st century will earn them a competitive edge in global markets. What various countries in Europe are doing in this respect will be examined.

The Protocol contains legally binding emissions targets for 36 industrialized countries. These countries are to reduce their collective emissions of six key greenhouse gases by at least 5% by 2008-2012, compared to 1990 levels. This first five-year target period is only a first step. While developing countries do not now have specific emissions targets, they too are committed under the 1992 Climate Change Convention to taking measures to limit emissions; the Protocol will open up new avenues for assisting them to do so. In addition to inspiring national action to cut emissions, the Protocol's entry into force will strengthen international cooperation through the early start-up of an international "emissions trading" regime enabling industrialized countries to buy and sell emissions credits amongst themselves; this market-based approach will improve the efficiency and cost-effectiveness of emissions cuts. the "*clean development mechanism*" (CDM), through which industrialized countries can promote sustainable development by financing emissions-reduction projects in developing countries in return for credit against their Kyoto targets cooperative projects under the system for "*joint implementation*", whereby one developed country can finance emission reductions in another developed country.

Developments in the power industry in Europe depend on expectations for future political, financial and technical conditions. Embedding of renewable energy sources is a quite challenging task, based on conditions defined by the Kyoto Protocol.

The trend in European power industry developments will be influenced by:

- Liberalization and globalization with the goal to open markets, not only for delivery of equipment but also to include new market players in the generation and transmission of the energy.
- Increasing environmental constraints (e.g. CO_2 reduction, regenerative power generation, and difficulties to get right of way for overhead lines) will influence the type and location of new generation and changes in the structure of power systems.
- Continuous increase of price for oil and gas can speed up the use of new generation technologies if they would be technically available.

In the deregulated environment, responsibilities for generation, transmission and distribution are separated. However, from technical point of view there are strong interdependencies among all the parts of power systems. Generation locations depend on the available primary energy sources (water, wind, etc.), mostly not close to the centers of power demand. The transmission system then has to transmit power over long distances. In case primary energy as gas or coal is available close to the load centers or it can be transported by other means (e.g. pipelines, shipping), generation can be placed close to the load, even in sub-transmission or distribution systems.

Financing of power plants plays an important role in the deregulated environment. Therefore payback times are an important factor in the decision for new power stations. Technologies with the shorter payback have economic advantages.

In the decades to come it can be expected that the main primary energy will still be gas, with declining use of coal. Studies show that the gas exploitation will increase for more than 4 times in the next 30 years. The renewable power generation (wind, solar and biomass) will increase considerably in some countries, especially in Europe; however, because of still high costs and the need for additional generation as running reserve, there are many on-going discussions on the feasibility of embedding large amounts of renewable energies within the existing grids.

New technologies as fuel cells are still in the early phase of the development. To be economical, the production costs have to be reduced considerably. This depends, however, on the progress in the development of new materials. The expectations for the economic break-through are therefore uncertain. In the next 30 years fuel cells will be used only for small ratings in distribution networks and will not play a major role in the power industry.

Development in the field of fusion to produce electric energy is just at the beginning with problems in the field of materials that have to resist very high temperatures. Its realization in the near future cannot be expected. It can, however, be possible that fusion generation will be built in 50 years or even later.

According to the expectation for increasing power demand in the next decades the existing systems in many industrialized countries, also in Europe, will be loaded by additional power of at least 60%, without the possibility to build a larger number of new overhead lines. The existing lines, in Europe with a relatively low voltage level of only 400 kV will therefore be loaded up to their thermal limits. The solution in densely populated areas will be to introduce more underground cables and preferably to use GIL (Gas insulated Lines) for bulk power transmission corridors, as GIL technology can transmit large amounts of power at reasonable costs through narrow rights of way. FACTS (Flexible AC Transmission Systems) technology could also help to improve the loading of power corridors. With the increasing load the short-circuit current will also further increase. Short-circuit current limiter solutions will be needed.

However, with the increasing complexity of power systems, the reliability of power supply will diminish as already shown by a number of large blackouts in Europe and America. Studies show that the probability for large blackouts is much higher than theoretically expected. The reason is that fault sequences leading to blackout do not result only from statistical failures. An essential role is played by human errors, insufficient maintenance and systematic errors in planning and operation, leading to cascading of the faults. These systematic errors cannot be completely avoided, because of too high complexity of the systems. Improvements can be made also by the use of HVDC. Back-to-back HVDC could separate parts of the interconnected systems to avoid widening of large disturbances throughout the system.

HVDC will further be increasingly used to transmit large power blocks from remote locations to the load centers.

The effective operation of large and complex power systems in many countries of Europe will ask for new modern control systems combined with new protection strategies. The goal of new control and protection will be to assure economic and reliable operation even under emerging conditions.

Power output of wind generation can vary fast in a wide range, depending on weather conditions. Hence, a sufficiently large amount of controlling power from the network is required to substitute the positive or negative deviation of actual wind power in feed to the scheduled wind power amount.

One possible solution is to use HVDC long distance transmission, integrated into a synchronous AC network to reinforce the interconnection of different parts of the system, when an increase of power exchange is requested without overloading weak links or bottlenecks in the existing grid. Such a situation is expected in the German network, when large amounts of renewable energy sources, e.g. wind parks, are connected to the northern parts of the grid. At present, a total amount of about 12 GW wind power has already been installed in Germany (out of 120 GW totally installed generation capacity). A further increase of up to 50 GW wind power capacities can be expected in the next decades, from which about 50% will be generated by off-shore wind parks in the north- and east-sea areas.

Both tasks, to transmit surplus power out of the northern wind generation area and to provide the controlling power from the generation in central and southern grid parts, would additionally load the existing network, thus leading to bottlenecks in the transmission system.

Loading in distribution systems will also increase leading to high current networks. In addition, decentralized power generation will be in larger extent connected to the distribution networks. The structure of distribution networks will therefore change from vertical oriented power in-feed to the mixed structure with part of power in-feed from the superposed power system and part delivered by own generation.

Distribution systems will operate in similar way as high voltage systems.

Because of high short-circuit currents and reliability reasons they will be separated into smaller systems interconnected by current limiters or DC back-to-back stations.

7.5 Acknowledgements

This Chapter has been prepared by Thomas J. Hammons (Chair International Practices for Energy Developments and Power Generation IEEE, University of Glasgow, UK), Yvonne Saßnick (Vattenfall Europe Transmission GmbH, Berlin, Germany) and Bernd Michael Buchholz (Vice President, Siemens AG, Erlangen, Germany). Contributors include: John Olav Tande (SINTEF Energy Research, Trondheim, Norway), Juan Manuel Rodríguez García, Fernando Soto Martos, and David Alvira Baeza (Red Eléctrica de España. Madrid, Spain), and Susana Bañares (Red Eléctrica Internacional., Madrid, Spain).

7.6 References

[1] S. Kohler. Energiewirtschaftliche Planung für die Netzintegration von Windenergieanlagen in Deutschland. VDE- Kongress 2004, ETG Fachtagung „Nachhaltige Energieversorgung", Berlin 18.-20. Oktober 2004

[2] U. Keussen. Folgen der Windenergieeinspeisung und Verteilung ihrer Lasten. FGE Tagung 2003, Aachen, 11.-12. September 2003

[3] Y. Sassnick et.al. Influence of Increased Wind Energy Infeed on the Transmission Network. CIGRE 2004, Paris, 29.August-3.September 2004

[4] REA* Generating Plants Connected to the High- and Extra-High Voltage Network, Guidelines (in Addition to the Grid Codes) for Renewable-Based Generating Plants Connection to and Parallel Operation on the High- and Extra-High Voltage Network. VDN, August 2004

[5] Bernd Michael Buchholz, Zbigniew A. Styczynski, Wilhelm Winter. Dynamic Simulation of Renewable Energy Sources and Requirements on Fault Ride Through Behavior. Proceedings of the IEEE PES General Meeting 2006 Montreal, Canada

[6] E.ON Grid Code-High and Extra High Voltage,. E.ON Netz GmbH, Bayreuth, 1 April 2006
[7] Eigenerzeugungsanlagen am Mittelspannungsnetz, VDEW Verlag, 2. Ausgabe 1998
[8] B. Buchholz et. al. Advanced Planning and Operation of Dispersed Generation ensuring Power Quality, Security and Efficiency in Distribution Systems. CIGRE 2004, Paris, 29.August-3.September 2004
[9] J.Scholtes, C. Schwaegerl. Energy Park KonWerl: Energy Management of a Decentralized Supply System. Concept and First Results. First International Conference on the Integration of Renewable Energy Sources and Distributed Energy Resources. Brussels, 1-3. December 2004.
[10] IEC 61400-21: Wind Turbine Generator Systems--Part 21: Measurement and Assessment of Power Quality Characteristics of Grid Connected Wind Turbines, Ed. 1.0, International Standard, 2001.
[11] Requirements for Wind Farms connected to the Transmission Grid. Second Edition. Document No 74174. Eltra (2000). http://www.eltra.dk/media/1030_12368.pdf. In Danish.
[12] M.T. Palsson, T. Toftevaag, K. Uhlen, J.O.G. Tande, "Large-Scale Wind Power Integration and Voltage Stability Limits in Regional Networks", 2002 IEEE-PES Summer Meeting, Proceedings.
[13] M.T. Palsson, T. Toftevaag, K. Uhlen, J.O.G. Tande (2003) Control Concepts to Enable Increased Wind Power Penetration. 2003 IEEE-PES Summer Meeting, Proceedings.
[14] J.O.G. Tande, K. Uhlen (2004) Cost Analysis Case Study of Grid Integration of Larger Wind Farms, Wind Engineering Volume 28, No. 3, 2004, pp 265–273.
[15] Royal Decree 436/2004
[16] Ministerio de Industria y Energía. Normas administrativas y técnicas para funcionamiento y conexión a las redes eléctricas de centrales hidroeléctricas de hasta 5.000 kVA y centrales de autogeneración eléctrica. B.O.E. 219, 12 September 1985.
[17] Impact of Wind Energy Generation on the Safety of the Electrical Transmission Network. Juan M Rodríguez , David Alvira red eléctrica de España. Domingo Beato, Ramón Iturbe, J. C. Cuadrado Empresarios Agrupados. Mariano Sanz, Andrés Llombart circe. Universidad de Zaragoza. José R Wilhelmi Universidad Politécnica de Madrid. Cigre 2004
[18] Condiciones Técnicas Aplicables a la Generación de Régimen Especial no Gestionable. Red Elé

8

Europe: Impact of Dispersed and Renewable Generation on Power System Structure

8.1 Introduction

In Europe the dependency on imported primary energy is increasing annually. As a countermeasure against this growing dependency, national programs inside the European Community are directed at increasing the share of renewable energy sources and the efficiency of power generation by cogeneration of heat and power (CHP). Targets have been set by the European Commission for each country to gain a sustainable electricity supply in the future.

Generally, the share of renewable energy sources has to be increased by 2010 from 14% to 22% and the share of CHP has to be doubled from 9% to 18%.

Today approximately 50 GW of wind power are operated in Europe, and about 50 % of it is located in Germany. Assuming that wind power production will grow primarily in the form of large wind farms feeding into the transmission grids with an additional 35 GW installed power by 2010, the dispersed generation based on CHP and small renewable sources shall achieve an additional growth to meet the mentioned goals.

The output of most of the renewable energy sources depends on meteorological conditions and the CHP output is driven by the demand for heat. The question arises, how can the power system be operated with such a large share of mostly non-dispatched power sources? How can the reserve power be limited, which is required for compensation of power fluctuations and ensuring a safe network operation?
Thus, it has become clear that advanced planning and energy management approaches have to be introduced to ensure that the existing high level of power quality will exist in the future as well.

In this context, the power system of the future might consist of a number of self-balancing distribution network areas. In each of these areas a significant share of the power demand will be covered by renewable and CHP generation. However, the power balance of these areas should be planable and dispatch able in such a way that the import or export of power from or into the higher-level network has to follow a schedule, which can be predicted with a high level of accuracy in advance.

As the result of this future set-up, the distribution networks will become active and have to provide contributions to such system services like active power balancing, reactive power control, islanded operation and black-start capability. These services have to be coordinated with the transmission system operators where the responsibility for system stability will be allocated in the future as well. On the other hand, large-scale integration of wind power at the transmission level combined with an international area for trading energy will lead to higher utilization of the transmission grids. Consequently, the transmission capability has to be strengthened and short-term congestions have to be managed in an efficient and innovative way.

8.1.1 New Challenges

Each of these trends creates new challenges for power system operation on all of its levels and requires the introduction of advanced and economic solutions concerning:

- Supervisory control for congestion management
- Real-time security assessment
- Coordinated centralized and decentralized energy management including the unit commitment based on predictions of fluctuating power sources, demand side and storage management
- Coordinated trade of energy and transmission capacity.

The new tasks require a significant growth of information exchange. Communication networks using the existing infrastructure with different communication technologies like radio channels, power line carrier, fiber optics or traditional telecommunication cables will be the means of exchange. International communication standards shall be applied to simplify the engineering and operation of these new types of communication networks.

Under these mentioned circumstances the interplay of transmission and distribution will reach a new quality.

8.2 Distributed Generation: Challenges and Possible Solutions

Distributed generation (DG), for the moment loosely defined as small-scale electricity generation, is a fairly new concept in electric energy markets, but the idea behind it is not new at all. In the early days of electricity generation, distributed generation was the rule, not the exception. The first power plants only supplied electric energy to customers connected to the '*microgrid*' in their vicinity. The first grids were DC based, and therefore, the supply voltage was limited, as was the distance covered between generator and consumer. Balancing supply and demand was partially done using local storage, i.e. batteries, directly coupled to the DC grid. Today, along with small-scale generation, local storage is also returning to the scene.

Later, technological evolutions, such as transformers, led to the emergence of AC grids, allowing for electric energy to be transported over longer distances, and economies of scale in electricity generation led to an increase in the power output of the generation units. All this resulted in increased convenience and lower per-unit costs. Large-scale interconnected electricity systems were constructed, consisting of meshed transmission and radially operated distribution grids, supplied by large central generation plants. Balancing supply and demand was done by the averaging effect of the combination of large amounts of instantaneously varying loads. The security of supply was guaranteed by the built-in redundancy. In fact, this interconnected high-voltage system made the economy of scale in generation possible, with the present 1.5 GW nuclear power plants as a final stage in the development. Storage is still present, with the best-known technology being pumped hydro plants.

In the last decade, technological innovations and a changing economic and regulatory environment resulted in a renewed interest for DG. This is confirmed by the IEA [1]. This chapter presents the technical challenges and possible solutions when large amounts of distributed generation are introduced.

8.2.1 Drivers for DG

The IEA identifies five major factors that contribute to the renewed interest in DG. These five factors can be grouped under two major driving forces, i.e. electricity market liberalization and environmental concerns. The developments in small-scale generation technologies have been around for a long time, but were as such not capable of pushing the "*economy of scale*" out of the system. Although it is sometimes indicated, it may be doubted that DG is capable of postponing, and is certainly not capable of avoiding, the development of new transmission lines, as, at the minimum, the grid has to be available as backup supply.

8.2.1.1 Liberalization of electricity markets

There is an increased interest from electricity suppliers in DG, because they see it as a tool that can help them fill in niches in the market, in which customers look for the best-suited electricity service. DG allows players in the electricity sector to respond in a flexible way to changing market conditions. In liberalized markets, it is important to adapt to the changing economic environment in the most flexible way. DG technologies in many cases provide flexibility because of their small sizes and assumed short construction lead times compared to most types of larger central power plants. However, the lead-time reduction is not always that evident. For instance, public resistance to wind energy and use of landfill gasses may be very high.

Many DG technologies are flexible in several respects: operation, size and expandability. Making use of DG allows a flexible reaction to electricity price evolutions. DG then serves as a hedge against these price fluctuations. Apparently, this is the major driver for the US demand for DG, i.e. using DG for continuous or peaking use (peak shaving). The energy efficiency is sometimes very debatable. In Europe, market demand for DG is, for the moment, driven by heating applications (through CHP), the introduction of renewable energies and potential efficiency improvements.

The second major driver of US demand for DG is quality of supply or reliability considerations. Reliability problems refer to sustained interruptions, being voltage drops to near zero (usually called outages). The liberalization of energy markets makes customers more aware of the value of a reliable electricity supply. In many European countries, the reliability level has been very high, although blackouts have occurred in recent years.

Customers do not really care about supply interruptions, as they do not feel it as a great risk. However, this may change in liberalized markets. A high reliability level implies high investment and maintenance costs for the network and generation infrastructure. Because of the incentives for cost-effectiveness that come from the introduction of competition in generation and actions from regulators aiming at short-term tariff reductions for network companies, it might be that reliability levels decrease. However, having a reliable power supply is very important for society as a whole, and industry specifically (chemicals, petroleum, refining, paper, metal, telecommunications, ...). Companies may find the grid reliability to be of an insufficient level and decide to invest in DG units in order to increase overall reliability of supply to the desired level.

Apart from voltage drops to near zero (reliability problems), one can also have smaller voltage deviations. The latter deviations are aspects of power quality. Power quality refers to the degree to which power characteristics align with the ideal sinusoidal voltage and current waveform, with current and voltage in balance [2]. Thus, strictly speaking, power quality encompasses reliability.

Insufficient power quality can be caused by failures and switching operations in the grid, mainly resulting in voltage dips, interruptions, and transients and by network disturbances from loads yielding flicker (fast voltage variations), harmonics, and phase imbalance. The nature of these disturbances is related to the 'short-circuit capacity', being a measure for the internal impedance in the grid, depending on its internal configuration (e.g. length of the lines, short-circuit capacity of generators and transformers) [3].

DG could partially serve as a substitute for investments in transmission and distribution capacity (demand for DG from T&D companies) or as a bypass for transmission and distribution costs (demand for DG from electricity customers). This is only possible to the extent that alternative primary fuels are locally available in sufficient quantities. For example, increased use of DG could result in new congestion problems in other networks, such as the natural gas distribution network.

Finally, DG can also contribute in the provision of ancillary services, including those necessary to maintain a sustained and stable grid operation of the customers. This may be the capability of the grid operator to generate active power on demand, for instance to stabilize a dropping frequency due to a sudden under capacity in generation or excess demand, or reactive power to support the voltage.

8.2.1.2 Environmental Concerns

At present, environmental policies are probably the major driving force for the demand for DG in Europe. Environmental regulations force players in the electricity market to look for cleaner energy solutions. Here, DG can also play a role, as it allows optimizing energy consumption of firms that have a large and constant demand for heat. Furthermore, most government policies aiming to promote the use of renewables also results in an increased impact of DG technologies, as renewables, except for large hydro and wind parks (certainly off-shore), have a decentralized nature.

8.2.2 Grid Protection and DG

Power can flow in a bi-directional way within a certain voltage level, but it usually flows uni-directionally from higher to lower voltage levels, i.e. from transmission to distribution grid. An increased share of DG units may induce power flows from low into medium-voltage grid. Thus, different protection schemes at both voltage levels may be required [4].

Safe operation and protection are to be guaranteed at all times. In addition, the protection system has to be sufficiently selective; in order to optimize reliability and availability of supplied power. This is less simple than it seems, since the fault current not only comes from the main power system grid in a unidirectional way, but also from the DG units, making detection far more complicated and the conventional hierarchy (selective) protection methods might fail. Therefore, a more 'active' protection system with some form of communication is required to keep up the required level of safety in the future.

The protection problems are illustrated by using a distribution system with five feeders in Figure 8.1. If a short circuit occurs at F2 or F3, the short-circuit current is supplied by the generators connected to this Feeder (G1 and G2), other DG units in adjacent feeders, and the main grid. If the contribution to the short-circuit current of G1 and G2 is large compared to that of the grid and the other Feeders, the current through the circuit breaker and fuse CB1 might be too low to operate in order to eliminate the short circuit in the feeder. On the other hand, if the contribution to the short-circuit current from generators in adjacent feeders is significant, healthy feeders (Feeder 4) might be disconnected before the faulty feeder is disconnected.

As long as islanding is not intended to backup a loss of mains, it should be avoided [5]. According to technical standards (e.g. IEEE 1547), DG must be automatically disconnected when faults or abnormal conditions occur, with the assumption that interconnection systems detect such conditions.

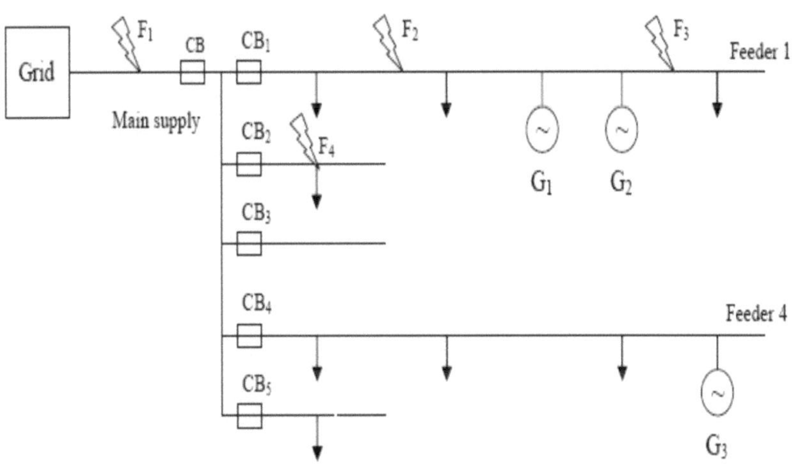

Fig. 8.1. Grid with Safety Problems due to High DG Penetration

In this way, conventional protection selectivity can be restored, guaranteeing person and equipment safety. In the future, when more DG is used, this requirement would reduce expected benefits of DG. To make optimal use of DG, unnecessary disconnection of DG should be avoided. Generators should be able to ride through minor disturbances [6].

DG flows can reduce the effectiveness of protection equipment. Customers wanting to operate in 'islanding' mode during an outage must take into account important technical (e.g. the capability to provide their own ancillary services) and safety considerations, such that no power is supplied to the grid during the time of the outage. Once the distribution grid is back in operation, the DG unit must be resynchronized with the grid voltage.

8.2.3 Voltage Quality and DG

Imbalances between demand and supply of electricity cause the system frequency to deviate from its rated 50/60 Hz value. These deviations should be kept within very narrow margins, since the proper functioning of many industrial and household applications depends on it. In economic terms, system frequency can be considered as a public good. As a consequence, the transmission grid operator is appointed to take care of the system frequency as well as of other services with a public good character that need to be provided.

The installation and connection of DG units are also likely to affect the system frequency. These units will free ride on the efforts of the transmission grid operator or the regulatory body to maintain system frequency. They will probably have to increase their efforts and have an impact on plants efficiency and emissions. Therefore, the connection of an increasing number of DG units should be carefully evaluated and planned upfront.

The relation between DG and power quality is an ambiguous one. On the one hand, many authors stress the beneficial effects of DG for power quality problems [1], including the potential positive effects of DG for voltage support and power factor corrections [4].

On the other hand, large-scale introduction of decentralized power generating units may lead to instability of the voltage profile: due to the bi-directional power flows and the complicated reactive power equilibrium arising when insufficient control is introduced, the voltage throughout the grid may fluctuate. Eventually an '*islanding*' situation may occur in which a local generator keeps a part of a disconnected grid energized leading to dangerous situations for the repair personnel coming in.

Others also stress the potential negative externalities on power quality, caused by the installation of DG capacity. According to [7], the impact on the local voltage level of DG connected to the distribution grid can be significant. The same reaction was noted through the CIRED questionnaire [8], where, next to the general impact on power quality, a rise in the voltage level in radial distribution systems is mentioned as one of the main technical connection issues of DG. The IEA [1] also mentions voltage control as an issue when DG is connected to the distribution grid. This does not need to be a problem when the grid operator faces difficulties with low voltages, since in that case the DG unit can contribute to the voltage support. But in other situations it can result in additional problems.

Small and medium-sized DG units often use asynchronous generators that are not capable of providing reactive power. Several options are available to solve this problem. On the other hand, DG-units with a power electronic interface are sometimes capable of delivering reactive power.

Some DG technologies (PV, fuel cells) produce direct current. Thus, these units must be connected to the grid via a DC–AC interface, which may contribute to higher harmonics. Special technologies are also required for systems producing a variable frequency AC voltage. Such power electronic interfaces have the disadvantage that they have virtually no 'inertia', which can be regarded as a small energy buffer capable to match fast changes in the power balance. Similar problems arise with variable wind speed machines [7].

8.2.4 Practical Distribution Network

An existing Belgian medium voltage distribution system segment has been used to study the power quality and voltage stability with different DG units (Figure 8.2). The system includes one transformer of 14 MVA, 70/10 kV and four cable feeders. The primary winding of the transformer is connected to the transmission grid and can be considered as an infinite node. Normal operation of the distribution system is in radial mode and the connections at node 111 with feeders 2, 3 and 4 are normally open.

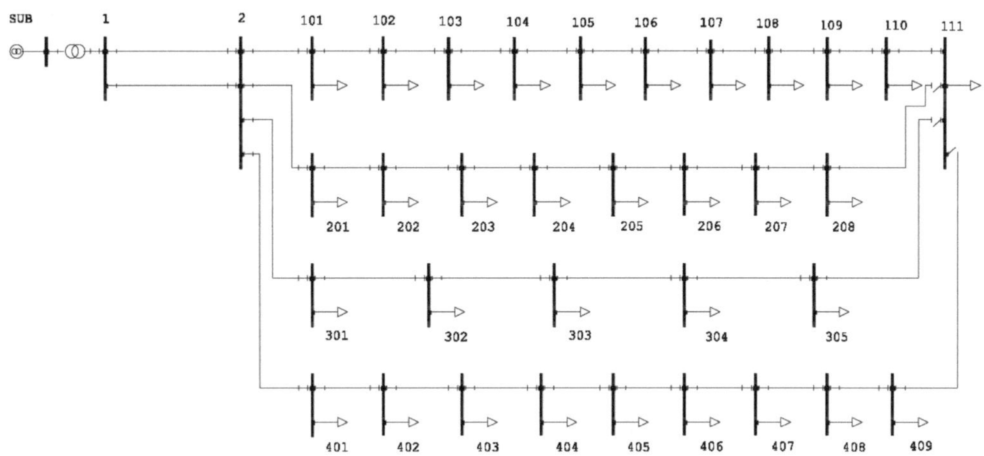

Fig. 8.2. Practical Distribution System

A DG unit is connected at node 406 of Feeder 4. The total load in the system is 9.92 MW, 4.9 MVAr. A synchronous and an induction generator are simulated with different power output. The synchronous generator is simulated at power factor 0.98 leading at 3 and 6 MW. The induction generator is simulated at power factor 0.95 lagging also at 3 and 6 MW. The power of the DG for both synchronous and induction generators raises the voltages of Feeder 4, compared to the base case without DG (Figure 8.3). For higher active and reactive power generation (synchronous 6 MW), an over voltage occurs at node 406 and its neighbors.

Figure 8.4 illustrates the voltage at node 406 with different power generation levels and power factors. Compared to the case where DG only injects active power or operates at the unity power factor, synchronous generators raise the voltage of the system faster due to reactive support. For induction generators, the voltage rise is slower and at a certain level of power generation, the voltage starts to decrease. This is due to the fact that induction generators need reactive power, yielding in a reduction of the voltage rise.

Through this study, it can be seen that the impact of induction generators is less than that of synchronous ones in terms of voltage rise (Figure 8.5). If an over voltage occurs with a synchronous generator, it has to operate under-excited and to absorb reactive power instead of injecting it.

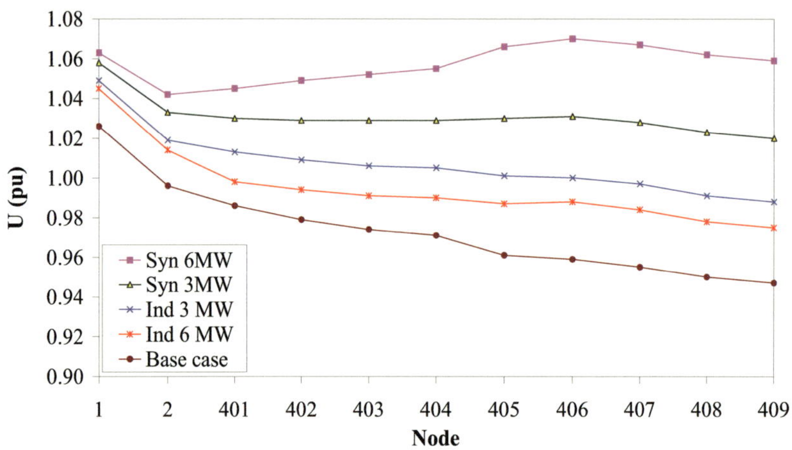

Fig. 8.3. Voltage Profile of Feeder 4 with DG Connected at Node 406

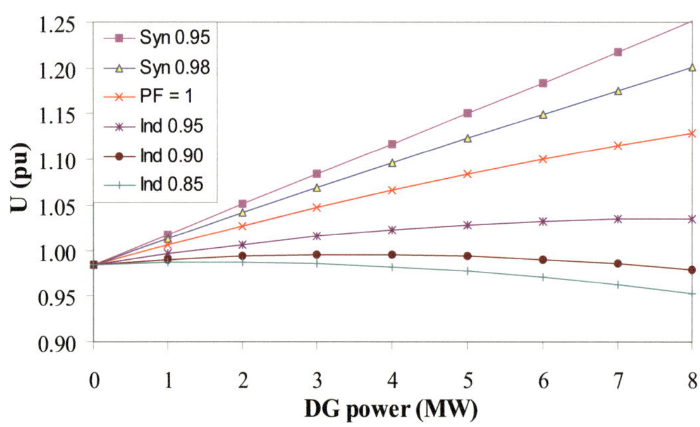

Fig. 8.4. Voltage at Node 406 with Different Power Factors

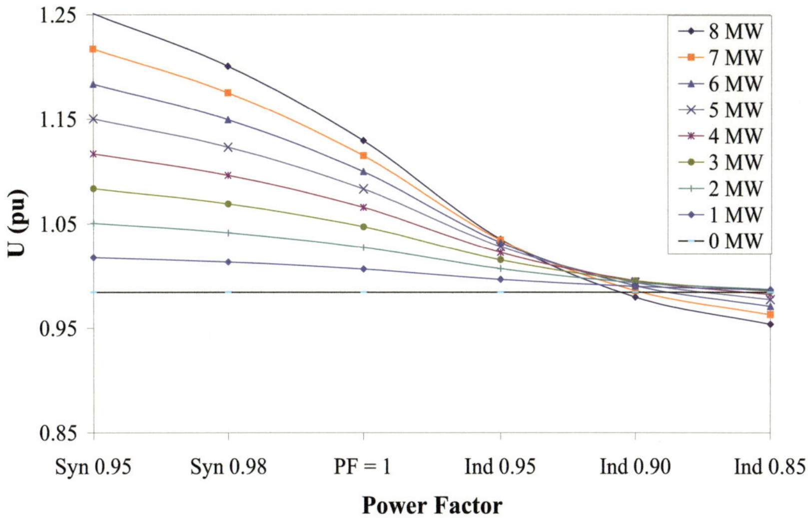

Fig. 8.5. Voltage at Node 406 with Different Power Generation Levels

In order to see the voltage fluctuation problem with DG, a photovoltaic (PV) system is used. The reactive power is produced by a capacitor of the inverter's grid filter and is almost constant. The PV system is treated as a PQ node with negative active power. The PV power is calculated from 5-s average irradiance data measured during one year in Leuven, Belgium. In this study, a PV array with 50 kW rated peak power is connected at node 304. Figure 8.6 shows the one-hour power output of the PV system at noon of a slightly cloudy summer day. In order to isolate the voltage fluctuation impact of PV from short-time load variation at individual nodes, the loads are assumed constant during the calculation. The total load in the system is 4.4 MW, 1.9 MVAr. In Figure 8.6, the voltage fluctuations correspond to the variations of injected active power of the PV system. At times when clouds cover the sun, the power generated can quickly drop by 60%, causing sudden variations in node voltages in the range of 0.1%. The installed capacity of PV in this study is rather low compared to the capacity of the distribution system and the loads, so the value of voltage fluctuation is limited. However, with a high connection density or the connection of a large PV system, the voltage fluctuation problem might become more severe.

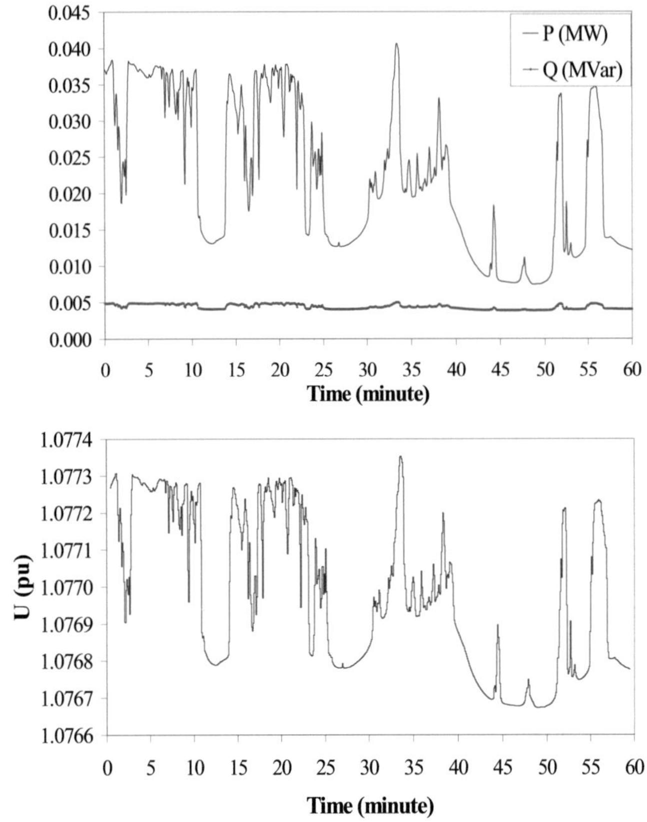

Fig. 8.6. Injected Power and Voltage at Node 304

A total DG capacity of 30% of the total system load is distributed equally over nodes 108, 204, and 406. The simulations have been carried out for induction and for synchronous generators. All operate at power factor 0.98 lagging. One of the 1-2 lines is opened during dynamic simulations at time t = 100 s. The distributed generators are connected at node 108, 204 and 406 with rated power 1 MW for both synchronous and induction generators.

The voltage dips are highest with constant power load characteristic and lowest with impedance load characteristic for both synchronous and induction generators (Figure 8.7 and Figure 8.8). With synchronous generators, after a short voltage dip, the voltage recovers close to the voltage level before the disturbance. For induction generators, the voltage does not recover due to the lack of reactive power support. There is not a big difference between a voltage dip in the base case and with DG connection, being around 1%. So the connection of DG in the distribution system does not affect dynamic voltage stability significantly. In most cases it reduces the voltage dip value.

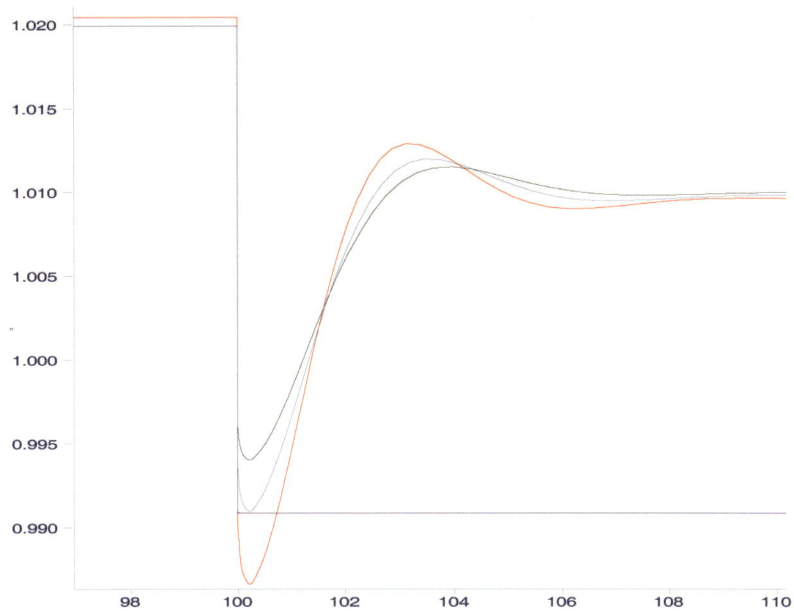

Fig. 8.7. Voltage Dip at Bus 2 with Synchronous Generator

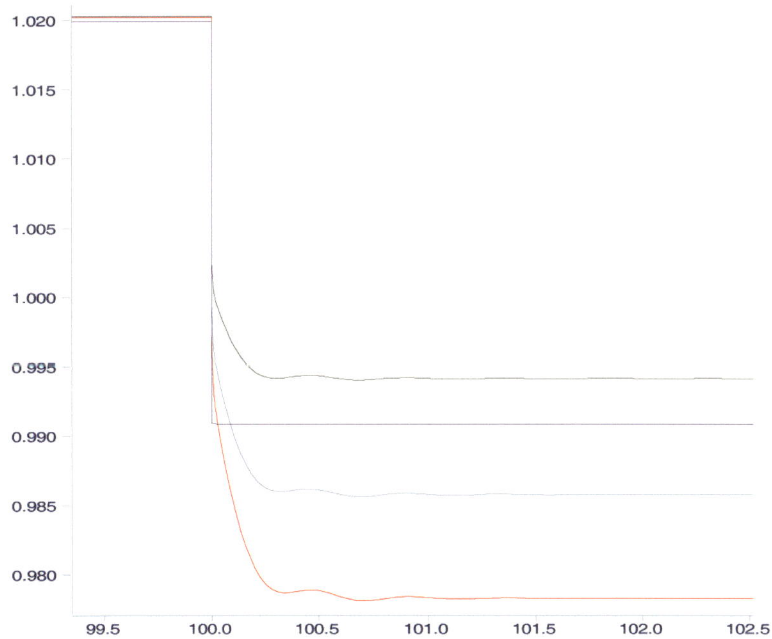

Fig. 8.8. Voltage Dip at Bus 2 with Induction Generator

In order to see the voltage dip problem when a DG starts up, an induction generator connected at node 108 with rated power of 3 MW is tested at lagging power factor of 0.9. When the induction generator starts up, it causes a transient and a voltage dip of up to 40%

in the system and lasts for several seconds (Figure 8.9). It is due to an initial magnetizing inrush transient and power transfer to bring the generator to its operating speed [9]. This results in a major problem for sensitive loads connected near the DG. If the distribution system is equipped with an under-voltage relay and the DG unit has islanding protection, the voltage dip may lead to an action of the protection relay resulting in an outage of the system. A soft-start circuit is required for large connected induction DG.

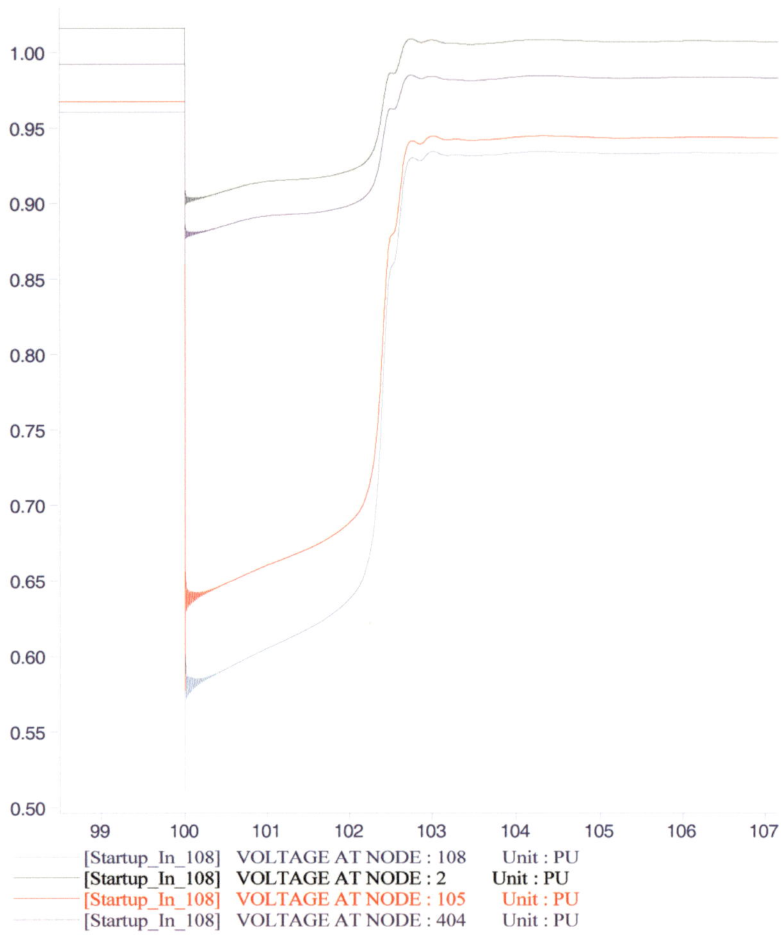

Fig. 8.9. Voltage Dip when an Induction Generator Starts Up

The voltage stability was studied for synchronous and induction generators with three cases of DG connection: a) one DG unit connected at node 108, b) one at node 2, c) DG units distributed in the system at nodes 108, 204, 406. The total load of the system is 9.92 MW, 4.9 MVAr, all impedances. The total installed capacity of the DG units in all cases is 3 MW. The voltage stability at node 111, at the end of feeder 1, was studied. DG units generally increase the voltage and support stability in the system (Figure 8.10 and Figure 8.11), however the connection point of the DG influences the voltage stability in the system. DG strongly supports the voltage at nearby nodes and has less impact on distant ones. This is also true for the other load characteristics. Compared to induction DG, the synchronous generator has a larger impact on the voltage stability because of its capability of reactive power injection. On the other hand, the influence of induction DG on voltage stability is not so different from the base case (without DG).

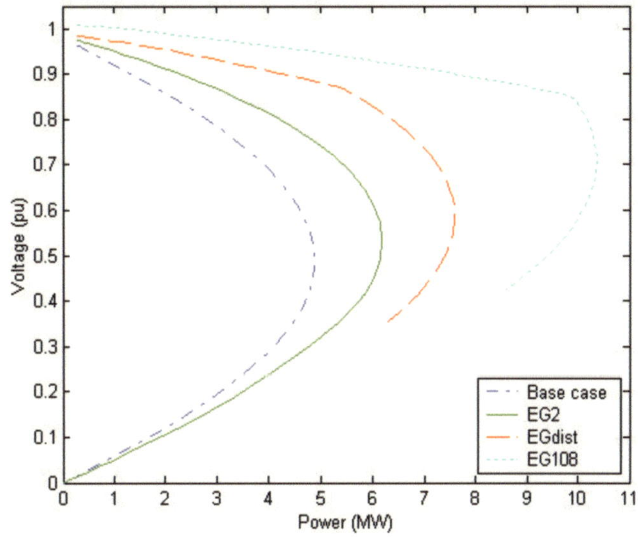

Fig. 8.10. Static Voltage Stability at Node 111 with a Synchronous Generator

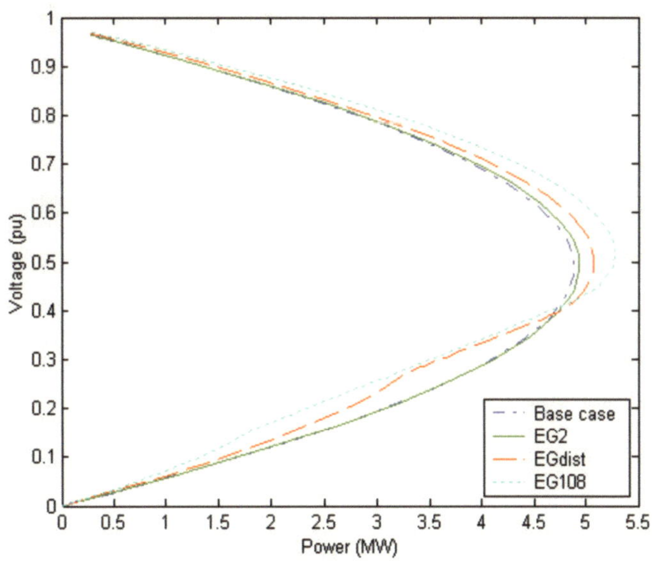

Fig. 8.11. Static Voltage Stability at Node 111 with an Induction Generator

8.2.5 Energy Security

In some discussions, energy security is linked to the diversification of primary energy supplies, while in others it is interpreted as the reliability of the electricity system. Under the first interpretation, energy security improves as the diversification of primary energy supplies increases. In this case, the advantages of DG are limited, as most technologies are directly or indirectly dependent on natural gas - with the exception of systems based on renewables.

Under the second interpretation, many authors think [1] that DG can contribute to reduce the risks and costs of blackouts. Here, DG is seen as an instrument that helps to reduce the private costs and risks of system failures for electricity customers. Others, like [8], claim that DG does not contribute to system security. On the contrary, it would have a negative effect. Such a negative impact on the system security occurs when the share of non-dispatch able generation capacity increases. Examples of such units are wind turbines, photovoltaic systems and cogeneration units closely tied to heat demand. The latter units cannot be centrally controlled because of the natural variability of their power supply. As a consequence, there is an increased need for regulating (backup) power.

8.2.6 General Summary

Section 8.2 began with the observed renewed interest in small-scale electricity generation. General elements of the drivers for this development are discussed both from the economic and environmental point of view. Small-scale generation is commonly called DG and we have tried to derive a consensus definition for this latter concept. It appears that there is no agreement on a precise definition as the concept encompasses many technologies and many applications in different environments. In our view, the best definition of DG that generally applies seems to be 'an electric power generation source that is connected directly to the distribution network or on the customer side of the meter'. Depending on the interest or background of the one confronted with this technology, additional limiting aspects might be considered. A further narrowing of this 'common denominator' definition might be necessary depending on the research questions that are being investigated. However, a general and broadly understandable description as proposed here, is required to allow communicating on this concept.

From a technical viewpoint, Section 8.2 discusses the impact on the protection and the safety of the grid. A lot of attention has been paid to the interaction of the DG units with the quality of the grid voltage. Both static and dynamic voltage analyses are used to demonstrate the interactions. The choice of generator type has a major influence: two types are distinguished, synchronous and induction; the impact of the power electronic converter that may be used was discussed. An actual grid was used for supporting the results obtained by simulations.

8.3 New Tasks Create New Solutions for Communication in Distribution Systems

8.3.1 Basic Principles and Tasks

In the environment of a growing share of dispersed and renewable generation the distribution networks will change from passive to active systems. In the power systems of the future distribution networks will also have to contribute to the system services in coordination with the transmission system. The idea of virtual power plants (VPP) will become reality where a number of dispersed and renewable generation units (partially with intermittent power output), storage units and controllable loads will be clustered and managed in such a way that the power exchange with the outer world can be scheduled and dispatched with a high level of accuracy. The decentralized energy management inside VPPs requires communication facilities that are mostly not applied in today's practice of the distribution system operation.

The efficiency of future communication networks at the distribution level requires some basic principles.

In contrast to the existing practice, where power generation is located on a rather concentrated area and, therefore, information and data is transferred on local networks or field busses, the supervisory control and dispatching of dispersed generation will be spread over a wide area. For economical reasons the pre-existing infrastructure has to be used; that also means the utilization of different communication channels like radio, fiber optics, power line carrier and telecommunication cables will be applied within one network as long as they are available in the environment.

The communication over the different physical layers has to be compliant to a common standard regarding data modeling and communication services. The main requirements for such a standard are:

- plug and play ability,
- possibilities for mapping to different physical layers,
- expandability of the data models and introduction of new models in accordance with the new and enhanced communication tasks.

Thus, if the communication network for dispatching the VPP covers a whole distribution network additional system services can be provided by the same network. Therefore, communication tasks for distribution networks of the future include:

- the contribution to the active power balancing through dispatch of power generation, storage and controllable loads in the framework of a VPP,
- the transfer of metered values as a support for the decentralized energy management and for billing,
- the provision of further system services like congestion management, reactive power and voltage control, fault location, network recovery after faults, islanded operation, black start capability etc.

The application of these ideas is investigated in the framework of the German "Network for Energy and Communication", a project sponsored by the German Ministry for Education and Research.

8.3.2 Case Study

The design of the communication network was investigated for a typical distribution network shown in Figure 8.12.

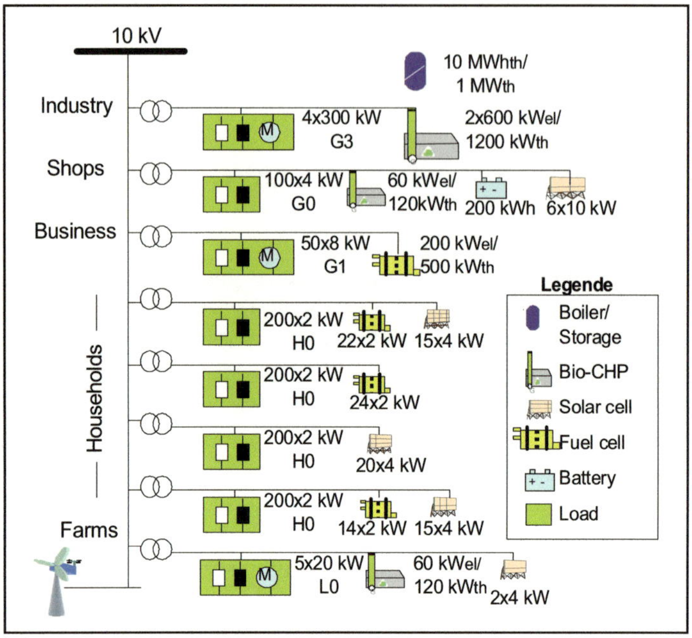

Fig. 8.12. Considered Distribution Network

Along a 10 kV feeder 8 ring main units supplying different types of low voltage consumers are connected. At the end of the feeder there is a further wind power plant.

The low voltage lines supply consumers with different load profiles in accordance with the German standard load profile types defined by the German Network Society (VDN): G3 - industry, G0 - shopping center, G1 - business center, H0 - households, L0 - rural farm. Various D&RES and storage units are located in the low voltage networks as shown in Figure 8.12. They provide their specific generation profile partially depending on weather conditions. Demand side management is planned with 12 x 20 kW in the industrial network, 10 x 2 kW in the shopping area and 40 x 2 kW in the business center.

For the distribution network described, the optimum communication network has to be designed in accordance with the following criteria:

- A maximum latency time is assigned to each class of information, e.g.

- Control with return information	2 s
- Alarm	1 s
- Event message	5 s
- Metered or measured value	2 s
- Power schedule (96 target values)	20 s

- The content and the classes of information exchange have to be defined for each active component of the network - loads, generators, storage units, substation equipment. The amount of data for communication varies by type. For example, only the metered value will be communicated every 15 minutes for non-controllable loads or photovoltaic units. On the other hand, the larger CHP plants provide 6 alarms, 24 event messages, 12 measured and 2 metered values, 6 controls, 2 target values as well as target profiles for active and reactive power.
- The volume of data transfer has to be defined in accordance with operational needs for worst case and normal scenarios. In the normal case the metered values of all components will be transferred in a 15 minutes interval. One time per day the target profiles of the generation units above 100 kW will be communicated. Furthermore, 40 target values, 20 event messages, and 10 controls will be communicated. In the worst-case scenario (e.g. voltage dip) each component will send a report with alarms and measured values, and this has to be performed within 5 s.
- The selection of the communication protocol defines the data volume for each data class. Chapter IV discusses special features of available IEC standards, in particular the application of IEC 61850.
- The selection of communication channels is based on their availability, a cost comparison of different alternatives and the baud rates providing the performance in worst case and normal scenarios.

The experience gained in initial pilot projects with VPPs [10], [11] underlined the need to apply communication protocols based on common standards for all channels used. Otherwise the engineering expenses will grow and the operation of the communication network will become inconvenient.

8.3.3 Communication Standards

The first international standards for digital communication in power systems were developed in the 1990s. These standards were limited regarding their 'plug and play' ability. Figure 8.13 gives an overview of the IEC standards for supervisory control in electric networks.

Only the latest standard IEC 61850 for communication in substations (published as standard in 2004) responds to the requirements of chapter II, topic 2.

The 'plug and play' - ability is reached by the detailed object modeling based on logical nodes (objects like circuit breaker or transformer etc.) and data (information like "status ON" or "Buchholz alarm" etc.) with the supplement of different attributes (like time stamps, validity information etc.) [12].

The mapping to different application layers was foreseen in the reference model of the standard in accordance with Figure 8.14.

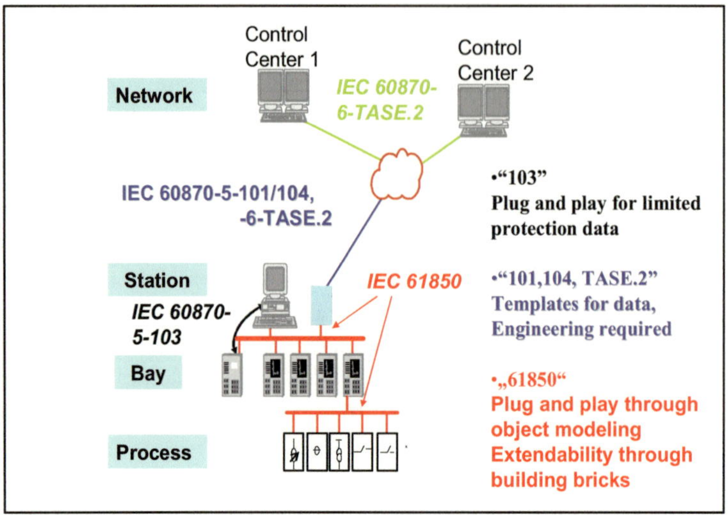

Fig. 8.13. IEC Standards for Communication in Electric Networks

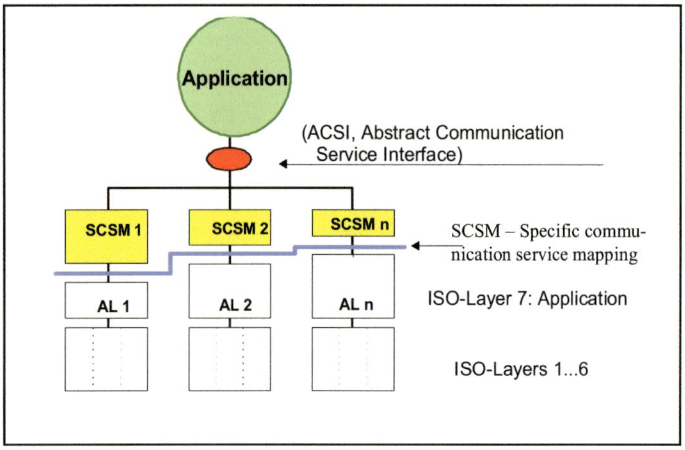

Fig. 8.14. Reference Model of IEC 61850

The abstract communication service model describes the data models and the services in an abstract form. The protocol requires the definition of all layers of the ISO/OSI model. IEC 61850 defines in Part 8.1 the 'Manufacturing Message System' (MMS) as the base for the first standard conform application layer (AL1) and Ethernet for the lower layers. The specific communication system mapping SCSM ensures the adaptation of the services and models to the layers selected. This way, IEC 61850 allows the adaptation of future communication methods to the core elements of the standard - the ACSI (described in the parts 7.1-7.4). Consequently, through the SCSM different link and physical layers can be applied.

Last but not least, the object models can be extended on demand. IEC 61850 defines the rules for building such extensions.

As a result of these features the standard IEC 61850 is suitable to serve as a general standard for all communication tasks in power systems. Therefore, the basic rules and models of IEC 61850 are inherited in the following subsequent standards:

- IEC 61400-25 for communication of wind power plants [13],
- IEC 62350 for communication of dispersed generation [14].

As a goal of the new standards it was declared that all existing services and models of IEC 61850 would be taken over, as defined and only necessary extensions will be added.

In accordance with Figure 8.12 there will be a need to communicate information from wind power plants, other D&RES and substation equipment over a common communication network. Consequently, the consistency of the data models used is mandatory.

The relevant IEC working groups of TC 57 (62350) and TC 88 (61400-25) are requested to ensure the consistency of all subsequent standards with IEC 61850. Otherwise there will be no acceptance of the new standards from both the power automation industry and the utilities.

8.3.4 Design of the Communication Network

IEC 61850 was analyzed regarding the size of telegrams for each data class. The results in Table 8.1 present the worst case, which means the maximum possible number of bytes. In practice the services of IEC 61850 create reports within a given time interval in which all changed information is to be embedded. Therefore, the net bytes will be much lower then stated. However, these figures build a good base for the communication network design. The design task consists of the distribution of the communication clients over the possible communication channels with minimum expenses and under the condition that the baud rates of the selected channels ensure the required performance in worst case and normal scenarios. A possible design of the communication network which meets the performance requirements and combines different physical channels is shown in Figure 8.15. The large CHP- plants of the industrial network play a significant role in the power balance of the distribution network and impact the energy tariff of the industrial plant. They are connected by a dedicated ISDN line that was available. The other generation and storage units in the shopping and business area as well as the access to weather forecast data (for load and renewable generation prediction) need only a dial up line. The wind power plant is connected via a radio channel with the aim to combine this kind of communication with the others.

The main load of communication is assigned to the 'Distribution Line Carrier' (DLC), which can reach baud rates higher than 300 kBd [15]. Over this channel the dispersed generation units in the household and rural networks communicate, the metered values of all loads are reported, the control commands for demand side management are sent out and the equipment in the substations is incorporated to provide a new class of distribution system management. For this network the installation of new communication lines was avoided.

Data class	Raw data array	Overhead Layer 7 (MMS)	Overhead Other layers	Overall
Status inform.	11	161	64	236
Control	14	1245	384	1643
Measured value	15	161	64	240
Metered value	15	161	64	240
Array (96 metered values)	1440	1320	128	2888
Target value	15	693	192	900
Schedule (96 target values)	480	388	128	996

Table 8.1. Telegram Size for Different Data Classes (Byte)

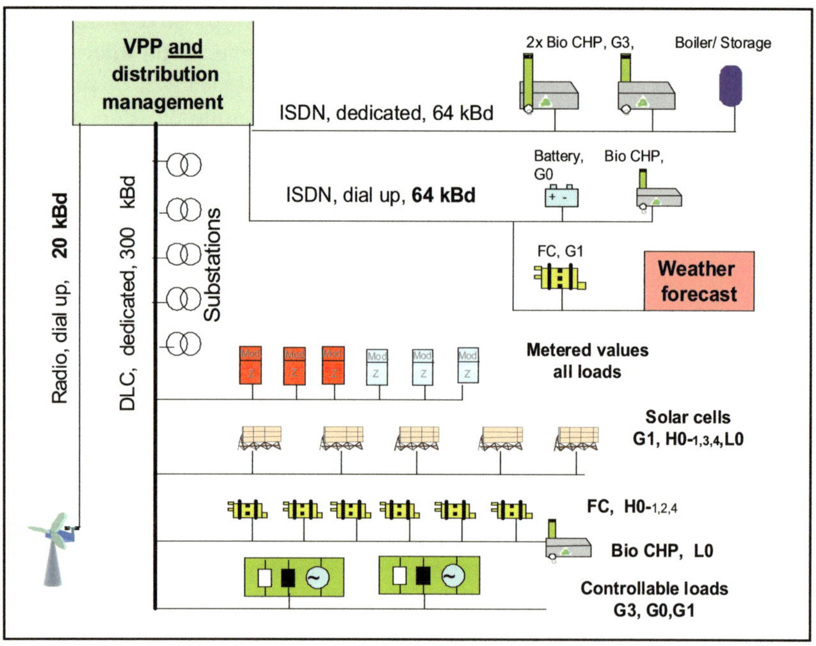

Fig. 8.15. Scheme of the Communication Network

8.3.5 Benefits for Other System Services

The availability of communication channels in the distribution level allows for the improvement of various system services. The example of a supply restoration after faults is demonstrated in Figure 8.16.

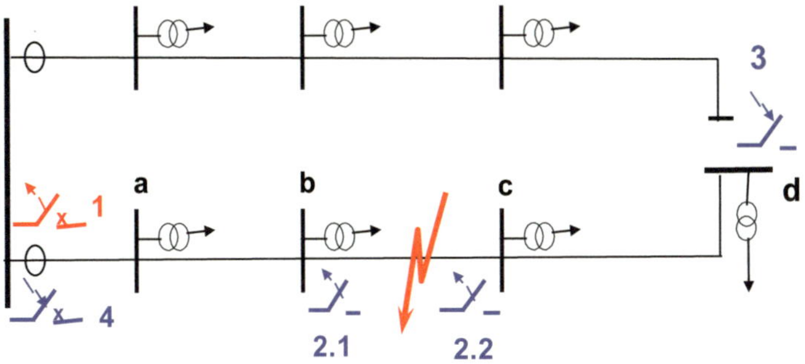

Fig. 8.16. Supply Restoration after Faults in an Open Loop

Preferably, distribution networks are operated with open loops. The loop in Figure 8.16 is disconnected in the ring main unit "d". In the event of a fault the protection of the feeding substation trips and the faulted feeder will be switched off. All ring main units (a-d) lose their supply (operation 1). Now, the maintenance staff allocates the faulted feeder part through driving along the feeder and reading the fault indicators in the ring main units. After localization the faulted part is disconnected by the switching operations 2. The restoration of supply is provided after that by the switching operations 3 and 4. The whole restoration procedure takes more than one hour on average. But, if communication channels are available in the ring main units the restoration procedure can be performed remotely. The restoration time will be shortened up to only a matter of minutes. Similar benefits can be demonstrated for other system services as well. Therefore, communication is a key to improve power quality.

In summary, the expected large-scale penetration of D&RES requires a new sharing of system services between the transmission and distribution levels. The distribution networks will become more active and communication networks have to be established for that purpose.

Setting up a cost efficient communication requires the use of existing communication channels and of standardized protocols. With the help of an example of a distribution system with different characteristics for load and generation, the design of the communication network was investigated. It was shown that IEC 61850 provides the required features to serve as a communication standard. However, the consistency of the subsequent standards IEC 61400-25 and IEC 62355 should be reached as a prerequisite for a broad acceptance in practice. Examples of actual deviations and inconsistencies are given. Furthermore, it was shown that the communication tasks of the example system can be performed by a combined communication network with dedicated and dial up communication channels on different physical media like ISDN, Distribution Line Carrier and radio transmission. The availability of communication channels in the distribution level benefits the management of system services and helps to improve the power quality.

8.4 Integrating Dispersed Generation into the Danish Power System

Denmark is electrically divided into two parts - western Denmark forms the northern part of the continental European synchronous area - and eastern Denmark constitutes the southern part of the Nordel synchronous area (Figure 8.17). The eastern and western Danish networks are planned to be connected by a High Voltage Direct Current (HVDC) link by the year 2009. As a link between the two synchronous areas, Denmark faces high-energy transits.

Since the early 1980s a huge amount of dispersed generation has been implemented into the grid - mainly in the continental European synchronous area part of Denmark, where today 23 % of the energy consumption is produced by wind turbines and about 32 % by CHP units. More than 50% of the total production capacity is implemented within local distribution grids, making control and forecasting of system operation very challenging.

Thus the transmission system operation requires careful planning as well as intelligent utilization of possibilities offered by the liberalized electricity market.

Energinet.dk is responsible for the secure and reliable operation of the power system (and natural gas), a well functioning energy market and for owning, operating and expanding the transmission infrastructure for electricity (and natural gas). Daily operation of a system with massive infeed from uncontrolled generation units is a challenging task and depends strongly on interconnections to neighboring countries and a well functioning international electricity market. The further - also international - growth of wind power capacity may lead to increasing demand for national security of supply as well as the implementation of an international market for ancillary services for the efficient utilization of the available resources.

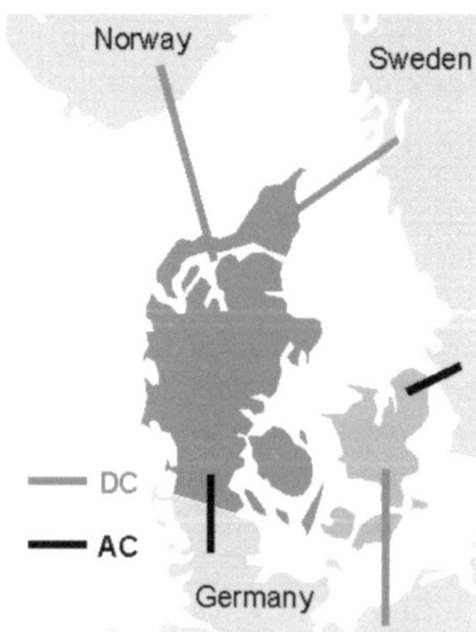

Fig. 8.17. Denmark between Two Synchronous Areas

8.4.1 System Overview

The transmission system in western Denmark is operated at 400 kV and 150 kV. To the south, it is connected to the continental European synchronous area via 400 kV, 220 kV and

150 kV AC-lines to Germany. To the north, it is connected to the Nordel synchronous area via HVDC links to Norway (1,000 MW) and Sweden (600 MW). The eastern Danish system is operated at 400 kV and 132 kV, respectively, as a meshed transmission system with AC connection to Sweden and HVDC connection to Germany.

Figure 8.17 and Figure 8.18 give the key figures of the Danish power system. The primary power plants are thermal units, fired by coal or gas. A significant part of today's installed capacity in the Danish system are decentralized units, such as wind turbines and combined heat and power (CHP) units, mostly connected to the distribution grid. This combination has resulted in a change of the classical hierarchical load flow structure - former passive networks have become active networks due to the changed load flow direction, especially on windy days.

In the western system the offshore wind farm Horns Rev A (HRA) with a rated power of 160 MW is connected to the 150 kV transmission system. The construction of the second offshore wind farm, Horns Rev B (HRB), with a rated power of 215 MW should be finished by the year 2009 [16].

In the eastern system another new big offshore wind plant with a rated power of 215 MW is planned to be operating in 2009 -2010.

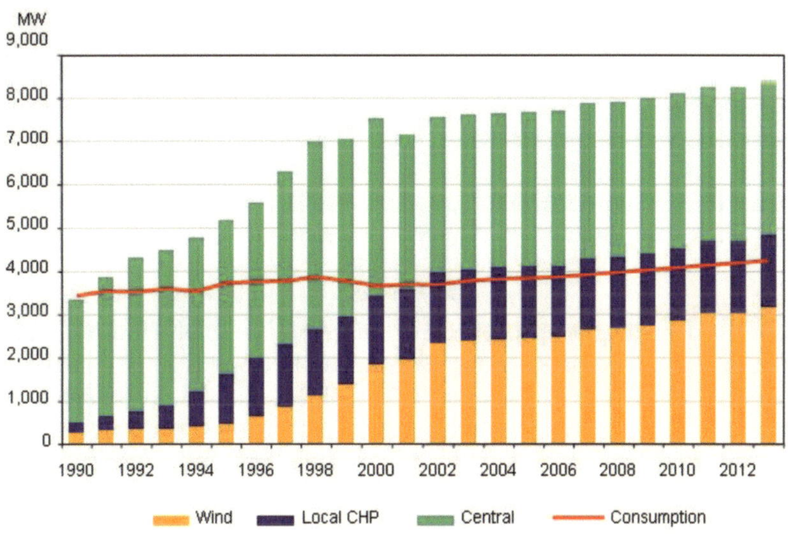

Top to bottom: Central, Local CHP, Wind. Line: Consumption

Fig. 8.18. Development of the Power Balance in Western Denmark

The significant share of wind power and CHP units has introduced several technical tasks to the system operator regarding the stable, safe and reliable operation of the transmission system, such as:

- Development of technical specifications for the grid connection of wind turbines that are based on prior experience; e.g. requirements like fault-ride-through capability of large offshore wind farms [17].
- Constant improvement of wind power forecasts.
- Long- and short-term balance for the Danish power system.
- Responsibility for voltage stability and power quality.
- Wind turbine modeling as part of the Danish power system model.
- Preparation of the system for the implementation of more wind power.

8.4.2 Wind Energy

Large offshore turbines usually are located close to each other and show significant correlation between their output powers. Experience from the operation of Horns Rev A (HRA) shows that power fluctuations within 10-min intervals can be remarkable high due to the concentration of wind power in a small area of about 20 km^2 [16]. The power gradients may reach values of 15 MW/min for this 160 MW wind farm resulting in changes of generated power from none up to the rated power within 10 to 15 minutes. Without control such power fluctuations may be introduced into the transmission system and even distributed to the neighboring transmission systems.

A control system has been developed which reduces this effect [16]. This is achieved by applying power gradient limits of the wind farm and by using secondary control of the primary power plants and, additionally, using fast power control of HVCD. The main target of keeping the power balance is to adjust power generation including power import and power consumption, including power export, as well as keeping the power exchange between western Denmark and the UCTE synchronous area at the planned level.

The high share of wind power within the system results sometimes in extreme requirements for system operation due to the power fluctuations mentioned above.

An impressing example is the hurricane on the 8th of January 2005 that crossed the whole area of Denmark resulting in a disconnection of nearly the total wind production (Figure 8.19). In this case, the system operator had to handle a record high imbalance between schedule and production of more than 1,700 MW. Until now a sufficient amount of regulating power has been available in the western Danish power system to compensate for the intense power fluctuations from HRA by applying the load-frequency controller (LFC) accessing the secondary control on the central power plants.

The second offshore wind farm HRB will be located very close to the existing wind farm HRA. An analysis showed that it might be critical to compensate for the additional power fluctuations using only the domestic regulating power 16]. A part of the power fluctuations will be reduced by the offshore wind farms` control themselves. In the analysis HRB was obliged to comply with the power gradient limit of +5 MW/min. Additionally, the use of the fast power control of the HVDC- connections will be necessary to keep the power balance in the western Danish power system. This requires applying the capacity of some of the HVDC- links to the regulating power to compensate for the wind farms` fast power fluctuations.

However, the LFC control is not capable of handling the power regulation that is required for the case shown in Figure 8.19.

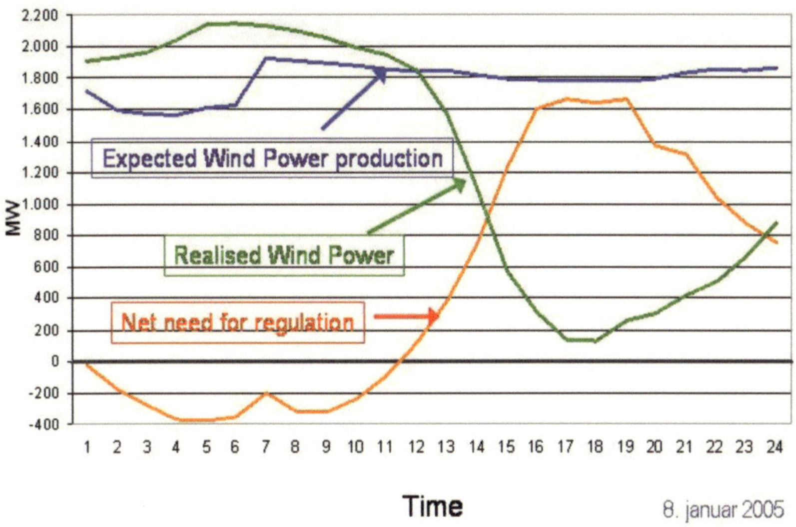

Fig. 8.19. Need for Regulation in Western Danish System during the Passing of a Wind Front

Improving forecasting systems is one of the possibilities to improve the power balance. Reliable wind forecasts are essential for power system operation in Denmark.

The planned active power from a wind farm is based on wind forecasts that are transferred to active power forecasts. The first active power forecast is made a day ahead, but can be updated during the day. The active power produced by wind farms is part of the power supplied from a group of power plants available to the Power Balance Responsible Player (PBRP). The PBRP controls the active power from this group of the power plants according to the latest power forecast in a way that complies with the planned total power production. Deviations between power forecast and the delivered total active power are injected into the transmission system and should therefore be minimized.

The aggregated western Danish wind power curve (Figure 8.20) has a very high power slope, resulting in a deviation of +320 MW for a +1 m/s wind velocity prediction fault appearing between wind speeds of 5 and 15 m/s. A relieving factor is the regional distribution of the wind turbines over the whole western Danish area.

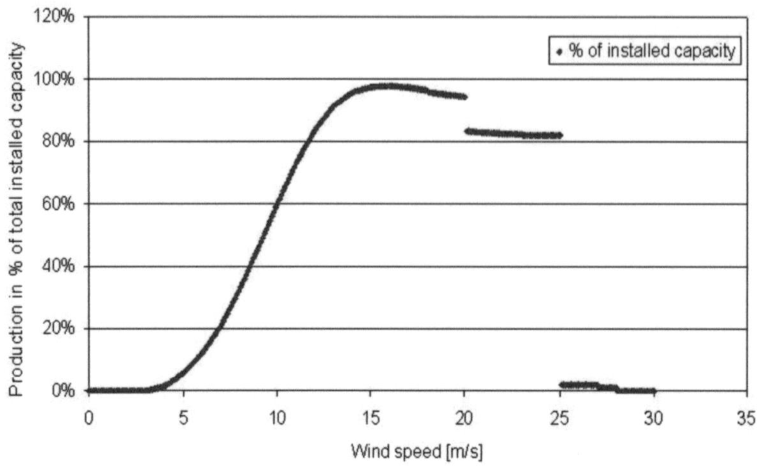

Fig. 8.20. Aggregated Wind Power Production Curve for Western Denmark

The wind forecast models have to be improved in several ways:

- Improvement of day-to-day forecasts because the amount of grid incorporated wind power is significant and still increasing (work in progress).
- Improvement of hour-by-hour forecasts: they have to comply with the power balances and planned operation of the power plants, planned power transits and consumption (work in progress).

8.4.2.1 MELTRA

In 2002, Energinet.dk funded a research project on ensemble forecasting at University College Cork (UCC), Ireland. In this context a real-time forecasting system called MELTRA was designed to meet specifically set requirements in Energinet.dk. It consists of 75 ensemble members and a graphics package for visualization of the forecasts (Figure 8.21).

MELTRA has undergone many changes since its first implementation. The upgraded 2005 system generates 3-day forecasts every hour and consists of around 6000 forecasts per day. Half of the forecasts are carried out as nested forecasts in higher resolution. The forecasts are converted into probabilities and, in combination with observations, provide the best possible forecasts of wind power. The MELTRA ensemble system is run on a 92 processor Linux cluster, which is believed to be a very cost-effective hardware solution. The resolution in the meteorological model is 45 km with a finer 5 km nested grid covering Denmark.

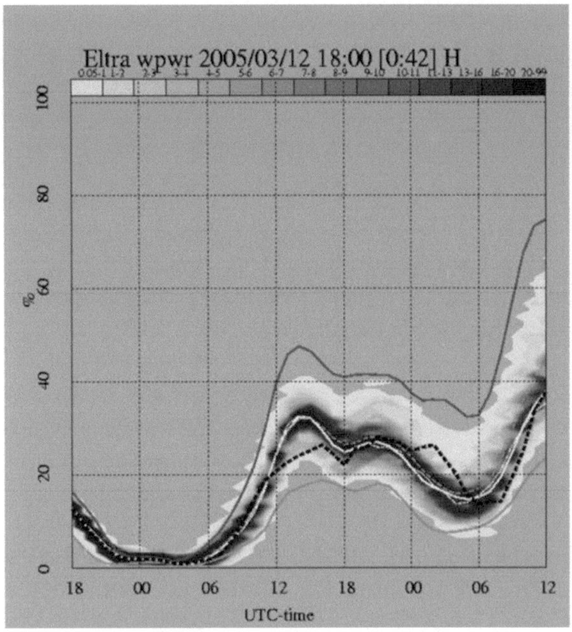

Grey Colors: Probability Distribution; White Curve: Average Prediction; Black Dotted Curve: Measured Production

Fig. 8.21. 48-hour Power Prediction for the Western Danish Area

The major benefits of the first year's real-time experience with the MELTRA system can be summarized as:

- Averaged over one year, the implemented ensemble technique has a potential of at least 20 % better forecasts of wind power compared to a single forecast.
- The ensemble technique is also better in predicting wind power for single sites or smaller areas than a single forecast [16].

8.4.2.2 Future requirements for further expansion of wind power

An increase of the share of wind energy up to 35 per cent of the Danish electricity demand by 2015 has been suggested and will probably be given serious consideration [18]. This requires focusing on regulation power that is available within the present frames. Several issues are under study at this time.

Utilization of domestic regulation power could be applied to further the development of price response mechanisms and better utilization of local scale CHP units introducing them to market terms.

The establishment of the planned Great-Belt connection between both Danish systems will allow for utilizing the regulating power control of both systems.

Further, the establishment of an offshore transmission system connecting the large offshore wind farms with the grids of Norway, Denmark, Germany and Holland may reduce the impact onto the Danish transmission system.

8.4.3 CHP Units

Since the energy crisis of the 1970s, small-scale CHP power plants have been established to supply local heating systems of small cities. Simultaneously industrial CHP units have been installed. This concept has been followed until today resulting in a high share of dispersed installed capacity, which is not as a matter of course available for power regulation and thus, does not contribute to system balance.

The distributed CHP-units' range in size is from a few kW up to 100 MW. Most of these units are gas turbines or gas engines. Traditionally the power production from these units depends on the heat demand, thus heat and electricity are strongly coupled. To eliminate this dependence, these units are equipped with heat storage tanks.

Most of the large thermal units are coal-fired CHP units that can extract steam for heat production. These units have an operating domain between 20 % and full power load without heat production. However, the operating domain for the power depends on the heat production - with higher heat production the minimum power load increases and the maximum power load decreases. According to the power station specifications [19], these thermal units have a regulating capability of 4 % of full load/minute in the operating domain from 50-90% and 2 % of full load/minute below 50 % and above 90 % load. Besides the normal regulating capabilities these units can disconnect the heat production and, for a short period, utilize the extracted steam for electricity generation.

Increasing security problems have led to a reconsideration of the traditional high degree of independence between TSOs and DSOs (distribution system operators).

A new control strategy shall include all local grids with DG into new responsibilities, such as control of reactive power, provision of data for security analyses, supervision of protection schemes at local CHP plants, updating under-frequency load shedding schemes and new restoration plans, including controlling dead start of local plants in emergency cases.

The implementation of such new responsibilities will require development of new control, communication and information systems. During normal operation all functions should be automatic. For emergency situations restoration plans have to be carefully prepared and trained. The targets concerning the systems redesign are:

- balance between supply and demand shall be ensured by sufficient available domestic resources
- operators need to have access to an improved knowledge of the actual system conditions, both locally and centrally
- efficient system control shall be available, especially during emergencies
- Black start capabilities using local generators shall be provided.

Presently, Energinet.dk is executing a cell controller pilot project (CCPP) defining a demonstration area of a real distribution network ("cell"), where a new concept implementing new communication systems and a new controller shall be implemented and tested according to the following ambitions[20]:

- in case of a regional emergency situation reaching the point of no return, the cell shall disconnect itself from the high voltage grid and transfer to island operation
- after a total system collapse, the cell has black-start ability to a state of island operation.

The CCPP aims to:

- gather information about feasibility and approaches to utility-scale microgrids
- develop requirements, specifications and preliminary solutions for a pilot implementation of the cell concept
- implement measurement and monitoring systems to gather and analyze data from the pilot area
- perform detailed design, development, implementation and testing of a selected pilot cell.

8.4.4 Aspects Concerning the Energy Market

The Nordic electricity market consists of several markets: the physical day-ahead market (Elspot), the hour-ahead (Elbas) trade and the real-time market for balance power (Figure 8.22).

The power plants find a Production Balance Responsible (PBR) to sell their energy production. The PBR sells the production either directly to the Nord Pool spot market or announces the capacity to Energinet.dk`s regulation power market. Energinet.dk transfers the regulation power bids to the Nordic TSOs Nordic Operational Information System (NOIS). In the NOIS a merit order list of the bids, visible to all TSOs, is composed. The present regulation measures are based on this list. Regulating power prices can differ in the event of network congestions, when several price areas have to be defined.

The residual market is a market for the production of energy that is not supplied by prioritized renewable generation. The commercial suppliers face a decreasing power demand leading to a decrease in the commercial production capacity's utilization and thereby a reduction in profit making opportunities.

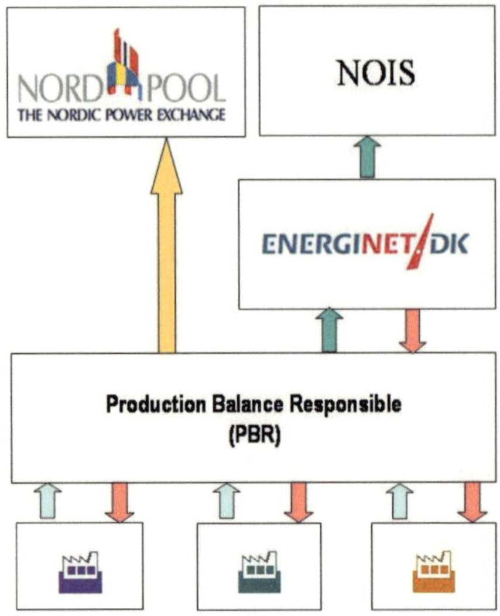

Fig. 8.22. Electricity Market Overview

8.4.4.1 SivaEl

The approach of defining the volume of the residual market is based on a fictitious west Danish 100 % thermal system with base-load and peak-load units [21]. The system is modeled in the simulation tool SIVAEL (simulation of heat and electricity), and the consequences of increased installation of wind power are analyzed by means of model simulations. The share of wind power is gradually increased from 0 % to 100 % coverage of the annual energy consumption. Two types of units are used: coal-fired base-load units and natural gas-fired gas turbines as peak-load units. Two assumptions are made; namely, base-load units are preferable when utilization times exceed 2,000 hours, whereas peak-load units are more profitable when utilization times are less than 2,000 hours. As for the calculations, the number of units and their distribution on base load or peak load are adjusted exogenously in the model in such a way that this criterion is observed.

A 100 % thermal west Danish system in 2025 with an annual consumption of about 26 TWh has been chosen as a basis in order to be able to relate the calculation results to something well known. Combined heat and power and international connections have been disregarded to maintain simplicity and generality - this means that the system must be able to make adjustments for variations in consumption and wind-power production.

The expansion of wind power is assumed to increase onshore and offshore in parallel. A maximum production of some 6 TWh onshore is assumed. Offshore, wind power production is some 20 TWh in the case of 100% share of wind power. Wind power production is included in the model as a time series based on wind-speed measurements offshore near Horns Rev and the island of Læsø and on wind-power production measurements from onshore wind turbines in Jutland and on Funen as well as from the offshore wind farm at Horns Rev.

SIVAEL solves the week-plan problem on an hourly basis and finds the optimum load dispatch with regard to start-stop, overhauls and outages. The optimum load occurs when the total variable costs are at a minimum.
Figure 8.23 shows the wind energy production, the share that can be sold immediately and the surplus electricity. It shows that the system can absorb about 30% of the wind power with no surplus electricity. On the other hand, the surplus grows substantially when the share of wind power is more than approximately 50%.

Following this idea, there will be two different residual markets: one for demand and one for overflow. The SIVAEL-Model is calculated for a share of 100 % wind power with a residual energy consumption of 8 TWh / year and a surplus energy of 8 TWh / year, thus the resulting residual market has an energy volume of 16 TWh and a capacity differential of about 9,000 MW. (Comparison: For a pure thermal system the volume of the electric energy market equals 26 TWh and the demand for capacity about 4,500 MW.)

In the future this business area can be cultivated by market players, e.g. by means of developing new products.

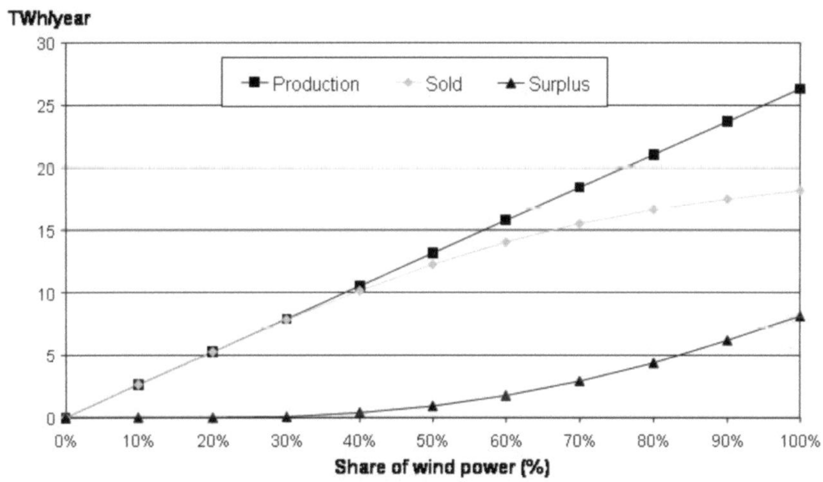

Fig. 8.23. Wind Power Production on an Annual Basis (TWh/year), the Share of Wind Power that Can Be Sold for the Assumed Consumption (TWh/year) and the Remaining Surplus

8.4.4.2 Demand response

The increasing share of wind energy has resulted in an increasing need for balance tools, which also may be located on the demand side. Demand response is defined as a short-term change in electricity consumption as a reaction to a market price signal [22]. The Nordel study [23].identifies demand response as both an alternative and a prerequisite for investments into new production capacity and recommends that all Nordic TSOs prepare action plans for developing demand response.

The TSO is responsible for maintaining the instantaneous balance between supply and demand for each control area. The TSO agrees with the supplier on the amount of power that has to be available at a certain time. If the reserve is activated it is financially compensated for according to the supplier's bid. Sometimes energy is very cheap - even free (Figure 8.24). It would be valuable to use this cheap energy rather than activating reserve energy that has to be paid for and simultaneously exporting the wind energy.

A further expansion of wind power capacity makes only sense if consumption is increased accordingly or thermal production can be reduced. Demand response manual reserves can be activated by suppliers or consumers, whereas up regulation means interrupted consumption and down regulation means extra consumption. If there is an unbalance in the system, either the production can be increased or the consumption decreased or vice versa - depending on the kind of unbalance. The smallest bid is 10 MW, and the price for being available as reserve power for the system operator can be between 27,000 EUR/MW/year and 67,000 EUR/MW/year for up regulation power and up to 20,000 EUR/MW/year for down regulation power. Thus, not only supply, but also electricity consumption should follow price signals. The former philosophy of influencing consumer behavior by means of time-tariffs or campaigns is substituted by new market products, which illustrate the market value of consumers` reaction and capitalize market gains. The system operator acts as a catalyst promoting the consumers` price flexibility. By this means utilization of cheap wind energy instead of valuable coal or oil shall be achieved. During Energinet.dk`s demonstration projects, for some big customers like such as an iron foundry, it has turned out to be economically efficient to install a parallel electricity based consumption system which is used during times of extremely low prices for wind energy.

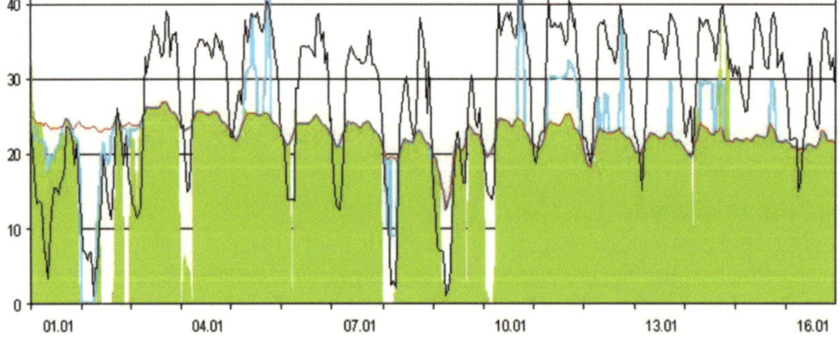

Fig. 8.24. Energy Prices in Denmark, Norway and at the EEX

In Denmark there is also a large technical potential for increased electricity consumption in district heating systems to substitute fossil fuels during periods of heavy wind production. Consequently, the substitution of primary resources is obtained and investments in non-economic peak load units can be avoided. The respective change of consumer behavior can be: moving the time of consumption to periods with lower prices; reducing or stopping consumption during periods when consumer benefit from using electricity does not exceed the price (possibly by means of substitution to another energy source); or increasing the consumption during times when the electricity price is lower than the marginal utility and the price of another energy source, e.g. during times of high wind production. This measure results in a smaller slope of the demand curve where, due to limited demand response, there may sometimes be no market clearing point found (Figure 8.25). An action plan has been made including 22 specific initiatives aiming at the development of demand response in the electricity market and all Nordic TSOs are cooperating on this topic [24].

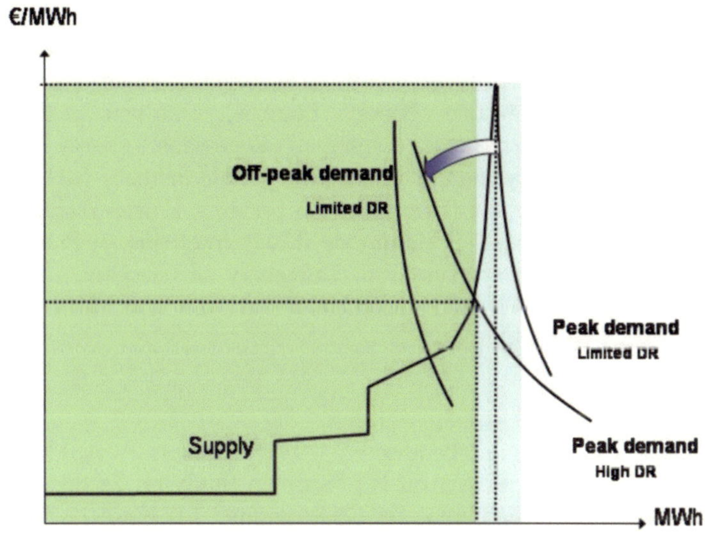

Fig. 8.25. Supply and Demand Curve for Different Elasticity Coefficients due to Grade of Demand Response

In summary, Section 8.4 has highlighted that the Danish system is facing various difficulties on several levels: Technically, a high share of dispersed generation challenges the transmission system operator who is responsible for reliability and security of supply and constantly has to balance supply and demand. This is additionally complicated by high transits passing through the system. Interconnections to neighboring countries are essential for the functioning of the system, and a further expansion of the network as well as the interconnections has to be planned carefully.

Referring to market requirements the Danish transmission system operator, being situated in two synchronous areas operating with different schedules, has to adapt to both systems and use the opportunities of the market to improve the national power balance situation by means of the real time market.

In Denmark a further wind energy expansion is expected, but it has been decided, that there will be a maximum limit for the price at which energy can be sold. Consequently, the future role of small-scale CHP units has to be newly defined aiming at better utilization through operation on market terms.

Also, the use of electricity is being re-discussed. A demand response project illustrated the potential of integrating the consumer into the well functioning of the market. For example, in times of high wind production it can be economically efficient to use electricity for district heating systems by using heat pumps or heat boilers.

8.5 Further Reading

Further reading on integrating dispersed renewable generation sources into European Grids is given in References [25].

8.6 Acknowledgement

This Chapter has been prepared by Zbigniew A. Styczynski (Head and Chair of Electric Power Networks and Renewable Energy Sources, Otto-von-Guericke University, Magdeburg, Germany and President, Center of Renewable Energy Saxonia Anhalt, Germany). Contributors include Johan Driesen and Ronnie Belmans (KU Leuven, Leuven, Belgium), Bernd Michael Buchholz (Director, PTD Services, Power Technologies, Siemens AG, Erlangen, Germany), Thomas J. Hammons (Chair International Practices for Energy Developments and Power Generation IEEE, University of Glasgow, UK), and Peter B. Eriksen, Antje G. Orths and Vladislav Akhmatov (Analysis and Methods, Energinet.dk, Fjordvejen, Fredericia, Denmark)

8.7 References

[1] IEA, Distributed Generation in Liberalised Electricity Markets, Paris, 128 pages, 2002.
[2] Eto J., Koomey J., Lehman B., Martin N., "Scoping Study on Trends in the Economic Value of Electricity Reliability to the US Economy," LBLN-47911, Berkeley,2001, 134 pages.
[3] Renner H., Fickert L., 1999. Costs and responsibility of power quality in the deregulated electricity market, Graz.
[4] Dondi P., Bayoumi D., Haederli C., Julian D., Suter M., "Network integration of distributed power generation," Journal of Power Sources, 106, 2002, pp.1–9.
[5] Woyte A., De Brabandere K., Van Dommelen D., Belmans R., Nijs J, "International harmonisation of grid connection guidelines: adequate requirements for the prevention of unintentional islanding," Progress in Photovoltaics: Research Applications, 2003, Vol.11, No.6, pp.407-424.
[6] Gatta F.M., Iliceto F., Lauria S. Masato P. "Behaviour of dispersed generation in distribution networks during system disturbances. Measures to prevent disconnection," Proceedings CIRED 2003, Barcelona, 12-15 May 2003.
[7] Ackermann T., Andersson G., Soder L., "Distributed generation: a definition," Electric Power Systems Research, 57, 2001, 195–204.
[8] CIRED, 1999: Dispersed generation, Preliminary report of CIRED working group WG04, June, p. 9+Appendix (p.30).

[9] Jenkins N., Allan R., Crossley P., Kirschen D., Strbac G., Embedded Generation, The Institute of Electrical Engineers, London, 2000

[10] B. Buchholz a.o. Advanced planning and operation of dispersed generation ensuring power quality, security and efficiency in distribution systems. CIGRE 2004, Paris, 29.August - 3.September 2004

[11] J. Scholtes, C. Schwaegerl. Energy Park KonWerl. Energy management of a decentralized Supply system. Concept and First results. First international conference on the integration of Renewable energy sources and Distributed energy resources. Brussels, 1.-3. December 2004

[12] IEC 61850 Part 1-10. Communication networks and systems in substations

[13] IEC 612400-25-2. Wind Turbines. Communication for monitoring and control of wind turbines. Part 25-2. Information models. IEC 88/214/CD

[14] IEC 62350. Communication systems for distributed energy resources. IEC 57/750/CD

[15] Bumiller, G., Sauter, T., Pratl, G. Treydl, A. Secure and reliable wide area power line communication for soft real- time applications within REMPLI. 2005 International Symposium on Power Line Communications and its Applications, Vancouver, April 6-8 2005

[16] V. Akhmatov; H. Abildgaard; J. Pedersen; P. B. Eriksen: "Integration of Offshore Wind Power into the Western Danish Power System" in Proc. 2005 Copenhagen Offshore Wind International conference and Exhibition, October 2005, Copenhagen, Denmark.

[17] Specifications TF 3.2.5, "Connection Requirements for Wind Turbines connected to voltages over 100 kV" (in Danish) Available: http://www.energinet.dk.

[18] P. B. Eriksen; Th. Ackermann; H. Abildgaard et. al.: "System Operation with High Wind Penetration", IEEE Power and Energy Magazine, vol 3 No. 5, pp 65-74, Nov. 2005.

[19] Power Station Specifications for Plants > 50 MW, Elsam, Denmark, SP92-230j, 16 pages + 3 pg annex, August 1998; Kraftværskspecifikationer for produktionsanlæg mellem 2 og 50 MW: Elsam, Denmark, SP92-017a, 16 sider + 5 sider bilag, september 1995 (in Danish).

[20] P. Lund, S. Cherian, T. Ackermann: "A Cell Controller for Autonomous Operation of a 60 kV Distribution Area" in Proc. 10th Kasseler Symposium Energie-Systemtechnik 2005, ISET, Kassel. pp. 66-85.

[21] J. Pedersen: "System and Market Changes in a Scenario of Increased Wind Power Production " in Proc. 2005 Copenhagen Offshore Wind International conference and Exhibition, October 2005, Copenhagen, Denmark.

[22] K. Behnke, S. Dupont Kristensen: "Nordel - Danish Action Plan for Demand response", Elkraft/ eltra, Nov. 2004 (intern document)

[23] ["Enhancing Efficient Functioning of the Nordic Electricity market", Nordel, Februar 2005. Available: http://www.Nordel.org.

[24] "Ensuring Balance between Demand and Supply in the Nordic Electricity Market", Nordel, 2004, Available: http://www.Nordel.org.

[25] T. J. Hammons: "Integrating Renewable Energy Sources into European Grids", International Journal of Electrical Power and Energy Systems, vol. 30, (8), 2008, pp. 462-475.

9

Status of Power Markets and Power Exchanges in Asia and Australia

Integration of electric power systems and power exchanges among countries, regions and companies is an objective tendency in world power industry development. The Asian region is rather promising in this respect since the sources of energy resources for electricity production are often very remote from the load centers. Besides, there are the so-called system effects from electric power systems integration that are beneficial for all the participants. The role of power exchanges increases still further under deregulated electricity markets particularly in terms of the possibilities to decrease the market prices of electricity.

The following viewpoints are discussed in this Chapter:

- Ideas of the different countries in Asia and Oceania of either the positive or negative role of power exchanges in a market environment;
- Estimations of potential limits in the power exchanges and substantiation of such limits if there are any;
- Concrete results of the studies on power exchanges in the feasibility studies of prospective projects of power exchanges.

9.1 Status of Reform and Power Exchange in India: Trading, Scheduling, and Real Time Operation Regional Grids

Though India opened up its power sector in nineties to private sector investment, initial impact was mainly in the form of generation addition and then with unbundling of generation, transmission and distribution, to some extent on the last segment also. Transmission as natural monopoly remains still under government-owned companies, both at central and state level, though right at the beginning of 1998 specifically it was opened to private enterprises to build, own and operate from point to point. With the open access in inter-state transmission to any distribution company, trader, generating company, captive plant or any permitted consumer as per November 2003 order[1] of Central Electricity Regulatory Commission (CERC) certain changes are, however, taking place. Under such circumstance changes in methodology of generation scheduling to meet demand are also inevitable to take into account this very aspect from time to time considering role of various participants in power market. However, at the same time aspect of system security vis-à-vis stability is given due importance in real time grid operation, as envisaged also under Electricity Act 2003[2].

9.1.1 Development of Indian Power System[3,4]

India has a federal structure with 28 States, 7 Union Territories and a Central Govt. Present installed capacity of India is 112 GW with 25% of hydro besides nuclear, gas, wind and conventional thermal plants. For the purpose of power system, the country was demarcated into five geo-political regions in the year 1964 and gradually different states within the region got integrated and by the 1980s five mature regional grids were under operation. In 1992 Eastern and Northeastern regions were interconnected. In 2002 the Northeast, East and West with a span of 2800 km. of synchronous grid became operational. There are four HVDC Back-to-Back stations of 500 / 1000 MW capacity each and three Bi-pole HVDC long lines for carrying bulk power. Indian power system also has multiple connections at different voltage levels with neighboring countries, like, Nepal and Bhutan. Cross border power exchanges are progressively increasing. There is wealth of experience regarding expansion of the grids and experience of operating large grids.

Resources are unevenly spread with hydrocarbon deposits in the East and Central parts of India and huge hydro potential in the Northeastern and Northern part of the Northern Grid. There is a promising availability of gas on the long coastal lines. The load growth has also been uneven with widely varying per capita income in different states. This calls for transfer of large blocks of power over long distances.

Central Electricity Authority, a statutory organization produces the national plans. Integrated resource planning approach is adopted. Transmission system expansion is coordinated for achieving a most optimal plan with least investment. Perspective plan and the long-term forecasting are also carried out by the Authority.

The Legislations on Electricity in India traversed a long distance and all the old act since1910 onwards have been merged and recast in the form of a consolidated Electricity Act 2003. Indian Electricity Grid Code (IEGC) and the State Electricity Grid Code (SEGC) are in place after public debate. The Regulators, Authorities and the state utilities are framing rules and Regulations. The Central Electricity Authority is developing metering codes.

Indian Electricity Act 2003 envisages Electricity Regulators at State level (State Electricity Regulatory Commission, SERC) to take care of intra-state affairs while the Central regulator (Central Electricity Regulatory Commission, CERC) to take care of inter-state matters. The tariffs, codes and directions on Open Access are now being issued by the Regulators in a fair and transparent way and the Government is distancing itself.

Transmission has been recognized as a separate activity in 1998 by the legislation. In line with the federal structure the Central Transmission Utility (CTU, at present Power Grid Corporation of India) and the State Transmission Utilities (STU, at present Transmission / Grid Company TRANCO or GRIDCO of the concerned state) have been created for coordinated development of the transmission segment. Transmission being a natural monopoly is a regulated entity and barred from trading as per the law. Transmission system in India has developed from 132/220 kV and now well-meshed 400 kV mature grid forms the backbone of Indian Grid. A rapid development is envisaged by the year 2012 matching with load growth and generation addition of 100,000 MW.

By and large the GENCO (Generating Company), TRANSCO and DISCO (Distribution Company), STU, CTU, SLDCs (State Load Dispatch Centers), RLDCs (Regional Load Dispatch Centers) and CERC structure has been followed while progressing with reforms and unbundling. There are variations in the models being adopted by different states. Some of the states have already privatized their distribution systems.

The Indian sub-continent with its vast geographical distances and diverse resources is struggling to achieve cost reduction through 'Economy of Scale'. The large size generators of 660 MW are being added as 500 MW sets have already stabilized and are dominating presently. For transferring large blocks of power, 765 kV transmission system has been envisaged overlaying 400 kV meshed network.

Private participation in generation by way of IPPs (Independent Power Producers) and Mega Power Projects supplemented with Government investment is envisaged. So far the transmission has been through the State / Central Government companies. Joint venture and IPTC (Independent Power Transmission Company) route have also been launched to attract private investment in the transmission segment. With unbundling and demarcated distribution companies, niche market is being created for private participants to enter into the field of Distribution. With Open Access, investment in captive power plants is likely to get a boost, as they would have access to enter the Indian power market.

765 kV transmission systems connecting the regions and the resource-rich areas and load centers would form a super highway for wheeling of power from source to sink. A massive capacity addition plan of 50,000 MW of hydro and 100,000 MW of thermal power has been launched and expected to yield result by the year 2012.

The variety of diversities between the different regions of India and its neighboring countries open a vast potential for coordinated expansion and operation to take care of time, season and resource diversity prevailing in the sub-continent. It would also enable to level the diversity caused by various uncertainties, like, investment, load growth, etc.

9.1.2 Grid Operation

The Indian Electricity Grid Code (IEGC) lays down rules, guidelines and standards to be followed by the various participants in the system to plan, develop, maintain and operate the power system in the most efficient, reliable and economic manner while facilitating healthy competition in the generation and supply of electricity. The IEGC covers roles of different organizations and their linkages, planning codes, connection conditions, operating codes, scheduling and dispatch codes, metering and management of the grid code.

The regional grids in India are operating as loose power pools in which the constituents have full autonomy and have the total responsibility for scheduling and dispatching their own resources, arranging any bilateral inter-change and regulating their drawl from the regional grid.

The Regional Load Dispatch Centers coordinate the entire activity of day-ahead scheduling. For the purpose of scheduling and settlement a day is divided into 96 blocks of 15 minutes each. The shared generation resources declare their availability and RLDCs communicate the entitlement to all the stakeholders. Based on the load-generation availability and economics, all the constituents furnish their requisition from the shared resources that are aggregated by the RLDCs and communicated to the shared generators. These are based on the long-term contracts and allocation normally done by the Central Government in consultation with the State Governments.

On day-to-day basis the utilities enter into bilateral agreements of different kinds. The SLDCs and RLDCs incorporate the same in the schedule provided there is no network congestion. In case of congestion the same are moderated by the SLDCs / RLDCs. There is an elaborate time line for the scheduling and dispatch procedure. Provisions also exist for revising and modifying the schedules by any of the participant in case of contingency for which at least six time blocks, i.e., one and a half hour notice is required. The scheduling is carried out through a web-based scheduler and all the revisions are posted on the web in a transparent way. At the end of the day the final schedule becomes the datum for calculation of 'Unscheduled Interchange' as well as payment of energy charges. In other words the schedules are Commitments / Contracts and payment of energy is decided and finalized based on the finally implemented schedules.

At present there are few Traders licensed to operate in the Inter-state Trading. Traders are given different categories of license depending on the volume of transactions and the Regulatory Commission assesses the financial capacity and other parameters of the Traders before issuance of license. Traders are to file the periodic Returns to the Regulator furnishing the details of the transactions.

The Open Access Regulations stipulated by the Commission and the Procedures framed by the Central Transmission Utility is followed by the Open Access customers, which are primarily the Traders. The Open Access Regulations enacted in 2003 are undergoing speedy refinements based on the experience of various stakeholders including the Operators. The volume of trading has grown phenomenally and huge number of transactions has already taken place.

Load Dispatch Centers have been declared as an apex body both at the State level and at Regional level, i.e., SLDCs and RLDCs. The National Load Dispatch Center (NLDC) has also been conceived to take care of inter-regional and cross boarder exchanges. Load Dispatch Centers are also barred from trading activity and are 'no profit no loss' centers with fees and charges being determined by the Regulators. Since both transmission and system operation are neutral to the market and barred from trading activity, at present Indian power system is having synergy with transmission and Grid operation. The Load Dispatch Centers both at Regional and State levels have been upgraded with state-of-the-art technology.

Each region acts as a pool. Control areas are demarcated with each state and shared generators being separate entity and there are many participants in the pool. After much debate and in consonance with the federal structure, India has opted for loose power pool and decentralized market.

The settlement system has undergone drastic evolution in recent years. A new scientific settlement system popularly known as 'Availability Based Tariff' (ABT) has been introduced in all the five regions in the country in a staggered fashion. The new mechanism has three parts, viz. capacity charges linked to the availability of the generation, scheduled energy charges based on the requisition and the schedules by the control areas.

The unique and the third component is termed as 'Unscheduled Interchange' (UI), which is deviation from the schedule and its pricing is linked to frequency. The UI mechanism has a self-healing property, brings in equilibrium and emulates all the properties of 'Non-Cooperation Game Theory' automatically. The mechanism while causing economy also complements reliability, yet maintaining the sovereignty of the utilities giving choice and freedom.

Besides long-term (25 years) and short-term day-ahead, spot / balancing market by way of UI mechanism where the prices are linked to the frequency has been created. The spot prices are linked to frequency that is said to be collectively controlled and effectively stabilized. It does not require elaborate calculations. Regulators tinker the UI vector from time to time in order to achieve economy and reliability by creating a pseudo competitor.

Unlike other Pools the Pricing Mechanism of Unscheduled Interchange, i.e., Schedule minus Actual is linked to frequency. The Central Regulator after public hearing and debate notifies the UI price curve. The slope of the curve, kinks, upper and lower ceilings are arrived at by the Regulator with a view to cause overall economy as well as quality in the grid.

The fundamental theory of equilibrium and the negative feedback has been adopted while deciding the UI price curve. As UI price is linked to frequency and as it is known that frequency deviations represent surplus and shortage situation, accordingly the UI price varies with real-time shortage or surplus. In other words, as the surplus emerges, frequency rises and the UI price starts coming down. Similarly, with shortage the frequency starts falling and the UI price rises. The participants in the Pool seeing the rise and fall take corrective action that acts as a negative feedback and dampens fluctuations and system reaches equilibrium. The beauty of the scheme is that the Pool price needs not be calculated. It is totally transparent. The mechanism encourages 'Merit Order Operation' in a distributed fashion as virtually all the generators compete with the prevailing UI price that keeps sliding.

The marked difference in the scheme expects the frequency to fluctuate in order to give a signal to the generators to adjust their output. A distributed optimization is effected. The mechanism is also akin to Non Co-operative Game Theory through which the best prices are achieved. The most interesting feature is that while economy is achieved the mechanism also compliments reliability. In UI Mechanism utilities while economically gaining also contribute to reliability. This makes the approach absolutely novel.

Trading of power has been recognized as a separate, distinct licensed activity by the legislation. There are variety of products being invented by the Traders and the prices are being discovered however with a benchmark of prevailing UI prices fixing the virtual roof and floor.

With the formulation of power pool, settlement system, trading, in 2003 Regulators introduced Open Access in the inter-state transmission. The plan is to progressively introduce Open Access for embedded and captive power plants. The Open Access has been primarily categorized as 'long-term' and 'short-term'. The detailed speaking orders with elaborate procedure has been put in place for calculation of transmission charges, obligation of losses, prioritization of allotment, etc.

The transmission development management is a coordinated activity and by and large there is not much of intra-regional congestion. However with increase in inter-regional flows congestion has started surfacing in the inter-regional links. Regulators have devised bidding procedure to take care of the congestion. Augmentation of Inter-Regional links capacity to 30,000 MW is envisaged by 2012.

9.1.3 Power Exchange

At present there is no formal 'Power Exchange' operating in India. However, the Buyers, Sellers and the Traders meet periodically in the various coordination meetings and deals are negotiated. Some of the constituents have also opted for tendering and bidding for power procurement through Traders in a competitive way. With continuous oversight by the Regulators the resource scheduling and turnover of power by trading is improving and causing economy to the sector while giving much desired choice to the utilities.

Basically the concept of a Power Exchange is that of a platform that enables market participants to go about their business of bidding, pricing, scheduling and settlement of transactions on a real-time basis. In the Indian context, Power Trading Corporation (PTC) of India, formed in the year 1999 in the public sector, was initially conceived as an intermediary with a primary focus on managing credit risk for the Mega Power Projects. However soon it recognized its larger mandate of creating a vibrant power market. The concept of an exchange gets subsumed in this mandate, and it came up with the statement of purpose as to be a frontrunner in developing a Power Market and striving to correct market distortions.

The frontrunner has conceived a roadmap for setting up a power exchange in the country. While shaping the concept, the frontrunner has the onus of visualizing the phasing of various activities and corresponding investments as also educating various market participants, existing and prospective, about the potential benefits. All this has to be dovetailed to the Indian context, with its own peculiarities and consequent capacity to absorb change.

With the initiative taken for the first time in 2001, few market participants took part utilizing the concept of exchanging surplus power with entities that have complementary deficits at a market determined rate. The structure of these transactions was simple, with seller entities supplying power on a round-the-clock basis for periods varying from a few months to one year to buyer entities. While seller entities benefited by the enhancement of cash flows due to better capacity utilization, the buyers got reliable supply at an economic, market-determined rate. At the same time, various linking entities in the supply chain like the CTU, STUs, RLDCs and SLDCs were able to make adjustments in their processes to allow these market determined exchange transactions to overlay existing long-term, bilateral transactions. The participants experienced the benefits of exploiting complementary surplus-deficit situations arising from an annual or seasonal time-epoch.

As market participants and the linking agencies gained confidence from the demonstration of success in these early transactions, more participants were initiated into the market. At the same time, the experience curve benefits started accruing to the participants and their power planning and operational processes became geared to take on shorter response time. At this stage, it was felt that the time was ripe to initiate services that exploited complementary demand-supply situations arising from shorter time-epochs, like even a one-day period. Therefore in 2002, new products were introduced that allowed flow of power for limited hours during a 24-hour period, like `Morning Peak', `Evening Peak', `Off-Peak' and various combinations like `18 Hours Supply'. As all participants benefited by utilizing these trading opportunities for shorter durations, many participants experienced the unique position of reversing roles from buyer to seller during the same 24-hour period. At the same time, implementation of Availability Based Tariff was started with the Western Region (WR), and PTC as Trader looked at opportunities arising from this situation. Therefore, 'ABT' Power arising from the need of the utilities in WR to balance schedules and optimize their revenues was sold to utilities in the Southern Region (SR) at a fixed rate (the regime in SR had until then not changed to ABT). This transaction, though small in terms of the volume traded, was a pre-cursor to `As-and-When-Available' power, a product evolved later in 2003 when all participants became subject to the ABT regime. During 2002, with the acquisition of long-term contracts for trading of power from Chukha and Kurichhu projects in Bhutan it has been possible to diversify the supply portfolio. The participants' confidence in the evolving market mechanism is perhaps best symbolized by the structuring of trading transactions for the sale of power from the 86 MW Malana HEP for periods ranging from one to three years, in effect making it the first plant in India to operate on the merchant power plant business model.

In the quest for greater efficiencies through a market based exchange mechanism, the 'As-and-When-Available' power as a product where sale and purchase is planned on a day-ahead basis in 2003 was introduced. At the same time, the Electricity Act was instituted, and it formalized a very important principle on which these transactions were structured, namely `Open Access' in transmission. Participants and transactions grew manifold, and about 30 participants were active in the market at any point of time during the year. Several transactions that involved use of transmission systems of four, and even all five-power regions of the country were structured successfully. Hence, new participants that came into the market were unique in their position. Some of them did not have significant sizes, but were in a position to relieve power system congestions, or help other participants in managing resources better because of the timing of the trading opportunities offered.

The challenge in moving ahead to the long-term scenario of an exchange is to reduce the response time from a `day-ahead' level to a real-time environment. While, even in the present scenario, it has been possible to effect transactions in periods as low as 6 hours from the user's request, a power exchange aims to reduces this window to the minimum time required by the System Operator. This part of the journey, possibly, presents the greatest challenge in terms of integrating the evolution process to an exchange-like platform to the changes taking place in the industry, post Electricity Act 2003. In order to meet this challenge a number of steps are contemplated that cover various aspects of the development of an exchange.

A successful exchange platform requires rewards for efficient operation to be accruing to participants equitably. This in turn, depends on the number and type of participants in the market. Efforts are being made for a further diversification of the market participants' base. As a result entry of prospective participants, like, Captive Generators, Co-generators, Industry Associations, energy intensive process industries, is expected into the market as Open Access customers. However, with all these there may be requirement of change in regulatory structures that would promote development of a market as well as wider participation in the market. This is a crucial component in the strategy towards implementation of an exchange.

The backbone of an exchange is an information repository. While all other components would require investment of time and buy-in of participants and linking entities, this is the key component that also requires a significant capital investment. The features of this information platform would include a single application that allows tracking of bid-to-bill (B2B) aspects of all transactions, enables validation at transaction and business levels, supports high volumes (typical range 500 to 5000 transactions in a year). The applications also have to be flexible in that they need to adapt to additional / specific markets. While initially the trading platform is envisaged more in the form of bulletin boards or an effective B2B platform, the same will evolve into a power exchange with real-time capabilities over a longer term. This is the correct approach to setting up a power exchange, as presently all the participants do not have online connectivity with even aggregate level scheduling and dispatch data available with the RLDCs.

Further to enable customers to adapt to an online environment in a phased manner, as a next step it is also necessary to initiate a Customer Relationship Management tool that would enable both availability of information at the market participants' desktops as well as allow capture of operational information. As the market matures, this would evolve into a B2B platform, or alternatively, integrate with an existing B2B platform. In fact the B2B platform would be a pre-cursor to a fully functional, real-time exchange.

Thus, with lot of changes taking place in the Indian power scenario, due to unbundling of generation, transmission and distribution and subsequent regulation providing Open Access of transmission, de facto Power Exchange is in place in grid operation. Scheduling of generation with Open Access of transmission system to meet demand is done keeping in mind trading of power involved on short-term and long-term basis. Under real time Grid Operation, system operates with the principle of Availability Based Tariff taking care of Unscheduled Interchange, of course subject to the constraint of system security and stability.

9.2 The Influence of Transmission on Further Development of Power Exchange in the Australian National Electricity Market

The national electricity market (NEM) is the market for the wholesale supply and purchase of electricity in five Australian states and territories - the Australian Capital Territory, New

South Wales, Queensland, South Australia, and Victoria - together with a regime of open access to the transmission and distribution networks in those states and territories[5]. Tasmania intends joining the market following completion of Basslink.

The market was launched on 13 December 1998.
The objectives of the national electricity market are that:

- the market should be competitive;
- customers should be able to choose which supplier (including generators and retailers) they will trade with;
- any person wishing to do so should be able to gain access to the interconnected transmission and distribution network;
- a person wishing to enter the market should not be treated more favorably or less favorably than if that person was already participating in the market;
- a particular energy source or technology should not be treated more favorably or less favorably than another energy source or technology; and
- the provisions regulating trading of electricity in the market should not treat intrastate trading more favorably or less favorably than interstate trading of electricity.

9.2.1 Description

The NEM is based on a single dispatch model. Generators are dispatched every five minutes on the basis of bid price in $/MWh. An ancillary services market also operates simultaneously with the energy market to maintain the level of operating reserve.

The NEM spans more than 4000km. The long distance transmission between load and generation centers can require restricting the dispatch of generators to avoid overloading transmission. Presently, the NEM is divided into several pricing regions. The market can thus be described as a limited nodal pricing model. The resulting dispatch rules accommodate the effects of marginal inter-regional loss factors and inter-regional operating limits. The control center dispatch software, based on a Linear Programming algorithm, minimizes the cost of meeting the load subject to many variables including generator upper and lower operating limits, inter-regional line limits, and minimum reserve levels.

The inter-regional transmission system has been augmented several times since market inception, including connection of Queensland in 2000 through the Direct link dc link and the QNI ac link, then in 2002 with the Murray link dc interconnector between Victoria and South Australia. In November 2005, the Tasmanian and Victorian systems was interconnected for the first time by a 300km undersea HVdc cable.

Figure 9.1 shows the connection arrangement including the nominal interconnector limitations between regions.

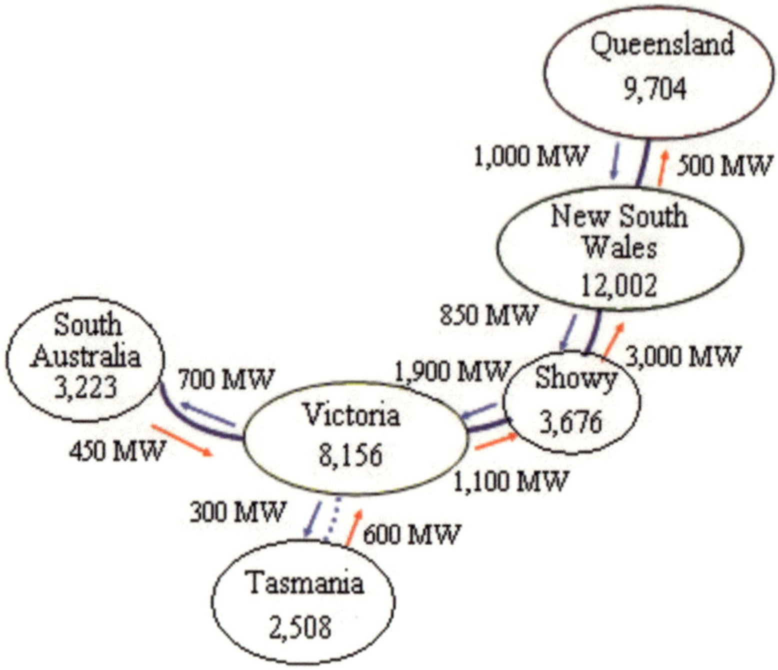

Figure 9.1. Generation and transfer capabilities (Summer 2004/05).

9.2.2 Issues

9.2.2.1 Generation utilization

Prior to and since market inception, the forced outage rates of generating plant have reduced substantially. This has allowed the margin between supply and demand to reduce without compromising reliability. The result of these efficiency gains has been to increase the dependence of reliable market operation on inter-regional interconnections.

The general reduction in the gap between annual installed capacity and annual half hourly peak demand over time is shown in Figure 9.2. While net generation of nearly 4000MW has been added since the start of the NEM due to market forces, the summer peak demand has increased by more than 6000MW. The data is for the mainland states only, but includes a net injection of equivalent generating capacity of 600MW from Bass link starting in 2005-06.

9.2.2.2 Resource Development

Growth, with accompanying pool price increases, is the main driver for the development of new generation.

At the same time, environmental factors are influencing the development of the market through restrictions on permitting plants that do not meet the highest levels of performance available for that fuel.

Several mechanisms are in place to encourage development of renewable or low emission generation within a market context:

- Commonwealth initiatives such as the 2% Mandatory Renewable Energy Target (MRET) scheme for additional renewable energy
- State initiatives such as the Queensland 13% gas scheme and the NSW Greenhouse Abatement Certificate (NGAC) scheme, both of which are aimed at enhancing efficiency, and development of lower emission plant such as gas.

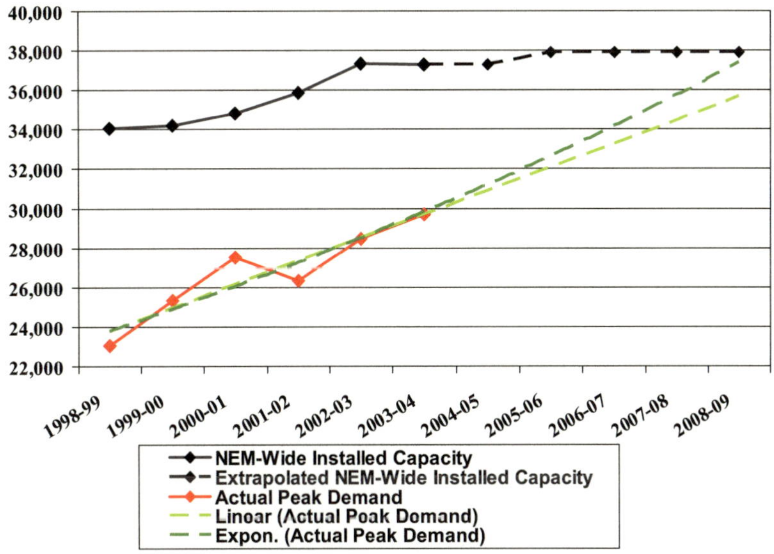

(a) Installed Capacity and Summer Peak Demand

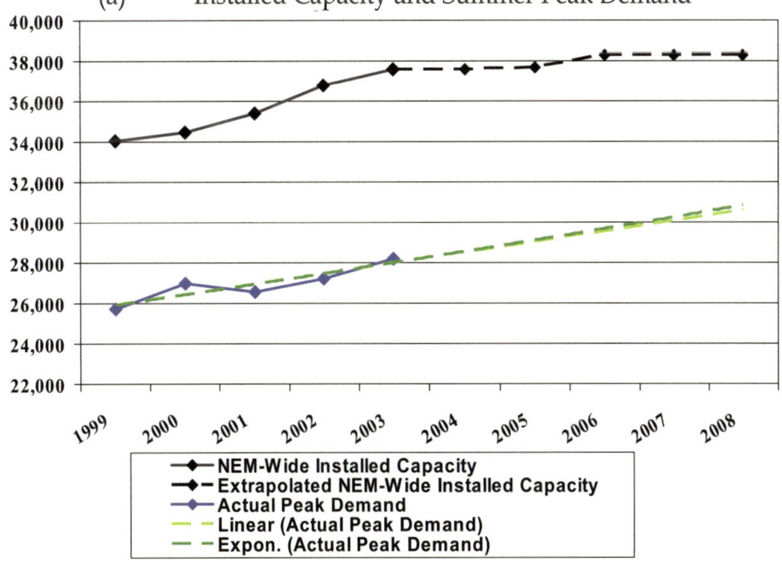

(b) Installed Capacity and Winter Peak Demand

Figure 9.2 Supply and demand

With the impending exhaustion of opportunities to gain further performance improvements from existing generation, many new plans are being developed by market participants to introduce new generation, predominantly gas, black coal, brown coal, wind and biomass. Wind resources dominate in the southern regions, while biomass tends to dominate in the northern regions.

Major gas pipelines are planned from Papua New Guinea and the Timor Sea, both north of Australia. If these projects eventuate they are of the scale to introduce major changes to the generation mix and utilization of transmission.

The Bass link project will add 2500MW of storage hydro generation to the existing mainland hydro (predominantly the 3676MW Snowy Mountains hydro), and introduce further dynamics to the hydrothermal interactions.

9.2.2.3 Transmission utilization

Flows are significantly different from the pre-market situation. While pre-NEM flows were dominated by issues within state boundaries, plus defined interchanges based on operating agreements, the market has expanded trading across interconnectors on the basis of bids by all generation including renewables.

This has tended to use transmission to a higher degree, resulting in increased incidence of operation at or near transmission limits.

The increasing incidence of transmission constraints resulting in price separation between regions with generation constrained off or down in the exporting region(s) has created the opportunity for further augmentation. However, since transmission is a regulated service, additional transmission augmentation has to be justified on the basis of reliability or cost reductions. Proposals are in place, however, to relax the conditions for regulated augmentations by accounting for other defined market benefits. This could open the way for expanded interconnections and/or new interconnections between existing regions.

Unregulated transmission can be built by a proponent seeking to recover its investment based on market revenues. So far, several such transmission links have been built, all based on HVdc technology and connecting regions with potential price differentials sufficient to justify the investments. One of these, Murray link, has since successfully applied for conversion to regulated status.

The potential for additional transmission connections to other regions and/or countries is reduced by the relatively small loads and long distances involved. The largest unconnected region is the 4000MW South West Interconnected System (SWIS) in Western Australia, which is approximately 2000km from the nearest connection point to the NEM. Since Western Australia has an abundance of energy available from the North West Shelf gas fields in the north of the state, and coal fields in the South West, there is no economic justification for a NEM-SWIS link at present.

Several smaller grids are operating in Australia - in North West Western Australia (the North West Interconnected System, of about 400MW, and several other unconnected generators with an additional several hundred MW), in the Northern Territory (approximately 400MW), and in North West Queensland (approximately 400MW). These may be linked to the NEM or to each other as part of anticipated resource developments, including gas pipelines from the Timor Sea and/or PNG.

9.2.3 Market Developments

The strict conditions associated with development of the NEM have resulted in a robust and successful electricity market. This market has succeeded in accommodating substantial load growth, while accommodating several thousand MW of new merchant plant of various types and sizes. There has also been withdrawal of some generation due to market forces. The success of the NEM can be put down to several factors, including:

- A single body/organization overseeing the market
- Multiple parties looking at system operation to look after commercial interests
- Careful specification and description of network limitations, resulting in a general increase in network utilization.
- expanded tools to improve analysis, which are used by market operators, participants and regulatory authorities.
- A uniformity of approach in the application of standards for connection and access, which previously did not exist.
- Transparency and governance strictly according to the National Electricity Code and National Electricity Law. High levels of transparency tend to ensure good governance.
- Active involvement at the political level as well as the industry level
- The high quality of staff in market operation
- The high level of horizontal and vertical disaggregation of the industry exposes the risks of markets and therefore drives commerciality. This tends to drive out the benefits in each part of the industry.
- The rules and regulations support and facilitate market operation.

9.2.4 The Way Forward

Since market inception, grid development has been minimal compared with the extensive changes to the operating regimes of existing generating plant and the new generating plant that has been developed.

This has resulted in increasing tension between regulated grid development and generation.

Planning and implementation of changes to the grid have been underway in several critical areas.

Grid companies will need sufficient incentives to consider augmentation to meet the imperatives of economic resource development in an environment that is likely to include further and stricter emissions limitations from generation sources.

9.3 Technical and Market System Effectiveness of Intersystem Power Exchanges in Russia

Historically the formation of large electric power systems and their interconnection into the Unified power system (UPS) of the country was related to the objective positive factors. Creation of Russia's UPS enhanced essentially the economic efficiency and reliability of power supply as compared to separately operating regional power systems. The influence of the system effect casts no doubt among energy specialists. A great number of publications have been dedicated to the analysis of different aspects of system efficiency of electric power systems.

This section deals with the notions of the potential effect of integration of electric power systems (EPS) and the effect realized in a market environment. The estimation technique is presented. The effect is estimated on the example of Russia's UPS.

9.3.1 Technical System Effect

The system effect in electric power industry is of a multi-factor character. Traditionally the following technology-based components of the intersystem effect have been set off at integration of power systems[6,7].

A "capacity" effect

- A decrease in demand for installed capacity of power plants by bringing into coincidence the load maxim, reducing the operating reserve, decreasing the reserves for routine maintenance;
- An increase in firm power of hydro power plants by raising the total firm power owing to asynchronous run off in different river basins and use of long-term regulation of water reservoirs to the benefits of neighboring power systems;
- A more complete use of commissioned capacity by decreasing the unused capacity in a large system.

A "structural "effect:

- Rationalization of power system structure by: using cheap (but economically inefficient in terms of transportation) energy resources at power plants with power transmission to neighboring systems; increasing the use of peak and free power of hydro power plants;
- A better use of hydropower in the high water years;
- An opportunity to construct power plants successively with the use of temporary surplus powers in the other power plants;
- Saving in the construction of electric networks for power supply to the areas of individual power systems connection.

A "frequency" effect

A Frequency Effect implies a lesser impact of an individual energy unit or a consumer in a large electric power system (EPS) on the system frequency as compared to a smaller system. The frequency effect allows the unit capacity of energy facilities to be chosen without constraints on the system requirements.

An "operation" effect
An Operation Effect implies a decrease in operating costs by optimizing the operating conditions of power plants in the integrated system, increasing the total density of load curves of power systems at integration, by widely using the cheap fuels.

An "environmental" effect supposes improvement of the environmental situation due to redistribution of power generation at power plants with its decrease in the areas with unfavorable environmental conditions.
All these components have objective material (technological) nature. However currently the assessment of these components only seems to be insufficient.

9.3.2 Market System Effect
At present the process of operation and expansion of power systems involves many subjects of relations in electric power industry: power companies, individual power plants, Governmental authorities (federal and regional), electricity consumers (production and agricultural consumers, transport sector and population). These subjects have different interests. For power companies as for the wholesale market subjects the main criterion is profit. The criteria of the governmental authorities include profitability of the sector (incomes to the budgets), impact of the electric power industry on the industrial production volumes, employment and living conditions of the population, environmental impact, energy security, etc. The consumers are interested in the level of electricity and heat tariffs, reliability and quality of power supply. In particular the decisions efficient from the viewpoint of a federal or "national economic" level can be unacceptable for the other subjects. Many decisions cannot be made until the interests of all the parties concerned are coordinated and the required compromise is achieved.

Thus, the traditional estimation of power systems efficiency with account for only technological factors corresponds currently only to its technically feasible limits. Let us call them technical system effects (TSE).

The fact that many subjects have different interests and affect the decision making on expansion and operation of power systems does not change the set of components of the effect but leads to variations in the set for different subjects of relations and to different estimations of the same components of the effect by different subjects of relations.

Consider the main factors that determine the system effects for different subjects of relations in a market environment (let us call them market system effects (MSE)) as applied to Russia's current electric power industry structure that consists of competing generating and selling companies, network companies as natural monopolies, electricity consumers.

The bids of generating companies for electricity supply to the wholesale markets build the function of a supply which is then related to the function of a demand for electricity on the part of selling companies and consumers. This relationship is used to determine the equilibrium price of electricity at the wholesale market. Profit, being the main criterion for the generating companies under competition will make them decrease the costs of electricity production by loading, first of all, the most efficient generating capacities. Thus, the market mechanisms will decrease the equilibrium electricity price at the wholesale market. This is possible at joint operation of generating companies in a system with no network constraints. Here the account should be taken of the constraints on participation of generating units in covering load curves and on provision of power supply reliability and power quality.

The relationship between the MSE and the above TSE components shows that the formation of equilibrium electricity price at the wholesale market involves realization of practically all the TSE components. However the extent of their realization is determined by the efficiency of the competing market mechanisms. Bearing in mind the fact that the ideal competition in electric power industry is practically unachievable due to a limited number of market subjects we can expect that the MSE will be smaller than the potential TSE.

Similar market mechanisms should operate at competition of selling companies at the consumer electricity markets that will result in realization of additional MSE components at this level.

Network companies (federal and distribution regional) play an auxiliary role in the considered market processes. They rend the required services on power transmission from suppliers to consumers, on provision of power supply reliability and power quality, thus being conducive to the enhancement of MSE as a result of electricity market operation.

It is necessary to emphasize that in a short-term context the competing mechanisms at the wholesale and consumer electricity markets may decrease the electricity prices even below the level which is determined by the complete realization of TSE, as a result of price bids of the generating companies below the electricity production cost with account for its components. However, in a long-term context this situation is fraught with negative consequences that may take the form of inadmissible reduction of reserve capacities, decrease in funds to maintain the equipment in service state, to update and replace it. This will result in disappearance of conditions for competition at the electricity markets. The trend may appear toward a sharp increase in the electricity prices that will call for their regulation.

The interests of consumers expressed through their main criteria imply the interest in efficient operation of the electricity markets, i.e. maximum realization of MSE and, thus decrease in the electricity prices.

The interests of the authorities are to a certain extent contradictory. For example, the electric power industry will be highly profitable at high profits of power companies and these high profits are possible at high electricity prices. At the same time the efficiency of the industrial production, the living conditions of the population and other interests call for decrease in these prices. However, on the whole the authorities are certainly interested in the efficient operation of electricity markets, i.e. in maximum MSE realization.

It should be noted that for the subjects of relations the real effect from the measures on intensification of power systems integration depends on the system of economic management in the country. The system of management to a considerable extent affects the redistribution of the real effect among the subjects of relations and can both foster and hinder the TSE realization.

9.3.3 Principles of Estimating the System Efficiency

When estimating the potential TSE we consider UPS of Russia as a technically and technologically single object disregarding the forms of property. In this context this estimation is objective and single-valued. Comparatively simply we determine the base for such estimation, i.e. the conditions the effect is estimated for. Here we should use the approaches and recommendations that were developed for the centrally managed electric power industry. In a certain sense the model of the centrally managed electric power industry facilitates the full realization of the potential TSE.

The TSE components for UPS as a whole were estimated, depending on the aspect of consideration, based either on the conditions that do not suppose realization of the effect (for example isolated operation of regional power systems) or on the existing level of the UPS integration.

The approaches to the estimation of MSE are less obvious. When analyzing UPS expansion for a long-term future they apparently should be maximum independent from the conditions and principles of legal-normative framework. We should orient to the situation when the legal-normative framework fosters the full realization of TSE. When considering short-term prospects it is necessary to take into account the active legal-normative framework.

For a qualitative estimation of the system effect it is necessary to use existing and develop new mathematical models for: estimation of individual TSE components; integral estimation of an effect (such models are necessary as the net present value is not a simple sum of individual components); estimation of effects for federal and regional authorities and energy consumers that include the assessment of budget efficiency, levels of electricity tariffs, level of employment, level of living conditions, etc.

A complex integral estimation of the TSE involves the optimization model of electric power system expansion "SOYUZ"[8-10].

9.3.4 Case Study

Below presented are the results of estimating the system efficiency of Russia's UPS expansion for the time horizon 2010-2030.

It is very complicated to fully estimate all the TSE components, therefore only the main ones were estimated: the "capacity" effect, excluding the effects of equipment maintenance optimization and use of asynchronous run off in the rivers of different water reservoirs; the "structural" effect disregarding the environmental components, the effect of successive construction and construction of transmission lines at the point of connection of adjacent power systems; the "operation" and "frequency" effects. The model "SOYUZ" was applied for the estimation.

In the analysis of MSE the changes in EPS profit, cost of electricity produced, customer profits, payments to the federal and regional budgets were evaluated from the standpoint of subjects of the wholesale market, authorities and customers.

Comparison was made by the variant that envisaged expansion of the UPS generating capacities without expansion of the intersystem electric network, whose transfer capabilities between interconnected EPSs (IPSs) remained at the existing level. The basic variant to be estimated was the variant of the UPS expansion that was recommended by design institutions with optimization of transfer capabilities of tie lines between IPSs.

UPS in the model was described by a 7-node network (by IPSs). Two time levels were modeled: the years 2010 and 2030.

Adaptation of the structure of generating capacities to different expansion conditions was foreseen by the necessary changes in capacity commissioning at condensing power plants. The main calculation results are presented in Table 9.1.

The Table shows that the expansion of intersystem tie lines over the time span to 2010 is effective. The discounted costs for the UPS operation and expansion in this variant are $340 million lower than in the variant without their expansion, the required installed capacity of power plants is 990 MW lower, the saving of investments makes up $1.8 billion, the annual expenditures decrease by $110 million per year.

Indices, million doll.	Efficiency	
	2010	2030
Discounted costs, total	340	2760
Including: Power plants	440	3080
Transmission lines	-110	-320
Capital investments, total	1810	12890
Including: Power plants	2510	14720
Transmission lines	-700	-1830
Installed capacity, MW	990	8740
Annual expenses	110	1220
Including: Power plants	130	1320
Transmission lines	-20	-100
Fuel costs	-50	-30
Payments to the budget	49	510
Net profit of power companies	88	900
Cost of electricity production	130	1320

Table 9.1 Basic components of the efficiency of UPS integration

Expansion of the intersystem tie lines "Siberia - Ural" and "North-West - Center" is the most effective. The effect of expansion of the intersystem tie lines depends primarily on the "capacity" components when connecting the IPS of Siberia due to coincidence of load maxim and power exchanges of IPSs in emergency situations; capacity commissioning at power plants in Siberia decreases with partial compensation for this decrease by the capacity commissioned in other systems and increased use of existing power plants.

Decrease of operating costs is determined mainly by the decrease in fixed costs as a result of decrease in the total installed capacity; the fuel costs even somewhat increase due to the change in the structure of fuel burnt. Decrease in the cost of electricity production is accompanied by the growing profit of power supply systems and growing payments to the budgets of different levels.

The calculated value of the effect over the time horizon to 2030 (see Table 1) is considerably higher (3-4 times) than to 2010. The effect value is much higher than the expenses for its realization by construction and maintenance of intersystem tie lines. One dollar invested in expansion of transmission lines saves $2.5. The relation between decrease in discounted costs and increase in costs for transmission lines is equal to 3.1. One dollar invested in expansion of transmission lines decreases the cost of electricity production by 19 cent, increases an annual profit in UPS by 12 cents, and brings 7 cents to the annual consolidated budget.

Thus, the studies performed confirm a high efficiency of measures on strengthening the integration of Russia's UPS. The estimates obtained are conservative enough, since they cover not all the components of the system effect. In particular, the possibilities for formation of UPS expansion conditions that are optimal in terms of gaining the highest system effect were not analyzed. The system effect within IPSs in terms of interaction between the power systems comprising it was not considered as well. Account for this internal effect can raise the obtained estimates by a factor of 1.5-2.0. Nonetheless, even such conservative estimates point to an absolute efficiency of EPS integration. It can be achieved by the proper implementation of the required principles of economic management and the provisions of the legal-normative-legal framework.

9.4 Deregulation of power systems in Asia: special considerations in developing countries

Power industry is undergoing restructuring throughout the world[11-17]. The traditional vertically monopolistic structure has been deregulated and replaced by gencos, transcos and discos with competitions introduced to gencos and discos in order to reach higher efficiency in electricity production and utilization. A lot of R&D work has been conducted on power markets[18-24]. However for different countries, the principal objectives and considerations of power system deregulation are different. In the developed countries, areas with higher electricity prices would like to introduce competitions so that electricity prices are able to reduce and social welfare can be maximized. In some countries, electric power supply sector is a regulated public utility, and the government would like to privatize the electric power industry so as to reduce government commitment and functions. In this regard, gencos, transcos and discos are formed with competition introduced to gencos and discos to stimulate their incentives in efficiency improvement. When efficiency grows up and electricity price goes down, it can in turn promote GDP growth. In the developing countries, the power system deregulation and power markets are to establish fair competition among gencos so as to encourage investments to power industry from various resources, such as foreign investments, local investments and IPPs etc.

The diversified resources attracted to power industry can well meet the requirement of fast growing electricity demand induced by blooming economy in the developing countries. This might be the main reason for Asian developing countries to promote deregulation in power systems.

Although power system deregulation is widely conducted in the world, it is not easy to successfully reach the initial targets set for the deregulation. Both experiences and lessons are obtained in practice. Considering that power industry is a significant infrastructure and in order to realize sustainable development of power systems at high efficiency, how to realize optimal system planning and reliable operation at acceptable electricity prices with qualifies service and how to transit to the market environment smoothly at lowest costs and lowest risks should be considered thoroughly in the market environment. This is actually an important task for the developing countries in Asia.

9.4.1 Considerations on Power System Deregulation in Developing Countries

9.4.1.1 Targets of deregulation

In Asian developing countries, power systems are usually owned by the state or provincial government before deregulation and operated as a vertical monopoly. Unified dispatch is applied to individual power grids with limited power exchange to external grids. Hence the interconnections among grids are relatively weak. Because of fast economic development, the electricity demand is growing rapidly and the power system expansion becomes a severe economic burden of the governments and a bottleneck of overall economy sustainable development. Investments from companies other than governments to power systems through approaches such as Build-Operation-Transfer (BOT) etc. are not so attractive. It is urgent to launch power system restructuring and deregulation and to establish power markets with fair competition so as to attract more investments from various resources to power industry. Moreover it is found that in some countries the monopolistic operation of power systems can also lead to low efficiency and even corruption for non-transparency and lack of surveillance in operation. It is clear that the power market is able to solve most of the problems appropriately.

Based on the facts mentioned above, the targets of power system deregulation and power market establishment in Asian developing countries can be outlined as follows with certain priority.

* To attract various investments to power systems through open-access and fair competition so that power supply can meet the fast growth of demand in developing countries.

* To reduce government commitment to power industry, which is a severe burden to the governments of developing countries concerning huge amount of investment to the infra-

structure construction; and to mitigate government functions in power industry so that the government can focus attention to significant issues of entire society.

* To encourage efficient electricity production and utilization through competition and electricity pricing mechanism. The energy resources can be saved. The production costs can be reduced and hopefully the electricity price can decrease. It can in turn increase social welfare and benefit to national GDP growth. Thus sustainable development of power industry can be realized.

* Through competition, the market-clearing price (MCP) of electricity can provide clear economic information for future power system development use; power supply services can be improved; gencos' profits may increase through higher operation efficiency; and the government tax return from power industry can be raised.

However the realization of the targets are not easy because power system deregulation involves the benefits of various parties and each party will try hard to protect its own interests and add influence to the power market policies and rules. To keep a large genco can make it possess noticeable market power to get more profits. To define a MCP with or without congestion management consideration will lead to totally different profit distributions. To organize a small provincial market or a large regional power market may bring about significantly different social welfare. In order to construct power markets with fairness, justice and transparency, thorough considerations should be made to various aspects, which will be discussed below.

9.4.1.2 Considerations on the scheme of power system deregulation

The overall power market scheme is usually considered by the government. It includes several aspects described below.

* According to the targets set for deregulation, it is very important for the government to consider carefully at the beginning on: how many markets should be organized throughout the country; who should be responsible to organize the power markets, how to organize gencos, transcos and discos; what are their functions; who is responsible for market operation and/or system operation; how to develop market rules; who is responsible for market surveillance; how to transit from current regulated market to future competitive market; etc. A timetable should also be developed for execution. It is clear that a lot of negotiations and government decisions should be made at this level in order to make deregulation process successful. International consultant might be helpful, however for the different economic and technical backgrounds, it is more important to rely on own conditions and experts to make final decision.

* After the fundamental decisions in launching the deregulation, the market rule development is a key step. A lot of lessons have been obtained in this aspect. For developing countries, copying power market rules of other countries or other markets such as natural gas etc. might lead to severe results. The market rules should be different from market to market with only principles unchanged. Therefore all market participants (suppliers, consumers, brokers, operators etc.) should have rights to join or examine rule development with consideration on all economic, technical and physical backgrounds, conditions and trends. The rule should try to be a perfect set on market organization, market functions, goods to be exchanged and services to be provided, sub-markets definition (futures, forward, option,

day-ahead, and balancing markets for electricity as well as the markets for transmission service and ancillary service, etc.), pricing mechanism, participants and relevant regulation, market operation and system operation rules under normal and emergency cases, settlement, surveillance, etc. The rule should be revised from time to time to meet new requests and evolve gradually.

* Another important task in power market operation is to develop a reliable computer supporting system to manage bilateral contract records, day-ahead market bidding and MCP - power selling/buying calculation, congestion management, and market information announcement etc.

It should be pointed out that different markets have different potential risks in operation, to which special attentions should be paid. The issue is discussed below.

9.4.1.3 Consideration on risks and their mitigations

To realize successful running of the power markets, a significant issue is to examine carefully the potential risks existing in a specific power market. The risks include natural resource types, amount and relevant risks, economic development in the market area, gencos and consumers' features, grid characteristics, potential market power, risk in power markets of single-buyer mode, special reliability and security issues in power system operation, power system planning/expansion in market environment, possibility of power exchange with adjacent power markets and relevant issues, and possible local government interference, etc.

For a power market operating very well for years, we should not say the market is perfect until it operates very well under extremely severe market conditions and can also keep power system sustainable and efficient development for a relatively longer time horizon.
Some risks, which are often appeared in the developing countries, will be discussed briefly in more detail below.

Unbalanced economic development within the market area. When a power market covers both rural area and industrial area, the power generation in rural area usually has lower electricity price than that in industrial area if there is no power exchange between the two areas or if special price policy is applied to the rural consumers. When they are in the same market, discriminating price or proper measures should be taken to protect rural area consumers. Otherwise, industrial area users would like to sign bilateral contract with gencos in rural area at lower prices and the electricity price in that area will rise and have negative impacts to rural users.

Hydraulic resource and hydrothermal dispatch in market environment. When a certain area has a higher hydraulic capacity ratio and/or the hydraulic resources should also be used for navigation, irrigation and flood control purposes controlled by the government agent, the risk of hydraulic power generation may occur. Basically there are several ways to dispatch hydropower: central dispatch with priority for hydropower taking care of peak load to maximize social welfare, thus all hydropower can get maximum benefits. However this might have negative impacts on thermal power generation profits especially when water resource is rich in certain years. However if hydraulic power plants bid in the same way as thermal power plants, the hydraulic resources might not be used efficiently. Besides some compensation should be provided to hydraulic power plants for their loss caused by diversified usages of water resources. Therefore certain rules should be developed to manage the case to make both thermal and hydraulic power companies have proper profits with incentives to make further investment to the power markets when load grows.

Market power. When a market has few power generation companies, the company with large market share is able to control the price and get high profits from it. Such market power is harmful to consumers and may reduce social welfare noticeably. Therefore market share and/or capacity limits should be applied to any gencos. Grid congestion is another source of market power. Improperly defined transmission rights or congestion management rules can be utilized by market participants and strengthen the locational market power. In peak load condition or when large units are under maintenance, some machines become 'must-run', the market power can be severe. In addition to well-defined market rules to manage congestion and mitigate market power, market power existence and severity should be supervised constantly. Market surveillance and certain penalty should be applied to solve the issue.

Reliable and secure power system operation under market environment. This is the most significant issue in deregulation. The power system deregulation should not sacrifice the system operation reliability and security. However the risk does exist. Lots of reasons can cause the issue, such as lack of reserve capacity, lack of reactive power supply and voltage control, poor organization of real-time power balancing market, poor emergency control and restorative control, wrong load forecasting and/or poor generation and transmission expansion planning which in turn leads to inadequacy of capacity, etc. In developing country, the grid construction may lag behind the generator installation and the resultant stressed network may cause severe problems under large and small disturbances.

Centralized dispatch is highly recommended in such case to make full use of system-wide resources to avoid system instability and collapse. In principle, the economic issue such as various transactions should be determined by market rules before real time operation. Even if the resource/service is used/provided in real-time, it should be scheduled beforehand regarding to time, price, quantity, and the way to use/provide. However secure and reliable system operation should be realized in real-time absolutely through central dispatch and control and the system operator should have the authority to make decision according to load demand, power quality and security requests etc. under normal and emergency conditions. The difference of system operation with or without power market existence is that in power markets the operator should dispatch/use the available resource according to bilateral contracts, day-ahead market, ancillary service market outcomes determined beforehand.

Considering former monopolistic operation condition, existing EMS and operation experience, it is recommended to have two computer systems for market operation and system operation respectively but with real-time link to each other for information exchange, software/database sharing and necessary mutual support. The same institute will manage both computer systems to realize cost-effective operation and well coordination. A system operator should get some training on market operation and vice versa so that he/she can understand the impacts of market environment on system operation very well.

Electricity pricing and demand elasticity. Pricing is a core issue in power markets. Bidding strategies are worked out based on pricing mechanism. For most power markets, MCP is used to determine day-ahead market price and electricity production amounts, which are then revised according to congestions if any. If the pay-as-bid principle is used in spot pricing, the ISO of a single buyer market can get significant benefits.

A disco, who buys electricity from the market and sell it to the end users, may suffer from fixed electricity price for end users. The load elasticity is very important in power markets. Market failure might happen when the energy supply price is extremely high while end user cannot see it.

Other merchandises in power markets should have their prices well defined as well. However at the beginning stage, the market rules should be simple and practical. Some services, such as reactive power supply and voltage control, reserve capacity, AGC etc. can be considered as a whole by the ISO and then distribute the cost to users as a lumped 'uplift'. The users' bill will not show the price for each service. This can be improved gradually when the technology is mature.

Smooth transition to the market environment. This is also an important point in deregulation. A violent transition should be avoided. Usually when a market has adequate generation capacity and the supply is a little greater than demand, it is a good time to start power markets. The price will be smooth in transition and market rules can be tested under normal conditions first. Operators can practice in the real market environment. In this case, reliable and secure system operation can usually be ensured. Otherwise risk will appear in the test operation of power markets.

Power system planning in market environment. Under market environments, generation and transmission network expansion are vital to power industry sustainable development. Accurate load and fuel price forecasting and correct estimation of profits are very important. Government policy and market rules in power system planning also play an important role in system expansion. In order to attract enough generation investments, a key point to developing countries, some rules with bias to gencos have to be applied. A typical price policy is two-part electricity price including capacity price and energy price, which can greatly reduce gencos' risks in capital return and encourage generation investments effectively. This price policy is now used in NE China regional power market. In the meantime, to let grid company (which may also be the operator of both power market and power system) have enough income as transmission network capital return and for future development, transmission service price can be set at a relatively higher level.

In power system planning, it is important to have an authorized institute to make centralized optimal planning with participation of market participants based on the information of economic development, load forecasting, energy resources available etc. Interested companies can then bid for power plant expansion and construction. However the transmission network expansion is a problem since it is difficult to get investment back in a short period through transmission service charge. The incentives to invest transmission line construction are therefore unclear. And the stressed network may lead to system instability under disturbances and even blackout under cascaded faults. Special measures and policies should be applied to solve the problem.

9.4.1.4 Other considerations

To ensure fairness, justice and transparency in market competition, government should have minimum interference to the markets. But special purpose committees can be organized to manage the overall power industry policy, energy planning and investment consideration, market surveillance, etc.

The market rules should also define duties, rights and benefits of individual market participants; include extraordinary operations of markets such as emergency stop, termination and re-start conditions and procedures.

In order to promote power exchange and obtain more economic benefits and social welfare, transaction between markets are highly encouraged which include hydrothermal coordination across the markets, emergency support to reduce reserve capacity cost, and peak-load-time-difference utilization, etc.

In certain area, special policies on electricity tariff are applied to special users, such as compensation to rural area users and special local industry, etc. This policy may exist at the beginning stage of power markets. It will be reduced and eliminated gradually when market is mature.

In general power system deregulation is a kind of 'system engineering', which is relevant to various aspects including economy, politics, society, engineering and technology. The success of power system deregulation will have great impacts on overall society and national economic development.

9.4.2 Power Markets in NE and NW China

China is a developing country in Asia with fast economic development in recent years. By the end of 2004, the total generation capacity reaches 447 GW, ranking at No. 2 in the world. Its electricity is still in short supply for the fast blooming economy.

Huge population, fast economic development and electricity demand growth cause severe burden of the government under traditional vertically monopolistic operation mode, in which the government is responsible for the investment of main power plants and transmission grids. In order to attract diversified resources to invest power industry, power system deregulation and power market establishment to ensure fair competition is the best and sustainable way to solve the problem.

The main steps in power system deregulation in China are as follows[25-28]. In Dec. 1998, the State Council of China announced to launch power industry reform in China. In the first stage, the restructuring was focus on forming independent power companies and grid companies. Five provincial power markets and Shanghai power market were built up and tested as pioneers. Competition was introduced to gencos and the markets were in the single buyer mode. However the provincial power markets only covered small areas, which strongly limited the power exchange across the provinces and the goal of optimal use of resources over a wide region could not be realized. In Feb. 2002, the State Council of China announced a new document of 'The Scheme of Power System Reform'. In this document, it is declared that to form regional power markets will be the future direction of power industry restructuring in China. In Dec. 2002, two independent grid companies were formed. They are S. China Grid Company and the State Grid Company. The latter includes 5 regional grid companies. They are Northeast, Northwest, N. China, E. China and Central China Grid Companies. The traditional government owned power plants were reformed into five generation-group companies with their individual shares in each regional power market less than 20%. In March 2003, the State Power Market Surveillance Commission (SPSC) started to work. Under the leadership of SPSC, Northeast and East China regional power markets were built up and committed to test operation in January and May of 2004 respectively. In the meantime all original provincial power markets were asked to stop operation. Other regional grid companies are actively making R&D to ensure well design and optimal operation of future regional power market and avoid any potential risks.

The power market in NE China and R&D work in NW China will be discussed below.

9.4.2.1 Northeast Regional Power Market in China

The Northeast region covers three provinces (Hei-long-jiang, Ji-lin and Liao-ning) of NE China and East of Inner-Mongolia with the population of 100M and generation capacity of 41GW (85% thermal power and 15% hydropower) at the end of 2003. The rich coalmines in the north and heavy load in the south cause a long transmission corridor from N. to S. of the region. In addition, a short power transmission corridor exists in mid-region from west (thermal power) and east (hydraulic power) to central part. The NE has been interconnected to N. China to sell their electricity.

From the Northeast Regional Power Market, we can see some rules which are typically defined based on developing country conditions and significant to secure operation and sustainable development.

* The NE Grid Company owns the regional transmission network and runs the regional power market and power system in parallel. It is also responsible for organizing generation planning and transmission planning according to the load forecasting although the Grid Company is not responsible for the construction of power plants.

* The NE Grid Company keeps main hydraulic power plants in hand (10% in capacity) for frequency control and peak load dispatch to avoid sky-high prices under heavily loading conditions.

* Two-part electricity pricing is used, in which capacity price and energy price are separate. The capacity price is defined by the State Tariff Department, while energy price is generated

through competition. This is good to encourage future investment in power generation and solve sunk-cost issue.

* The market in its first stage adopts single buyer mode. For long-term contract, pay-as-bid rule is applied for settlement, which can protect the benefits of the Grid Company and consumers. For day-ahead spot market, MCP is used. In general major energy transactions are through forward contract (round 80-90% of total power exchange); only a small amount is exchanged through spot market. Besides the generation option market is under organization to reduce gencos' risks.

* All transmission and ancillary services are centrally dispatched and the cost is distributed to consumers as a 'transmission cost' in average cost nature, which is also examined by the state Tariff Department. In order to avoid high service price, each power plant should provide certain reactive power and reserve power and join voltage and frequency regulation without payment. Later on a compensation method will be worked out.

* Bilateral transaction is limited to large consumers and power plants as test right now. It will be gradually increase. Currently only machines at or above 200MW are in the market. Later on it can be generators at or above 100MW.

* Ancillary service market is under consideration. Competition among discos and the transition of the market from single buyer to whole sale modes are also under consideration. But these will commit only when the current market operates well and experiences have been cumulated enough.

9.4.3 R&D for Future Northwest Regional Power Market

The Northwest Grid covers four provinces or autonomous region. They are Shan-xi, Gan-su, Ning-xia and Qing-hai. Its area is about one seventh of China. By the end of 2003, the total generation capacity was 20 GW with 65% thermal and 35% hydraulic power. However the economic development is quite unbalanced in the region, relatively rich in Shan-xi and poor in Qing-hai. Hence the electricity prices are quite different in the two areas. The upper Yellow river provides rich hydraulic resources. However the request of nevegation, irrigation and floor protection, which is controlled by certain official organization, may significantly reduce the profits of hydraulic power plants and bring about difficulties in day-to-day dispatch. Another issue is caused by larger area and lower population density, which means more transmission system investment. The western area hydraulic power and eastern area thermal power can be dispatched properly to yield most economic benefits. The peak load time difference between western and eastern areas also asks for large amount of power transfer between east and west parts. However currently relative weak network may lead to severe congestion and relevant market power if power market is committed. The region is going to interconnect to N. China and Central China Grids to sell electricity to the two regions.

Based on the facts mentioned above, NW Grid Company is actively conducting R&D towards future power market operation. Three R&D projects have been implemented with the help of universities. The three projects covered the following three aspects respectively (i) overall market structure, market operation mode and market rule etc.; (ii) pricing mechanism and grid average transmission cost estimation; and (iii) potential risks and their mitigations in the NW Regional Power Market. The last project was jointly implemented by the University of Hong Kong and NW Power Grid Company.

All the R&D projects examined the characteristics of the Grid thoroughly and made valuable qualitative and quantitative suggestions on future power market operation. Experiences and lessons from world power market practices are also considered in-depth, which laid a solid foundation for future successful operation of the power market and secure and reliable operation of the power system.

9.5 East China Power Market Development and TRIAL OPERATION

East China Power Grid (ECPG) covers the geographical areas of Shanghai Municipality, Jiangsu, Zhejiang, Anhui and Fujian Provinces. By the end of 2003, the total installed capacity in East China Power Grid had reached 81,300MW, with total annual electric energy consumption of 453TWh. There are 45 substations and power plants at 500KV level. The length of 500KV AC transmission lines had reached 7460 Km. 500KV DC transmission lines connect the East China Power Grid with Central China Power System where Three Gorge Power Station, the largest in the globe, is located. The large scale East China Power Grid offers an excellent platform for the trading of energy-related commodities.

The region East China Power Grid covers is one of the fastest economically growing and most prosperous regions in China, with a two digit annual GDP increases in recent years. The population accounts for 1/6 of China's, consuming 1/5 of electric energy in China, and having 1/3 of China's total economic production. The average per capital GDP in this region is over 1.7 times of domestic average level. The continuously fast economic growth has triggered two-digit increases in both energy consumption and peak load for some consecutive years. The fast economic growth, together with the dynamic market activities in other sectors and relatively advanced market management experience, provide a rather suitable external environment for the development of East China Power Market (ECPM).

Before industry restructuring, East China Power Grid, the parental company of four provincial grid companies, had a long tradition of executing inter-provincial energy transactions. These transactions range from centrally planned proportional allotment, residual energy transactions, to bi-lateral contracts. In general, provincial grid companies participate in such transactions while individual generators do not. In particular, the bi-lateral trading among the provinces is booming in these years. The bi-lateral trading energy has reached 4.855 TWh in 2001, 11.836TWh in 2002 and 25.594 TWh in 2003, with an average increase rate 129.6%. From a historical viewpoint, these inter-provincial energy trading laid a solid foundation for the development of the East China Power Market.

During the period from 2000 to 2003, a number of provinces in east China region have opened free electricity markets in a provincial level. These markets were criticized because they offer competition only in provincial level. This motivated the idea of inter-provincial markets, which is of much the same flavor of Regional Transmission Organization (RTO) in USA.

Traditional wisdom suggests that the construction and development of the regional power market is a process of learning both international experiences and lessons that have been accumulated over times. It is also a process of reaching agreement and increasing awareness of the market initiatives. From June 2003, when the East China Power Grid was selected as one of the regional power market pilots, to May 18th, 2004 when trial market operation started*, it took almost one year to complete the initial market preparation and construction. The major milestones can be listed below:

- In March 2002, China government issued Power Industry Institutional Reform Plan (Policy No. 5), which outlines the direction of China's Power Industry reform. Generation business has been separated from transmission business and five major generation companies have been formed. The establishment of the regional power market is given the top priority.
- In June 2003, State Electricity Regulatory Commission (SERC) presided a meeting in Shanghai and clearly addressed the intention of establishing a pilot power market in East China.
- According to the Policy No. 5, East China Grid Company Limited (ECG) was established on September 28, 2003, which bears leading responsibility for developing ECPM.
- In November 2003, SERC issued the ECPM Pilot Plan, which is the blue print of the power market development.
- In April 2004, SERC issued the ECPM Codes.
- On May 18th, 2004, ECPM trial operation was commenced. This is the largest regional power market pilot in China.

9.5.1 The Development of the Market

9.5.1.1 Guidelines

The Policy No. 5 guides the development of ECPM. The market development has to respect the unique characteristics of power industry, and international experiences are exploited. It starts with the separation of generation from transmission functions, which aims at optimal resources allocation by introducing competition into wholesale market. The market is expected to facilitate a sustainable and healthy development for power industry in East China Region to satisfy the requirement of regional economic growth as well as the needs of people's life.
The objectives of ECPM Development include:

♦ Introducing competition, breaking down market barriers and achieving optimal resource allocation in the East China region;
♦ Encouraging power industry to improve management, efficiency and service standard etc;
♦ Establishing a unified, open, competitive and stable regional power market under government regulation.

9.5.1.2 The principles of market design

The institutional structure of ECPM is judiciously chosen taking into account the characteristics of energy resources and consumption types within the region, technical requirements, the current institutional structure, and local economic conditions[29].

---

* In the language of Chinese power community, "trial operation" is the first step towards a competitive market, followed by "pre-operation", and "official operation". In trial operation, market dispatch is performed based on bids, settlement is performed but financial transaction is not executed. In "pre-operation", financial transaction is executed but the market can be suspended if certain condition (say, extremely hot weather) is met. In "official operation", market can not be suspended in general.

In the pre-restructuring model, the buyers and sellers are limited only to the provincial power companies serving as either single buyers or single sellers who trade for surplus or deficit of their own provinces. In the new market, all the market entities are competing on the same regional market platform. This design is expected to produce a better outcome in terms of resource allocation over a wider area.

In the preliminary stage of the ECPM, five buyers (four provincial power companies and Shanghai Municipal Power Companies), 171 sellers (conventional coal-fueled generators with capacity over 100MW) will participate in the centrally controlled regional power market.

A 500KV transmission network that connects the provinces and Shanghai Municipality has formed, which facilitates the inter-provincial energy trading. The separation of generation from transmission was largely completed by 2004. Shanghai Municipality and Zhejiang Province who managed province-based power market pilots have accumulated some experiences for the development of the regional market.

9.5.1.3 The stage objectives

In accordance with the principles of integrated design and phase-in implementation, ECPM development shall be implemented in three stages.

- Stage one objective: Let most of the generation companies participate the competitive market, and a portion of the energy be centrally traded in the ECPM platform. Investigate to allow large customers to directly purchase from generation companies in a controlled fashion. Establish legal and regulatory framework for the operation of ECPM. Reform and improve pricing mechanisms for electricity. Establish an ECPM coordinated by the region and its provinces.
- Stage two objectives: Let gradually more generation companies participate in the market, thus increase the level of competition; start bi-lateral trade between large customers, independent distributors and generation companies, set up ancillary services market and transmission rights market, increase trading products; Establish a well functioning market regulation framework and a market under unified operational control.
- Stage three objectives: Introduce competition into retail segment of the industry, and allow all eligible customers directly participate in the market competition with more flexible trading arrangements and products. Establish financial market for energy and form a unified, open, competitive and orderly ECPM under government regulation.

9.5.1.4 The trading arrangements

At initial stages of market development, market trading is a combination of contract trading and physical trading. Contract trading will be the major trading pattern, supplemented by physical trading. A small portion of the energy generated by generators is traded in monthly market and day-ahead market, the remaining energy is sold to network companies through annul contracts for difference (CfDs) and physical contract. Ancillary service is not included in this market at this stage. Non-market generators are expected to sign contracts with network companies under the guidance of SERC.

In the short term, ECPM trading includes annual contract, monthly contract, day-ahead market and real time balancing mechanism. With the development of the market, other forms of trading could be introduced such as quarterly competition and weekly competition. At initial stages of market development, approximately 85% of energy is determined by annual contract; the remaining 15% or so is traded in the market. Depending on the progress of market development, the portion of energy exposed to market competition can be increased.

9.5.1.5 Bidding and market clearing mechanism of monthly market

The monthly market of ECPM, the first market to go under trial operation, was divided into two segments. One is for the peak hours trading with a price cap 482 RMB/MWh, the other is for the off-peak time with a price cap 321 RMB/MWh. Both markets follow the familiar market clearing mechanism while transmission constraints are taking into consideration. Mathematically, the market clearing is represented as the following linear program in which the mathematical symbols are self-explanatory:

$$\text{Max} \sum_{p \in Prv} \sum_{i \in PrvBlock_p} \text{PrvPrice}_{p,i} \times \text{PrvBid}_{p,i} - \sum_{g \in Gen} \sum_{j \in GenBlock_g} \text{GenPrice}_{g,j} \times \text{GenOffer}_{g,j} \Big/ \text{GLF}_{g,j}$$

$$S.T. \sum_{p \in Prv} \sum_{i \in PrvBlock_p} \text{PrvBid}_{p,i} - \sum_{g \in Gen} \sum_{j \in GenBlock_g} \text{GenOffer}_{g,j} = 0$$

$$0 \le \text{PrvBid}_{p,i} \le \text{PrvBidMax}_{p,i}$$

$$0 \le \text{GenOffer}_{g,i} \le \text{GenOfferMax}_{g,i}$$

$$\sum_{i \in PrvBlock_p} \text{PrvBid}_{p,i} - \sum_{\substack{j \in GenBlock_g \\ g \in GenPrv_p}} \text{GenOffer}_{g,j} < \text{PrvInMax}_p$$

$$\sum_{\substack{j \in GenBlock_g \\ g \in GenPrv_p}} \text{GenOffer}_{g,j} - \sum_{i \in PrvBlock_p} \text{PrvBid}_{p,i} < \text{PrvOutMax}_p$$

Implicit in the above formula is a zonal transmission model; each province (Prv) is considered as a transmission zone. The duals of the above linear program form the zonal marginal prices [30].

9.5.2 The trial operation of the market

According to a SERC order, ECPM trial operation, which started on May 18th, 2004, only involves monthly competition. The monthly competition simulates the following processes: generator unit registration, bidding, price ranking and information publishing.

In this trial operation, both supplier and buyer should bid and offer for their intended energy. The price and volume that will be cleared in the monthly market is determined by the bidding and offering curve from both generators and provincial buyers respectively, although, it will not be settled during the trial operation period.

However, the provisions of the ECPM Code on market administration and network operation shall be strictly implemented even during the trial operation period in order for the safer operation during simulation stage. Market administration comprises market participant management, unit registration, market suspension, etc.

9.5.2.1 Simulation results

Results of June 2004

ECPM cleared at total electricity 4.503 TWh with an average price 369.5 RMB/MWh in June 2004. Respectively, in peak load time the cleared quantity is 3.326 TWh, accounting for 73.86%, the cleared price is 406.8 RMB/MWh; in off-peak load time the cleared quantity is 1.177 TWh, accounting for26.14%, and the cleared price is 273.1 RMB/MWh (see Table 9.2 and Table 9.3 below).

Unit: RMB/MWh; MWh

Provinces	Clear Price	Supply	Demand	Export
Shanghai	406.8	538.020	143.220	394.800
Jiangsu	406.8	11895.460	1621.200	274.260
Zhejiang	406.8	248.220	1275.960	-1027.740
Anhui	389.0	568.680	0	568.680
Fujian	406.8	75.600	285.600	-210.000
Aggregation		3325.980	3325.980	0

Table 9.2 The simulation results of peak load time in June 2004

Unit: RMB/MWh; MWh

Provinces	Clear Price	Supply	Demand	Export
Shanghai	273,1	266.100	600	265.500
Jiangsu	273,1	292.200	0	292.200
Zhejiang	273,1	132.000	708.300	-576.300
Anhui	272.2	434.100	277.800	156.300
Fujian	273.1	52.800	190.500	-137.700
Aggregation		1.177.200	1177.200	0

Table 9.3 The simulation results of off-peak load time in June 2004

The market clearing of on peak and off-peak hours are further illustrated in Figure 9.3 and Figure 9.4.

Results of July 2004

ECPM cleared at total electricity 6.820 TWh with an average price 405.4 RMB/MWh in July 2004. Respectively, in peak load time the cleared quantity is 4.849 TWh, accounting for 71.09%, the cleared price is 468 RMB/MWh; in off-peak load time the cleared quantity is 1.972 TWh, accounting for28.91%, and the cleared price is 289.3 RMB/MWh (see Table 9.4 and Table 9.5 below).

Unit: RMB/MWh; MWh

Provinces	Clear Price	Supply	Demand	Export
Shanghai	468.0	1014.258	954.366	59.892
Jiangsu	449.0	2139.620	2070.180	69.440
Zhejiang	468.0	607.600	1059.828	-452.228
Anhui	416.4	687.890	186.186	501.704
Fujian	468.0	399.280	578.088	-178.808
Aggregation		4848.648	4848.648	0

Table 9.4 The simulation results of peak load time in June 2004

Unit: RMB/MWh; MWh

Provinces	Clear Price	Supply	Demand	Export
Shanghai	289.3	865.520	0	865.520
Jiangsu	289.3	458.490	193.130	265.360
Zhejiang	289.3	244.280	1399.030	-1154.750
Anhui	289.3	288.610	199.950	88.660
Fujian	289.3	115.010	179.800	-64.790
Aggregation		1971.910	1971.910	0

Table 9.5 The simulation results of off-peak load time in Jule 2004

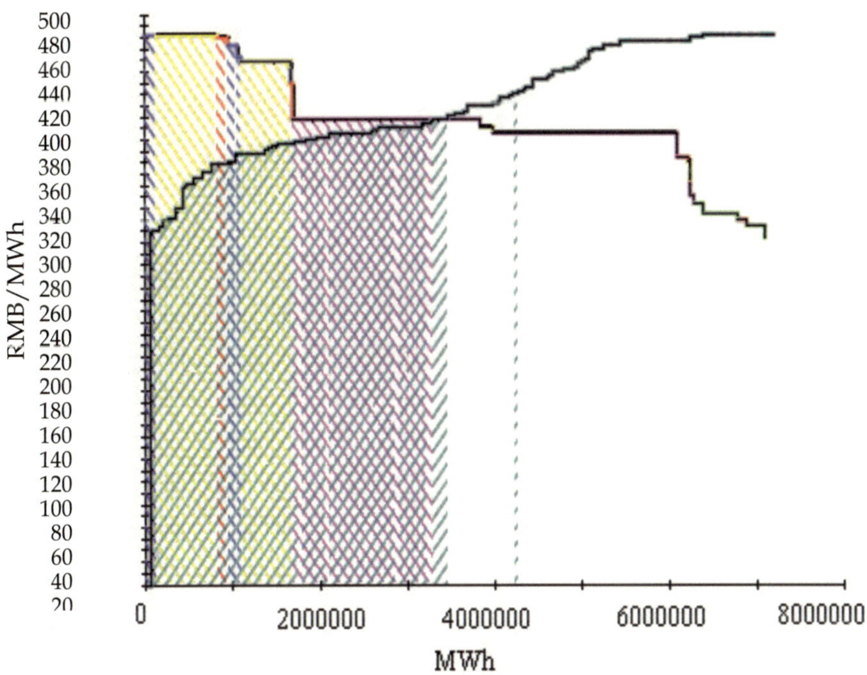

Figure 9.3 The merit curve of peak load time in June 2004

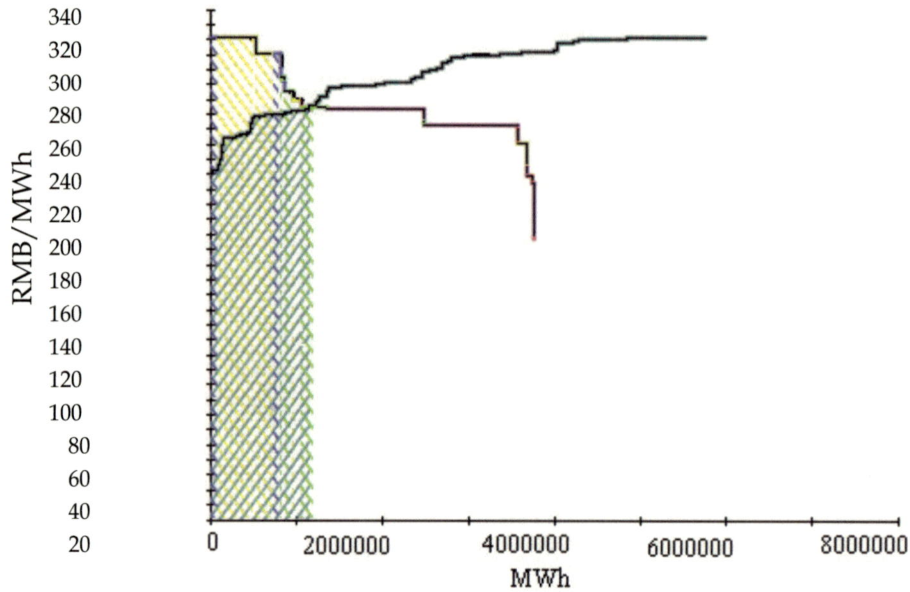

Figure 9.4 The merit curve of off-peak load in June 2004

The market clearing of on-peak and off-peak hours are further illustrated in Figure 9.5 and Figure 9.6.

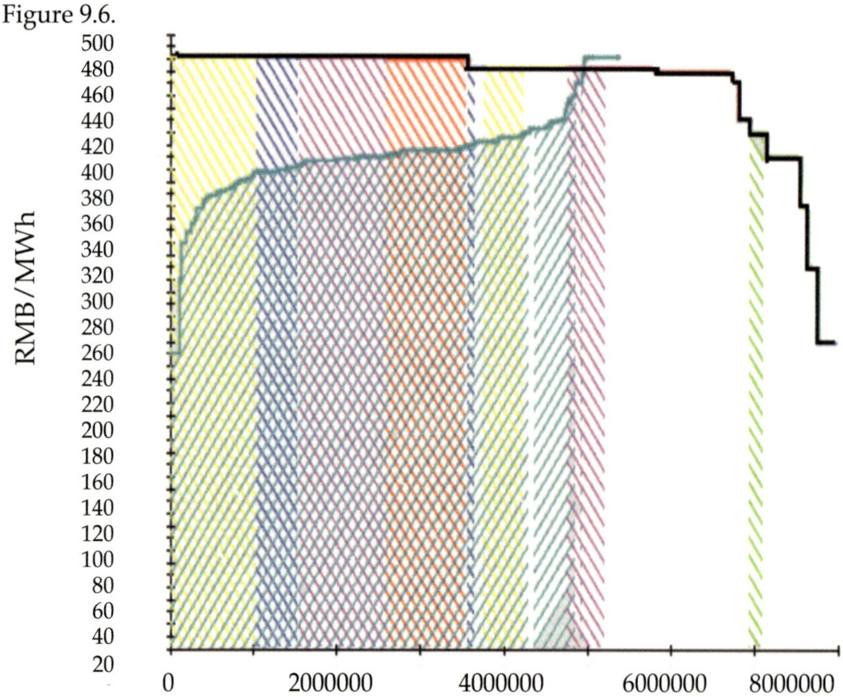

Figure 9.5 The Merit curves of Peak Load Time in July 2004

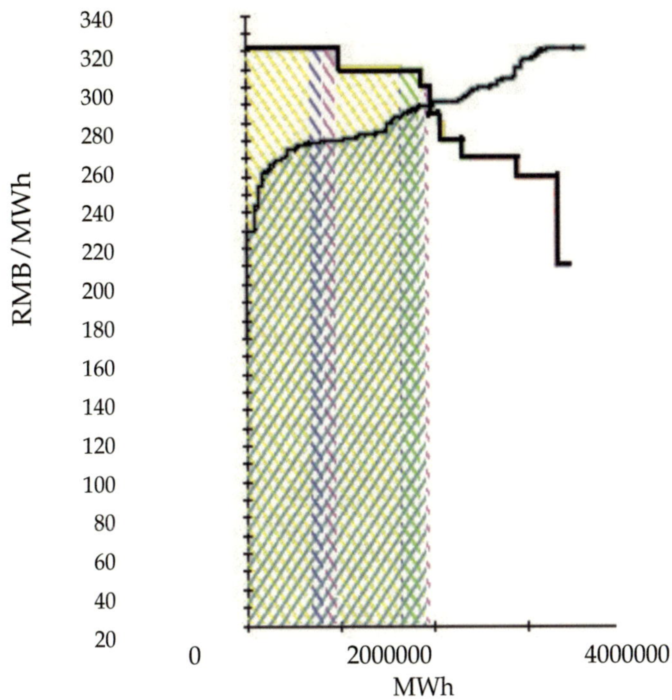

Figure 9.6 The Merit Curves of Off-peak Load in July 2004

9.5.2.2 Results analysis

The significance of the regional power market lies in optimizing resource allocation over a wider area, compared to the previous provincial power market. The problem is how to quantify the significance. In what follows we give a preliminary solution to this question.

The significance can be represented by the market efficiency. According to the principle of economics [31] the market efficiency can be quantified by the total surplus, which is the sum of consumer's surplus and producer's surplus. In the regional power market, the Tie Line Flows among the provinces has a significant impact over the market efficiency. Assuming the transfer limits of the Tie Line Flows are zero, actual market flows and infinite respectively, the efficiency of the provincial market, actual regional market and ideal regional market can be calculated, the results are included in Table 9.6.

From Table 9.6, it can be seen that the market efficiency of the actual regional market is 774 million RMB, while the province's is 574 million RMB. The regional power market proved to be maximizing the market efficiency. Figure 9.7 further illustrates the relationship between tie-line transfer capacities and market efficiency.

Unit: million RMB

	Provincial	Regional	Regional (ideal)
Off-peak, June	23	52	52
Peak, June	149	232	235
Off-peak, July	44	93	93
Peak, July	357	397	411
Aggregation	574	774	791

Table 9.6 The regional market efficiency compared with the provinces

Certainly, provincial markets with zero tie-line capacities are rather idealized markets. The transactions exist among provincial markets help to improve (provincial) market efficiency. However, the difference between an inter-provincial market and a group of provincial markets with transactions is fundamental.

Another issue that is receiving attention is the simulated deficit of network companies. As a result of the tight supply situation, the network companies' payment in the wholesale market has increased on a large scale. The analysis on trial operation revealed that the increased payment is 171 million RMB in June, and 334 million RMB in July respectively. The total increased payment, 505 million RMB, is far beyond the network companies' budget. This tendency has been one of the major concerns of the regional market.

We complete this section by summarizing the trial operation of the monthly market in Figure 9.8, where ISO means Independent System Operator.

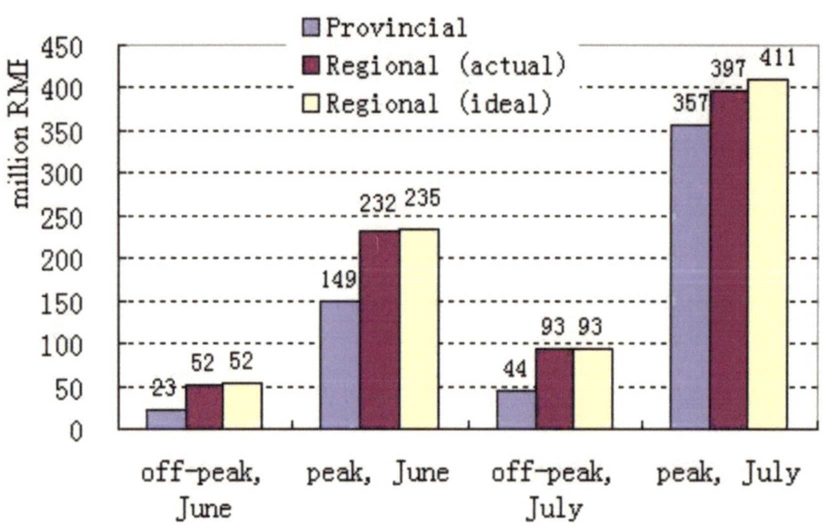

Figure 9.7 Tie line flows affecting market efficiency.

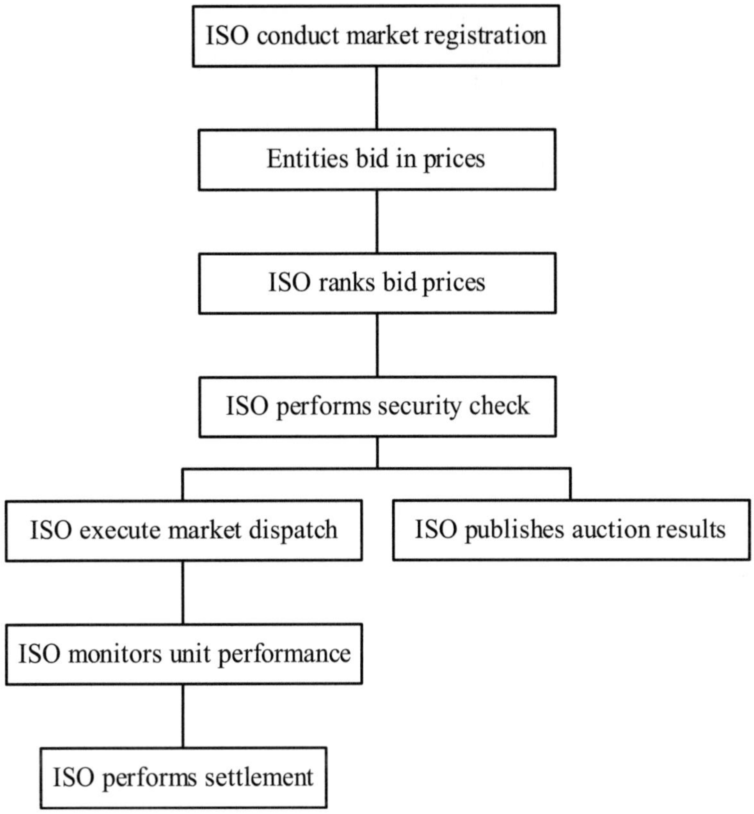

Figure 9.8 The Monthly Market Operation.

9.5.3 Lessons and Recommendations

Despite of the success of the trial operation, there are lessons and recommendations worth mentioning. They are summarized in this section.

9.5.3.1 Regulations and policies need to be improved

The ECPM pilot is developed and operated in a trial fashion under the condition of lacking relevant regulations and policies. With the pilot work going, however, improving market regulations and policies becomes extremely important. The Electrical Regulatory Ordinance has not come forth, the electricity transmission tariff has not been established, and the retail prices has not been linked with the wholesale price. The immaturity of legal environment has limited the success of ECPM pilot.

9.5.3.2 The supply-demand gap threatens the safety of the power system

The East China Power Grid has a wide supply-demand gap for the time being. The Month-ahead market has affected the electricity balance of the provinces, and changed the Demand Side Management of the region. Those provinces, which failed to purchase energy in market, will lead to heavier pressure on their power supply responsibility. This kind of pressure will threaten the safety of network operation because it makes the buyer more irrational also brings the native provincial governments more challenges. A favorable macroscopic environment with the understanding and support of the whole society is necessary for changing the power industry from the planned economy to the market economy.

9.5.3.3 The tight supply situation results in market price rising rapidly

Opening the wholesale market under the tight supply situation, will certainly lead to the market clear price rising rapidly. This has been proved by the trial operation of the monthly ECPM. It was forecasted that the peak supply gap would reach 18,000MW in East China Power Grid in summer 2004, further more, the severe condition would continue to develop until 2005. The price rising reflects the gap between supply and demand, but the fixed retail price makes the final consumer not be aware of the market signal, and on the other hand, makes the Grid Company take much financial risks. This is just the story of Pacific Gas and Electricity of USA in 2000 market crisis.

9.5.3.4 Risk mitigation methods need to be studied

A free market is just like a sword with two blades. One blade optimizes the allocation of resource, while the other brings enormous risks. Under the environment with incomplete regulation and intense situation between supply and demand, the power market risk is magnified and focused on transmission utilities. In the first two months of trial operation in ECPM, the increased payment of the transmission utilities has been 5.05 hundred million in total, which has broken the balance between generation and transmission presently and has gone beyond the finance capacity of the transmission utilities. The risk mitigation methods need to be studied deeply.

9.5.4 Concluding Remarks

As one of the pilots of regional power market, ECPM has renovated the provincial power markets to the regional platform with uniform codes, realized the inter-provincial competition under the multiple control area model, and optimized the resources allocation within a wider area. Despite of the institutional complexities, the monthly market competition, which adopts a two-way bidding mechanism, maximized the market efficiency and enhanced the social welfare.

The trial operation of the regional market has converted administrative relationships between the market participants into contractual relationships. The responsibilities and obligations of the market participants are declared by the market codes definitely. At the same time, the regional market clearing prices as signals indicate investments on the generation and transmission sectors. In addition, the ECPM model offered a good example for further restructuring of China power industry.

Although the ECPM faces a lot of difficulties and disadvantageous situations, it is believed that market reform will bring positive changes to social welfare and should be pursued down the road. To this end, international experiences such as those of Australia, North America are exploited. It is likely that the concept of pool-based market, widely accepted in the above countries, will be followed in the future market design. Although the suitability of Standard Market Design[32] remains skeptical, a day-ahead market with zonal marginal prices calculated centrally is envisioned.

Another question that ECPM faces is whether or not it is feasible to centrally manage a market with rapidly increasing installed capacity. The total installed capacity of the market will exceed 100,000MW in 2011. An extra-large scale market presents challenges to fundamental control center functions such as but not limited to security analysis, real-time dispatch, unit commitment, ancillary service provision, and AGC. This is in fact a question of common interest in the power community[33].

9.6 Status and Perspective of Electric Power Industry in Korea

In recent years, the electric power industry in which monopoly had been taken for granted, has experienced drastic shifts. These changes have triggered active movements toward competition through industrial restructuring and liberalization. In step with this trend, Korea has gradually carried out privatization in the state-monopolized industry. Our objective is not only to overcome management inefficiency but also to restructure it in a way to more effectively befit the market competition. To begin, the generation sector has been divided into six separate segments according to the similarity of fuel composition and generation capacity. Also, the Korea Electricity Commission (KOREC) was established as a regulatory body and the Korea Power Exchange (KPX) was opened in 2001.

9.6.1 Korea Power Exchange

FOR decades, the electric power industry of Korea had been monopolized by the state-run Korea Electric Power Corporation (KEPCO). However, deregulation of the industry became indispensable so as to establish a competitive electricity market in which electricity can be traded as a commodity. In this context, a market in which electricity can be bought and sold has formed and competition among generation companies has commenced since April 2001.

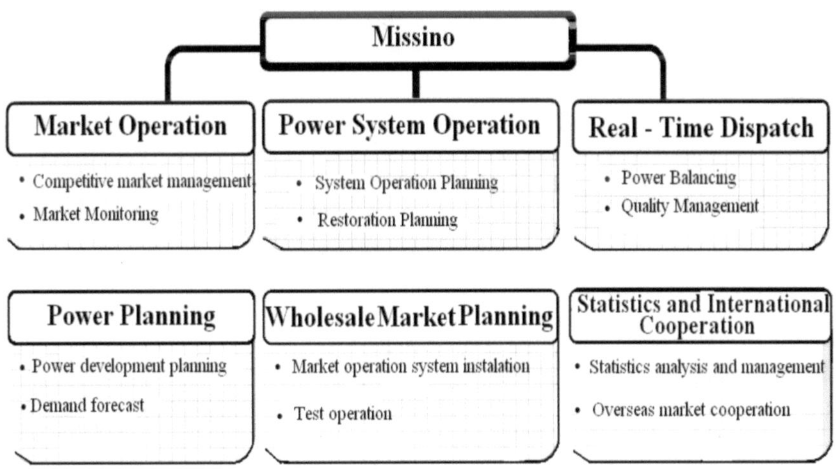

Figure 9.9 KPX's main functions

The Korea Power Exchange (KPX) was introduced on April 4, 2001 as a non-profit independent organization in accordance with Article 35 of the Electricity Business Law to perform main functions, that is, market and system operations, settlement, revision of market rules, administration for regular and associate members, load forecast and establishment of a power development plan, and so on. The KPX's mission can be briefly described in Figure 9.9.

The futuristic reform of the electricity market will be addressed as a transition from the generation competition market to the wholesale competition and finally the retail competition market[34]. Accordingly, the structure will be changed from Cost-Based Pool (CBP) to Price-Based Pool (PBP) and eventually to Two-Way Bidding Pool (TWBP) in Figure 9.10.

At the generation competition stage, the principles of reform include the adoption of the free competitive environment among the subsidiary generation companies, separation of the distribution business, privatization, establishment of independent regulatory authorities and the bilateral power exchanges available only to large customers. At the wholesale competition stage, the distribution and the sales business will be monopolized on a regional basis and the consumers' right in choosing the electricity supplier will be gradually elevated. In the long run, at the retail competition stage, the regional monopoly of distribution business and sales companies will not be prevented until 2009. There will be new sales companies such as the consumer associations and the expert power sales companies, which will reflect prospects from consumer-oriented viewpoint. The application of PBP and TWBP will be timely made subject to the change of the situation in the market structure.

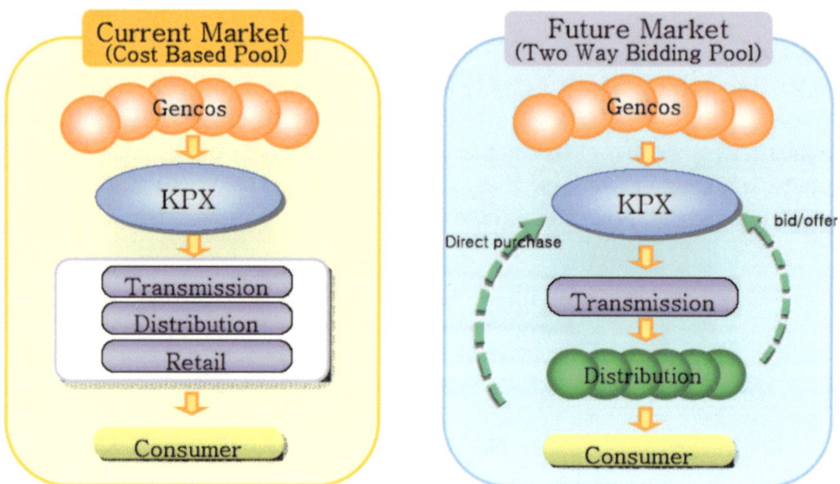

Figure 9.10 CBP vs. TWBP.

9.6.2 Overview of Korean Electricity Industry

Due to the increasing demand of electricity with a growth rate of over 7.5% per annum, the industry had the total electricity generating capacity of 56,053 (MW) in 2003. In 2015, the capacity will be approximately 1.5 times as much as current capacity, which will come to 77,023 (MW) and be expected to sufficiently meet the forecast load-demand. Figure 9.11 shows the installed capacity according to the source of power and individual generating companies, respectively.

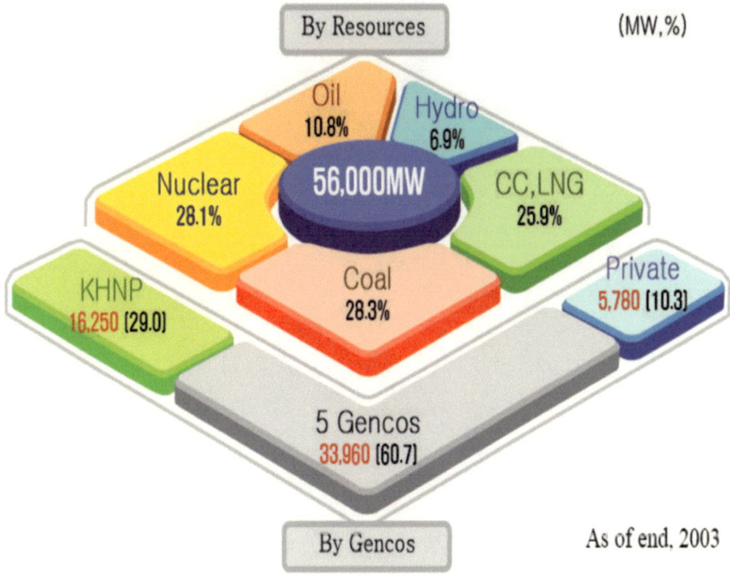

Figure 9.11 Installed generating capacity according to resources and GENCOs

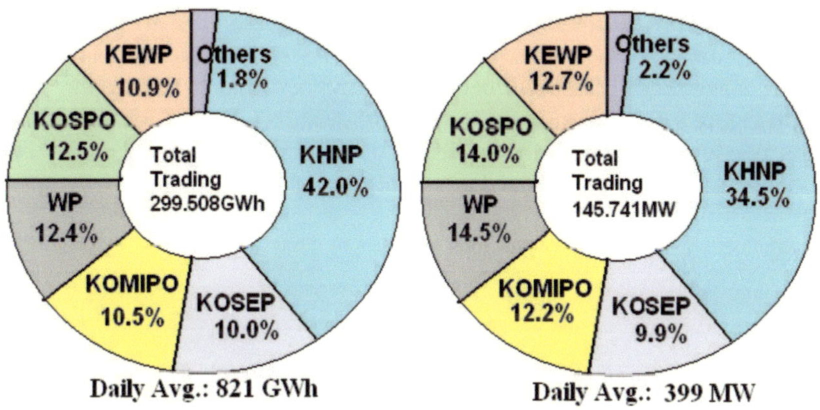

Figure 9.12 Daily average trading volume.

In April 2001, KEPCO's generation sector was split into six power generation subsidiaries including one big hydro and nuclear power company (Korea Hydro & Nuclear Power company, KHNP) and five thermal power companies (Korea East-West Power Co., LTD., Korea South-East Power Co., LTD., Korea Midland Power Co., LTD., Korea Southern Power Co., LTD., Korea Western Power Co., LTD.) on an equal basis in terms of commercial and technical aspects. Particularly, KHNP will remain as a public entity in consideration of nuclear safety, the characteristics of power supply and demand, nuclear power development, and the capability of building new plants, while the other five power generation companies, consisting of fossil and pump storage power plants, will be privatized in the near future. Figure 9.12 depicts the daily average trading volume and we have seen a 15 percent increase of the trading volume in 2003 compared to the same period in last year, as shown in Figure 9.13.

Figure 9.13 Total trading amounts from 2001 to 2003.

During the generation competition, the power plants under construction were built as planned to keep the capacity reserve margin above 15% by 2005. At the same time, the Demand Side Management (DSM) will be strongly implemented to reduce the peak load as illustrated in Figure 9.14. To secure the investment resources needed for activating the DSM program as previously planned, the collection of "Electricity Supply Industry Foundation Funds" for DSM program is legitimized pursuant to article 49 of the Electricity Business Act.

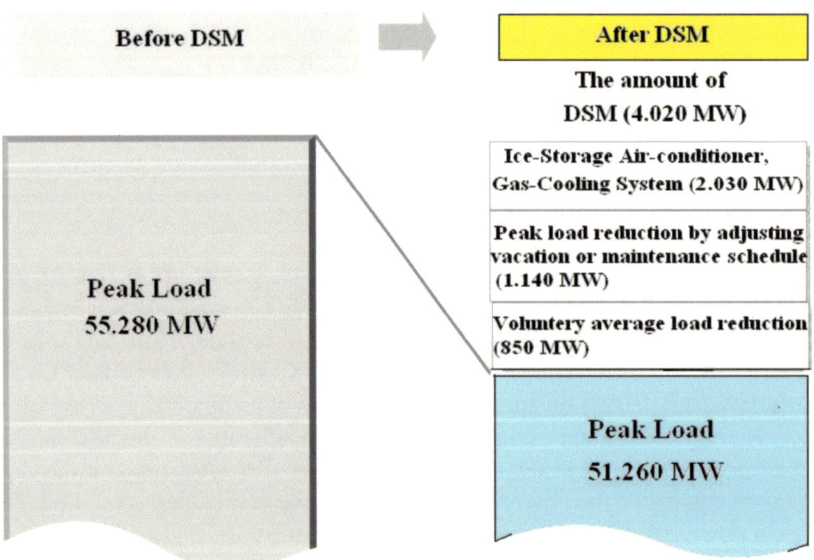

Figure 9.14 Positive effect of demand side management program

In Korea, the seasonal load-demand pattern can be characterized as follows (See Figure 9.15):

- Summer: annual peak load (12:00 ~ 13:00) due to cooling load
- Winter: peak load (23:00) due to heating load
- Spring: the lowest load of a year without consuming neither cooling nor heating loads Besides, Figure 9.16 and Figure 9.17 represent the summer peak load (47,385 MW on August 22, 2003) and the winter peak load (46,387 MW on February 5, 2004), respectively.

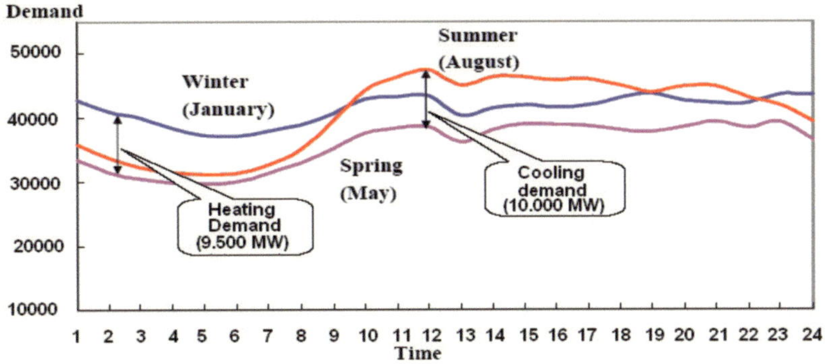

Figure 9.15 Seasonal load-demand pattern

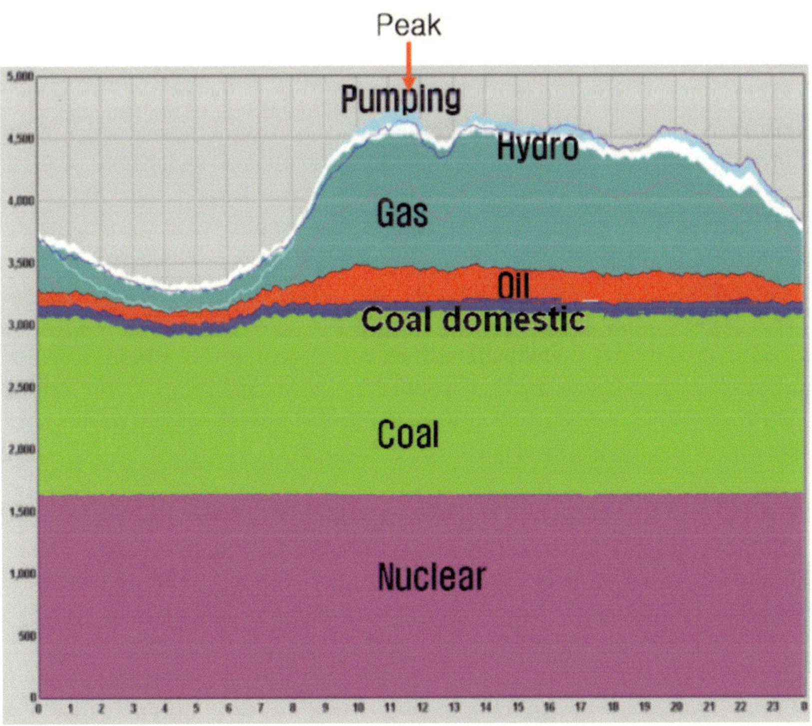

Figure 9.16 Hourly load curve for summer peak load on August 22, 2003.

Table 9.7 summarizes the transmission and distribution facilities in Korea[35]. The transmission lines including both overhead and underground have a length of 28,260 (km) and the installed transformers totaled 1,672 in 2003. In addition, the distribution lines run radially for a length of 376,454 (km).

Classification		2000	2001	2002	2003
	765 kV	595	662	662	662
	345 kV	7.281	7.345	7.497	7.740
Transmission	154 kV	16.747	17.576	18.144	18.595
line	664kV below	1.727	1.540	1.402	1,031
(c-km)	DC180kV	232	232	232	232
	Total	25.582	27.355	27.937	28.260
Transformer	765 kV	-	1.110	7.110	7.110
capacity	345 KV	53.115	63.577	69.078	75,660
(MVA)	154 KV	70,886	78.119	83.364	89.228
	66 KV below	1,699	1.473	1.286	1.068
	Total	125.700	144.279	160.838	173.066
Distribution	Route				
Facilities	Length	351.264	358.328	366.938	376.454
(km,	Supporter	6.439	6.695	6.875	7.171
1.000Set, EA)	Transformer	1308.947	1428,510	1546.088	1618.889

Table 9.7 Facilities of transmission and distribution in KOREA

9.6.3 Measures in Power System Operations

Figure 9.18 is a schematic showing six routes connecting metropolitan regions and others as well as a large amount of real power flows through the designated "flowgates"[36]. More than 40% of system load is in the metropolitan region, while the majority of generation is in the non-metropolitan regions. Further, most generating units with low generation costs are scattered all over the non-metropolitan regions.

For the purpose of economic benefits, therefore, real power generation in non-metropolitan regions increases in parallel with the consumption level, resulting in the power transfer from the south and central parts of the Korean electric power system to the northwestern part through one of the most critical corridors of the grid.

Even more striking is the concept of transfer capability that would be eventually bounded by applicable line ratings, reactive support, and dynamic limitations because greater volume of power flows into a region in normal states can give rise to cascading failures in the N-1 steady-state security criteria[37, 38]. After privatization of generators, power system engineers in Korea emphasize that the trend of heavier real power flows into the metropolitan region will continue or become profound, and that the constraint of the interface flows will be vital to our national-interest transmission bottlenecks, leading to congestion that significantly decreases reliability, restricts competition, enhances opportunities for suppliers to exploit market power, increases prices to customers, and increases infrastructure vulnerabilities.

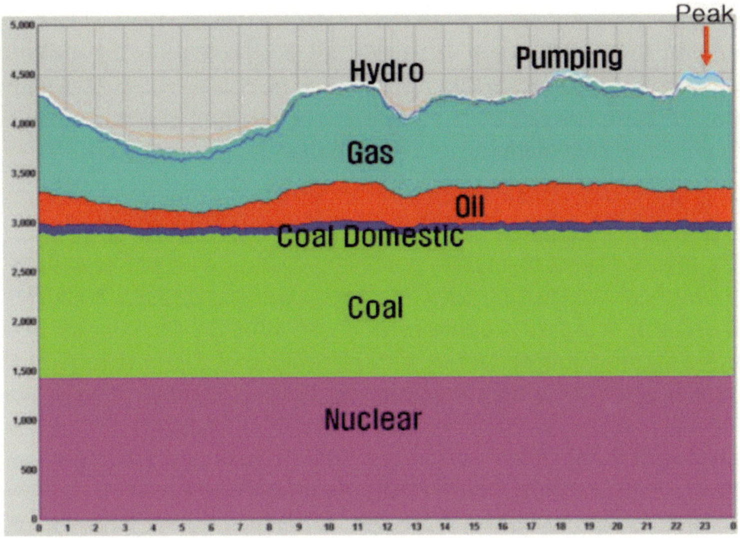

Figure 9.17 Hourly load curve for winter peak load on February 5, 2004.

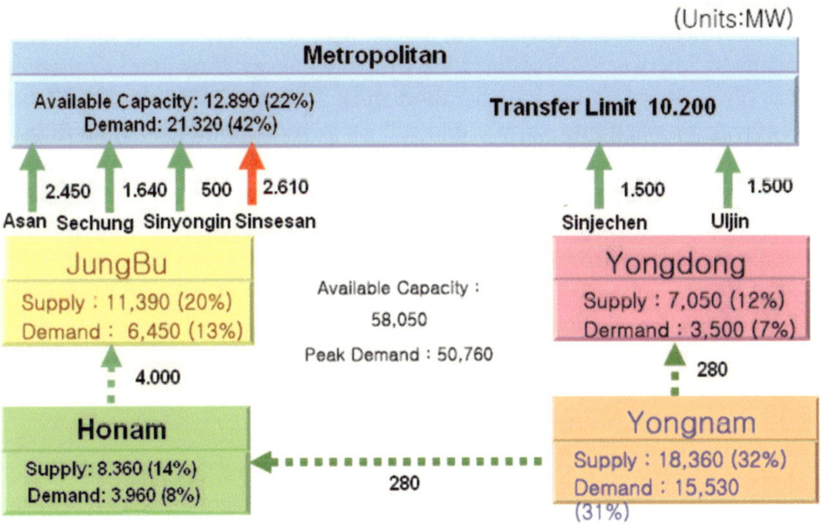

Figure 9.18 Total transfer capability in tie lines between metropolitan region and adjacent regions.

Typically the transmission network planning approach includes a set of fundamentals, some realistic events, under which the system must be able to operate and specified consequences that are accepted under the operation[39]. As the electricity sector is getting more and more liberalized, a number of questions have been raised regarding the grid planning, e.g., does the market opening require network reinforcement and can the market requirements be an argument for that reinforcement? The network planning approach now involves a set of additional parameters like market prices, transmission pricing, and investment policies.

Thus the transition from monopoly to an open electricity market is a global process, which has been going on for several years. In an overall perspective the open electricity market means liberalizing the sector to create competition in power generation and supply. The introduction of the competitive electricity market has resulted in new frameworks and considerations in power system planning and operations.

9.7 Outlook for Power Exchange between Russia, DPRK and ROK

Since the 1990s, many papers have been published dealing with power system interconnection between Northeast Asian countries. Electricity trading through NEAREST offers mutual benefits, and can be a good countermeasure to solve the environmental and technical problems caused by the independent system operations of each country. Power exchange between countries contributes the infrastructure to open trading markets, while interconnected systems between NEA countries will have more technical and economic advantages when compared with independent system operation conditions. However, this power system interconnection could not become a reality until now due to social, economic and political regime differences. Basically, the ROK, the DPRK and Russia have the most powerful potential in NEAREST, when their status and future prospects are considered. These three countries have different situations and backgrounds on power system interconnection from technical, economic and political viewpoints. The ROK power system is an island, having been isolated from the DPRK network in 1945. Also, the ROK is very poor in natural resources and must import 97.4% of the total primary energy consumed domestically. Also, the ROK has difficulties relating to generation sites. Since the 1980s, the DPRK has suffered from a deficiency of electricity supply and wants to be supported by the ROK. After the summit between the DPRK and the ROK in 2000, the DPRK has requested electricity support with a short-term capacity of 500MW, and a long-term capacity of 2,000MW from the ROK government. Conversely, East Russia, FER (Far East Russia) and ES (East Siberia), have plenty of coal, gas and hydro resources. Also, Russia has surplus power plants and generation potential due to the economic decline since 1990. Russia has plenty of power export potential. Therefore, this section evaluates the prospect of power exchange considering future demand/surplus supply plans and exchangeable power in technical and economic aspects.

9.7.1 Power interconnection scenarios for "RFE – DPRK - ROK"

Many scenarios on NEAREST have been published by institutes working on power interconnection topics, including as ESI, KERI, and others[40-42]. Most of these scenario analyses, however, have simply estimated the rough parameters of interconnection scenarios, including voltage level, capacity, and line length of inter-ties. The basic contents and concepts covered by these scenario analyses have been largely similar to each other. The main scenarios either under discussion, or currently being studied, are as follows.

9.7.1.1 Potential local interconnections under discussion

Russia has a plan to interconnect its power grid with that of the DPRK. This interconnection might ultimately be extended to the ROK. A number of problems, however, including obtaining financing, pose significant barriers to this project. Table 9.8 describes the general ratings of this interconnection plan. This project will include a 380km, ±500 kV DC line between VLADIVOSTOK and CHEONG-

JIN. This interconnected line will be operated at 220 kV AC during the first stage of the project, and will be changed to 500 kV AC operation after the 500 kV line between "CHUGUEVKA-NAHODKA-VLADIVOSTOK" is put into operation. In its final stage, the line would be modified as a ±500 kV HVDC line in the future.

Power volume to be transmitted (mln. kWh)	1500 - 2500
Load to be transmitted (MW)	300 - 500
Frequency (Hz)	50
Voltage (kV)	220/500
Length of line in Russian territory (km)	250
Length of line in DPRK territory (km)	130
Cost of construction (mln. USD)	160 - 180
Period of construction (years)	3 - 4
Period of investment repayment (years)	8 - 10

Table 9.8 Overview of interconnected system between FER-DPRK

Also, the ROK and the DPRK are seeking to develop an industrial complex at GAESUNG, near the shared border of the two countries (but inside the DPRK). The required electricity for the GAESUNG industrial complex might be supplied by the ROK. This project is utterly dependent on the political situation between the two parties. At the first stage, the ROK and the DPRK agreed to construct two distribution circuits rated 2×10MW, which are now under construction. Finally, the basic rating of transmission line supplying electricity for the GAESUNG industrial complex is 154Kv, 200 MW with a length of 40km.

9.7.1.2 New scenarios including KEDO N/P

Basically, KERI investigated new six interconnection scenarios for the "RFE-DPRK-ROK" interconnection[40]. "Russia-DPRK-ROK" interconnection can present various scenarios according to the following factors and hypotheses.

i. Whether KEDO nuclear power plant is included in NEAREST or not.
ii. Accomplishment of "VLADIVOSTOK-CHEONGJIN" local interconnected system under disconnection to the DPRK system, and the future possibility of re-connection to the DPRK system of CHEONGJIN load.
iii. Power supply plan to GAESUNG industrial complex under disconnection to the DPRK system and future possibility on re-connection to the DPRK system.
iv. Capacity and voltage of the interconnected system.

For example, in order to include the KEDO N/P in a power interconnection network, we can consider the interconnection route "VLADIVOSTOK-SINPO" as a tentative hypothesis. This scenario is somewhat different from the existing scenario for a "VLADIVOSTOK-CHEONGJIN" interconnection that is under discussion between Russia and the DPRK. The "VLADIVOSTOK-SINPO" scenario could be one of the alternatives for the effective utilization of the KEDO N/P. If this scenario is implemented, after the commissioning of KEDO N/P, by means of the interconnection the DPRK can earn revenues by trading seasonal surplus electricity, or can be supported with electricity imports at times of seasonal shortages of electricity. This implies that all of the interconnected countries in this scenario can reap benefits by trading seasonal surplus electricity.

9.7.2 Estimated prospective export/import potential

9.7.2.1 Power industry of the ROK

Table 9.9 describes the present status and future projections for installed generating capacity in the ROK according to the 1st power supply/demand plan after restructuring. The installed capacity is expected to rise to 77,024MW by 2015. In terms of the plant mix, the share of oil and coal plants are projected to decrease over the next 12 years, while the share of nuclear capacity is projected to increase.

Table 9.10 describes the present and future total electricity production in the ROK. As shown in this table, the expectation is that the total generation portion provided by nuclear power plants will rise slightly in the future. In contrast, the fraction of generation provided by thermal plants such as coal- and oil-fired units will decrease.

Year	Nuclear	Coal	Gas	Oil	Hydro	SUM
2002	15716 (29.2%)	15931 (29.6%)	13618 (25.3%)	4660 (8.7%)	3876 (7.2%)	53801
2005	17716 (28.6%)	18165 (29.4%)	16814 (27.2%)	4667 (7.5%)	4485 (7.3%)	61847
2010	23116 (29.3%)	24265 (30.7%)	20437 (25.9%)	4817 (6.1%)	6385 (8.1%)	72635
2015	26637 (34.6%)	22240 (28.9%)	19550 (25.4%)	2212 (2.9%)	6385 (8.3%)	77024

Table 9.9 Present and future projected generating capacity in the ROK (MW)

Year	Nuclear	Coal	Gas	Oil	Hydro	Etc.	SUM
2002	122.8 (43.2%)	117.9 (41.5%)	29.7 (10.4%)	26.7 (2.8%)	6.0 (2.1%)	- (0.0%)	344.8
2005	134.1 (40.6%)	132.7 (40.2%)	45.6 (13.8%)	24.8 (2.9%)	6.7 (2.0%)	1.4 (0.4%)	399.0
2010	166.7 (42.1%)	175.2 (44.3%)	26.5 (6.7%)	17.9 (4.5%)	8.5 (2.2%)	1.0 (0.3%)	435.0
2015	210.3 (47.2%)	165.4 (37.1%)	49.0 (11.0%)	12.0 (2.7%)	9.3 (2.1%)	-(0.0%)	445.9

Table 9.10 Present and future projected electricity production in the ROK (TWh)

Although the projections shown in Table 9.9 indicate that nuclear power's share of future ROK installed capacity and electricity production are expected to be higher than they are at present, it should be noted that these projections should be considered just as long-term targets. Factors such as the shortage of land in the ROK suitable for nuclear plant construction, and public resistance to building power plants, especially nuclear plants (the "NIMBY", or "not in my back yard" movement) will likely make these targets difficult to achieve. As a result of the "NIMBY" movement in the ROK, and the public fear of atomic energy, construction of new nuclear power plants faces difficulties. Furthermore, building thermal power plants fueled with coal, oil and gas is problematic because of the constraints on GHG emissions specified under the Kyoto protocol. Therefore, as a matter of government policy, it is necessary to establish a future general plan and countermeasures that will help to assure that future electricity demand is met, while still reducing GHG emissions.

9.7.2.2 Power industry of DPRK

Even though we have some DPRK power industry and power system data, most of the DPRK data is quite uncertain[43]. The DPRK had been suffering from electricity deficiency since the 1980s and most of its hydro/thermal plants are out of date. Because of this, the DPRK had not published formal statistics since the late 1990s, so we could not use existing outdated formal statistics when evaluating the prospect of the DPRK power balance. We could only estimate and treat the DPRK system as a black box.

9.7.2.3 RFE power balance and export potential

A study of the power export potential of East Russia (ER), including East Siberia (ES) and Russian Far East (RFE), up to 2020 was done in [44]. In Tables 9.11-9.13, min/max value is based on the future minimum/maximum domestic demand. Three categories of power export potential are identified. The first one is power that can be additionally generated by existing power plants up to 2005. The second category of power export includes power from power plants that can be additionally generated during the summer season. The third category of power export potential includes power generation from power plants that should be additionally constructed in ER for export purposes.

Tables 9.11, 9.12 indicate power balances for the RFE interconnected power system compiled using data prepared by ESI for NEAREST DB. Hydropower capacity is supposed to be significantly developed in the RFE, according to power balances in Tables 9.11, 9.12. Bureyskaya HPP, with total capacity of 2000 MW (6×333 MW) and average yearly generation of 7.1 TWh, is constructed, with a third unit phased in by the end of 2004. Three more units were planned by 2009. Nizhne-Bureyskaya HPP, with total capacity of 428 MW (4×107 MW) and average yearly generation of 1.6 TWh, is the second stage of the Bureysk cascade of HPPs. It is supposed to be completed by 2010. Cascade of Nizhnezeysk HPPs, of an installed capacity and average power generation of 349 MW and 2,12 TWh/year respectively, will be completed in the period 2010-2012. Additionally Urgalsk HPP-1, with a power generation of 600 MW and 1.8 TWh/year, and Dalnerechensk hydropower complex, with a generation capacity of 595 MW and 1.4 TWh/year, are supposed to be introduced by 2015-2020, depending on scenarios of rates of electricity consumption growth in the RFE. Steam TPPs are not supposed to be developed in the RFE. In fact, they are planned to retire, and new steam TPP capacity is not to be commissioned. Development of co-generation TPPs is mainly determined by the demand of heat consumers.

Capacity, peak load and transfer	2001	2005		2010		2015		2020	
		Min	Max	Min	Max	Min	Max	Min	Max
Hydro	1.33	2.2	2.2	4.0	4.0	4.7	5.3	4.7	5.3
Steam turbine	2.61	2.5	2.5	2.4	2.4	2.5	2.5	1.6	1.6
Co-generation	3.17	3.5	3.5	3.6	3.8	3.8	4.3	5.4	5.4
Nuclear	-	-	-	-	-	0.6	0.6	1.3	1.3
Total capaity	7.11	8.2	8.2	10.0	10.2	11.6	12.7	13.0	13.6
Peak load	4.74	5.33	5.77	5.80	6.74	6.33	7.96	6.93	8.85
Power tranfer to adjacent regions	0.04	0.32	0.35	0.85	0.85	0.85	0.85	0.85	0.85
Peak load and power transfer	4.78	5.65	6.12	6.65	7.59	7.18	8.81	7.78	9.7
Capacity reserve rate, %	48.7	45.1	34.0	50.4	34.4	61.6	44.2	67.1	40.2

Table 9.11 Capacity balance for RFE IPS, GW

Power generation, electricity consumption and transfer	2001	2005		2010		2015		2020	
		Min	Max	Min	Max	Min	Max	Min	Max
Hydro	4.85	8.9	8.9	13.7	13.7	16.3	16.3	17.0	19.0
Steam turbine	6.05	6.8	6.8	5.2	6.6	4.9	7.7	1.2	3.0
Co-generation	14.6	14.5	16.6	17.1	19.9	17.7	22.8	18.9	21.6
Nuclear	-	-	-	-	-	-	-	3.8	7.8
Total genertion	25.5	30.2	32.3	36.0	40.2	38.9	46.8	40.9	51.4
Electricity consumption	25.2	28.5	30.6	31.5	35.7	34.4	42.3	37.4	46.9
Electricity transfer to adjacent regions	0.29	1.7	1.7	4.5	4.5	4.5	4.5	4.5	4.5
Electricity consumption and transfer	25.5	30.2	32.3	36.0	40.2	38.9	46.8	41.9	51.4

Table 9.12 Electricity balance for RFE IPS, TWh/year

As can be seen from Table 9.13, power export potential, which does not require additional capacity commissioning (apart from that required for meeting domestic power loads), and, therefore, additional investment, can be quite sufficient, exceeding 4 GW of capacity in summer, and 2 GW in winter, and 16-18 TWh/year of power generation in the beginning of the period under consideration. At the end of the considered period, export potential declines to about 2.5-3.0 GW of capacity in summer only (because of exhausting existing excessive capacity), and 5-6 TWh/year of power generation.

Potential		2005	2010		2015		2020	
			Min	Max	Min	Max	Min	Max
Capacity, GW	Winter	2.4	1.2	0.4	0.7	0.0	0.0	0.0
	Summer	4.3	3.3	2.8	3.0	2.9	2.5	3.2
Power generation, TWh		18.1	11.5	7.4	8.6	5.8	5.0	6.4

Table 9.13 Total RFE Power Export Potential of Existing Plants

	Power plants	Installed capacity, GW	Annual average generation, TWh	Years of commissioning
Hydro	Bureysk (together with Nizhne-reysk)	2.428	8.7	By 2010
	Cascade of Nizhnezeysk	0.349	2.12	By 2015
	Dalnerechensk	0.595	1.4	By 2015
	Urgalsk-1	0.6	1.8	By 2015
	Subtotal	3.972	14.02	-
Nuclear	Primorye	1.3	9.75	By 2020
				-
Total		5.27	23.77	-

Table 9.14 Power plant capacities to be commissioned in RFE by 2020

	Power plants	Installed capacity, GW	Average yearly generation, TWh
Hydro	Urgalsk-1	0-0.6	0-1.8
	Gilyuisk	0.38	1.15
	South Yakutian hydropower complex, including:	5.0	23.45
	Cascade of Sredne-Uchursk and Uchursk HPPs	3.7	17.2
	Cascade of Idjeksk and Timtonsk HPPs	1.3	6.25
	Khingansk	1.2	5.8
	Subtotal	6.58-7.18	30.4-32.2
Thermal	Sakhalin (Gas)	4.0	26.0
	Sakhalin (Coal)	2.0	13.0
	Urgalsk (Coal)	1.2	7.5
	Subtotal	7.2	46.5
Nuclear	Far East	2.5	18
Total		16.28-16.88	94.9-96.7

Table 9.15 Power Plant Capacities to be commissioned in RFE after 2020

Table 9.14 and Table 9.15 shows prospective power plants, which can be constructed within (or close to) the area of the RFE IPS in and beyond 2020. As can be seen from Table 9.15, the total power potential of the third category can exceed 16 GW and 95 TWh/year. In addition to this potential, construction of the Tugursk tidal power plant, with a capacity of nearly 7 GW and a yearly power generation of 17 TWh, can be possible beyond 2025-2030.

9.7.3 Admissible Interconnected Capacity in Technical Viewpoints

9.7.3.1 Evaluation of maximum exchangeable power

An evaluation of maximum exchangeable power was performed by KERI[45, 46]. It can be evaluated by taking into account the following technical aspects, such as ROW (Right of Way) and system constraints. ROW constraint means the geographical constraints that the interconnected line should pass through. Also, system constraints include technical problems, such as load flow and stability analysis. The study results of technical aspects are as follows.

ROW constraint: Considering the geographical situation between Russia and the Korean peninsula, a two-bipole system having a capacity of 7 GW can be built.

Load flow analysis: There is no violation of overload and voltage in a steady state up to 7 GW of inflow power. However, in N-1 contingency, some violations happen as the inflow power exceeds 4 GW. Therefore, 4 GW seems to be the maximum exchangeable power.

Dynamic analysis: The power system frequency of the ROK can keep the standard when losing 2 GW of power. However, loss of more than 3 GW of power makes frequency violate the standard. Considering a one-bipole trip, 4 GW is the maximum exchangeable power.

Finally, we can say that 4 GW of power exchange is the maximum exchangeable power from a technical viewpoint between Russia and the ROK at present status, and this result could satisfy the security points.

9.7.3.2 Evaluation of minimum exchangeable power

Minimum exchangeable power is evaluated through a comparison of total costs and benefits of the interconnected line during its life cycle span of 30 years. The total cost of interconnected lines, life cycle costs, consist of initial investment and operating costs. Initial costs include the construction cost of transmission lines and converter stations, operating costs means the maintenance costs of transmission lines and converter stations. The benefit of interconnection comes from the electricity tariff difference between the ROK and Russia. The electricity tariff difference in 2000 was $0.0383/kWh, but this difference has been getting decreased because the annual rate of increase for electricity tariffs in Russia will be higher than that of the ROK. Table 9.16 shows the total cost and benefits of interconnected lines. If 1 GW or 2 GW of power is exchanged between the ROK and Russia, the total cost is much more than the accrued benefits, a situation that cannot assure an economic advantage. However, more than 3 GW of exchange power can guarantee the interconnection project will be in the black. Therefore, we can propose that minimum exchangeable power, from an economic viewpoint, will be 3 GW.

Exchange power	Cost (billion $)	Benefit (billion $)
1GW	4.13	3.16
2GW	6.60	6.33
3GW	7.82	9.49
4GW	10.56	12.65

Table 9.16 Total cost and benefits

Benefits are affected by a decrease in the rate of electricity tariff differences between the ROK and Russia. The lower the decreasing rate is, the more we can expect benefits. Figure 9.19 shows the sensitivity of benefits with variations of the decrease rate. In this figure, the horizontal axis is the decrease rate and vertical axis shows benefits. In the case of 1GW of exchange power, the benefit is $5.24billion, with a 1% decreasing rate, but the benefit is reduced to $2.11billion with a 9% decreasing rate. With a 5% decreasing rate, more than 3GW of exchange power is needed to assure economic feasibility. More than 1GW of exchange power, with a 1% decreasing rate makes the interconnection project beneficial, but if decreasing rate increases over 7%, the cost is larger than the benefit with 1 GW to 4 GW of exchange power. Figure 9.20 shows the Benefit/Cost ratio with a 5% decreasing rate. In this figure, the horizontal axis means exchange power and vertical axis means B/C ratio. As exchange power grows, B/C ratio also increases up to 3 GW. However, B/C ratio decreases from more than 4 GW, as shown in Figure 9.20. So, we can say that ranging from 3 GW to 4 GW is a more reasonable exchange power in economic terms.

As a result, the minimum exchangeable power is about 3GW, and optimal exchangeable power range, considering technical and economic viewpoints, is expected to 3~4 GW.

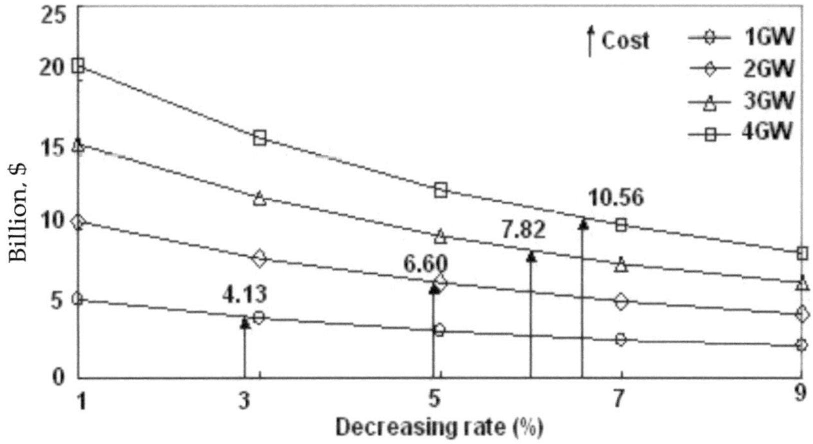

Figure 9.19 Sensitivity of benefit to variations in decreasing rate.

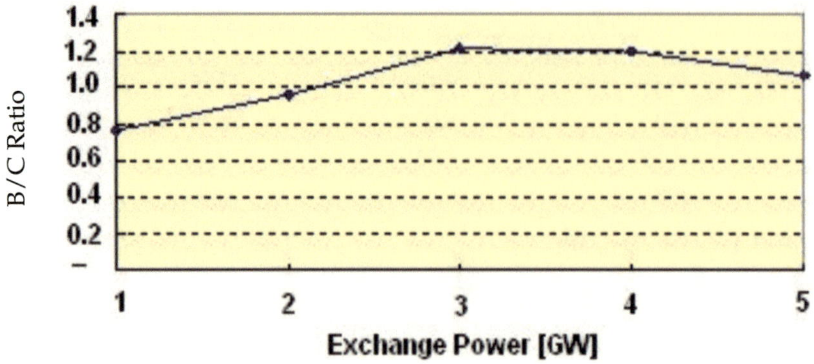

Figure 9.20 B/C Ratio with a 5% decreasing rate.

Thus, above study examines the future outlook of exchange power between the ROK, the DPRK and RFE from technical and economic viewpoints. The main results of this study on power system interconnection are as follows.

1. Excessive capacity and power generation for the RFE system was estimated in the paper. Power export potential, which does not require additional capacity commissioning and, therefore, additional investment, can be quite sufficient exceeding 4 GW of capacity in summer, 2 GW of capacity in winter, and 16-18 TWh/year of power generation at the beginning of the considered period. At the end of the considered period, the export potential declines to about 2.5-3.0 GW of capacity only in summer (because of exhausting existing excessive capacity) and 5-6 TWh/year of power generation. The total power exports potential, including new commissioning plants can exceed 16 GW and 95 TWh/year. In addition to this potential, construction of the Tugursk tidal power plant, with a capacity of nearly 7 GW and yearly power generation of 17 TWh, can be possible beyond 2025-2030.

2. The maximum acceptable exchange power between Russia and the ROK at present status, from a technical viewpoint, is 4 GW and this result could satisfy security points. In addition to maximum exchangeable power, the minimum exchangeable power, by comparing total costs and benefits of interconnected lines, is evaluated at 3GW. At this time, we can say that the range of 3 GW to 4 GW seems to be a reasonable power exchange level between the ROK and RFE systems.

3. This study is based on a hypothesis, and research concepts, not on practical engineering projects. Therefore, more detailed engineering work from the technical and economic viewpoints are required for the realization of NEAREST. Above all, we could not estimate the prospect for the DPRK system because we have no accurate DPRK power industry data and, consequently, the exact details are uncertain.

9.8 Northeast Asia Interconnection, and Power Flow Considering Seasonal Load Patterns

Economical and technical considerations are usually the underlying factors for interconnecting electric power systems. Among some of the benefits that may be realized are plant capacity savings, interchange due to diversity, emergency power interchange, spinning reserve savings. Development of such ties in the future can result in more effective utilization of power stations installed capacities, fuel economy, to improvement of ecological situation in a region.

However, the planning of interconnection is a demanding task and needs to meet a wide range of technical aspects. The interconnection of the power systems among North-East Asian countries (Russia, China, Mongolia, Japan, and Korea) has been proposed on numerous occasions, but little progress has been made due to the complicated political issues and economical problems involved. Interstate electrical ties of power systems of the Northeast Asia countries now practically are not developed. Now, the necessity for this power system interconnection is increasingly being felt due to the benefit of each country. Because of these reasons, Korea peninsula takes the role connect a bridge between different areas of Northeast Asia, such as Russia, Mongolia, China, and Japan[47-51]. The problem of utilizing 2,000MW power output after the successful construction for the Sinpo nuclear power plant in future has been studied, and a 765 kV HVAC interconnection between South Korea and North Korea has been discussed with several papers[52-58].

In South Korea, the potential increase in power demand is higher than that of any other country. The metropolitan area situated in the central parts consumed nearly 43% of the total electricity generated, and the southeast area consumed about 33%.

However, most of the large-scale power plants have been constructed in the southern part of South Korea. Consequently, the existing power grid includes multiple routes designed to supply the metropolitan area so that, by and large, the direction of power flow is toward the north. The future substitutes are to relieving the problems of power imbalance and the shortage of power in the Seoul metropolitan areas in South Korea and the Pyongyang metropolitan areas in North Korea.

In this Section, we present various scenarios and the accompanying power flow analyses considering on seasonal load patterns, in order to provide the interconnection of the electric power grids. A distribution map of the projected power flow will be drawn by the results of simulations performed using the PSS/E tool.

9.8.1 Power System Status and Seasonal Load Patterns in Northeast Asia

In this Subsection, we will explain the general characteristics and the seasonal load patterns of the existing power systems used in South Korea, North Korea, Russia, China, and Japan[59-66].

9.8.1.1 Power system and seasonal load patterns in South Korea

The South Korean electricity generation system can be divided into 7 geographical areas that take geographical boundaries into account. The transmission voltages used are 345kV for the major networks, and 154kV or 66kV for the local systems. Most 66kV lines are now either being removed or replaced by higher voltage lines. Power system on Jeju Island is now connected to the mainland via a 100km-long submarine transmission system, comprised of HVDC (High Voltage Direct Current) cables. Because the power demand is increasing rapidly in the metropolitan area, 765kV facilities are in the process of being constructed and now come into operation in order to provide a stable large-scale power transmission between the large power generation plants and the areas where the consumers are located. Figure 9.21 represent the load curve for day and the load curve for month in South Korea.

Table 9.17 shows the current status of KEPCO's transmission grid facilities at the end of 2001. Table 9.18 represents a mid-to-long term forecast in demand and supply. Table 9.19 shows a power capacity of 6 generating companies in South Korea, 2002. (The bellow data had obtained from KEPCO in Korea) Figure 9.22 represents a load demand and a generating facility capacity for districts.

9.8.1.2 Power system and seasonal load patterns in North Korea

Figure 9.23 represents the load curve for day and the load curve for month with the assumed material in North Korea. As shown in bellow Figure, the pattern of a curve has a flat and small variation.

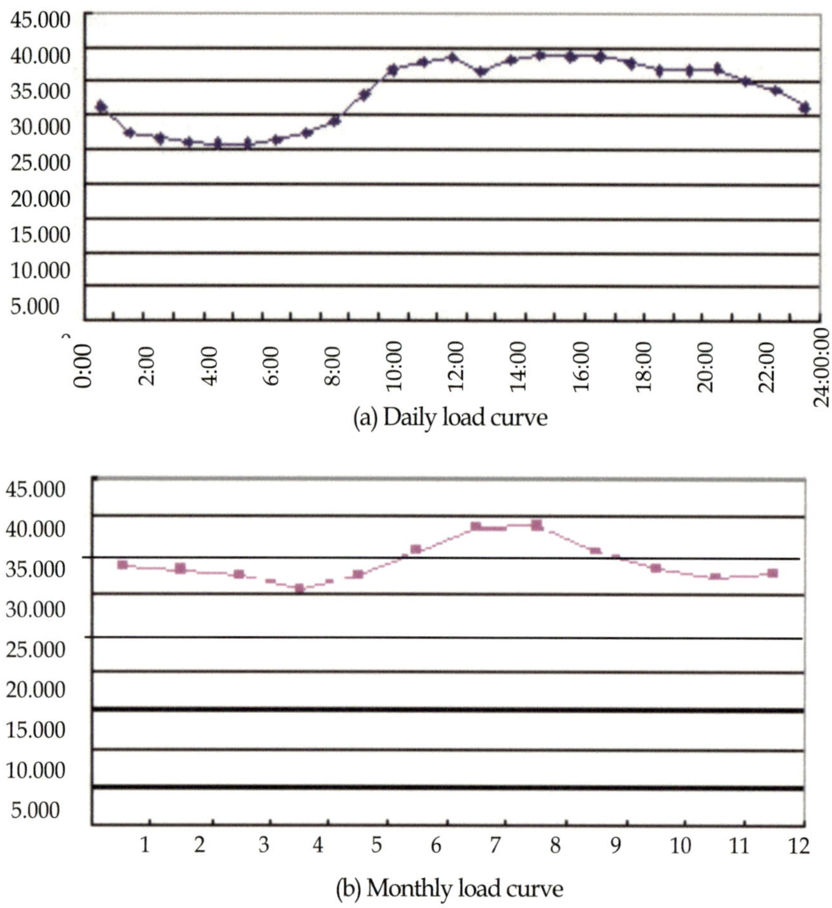

Figure 9.21 South Korea load curves for day and for month.

(At the end of 2001)

	Transmission Facilities				Substation Facilities	
	Circuit length (C-km)			Support (ea)	Number of substation (ea)	Transformer capacity (MVA)
	Ovehead	Underground	Total			
765 kV	662	-	662	666	1	1,110
345 kV	7,234	111	7,345	9,914	65	63,577
180 kV(HVDC)	30	202	232	553	-	-
154 kV	16,111	1,465	17,576	24,581	449	78,119
66 kV	1,531	9	1,540	7,112	25	1,225
22 kV	-	-	-	-	9	248
Total	**24,037**	**1,778**	**25,815**	**42,826**	**540**	**144,279**

Table 9.17 Current status of KEPCO's transmission grid facilities

Year	Peak Demand [MW]	Installed Capacity [MW, as of year end] (%)						Capacity Margin [%]
		Nuclear	Coal	LNG	Oil	Hydro	Total	
2001 (Record)	43,130	13,720 (27.0)	15,530 (30.5)	12,870 (25.3)	4,870 (9.6)	3,880 (7.6)	50,860 (100)	15.1
2005	51,860	17,720 (28.6)	18,170 (29.3)	16,810 (27.2)	4,670 (7.6)	4,490 (7.3)	61,850 (100)	16.8
2010	60,620	23,120 (29.2)	24,270 (30.7)	20,440 (25.9)	4,820 (6.1)	6,390 (8.1)	79,020 (100)	25.1

Table 9.18 Mid-to-long term forecast in demand and supply

Company	Base (MW)	Middle (MW)	Peak (MW)	Total (MW)
KOSEPCO	3,565	500	1,500	5,565
KOMIPO	3,400	0	3,337	6,737
KOWEPO	3,066	1,400	2,880	7,346
KOSPO	3,000	400	2,200	5,600
KEWESPO	2,900	1,800	2,800	7,500
KHNP	15,715	0	528	16,243
OTHERS	0	58	4,186	4,244
TOTAL	31,646	4,158	17,431	53,235
%	**59.5**	**7.8**	**32.7**	**100**

Table 9.19 Power capacity for generation companies in South Korea, 2002

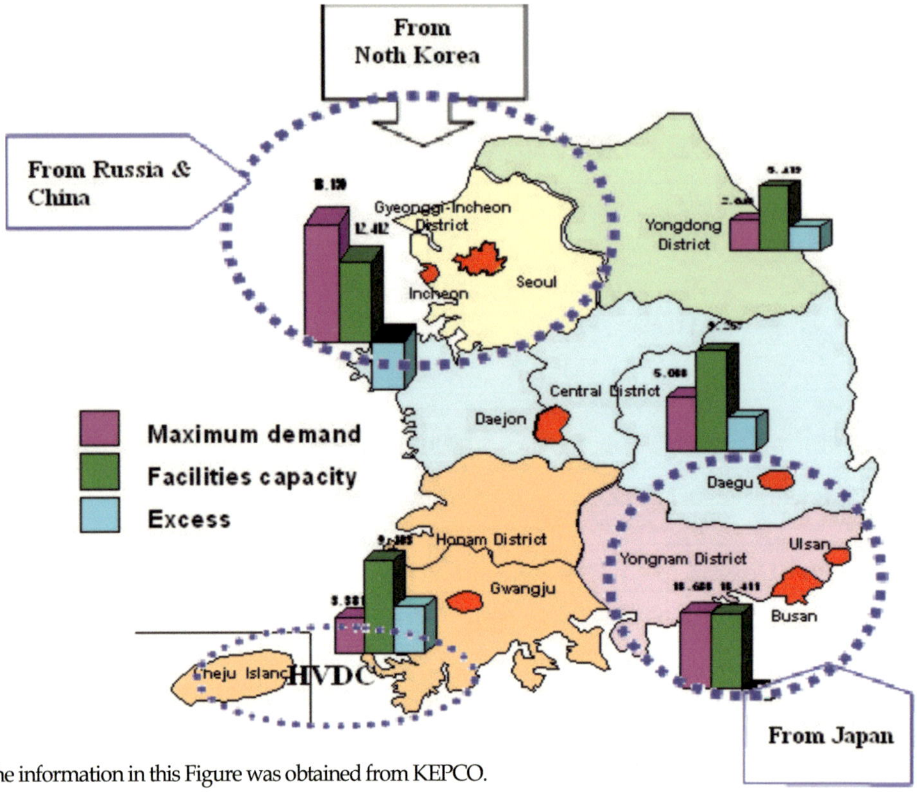

* The information in this Figure was obtained from KEPCO.

Figure 9.22 Demand and facility capacity by regions

At present, the data about transmission system of North Korea are insufficient and are not arranged well. There are only a little data from Russia, UN, CIA, the Korean Board of Unification, etc. Accordingly, the previous researches of interconnection in the Korean Peninsula have just focused on the analyses of the present data and scenarios. This study assumes that the power system in North Korea is divided into 5 areas. The power system in North Korea is smaller than that in South Korea. Most of the hydroelectric power plants are located in the hilly region of the northern areas in North Korea and most of the thermoelectric power plants are located in the metropolitan area. Moreover, power capacity in North Korea has been estimated to be approximately 7,000MW. Currently, it is known that transmission line voltage is composed of 110kV and 220kV.

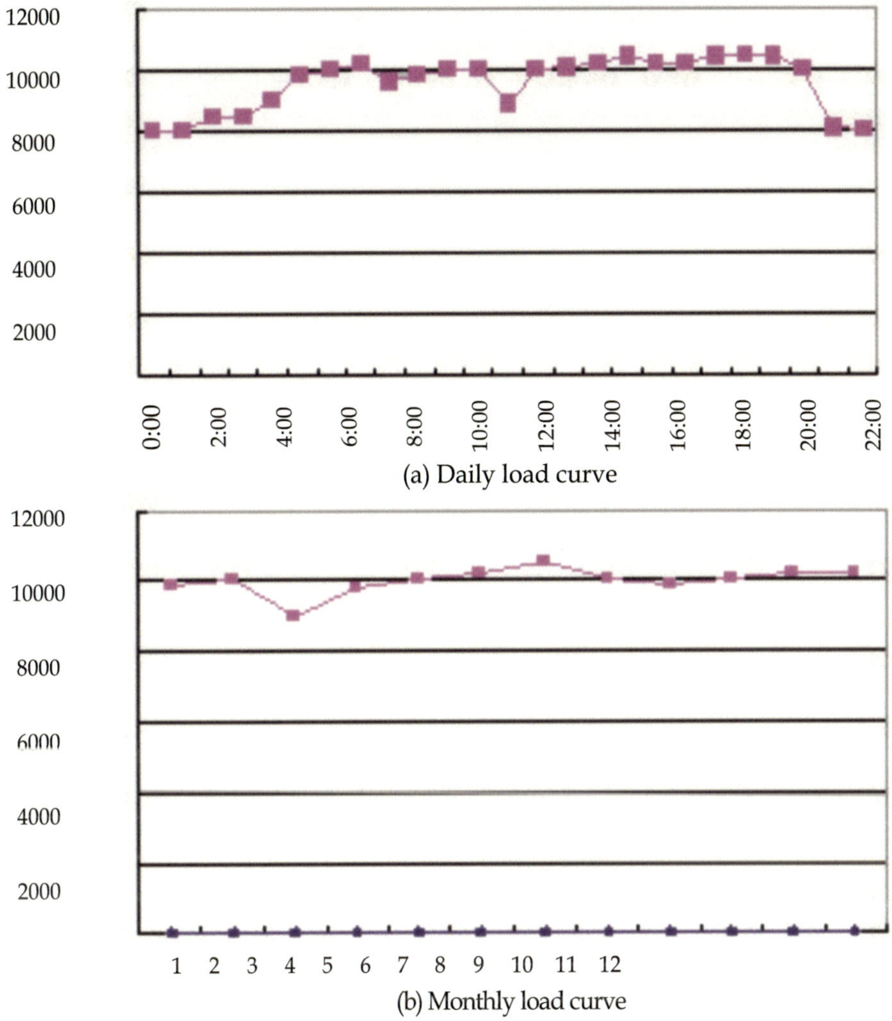

Figure 9.23 North Korea load curves for day and month (Assumed Material)

9.8.1.3 Power system and seasonal load patterns in Far East Russia

The above data had been obtained from SEI in Russia. Table 9.20 represents a present seasonal data of power in Russia (2001). Table 9.21 is a present seasonal data of power in East Siberia (2001). Table 9.22 shows a present seasonal data of power in Russian Far East (2001).

Type		Present seasonal data				Year
		Spring	Summer	Autumn	Winter	
Hydro	Hydro Pumped-storage power	45.3	48.0	41.7	40.9	175.9
Nuclear		33.3	27.7	36.8	39.1	136.9
Thermal		140.9	105.2	146.5	185.9	578.5
Including	Conventional steam turbine	56.9	46.2	64.4	80.7	248.3
	Co-generation	83.4	58.6	81.6	104.5	328.0
	Renewable energy	-	-	-	-	-
	Total	**219.5**	**180.9**	**225.0**	**265.9**	**891.3**

Table 9.20 Present seasonal data of power in Russia (2001, TWh)

Type		Present seasonal Data				Year
		Spring	Summer	Autumn	Winter	
Hydro	Hydro Pumped-storage power	22.0	26.4	24.2	22.3	94.9
Nuclear		-	-	-	-	-
Thermal		9.9	3.9	8.7	14.3	36.8
Including	Conventional steam turbine	5.1	1.0	4.1	8.4	18.6
	Co-generation	4.8	2.9	4.6	5.9	18.2
	Renewable energy	-	-	-	-	-
	Total	**31.9**	**30.3**	**32.9**	**36.6**	**131.7**

Table 9.21 Present seasonal data of power in East Siberia (2001, TWh)

Unified Power System (UPS) of Russian East provides with the electric power the most inhabited and industrially developed regions of the Russian Far East. UPS of Russian East consist of seven large regional electric power systems: Amur, Far East, Kamchatka, Magadan, Sakhalin, Khabarovsk and Yakutsk. Now the Amur, Khabarovsk and Far East electric power systems are united on parallel operation, in parallel with them the southern part of the Yakut electric power system is working also. The maximum of electric loading in UPS falls at winter and makes about 5.8 GW (based on the data for 2001). The minimum of electric loadings makes approximately half from a maximum and falls at the summer period. The maximum of in UPS was in 1990 and made approximately 30 billion kWh. In 2000 value of electrical energy consumption has made approximately 24 billion kWh, in 2001 this value has made 25.5 billion kWh. It was planned, that by 2005 consumption will make about 28.7 billion kWh by 2010 - 32 billion kWh, and by 2025 will make about 50 billion kWh.

Type		Present seasonal Data				Year
		Spring	Summer	Autumn	Winter	
Hydro	Hydro Pumped-storage power	1.13	0.98	0.97	1.77	4.85
Nuclear		-	-	-	-	-
Thermal		5.29	3.57	5.04	6.75	20.65
Including	Conventional steam turbine	1.54	1.27	1.52	1.72	6.05
	Co-generation	3.75	2.30	3.52	5.03	14.60
	Renewable energy	-	-	-	-	-
Total		**6.42**	**4.55**	**6.01**	**8.52**	**25.50**

Table 9.22 Present seasonal data of power in Russian Far East (2001, TWh)

The current consumption is distributed non-uniformly. More than 40 % of the electric power is consumed in the Far East electric power system. The rest of 60% are distributed between the Khabarovsk, Amur and Yakut electric power systems. Backbone electrical network of the UPS consist of 220 and 500 kV transmission lines. General extent of 500 kV lines makes about 2000 km The total installed capacity of power stations (nuclear, thermal and hydro) make about 11 GW[59]. Figure 9.24 represents the HVDC interconnection lines in Siberia and Far East Russia[50].

9.8.1.4 Power system status in North East China

Figure 9.25 represents the seven regions and power consumption map in China. This Figure was obtained from EPRI in China.

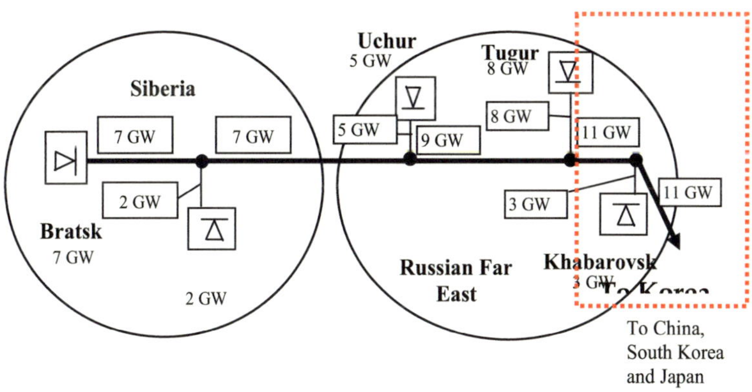

Figure 9.24 HVDC Interconnection Lines in Siberia and Far East Russia

This map shows an overview of the different regional grid systems within China, showing year 2002 generating capacities and outputs in each region, as well as indicating interconnections between regional grids. In China, Liaoning's power network covering the 147,500 square kilometers of land is a modern power network with long history and full of vigor.

Liaoning province is the power load center in Northeast China. It has one 500kV line and six 220kV lines to connect with the power network in Jilin province. It also has two 500kV lines and one 220kV line to connect with eastern part of an Inner Mongolia. By the end of 2000, the total installed capacity in Liaoning province was 15,185MW (hydro power: 1,156MW; thermal power: 12,559MW). The total installed capacity of the wholly-owned and holding power generation plants of Liaoning Electric Power Co., Ltd. is 2,854MW (hydro power: 456MW; thermal power: 2,398MW) and takes up 18.8% of the total installed capacity of the whole province. The independent power generation company has a total installed capacity of 10,861MW (hydro power: 488MW; thermal power: 10,373MW) and takes up 71.5%. The local self-supply power plants have a total installed capacity of 3,006MW, taking up 19.8%. The installed capacity of the plant at Sino-Korean boundary river is 545MW, taking up 3.6%.

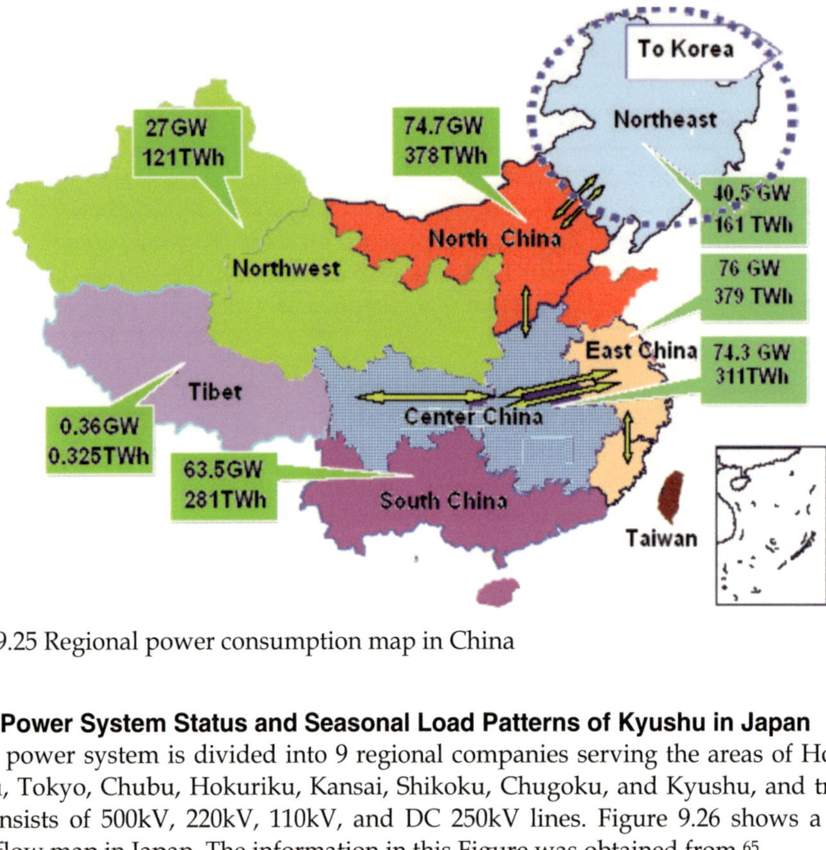

Figure 9.25 Regional power consumption map in China

9.8.1.5 Power System Status and Seasonal Load Patterns of Kyushu in Japan

Japan's power system is divided into 9 regional companies serving the areas of Hokkaido, Tohoku, Tokyo, Chubu, Hokuriku, Kansai, Shikoku, Chugoku, and Kyushu, and transmission consists of 500kV, 220kV, 110kV, and DC 250kV lines. Figure 9.26 shows a cascade power flow map in Japan. The information in this Figure was obtained from [65].

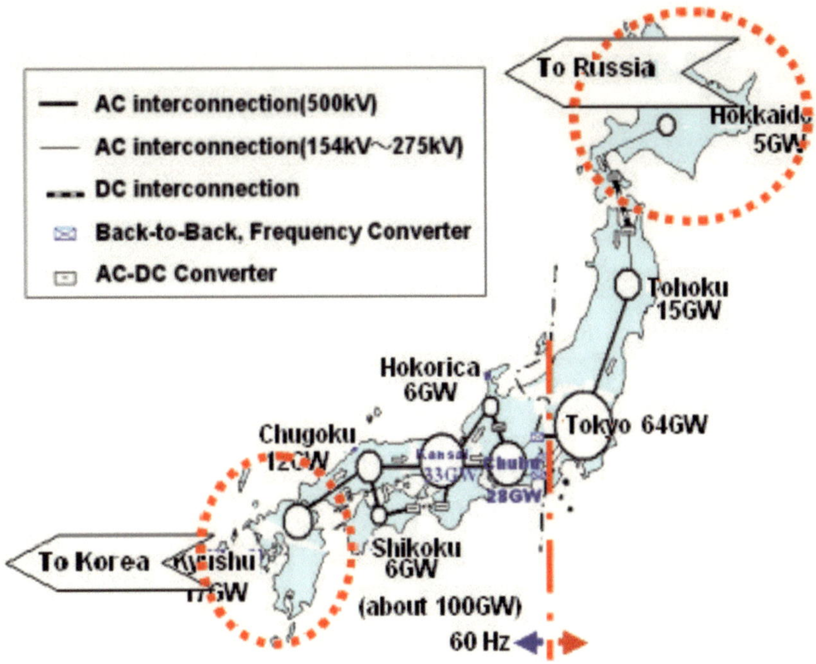

Figure 9.26 Cascade power flow map in Japan

The frequency used is 60Hz in the western part and 50Hz in the eastern part of the country. According to statistics published in 2001, the total generating capacity of the nine power companies is 33,765MW due to hydropower, 118,112MW due to thermal power, and 42,300MW due to nuclear power. The total capacity is therefore 194,177MW.

Kyushu's infrastructure is composed of nuclear, thermal, hydro, and geothermal power generating plants. In Kyushu region of Japan, 2001, summer peak has 16,743[MW], and winter peak has 12,961[MW]. The nuclear power plants are located both in the southwest coastal region and at the furthermost tip of Kyushu's northwest coast. The thermal power plants are located mainly on Kyushu's northeast and the northwest coasts. The hydro power plants are randomly distributed within the north and south central regions. The geothermal power plants are located in the north and south central regions. Among these regions, Kyushu has a total land area of 42,163 km^2 and is located in the southernmost part of Japan. The generating capacity of Kyushu's Electric Power Company is approximately 30,200MW. The backbone of its transmission system consists of 500kV, 220kV, and some 110kV lines.

9.8.2 Assumed Possible Interconnection Scenarios in North East Asia

Several cases of maps are drawn according to the assumed scenario in Figure 9.27, which has possible scenarios among Russia, China, North Korea, South Korea and Japan.

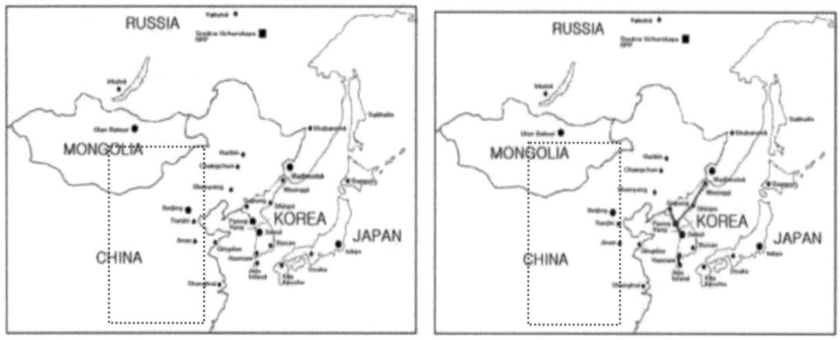

(a) Separation for North Korea and South. (b) North Korea-South Korea

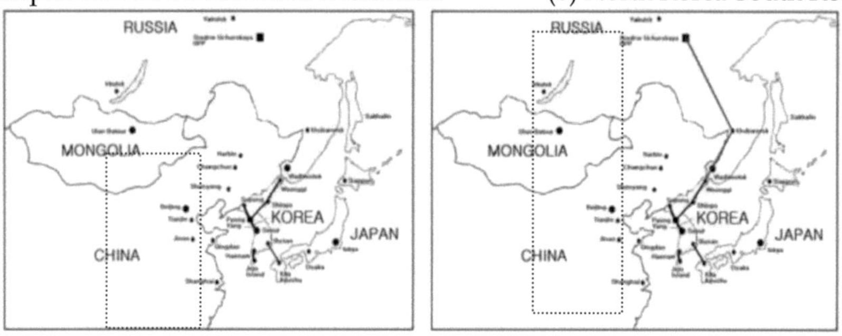

(c) North Korea-South Korea-Japan. (d) Russia-North Korea-South Korea-Japan

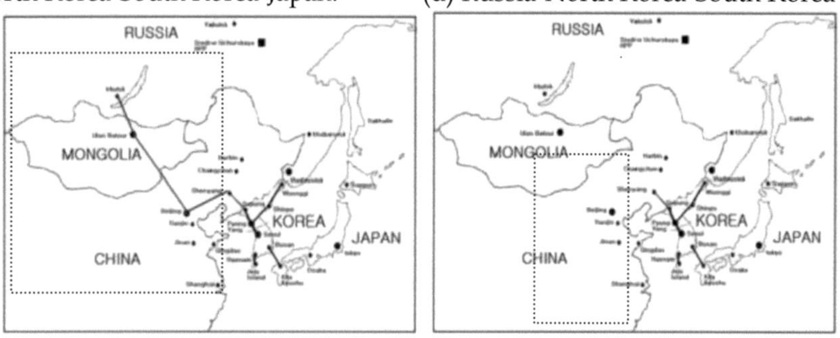

(e) Russia-Mongo-China-South Korea-Japan. (f) China-North Korea-South Korea-Japan

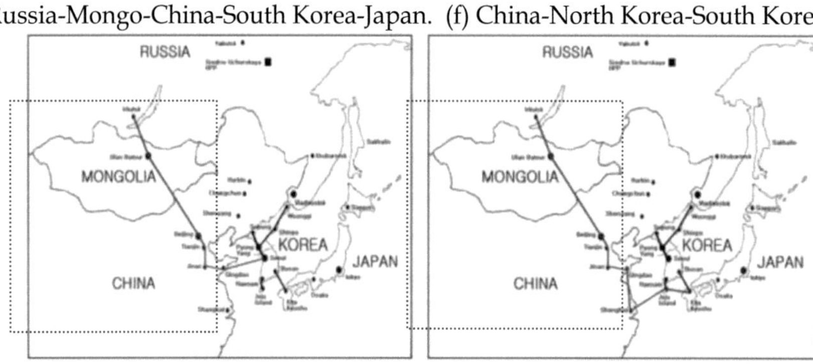

(g) Russia-Mongo-China-South Korea-Japan. (h) Russia-Mongo-China-South Korea-Japan

Figure 9.27 Possible scenarios among Russia, China, North Korea, South Korea and Japan

9.8.3 Assumed Seasonal Power exchange Quantity for Power Flow Calculation

Table 9.23 represents the assumed peak load data for summer and winter in South Korea, 2005. To simulation the PSS/E package, the load was decreased with 2,000MW in summer season and decreased with 1,000MW in winter season. Table 9.24 has the assumed peak data for summer and winter in North Korea, 2005. All the load and supply patterns were assumed with constant quantity. Table 9.25 is the assumed peak data for summer and winter at Kyushu in Japan, 2001. Table 9.26 has the assumed export power for summer and winter in Far East Russia. Table 9.27 represents the assumed export power for summer and winter in North East China.

Thus, the purpose of this Section was to execute a power flow analysis considering seasonal load patterns for the increase or for the decrease of a reserve power for the future power shortages faced by the metropolitan areas or by the southeastern area of the South Korea in North-East Asia. Several cases were considered as follows:

- Securing South Korea's power reserve by a power interchange considering seasonal effects in North East Asia countries.
- Drawing possible scenarios and power flow maps for relieving the power shortages faced by the metropolitan areas and southeastern area in Korean Peninsula.
- Considering seasonal load patterns and studying power flow for the interconnection with 2,000MW in Far-East Russia or in Northeast China, and 1,000MW in Japan to utilizing remote power sources.

The preliminary considerations above consist only of a scenario-based power flow analysis included with seasonal load patterns; however, the results of this research may be referred to the government for use in the establishment of a future construction plan for the power system in South Korea. Moreover, these may be expecting to improve political and economical relationships in North East Asia countries.

Seasons	Generation [MW]	Load [MW]	Receive Power [MW]
Summer peak	51857.8	51,090.4	2,000+1,000
Winter peak	41,857.8	41,090.4	1,000+500

Table 9.23 Assumed peak data for summer and winter in South Korea, 2005

Seasons	Generation [MW]	Load [MW]	Transmission P [MW]
Summer peak	9,000	9,000	-
Winter peak	9,000	9,000	-

Table 9.24 Assumed peak data for summer and winter in North Korea, 2005

Seasons	Generation [MW]	Load [MW]	Transmission Power (Japan → Korea)
Summer peak	17,743	16,743	1,000
Winter peak	13,461	12,961	500

Table 9.25 Assumed peak data for summer and winter at Kyushu in Japan, 2001

Seasons	Generation [MW]	Load [MW]	Transmission Power (Russia → Korea)
Summer peak	2,000	0	2,000
Winter peak	1,000	0	1,000

Table 9.26 Assumed export power for summer and winter in Far East Russia

Seasons	Generation [MW]	Load [MW]	Transmission Power (China → Korea)
Summer peak	2,000	0	2,000
Winter peak	1,000	0	1,000

Table 9.27 Assumed export power for summer and winter in North east China

9.9 Acknowledgements

This Chapter has been prepared by Nikolai I. Voropai, Professor, Corresponding Member of RAS, Director of Energy Systems Institute, Irkutsk, Russia. Contributors include colleagues at the Institute and Members of the IEEE PES W.G. on Asian and Australian Electricity Infrastructure.

9.10 References

[1]. *Open Access in Inter-State Transmission,* Central Electricity Regulatory Commission, New Delhi, India, Nov. 2003.

[2]. *Electricity Act 2003,* Ministry of Power, Government of India, New Delhi, India, June 2003.

[3]. Mukhopadhyay, S., "Interconnection of Power Grids in South Asia", *Proc. 2003 IEEE PES General Meeting,* Toronto, Ontario, Canada.

[4]. Mukhopadhyay, S., "Power Generation and Transmission Planning in India - Methodology, Problems and Investments", *Proc. 2004 IEEE PES General Meeting,* Denver, Colorado, USA.

[5]. National Electricity Code Administrator website. www.neca.com.au

[6]. Ershevich, V.V., Antimenko, Yu.A., "Efficiency of the Unified Electric Power System Operation on the Territory of the Former USSR", *Izv. RAN. Energetika,* 1993, No. 1 (in Russian).

[7]. Voropai, N.I., Ershevich, V.V., Rudenko, Yu.N., *Development of the International Interconnections - the Way to Creation of the Global Power System,* Irkutsk: SEI SB RAS, 1995, Vol. 10 (in Russian).

[8]. Belyev, L.S., Voizekhovskaya, G.V., Saveliev, V.A., *A System Approach to Power System Development Management,* Novosibirsk: Nauka, 1980 (in Russian).

[9]. Belyaev, L.S., Kononov, Yu.D., Makarov, A.A., "Methods and Models for Optimization of Energy Systems Development", *Soviet Experience Review of Energy Models.,* Laxenburg: IIASA, 1976, No. 3.

[10]. Voropai, N.I., Trufanov, V.V., Selifanov, V.V., Sheveleva, G.I., "Modeling of Power Systems Expansion and Estimation of System Efficiency of Their Integration in the Liberalized Environment", *Proc. CIGRE 2004 Session,* Rep. C1-103.

[11]. http://www.nemmco.com.au/
[12]. http://www.pjm.com
[13]. http://www.caiso.com
[14]. http://www.nordpool.com
[15]. http://www.elecpool.com
[16]. http://www.open.gov.uk/offer/offer.htm
[17]. http://www.ngc.co.uk
[18]. Schweppe, F. C., et al, *Spot Pricing of Electricity*, Kluwer Academic Publisher, 1988.
[19]. Chao, H. P., Huntington, H. G., *Designing Competitive Electricity Markets*, Kluwer Academic Publisher, 1998.
[20]. Ilic, M., Galiana, F., Fink, L., *Power Systems Restructuring, Engineering and Economics*, Kluwer Academic Publisher, 1998.
[21]. Cameron, L., "Transmission Investment: Obstacles to a Market Approach", *The Electricity Journal*, 2001, Vol. 14, No. 2.
[22]. Kahn, E. P., "Numerical Techniques for Analyzing Market Power in Electricity" , *The Electricity Journal* , 1998 , Vol. 11, No. 6.
[23]. Oren, S.S., Ross, A.M., "Economic Congestion Relief Across Multiple Regions Requires Tradable Physical Flow-Gate Rights", *IEEE Trans. on PWRS*, 2002, Vol. 17, No. 1.
[24]. Wu, F. F., Ni, Y., Wei, P., "Power Transfer Allocation for Open Access Using Graph Theory: Fundamentals and Applications in Systems without Loopflow", *IEEE Trans. on PWRS*, 2000,Vol. 15, No. 3.
[25]. *State Power Information Network*, http:// www.sp.com.cn
[26]. *Electric Power Info.-Net of China*, http://www.zdxw.com.cn
[27]. Association of the Chinese Electric Power Enterprises, http://www.cec. org.cn
[28]. *Electric Power News Net of China*, http://www.cepn.sp.com.cn
[29]. East China Power Market Steering Committee Office, "East China Power Market Pilot Work Documents", No. 18-19, 2004.
[30]. Gan, D., Bourcier, D. V., "Locational Market Power Screen and Congestion Management: Experience and Suggestions", *IEEE Transactions on Power Systems*, 2002, Vol. 17, No. 1.
[31]. Mas-Colell, A., Whinston, M. D., Green, J. R., *Microeconomic Theory*, Oxford University Press, Oxford, UK, 1995.
[32]. Federal Energy Regulation Council (FERC), "Working Paper on Standardized Transmission Service and Wholesale Electricity Market Design", March 16, 2002. http://www.ferc.fed.gov
[33]. LECG, LLC, Kema Consulting, Inc, "Feasibility Study for a Combined Day-Ahead Electricity Market in the Northeast", 2nd Draft Report, Albany, April 26, 2001.
[34]. Hunt, S. , Shuttleworth, G., *Competition and Choice in Electricity*, New York, Wiley, 1997.
[35]. http://www.kpx.or.kr
[36]. Hur, D., "Determination of Transmission Transfer Capability Using Distributed Contingency-Constrained Optimal Power Flow and P-V Analysis," Ph.D. dissertation, School of Elect. Eng., Seoul Nat. Univ., Korea, 2004.
[37]. Hur, D., Park, J. K, Kim, B. H., "Application of Distributed Optimal Power Flow to Power System Security Assessment," *Electr. Power Components Syst.*, 2003, Vol. 31, No.1.

[38]. Hur, D., Park, J. K., Kim, B. H., "An Efficient Methodology for Security Assessment of Power Systems based on Distributed Optimal Power Flow," *Eur. Trans. Elect. Power*, 2003, Vol. 13, No. 3.
[39]. *Network Planning in a Deregulated Environment*, CIGRE WG 37-30, Feb. 2003.
[40]. Park, D.et al, "1st Report for Infrastructure on NEAREST Project", KERI, Tech. Report. Nov. 2003.
[41]. Park, D., Podkovalnikov S., "Analysis of Scenarios for Potential Power System Interconnections in Northeast Asia", AEC Conference, September 2004.
[42]. Belyaev, L., Podkovalnikov, S., "An Approach to and Results of Effectiveness Assessment of Inter-Tie RFE - DPRK - ROK>", *AEC Conf.*, Irkutsk, Russia, September 21-22, 2000.
[43]. "A Preliminary Study of the Power System Interconnections in Northeast Asia Countries", KERI, KEPCO, Tech. Rep, 2000.
[44]. Podkovalnikov, S., "East Siberia and Rusian Far East Estimated Prospective Export Potentials", KERI-ESI Tech. Report, Nov. 2004.
[45]. Yoon, J., et al, "Maximum Exchange Power Between Russia and Republic of Korea", *AEC - 2005 Conf.*, Irkutsk, Russia, September 13-17, 2004.
[46]. Yoon, J., et al, "Economic Analysis Methodology of Power System Interconnections Considering Conventional Economic Benefits and Environmental Effects", *ICEE Conf.*, July 20-24, 2002, Xian, China.
[47]. Voropai, N.I., Kononov, Y.D. and Saneev, B.G., "Prerequisites and Directions of Energy Integration in North-Eastern Asia", *Proceeding of International Conference*, Irkutsk, Russia, September 22-26, 1998.
[48]. Park, D. W., Hwang, C., Na, K. Y, Kim, I. S., "The Status Quo and Prospects of Korea Power System", *Proc. of Int. Conf.*, Irkutsk, Russia, September 22-26, 1998.
[49]. Long, W. F., Stovall ,J. P., "Comparison of Costs and Benefits for DC and AC Transmission", *CIGRE Symposium on DC and AC Transmission Interaction and Comparisons*, Boston, USA, June 13-17, 1987.
[50]. Belyaev, L. S., Chudinova, L. Yu., Koshchceev, L. A., Podkovalnikov, S. V., Savelyev, V. A., Voropai, N. I., "The High Direct Current Bus Siberia-Russian Far East", *Proc. of Int. Conf. ECNEA-2002 (3rd)*, Irkutsk, Russia, September 9-13, 2002.
[51]. Samorodov, G., Krasilnikova, T., Zilberman, S., Iatsenko, R., Kobylin, V., Drujinin, A., "Consideration on Technical-Economic and Reliability Performance of the Transmission System from South-Yakutia Hydro Power Complex to Korea", *Proc. of Int. Conf. ECNEA-2002 (3rd)*, Irkutsk, Russia, September 9-13, 2002.
[52]. Shin ,J. R, Kim, B. S., Choi ,Y. J., "Power System Linkage between South and North in Korean Peninsula: A Proposal with Supposed Situation", *Proc. of ICEE 2001*, Xian, China, July 23-27, 2001.
[53]. Jang ,Y. J., Lee, S. S., Park ,J. K., Kim, K. H., "Scenarios based Power Flow Analysis for the Interconnection of Power Systems between South and North Korea", *Proc. of ICEE 2001*, Xian, China, July 23-27, 2001.
[54]. Nahm, J. I., "Electric Power Supply in Korea & The KEDO Project", Tech. Report, 2002.
[55]. Lee ,S. S., Jang, G. S., Park, J. K., Honma,T., Minakawa ,T., "Scenario and Power Flow Analysis for 765kV Interconnection between South and North Korea", *Proc. of ICEE 2002*, Jeju, Korea, July 13-17, 2002.

[56]. Lee, S. S., Park, J. K., "765kV Interconnection Scenarios and Power Flow Analysis in Korean Peninsula", *Proc. of Int. Conf. ECNEA-2002 (3rd)*, Irkutsk, Russia, September 9-13, 2002.

[57]. Lee, S. S., Park, J. K., Moon, S. I., "Power System Interconnection Scenario and Analysis Between Korean Peninsula and Japan", *IEEE 2003 General Meeting*, Toronto, Canada, July 2-6, 2003.

[58]. Lee, S. S., Park, J. K., Moon, S. I., Moon ,J. F., Kim J. C., Kim, S. K., Kim, H. Y., "North-East Asia Interconnection Scenario Map, and Power Reserve Strategy in South Korea", *IEEE 2004 General Meeting*, Denver, USA, June 24-28, 2004.

[59]. Gerasimov, A. S., Koshcheev, L. A., "Russia - Korea Interstate Electrical Tie", *Proc. of Int. Conf. ECNEA-2004 (4rd)*, Irkutsk, Russia, September 13-17, 2002.

[60]. 2001' Report, (Korea Power Exchange, Tech Report, 2001.

[61]. 2000' Annual Report, Liaoning Electric Power Co., LTD.

[62]. Manual of Central Load Dispatching Center, Kyushu Electric Power Co., Inc , 2001.

[63]. Annual Report of Central Dispatch Center of Russia, Technical Part, 2001 (in Russian).

[64]. Annual Report of Regional Dispatch Center of RFE, 2001(in Russian).

[65]. Takeshi, T., "Liberalization of Electricity Market in Japan", *Korea-Japan Symp.*, Seoul, Korea, July 2-6, 2002.

[66]. KERI, IEA, APERC, Vostokenergo, International Symposium on NEAREST, Vladivostok, Russia, August 24-27, 2004.

10

Power Generation in Southern Africa: Energy Trading and the Southern African Power Pool

This Chapter reviews power generation and energy trading arrangements that exist in southern Africa. The Chapter also considers the operations and workings of the Southern African Power Pool. The Southern African Power Pool (SAPP) was created in April 1995 through the Southern African Development Community (SADC) treaty that was signed to optimize the use of available energy resources amongst the countries in the region and support one another during emergencies. At the time of creation, the SADC governments agreed to allow their national power utilities to enter into the necessary agreements that regulate the establishment and operation of the SAPP. SAPP membership was therefore restricted to national power utilities of the SADC member states as stipulated in the Inter-Governmental Memorandum of Understanding (IGMOU). In the Revised IGMOU of 23 February 2006, SAPP membership was extended to include other Electricity Supply Enterprises within the SADC region.

10.1 Structure and Governing Documents

There are four legal documents covering the rights and obligations of the SAPP members and participants:

(i.) *Inter-governmental memorandum of understanding* (IGMOU) that grants permission for the utilities to participate in the SAPP and enter into contracts, and guarantees the financial and technical performance of the power utilities. The original document was signed in 1995 by SADC members, excluding the Democratic Republic of Congo (DRC), Madagascar, Mauritius and Seychelles. All the SADC countries, with the exception of Madagascar, Mauritius and Seychelles, signed the Revised IGMOU on 23 February 2006.

(ii.) *Inter-utility memorandum of understanding (IUMOU)* between participants, defining ownership of assets and other rights, e.g. provision for change in status from participating to operating member. The Revised IUMOU was signed by all the SAPP member utilities on 25 April 2007 in Harare, Zimbabwe, with the exception of SNEL of the DRC and TANESCO of Tanzania. TANESCO signed the Revised IUMOU in February 2008 and SNEL in April 2008. The Revised IUMOU has defined a new structure for the management and operations of the SAPP.

(iii.) *Agreement between operating members (ABOM)*, which determines the interaction between the utilities with respect to operating responsibilities under normal and emergency conditions. Operating Members only, i.e., members whose transmission system is interconnected to the SAPP grid signed this document. The document is currently under review and when completed would be signed by all Operating Members.

(iv.) *Operating guidelines (OG)*, which defines the sharing of costs and functional responsibilities for plant operation and maintenance including safety rules.

The basis for the SAPP as defined in the Revised IGMOU is the need for all participants to:

(a) Co-ordinate and co-operate in the planning and operation of their systems to minimize costs while maintaining reliability, autonomy and self-sufficiency to the degree they desire;
(b) Fully recover their costs and share equitably in the resulting benefits, including reductions in required generating capacity, reductions in fuel costs and improved use of hydroelectric energy; and
(c) Co-ordinate and co-operate in the planning, development and operation of a regional electricity market based on the requirements of SADC Member States.

In order to carry out the vision of the SAPP, a Coordination Center was established in Harare, Zimbabwe, in February 2000 to act as a focal point for all the SAPP activities. A Host Country Agreement (HCA) was afterwards signed between the Government of Zimbabwe and SAPP on 13 March 2006 giving the SAPP Coordination Center a Diplomatic Status. Also a Memorandum of Understanding between SAPP and the Regional Electricity Regulators Association (RERA) on liaison and interaction between the two parties was entered into in April 2007.

The structure of the SAPP is shown in Figure 10.1.

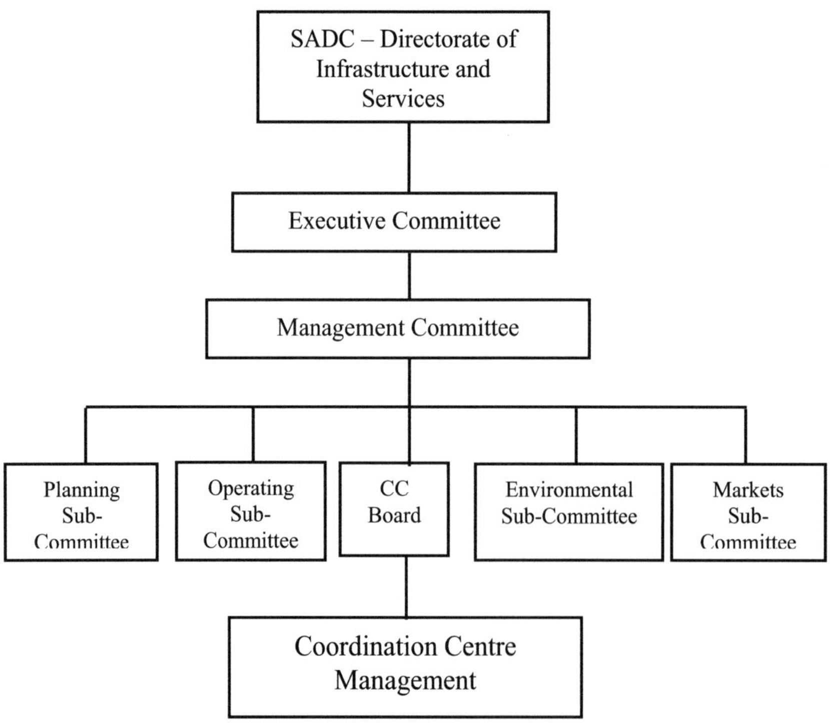

Fig. 10.1 Reporting Structure of the SAPP

The *SADC Government Ministers and Officials* are responsible for policy matters normally under their control within the national administrative and legislative mechanisms regulating the relations between the Government and the national power utility.

The *Chief Executives* of the members and a representative from the SADC Secretariat form the *Executive Committee.* The Executive Committee will refer matters such as requests for membership by non-SADC countries and major policy issues that may arise to the SADC Ad Hoc Committee of Energy Ministers. A country with more than one member utility would need to designate one utility to represent it on the Executive Committee.

The *Management Committee* oversees and decides on the recommendations of the Sub-Committees and the Coordination Center Board.

The *Operating Sub- Committees* consist of representatives from those power utilities already interconnected and exchange power on a major scale (Operating Members), presently 9 countries (Botswana, South Africa, Zambia, Zimbabwe, Democratic Republic of Congo, Lesotho, Mozambique, Namibia and Swaziland). The duties of the sub-committee include the establishment and updating of methods and standards to measure technical performance, operating procedures including operating reserve obligations

The *Planning Sub-Committee* establishes and updates common planning and reliability standards, review integrated generation and transmission plans, evaluate software and other planning tools, determine transfer capability between systems etc.

The *Environmental Sub-Committee* consists of appointed representatives from each Operating Member. The committee develops Environmental Guidelines for SAPP; liaise with Governments to keep abreast of world and regional matters relating to air quality, water quality, land use and other environmental issues. Where Governments have in place related Environmental Organizations, the Committee has to liaise with them to assist one another on specific issues.

The *Markets Sub-Committee* is responsible for the design and continued development of the electricity market in the region and determines criteria to authorize this trade.

All the *Sub-Committees* consist of a maximum of two representatives per Member who are of sufficient seniority in their own organization to make all relevant decisions.

The *Coordination Center* reports to a Co-ordination Center Board consisting of a maximum of two representatives of each National Power Utility (i.e. the signatories of the IUMOU).

10.1.1 SAPP Vision

The vision of the SAPP is to facilitate the development of competitive electricity market where an end user within the SADC region ultimately has possibility of choosing the preferred supplier of electrical energy. To promote the vision and change it into a reality, SAPP is about to change from a cooperative pool to a competitive power market trading both physical and financial contracts. The challenge for SAPP will be to manage all the difficulties and uncertainties envisaged to emerge during the transition period from administrating a corporative market to the geographical biggest competitive pool in the World. At the same time as the transition is taking place, the SAPP has run out of generation surplus capacity resulting in load shedding in a number of member countries.

10.1.2 SAPP Objectives

The SAPP objectives are:

- To provide a forum for the development of a world class, robust, safe, efficient, reliable and stable interconnected electrical system in the region.
- Harmonise inter-utility relationships.
- Co-ordinate the development of common regional standards on quality of supply; measurement and monitoring of systems performance; enforcement of standards, and facilitate the development of regional expertise through training programs and research.

10.1.3 SAPP Mission, Strategy and Values

Mission

The Mission of SAPP is to provide the least cost, environmentally friendly and affordable energy and increase accessibility to rural communities.

The Strategy

In its operation the SAPP aims at being the most preferred region for investment for value for money by energy intensive users.

The Values

- Respect for others and develop mutual trust
- Honesty, complete fairness and integrity in dealing with issues
- Selfless discharge of duties
- Full accountability to the organization and its stakeholders
- Encourage openness and objectivity.

10.1.4 SAPP Coordination Center

The SAPP Coordination Center was established in Harare, Zimbabwe, at the beginning of the year 2000. The Center represents a focal point of SAPP and a staff to further its vision and technical challenges. In addition to the Manager, a total of seven (7) support staff in fields of Finance, Information Technology, Environment and Secretarial are presently employed at the Coordination Center. The functions of the SAPP Coordination Center are to:

- Implement the SAPP objectives; provide a focal point for SAPP activities; facilitate the implementation of a competitive electricity market in the SADC region;
- Monitor the operations of SAPP transactions between the members;
- Carry out technical studies on the power pool to evaluate the impact of future projects on the operation of the pool;

- Coordinate the training of members of staff to improve the region's knowledge of power pool operations; and
- Provide power pool statistics and maintaining a pool database for planning and development.

A website was developed as a means for SAPP to communicate with the world and inform interested persons of its activities. The Coordination Center also acts as a secretariat for the various SAPP committees and its sub-committees.

The twelve members of SAPP fund the activities of the Coordination Center through an annual subscription fund. The Coordination Center makes a budget and this is presented to the Coordination Center Board for approval. The Coordination Center Board is made up of senior managers of utility representatives and one of their functions is to oversee the activities of the Coordination Center including the approval of the budget. This budget is used to pay for staff salaries and other SAPP operational costs.

Internationally reputable auditors have been appointed to audit the SAPP Coordination Center finances periodically. The audited financial report is then distributed to members and is also published as part of the SAPP

10.1.5 SAPP Membership

The governance and membership of the SAPP was derived from the desire for economic co-operation and integration, equitable sharing of resources and support of one another in times of crisis under the SADC protocol. The environment under which the power pool now operates, and the ongoing development of a competitive market, will significantly change the basis for the operation of the SAPP. The Pool has therefore recently reviewed its governance and membership in order to achieve a competitive market including giving access for an increased number of participants.

Full Name of Utility	*Status*	*Abbreviation*	*Country*
Botswana Power Corporation	OP	BPC	Botswana
Electricidade de Moçambique	OP	EDM	Mozambique
Electricity Supply Commission of Malawi	NP	ESCOM	Malawi
Empresa Nacional de Electricidade	NP	ENE	Angola
Eskom	OP	Eskom	RSA
Lesotho Electricity Corporation	OP	LEC	Lesotho
NamPower	OP	NamPower	Namibia
Societe Nationale d'Electricite	OP	SNEL	DRC
Swaziland Electricity Board	OP	SEB	Swaziland
Tanzania Electricity Supply Company Ltd	NP	TANESCO	Tanzania
ZESCO Limited	OP	ZESCO	Zambia
Zimbabwe Electricity Supply Authority	OP	ZESA	Zimbabwe

OP = Operating member NP = Non-Operating member

Table 10.1 SAPP Membership

SAPP membership is as per the latest revision of the IUMOU open to national power utilities and other Electricity Supply Enterprises (Power Utility, Independent Power

Producer, Independent Transmission Company and/or Service Provider for the electricity market), from SADC member countries. There are currently twelve SAPP members as indicated in Table 10.1, nine operating members and three non-operating members.

10.2 Sapp Achievements

From the time that the SAPP was created in 1995, the following achievements have been made:

10.2.1 Coordination Center

The official opening of the SAPP Co-ordination Center in Harare on the 18th of November 2002 was marked as a great success. The Guest of Honor was the Minister of Petroleum of Angola: The Honorable, José Maria Botelho de Vasconcelos.

10.2.2 Documentation Review and SAPP Restructuring

The signing of the Revised Inter-Governmental Memorandum of Understanding (IGMOU) by the Ministers responsible for energy in the SADC region in Gaborone, Botswana, on 23 February 2006, was the beginning of the restructuring of the SAPP. The Chief Executives of the SAPP Member Utilities then signed the Revised Inter-Utility Memorandum of Understanding (IUMOU) on 25 April 2007 in Harare, Zimbabwe. Therefore, other Electricity Supply Enterprises (Power Utility, Independent Power Producer, Independent Transmission Company and/or Service Provider for the electricity market), from SADC member countries can now join the SAPP.

10.2.3 Cooperation with the Regional Electricity Regulatory Association (RERA)

The resolution of the SAPP-RERA relationship and the signing of the SAPP-RERA Memorandum of Understanding on 25 April 2007 in Harare, Zimbabwe. This is a cooperation agreement that will allow the two institutions to work together and cooperate for the common good of the SADC region.

10.2.4 Transmission wheeling charges and losses

The SAPP adopted a scientific method for the determination of transmission wheeling charges. The new transmission wheeling charges were implemented over a three-year period starting from the 1st of January 2003. In the same year, the SAPP also approved the enforcement of Article 11.3.3 of the Agreement between Operating Members on transmission losses.

10.2.5 Development of a competitive electricity market

- In April 2001, the SAPP started the short-term energy market (STEM) as a precursor to a full competitive market. At the time of this publication, there are eight participants on the STEM from an initial number of two at the start of the market in April 2001.
- The development of the competitive electricity market started in January 2004 when an Agreement between the Government of Norway and SAPP provided SAPP with a grant to the amount of NOK 35 million for this purpose. The SAPP is currently testing the day-

ahead market-trading platform that has been developed by NordPool. The SAPP Executive Committee will determine the date for the market opening. The recommendations of the Management Committee are to wait until governance issues are resolved within the SAPP. It was expected that the opening would take place towards the end of 2007.

- In order to assure a proper development and operation of a competitive electricity market, the SAPP has developed long-term transmission pricing policies and implementation procedures and an ancillary services market. SAPP and Sida signed an agreement in July 2004 covering financial assistance to provide the necessary consultancy services for this and an English company, Power Planning Associates (PPA) was assigned to carry out the task.

10.2.6 Completed transmission projects
The following transmission lines have been commissioned:

- The 400kV Matimba (South Africa) - Insukamini (Zimbabwe) interconnector linking Eskom of South Africa and ZESA of Zimbabwe in 1995.
- BPC Phokoje substation was tapped into the Matimba line to allow Botswana's tapping into the SAPP grid at 400kV in 1998.
- The 330kV Mozambique-Zimbabwe interconnector was commissioned in 1997.
- The restoration of the 533kV DC lines between Cahora Bassa in Mozambique and Apollo substation in South Africa was completed in 1998.
- The 400kV line between Aggeneis in South Africa and Kookerboom in Namibia in 2001.
- The 400kV line between Arnot in South Africa and Maputo in Mozambique in 2001.
- The 400kV line between Camden in South Africa via Edwaleni in Swaziland to Maputo in Mozambique in 2000.
- The 220kV Livingstone (Zambia)-Katima Mulilo (Namibia) interconnector was commissioned in 2006.

10.2.7 Establishment of Westcor
The establishment and launching of the Western Power Corridor (Westcor) in April 2002 to develop the hydropower generation resources in the DRC, Angola and Namibia; and the transmission links from the DRC via Angola, Namibia, Botswana to South Africa, including a telecommunication network has been a great welcome to the region. A Project Office was opened in May 2006 in Gaborone, Botswana.

10.2.8 Environmental Guidelines
The SAPP has completed and approved the following environmental guidelines:

- Environmental Impact assessment (EIA) Guidelines for Transmission Lines
- Environmental Impact assessment (EIA) Guidelines for Thermal Power Plants
- Guidelines on the Management of Oil Spills
- Guidelines for the Safe Control, Processing, Storing, Removing and Handling of Asbestos Containing Material

- Guidelines for Management and Control of Electricity Infrastructure with regard to Animal Interaction.

10.2.9 Other Completed Projects

The other completed projects include the following:

- Completion of the SAPP Pool Plan in 2001. In 2006, the SAPP received a World Bank grant to review the Pool Plan and the Revised Pool Plan was completed in November 2007.
- In 2001, the SAPP received a World Bank grant to conduct a telecommunications study on how best to link the three control areas. The recommendations of the study were to use a VSAT solution in the short-term and fiber in the long-term. The SAPP has now completed the implementation of a VSAT solution and the project has been commissioned.
- Frequency relaxation project was completed in 2003. The SAPP relaxed the operating frequency from 50 +/-0.05 Hz to 50 +/-0.15 Hz. The new frequency bands were implemented from January 2003.

10.3 Energy Trading

10.3.1 Bilateral Contracts

Based on the current SAPP Inter-Governmental Memorandum of Understanding, the general arrangement for electricity trading in the SAPP is for the national power utilities to engage into long-term and short-term bilateral contracts for the sourcing and consumption of electrical energy. Thus, the intergovernmental agreements and the bilateral contracts between the utilities form the basis and foundation for cross border electricity trading in the SADC region. The routine activities that follow include scheduling, settlements, monitoring of the quality of supply and detailed investigations are conducted into inadvertent energy flows and major power system faults and disturbances [1].

The prices for the bi-lateral energy contracts are negotiated between the buyer and the seller. The pricing structure for bilateral contracts is diverse with some contracts having capacity and energy rates which take cognizance of the time of use, peak or off peak. Other contracts have flat energy rates.

Bilateral agreements provide for the assurance of security of supply but are not flexible to accommodate varying demand profiles and varying prices. To explore further the benefits thereof, the sourcing and scheduling of electrical energy closer to the time of dispatch, the SAPP developed the short-term energy market (STEM) as one option for sourcing and securing supplies closer to real time dispatch. STEM has been designed to specifically mimic a real time dispatch.

Figure 10.2 shows the bilateral agreements in force from 2005.

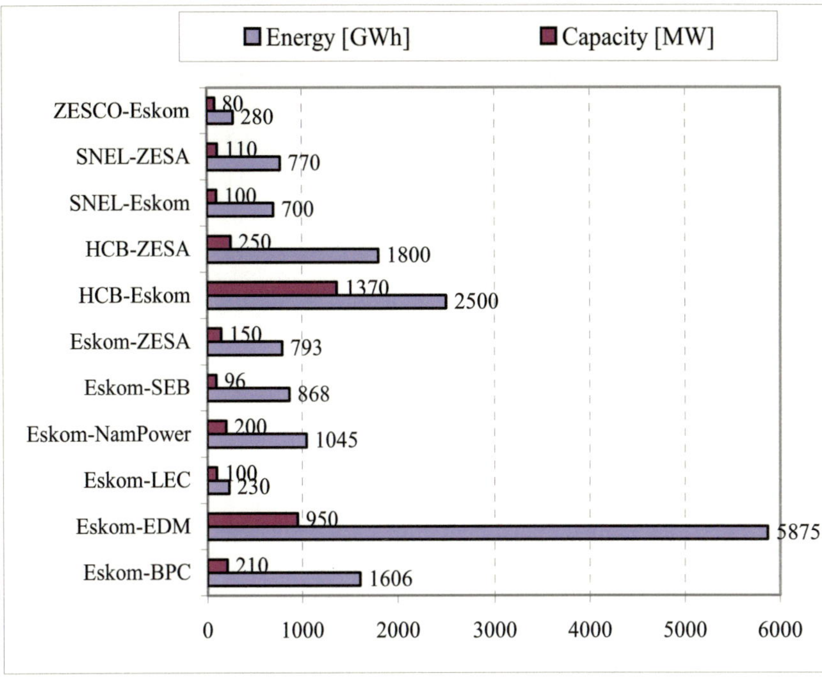

HCB hydro supply: 1,770MW, Eskom thermal supply: 1,706MW

Fig. 10.2 The 2005 Bilateral Contracts in SAPP

10.3.2 The Short-term Energy Market

The goal of standard market design is to establish an efficient and robustly competitive wholesale electricity marketplace for the benefit of consumers. This could be done through the development of consistent market mechanisms and efficient price signals for the procurement and reliable transmission of electricity combined with the assurance of fair and open access to the transmission system [3]. For the short-term energy market (STEM) design, the following criteria were used [2,3,4]:

i.) *Transmission rights*
Long and short-term bilateral contracts between participants were given priority over STEM contracts for transmission on the SAPP interconnectors. All the STEM contracts are subject to the transfer constraints as verified by the SAPP Co-ordination Center.

ii.) *Security requirements*
Participants are required to lodge sufficient security deposit with the Co-ordination Center before trading commences and separate security is required for each energy contract.

iii.) *Settlement*
Participants have the full obligation to pay for the energy traded and the associated energy costs. The settlement amounts are based on the invoices and are payable into the Co-ordination Center's clearing account. It is the responsibility of the Participants (buyers) to ensure that sufficient funds are paid into the clearing account for the Co-ordination Center to effect payment to the respective Participants (sellers).

iv.) *Currency of trade*
The choice of currency is either the United States Dollar or the South Africa Rand dependent on the agreement between the buyer and the seller.

v.) *Allocation method*
The allocation of available quantities based on the available transmission capability is by fair competitive bidding with equal sharing of available quantities to the buyers.

vi.) *Firm contracts*
Once contracted, the quantities and the prices are firm and fixed. There are currently three energy contracts that have been promoted in the STEM as follows; monthly, weekly and daily contracts. Daily contracts have been most consistent and have been greatly used by participants.

Table 10.2 summarizes the daily trading routine in the STEM. It is important to note that the period for submission of bids and offers close simultaneously.

At 08:30 HRS, a day before trading – The Center publishes the exchange rate between the United States Dollar and the South African Rand.
Any time before 09:00 HRS, a day before trading - Participants submit bids and offers to the Co-ordination Center for future daily contracts.
At 10:00 HRS, a day before trading - The market closes and the Co-ordination Center matches bids and offers for any future trading day;
At 14:00 HRS, a day before trading - The Co-ordination Center publishes the results to all Participants.
At 15:00 HRS, a day before trading – Participants may enter into post-STEM contracts and inform the Coordination Center accordingly.

Table 10.2 Daily Trading Routine in the STEM

For the period from 1 April 2005 to 31 March 2006, corresponding to the SAPP Coordination Center fiscal period, the power supply on the short-term energy market (STEM) was 423-GWh and the corresponding demand was 3,700-GWh. The traded energy was 178-GWh at an average cost of 0.96 USc/kWh. For a similar period from 1 April 2006 to 31 March 2007, supply and demand figures were 377-GWh and 1,118-GWh, respectively. The energy traded was 226-GWh at an average cost of 1.38 KWh/kWh. This period recorded an increase in the cost of energy, but with a much lower power demand [see Figure 10.3].

The total energy sales for the period from 1 April 2005 to 31 March 2006 was *US$2.2 million* and the corresponding sales for the period from 1 April 2006 to 31 March 2007 was *US$3.1 million,* Figure 10.4. Though the same quantity of energy was traded during both periods, it is seen that the cost of energy in the 2006 period had increased due to reduced power supply on the market.

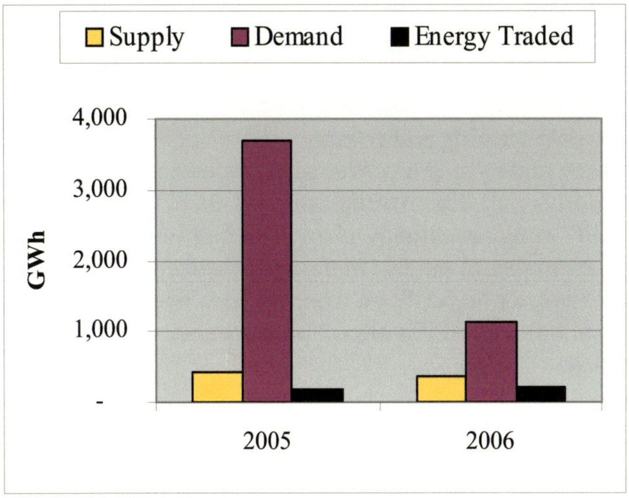

Fig. 10.3 Energy Trading Summary
(1 April to 31 March of the following year)

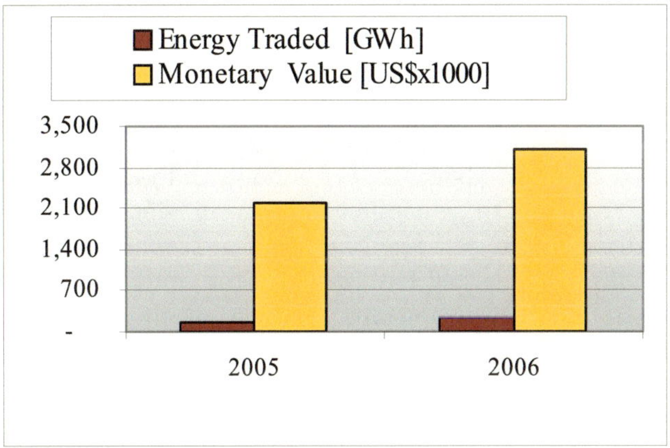

Fig. 10.4 Energy trade versus monetary value
(1 April to 31 March of following year)

The development of the competitive electricity market started in January 2004 when an Agreement between the Government of Norway and SAPP provided SAPP with a grant to the amount of NOK 35 million for this purpose. The competitive market will replace STEM. STEM was developed as a precursor to a full competitive market. The experience derived from STEM operations has formed the basis for the development and implementation of a full competitive electricity market for the SADC region [5,6].

10.4 Regional Challenges

Despite the stated achievements, the SAPP is still faced with the following key challenges that lie ahead as follows:

1) **Electricity sector restructuring and reforms**
   SAPP Members are undergoing a power sector reform process and the restructuring is taking various forms [4]. The restructuring of SAPP members will mean that the members of SAPP would eventually change as more utilities are unbundled by their governments. The number of players in the SAPP is likely to increase as a result and this will have a major impact on SAPP membership and operations. Whilst SAPP members are being restructured, the SAPP is also making a transition from a cooperative pool into a competitive pool.
2) **Electrification**
   Electrification and particularly rural electrification is the cornerstone for economic integration and development. The level of electrification for most SAPP member countries is less than 30% meaning that a lot of people have no access to clean energy. The challenge is to increase access to modern energy services and delivery.
3) **Human resource capacity and impact of HIV/AIDS on the utilities**
   This is a regional problem and it is affecting the operations of member utilities. More and more educated and trained people are dying as a result and replacing them is at great cost to members and the region as a whole.
4) **Diminishing generation surplus capacity**
   The biggest challenge that the SADC region is facing is the diminishing generation surplus capacity. In the last ten to fifteen years, power demand in the SADC region has been increasing at a rate of about 3% per annum. Unfortunately, there have been no corresponding investments in generation and transmission infrastructure to match the increase in the demand and as a result, generation surplus reserve capacity has been diminishing steadily over the past few years [7]. The continued diminishing generation surplus capacity in the SADC region would have a negative impact on the economies of the region and potential investors would be frightened.

The rise in the regional power demand has been caused by the following identified factors:

- Economic expansion in member states requiring more power to supply the new industries,
- Increase in population of most SADC member states coupled with increased electrification programs,
- Non-economic tariffs in most member states that do not support re-investments in power generation, but allow large energy intensive users to come to the SADC region and set up their operations, and
- No significant capital injection into generation and transmission projects from either the private or the public sector.

The total installed capacity in countries included in SAPP is about 55,000MW [see Table 10.3], but the available capacity is only 47,000 MW due to technical limitations. The dependable capacity is further reduced to 43,000 MW as the available hydro capacity varies depending on season and other constraints. The Peak Demand in 2006 was 42,000 MW resulting in load shedding in rather extensive parts of the region. Bearing in mind that there is a need for continuous reserves of above 4000 MW not included in these figures it goes without saying that the regional deficit situation is becoming a severe challenge for the utilities. From experiences globally such challenges are best met through a strictly formalized regional power cooperation, e.g. through power pools like SAPP.

No.	*Country*	*Utility*	*Installed Capacity [MW]*	*Available Capacity [MW]*
1	Angola	ENE	1,128	943
2	Botswana	BPC	132	90
3	DRC	SNEL	2,442	1,170
4	Lesotho	LEC	72	70
5	Malawi	ESCOM	302	246
6	Mozambique	EDM	233	174
		HCB	2,250	2,075
7	Namibia	NamPower	393	360
8	South Africa	Eskom	43,061	38,764
9	Swaziland	SEB	71	70
10	Tanzania	TANESCO	1,186	780
11	Zambia	ZESCO	1,737	1,200
12	Zimbabwe	ZESA	2,045	1,125
	Interconnected SAPP		**52,416**	**45,098**
	Total SAPP		**55,052**	**47,067**

Table 10.3 SAPP Installed and Available Capacity

In the period 2004-2006 a total of 1140 MW installed capacity was commissioned consisting of both constructions of new plants and upgrading of existing plants. In 2007 a further capacity of 1450 MW was installed, mainly in South Africa. Existing plans for the period 2007-2010 indicate rehabilitation and short-term generation projects of approximately 13,500 MW if sufficient funding resources are made available.

Even with an optimistic implementation rate for generation projects, the existing growth rate in electrical energy use of 3.6 % p.a. or between 1000 - 1500 MW per year will imply clear risks of further load shedding in parts of the region. Rather extensive Demand Side Management initiatives are consequently required and some have already been initiated, in particular in South Africa, with positive results.

A survey carried out in 2006 by the SAPP Coordination Center [8] reviewed that all the SAPP Member Utilities registered a positive growth in power demand during the period from 2001 to 2005 mainly due to the increase in economic activities in their countries. The Utilities' peak demand occurred almost at the same time and there was basically no load diversity in the interconnected SAPP grid and no benefits of time differences in the region.

From Figure 10.5, it is seen that the non-coincidental peak demand in the SAPP during the winter of 2006 was about 42,000MW against an available capacity of 45,000MW. Accordingly to the SAPP agreements, SAPP Members are required to carry a generation reserve margin of about 10.2%. This means that the maximum peak that the SAPP should reach is 40,400MW (i.e. 45,000MW available capacity less 10.2%). Therefore, the recorded 2006 peak should be set as the maximum peak that the SAPP could achieve with the available capacity. Unfortunately, the load is still increasing and the generation capacity is static indicating that the maximum peak in the coming years will rise beyond the stipulated limit. This is a demonstration of the diminishing generation surplus capacity that the region is now experiencing and should be reversed. Figure 10.5 also shows that in 2007, the SAPP peak demand equalled the available generation capacity and the region had not much reserve capacity to fall back on. Figure 10.6 also confirms what is highlighted in Figure 10.5 that the region runs out of generation surplus capacity in 2007. Figure 10.6 indicates the reserve capacity position in the SAPP if no new generation capacity is built in the next few years. In 1998, the SAPP had generation reserve capacity of over 11,000MW i.e., about 24%. Over the years, the generation reserve has been diminishing steadily due to the reasons given above and is expected to continue reducing unless new investment in generation infrastructure is done.

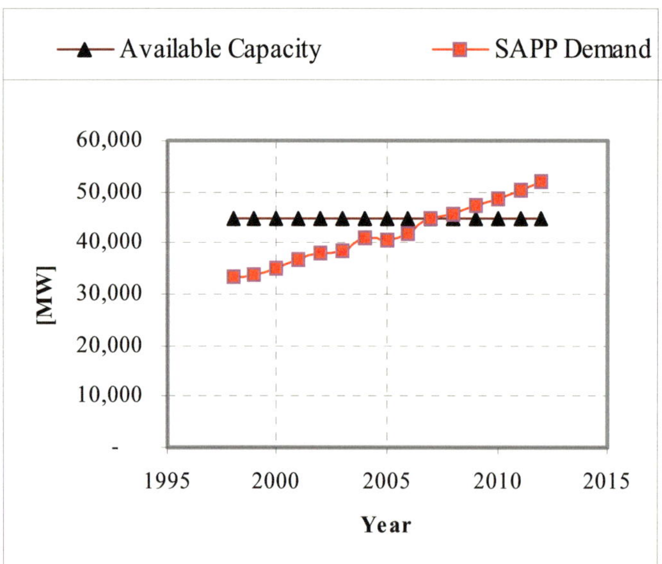

Fig. 10.5 Historic and forecast peak demand growth (1998 - 2012)

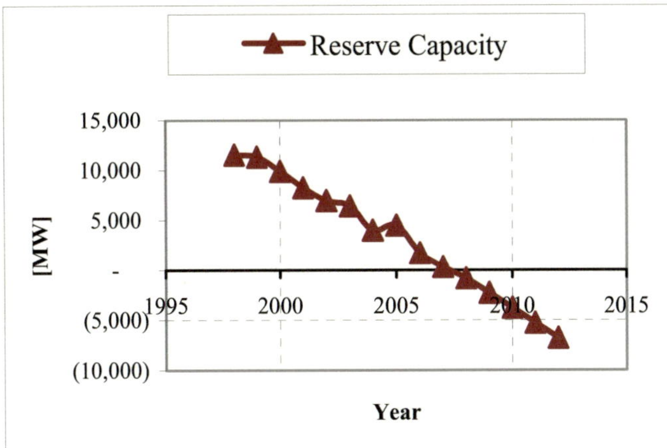

Fig. 10.6 SAPP Reserve Capacity profile: 1998 to 2012

In order to reverse the diminishing generation surplus capacity and to avert an impending energy crisis in the SADC region, the SAPP has put in place the following measures:

a) Implementing Priority Projects

The SAPP has formulated Priority Project Listing, which is expected to act as a project investment guideline to Investors, the Public and the Private Sector. The agreed SAPP Priority projects are as follows:

1) *Rehabilitation and associated infrastructure projects:* These are currently in progress and most of them are under construction. Once completed, they will add 3,200MW of power to the SADC grid. The estimated cost is around US$1.4 billion.
2) *Short-term generation projects:* These projects are expected to be completed in 2010. Feasibility studies and environmental impact assessments on the projects have been completed. Some of the projects have secured funding and for those with no funding available, the SAPP is sourcing funds via different initiatives. Once completed, the short-term generation projects will add about 4,200 MW to the grid at a cost of approximately US$3.8 billion.
3) *Transmission projects:* There aim is to interconnect the three non-operating members of the SAPP (Angola, Malawi and Tanzania) to the SAPP grid. The other and mainly internal transmission projects are aimed at relieving congestion on the SAPP grid and evacuation of power from the generating stations to the load centers. The north-south congestion program that was started is aimed at relieving congestion on the SAPP transmission grid between the north and the south, and also promotes and facilitates trade amongst SAPP member countries.
4) *Medium to long-term generation projects*: These are meant to supply power to the SADC region in the medium to long-term. Notable among them is the Western Power Corridor Project, Westcor, which is expected to move 3,500-4,000MW of power from Inga-3 in the DRC to southern Africa and to pick up 6,500MW of generation at Kwanza River in Angola.

b) Marketing of priority projects
The SAPP and NEPAD are working with the Ministers responsible for energy in the SADC region to market the priority projects and to attract funding for the short and long-term generation and transmission projects. A SADC Regional Electricity Investment Conference (REIC) was held in Namibia in September 2005 aimed at attracting investors into the SADC power sector. A follow-up conference was planned in the following year.

c) Energy Regulation and Tariffs
The Ministers responsible for energy in the SADC region pledged to address regulation, implement cost reflective tariffs and adopt regulatory principles that would enhance those tariffs. Political support from the SADC governments is essential for cost reflective tariffs to be implemented. A tariff study has been initiated by SAPP. The objective of the study is to review the tariff setting principles used by SADC governments and their national power utilities and to compare them with best practices from around the world. The study will also review the issues surrounding tariff settings and electricity pricing including the role of the regulator in those countries with a regulator and the importance of having a regulator in some cases.

In order for the SAPP to complete the projects in progress and those under rehabilitation and to implement the short and long-term projects, an estimated total of US$43 billion would be required, as indicated in Table 10.4.

SAPP Generation Projects	*Capacity [MW]*	*Estimated Cost [US$ Million]*	*Period of Implementation*
In Progress	3,211	1,410	2005 - 2007
Rehabilitation	1,048	523	2007 - 2010
Short-term (New Build)	4,217	3,830	2005 - 2010
Long-term (New Build)	43,542	37,585	2011 - 2020
Total Planned Capacity	52,018	43,348	

Table 10.4 Cost and Timing of the SAPP Projects

In South Africa for example, in order to deliver the required capacity, Eskom plans to spend over R97 billion (about US$14 billion) over a 5-year period in capacity expansion including rehabilitation. The return to service of the three-mothballed power stations Camden, Grootvlei and Komati were completed before the end of July 2007. Major capacity expansion in South Africa will include new coal fired base load stations, new pumped storage technology, open cycle gas turbines (at Atlantis and Mossel Bay), nuclear, and the associated transmission lines. The open cycle gas turbines at Atlantis and Mossel Bay was completed before the end of April 2007.

The SAPP is currently faced with the challenge of a diminishing generation surplus capacity. The continued diminishing generation surplus capacity will have a negative impact on the economies of the SADC region if it is not reversed. In order to reverse the diminishing generation surplus capacity, the SAPP has developed a program of implementing the priority generation and transmission projects so as to avert an energy crisis. The success to the implementation of this program is key to the development of the region, noting that energy is the cornerstone of development.

10.5 Acknowledgements

Lawrence Musaba, Coordination Center Manager, Southern African Power Pool, Harare, Zimbabwe and Pathmanathan Naidoo, Senior General Manager (Special Projects), Office of the Chief Executive, Eskom, Johannesburg, South Africa, and Chief Executive, The Western Power Corridor Company Ltd., Gaborone, Botswana; has prepared and coordinated this Chapter. Contributors include Alison Chikova (System Studies Supervisor, Southern African Power Pool, Harare, Zimbabwe), Thomas J. Hammons (Chair International Practices for Energy Development and Power Generation IEEE, University of Glasgow, UK) and colleagues at Eskom and the Southern Power Pool, Zimbabwe.

10.6 References

1. Musaba, L., *Energy Trading in SAPP,* Energize - Power Journal of the South African Institute of Electrical Engineers, July 2003.
2. Musaba, L., *Electricity Trading in SAPP and the Short-Term Energy Market,* Conference proceedings of the SADC Energy Investment Conference and Exhibition, Victoria Falls, Zimbabwe, 2-6 Sept. 2001.
3. *Principles on Standard Market Design.* Edison Electric Institute. July 2002.
4. Musaba, L., *Power Sector Reforms in SAPP Member Countries,* Proceedings of the Power Africa Summit 2004, Midrand, South Africa, 7-10 June 2004.
5. Musaba, L., Naidoo, P., Chikova, A., *Towards Developing a Competitive Market for Regional Electricity Cross-Border Trading: The Case for the Southern African Power Pool,* IEEE PES Annual General Meeting, Denver, USA, June 2004, Conf. Record.
6. Musaba, L., *Cross-border trading with the Southern African Power Pool,* IEE Power Engineer Journal, pp.30-33, April/May 2004.
7. Musaba, L., Chikova, A., Naidoo, P., *A continent in crisis - Effective Planning for Security and Stability of Regional Power Supply,* Proceedings of the Pan African Power Congress 2007, 18-20 April 2007, Johannesburg, South Africa.
8. *Report on Performance of SAPP Members 2001-2005,* Sept. 2006.

11

Electricity Infrastructure in Asian Region and Energy Security Problems

11.1 Introduction

The Asian region is a promising region for creation of a joint interstate electricity infrastructure. It embraces the industrially developed (Japan, Republic of Korea) and intensively developing (China, India, Vietnam) countries. The distribution of and demand for the fuel resources for electricity production do not coincide geographically. In the last years a number of bilateral studies (Russia-Japan, Russia-Republic of Korea, Russia-China) have been carried out. The studies have dealt with the potential options of interstate electric ties in East Asia. There are also a number of conceptual studies on potential interstate electricity infrastructure in East Asia (Russia, Japan, China, Republic of Korea, KPDR, Mongolia), South Asia and Oceania (Vietnam, Thailand, Myanmar, Indonesia, Philippines, Kampuchea), India with neighboring countries. All these studies form a basis for interstate cooperation in the sphere of the electric power industry and for the increase of interstate electricity exchanges, which is beneficial for all the participants.

At the same time each country encounters the problem of energy security. Energy security is the immunity of the country, its economy, society and citizens from the threats of impossibility to fully meet their energy demands by economically accessible energy resources of an admissible quality from the threats of energy supply interruption. An important aspect of energy security in any country is its energy independence that can be provided by diversified external sources of energy resources, including electric power.

This Chapter deals with the answers to the following questions:

- What are the views of different countries in the Asian region of interstate electricity infrastructure development?
- Is the problem of energy security a limiting factor for interstate electricity infrastructure development in the Asian region?
- What are the views of different countries in the region of the impact of this limitation?
- Do the growing interstate electricity exchanges in the Asian region play an important role for increasing energy independence of the countries in terms of other energy resources?

11.2 Energy Security as a Factor of the Common Energy Cooperation in East Asia

Energy Security (ES) implies here protect ability of a person, society, state, economy from threats of deficiency in meeting their energy demands by economically accessible fuel resources (FR) of an acceptable quality, from threats of disturbances of stable and uninterrupted power supply. The indicated state of protect ability corresponds to the full meeting of the reasonable demands under normal conditions and to the guaranteed meeting of the minimum required demands under extreme conditions. In so doing, the reliable and guaranteed external supply of energy carriers is the most important component of ES for the countries and regions with insignificant reserves of local FR (for instance, Japan, Central and Volgo-Vyatsky areas of Russia); energy independence: capability to manage with local resources at a loss/decrease of external supplies - for the countries and regions with a middle level of provision (USA, UK, Povolzhie and Ural areas of Russia, etc.); for relatively well provided territories (Russia, Canada, Turkmenistan, Iran, West Siberia, etc.) the following four factors which also play an important role for the mentioned above territories are decisive in the ES support:

a) Capability of the fuel and energy complex (FEC) to supply energy carriers continuously and in sufficient volume, creating energy prerequisites for stable operation and progressive development of the economy and maintenance of the adequate living level of citizens;

b) Capability of consumers to efficiently consume energy, to limit energy demand, decreasing it, thus to prevent energy balance tension and deficiency;

c) Balanced FR supply and demand with regard to economically feasible export and import of FR;

d) Favorable socio-political, legal, economic and international conditions for realization of the above capabilities by FR producers and consumers.

It seems to be a common practice that ES is considered an important component of economic security being in its turn one of the basic constituents of the national security of each country.

Analysis of the up-to-date state and trends in the development of FEC and its industries as well as conditions of their development and functioning has allowed one to reveal a wide range of threats to energy security of Russia. These threats grouped in 6 classes are presented in Table 11.1 (in more detail about the subject-matter, problems of ES and threats to ES (see [1-6]).

The most important economic threats are sharp shortage of investment resources, entailing insufficient volumes of new capacities put into operation (3-5 times less than the minimum required ones), reconstruction and reequipping in the FEC industries; non-compensated retirement of capacities; sharp decrease in prospecting and, hence, a delay between increase in explored reserves, in particular concerning oil, and production volumes; operation of inefficient and deeply worn equipment. Mitigation of these threats to ES is one of the basic incentives to activate strategic efforts of Russia for formation of the common energy spaces in Europe and East Asia and its integration in them.

Russia is still the major energy country (Table 11.2) and its FEC is the most important component of the economy (Table 11.3) [7,8].

Russia can fully meet its demands for fuel resources and supply their considerable amount for export. Particularly it concerns oil and gas. Table 11.4 presents corresponding data for the pre-crisis 1990 year. Though the internal consumption, production of energy carriers and their export to some extent be considerably reduced due to the economic crisis, still the situation has not changed qualitatively. In 2003 Russia exported 224 million t of oil; 189 billion m^3 of gas; about 78 million t of oil products; about 66 million t of coal and 12 billion kWh of electric power. However, this export (as well as the energy co-operation on the whole) is oriented mainly to the countries of West and Central Europe and the European countries of CIS (Ukraine, Belarus, etc.).

Group of Threats	Threats
Economic	• Shortage of investments. • Uncontrolled energy use by the economy. • Monopolization of energy markets. • Price disproportions. • Low technical level of energy equipment. • Weak energy-transport ties with energy imbalances of regions. • Excessive concentration and centralization in energy. • Weak diversification of energy supply. • Insufficient amount of energy reserves and stocks. • Leading growth of FR demand.
Socio-political	• Labor conflicts, strikes. • Political, ethnic conflicts, terrorist acts. • Ecological extremism. • Limitation of free flows of energy goods between the regions. • Conflicts between different levels of authorities, separatism. • Low qualification, discipline of personnel, carelessness. • Criminalization in energy sphere.
External economic and external political	• High dependence of FEC on imported equipment and materials, disruption of delivery. • Dependence of energy supply to individual regions on external FR supplies. • Discrimination measures against the Russian FR. • Critical dependence of the Russian FR exports on conditions of their transport.
Technogenic	• Ageing, wear of equipment. • Accidents, explosions, fires on the FEC objects. • Accidents, explosions, fires on the objects of other branches connected with the FEC objects.
Natural	• Natural disasters (earthquakes, storms, etc.) • Severe winters. • Long-term low waters, particularly multiyear ones, on the rivers with hydro power plants.
Managerial-legal	• Inefficient economic and social policies, mistakes in their realization. • Weak mechanisms of antimonopoly policy and regulation of natural monopolies. • Incompleteness and imperfection of energy legislation. • Weak state control. • Excessive state interference. • Inefficient energy saving policy. • Poor management quality of enterprises, companies, corporations.

Table 11.1 Main Threats to Energy Security of Russia

Fuel Resource	Fraction in World Explored Reserves, %	Fraction of World Production, %
Oil	13	11
Gas	33	24
Coal	20	6

Table 11.2 Role of Russia in the World Energy (2003)

Unit	Percent
In industrial production	31-30
In receipts to federal budget	46-42
In export	54-60
In production investments	26-31
In number of production personnel	13-14

Table 11. 3 Share of Fuel and Energy Complex in the Structure of the Russian Economy in 2002-2003, %

There are two arguments in favor of diversification of the Russian external energy ties by their extension in the eastern direction. The first consists in the fact that the eastern territories of Russia (East Siberia and Far East) possess a sufficient energy potential for development of FR export to the East Asia countries. Whereas West Siberia meets local oil and gas demands, the main demands of European Russia and also exports these energy resources to the West, the East-Siberian and Far-Eastern oil/gas complexes under formation as well as the electric utility industry of these regions based on hydro energy and coal can both meet the local needs (and transmit a part of electric power from Kansk-Achinsk Fuel and Energy Complex to the West) and develop the eastern direction of the Russian energy policy.

Energy Carrier	Production	Import to Russia	Export from Russia	Net export
Oil, including gas condensate, million t	516.3	16.8	210.2	193.4
Natural gas, billion m^3	640.2	31.3	212.0	180.7
Coal, million t	396.3	43.2	53.6	10.4
million .c.e		28.3	35.2	6.9
Basic oil products (diesel fuel, motor and avia-tion gasoline, engine fuel, fur-nace and marine residual oil), million t	239.3	10.1	65.8	55.7
Electric power, billion kWh	1082.2	...	...	8.4

Table 11.4 Some Indices of the Russian Fuel Balance for 1990 [9]

Two considered eastern regions of Russia (ERR) - East Siberia and Far East with a territory (10.3 million km^2) making up 60% of the whole country and population of 16.7 million people that produced 13% of GDP of Russia in 1995 possess the major reserves of natural energy resources. The initial potential oil resources of ERR within the shelf of the Far-Eastern and arctic seas are estimated approximately at 17.8 billion t; and those of natural gas are estimated approximately at about 56 trillion m³. However these resources are mainly the forecasted ones whose exploration extent is low. As for the reserves in commercial categories, the fraction of ERR for oil is more than 17.5% of that for the whole of Russia (basic reserves are in West Siberia); and natural (free) gas more than 20%, respectively [10]. In magnitude the gas reserves of ERR are about 5 trillion m³ according to [10], the oil reserves (only in East Siberia) are about 1 billion t according to [7]. Based on these reserves potential oil production in ERR in 2010 is estimated at about 34-42 million t (with active participation of foreign investors). In the more remote future it can reach 70-75 million t/year, the export resources will be about 40 million t. Gas production in 2010 in ERR is estimated at 30-60 billion m³, including about 20 billion m³ on the Sakhalin shelf. Export potentialities of ERR in the more distant future are estimated at 50 billion m³ /year [7].

Eastern regions of Russia possess the largest explored balance resources of hard and brown coals, more than 17 billion t. The economically efficient hydro power potential of ERR is 75% for the whole of Russia, i.e., more than 640 billion kWh, more than 135 billion kWh (33% in East Siberia and 6% in Far East), including the hydro power plants under construction have been realized. Possible scales of electric power production in ERR at the level of 2010 reach 255-260 billion kWh/year at local consumption of up to 230 billion kWh (about 180 billion kWh in 1990) [7]. The difference forms an essential part of the export potential.

The second factor of changing the priorities in the external energy policy of Russia is a growing role of the Asia-Pacific region, in particular of the East Asia countries, deficient in energy resources. There is a stable tendency in the world economy to turn this region into the most important center of the world economy.

№	Project	Capacity, million t/year	Length, km	Diameter, mm	Investment, US$ million	Project participants
1	Taishet- Perevoz-naya bay	50	3885	1020-1220	5817	Transneft
	Including: Taishet -Skovoro-dino	80	2047	1020-1220	3430	Transneft
2	Sakhalin-1 (De Kastri-Kom-somolsk-on-Amur)	12.5	207	500	500	Exxon Neftegaz Rosneft Rosneft-Sakhalinmorneftegaz ONGC SODECO
2	Sakhalin-2 (Yuzhno-sakhalinsk-Prigorodny)	10	800	500	1000	Sakhalin Energy Investment Company, Ltd Marathon Mitsui Shell Mitsubishi

Table 11.5 Prospective Projects on Construction of Main oil Pipelines in East Siberia and the Far East with Penetration to Foreign oil Markets

Import of energy resources is crucial for Japan and South Korea, since these countries have practically no natural fuel resources of their own. However, during the recent years Japan has been intensively searching for alternative gas sources, which could at least partially reduce dependence of the country on foreign supplies. In particular, the program for prospecting and development of the gas hydrate shelf fields near the sea coast of Japan has been elaborated. By now 12 hydrate-bearing areas of the shelf, containing about 6 trillion m^3 of methane [11] has been explored. But the commercial exploitation of the gas hydrate fields is a matter of the distant future. China cfn be is important importer of fuel and energy resources in nearest future.

Russia also pursues its economic interest in East Asia region. In this connection in different time joint discussions and work on a number of energy projects "Russia-East Asia countries", which are at different stages of realization, were started. Some characteristics of these projects are given in Table 11.5.

Having considered the subject matter and main incentives of the Russia-East Asia energy export-import co-operation it is necessary to estimate it in terms of energy security (of Russia and its partners) and partially in terms of economic and ecological security.

Positive arguments for the East Asia countries consist in the following:

1. Provision of economically beneficial balance of their energy supply.

2. Provision of diversification of sources of hydrocarbons supplies: Russia, objectively interested in stable East-Asia market for its oil and particularly natural gas joins the conventional sources - countries of Persian Gulf, Africa and partly Southeast Asia and Australia (gas). In so doing the Russian sources themselves are also diversified (Sakhalin, Irkutsk region, Yakutia).

3. Improvement of the structure of fuel balances of the East Asia countries by using environmentally clean fuel (gas) of higher quality and supplying "clean" (for these countries) electric power as an alternative to its production by the coal-fired thermal power plants of their own and on the whole as an alternative to development of coal technologies. Though this idea is formulated in terms of ecological security, it is important for energy security: firstly, the latter implies both the quantitative meeting of demand and acceptability of energy carrier quality (see above the ES definition); secondly, the considered improvement of the fuel balance structure reduces any energy feelings and movements in the society, ecological extremism - one of the major threats to ES (see Table 11.1).

4. Extension of the possibilities for the East-Asian companies to penetrate into the Russian markets of investments, equipment, technologies and other goods and services. Generally speaking, this argument testifies to the advantages of the considered projects in terms of economic security of the East Asia countries. At the same time such penetration favors commercial success and development of energy machine building, R&D works in energy and associated spheres of activities of these countries, which is obviously important for strengthening of their ES.

5. Improvement of the EPS reliability of the East Asia countries, reliability of their power supply and achievement of the other known "system" effects due to interconnection of

power system at implementation of electric power projects (Table 11.6). Realization of these projects will lead to replacement of some capacities of power plants, particularly thermal ones, in Japan, South Korea, China at the expense of construction of power plants in East Siberia and Far East, rich in hydro resources. Correspondingly the environmental situation in the East Asia countries is improved.

The considered projects of EPS interconnection allow the countries-importers of FR to diversify their import at the expense of electricity, since electric power supplies can replace the shortfalls in supplies of some fuels. It undoubtedly strengthens their energy security.

We believe that the objections to the considered projects and co-operation from the ES viewpoint of East Asia countries are the following:

1. There is a danger of either monopoly position of Russia in FR supplies to the market of any country or its extremely large fraction enabling one to use these supplies as an instrument of economic or political pressure. But as is seen, in particular from comparison of the data in Tables 11.5 and 11.6 this situation is hardly probable.

2. The interruptions of the continuous FR supplies due to insufficient political and social stability in Russia are possible. However, one can suppose with a high degree of certainty that by the beginning of the period of supplies under the considered projects the situation in Russia will become more stable. Besides, the regions of conventional FR supplies to the East Asia countries do not belong at all to the politically stable ones, taking into account growing fundamentalist tendencies there, interstate and interethnic conflicts, etc.

Let us consider positive arguments in terms of ES of Russia and its East- Siberian and Far-Eastern regions.

1. The basic argument which has been already mentioned is overcoming (mitigation) of the threat (to ES) of deficient investments, possibility to get direct investments, credits for development of the Russian, first of all the East-Siberian and Far-Eastern energy resources in the amounts, sufficient for their export to East Asia, their supply to the internal market and for meeting the local demand for hydrocarbons, including creation of the appropriate transport and other infrastructures. Besides, the expected investments and revenues from the export should and can be used for updating and reequipping the FEC industries and enterprises, as well as other branches of the economy (especially in frames of energy saving programs), and for solving social problems in restructurisation of the energy sector in the eastern areas of the country.

2. The possibilities for introduction of efficient technologies and equipment for solution of the problems indicated in item 1, for updating and reconstruction of the production capacities of FEC as a whole are extended and hence, the threat of its low technological level and deep wear is reduced (Table 11.1).

3. A new stable market of the Russian FR is being formed. This fact is directly connected with the financial and external economic security of Russia contributing indirectly to provision of its ES.

4. Electric power effect, i.e. mutual effect due to interconnection of power systems that is similar to the described one for the ES of the East Asia countries (item 5) is attained.

5. Development of the economic (energy) co-operation in the region, determined by performance of the considered export projects will promote implementation of more progressive forms of co-operation. Though this argument concerns first of all the economic security; support of the Russian energy machine building, design and construction organizations, associated with such projects is of great importance for ES of Russia.

6. Realization of export projects reduces the social tension in Russia, especially in its eastern regions which is an essential threat to ES, since it allows one to raise the level of employment for the population, particularly the skilled workers owing to construction of energy objects and creation of the corresponding infrastructure, their servicing, production of materials and equipment, etc.

Finally, the negative arguments (objections) in terms of ES of Russia can be formulated as follows:

1. The most frequently expressed objection consists in the fact that the currently considered gas projects take into account the local gas needs insufficiently. In particular similar remarks concern the Sakhalin projects, since they mainly solve energy problems of the countries-investors and on the territory of Russia meet local demands only and do not lead to a radical change to better in energy supply of the south of Far East.

2. The threat of premature depletion of highly efficient fields of non-renewable natural resources (oil and gas), a failure to preserve them for future generations, and, hence, potential weakening of ES of Russia and particularly of its eastern regions in the middle of the 21st century seems to be serious enough. In fact, it concerns commensuration of the today's effects (including those from the ES view point) and future losses.

3. An extreme influence of foreign owners of the Russian energy enterprises on the decisions made by Russia in the energy sphere and on utilization of the strategic resources on the whole can form a definite threat while realizing export projects. This and partially two previous threats can be overcome by the thoroughly developed legislation, comprehensive substantiation of the corresponding agreements, state participation or state control at the regional and municipal levels, in the process of preparation and realization of agreements, maximum possible publicity and public control in the given sphere.

4. Export of foreign technologies and equipment within the framework of energy co-operation leads to a great dependence of Russian Federation on supply of spare parts from abroad. Licensing and arrangement of production of extremely important spare parts and units by the Russian industry including joint ventures, which are envisaged in the corresponding agreements, could contribute to elimination of this threat.

On the whole it should be pointed out that the enumerated aspects, particularly those of the negative influence on ES are taken into account insufficiently in the currently considered projects. Elimination of this drawback is still an urgent problem to be solved.

11.3 Energy Security in the Asia-Pacific Region

Energy security may be achieved when a state is able to minimize vulnerability to resource supply disruptions, access reliably energy at reasonable and/or market-driven prices, and consume resources that least damage the environment and/or promote sustainable development. By extension of this broad definition, energy becomes a security concern when states are denied access—whether it is the actual resource itself or by way of volatile and/or unfair pricing. There exists a vast literature on the history, politics, and economics of resource consumption in the world, and how states may act in an effort to secure their needs. Relatively little is written, however, on understanding how, why, and whether energy—as the unit of analysis—triggers competition or cooperation among states at the national security level.

By 2010, energy use in developing Asia (including China and India, but excluding Japan, Australia, and New Zealand) is projected to surpass consumption of all of North America; by 2020 it is expected to exceed North American consumption by more than 36%. The Asia-Pacific region will consume more than half of the world's energy supply, and will emerge as the dominant energy consumer by the early next century. Some scholars argue that the region's growing energy needs have led to new strategic relations with other parts of the world, especially the Middle East, and have raised new questions about the reliability of the international market system in providing predictable and affordable access to energy resources.

What are the pressing energy issues of the Asia-Pacific region? Do energy needs pose new challenges to Asia-Pacific security? How important is energy as a source of tension between states, or are energy security matters considered "low politics" which have a higher tendency toward resolution rather than conflict?

In an attempt to explore these and other questions, the Asia-Pacific Center for Security Studies held a one-day seminar on regional energy security on January 15, 1999. The specific purposes of the seminar were to assess the current and future energy outlook for the region, identify the salient factors that influence energy security, and evaluate whether new aspects of the energy scenario raise new security challenges. The seminar was organized into four sessions: "Outlook on Energy Supply and Demand in Asia," "Energy Constraints and International Politics on Key Asian States" (two case studies on China and India), "Comparative Perspectives on the Politics of Energy Resource Management in the Asia-Pacific Region," and "Concluding Discussions: Lessons Learned." This seminar report draws on the papers, presentations, and subsequent discussions held during the one-day proceedings. It also incorporates the broader literature on the subject as it relates to matters discussed at the seminar.

In short, discussions led to a general conclusion that the energy security debate of the nineties is less about energy, per se. Unlike the energy security debates of the 1970s and 1980s, which focused on supply shortages as a source of conflict and competition, rivalry and competition over energy itself is a non-issue in today's debate. In fact, some would go so far as to argue that the common challenge of greater external reliance on energy supplies among Asian states would create incentives to cooperate, not compete. The security concerns, rather, arise from issues related to the access, transportation, and reasonable prices and distribution of energy, specifically, the changing patterns in trade, a greater reliance on the Middle East for oil and thus a greater reliance on open access to sea-lanes, and shifting strategic relationships.

Overall, discussions were not pessimistic about the future of Asia's energy security. The general sentiment seemed to be that the identified problems were not entirely unavoidable if countries applied a constructive use of diplomacy during tensions; shaped national policies based on facts - not perceptions - regarding resource potential in certain parts of the region; and made commitments to utilize relationships already in place, such as bilateral alliances, to explore energy security questions.

Section 11.3 presents an overview of the energy security in the Asia-Pacific region [12-20].

11.3.1 Why Energy Security in Asia

The view that increasing competition for energy resources, a consequence of increasing Asian economic growth, is producing growing insecurity in the Asia-Pacific region is best proffered by Kent Calder in Pacific Defense, his 1996 analysis of the US role in the future of Asia.

Calder argues that economic growth gives Asian nations the resources to strengthen their military might, but that it also results in rising energy demands, and the resulting need to secure stable energy supplies in competition with one's neighbors increases global insecurity and a region-wide arms buildup.

Petroleum, coal, and natural gas continue to be in insufficient supply in Asia, which provides only 11 percent of global oil production and 4.5 percent of reserves. Japan, with half the region's economic output, remains 95 percent dependent on oil imports. The growing Chinese economy's hunger for energy will soon make that country a net oil importer despite its status as the top supplier (with Indonesia) of energy in Asia. And increasing demand among other countries in the region will intensify competition for oil supplies and raise insecurity about neighbors' plans to ensure a supply of energy.

More important than rising Asian dependence upon Middle East oil-producing nations per se (an East-West center study estimates that Asia's share of oil imports from the Middle East will rise from 70 percent in 1993 to 95 percent in 2010) is the tension surrounding reliability of access to shipping lanes from the Middle East.

Asia Pacific region use increased with the strongest growth (6.3 percent) in 2003. Among fossil fuels, coal grew fastest in 2003, with an increase of 6.9 percent, largely due to a reported increase of more than 15 percent in China. Chinese oil demand has also doubled over the past 10 years, leading BP's chief executive to conclude in his foreword that China "will be a major influence on the world energy scene from now on."

The approaches to the Strait of Malacca (for smaller tankers) and the Lombok and Makassar Straits in Indonesia (for larger tankers) are surrounded by Southeast Asian nations (Malaysia, Indonesia, and Singapore), which control those straits, and adjacent waters with increasing naval might. Calder maintains that China's strengthening naval presence and territorial claims to waters of the South China Seas, reflecting its own desire to secure shipping lanes for its energy supply and trading routes, will likely further heighten tension in the waters of Southeast Asia.

One solution to the energy demand crisis in Northeast Asia is nuclear energy. However, growing civilian nuclear power programs raise the risk of diversion of nuclear materials for military purposes, as is widely feared in North Korea. Northeast Asia includes three nuclear weapons states (the U.S., Russia, and China), and Japan and South Korea maintain large and growing civilian nuclear programs, which further contribute to anxieties in the region.

11.3.2 A Region at Risk: The Asia-Pacific

The dramatic geopolitical shifts stemming from the end of the Cold War and the global war on terrorism in the wake of September 11th have resulted in an abrupt restructuring of the traditionally bipolar system of global governance that has served as the norm for 20th century. Of all the regions subject to the repercussions of this new geopolitical landscape, the Asia Pacific region has emerged as one of the key arenas. A convergence of new factors, ranging from the threats posed by Al Qaeda to the sweeping engagement of the U.S. military throughout the region, has endowed the region with a significantly enhanced strategic importance.

The implications for the Asia-Pacific region from within this new prism of global geopolitics and a greater reliance on military security have also been deepened by several underlying characteristics. Specifically, the Asia-Pacific has seen a pattern of increasing insecurity in recent years that has exposed the absence of any regional institution capable of forging common and cooperative security. This pattern of mounting threats has been marked by three escalating crises: the Taiwan Straits crisis in 1996, the Asian financial crisis of 1997-1999 and the recent North Korean nuclear crisis. There is also a danger of a fourth crisis, involving Chinese frustration with the intricacies of Taiwan's political ambitions.

This absence of a governing regional structure has only exacerbated the region's vulnerability within a new post-Cold War/post-September 11th threat matrix. Although there has been some attempt to address this regional insecurity through existing regional organizations such as the Association of Southeast Asian Nations (ASEAN), the regional states still lack the political will, military capability and experience to adequately enforce security in any significant multilateral approach. And as the only substantive security architecture in the region is limited to the web of bilateral security treaties centered on the United States, there is a serious need for a new security regionalism. Such an effort can link Asian-Pacific economic cooperation to a regional security process and also build on the regional powers of Australia, Japan and South Korea, each of which have been recently "deputized" by the United States. Therefore, energy security may offer the most effective avenue toward this "securities regionalism," especially given the genuine level of cooperation and shared interests in seeking adequate and secure supplies of energy. Such a need for regionalized security is also reflected in the less visible security challenges facing the Asia-Pacific region. These security problems are concentrated in the core of the region, in the very foundations of the still incomplete state- and nation-building process, and stem from the fragility and weakness of these states. Coupled with the economic, social and environmental issues in the region, the complexity of these threats requires a multilateral, yet regionally based approach.

11.3.3 The Economics of Energy Security

In terms of pure economics, the outlook for energy security in the Asia-Pacific looks particularly troubling, with rising levels of oil consumption and an even stronger rise in demand, Figure 11.1. Some experts, such as Ji Guoxing of the Shanghai Institute of International Strategy Studies, contend that the Asia-Pacific region's dependence on Middle Eastern oil may exceed 90% by 2010. While oil fields in Russian Siberia and Central Asia do offer some short-term energy relief, the lack of existing infrastructure to facilitate the transport of this oil poses costly political and economic challenges of their own.

Aside from the dependence on imports from the Middle East, there is also a danger of tension stemming from such an oil shortage within the Asia-Pacific region itself. The growing demand for energy may strain relations between such important regional actors as China and Japan, for example, which may then engender a set of new destabilizing regional or international conflicts. But an even more immediate problem is the effect of oil market volatility on the region, with the sharp rise in oil prices putting particular pressure on the currencies of some crude importing emerging market countries and the dangers of soaring current account deficits and weaker economic growth. This also threatens to impact the record of growth that has served as the driving force for Asian stability and development since the end of World War II. And while Asia is seen as the most affected region, the surge in oil prices also threatens other struggling oil importers.

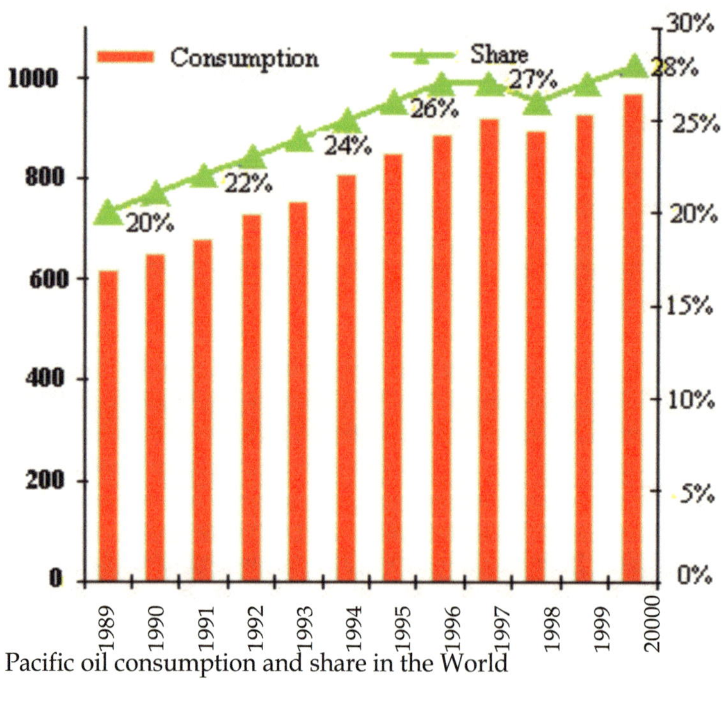

Fig. 11.1 Asia Pacific oil consumption and share in the World

11.3.4 Regional Energy Security in the Asia-Pacific

Energy security in the Asia-Pacific remains a complex and multifaceted challenge, with four main strategic issues mandating coordinated action:

1) Measures are needed to reduce Asian dependence on fossil fuel or to secure an adequate alternative supply to meet rising demand,
2) The need to address the environmental impact of the region's energy structure, as seen by the environmental repercussions from the heavy coal use in Chinese industries, for example,
3) The necessity for ensuring nuclear security in the face of regional ambitions to expand nuclear power, and
4) Specific policies to improve the vulnerable regional energy infrastructure and transportation networks, as well as safeguarding vital sea-lanes and "chokepoints."

As demonstrated by the set of four strategic priorities areas listed above, regional energy security in the Asia-Pacific requires a multilateral approach. There is a potential for regional cooperation, stemming from the convergence of national interests in the face of recent transnational threats. Much of these shared interests and threats have only been revealed in the aftermath of September 11 and the ensuing global "war on terrorism."

To date, the regional approach to Asia-Pacific energy security has been focused on petroleum security, conservation and the search for alternative fuels. Specific examples of regional cooperation are largely through the Association of Southeast Asian Nations (ASEAN), and include a Petroleum Security Agreement, requiring ASEAN member states to provide crude oil and/or petroleum products for countries in short supply. Studies for a Trans-ASEAN Gas Transmission System and an ASEAN Power Grid have also been initiated aimed at ensuring a reliable supply of energy to the region, with some notable progress to date related to cooperation in natural gas use and energy management.

Regional energy security was formalized as a priority issue at an Asia-Pacific Economic Cooperation (APEC) Energy Security Initiative Workshop on "Elements of Energy Security Policy in the Context of Petroleum," held in Bangkok, Thailand in September 2001. Dr. Piyasavasti Amranand, the Secretary General of Thailand's National Energy Policy Office (NEPO), reported to the APEC workshop that the current imbalance between reserves, production, and consumption of oil within the region has elevated oil security as a major concern for APEC officials. Amranand stated that the total reserves in the APEC region are far less than regional demand, exacerbating the regional dependence on oil imports, especially from the Middle East, therefore, made energy security a key element in establishing economic development policies.

Thailand has long been sharing information with the Asia-Pacific Energy Research Center (APERC) and other research centers, such as the ASEAN Center for Energy (ACE), and has also implemented other measures that have substantially enhanced the energy security of the country. Strategic oil stockpiling by the Thai private sector is one of the measures, but there is an inadequate government role in developing a state-owned stockpile.

The 2001 APEC workshop also recognized the security of tanker traffic as a main concern. In an address to the workshop, APERC President Tatsuo Masuda explained that the combination of vulnerable transport from the Middle East and West Africa with the fact that tankers are getting smaller, while the number of tankers crossing the Indian Ocean to Asia triples or quadruples, necessitates a reduction of the risks posed by tanker traffic. Masuda specifically pointed to the need for pipeline infrastructure projects connecting Russia, China, Korea, and Japan, as a means by which to reduce this risk.

11.3.5 Building Energy-Strategic Relationships

11.3.5.1 China

China's interest in expanding its resource linkages with Central Asia, Russia, and the Middle East will be better integrated. The government plans to promote market penetration into these areas in an effort to secure oil supplies. Given the political environment of the Persian Gulf, China plans to choose and develop better relations with "niche markets" such as Iran and Iraq while maintaining traditional relationships with other markets such as Sudan and Nigeria. The Chinese government has also enhanced its relations with Arab producers in the Gulf and North Africa, where approximately 36% of the world's oil reserves are located. In order to gain access to the Arab market, Chinese oil construction services and technical support units have been expanded in Kuwait, Iraq and several other Arab countries. At the same time, Arab producers are encouraged to enter Chinese offshore upstream and downstream projects. In the future, China plans to establish new linkages with Arab producers, which comprise sea shipments, land pipelines, and other investments.

China's past reluctance to see Russian resource development in Eastern Siberia has changed with its growing interest in the resources of Eastern Siberia and the Russian Far East region. Beijing expects gas imports from Eastern Siberia to double by 2010. Efforts toward building a strategic partnership between China and Russia have led to signed commitments in 1996 for major oil and gas pipelines projects with over US$20 billion in investment requirements. Sino-Russian relations remain complicated, however, by political and diplomatic uncertainties. It is expressed concern that a Sino-Russian confrontation could emerge in the long-term as pipeline projects routed east-west from Central Asia to China may undermine Russia's historical authority and control over resources in the region. Others raised the specter of possible competition among the United States, EU interests, China, and Russia over control of resources in the Central Asia/Caspian Basin area.

11.3.5.2 India

Coal and oil constitute India's primary energy sources. Figures for 1997 indicate that the share of coal in total primary energy consumption was about 56.2% and the share of oil was about 32%, making up almost 90% of India's total energy needs. Energy consumption has kept up with the pace of economic growth of about 6% since the post-reform period beginning in 1991. After the oil shocks of 1973, India's energy sector became heavy controlled by the state. A number of foreign companies were nationalized, and a number of public sector undertakings started in the coal, oil, and electricity sectors. However, rapidly growing energy demand outpaced the public sector's ability to provide adequate supplies. In an effort to attract foreign investment in the energy sector, the Indian government in 1991 began loosening state control over the energy sector by implementing phased programs for deregulating coal, oil and gas prices by 2002-2003. Progress in India's liberalization of its energy sector has been mixed, however. Multinational corporations still consider the transaction costs in India's energy sector to be too high.

Coal and oil remain India's primary energy source. Coal consumption has steadily increased in the last decade; India used 283 million metric tons (MMT) of coal in 1997/1998, or 6.5% of the world's total consumption of coal. India's coal usage is expected to double from 405 MMT per year in 2001 to over 800 MMT by 2010. While India will continue to rely heavily on coal, its consumption of oil will steadily rise. India's oil demand currently exceeds 1.75 million b/d, and is the fourth largest oil consumer in the region after Japan (5.78 million b/d), China (4.01 million b/d), and South Korea (2.25 million b/d). Similar to energy patterns in the rest of Asia, India will consume a predominant amount of oil in the medium to long term. India imports about 700,000 barrels of the 1.7 million barrels it currently consumes per day. Imports are likely to increase to 1.5 million b/d by 2010, most from Russia, Iran, Saudi Arabia, Iraq, and the United Arab Emirates (UAE). The bulk of its oil imports will continue to come from the Middle East.

Natural gas is a distant third in terms of India's current energy use, accounting for only an 8.5% share of the country's primary energy needs. Natural gas, however, could become a potentially important resource in the future. For instance, it was estimated that India faced a shortfall of 50,000 megawatts in power generation by 2000, but the country will continue to turn to traditional options to fuel its power stations, namely coal, oil, and hydroelectric power. LNG is not more amply used in India for two reasons: it is a scarce domestic commodity and its transportation methods over the mainland are poor. Forecasts of growth in demand suggest that 20 25 million tons per year of LNG could be imported into India by 2010, limited mainly by the lack of import terminals to receive LNG shipments. A large factor in predicting growing future consumption of LNG, however, is India's close proximity to some of the world's largest gas reserves, namely in Yemen, Qatar, Oman, Yemen, Indonesia, and Malaysia. Any growth in India's share of LNG consumption will depend on the coordination of various agencies in India such as the state maritime boards, the Finance Ministry, port authorities, the Ministry of Petroleum, the Ministry of Power, the state electricity boards, national and state grids and India's financial institutions.

Given India's energy interests in the Middle East, India has often taken positions contrary to the United States on Middle East issues. For instance, U.S. sanctions have delayed initiatives to import gas and oil from Iran and Iraq; but it is in India's interest to support the lifting of sanctions. Longstanding difficult relations with its neighbors also complicate India's energy security. For instance, the Oman-India deep-sea pipeline project, which needed to avoid the territorial waters of Pakistan, required the pipeline to be laid out at technologically infeasible depths. Another gas project, which would deliver gas from Iran through Pakistan to India, was cancelled because Pakistan would not permit pipelines to be laid on its territory or through the EEZ. Improvement in India-Pakistan relations is critical if huge gas reserves in the Middle East and Central Asia are to reach either country. However, persistent problems in this bilateral relationship have led India to adopt more costly alternatives for importing natural gas.

India expects no major or sudden changes in its energy needs in the near future, assuming steady economic growth and successful completion of its liberalization program in its energy sector. However, given India's increasing dependence on foreign sources for energy resources such as oil and gas, the government is concerned about several potential developments that could affect India's access to resources. Political instability in the Middle East, possible conflict between India and Pakistan, or potential tensions with China are all liable to negatively impact India's access to foreign supplies. Given the political uncertainties, India seeks an adequate emergency response capability to possible disruptions in energy supplies, a well-functioning international oil market, and a regional forum for cooperation on energy matters. India supports the creation of a regional version of the International Energy Agency (IEA) with the specific goal of improving states' emergency response capabilities to possible future disruptions in supply. Views were mixed about whether an energy-specific regional organization would be feasible; it is argued that such cooperation seemed ambitious given that more immediate inter-state security issues remained unresolved between India and its neighbors. Situated adjacent to the world's largest oil reserves in West Asia and along the route of the bulk of international oil trade moving from the Persian Gulf to East Asia and the Pacific, however, India is well placed to play a key leadership role in enhancing the "collective energy security" of the Indian Ocean Rim (IOR).

11.3.5.3 Japan

Japan, the world's second largest – and Asia's most powerful – economy, remains highly dependent on foreign suppliers for its energy resources. Japan's primary energy sources today are oil, coal, and gas. Although Japan was heavily dependent on oil during the 1970s – its share of oil consumption exceeded 70% at times during this decade – it pioneered trade in liquefied natural gas (LNG), sharply increasing the share of natural gas in the country's primary energy supply from 5% in 1980 to 12% in 1998. Today, Japan is the world's largest importer of LNG, accounting for 61.2% of total global LNG imports in 1996. Despite the rise in LNG consumption, Japan's primary energy source is oil, which accounts for about 53% of its total energy needs, followed by coal (18%), nuclear energy (16%), natural gas (12%), and hydroelectric fuel (2%). Japan's energy mix remains heavily oil-dependent because it possesses only incidental indigenous fossil fuel reserves and production. Its heavy reliance on oil is due to direct burning of crude for power generation. In terms of the outlook for Japan's resource needs, the share of oil is expected to decrease while that of natural gas will increase, with the role of nuclear power remaining uncertain. Although electricity generation will be increasingly met by LNG imports, Japan's consumption of oil is predicted to rise significantly, and a much greater amount of the resource will travel by sea, primarily from the Middle East. While still heavily reliant on oil, Japan in recent years has dramatically reduced its dependency from 77.4% in 1973 to 55.8% in 1995. Nevertheless, Japan remains concerned about the safety of shipping lanes.

11.3.6 Traditional and Newly Emerging Regional Security Concerns

Access to energy sources is a critical security issue for the Asia-Pacific region given the structure of its energy needs and expected future consumption patterns. Governments and security professionals continue to wrestle with traditional energy security concerns such as safe access to sea-lanes, reliable transportation, territorial conflicts, and attendant environmental security issues such as pollution. New sources of energy, such as natural gas, a more sophisticated and integrated energy market, and newly emerging strategic relationships have introduced new energy security considerations in the Asia-Pacific.

11.3.7 General Notes

1) Reducing dependency on oil has been a top priority in Japan's energy policy since the first oil shock in 1973. As a result, the amount of imported oil as a percentage of Japan's energy supply dropped from 78% in 1973, to 51% in 2001. However, the percentage of oil imports from the Middle East as a percentage of total oil imports has recently been on the rise and reached 86% in 2001.

2) According to the IEA's World Energy Outlook 2002, oil imports for developing countries in Asia are expected to increase dramatically from 4.9 Mbd (42% of demand) in 2000, to 24 Mbd (83% of demand) in 2030. In particular, net oil imports for China alone are expected to jump from 1.7 Mbd (35% of demand) in 2000, to 10 Mbd (83% of demand) in 2030.

3) The Japanese government has formulated a comprehensive policy, called the Hiranuma Initiative, which is aimed at maintaining energy stability in the Asian region. This policy was presented at a meeting of energy ministers from Japan, Korea, China, and the ASEAN nations at the IEF forum in Osaka, and was approved by all participants. The main points of this policy are (1) to promote cooperation in the development of natural gas resources in the Asia region, (2) to exchange information in emergencies, and (3) to cooperate with Asian countries in price negotiations with oil-producing nations.

4) KEDO is an international organization which was founded in 1995 based on the Agreed Framework between the US and North Korea. Since North Korea's admission that it had continued its nuclear weapons development program, the effectiveness of the framework itself has been called into question.

5) Japan's petroleum product tax per kiloliter is ¥1200 for gasoline, ¥570 for kerosene, ¥1270 for light oil, ¥2400 for low-sulfur crude oil, and ¥3410 for high-sulfur crude oil. Naphtha is tax-exempt. In tandem with a 1972 policy of deregulation of fuel oil imports, a proportional tax was introduced to support refining near areas of consumption, and this high secondary tax is still in effect today. On imported fuel oil priced at ¥20,000 per kiloliter, the tax amounts to between 12% and 17%. In contrast, Korea imposes a uniform tax on all petroleum products of 7% of the import price – except on naphtha, for which the taxation rate is 1%.

Therefore, the imperative for energy security in such vulnerable strategic regions as the Asia-Pacific is paramount for global stability and development. The priority of this challenge for the Asia-Pacific region is also no accident, as it is the world's fastest growing energy consumer, with projected demand to steadily surpass other regions for some time. But it remains to be seen whether this troubled region will be able to forge a collective and cooperative approach in the wake of the daunting challenges and demands posed by the global "war on terrorism" and an increasingly destabilizing unipolar world.

11.4 Prospects of Electricity Infrastructure in East Asia

The problem of forming interstate electric ties and interconnecting electric power systems of countries and regions of East Asia (EA), including Siberia and Far East of Russia, China, Mongolia, Democratic People's Republic of Korea (DPRK), Republic of Korea (ROK) and Japan has attracted ever-greater attention in recent years. There are favorable preconditions for electric power cooperation and creation of power interconnection in EA. First of all these are: a) uneven distribution of fuel and energy resources (in particular hydro energy) on the territory; b) substantial difference in tariffs for electricity in various power systems; c) different seasons and hours of annual load maxims in power systems.

Taking into account these factors economic effectiveness and prospects of forming interstate electric ties (ISETs) and interconnection of power systems in East Asia have been studied. The results of the studies are presented below.

11.4.1 Creation and Development of Common Electric Power Space of EA Countries

The concept of common electric power space (CEPS) is the basic one for studying interstate electric ties and interconnection of electric power systems (EPS) in EA. The following definition is suggested. *CEPS is a territory interconnected by electric ties and contract agreements for mutually beneficial exchange (trade) of electric power (and fuel for power plants).* It is kept in mind that ISETs and transmission lines in each country create technical infrastructure of CEPS and interstate agreements and internal legal acts are economic, financial and legal components of CEPS. Figure 11.2 presents block diagram of potential ties, forming CEPS. Formation of common electric power space in East Asia is aimed at creation of favorable conditions for:

1. Free and mutually beneficial export-import of electricity and power;
2. Use of effects of interconnecting national and regional electric power systems.

In this connection CEPS formation is reasonable, on the one hand, owing to different natural-climatic and economic conditions of EA countries and, on the other hand, owing to substantial energy and economic effects that can be achieved by interconnecting power systems of different countries.

Thus, Japan and ROK are insufficiently provided with fuel and energy resources and the cost of fuel and energy is high in these countries. Russia, China and DPRK are provided with resources much better but are far behind in capabilities to finance energy development. Besides, China has high rates of economy development and growth of electricity demand, which causes power supply problems. All this makes export-import of electric power potentially expedient, first of all, export from Russia to the other countries of the region.

As for the effect of interconnecting power systems, it can be particularly great in the considered region owing to different seasons (and hours of day) of annual load maxims of consumers (based on these maxims the total required installed capacities of power plants and their commissioning are determined). In Russia, North EPSs of China, DPRK and Mongolia annual load maximum is in winter in the evening hours, and in Japan and ROK - in summer in the daytime. A detailed description of effects of interconnecting EPSs with different seasons of load maxims is presented in [21]. Their brief description is presented below.

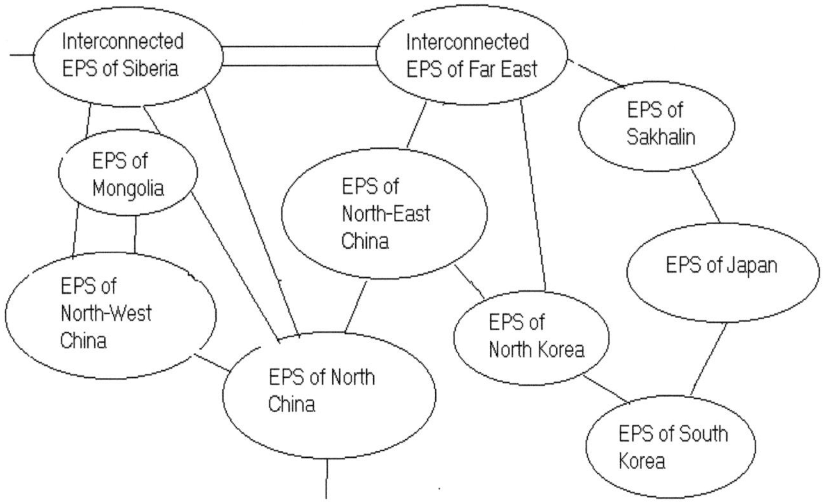

Fig. 11.2 Scheme of ISETs in CEPS of EA countries

In summer period of electric load decrease, thermal power plants (TPPs) of EPS with winter maximum of consumer electric load can be additionally loaded. This additional generation can be transmitted to power systems with summer consumer load maximum replacing semi-peak TPPs from the balance of capacities there. In winter period, to the contrary, TPPs of power system with summer load maximum can be additionally loaded and their generation will be transmitted to EPS with winter maximum of consumer load. Thus, installed capacity of power plants in power systems with different season maxims can be decreased only by commissioning interstate electric ties connecting them. Thus, this decrease in each EPS will be approximately equal to an ISET transfer capability.

The effect of interconnecting EPSs with different seasons of load maxims can be used when commissioning new capacities. Capacity of a new power plant, commissioned in the interconnection of power systems with different seasons of load maxims can be used in balances of capacities of both power systems (in one - in summer, in the other - in winter). Here each kW of capacity of this plant will substitute up to 2 kW of installed capacities of other power plants. This effect can be used when constructing any types of power plants - hydraulic, thermal burning fossil fuel, nuclear, tidal (if their power supply is regulated).

Common electric power space should be created based on the following (so far preliminarily formulated) principles and required conditions:

1. Attaining energy-economic and environmental efficiency of ISETs, to be constructed within CEPS, for all countries-participants.
2. Maintenance of energy security and accepted levels of power supply reliability of all the countries, comprising CEPS.
3. Joint elaboration by countries-participants of legal, prescriptive and methodological grounds for creation, operation and development of CEPS.

4. Formation of bilateral and multilateral commissions (or other bodies) to consider and solve the problems of coordinated development and operation of CEPS.
5. Conclusion of bilateral and multilateral agreements on conditions and guaranties of power supply and exchange between the countries.

Since the interstate electric ties existing in the region are weak, formation of EA CEPS will start from the scratch, in fact. Obviously it will proceed by stages, which can be set presumably only.

At the *first stage*, as one can suppose the ties within Russia, Korea, China, Mongolia and Japan will be getting stronger. Then, ISETs connecting East Russia with other EA countries are expected to be in place.

On the whole the first stage of CEPS formation will be characterized by the bilateral solving the problems of construction and control of power flows for each individual ISET. In fact, at the given stage one cannot speak of electric power space in EA countries in the full sense of this concept.

The *second stage* of CEPS formation will start after the constructed ISETs begin to noticeably affect energy balances and operating conditions of the interconnected EPSs. Here the power flows on individual ISETs can affect operating conditions of several EPSs, shown in Figure 11.2. There can be transit flows via some countries (for instance, via DPRK from Russia or China to Republic of Korea), etc. This will require ISETs construction, effects given by them and regimes of power flows to be coordinated at the level of several countries, and later, probably at the level of the whole interconnection.

11.4.2 Estimation of Prospective Electricity Demands of NEA Countries and Free Volumes of Their Electricity Markets

Electricity demand of NEA countries was estimated for a time span to 2010-2020. The calculations were done based on the information on levels and rates of power consumption in the countries of the region, presented in [22-25]. Taking into account a substantial uncertainty of the information the electricity consumption estimates were set by a range.

Free volume of electricity market is considered to mean the part of prospective electricity consumption that is not covered by generation of existing and predetermined power plants. Predetermined plants are those being constructed or whose construction is decided.

Electricity demand and market volume for China and Japan were so far determined for the "Northern" territories only. The "Northern" territories of China include the provinces served by power systems of Northeast, North, North-West (including Xinjiang autonomous region) and power systems of Shandong province. The "Northern" territories of Japan include prefectures served by power systems of Hokkaido, Tohoku and Tokyo, i.e. the whole 50 Hz zone of Japanese national electric power system. Electricity demand for Russia was determined for its East territories only, including East part of Interconnected EPS (IEPS) of Siberia, IEPS of Russian Far East (RFE) and EPS of Sakhalin. The volume of East-Russian electricity market was not calculated since, due to available and predetermined power plants, Siberia and RFE have surplus capacities for a time span to 2010-2020.

Calculations were done based on annual rates of electricity consumption growth assumed from [22-25], etc. and presented in Table 11.6. The presented values were rounded off with a precision of 0.5%.

In Table 11.7 numerator presents obtained estimates of prospective electricity demand and denominator - estimates of free volume of electricity markets through EA countries. The largest fraction of electricity demand falls on China. In 2000 its fraction is about 40% and by 2020 it may reach nearly 60%. Electricity demand of Japan, Republic of Korea and East Russia is also substantial. The fraction of DPRK and Mongolia is much smaller.

As is seen from Table 11.7, by 2010 free volume of the electric power markets in the region increases reaching 300-550 TWh/year and by 2020 it increases several times more. Like the case with electricity demand, the main fraction of market volume belongs to China. However it is much greater, being about 80-90%. The role of Japan and ROK is less significant. These countries may as well have free volumes of electricity market, reaching several hundreds of TWh/year in 2020.

Countries	2000-2010	2010-2020
China	6.5-7.0	5.0-6.5
Japan	1.5-3.0	1.0-1.5
Republic of Korea	3.5-4.5	2.0
DPRK	2.5	2.5-3.5
Mongolia	2.5	2.5-3.5
Eastern Russia	2.5-3.5	3.5-4.5

Table 11.6 Annual Rates of Electricity Demand Growth of NEA Countries, %

Countries	2010	2020
North China	$\frac{910-1050}{280-400}$	$\frac{1550-1950}{920-1300}$
North Japan	$\frac{455-560}{10-105}$	$\frac{495-660}{50-205}$
Republic of Korea	$\frac{310-345}{0-35}$	$\frac{370-415}{60-105}$
DPRK	$\frac{35}{5-10}$	$\frac{40}{10-15}$
Mongolia	$\frac{3.5}{1}$	$\frac{4}{1.5}$
East Russia	150-160	180-200
TOTAL	$\frac{1865-2155}{295-550}$	$\frac{2640-3270}{1040-1625}$

Table 11.7 Electricity Demand and Free Volume of Electricity Markets in the NEA Countries, TWh/year

11.4.3 Export Electric Power Projects of East Russia

At the expected rates of electricity demand growth on the territory served by IEPSs of Siberia and RFE there will be underused electric power generation even at a limited number of new generating capacities to be commissioned in the nearest future. The magnitude of the generation determines minimum potentialities of electric power export to EA countries, which requires construction of interstate transmission lines only.

At the same time export potential of Eastern IEPSs of Russia can be substantially increased by development and implementation of special export projects envisaging construction of electric power sources jointly with transmission lines.

Russian research and design organizations have studied the prospects of developing interstate electric ties between EPSs of East Russia and EA countries [26,27], etc. As a result potential directions of such ties were revealed and pre-feasibility studies of individual ISETs were made. Such export electric power projects are given below.

1. ISET "Bratsk-Beijing", length – 2600 km, voltage ± 600 kV, transfer capability – 3 GW, transmitting electricity – 18 TWh/year, cost – $ 1.5 Bln.
2. "Bureysk Hydro – Kharbin", 700 km, ± 400 kV, 1 GW, 3 TWh/year, $ 2 Bln. Cost for the transmission itself is about $ 250 Mln. The rest cost is for completing the hydropower plant construction.
3. "RFE – DPRK – ROK", 1100 km (additionally, transfer capability of 700 km of bulk power transmission lines of RFE power systems are needed to be enlarged), ± 500 kV, 4 GW on the section "RFE – DPRK" and 8 GW on the section "DPRK– ROK", 7 TWh/year, $ 2 Bln. – cost for ISET and additionally $ 2.8 Bln. – cost for Primorye Nuclear power plant.
4. "Sakhalin – Japan", 470 km, ± 500 kV, 4 GW, 23 TWh/year, $ 6.7 Bln (including $ 2.6 Bln. – cost of transmission and $ 4.1 Bln. – cost of export gas fired power plant on Sakhalin).
5. "RFE Nuclear – China – ROK", 2300 km, ± 500 kV, 2.5 GW, 18 TWh/year, $ 3 Bln. – cost of transmission and additionally $ 4 Bln. – cost of the nuclear power plant.
6. "Uchursk Hydro – China – ROK", 3500 km, ± 500 kV, 3.5 GW, 17 TWh/year, $ 4.5 Bln. – cost of transmission and additionally $ 6 Bln. – cost of the hydropower plant.

Thus, as a conclusion, we can say following:
1) Power cooperation of EA countries with formation of common electric power space and power interconnection will give substantial energy-economic effects to the countries-participants.
2) Creation of interstate power interconnection in EA opens a great market of new electric power technologies, in particular, on DC power transmission.
3) Development of methodology for assessment of ISETs economic effectiveness is required, in particular from the viewpoint of investors and potential owners.
4) Cooperation of efforts of research, design and other concerned power and financial organizations and state bodies is required to carry out further studies of interconnecting power systems of EA countries.

11.5 Assessment of Energy Supply Systems with an Energy Infrastructure Model for Asia

While energy demands in China, Southeast Asia and East Asia are projected to grow substantially over the coming decades, the development and exploitation of energy resources in Asia including East Siberia and the Russian Far East have attracted considerable attentions. It has become increasingly important to answer the question of what energy infrastructure, such as long distance natural gas pipelines and international electricity networks, should be constructed in the region, and the question of how their energy demands should be satisfied economically, securely and environmentally benignly.

Let us see briefly the outlook of primary energy supplies in Asia. Coal is an abundant and broadly distributed fossil fuel in Asia, and is expected to continue to be a major energy resource. Although the price of coal per unit calorific value has been relatively inexpensive, the growing demands of coal will not be met without the extensive development of its transportation infrastructures such as railroads and bulk carriers. In the case of crude oil, the amount of its resource in the region is not plentiful as that of coal, and is unevenly distributed. Oil supplies for Asia will continue to be increasingly dependent upon the Middle East, and such over-dependency of oil procurement on the single geopolitical region may potentially aggravate the energy securities of the countries in Asia. Natural gas is a clean and high quality fuel. It generates less CO_2 than any other fossil fuels on a per calorie basis. From the viewpoint of environmental protection, natural gas is the best substitute for oil and coal. However, enormous capital investment for its transportation infrastructure of liquefied natural gas tankers, liquefaction and re-gasification facilities, as well as, extensive pipeline networks in Asia will be required in order to increase its share in total primary energy supply.

In response to the above questions, the purpose of the study is to obtain insights into the optimal future configuration and operation of Asian energy infrastructure in a long run, and also potential roles of emerging energy related new technologies. For this purpose, the author's research group has been developing a large-scale global energy system model, which minimizes inter-temporally the sum of the discounted total energy system up until the year of 2100 with a dynamic linear-programming technique [28,29].

11.5.1 Global Energy Infrastructure Model

11.5.1.1 Geographical Coverage and Transportation Network

The geographical coverage of the energy model is the whole world. The world is geographically divided into 54 regions in order to express the detailed regional characteristics of socio-economic and geographical conditions.

As seen in Figure 11.3, the model assumes a transportation infrastructure network of 82 nodes.

The bright circle markers in the figure show the geographical location of 54 city nodes that represent energy consuming areas of the respective divided world regions. The dark rectangle markers indicate the location of 28 energy production nodes that were added to the network in order to express remote fuel production sites far from major cities.

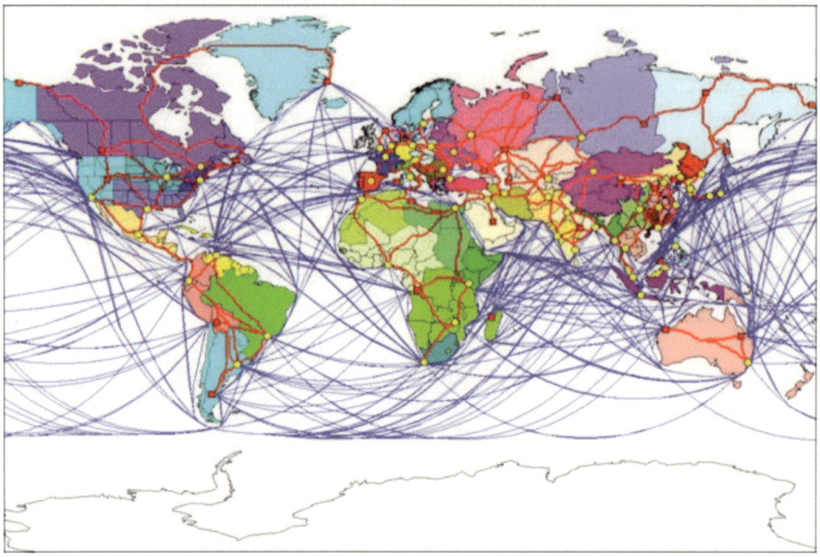

Fig. 11.3 World division framework and transportation network

The nodes are connected with plausible land and/or ocean transportation routes. The model takes account of transportations of coal, oil, natural gas, H_2, synthetic liquid fuels, captured CO_2 and electricity. As specific measures for transportation, the model assumes freight trains, on-shore and/or offshore pipelines, overhead power transmission lines, submarine power cables and various types of ships. The specific capacity and operation of each transportation route is determined as the result of minimization of the total energy system cost through linear programming.

11.5.1.2 System Structure of the Energy Model

Figure 11.4 indicates the assumed possible energy flow at each node in this energy model. Fossil fuels and biomass gasification, methane synthesis, DME synthesis, methanol synthesis, indirect coal liquefaction, H_2 production and electric power generation are considered as technological options for energy conversion. An elaborate integration of these conversion plants together with CO_2 capture facilities provides for a large of low carbon-intensive fuels with little additional CO_2 emissions from their conversion processes. Such an integrated energy system can be expected to contribute to remarkable reductions in CO_2 emissions from end-use sectors.

With respect to electricity generation sectors, the model explicitly takes into account daily load curves expressed simply with three time periods (morning, afternoon and evening) by season (summer, winter and intermediate), so as to determine how each type of power plant will be operated in accordance with diurnal and seasonal variation of electricity demands and renewable supplies. This is because the capacity factors of electric power plants are supposed to have a large influence on their economic characteristics.

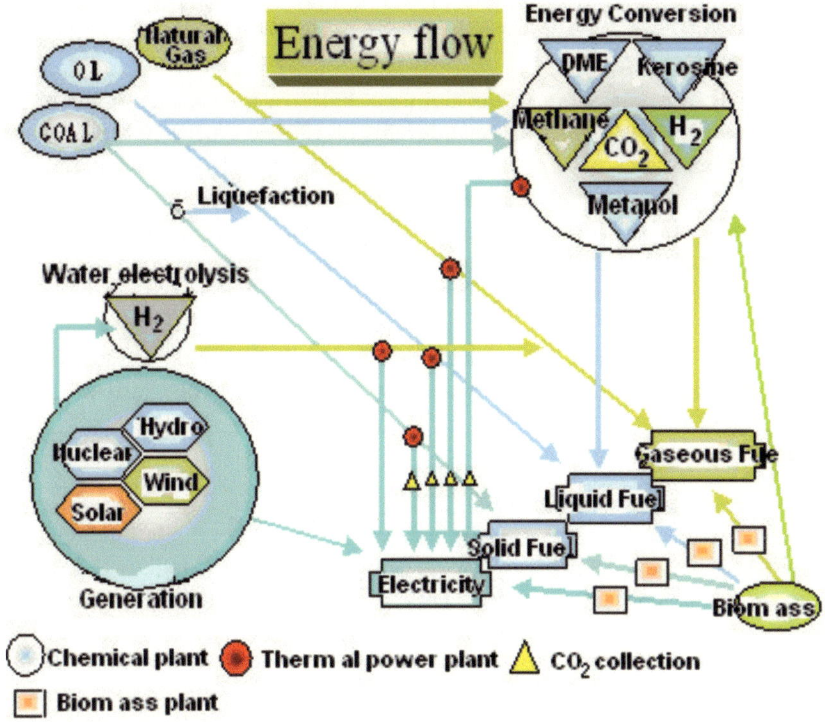

Fig. 11.4 Configuration of the energy system of each node of the model

One of the notable features of the model is that it can explicitly analyze the roles of processes of CO_2 capture and storage in the energy system. As specific measures for CO_2 capture, the model takes into account both chemical absorption from flue gas of thermal power plants and physical adsorption from the output gases of fossil fuel reforming processes. There are two major methods for CO_2 storage: geological storage and ocean storage. Geological storage is classified into three types: 1) injection of CO_2 into oil wells for enhanced oil recovery (EOR) operation; 2) storage of CO_2 in depleted natural gas wells; and 3) storage of CO_2 in aquifers. The model takes account of all the CO_2 capture and storage technologies, and can assess their future potentials by node in the model.

11.5.1.3 Mathematical Formulation

The model built here is mathematically formulated as a multi-period inter-temporal linear optimization problem with linear inequality and equality constraints. The constraints represent supply and demand balances of each type of energy carriers by node, energy and CO_2 balances in various energy conversion processes, and state equations for several inter-temporal dynamics, such as the depletions of fossil fuel resources and geological CO_2 storage reservoirs' capacities, the vintage structures of various facilities in the energy system and so forth. The objective function of the problem is defined as the sum of the discounted total energy system costs distributed over time, which include fuel production costs, amortized capital costs, maintenance and operation costs, energy transportation costs, CO_2 capture and storage costs, and energy saving costs measured as the losses of consumer surpluses. The supply cost curves of fossil fuels by node are expressed as step-wise linear functions with respect to their amounts of cumulative productions.

The model seeks the optimal regional development paths of the energy-related infrastructure for the years from 2000 through 2100, at intervals of 10 years, using a linear-programming technique. The model, therefore, does not take into account any nonlinear effects, such as economies-of-scale with respect to unit construction costs of various facilities, especially those of pipelines. Furthermore, for simplicity, all the variables in the model are treated as continuous real numbers, although some of them, such as those expressing the number of tankers, should be indeed treated as discrete integer numbers in the real world.

The number of the variables of the model is about one million, and the problem is solved with an interior point method of linear programming. This research group uses commercial software of CPLEX.

11.5.1.4 Reference Energy Demand Scenario

The final consumption sector of the model is disaggregated into the following four types of secondary energy carriers: 1) gaseous fuel, 2) liquid fuel, 3) solid fuel, and 4) electricity. In the case of electricity consumption, as mentioned before, the model explicitly takes into account daily load duration curves by season. The future energy consumption in the model is exogenously given as reference scenarios by energy carrier type, by node of the network and by year.

The future energy demand scenarios in this model are based on SRES B2 scenario that was made by IPCC. The amounts of energy consumption by node were principally adjusted by demographic data on geographical distribution of national populations in the future.

11.5.2 Simulation Results of the Model

This Subsection presents some of the simulation results of the energy infrastructure model. This study assumes two policy cases for simulation of the model, i.e., a reference case (REF case) and a controlled case (CON case). In CON case, the atmospheric CO_2 concentrations until the year 2100 are limited below 550 ppm, and additionally the annual CO_2 emissions of each Annex1 country of Kyoto Protocol are assumed to be reduced less than 20% of their respect emission levels of 1990 by 2050 and thereafter. No emission trades of greenhouse gases were assumed in the simulation.

11.5.2.1 Reference Case Results

The profile of the world primary energy production in REF Case is shown in Figure 11.5. When no CO_2 regulation exists, the share of coal in primary energy supply is very large because of its low price, and natural gas including unconventional gas is to become the second most important primary energy source. World electricity generation is shown in Figure 11.6. The share of coal-fired generation is substantially large. Photovoltaics will begin to be used practically after 2050, partly because in this model, the unit capital costs of solar cells are expected to be reduced by 3.6% per annum until 2050.

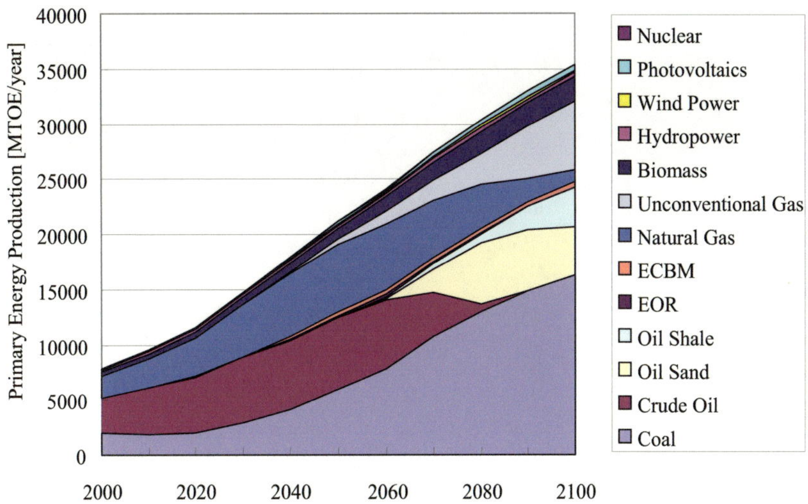

Top to bottom: Nuclear, Photovoltaics, Wind Power, Hydropower, Biomass, Unconventional Gas, Natural Gas, ECBM, EOR, Oil Shale, Oil Sand, Crude Oil, Coal

Fig. 11.5 World primary energy production in REF case

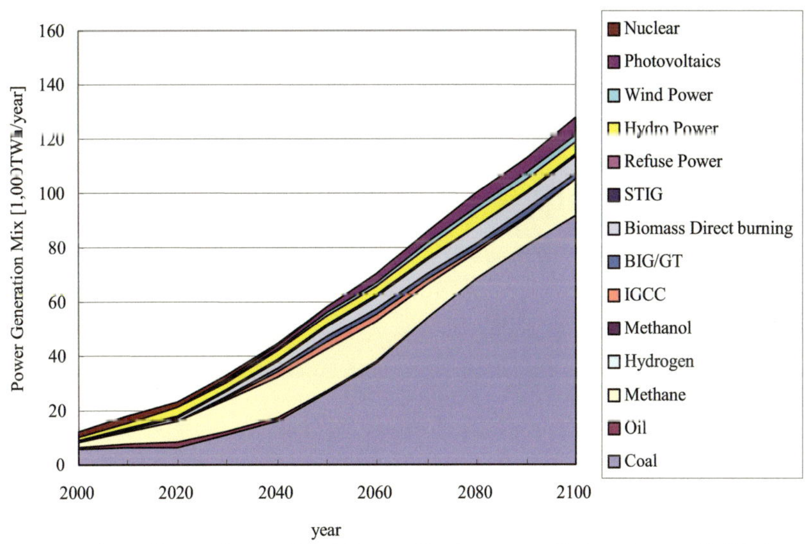

Top to bottom: Nuclear, Photovoltaics, Wind Power, Hydropower, Refuse Power, STIG, Biomass-Direct burning, BIG/GT, IGCC, Methanol, Hydrogen, Methane, Oil, Coal

Fig. 11.6 World power generation mixes in REF case

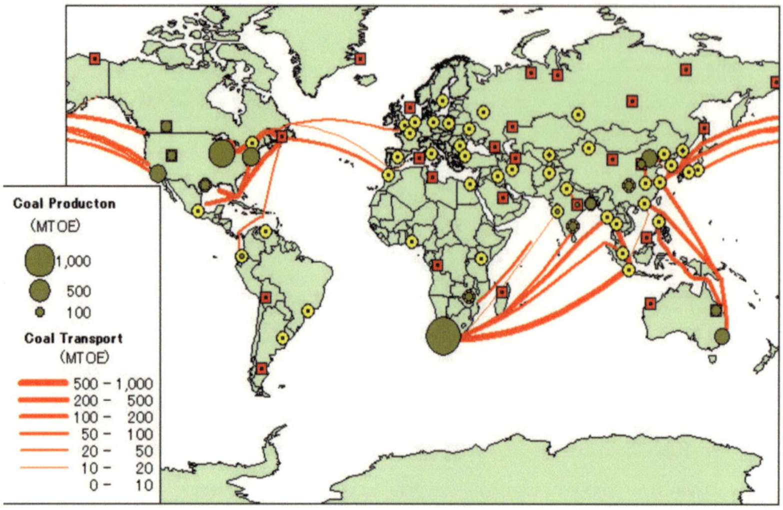

Fig. 11.7 Coal production and transportation of 2050 in REF case

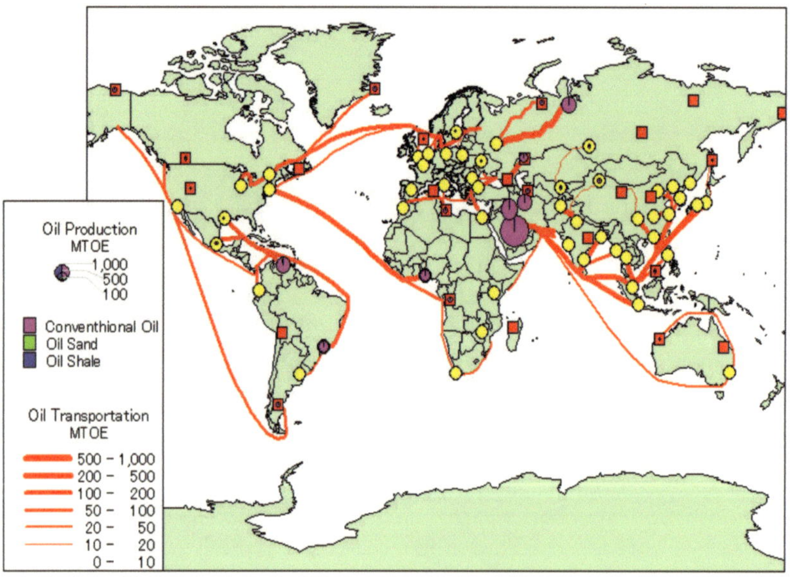

Fig. 11.8 Oil production and transportation of 2050 in REF case

The calculated global patterns of coal, oil and natural gas productions and transportations for the year 2050 are shown in Figures 11.7, 11.8 and 11.9, respectively. The cost of coal land transport by rail is rather expensive, and most of coal for international market trades is therefore transported by ship. North America and South Africa are expected to be large coal producing regions. In the case of oil, its costs for both land and ocean transportations are relatively low. This means that one single world region can provide oil resources economically for the rest of the world via its international market trades. The Middle East is therefore expected to continue to be a major oil-exporting region over the century. As for natural gas, its transportation cost is relatively high, and its international markets are divided into several local markets by continent. An increased reliance on natural gas would provide countries with more geographically diversified energy supply structures, thus improving the securities for their energy procurement.

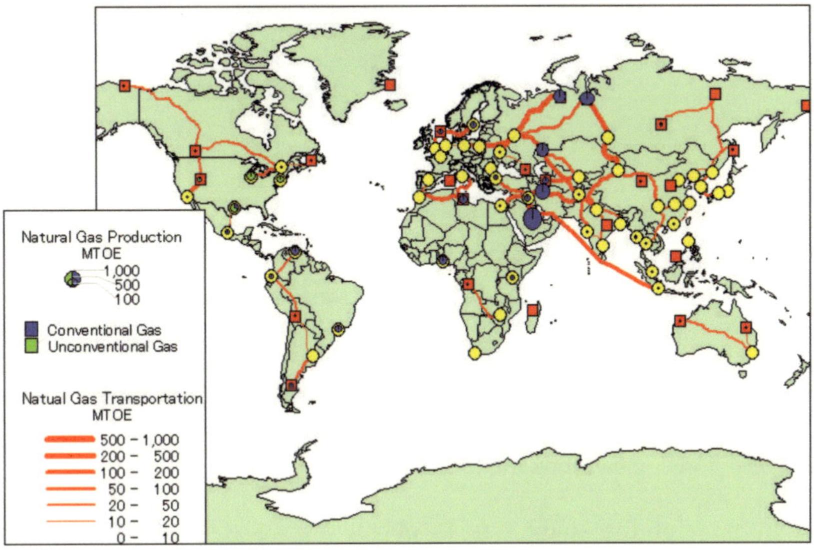

Fig. 11.9 Natural gas production and transportation of 2050 in REF case

Asia. In REF case, the necessity of the development of region-wide electricity grids among Northeast Asian countries does not seem apparent in the model result. The model indicates that transporting those fuels by rail or pipeline and generating electricity close to the energy consuming cities seems to be more economical than generating electricity at mine mouth or wellhead and transmitting electricity for a long distance by power transmission line. Obviously, the results are highly dependent upon the assumption about the relative cost competitiveness of power transmission lines against other types of energy transporting measures.

11.5.2.2 Controlled Case Results

In CON case, energy conservation, fuel switching, and CO_2 capture and storage leads to significant CO_2 emission reduction. The fuel switching means changing high carbon content fuels, like coal, to less carbonintensive fuels, like natural gas. World electric generation in Regulation Case is shown in Figure 11.10. The share of coal-fired generation is reduced and the share of natural gas-fired generation is increasing. Additionally coal-fired generation is changed to IGCC that enables us to conduct CO_2 capture more efficiently.

In CON case, the extensive international power transmission network around Japan can be seen in the figure for the year 2050. A very stringent CO_2 emission control policy may enhance significantly the economic viability of imported electricity from Sakhalin Island and the Korean Peninsula for Japanese power market.

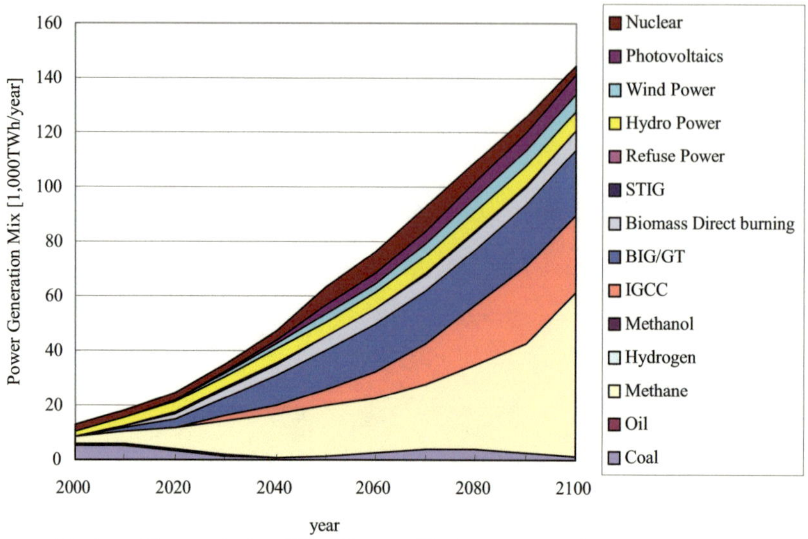

Top to bottom: Nuclear, Photovoltaics, Wind Power, Hydropower, Refuse Power, STIG, Biomass-Direct burning, BIG/GT, IGCC, Methanol, Hydrogen, Methane, Oil, Coal

Fig. 11.10 World power generation mixes in CON case

Thus, the purpose of the study was to obtain the insights into the possible future configuration and operation of energy infrastructure in Asia where energy demands are rapidly growing. This study presented the outline of the energy system model built in the study, and showed a part of the results obtained. Bearing in mind the considerable uncertainties as to various assumptions made in the model, the simulation results indicate that the economic validity of the development of region-wide electricity grids among Northeast Asian countries does not necessarily seem obvious in REF case. The current energy model cannot take account of the influences of energy security issues explicitly. It is necessary to conduct extensive sensitivity analyses to derive a general conclusion to the topic.

This study has not yet been completed, and the following research topics are to be incorporated in future studies, i.e. further improvement of the accuracy of the data on fossil fuel resources and production costs, sensitivity analyses of future energy demand scenarios, and consideration of nonlinear effects of infrastructure, such as economies-of-scale.

11.6 China Power Grid and its Future Development

Over decades, the China power grid has experienced a long developing process, from individual-provincial networks to across provincial power networks and to the current regional-interconnected power grid. This general development process of the electrical system has been driven by the pursuit of optimal utilization of energy resources on a nationwide scale. This is also impetus to future development of the China power grid. A feasible power network configuration is basic for secure and stable system operation, and power network planning has always emphasized the overall sense and combination of short- and long-term considerations. In the planning of a nationwide power grid, principles are adhered to, which are high security of the power grid, unification and entirety of planning, combination of short-term and long-term planning, moderately advanced construction, and advanced technology. In this Section, from the point of power system planning, future development of the China power grid is introduced; the planning of a nationwide power grid, and related problems are discussed based on current power grid structures [30].

11.6.1 The Current Situation of China Power Grid

Up to 2003, total installed capacity in China was about 391,400MW with yearly generation of 1.9×10^{12}kWh. Six regional power grids were established as shown in Figure 11.11. They are: East-North China power grid, North China power grid, West-North China power grid, East China power grid, Central China power and South China power grid. The main frame of the power network in East-North China, North China, Central China, East China and South China power grid is 500kV and 220kV transmission lines except that in West-North China power grid it is 750kV and 330kV transmission lines. Up to the year 2003, total length of transmission lines above 220kV was about 2.1×10^{5}km and total capacity of transformers above 220kV was about 588GVA.

Currently, two 500kV AC lines, forming a large synchronous interconnected grid, interconnect the East-North China and North China power grids and there are two ±500kV HVDC transmission lines that have combined transmission capacity of 4200MW from the Central China power grid to the East China power grid.

Fig. 11.11 Diagram of china power grid and energy resources distributions

In recent years, unexpectedly rapid economic developments have caused a power supply crisis in China. The crisis was exacerbated by coal supply shortages and drought together with high summer temperatures, and the crisis induced extended rolling brownouts and blackouts. Under this situation, construction of new generation capacity and transmission facilities became urgent and necessary to resolve the power supply crisis within two or three years. With respect to planning of the future power grid, optimal and maximum utilization of energy resources are being given special attention.

11.6.2 Planning of Nationwide Interconnected Power Grid

China is abundant in coal and hydro resources, but 80% of coal reserves are in the north and northwest China, and over 80% of hydro resources are in the west region. The power consumption of the east and coastal region is over 60% while the coal resources there are only 10% and hydro resources are only 14%. The uneven distribution of energy resources and its economic location has required transmitting power from the west to east, which is called *"West-to-East Electricity Transmitting"*. A brief indication of the hydro and coal distribution in China is also shown in Figure 11.11.

The nationwide interconnected power grid mainly comprises three west-to-east electricity-transmitting channels, termed the North Channel, the Central Channel and the South Channel. The North Channel consists of two parts. The first is to exploit coal resources in west Inner Mongolia and Shanxi and build AC lines to transmit electrical power to Beijing, Tianjin, Hebei and Shangdong. The second is to exploit coal resources in north Shanxi, hydropower and thermal power in west-north China and to transmit power to Beijing, Tianjin, Hebei and Shangdong by AC lines or DC lines. The central channel is to develop large hydropower stations along Jinsha River, Yalong River and Dadu River and build AC and DC lines to transmit power to East China, Central China and South China. The south channel is to exploit hydro and coal power in Yunnan and Guizhou, and transmit the electricity to Guangdong and Guangxi through AC or DC lines. From electricity demand forecasts and analyses, the total amount of power to be transmitted from west to east will be about 58,000MW, and will reach about 130,000MW in year 2020.

In respect to power transmission between the different regional power grids, a nationwide interconnected power grid would be established by year 2010. By 2020, interconnection between the regional power grids would be strengthened, and a higher voltage power grid will start to be built. Comprehensive planning and research in the structure of a nationwide power grid has been undertaken in the past few years. On deciding the number and scale of the synchronous grids, the following technical problems are under consideration: influence to transfer ability of key transmitting paths, ability of the system to withstand severe system faults, transient stability of the whole system, and adaptability to future development. Another consideration for the scale of the synchronous grid is to avoid *"strong DC and week AC interconnection"*. That means avoiding the amount of transmitted power through HVDC interconnection lines if it is larger than that through AC interconnection lines between two regional power grids. Large scale loop structure must also be avoided, because in emergency cases, the loss of HVDC lines in *"strong DC and weak AC interconnection"* structures or main AC interconnection lines in loop structure will cause a large amount of power transfer that would threaten the receiving system. In the worst case, this could induce blackout in a large area.

Regional power grids with a large amount of power transmitted to other areas by HVDC lines are therefore suggested in creation of independent synchronous systems.

Considering feasibility of technology and the economy, a structure with 4 to 5 synchronous power grids in China is acceptable. By year 2010, the East-North China, North China and East China power grids will be interconnected by AC lines and form a large synchronous power grid.

The Central China, West-North China and South China power grids are independent synchronous power grids, and are interconnected with other synchronous power grids by HVDC lines. By the year 2020, the basic structure of a nationwide power grid will be similar; the large synchronous power grids including East-North China, North China and East China may be divided into two synchronous grids or not changed, which will be decided by development of the inner network in each region. The possibility of converting AC interconnections to Back-to-Back DC connections exists between the East-North China and North China power grid and the North China and East China power grid. Under this basic structure of a nationwide power grid, the transient stability, power transfer ability, and power supply reliability of the whole system is improved.

11.6.3 Specific Problems Concerned in National-Wide Power Grid

As introduced above, in near future, the relationship between different regional grids will become more complex and tight. A change in one area may influence system stability in a large area. System simulations are being undertaken comprehensively to study stability of a nationwide power grid for year 2010 and year 2020. The simulation results show that for the power grid structure of year 2010, the system may remain stable under "n-1" contingencies. Further, the system may remain stable under "n-2" contingencies by applying measures such as disconnecting machines, bipole converter blocking on HVDC lines, and loss of two main AC transmission lines. In year 2020, the inner structure of each regional grid and the connection between synchronous grids will be strengthened, which leads to a result that the stability level of the whole system increases a lot. For the power grid structure of year 2020, the system may remain stable under almost all "n-2" contingencies.

Besides ability of the system to remain in synchronism in nationwide power grid planning, several other specific problems are being examined.

11.6.3.1 Low Frequency Oscillation

Low frequency oscillation is one of the important stability problems in weak-connected power systems. If the oscillation is strong, the transient transmission ability of the transmission lines or generators will be decreased. In the worst case, the reliability of the whole system will be impaired. From the calculations, some modes of low frequency oscillations are found. In the large synchronous power grid, the East-North China and North China and East China power grid, East-North China region and part of North China region have low frequency oscillation at about 0.19Hz relative to East China region and inner part of East China regional system.

And in the North China regional system, there are low frequency oscillations between different parts of system. In the Central China power grid, the oscillation is at about 0.23 to

0.25Hz between the west and east part of the system. Because the damping capability is acceptable and sufficient to restrain the oscillation, the problem is not very severe, but for reliability and stability of system operation, installation of PSS for generators is emphasized. These are currently installed for most of the generators but not enabled.

11.6.3.2 Stability of Receiving Systems

Pear River Delta, Yangtze River Delta and Beijing-Tianjin-Hebei area are economical-developed districts. Hence, they are load centers, the "*East*" in the "*West to East Electricity Transmitting*" project. The proportion of imported power to maximum load is relatively high. Take Beijing-Tianjin-Heibei load center for example; in year 2010, the proportion of imported power to the maximum load is about 36%, and that in year 2020 it is near 50%. With such a large amount of imported power, influences to stability of the receiving systems become complex and difficult to control. Specific research on this topic is taking place and some guidelines are being drawn for the planning of the receiving systems.

(a) Optimizing of receiving-end network structure: Most load centers in China have ring network structure to ensure reliability of power supply. With increase of imported power, requirements for a feasible and rational receiving-end system become higher. In the planning stage, the length of the ring and the layout of substations should be considered from a longer-point of view to have good adaptability to long-term development. The network structure should be simple and clear, and optimizing adjustments should be kept with the whole power grid development.

(b) The transmission paths to the receiving system should be independent and dispersed, if conditions allow, and the amount of transmitted power to the load center in each transmission path should be limited to a certain percent, which is 10~15% in China. So that, in the most severe case of losing the whole transmission path, loss of transmitted power would not cause stability problems in the receiving system.

(c) Receiving systems should have enough reacting power support, including static reactive power and dynamic reactive power. As known, lack of reactive power support is one of the most important reasons for voltage instability; it has been given more and more attention in recent years. Since reactive power is not generally transmitted during normal operation, the receiving systems have to supply enough reactive power to correspond to their imported active power. Capacitors and generators are traditional means for static and dynamic reactive power support. It is required that newly added generators must follow the so called "*Layered and zoned*" principle in China, that is to ensure local dynamic power in receiving systems, and the power factor of new generation should be 0.85. Besides, FACTS devices as an advanced technology for power system stability are highly recommended in the future power system.

11.6.3.3 Security of Multi-in Feed HVDC Systems

HVDC technology is being applied widely in the "*West-to-East Electricity Transmitting*" project, mainly for large hydro power stations to transfer power to load centers. As planning, in year 2020, there would be about 22 HVDC transmission projects (±500kV and

±600kV) in China. For the receiving-end, there would be several HVDC converter stations in relatively small areas. The high density of DC converter stations will influence operating reliability in converters. For example, 8 DC converter stations are planned in the east China area, mainly around Shanghai. Because the interrelationships between multiple converters may induce failure in commutation and sub synchronous oscillation (that have appeared in some HVDC systems in the world), security of multi-infeed HVDC systems is being given great attention and study, and detailed electromagnetic transient simulations are required.

11.6.4 Future Development of China Power Grid

Currently, the study of ultra high voltage AC transmission system (1000kV or 1150kV) and DC transmission line (±750kV or ±800kV) are initiated for future power grid development. The higher voltage power grid based on the 500kV nationwide power grid will be tentatively built near year 2020 as a long-term plan.

11.7 Acknowledgements

Nikolai I. Voropai, Professor, Corresponding Member of RAS, Director of Energy Systems Institute, Irkutsk, Russia has prepared and coordinated this Chapter. Contributors include colleagues at the Institute and Members of the IEEE PES W.G. on Asian and Australian Electricity Infrastructure.

11.8 References

[1]. Voropai, N.I., Klimenko, S.M., Krivorutsky, L.D., Pyatkova, N.I., Rudenko, Yu.N., Senderov, S.M., Slavin, G.B., Cheltsov, M.B., "Some Problems of Energy Security of Russia and Its Regions", *Energetika Rossii v perekhodny period: problemy i nauchniye osnovy razvitiya i upravleniya*. Novosibirsk, Nauka, 1996 (in Russian).

[2]. Lobov, O.I., "Energy Security Problems of Russia and the Irterconnection with Energy Security of Europe" *Energiya: Eknomika, tekhnika, ekologiya*, 1996, No.1 (in Russian).

[3]. Makarov, A.A., "Energy Security Problems of Russia, "*Ekonomika i upravleniye neftegazovoi promyshlennosti,* 1997, No.3 (in Russian).

[4]. Bushuev, V.V., Mastepanov, A.M., Rodionov, P.I., "Energy security of Russia",*Gazovaya promyshlennost*, 1997, No.8 (in Russian).

[5]. Voropai, N.I., Klimenko, S.M., Krivorutsky, L.D., Massel, L.V., Pyatkova, N.I., Senderov, S.M., Slavin, G.B., Cheltsov, M.B., *Energy Security of Russia (Introduction into the Pproplem):* Preprint No.1, Irkutsk, SEI SO RAN, 1997 (in Rusian).

[6]. Bushuev, V.V., Voropai, N.I., Mastepanov, A.M., Shafranik, Yu.K., e.a., *Energy Security of Russia*, Novosibirsk, Nauka, 1998 (in Russian).

[7]. Mastepanov, A.M., *Regional and External Economic Aspects of Energy Policy of Russia*, Moskva, VNIIOENG, 1997 (in Russian).

[8]. *Energy strategy of Russia up to 2020*, Moscow, Inst.of En.Strat.2003 (in Russian)

[9]. Kudinov, Yu.S., *Economic Problems of Development of the Fuel and Energy Complex of Russian Federation*. Part 1. Moskow, Minprirody, 1996 (in Russian).

[10]. Dmitrievsky, A.N., Shakhnovsky, I.M., "Resource basis of the Oil and Gas Industry of Russia", *TEK*, 1998, No.1-2 (in Russian).

[11]. Yakushev, V.S., "Is Japan the Major Natural Gas exporter of the 21 century? ", *Gazovaya promyshlennost*, 1998, No. 1 (in Russian).

[12]. Calder, K. E., Pacific Defense: Arms, Energy, and America's Future in Asia, New York: William Morrow, 1996.
[13]. Luft, G., Korin, A., "Terror's Next Target," The Journal of International Security Affairs, December 2003.
[14]. Clover, C., Fifield, A., "More to Oil Shocks than Middle East," Financial Times, July 29, 2004.
[15]. Priddle, R., "Living in Interesting Times: Energy Security," Asia Energy Security Seminar, Tokyo, Japan, March 4, 2002.
[16]. Asia Pacific Energy Research Centre (APERC), Energy Security Initiative: Emergency Oil Stocks as an Option to Respond to Oil Supply Disruptions, APERC Background Report, Tokyo, Japan, 2002.
[17]. Martin, W. F., et al, Maintaining Energy Security in a Global Context, New York: The Trilateral Commission, 1996.
[18]. http://www.eia.doe.gov/oiaf/ieo98/world.html.
[19]. http://www.bp.com.
[20]. http://www.apcss.org.
[21]. Belyaev, L.S., Kovalev, G.F., Podkovalnikov, S.V.,"Effectiveness of Interstate Electric Ties in the Northern Part of Pacific Region", Izvestiya RAN, Energetika. 1997, № 6 (In Russian).
[22]. Gao Yan, "To Forward China Power Industry with Great Opportunity and Expectation to the 21st Century", *Electra,* 1998, No.180.
[23]. *International Energy Outlook.* 1999. IEA, Office of Integrated Analysis and Forecasting US DOE. Washington DC.
[24]. *Electricity Review of Japan.* The Federation of Electric Power Companies, 1998.
[25]. *Annual Report,* Korea Electric Power Corporation, 1999.
[26]. *Eastern Energy Policy of Russia and Problems of Integration in Energy Space of Asian-Pacific Region.* Proc. of Int. Conf., Irkutsk, September 22-26, 1998 Irkutsk, Russia.
[27]. Belyaev, L.S., Chudinova, L.Yu., Khamisov, O.V., Kovalev, G.F., Lebedeva, L.M., Podkovalnikov, S.V., Savelyev, V.A., "Studies of Interstate Electric Ties in Northeast Asia ", *Int. Journal of Global Energy Issues.* 2002, Vol.17, No. 2.
[28]. Fujii, Yasumasa, Fukushima, Ryo, Yamaji, Kenji, "Analysis of the Optimal Configuration of Energy Transportation Infrastructure in Asia with a Linear Programming Energy System Model" *Int. Journal Global Energy Issues,* 2002, Vol.18, No.1.
[29]. Fujii, Yasumasa, Yamaji, Kenji, "Assessment of technological options in the global energy system for limiting the atmospheric CO_2 concentration", *Environmental, Economics and Policy Studies,* 1998, No.1.
[30]. Work Group of National-Wide Power Grid Planning, *Planning of National-Wide Power Grid in China,* 2003, Vol. I-VII.

12

Integrated Natural Gas-Electricity Resource Adequacy Planning In Latin America

Natural gas (NG) is considered as one of the most promising sources to supply the world energy demand, with a consumption expanding at a very accelerated pace. The largest use still is for industrial heating. The second largest use is for electric power generation, which experienced a strong growth after the development of combined-cycle generation technology (CC-NG) in the 1980s. Besides efficient, CC-NG is competitive in modules quite smaller than those of other technologies, such as coal. This has contributed to foster the implementation of power plants based on CC-NG in electricity markets worldwide and created interdependency between the electricity and the gas sectors.

Latin America boasts natural gas reserves and high-growth energy markets. The need to diversify away from heavy investments in hydropower and expensive oil has driven many countries to promote the use of natural gas, especially for power generation. This was facilitated given the abundant reserves of gas in several countries in the region, particularly Venezuela, Argentina and Bolivia, and their interconnection with other markets. These developments were coupled with additional challenges, such as (i) the competition between hydro and thermal generation in a heavily hydro balanced region, (ii) the building up and later breaking of cross-country natural gas agreements, (iii) the competition between natural gas and other resources for power generation and electric transmission, and (iv) the development of the natural gas industry in an environment where its requirements are very volatile due to the randomness of hydro inflows.

More recently, liquefied natural gas (LNG) started to be considered an option to ensure the adequacy of natural gas supply for power generation. Brazil and Chile are leading the implementation process of regasification facilities. However, the region has also potential to become an exporter of LNG in the medium-term once the potential gas reserves that require deep drilling become commercially available.

This chapter addresses natural gas-electricity resource adequacy expansion and planning in Latin America. Five "*case studies*" were chosen for the analysis: an individual analysis of the developments of natural gas in four countries (Brazil, Chile, Colombia and Mexico), and power and natural gas integration in the Southern Cone. The emphasis is on the institutional and operational arrangements adopted in each country, and the competition between electricity transmission and natural gas pipelines. The success/difficulties observed in handling recent conflicts in the region that arose from natural gas supply difficulties are also provided. A section devoted to analyze the introduction of LNG in the region is also presented.

12.1 Introduction

Latin America has been in recent years one of the most intensive regions for natural gas and electricity development [1]. The region is very hydropower dependent (about 57% of the region's installed capacity is hydro) and the need to diversify away from heavy investments in hydropower and oil is driving many countries to promote use of natural gas, especially for power generation. Examples of these developments are in Brazil, Chile and Colombia. Other countries, such as Mexico, take advantage of natural gas to displace oil-fired generation. The countries of the region have great diversity in size, electrical installed capacity, electrical power demand, and electrical transmission/natural gas network characteristics (level of meshing and geographical extension). Figure 12.1 shows the share of hydro and thermal power and the installed capacity in each country in the region.

Hydropower reserves are still high and the continent boasts abundant natural gas reserves and high-growth energy markets, as shown in Figure 12.2.

Total Installed Capacity (2001): 221 GW (57%H, 43%T)

Fig. 12.1. Installed Capacity in Latin America: Brazil, Mexico, Chile, Argentina, and Colombia, etc

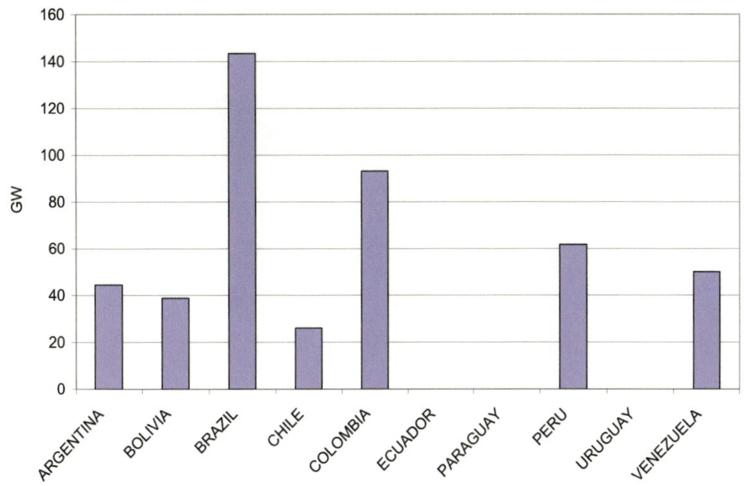

source: OLADE (Organización Latinoamericana de Energia), 2002

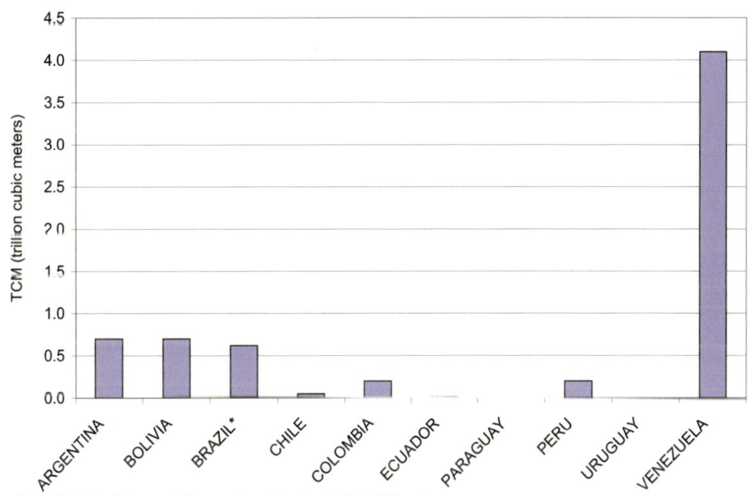

**Brazil includes the recent discoveries at Santos Basin (~419bcm)*

Fig. 12.2. Hydro Potential and Natural Gas Reserves in Latin America

The economic reforms have opened to private investors a number of sectors previously reserved to the state. This has led the region to develop an infrastructure of electricity and natural gas pipelines, both in each country separately as well as cross-border energy interconnections. These interconnections consist of basically cross-border electricity transmission links, power projects at the border (mainly hydro plants jointly owned by different countries) and cross-border natural gas pipelines. Figure 12.3 shows the main cross-border energy interconnections in the Southern part of Latin America.

Because the regional infrastructure is still developing, heavy investments in both generation and transmission investments are required. In those countries where hydropower is an expansion option, it is also necessary to determine the most economic trade-off between cheaper distant hydro, with higher electricity transmission costs, and more expensive "*local*" gas or coal fired thermal generation, with lower electricity network costs plus the cost of the correspondent gas pipeline.

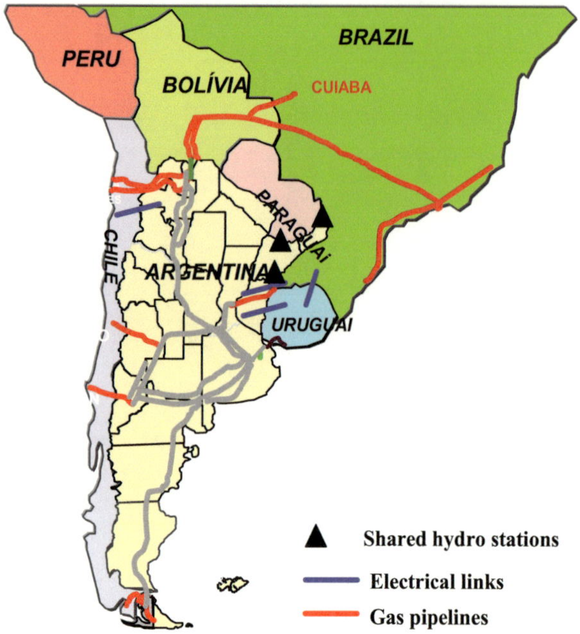

Fig. 12.3. Southern Cone: Electricity and Gas Cross-Border Interconnections

Another important issue in Latin America is the multi-country electricity-gas markets. These are a natural evolution to the existing "official" cross-border interconnections, which were originally established by the countries' governments for sharing reserves and carrying out limited economic interchanges. A significant development took place, as Bolivia and Argentina became abundant natural gas suppliers for Brazil and Chile, providing a fuel that was economically and environmentally attractive for the receivers. The energy and gas links were originally built as private initiatives and were mainly carried out under a local (not regional) framework and without integrated planning and were thus decoupled from each other. Recent episodes in the electricity and gas sectors in the Southern Cone, both in terms of crisis (for example, natural gas supply difficulties in Argentina that directly affected the Chilean and the Brazilian markets) and in terms of new opportunities (such as new natural gas sites and new gas pipelines) have brought up and restored the opportunity to work towards robust energy regional integration. Nevertheless, the environment is one of mutual mistrust among countries to build back any dependence of energy supply. As explained next, the difficulties faced by Chile after a unilateral cut of gas transfers from Argentina, breaking mutual government agreements, will make it very difficult to formulate any energy integration between those two countries.

12.2 Electricity and Gas Deregulation

The electricity and gas industries in the region experienced deregulation processes in several countries in the region, which made them evolve from centrally State controlled industries to privately driven ones, as illustrated in Figures 12.4a and 12.4b and explained in Chapter 17. Both Chile and Argentina formulated open markets in both sectors, creating market competitive conditions where feasible. A completely open market driven sector is that of the gas distribution to final consumers in Chile, where no prices are fixed by the government, as it is understood that competition naturally arises among different fuels. For example, natural gas can be replaced by bottled liquefied gas, so that neither is regulated, and price balances are reached when several suppliers compete for the market.

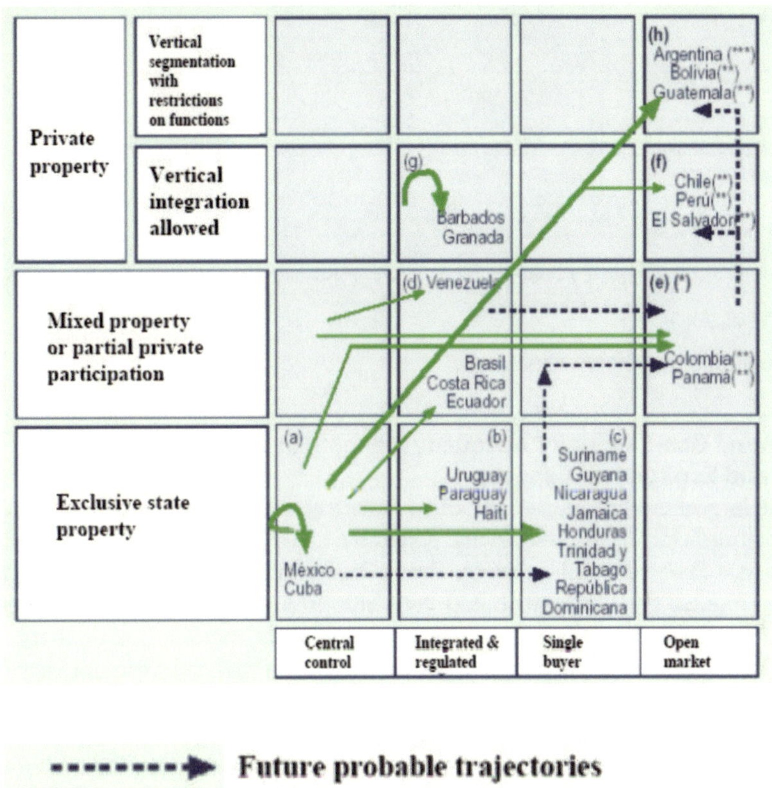

Source: Cepal, 2003 [13])

Fig. 12.4a. Electricity Sector Reform Process

These coherent processes in both electricity and gas have facilitated their interaction and competition, stimulating hybrid solutions, which used the technical and economic advantages of both. For instance, private investors can transmit energy as natural gas through pipelines or as electricity through transmission lines, either generating electricity at the gas field end or at the consumption end. Additionally, regulations facilitated this competition by forcing open access schemes to monolithic transport facilities that were built in several countries in the region. Figures 12.4a and 12.4b describe in general the deregulation process of the electricity and gas industries in the region.

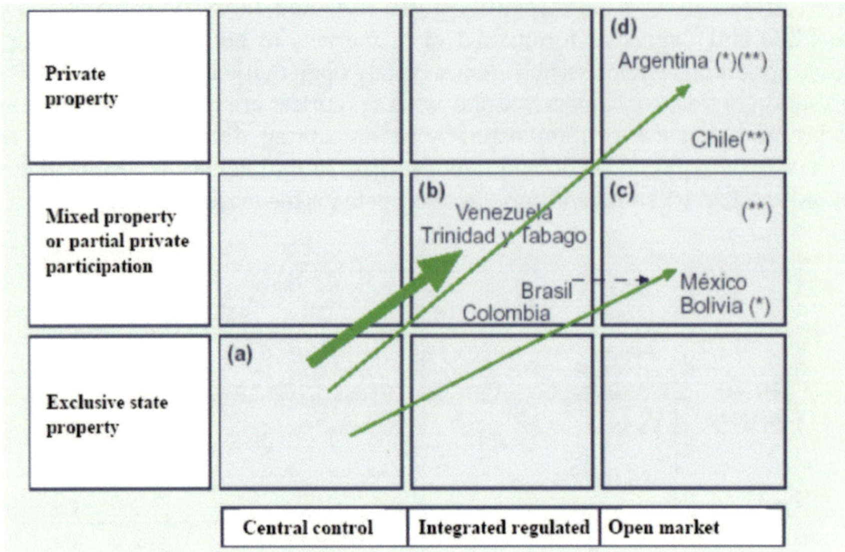

(Source: Copal, 2003 [13])

Fig. 12.4b. Gas Sector Reform Process

12.3 Integrated Gas-Electricity Adequacy Planning In Brazil: Technical and Economical Aspects

Brazil is the largest energy market in South America, accounting for 40% of the continent's energy consumption. On the electricity generation side, the country is hydro-dominated. Despite its gas reserves and imports, Brazil has a relatively undeveloped gas market. Historically, natural gas has contributed very little to Brazil's energy mix. The country has little or no need for space heating; hence there is little market potential for gas in the residential and commercial sectors, and local distribution networks are not very developed. As a result, gas consumption in the country is concentrated in the energy-intensive industries that replaced oil derivatives and electricity use by natural gas. Although the natural gas demand for industrial/vehicle use has been growing at relative high rates, this demand growth solely is unlikely to justify large investments in gas production and transportation. This implies that, at the moment, the power sector is the largest potential market for natural gas, which can provide the necessary anchor to spur production and infrastructure investments in this sector.

However, development of the natural gas industry in an environment where its requirements are very volatile due to the randomness of inflows is a key issue in Brazil. This immediate dependence on gas consumption from power generation creates special challenges for the country in the electricity-gas integration. Theses challenges will now be briefly discussed.

12.3.1 The Brazilian Electricity and Natural Gas Sectors

The Brazilian interconnected power system had, in 2007, a total installed capacity of about 100 GW. The system is hydro-dominated: 85% of the 100 GW installed capacity and more than 90% of the 400 TWh energy production comes from hydropower. Some plants have large reservoirs, capable of multi-year regulation. Thermal generation includes nuclear, natural gas, coal and diesel plants. The area supplied by the system is served by 80,000 km of a meshed transmission network. The main direct international interconnections are the back-to-back links with Argentina, with a maximum transfer capacity of 2,200 MW. Power sector reform with emphasis on privatization and competition was initiated in Brazil in 1996. The reform process was disrupted in mid-implementation by a severe energy rationing that took place along 9 months in 2001 - 2002.

Following some regulatory changes that occurred after rationing, a new model for the power sector was launched in 2004. The main objectives of this new model are to provide more secure conditions for investors and assure security of electricity supply. A review and assessment of this power sector model are provided in greater detail in [2,3] and in Chapter 17 of this book. Its main highlights are: (a) every load (regulated plus deregulated consumers) must be covered 100% by bilateral financial contracts at all times and all contracts must be "backed" by firm energy certificates; and (b) Discos must contract their energy through public Power Purchase Agreements (PPA) auctions, with standardized rules and contracts of different maturities.

The objective of rule (a) is to enforce security of supply by a contract obligation with physical coverage. The need to cover with contracts load growth will drive the system's expansion, and if the system is 100% contracted and contracts have physical coverage, then supply reliability is assured within the 'supply risk' defined in the calculation of the physical coverage capacity of each plant (firm energy in case of hydro). In turn, the objective of rule (b) is to promote the most efficient purchase mechanism for regulated consumers and provide secure conditions for investors. Regulated auctions for new energy are carried out on a least-cost basis three or five years in advance of delivery. This is to give enough time to build the project. Long-term PPA (15 to 30 years) is offered to investors and can be used as collateral with banks to obtain financing. In addition, all competing projects in the auction must have a prior environmental license in order to reduce environmental risk in the future. Energy supply auctions have been constantly carried out in the country since 2004. The first auction to contract new energy was carried out in 2005 and was of great interest to international investors looking to South America's energy market. Candidate suppliers included a wide variety of technologies, comprising new hydro projects, gas, coal and oil-fired plants, sugarcane biomass and international interconnections. Overall, the auctions until now contracted some 10 GW including a mix of technology from all candidate suppliers.

On the natural gas side, Brazil has proven gas reserves (are estimated at 220 billion cubic meters (bcm)), accounting for 4% of South America's total proven reserves. Despite this low level of reserves, Brazil is thought to have substantial potential for new gas resources. In particular, the discovery of a large offshore natural gas field (Santos basin) was announced in 2004. Even though this field requires very deep sea drilling (4000 m), its reserves are estimated at 420 bcm, which can double the level of the current reserves. National gas production (available to market) is in the range of 27 MMm^3/day. Production in the recent discovered Santos field are expected to increase this figure in about 20~25 MMm^3/day when it starts operating (current forecast is 2011). Since 1999 Brazil has been importing gas from Bolivia through the "Gasbol" pipeline. It is the largest capacity pipeline in Latin America, with 30 MMm^3/day, built by private investors. Imports in 2005 were in the range of 26 MMm^3/day. Brazil used to import gas (2.8 MMm^3/d) from Argentina to supply a 600 MW thermal plant on the Brazilian side of the border between the countries. However, as will be discussed later, Argentina has struggled to meet its own domestic gas needs and has started cutting exports to Chile and Brazil.

Without considering gas for power use, natural gas consumption in 2007 was in the range of 48 MMm^3/day. Most of the gas is used in the industrial sector. Because there is virtually no need for space heating in Brazil, gas use for the residential and commercial sectors remain limited to cooking and water heating. The use of gas for transport has been increasing. This is mostly encouraged by the competitive price of compressed natural gas (about half the price of gasoline when driving the same distance). There has been a strong growth outlook over recent years, where industrial and transportation sectors have been the main growth areas (motivated by government policy and increase in oil prices).

Since the 1990s Brazil has been calling for a larger share of thermal capacity to be fuelled mainly by natural gas. This is to reduce dependence on hydroelectricity and to boost natural gas demand. However, little happened until 1999-2000, where under imminence of the energy supply crisis, a program for an incentive for thermal generation was launched. This program resulted in the construction of about 7000 MW of gas-fired plants by 2005 (out of a planned 14,000 MW capacity). This corresponded to maximum gas consumption in the range of 35 MMm^3/day, about the same amount as the entire 'non-power' gas demand. In addition, thermal plants' dispatch depends on the hydrology: if the system is 'wet', the entire electricity load can be met with hydro generation alone.

As Brazil is a country the size of a continent, several distinct gas markets can be expected to develop, each characterized by its own supply sources, demand centers and transportation networks. Today three natural gas markets can be distinguished in Brazil: the largest and most developed system by far comprises the South, Southeast and Central-West regions. Coastal cities from the Northeast form the country's second natural gas system. The third system, with abundant reserves still to be developed, is the Amazon region, located North in the country. Integration between the Northeast and Southeast is planned.

Figure 12.5 shows the main (cross-regions) natural gas and electricity transportation networks. It can be seen that the gas network is still developing its infrastructure when compared to the existing electricity network.

As opposed to the power sector, Brazil currently does not yet have a clear policy or guidelines concerning the gas sector. The law that liberalized the petroleum sector in 1997 treats gas as a by-product of oil. The Brazilian government worked on a new law in 2008 for the gas sector.

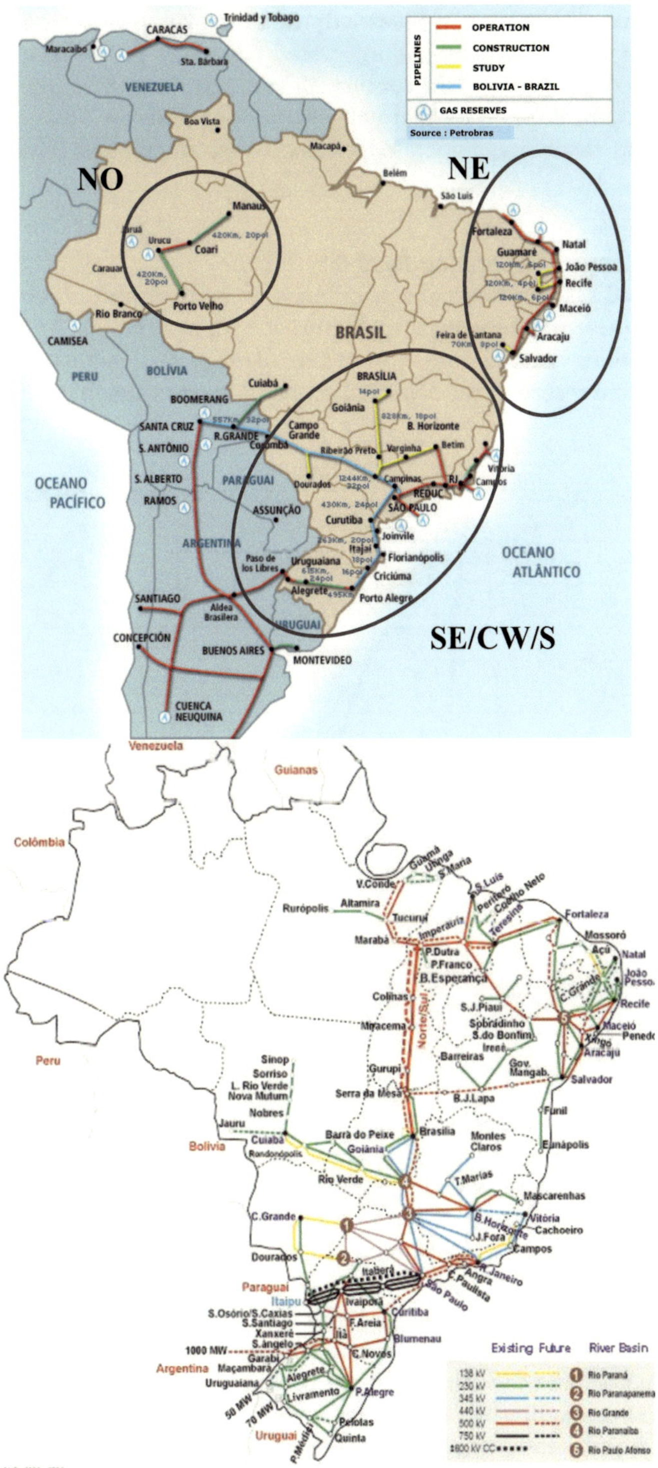

Fig. 12.5. Main Natural Gas and Electricity Transportation Network

12.3.2 Brazil's Main Challenges in Electricity-Gas Integrated Adequacy Planning

The main challenges in electricity-gas integrated adequacy planning in Brazil are those related to the high dependence of the country on hydropower and in turn the reliance of the gas market on gas consumption for thermal generation to develop. This section details three of them: operating flexibility, integrated electricity-gas operations planning and flexible LNG supply.

12.3.2.1 The challenge of operating flexibility

In hydro-dominated systems thermal generation is generally useful as backup for periods of low rainfall. This means that the existing thermal plants may be idle in periods of high (or average) precipitation, which occurs most of time. This pattern is illustrated in Figure 12.6, which shows a recent record of the observed energy market prices in the Brazilian Southeast system.

Fig.12. 6. - Historical Monthly Short-Run Marginal Costs

Figure 12.6 shows that the system marginal cost is very volatile, leading to very "sporadic" dispatch of thermal generation (associated to bad inflow conditions). The reason for this behavior is that predominantly hydro systems are designed to ensure load supply under adverse hydrological conditions, which occur very infrequently. Hence, for most of the time there are temporary energy surpluses, which result in very low market prices and no need of thermal dispatch. In turn, an occurrence of a dry period usually calls for the dispatch of all thermal plants "*at the same time*", which in turn calls for a robust pipeline network capable of meeting this "*volatile*" gas demand. One of the consequences of this "*feast or famine*" price characteristic is that it creates a very "*volatile*" gas demand from power generation. Since it is not economical to build production and transportation infrastructure to be idle for most of the time, this "*irregular*" consumption pattern from power generation creates a complex problem for investment decisions in new gas fields and in new pipelines that may be either excessive or insufficient depending on hydrological conditions. Take or pay and ship or pay clauses on the gas contracts between gas producers and thermal plant owners are mandatory to alleviate part of the financial uncertainty: these clauses are used to stabilize, from the gas producer point of view, the irregular cash flow that would arise from market operation of the power plants. On the other hand, these clauses decrease competitiveness of projects since thermal owners are paying a fixed price for gas independently of use. Thus, hydropower or flexible oil-fired and coal tend to be cheaper expansion supply options.

One alternative to alleviate these fixed expenses is to allow the pre-purchased gas to be resold in a secondary market whenever the gas-fired plants are not dispatched. In this scheme, flexible contracts would be offered (e.g. auctioned) to industrial consumers, who in turn would switch to an alternate fuel when the gas supply is interrupted due to thermal dispatch. The revenues obtained with the flexible contracts could be used as an extra income to reduce the fixed-costs of thermal plants. In addition, two consumers would be sharing the same gas infrastructure, thus optimizing investment needs.

12.3.2.2 Integrated Electricity-Gas Operations Planning

In Brazil, the system operator undertakes power system dispatch centrally. The system operator acts as if all plants belonged to the same owner. Hydro plants are dispatched based on their expected opportunity costs ("*water values*"), which are computed by a multi-stage stochastic optimization hydrothermal scheduling model that takes into account a detailed representation of hydro plant operation and inflow uncertainties [4]. Traditional hydro scheduling models used for system dispatch take into account a detailed representation of the power system (including electricity network), but do not take into account representation of constraints of infrastructure (production and transportation) of the natural gas sector. In other words, the approach assumes no constraints in the gas sector. This decoupling may imply dispatch results for the power sector that can be dangerously "*optimistic*". This is because the model may consider thermal dispatches that will be "*infeasible*" due to gas production or transportation constraints. For example, in January 2004, a shortage of hydropower in Northeast Brazil implied the System Operator to dispatch existing gas-fired resources in the region. However, only 30% of the gas-fired capacity installed in that region was able to generate due to gas production and transportation constraints, which were not "*seen*" by the hydro scheduling model. Therefore, an integrated modeling of the gas sector (production and transportation constraints) in operations of the power sector is one of the main challenges—and was analyzed in [5].

12.3.2.3 Integration of flexible LNG supply

In end-2006 a "*dispatch test*" performed by Aneel (power sector regulator) in the gas thermal power plants disclosed that the concern about gas supply and demand balance was actually legitimate, because about 50% of the tested capacity in the South/ Southeast-Center West Regions did not manage to produce energy due to fuel deficiency. The reason for the lack of fuel was two fold: (i) lack of local gas supply (non-power consumption grew at a faster pace than local gas supply) and (ii) difficulties with Bolivisn and Argentinean gas imports. In particular, the latter cut gas supply to Brazil to prioritize the supply of its own market. This has affected directly a 2,200 MW interconnection with Brazil and a 600 MW gas-fired plant.

In an effort to increase the natural gas supply in the country and to reduce the dependence of its neighbors, Petrobras announced in 2006 the construction of re-gasification stations, so as to import LNG, from 2009, to the Southeast and Northeast Regions. These gas imports would come from LNG exporters such as Trinidad & Tobago and Nigeria and Petrobras decided to implement mobile floating storage re-gasification units (FSRU).

Since the scheduling of LNG ships must be done some time before the actual need, observe that there is an uncertainty relating LNG supply and thermal needs, highly influenced by hydrology: if scheduled well beforehand, better gas prices can be obtained but higher is the uncertainty of hydrology on thermal dispatch; if scheduled close to the delivery, the uncertainty on hydrology is smaller but the gas price might be higher. Therefore, once more, it becomes necessary to develop flexible supply and demand options. Concepts discussed include: flexible (interruptible) gas contracts for the industry and storage options such as the use of hydro reservoir capacity. For example, delivered gas but not used could be stored as water in hydro reservoirs for future use (as a "call" option). These issues will be discussed in more detail in section 12.7.

12.4 Chile: Uncertainty in Natural Gas Supply

Chile was a pioneer worldwide in liberalizing the electrical generation segment in 1982, introducing a competitive and private market, where the entrance of new agents depends on the economic signals that the investors gather from the market. Therefore, in Chile the decision as what are the technologies to be developed essentially relies in private investment evaluation. The government is solely limited to generate the conditions so that it is possible to reach economic efficiency.

The process of liberalization and deregulation of the electricity market was companioned by the privatization of the existing state owned electrical companies. Currently, the governments influence in the sector is limited mainly to regulation functions, indicative expansion planning and to the fixation of the electrical tariffs for regulated clients. Historically, objectives as the diversification of the power matrix and the environmental sustainability has conformed a secondary level. The development of the generation segment has occurred in a frame of a technological neutrality as far as the technologies and fuels used, having all types of energy sources to compete in similar conditions of quality and price. Chile is a country with limited energy resources other than its hydro reserves in the Andes. Its own oil provides less than 10% of the country's needs and its coal is of poor quality, so imported coal has to be used for electric generation. Hydroelectric generation has developed using most of the low investment cost resources in the central part of the country, and remaining significant reserves are over 2,000 km south of the main load [7]. Although at present the Central Interconnected System (SIC) has a generation park with presence of different power generation technologies, 15 years ago supply presented a markedly hydroelectric component, with a participation of 78% in the total installed capacity. Figure 12.7 shows the evolution of the generation capacity in the SIC.

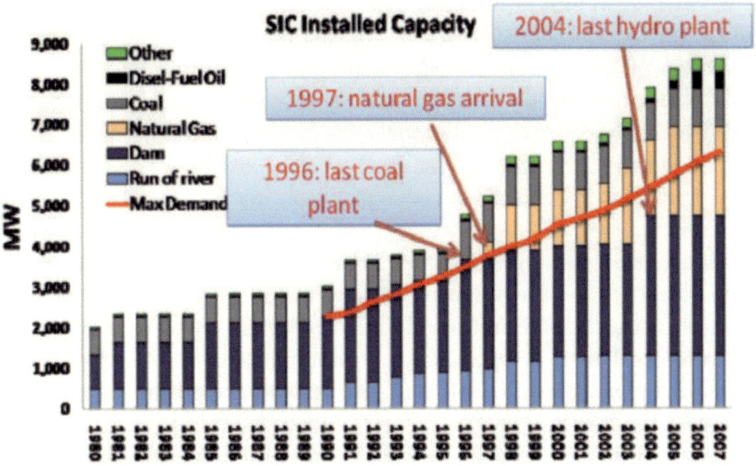

Fig. 12.7. Installed Capacity Evolution in Central Interconnected System

The conditions created for market competition in electricity drove private investors to look for more economic technologies and fuels. Argentinean gas arose as an attractive abundant economic alternative, coupled to the availability of the low investment required by efficient combined-cycle generation technologies. With the private interest to use the gas and the support of both the Argentinean and Chilean governments, an energy integration protocol was signed in 1995 between both countries. Under that protocol, both governments agreed to establish the necessary regulations to allow freedom of trade, export, import and transportation of natural gas. Private investors invested heavily in several pipelines that crossed the Andes and defined an energy supply path that would rely heavily on combined cycle generation. The protocol worked very well and Chile fully relied on Argentina to provide the necessary energy required to sustain its important economic growth. Gas exports grew steadily through four international pipelines (Figure 12.8). The petrochemical industry and the thermoelectric generation became the main users of natural gas (Figure 12.8).

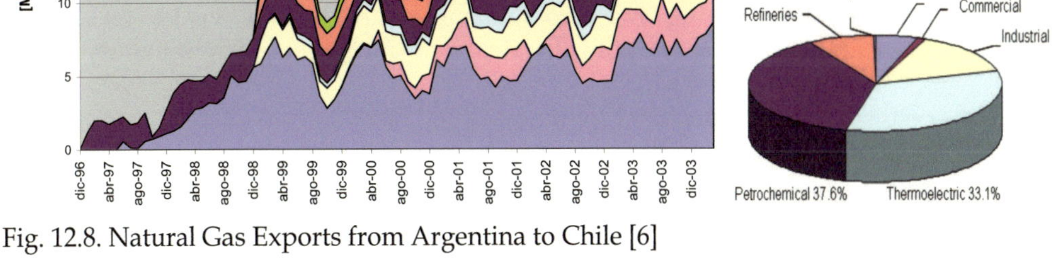

Fig. 12.8. Natural Gas Exports from Argentina to Chile [6]

The arrival of this economic fuel and the efficient generation technologies meant a significant reduction in electricity prices in the main central interconnected system. The energy spot price in the 1994 -2007 period is shown in Figure 12.9 together with the energy generation composition between thermal and hydro sources. A more balanced energy matrix was achieved with the arrival of natural gas from Argentina, but still very dependent on hydrological conditions, as can be seen in the 1997-1999 period, where high spot prices and supply restriction were faced as the consequence of intense droughts in the country.

However, when a macroeconomic economic crisis started affecting Argentina in 2001, the country changed dramatically its government policies, among them those relayed to energy, electricity and gas included. State returned to control energy prices, freezing them and not relating them to the real costs of production. Argentina natural gas prices were reduced to one third of their previous levels (due to a severe devaluation of the Argentinean peso) and this led to an escalating demand, not necessarily backed by investment in the exploration of new gas fields neither in new pipelines. This led to problems with its gas supply. Then, unilaterally, the Argentinean government, violating its agreements with Chile, in April 2004 decided to reduce gas exports to Chile. The Chilean power sector since the electricity deregulation process started has had several opportunities to have the strengths and weaknesses as a market model tested, and this was a particular major blow [6]. Figure 12.10 shows Argentinean gas import restrictions in the 2004-2008 periods.

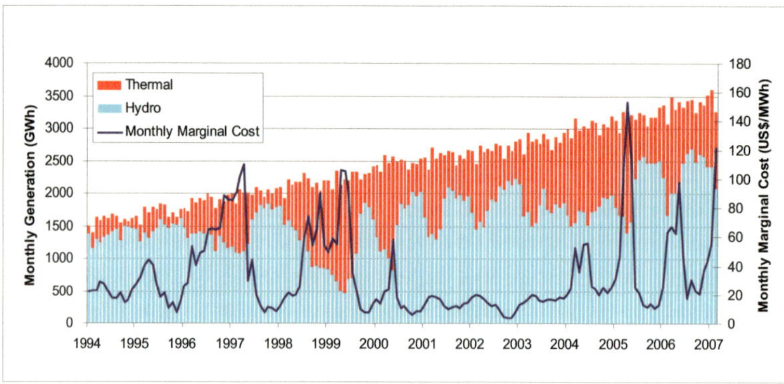

Fig. 12.9. SIC Hydrothermal Generation Structure and Marginal Cost

The non compliance by the Argentinean government of its international agreements, deciding to favor national supply, was not only detrimental to the consumers of Chile, but also to the neighboring countries of Uruguay and Brazil. Bolivia has significant natural gas resources and it is increasing exports to Brazil and Argentina, helping the later to diminish its crisis. However, given its long-term border disputes with Chile (Bolivia lost its access to the Pacific in a 19th century war with Chile), it denies the fuel to the adjacent country.

Chile was neither aware nor prepared for the surfacing conditions. As a demonstration, the National Energy Commission, in its indicative plan of April 2004, projected the building of seven combined cycle natural gas plants in the next decade, all fed by pipelines from Argentina. Mainly expansions of existing electric transmission corridors were included in that plan. Major new hydro plants and interconnections with other systems were postponed until 2010 or later, gas continued to be the major driver of expansion in a market with demand growing around 7% per year. With the crisis developing, the October 2004 indicative plan introduced radical changes to the government view of energy supply expansion. Only one combined cycle plant based on Argentinean gas was considered for 2007. The government decided to bet on LNG as the alternative and defined a project to build the necessary installations to import it from abroad.

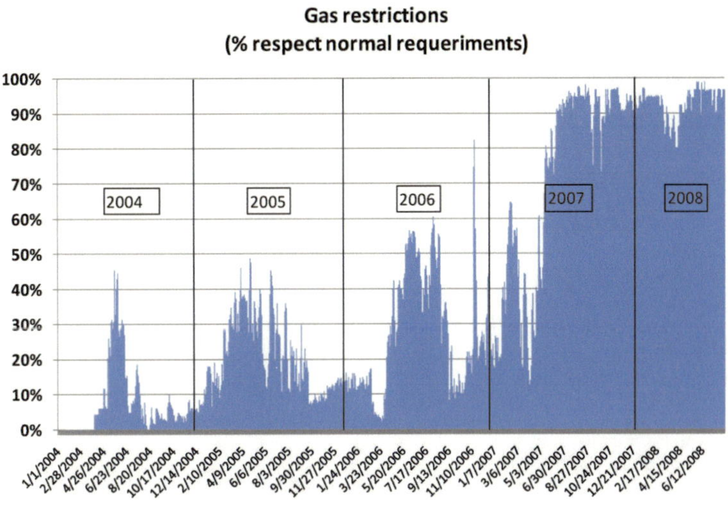

Fig. 12.10. Argentinean Gas Import Restrictions

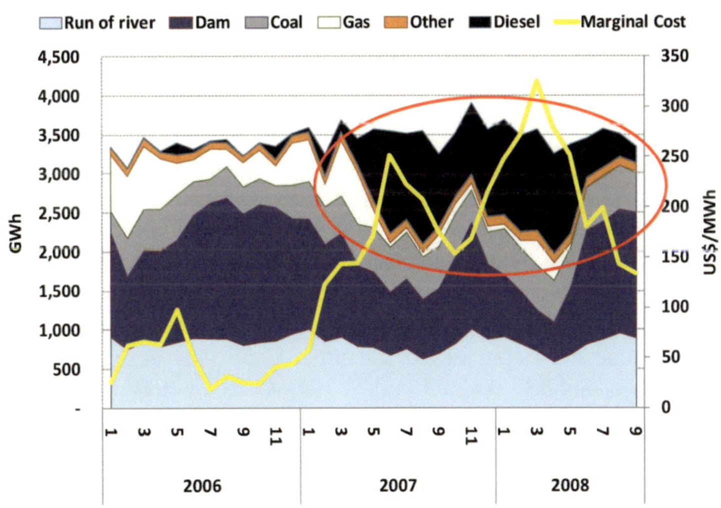

Fig. 12.11. Energy Generation by Source and Marginal Cost in SIC

As a consequence of the growing natural gas restrictions from Argentina, which reached 100 % as of 2008, combined cycle plants replaced the suspended fuel with diesel, facing fuel costs that were several times higher, as can been seen in Figure 12.11, were energy generation by source and spot market prices are showed.

But in the deregulated privatized Chilean power market, where private capital is the one making investment decisions, there is little space for the government to act, unless changes to laws are introduced. The electricity price scheme relies essentially on market competitive forces, with only part of it, prices for small consumers (under 500 kW), being regulated.

In retrospect, it is the cost of different generation technologies that will drive development. And the comparison has to be centered on the particular geographic conditions and infrastructure development of Chile. A recent comparison [6] was made for the Chilean investment environment (local cost of capital taken into consideration); where LNG combined cycle plants compete with circulating fluidized-bed boilers fuelled by coal. Cost of different generation technologies are stated in Table 12.1.

Tecnologies	Investment (US$/kW)	Average generation cost (US$/MWh)	Fuel cost (US$/mBTU)
Reservoirs (400 MW)	1100	25.5	0.0
Run of river hydro (400 MW)	1250	28.7	0.0
Combined cycle natural gas (394 MW)	530	31.6	2.8
Coal circulating fluidized bed (250 MW)	1270	44.5	1.8
Combined cycle liquefied natural gas (394 MW)	530	45.1	4.7
Combined cycle diesel (394 MW)	550	66.8	7.9
Gas turbine (120 MW)	430	103.5	7.9

Table 12.1. Cost of Different Generation Technologies, Chile, 2004 [6]

However, an essential question troubled investors at the time of the crisis, what if gas supply from Argentina returns to normal? If a decision was made, for example, to contract a coal investment, and cheaper natural gas starts re flowing without restrictions from Argentina, who will make the cost? High financial exposures may arise depending on decisions made. As indicated, in a deregulated environment, where private investors are the ones that decide expansion, the government is uneasy when faced with an uncertain energy supply that may seriously hurt economic development. At the time, Chile was returning to high economic growth rates, therefore, nobody wanted energy deficits to shadow economic development. Thus, the government looked for alternatives. Capacity payment regulations were modified to better take into account unreliable gas supply. A gas "*drought*" concept was introduced that de-rates combined cycle plants that do not have alternative fuel arrangements, and therefore reduces their capacity payments. Another alternative was considered, but discarded, was to limit by law the dependence on foreign fuels to a certain percentage of national consumption; the idea was that imports from a particular country should not exceed a certain value (a similar concept is used in Spain). Critics of this alternative argued it represented a State intervention that would imply higher long-term energy costs for the country as a whole. The LNG in Chile is discussed in more details in section 12.7.

In order to incentive investments in new generation capacity, the government went ahead and introduced a major regulation change in 2005 by introducing energy auctions for long term contracts to supply regulated end users. These auctions are based on up to 15 year contracts to be signed with distribution companies, at prices set in a bidding process, as explained in depth in Chapter 17. Through this major reform, risk could be reduced and new investment started flowing, as it can be seen in Table 12.2. Because of the lack of confidence in regional energy suppliers, the energy matrix shifted to local hydro development and coal fired plants, leaving the LNG to a backup role.

Year	MW			
	Coal	Gas	Diesel	Hydro
2008	-	-	373	-
2009	139	240	232	155
2010	924	-	-	172
2011	482	-	-	327
2012	445	-	-	553
2013	-	-	-	705
2014	250	-	-	660
2015	-	-	-	-
2016	300	-	-	500
2017	300	-	-	-
TOTAL	2.840	240	605	3.072

Table 12.2. New Investment in Power Generation in Chile, 2008 (NEC, April 2008)

Along the disputes that arose on the Argentinean government not to comply with the international agreements on gas supply, the Chilean government chose not to go into an international litigation, believing that a permanent political negotiation would be more fruitful in the short term solution of the gas supply problem. This was also supported by observing the long disputes that were taking place between Argentina and its foreign lenders and with foreign companies and investors that had their interested affected by the crisis.

12.5 Mexico: Growing Interactions Between Mexican Gas Markets And Electricity System Planning

The Mexican gas and electricity sectors have followed a non-synchronized agenda of reforms [8]. Some of the reforms in the gas sector have not been completed and the more profound reforms in the electricity sector have remained idle or without consensus for recent years. The interactions among the two systems, one -the gas sector- with a considerable degree of openness and the other, basically vertically integrated, have not avoided their growing interactions due to new technological and cost development in the gas markets. These interactions have direct impacts in electricity system resource planning. The increasing interactions between the two systems pose important questions. Among them is the need for new planning tools that represent important opportunities for research. On the energy side, of increasing importance for Mexico for the reliable and cost effective supply of gas to the electricity systems is LNG. This represents a clear alternative to continental gas supply to Mexico's electricity systems. The dynamics of LNG markets have also had an effect on traditional electricity system planning where more complex tolls for system planning may be required.

12.5.1 Gas Supply Demand for Electricity Production

Electricity expansion planning in Mexico indicates that least-cost expansion planning of the system will continue to rely in combined cycle plants for the next ten years (Figure 12.12). This has been the case in the last decade.

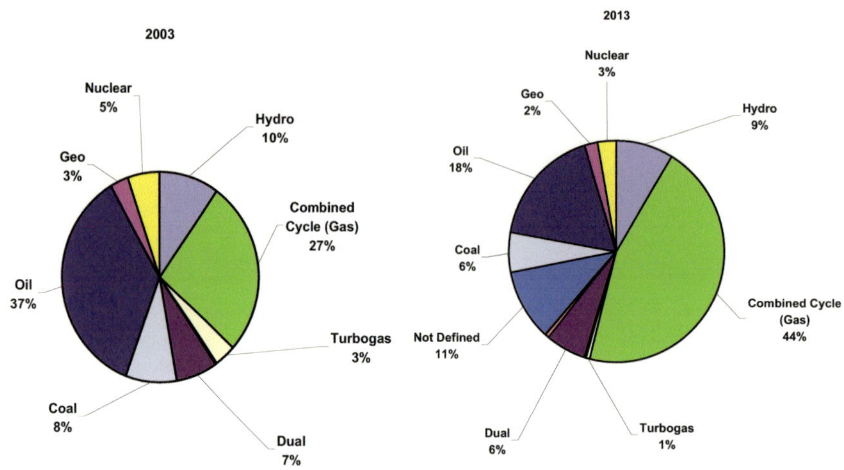

Fig. 12.12. Electricity Generation Installed Capacity Shares by Fuel Type, Actual (2003) and Planned (2013)

The share of gas as fuel for electricity supply will grow from 27% to 44 % of total electricity production from 2003 to 2013. The increasing extension of the national gas pipeline system and its connection to the US market and the growing worldwide Liquefied Natural Gas Market have resulted in interesting interaction among the traditional planning of an almost vertical integrated electricity utility and a more open and mature market for natural gas. Gas consumption for electricity generation (MM cubic feet/day) and planned LNG installations in Mexico is indicated in Figure 12.13.

12.5.2 Gas/Electricity Network Interactions

A specific project for electricity generation called Tamazunchale consisting of a large combined cycle plant of around 1000 MW required to supply the central region of Mexico was identified by the classic cost-minimization approach that is used for the electricity system expansion planning. Current models did not capture the fact that the territorial sitting of the plant had different alternatives that would require either: (i) the sitting of the plant beside an existing gas pipeline with the need of a new transmission line to connect the plant, or (ii) the sitting of the plant beside an existing transmission line with the need of a new gas pipeline to transport gas supply to the plant. The decision of sitting was left to the investors (i.e. to the market) in a bidding process that asked for a 1046 MW combined cycle plant with two different sitting options. Therefore, one important issue was how traditional vertical integrated planning interacted with a bidding (market) mechanism that asked for a long-term contract for electricity supply with two alternate delivery points. This issue had to be resolved for the interaction between the Gas/Electricity transport choices for the project.

12.5.3 LNG/Electricity Expansion Interactions

The increasing consumption of gas in Mexico for electricity production along with the lesser than expected national growth of the internal production of gas indicates that import of gas from the US (Texan or Californian) market through the national pipeline system will still be an alternative to a secure gas supply, although not at a competitive price. However, the increasing maturation of the LNG market worldwide makes this alternative an even less cost-effective alternative for supply of gas if certain considerations are made in the electricity generation expansion plans. Therefore, a basic challenge for Mexico is to incorporate the dynamics of LNG markets in traditional expansion models in order to better capture the costs and benefits of LNG as a supply source for the country instead of using pipeline gas from US markets through the national system.

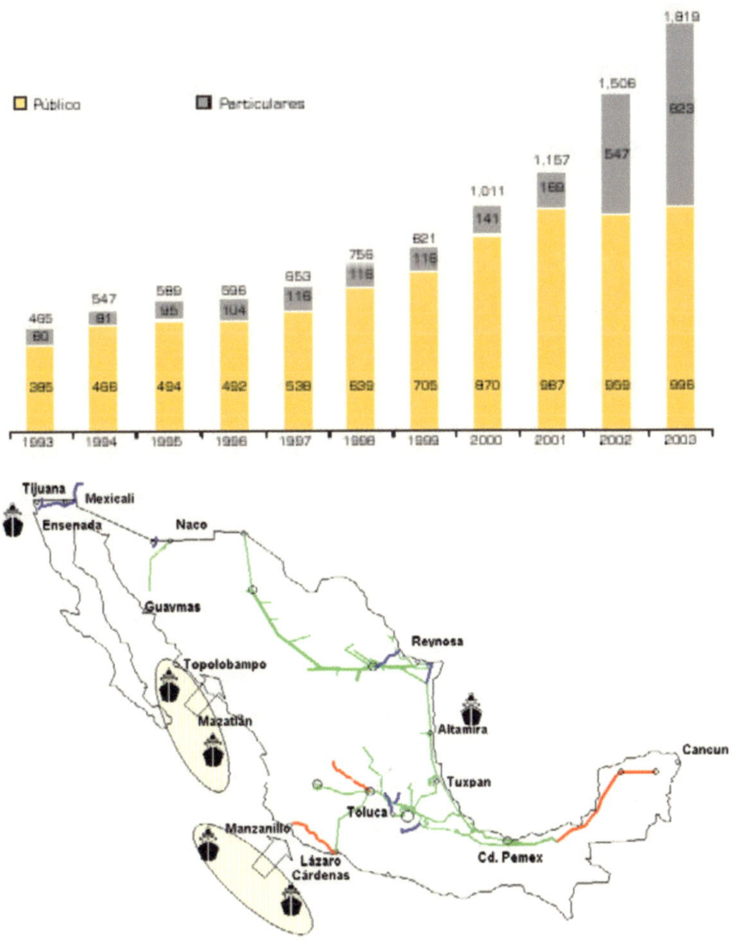

Fig. 12.13. Gas Consumption for Electricity Generation (MM cubic feet/day) and Planned LNG Installations

12.6 Natural Gas and Electricity Market Issues in Colombia

Colombia has numerous primary energy resources: oil and associated natural gas in the Interior region of the country, free natural gas in the Atlantic Coast region, hydroelectric resources mainly in the Andean Mountains and extensive coal deposits both in the Atlantic Coast and the Interior regions. Hydroelectricity is used to serve around 65% of the electricity market; the remaining 35% is served by coal and natural gas fired plants. Natural gas is also used in oil refining, industrial, residential, commercial and transportation. As in Brazil, development of the natural gas industry in an environment where its requirements are very volatile due to randomness of river discharges is a key issue in the Colombian energy sector.

Development of the natural gas industry in Colombia is recent. Although there were local natural gas uses since the 1950s, its massive utilization started in the middle of the 1970s in the Atlantic Coast region with the utilization of free natural gas reserves located in the region. In the middle of the 1980s a Government plan accelerated natural gas service extension towards urban centers. Later on, in the 1990s, another incentive plan was implemented. Its main component was for gas transportation infrastructure. It is in operation today connecting the gas fields with main consumption centers. The above actions have been complemented with an increase of natural gas reserves due to new findings in the Interior, the start of a new regulatory framework for the natural gas market, and by the dynamics of new natural gas demands. In particular, since the start of this Plan, 3010 MW of new gas fired plants have been installed, representing 22% of the total power capacity in the country.

Demand for natural gas in Colombia has been growing significantly, subject to volatility due to gas consumption for thermoelectricity that in 1998 reached an annual average of 304 MBTU/day. Natural gas consumption in Colombia rose to 589 MBTU/day in 2003, of which 181 MBTU/day was for electricity generation. Average supply of natural gas in Colombia during 2003 was 595 MBTU/day, 478 MBTU/day of it produced in the Atlantic Coast fields. It is expected that an interconnection gas pipeline with Venezuela will start operation in 2007. This will enable natural gas exports to the country for several years and, eventually, will allow future natural gas imports. This interconnection would enlarge the Colombian gas market, enabling international natural gas traders to develop the Colombian natural gas reserves.

The gas supply, transportation and supply outlook in Columbia is indicated in Figure 12.14.

Natural gas demand for electricity generation in the country is subject to large volatility. It is highly seasonal due to the nature of the Colombian power system that has a large hydroelectric component. River discharges are substantially affected by the El Niño phenomenon. Its occurrence implies large thermoelectric use to compensate for the decrease in hydroelectric generation. Guerrilla attacks to the transmission infrastructure are another source of uncertainty in demand for natural gas since it forces thermal generation in some areas that do not have hydroelectric resources.

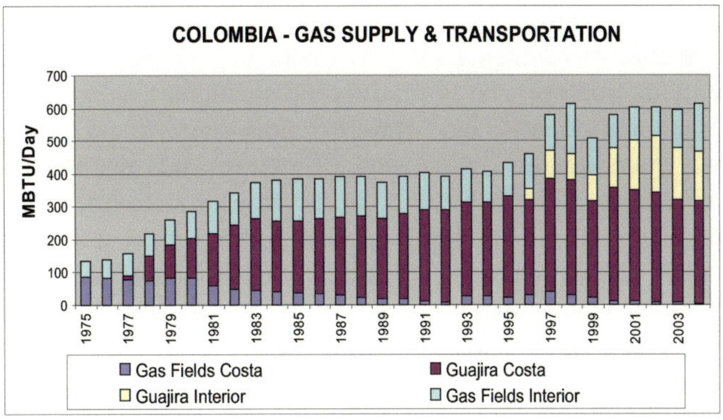

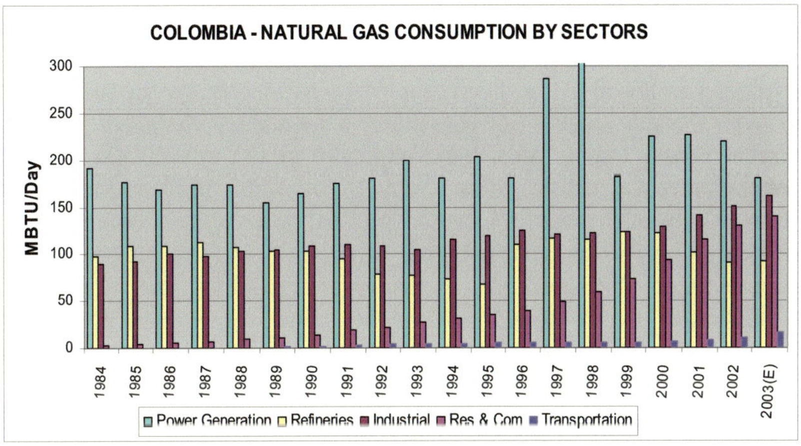

Fig. 12.14. Gas Supply, Transportation and Supply Outlook

There has been a relevant investment from state and private companies in recent years to connect main production gas fields to the principal consumer centers around the country through the construction of new gas pipeline grids. Estimates of natural gas demand in Colombia in sectors different from electricity generation assume that the Atlantic Coast regions have the largest and most developed markets. Under such assumptions, the highest demand increases would occur in the Colombian Interior region. This is a result of natural gas penetration that would occur in the residential, industrial and transportation sectors.

The forecasted natural gas demand in the industrial sector has been influenced by strict environmental regulation on emissions since year 2000. Environmentally aggressive fuels have been substituted by natural gas in the sector.

Natural Gas and Electricity markets have strong links in Colombia and there are several issues related to the interaction between them [8]. These include:

a) Capacity Charges: The large hydroelectric component of the installed capacity in Colombia implies that some of the natural gas fired plants have very low dispatch probability but are required to guarantee supply reliability. The main issue related to this is the design of an appropriate capacity charge mechanism to create financial incentives for the installation, operation and maintenance of these types of plants without creating economic adverse distortions.

b) Power transmission and gas transportation charges: Achievement of optimal integrated operation and expansion of power and gas transportation systems require correct incentives given by an appropriate scheme of regulated charges. Colombia has a simplified stamp and deep connection charge scheme for power transmission while complex distance related charges are applied to gas transportation. This creates perverse incentives to integrated power-gas system optimal operation and expansion. In addition, volatility in gas demand (from randomness of hydroelectric generation) constitutes a challenge.

c) Natural Gas vs. Electricity Markets: The Colombian electricity market is a price bid based highly competitive market with more than 30 generators participating while the Colombian natural gas market is reduced to a few participants requiring regulated wellhead prices. Even though the regulatory agency has given the signal to open the gas market this constitutes a regulatory challenge given the related market power issues. Also, the complexity of the natural gas based electricity generation cost structure within a main hydroelectric bid based market constitutes an issue to be addressed for incentive optimal power system operation.

d) Market surveillance: International experience of bid based power markets demonstrates the need of a market surveillance mechanism to prevent inefficiencies due to market power actions and to guarantee appropriate market development. In the Colombian case, inclusion of the gas market in the surveillance scheme is a critical issue that needs a solution.

12.7 LNG in South America

As discussed previously, LNG is increasingly at the heart of energy policymaking in South America. The rationale behind LNG projects varies among countries and sometimes within the same country. However, there are three main drivers behind LNG import and export projects in South America.

a) Gas imbalances: the first reason for importing or exporting LNG is related to the region's natural gas balance: there are countries or sub regions with gas surpluses and others with deficits. Brazil, for example, has a growing potential natural gas market and still not enough gas production. Given the large distances and the geographical obstacles, it is not always possible or economical to export or import pipeline gas. LNG imports are being sought as a way to increase gas supply. On the other hand, countries with abundant gas resources, such as Peru and Venezuela, are looking at LNG exports as a way to market their natural gas and monetize their reserves;

b) Security: the second reason is geopolitical and is related to energy security and the diversification of natural gas supplies and markets. In Brazil and Chile imports from neighboring countries have proven to be unreliable and further dependence on supply from a single country is deemed to be undesirable. LNG might become a way to diversify gas supply and some bargaining power in the discussion with regional suppliers. Similarly, Peru could export gas regionally by pipeline, but the LNG export option is considered less politically charged than pipeline.

c) Flexibility of gas supply: the third reason for LNG imports is related to the nature of gas demand and a growing need for flexibility in gas supply. Because of the hydro predominance in the region, gas-fired dispatch is very much volatile and flexibility is an attractive attribute. However, flexibility comes at a price and it remains to be seen whether LNG is a cost-effective way of achieving supply flexibility. Specifically, in Brazil a large portion of gas demand is linked to the power sector and is highly variable because of the country's dependence on hydropower. LNG imports are deemed to provide more flexibility at a lower cost than building large pipelines.

This section analyzes the introduction of LNG in Chile and in Brazil.

12.7.1 Main Challenges for LNG in Chile

As discussed in section 12.4, since 2004 Argentina has struggled to meet its own domestic gas needs and has started cutting exports to Chile. Total annual exports to Chile have been falling since 2005 and cuts started to be frequent and recently (2007) have reached as high as 95 percent of committed volumes on several occasions, as shown in Figure 12.10. Restrictions have affected mainly the thermal power sector and the industrial sector, forcing power plants and industrial consumers to switch to costlier fuels.

In response, Chile has launched a program to import LNG not only to supply additional gas demand but also to replace decreasing Argentine exports. An LNG terminal is being constructed in Quintero, Central Chile. Figure 12.15 shows the terminal's location. Its construction is well advanced; the terminal started partial operations in second quarter 2009, with full-scale operation by late 2010.

A pool of off takers including State owned oil company ENAP, power generator Endesa Chile, and gas distributor Metrogas was created. In early 2006 the pool selected UK gas company BG Group both to supply LNG and to construct the terminal. Off takers have already contracted 6 MMcm per day of regasification capacity (final capacity could be as high as 12 MMcm per day). Other off takers (mainly power plants) is expected to soak up the additional capacity. The plant is being constructed with a possible expansion in mind (a third tank would bring capacity to 20 MMcm per day).

Plans for another LNG regasification terminal in northern Chile have also been announced, led by Codelco, the State owned copper mining company. This system is much more dependent on gas. About 58% of capacity is gas fired, as the region has none of the hydro potential of the center and south. There are no connections between the SING and the SIC power grids, nor are there any connections between the respective gas networks. The mining companies are the main off takers of gas-based electricity in the north. However, in this region LNG would face a direct competition from coal imports and coal-based power generation.

There is yet no indication of the price at which GNL Chile will buy the LNG but it is certain to be much higher than the current import price from Argentina yet lower than the price of oil products (mainly diesel oil) currently used to replace missing gas.
LNG's competitiveness with other fuels and sources of power will be critical for the development of LNG imports. Chilean gas consumers may agree to pay a premium for supply security, given the risk embedded in Argentine gas imports. However, as much of the gas is used in power generation, LNG will need to be competitive with other fuel sources (such as coal, hydro, etc). Investors in the power sector are betting that coal will be more competitive than LNG and are already building new power plants based on that fuel, with LNG being considered to play a backup function, for existing combined cycle plants, rather than a basis for generation expansion.

It is important to notice that LNG installations are being developed essentially with the government driving the initiative, in one case through the State owned oil company ENAP and in the second through the also State owned mining company Codelco. In a liberalized market like the Chilean one, this has been justified on political grounds, on the interest of the government to secure energy supply, making the country independent from Argentina.

Fig. 12.15 Quintero's Terminal Location

12.7.2 Main Challenges for LNG in Brazil

The question of natural gas supply for thermal generation has been the object of concern by the authorities ever since the conception of the new model for the Electrical Sector. As discussed in section 12.3, Petrobras announced recently (2006) the construction of re-gasification stations, so as to import liquefied natural gas (LNG), from 2009, to the Southeast and Northeast Regions, in order to increase the natural gas supply in the country.

12.7.2.1 The business model: LNG flexible supply

The introduction of LNG is observed with interest by the electrical sector, for three main reasons: (i) to diversify gas supply sources, (ii) a contract market with shorter ranges and greater flexibility has been emerging. This way, ships for LNG delivery may be contracted according to consumption needs and, thus, have the potential for rendering flexible the natural gas supply to thermal power plants and other clients; and (iii) it is possible to build thermoelectric plants located relatively close to the major LNG delivery ports, thus avoiding investment (fixed costs) in gas pipelines.

In this manner, the final cost to the consumer of thermal energy produced from LNG may become more attractive. This because the flexible supply of gas provided by LNG permits thermal power plants to be operated in the mode of complementing hydroelectric production and, therefore, fossil fuel to be saved. As discussed in [5], the final consequence of this operation is the reduction of energy cost to the consumer. Actually, Petrobras announced its intention of contracting LNG to supply the Brazilian market in a flexible manner.

The business model to procure flexible LNG contracts is innovative and very challenging given the LNG volumes at stake and the current tightness of the LNG international market. The idea is to take advantage of the recently developed short-term LNG market and to sign a contract with flexibility clauses. This could be an option contract whereby an LNG provider to US market would divert ships to Brazil at Petrobras's convenience.

12.7.2.2 Challenges for LNG supply

Nevertheless, although LNG may provide flexibility in gas supply to thermal power plants, it has one important characteristic: its price (as a commodity) strongly depends on how much in advance its order is placed. For example, a LNG order placed one year in advance can normally have a fixed price, since the vendor has the possibility of contracting adequate hedges against the oscillations of the strongly uncertain and volatile international prices. On the other hand, a LNG order placed just a few weeks in advance has a price above that of usual references, associated to the opportunity cost of displacing this gas with respect to its destination market, and increased by an "urgency rate". For instance, a LNG request for "next month" may involve the displacement of a ship intended for the United States market which has a reference price corresponding to that associated to Henry Hub. In this case, the price for the Brazilian market would be, at least, the opportunity cost of this gas (Henry Hub price) increased by a spread (e.g., 10%).

In this context, an important decision problem for the LNG buyer consists in determining, each year, the shipping schedule so as to fulfill gas demand and to minimize its purchase price. This problem becomes more complex on account of the features of the electrical sector's natural gas consumption, which is potentially high and has a strong uncertainty component, as the National System Operator has the prerogative of setting thermal plants in motion without advance notice.

At first sight, the only way to solve this conflict between anticipation of fuel order and uncertainty as to the moment of thermal plants dispatch would be the construction of physical reservoirs for LNG storage. However, the cost of these reservoirs would be very high, if the gas storage capacity were sufficient to cover the period of thermal plants operation, which could last some months. It is at this point that the concept of a virtual reservoir appears: instead of storing gas in a physical reservoir, in order to generate later electric energy, one possibility would be to pre-generate this electric energy as soon as the previously programmed LNG shipments arrive, and to store this energy in the form of water in the system hydro plants reservoirs, as energy credits for the future use by thermal power plants. This way, the dispatch needs would be matched to the LNG supply logic. The concept of virtual reservoir was recently introduced in the Brazilian market rules.

12.7.2.3 Virtual gas storage: gas stored in hydro reservoirs

As described above, the expectation of a LNG order for gas to be used in thermal dispatch may be frustrated by the occurrence of a more favorable hydrology than that expected. In this case, the requested natural gas would not be needed after the arrival of the liquefied gas carrier ships at the re-gasification stations. Symmetrically, a less favorable hydrology than that expected could lead to the need of an "*immediate*" thermal dispatch, not allowing sufficient time for the arrival of the ship carrying the required fuel.

An interesting mechanism to relieve this problem can be found in the very physical characteristic of the Brazilian hydroelectric system: the presence of reservoirs with large storage capacities provides a storage flexibility which could be used by thermal power plants to store as equivalent water, through a "*forced dispatch*", the delivered natural gas that otherwise would not be used. In this case, the thermal power plants would retain a credit of natural gas stored in the hydro plants reservoirs in the form of water, meaning that hydroelectric storage could be used as a buffer by thermal plants so as to permit the storage of non-utilized natural gas.

The following steps describe a simplified version of the virtual reservoir scheme:

(1) Assume that a ship has just arrived, carrying sufficient LNG to supply 2 GW avg of thermal generation for one week. Assume, also, that the ISO announced that it intends to dispatch 50 GW avg of hydroelectric plants next week.
(2) The thermal power plant notifies ISO that it intends to pre-generate 2 GW avg; ISO reschedules hydroelectric plants generation to 48 GW avg, so as to accommodate thermal plant pre-generation.
(3) ONS records in the accounts the reservoirs storage reduction as if hydro plants had actually generated the scheduled 50 GW. In other words, the physical volume of the water stored in the reservoirs will be greater than the accounted stored volume.

(4) The difference between physical and accounted storage (corresponding to the pre-generated 2 GW avg) is credited to the thermal plant as an energy option (*"call option"*) that may be actuated at any moment.
(5) Finally, assume that some time later ISO announces that it intends to dispatch 48 GW avg of hydroelectric energy and 2 GW avg of thermoelectric energy. As mentioned above, the thermal plant may decide to generate physically (if, by a coincidence, a new LNG ship happens to have just arrived) or to apply the option of using the stored energy. In the latter case, the thermal plant follows a procedure inverse to that of item (2): it notifies ISO that it is going to utilize its stored energy, and ISO reschedules the hydroelectric generation to 50 GW avg.

The great risk for the thermal producer in this arrangement is that of water spillage from the physical reservoir: in this case, *"accounted"* hydroelectric energy will be spilled before the "physical" energy.

Of course, the procedure to be implemented involves more complex aspects, not addressed in this Chapter, such as transmission restrictions, storage management for the various hydroelectric plants, and compatibility with the mechanism of energy reallocation, among others. Yet, in brief, virtual storage utilization permits, through a swap operation, to accommodate the need to order LNG without affecting the system optimum policy and operation, thus favoring the ingress of flexible gas supply and the possibility of preparing strategies for its cost reduction.

12.7.3 Virtual Gas Storage and Smart Electricity-Gas Swaps

Finally, the introduction of flexible LNG supply in the region can bring up several opportunities to integrate the electricity and gas markets in the region. This is because energy swaps with LNG are much more economical than the proposed point-to-point pipelines. An example of gas-electricity integration is the so-called *"gas exports from Brazil to Chile without gas or pipelines"*. Essentially, Chile purchases 2000 MW of electricity from Brazil, for delivery to Argentina (via the Brazil-Argentina DC link). The power from Brazil now displaces 2000 MW of gas-fired thermal generation in Argentina, which frees up 10 MM^3/day of natural gas supply, which is (finally) shipped to Chile.

Another example is the use of LNG against the proposed *"Southern Gas Pipeline"*, from Venezuela to Brazil and Argentina. A more rational solution would be to send LNG from Venezuela to the Northeast region of Brazil, thus decreasing the need to send gas from the Brazilian Southeastern region to the Northeast. The surplus production is then sent by LNG to Montevideo, and from there through an existing pipeline to Buenos Aires.

Many other possibilities can be designed but, in essence, LNG brings opportunities for intelligent and economic integration of the regional energy market.

12.8 Power and Natural Gas Integration in the Southern Cone – Past, Present and Future

Regional power integration in the Southern Cone of Latin America had its inception before any political and economic partnership projects [10,11]. It exhibits a wealthy history of shared undertakings and a variety of physical links and exchanges. In its early stages, a characteristic of the way regional power integration evolved in this region was the development of bi-national hydro plants. This development gave rise to a parallel integration of the very high voltage networks existing in the region and to the implementation of a large exchange capacity, which has not always been properly utilized. In the 1990s, as a consequence of the growing trend toward development of a regional block, Power and Natural Gas Integration Protocols were signed within the Southern Cone, in parallel with market reform measures. At this point, the challenge was to integrate a supra-national regulatory framework structuring and promoting the development of mainly private investment projects with the prospective integration and liberalization of gas and power trade. In this context, high capacity works were implemented in the power sector as private undertakings, such as the 2200 MW Brazil--Argentina connection. Natural gas connections were also implemented between Argentina, Bolivia, Brazil and Chile. In addition, integrated projects involving gas exports and power generation were also developed, as shown in Figure 12.3.

The regional integration process was ultimately adapted to the primary resource matrix available in each country, with increasing expectations as regards satisfying local demand with foreign supplies. As discussed in Section 12.4, Chile undertook a program involving change of its power supply on the basis of gas imported from Argentina. A similar situation, but to a lesser extent, arose in Brazil with Bolivian gas.

This scheme was geared toward full utilization of existing network capacities and the generation of new links. The coexistence of firm exchanges (based on long-term contracts) and spot exchanges were not conflictive, as the market operated on the basis of capacity surplus. The full utilization of internal power and gas network capacities led the systems to a border situation where the interaction between natural gas and power (a characteristic feature of this new stage) took on a dominant role in the rationale of system development. Towards 2002, when the whole system suffered the shock of the Argentine crisis, the regional system, without exhibiting features of an open market, already showed the following traits: (i) Long term gas operations: exports from Argentina to Chile and Brazil; exports from Bolivia to Brazil; (ii) Long term power operations: capacity and energy exports from Argentina to Brazil; (iii) exports from bi-national entities (hydro plants) from Paraguay to Argentina and Brazil; (iv) spot operations: exchanges at bi-national power stations.

The integration scenario has shown some signs of stagnation since 2003, especially in view of the relative isolation of individual plans and a stronger emphasis on self-sufficiency at the national level. Energy independence has become a goal in a region where there are still no international legal frameworks that support integration processes not to be altered at mid road, as it did happen between Argentina and Chile and between Bolivia and Brazil.

12.8.1 Regulatory and Commercial Situation

During the last few years, the pace of reforms has slowed down at the international level, and market organization at national level is undergoing active reviews. Without having fully retreated from the systems implemented in the 1990s, transition periods are under way both in Argentina and Brazil, with a higher degree of participation by the State in sector management.
An important area affected by these changes was the integration of the markets at regional level: the regulatory frameworks governing interconnections have proven to be inadequate, despite the many protocols and agreements in force. In a context of strong national debates, protectionist or isolationist schemes imposing restrictions on compliance with contractual conditions have been retaken. It is as if the contracts freely entered into by private parties lacked a smooth relationship with the guarantee of supply in each country.

An aspect contributing to the integration is progress made as regards operating regimes and the coordination of load dispatches and network usage, all of which was facilitated by the long working experience with interconnected systems. It is true that competition has taken place with respect to firm and uninterruptible access to the networks. The role of distribution between the public and private sectors is on hold. Although the high rate of privatizations that characterized the 1990s has slowed down, no significant re-nationalizations have taken place. In Argentina, Chile and Brazil this has resulted in a mixed system sporting a wholesale market with significant private participation.

Reviews have focused mainly on the search for more effective regulation and control and on the adjustment of the pricing systems both at the wholesale and retail levels. This is to ensure efficient, low-cost procedures that, in turn, make the financing of any required investments feasible. In this sense, a review is being made of the role of the capacity and energy supply contracts with distributors, traders and large consumers and their relation with the spot pricing systems.

12.8.2 Southern Cone Integration Issues

Regional energy integration is the key to development. It is a project dating back quite a few years and in full development. However, at present there is a need to guarantee stable rules of the game and dispute settlement mechanisms based on agreements made at the highest political level. Today, there are a large number of outstanding issues related to integration in the Political, Institutional and Regulatory Areas. Examples of these issues include:

a) Guidelines for the future of economic integration and regional policies. The complementary and alternative political and economic integration processes include and determine infrastructure and services integration projects. Within this supra-sectarian framework, some noteworthy aspects are homogeneous tax treatment and the stabilization of exports and import authorization regulations.
b) Adaptations of existing energy integration protocols under the light of recent events (crises of the power and gas contracts between Argentina, Chile, Brazil, Bolivia, etc.). There is a need for higher-hierarchy multinational agreements with a larger degree of flexibility in order to adapt to particular situations that may affect

performance. To align policies and regulations among the various countries is an important step that would encourage spot and long-term exchanges.

c) Fostering the stabilization of mechanisms aimed at establishing price benchmarks for exchanges and eliminating circumstantial distortions.
d) The tendency to integrate open and competitive markets with long-term contracts and spot exchanges should be maintained, since such markets allow minimizing supply costs in the long-term. For this purpose, it is essential to develop effective non-discriminatory treatment mechanisms for demand and local and foreign supply, within the framework of liberalization and regional trade opening.
e) At present, capital market conditions are not positive for the sector. This causes delays in expansion projects. An integrated activity could increase fund availability for the various types of works: hydro stations, thermal power stations, power and natural gas transport, etc.
f) Creation of flexibilities and integrated electricity-gas swaps in the region using the existing infrastructure. For example, Brazil could export electricity to Argentina, thus displacing gas-fired generation and freeing more gas to be exported from Argentina to Chile. These types of arrangements should become common in the region.

Regional integration should not only include but also advance beyond infrastructure connections and individual exchanges. Ideally, free, long-term and spot exchange markets should be created between regional producers and consumers, with due safeguards against crises or emergencies. Regional integration is not just one more option; it is an obligation that must be undertaken to reduce social and environmental costs in the region. For this purpose, commitments at the highest level and stable national and international policies are required, to promote investment and efficient operation by adequately distributing the roles between the public and private sectors.

12.9 Conclusions

The primary challenge for Latin American countries is to ensure sufficient capacity and investment to serve reliably their growing economies. The region has emerged as one of the most dynamic areas for natural gas and electricity developments. In this sense, each country has adopted a different scheme to achieve the target of electricity and gas supply adequacy. Over recent years, these different schemes have had positive and negative repercussions. Among the countries analyzed, the different schemes, the degrees of market evolution, and market opening have resulted in active electricity markets (in Brazil, Chile, Argentina, Colombia), and gas markets (in Argentina, and Colombia). No country has been able to develop an active integrated electricity-gas market. Resource adequacy planning has always been carried out separately and characterized by the particularities of each country.

The high dependence of some countries such as Brazil and Colombia on hydropower creates challenges for the smooth insertion of gas-fired generation. Countries like Chile are facing the challenge of "*gas supply under uncertainty*", since the so far stable gas import contracts with Argentina have turned out to be "uncertain". A promising issue in the region is multi-country electricity markets. These are a natural evolution to the existing "*official*" international interconnections, which in turn were originally established by the countries' governments for sharing reserves and carrying out limited economic interchanges. The creation of a regional market is a natural step towards economic efficiency and economic growth, but important aspects still remain to be discussed, such as the compatibility of regulatory frameworks, tax systems, the political stability of long-term contracts, and need to harmonize supply adequacy actions in the region.

More recently, LNG has emerged as an attractive option. However, South America is a latecomer to the LNG business. Other regions and countries have already incorporated this external natural gas supply source in their portfolios for many years. However, some opportunities could arise from this late arrival. In particular, the evolving rules of the global LNG market could allow for more flexible supply. This, in turn, brings opportunities for intelligent and economic integration of the regional energy market. The energy swaps with LNG are much more economical than the proposed point-to-point pipelines. An example of gas-electricity integration is the so-called "*gas exports from Brazil to Chile without gas or pipelines*". Essentially, Chile would purchase 2000 MW of electricity from Brazil, for delivery to Argentina (via the existing 2,000 MW Brazil-Argentina DC link). The power from Brazil would displace 2000 MW of gas-fired thermal generation in Argentina, which would free up 10 MM^3/day of natural gas supply, which would be (finally) shipped to Chile.

Finally, the ultimate amount of LNG imported will depend crucially on the development of the natural gas reserves in the region. The region has significant reserves and the challenge is how to monetize them and serve the regional and sub regional markets. The situation varies widely among LNG importers: there are countries with growing potential natural gas reserves (Mexico), which was not discussed in this Chapter; those with very little potential (Chile) and those with substantial reserves but still not enough to supply their large market potential (Brazil). The result will likely be a mix of and local/regional gas with LNG playing a smaller, but still important role in balancing supply and demand.

12.10 Acknowledgements

This Chapter has been compiled by Dr. Luiz A. Barroso, PSR, Rio de Janeiro, Brazil; Chair of the IEEE PES W.G. on Latin America Infrastructure; Bernardo Bezerra, PSR, Rio de Janeiro, Brazil, Sebastián Mocarquer, Systep, Chile and Dr. Hugh Rudnick, Pontificia Universidad Catolica de Chile, Chile. Contributing authors include B. Flach (PSR, Brazil), M. V. Pereira (PSR, Brazil), R. Kelman (PSR, Brazil), R. Moreno (Systep, Chile), M. Madrigal (Worldbank, USA), G. Arroyo (CFE, Mexico), J. Mejía and A. Brugman (Colombia) and L. Sbértoli (Sigla, Argentina). The Chapter is primarily based on an up-date of the papers presented at the Panel Session on "*Integrated Electricity and gas Resource Adequacy Planning in Latin America*" at the IEEE-PES 2005 General Meeting (GM2005) in San Francisco ([6-11]).

12.11 References

[1] IEA - South American Gas - Daring to Tap the Bounty, IEA Press, 2003.

[2] M. Pereira, L. A. Barroso and J. Rosenblatt "Supply adequacy in the Brazilian power market" Proceedings of the IEEE General Meeting, Denver, 2004 - Available at http://www.psr-inc.com

[3] H. Rudnick, L.A. Barroso, C. Skerk, and A. Blanco. "South American Reform Lessons - Twenty Years of Restructuring and Reform in Argentina, Brazil and Chile". IEEE Power and Energy Magazine, Vol. 3, (4) July/August 2005, pp. 49-59.
[4] M.V.Pereira, N. Campodónico, and R. Kelman, "Long-term hydro scheduling based on stochastic models", Proceedings of EPSOM Conference, Zurich, 1998 - Available at http://www.psr-inc.com
[5] B. Bezerra, R. Kelman, L.A. Barroso, B. Flack, M.L. Latorre, N. Campodonico and M.V. Pereira. "Integrated Electricity-Gas Operations Planning in Hydrothermal Systems". Proceedings of the X Symposium of Specialists in Electric Operational and Expansion Planning, Brazil, 2006, pp. 1-7.
[6] H. Rudnick, "Electricity Generation and Transmission Expansion under Uncertainty in Natural Gas". Proceedings of IEEE 2005 PES General Meeting, San Francisco, 2005, paper 05GM1094, pp. 1-2.
[7] H. Rudnick, "Energy Risk in Latin America: the Growing Challenges". Keynote paper, Proceedings of the International Conference on Energy Trading and Risk Management, November 2005, IEE, ISBN 9780863415807G.
[8] J.M. Mejía and A. Brugman, "Natural Gas and Electricity Market Issues in Colombia". Proceedings of IEEE 2005 PES General Meeting, San Francisco, 2005, paper 05GM0311, pp. 1-4.T J Hammons and J S McConnach. Proposed Standard for the Quantification of CO2 Emission Credits, Electric Power Components and Systems, Taylor & Francis, Vol. 33, (1), pp. 39-58, 2005. L.
[9] L. Sbertoli, "Power and Natural Gas Integration in the Southern Cone: Past, Present and Future". Proceedings of IEEE 2005 PES General Meeting, San Francisco, 2005, paper 05GM0310, pp.1-4.
[10] M. Tavares, "The Role of Natural Gas as an Instrument for the Energy Integration in Latin America". Proceedings of IEEE 2005 PES General Meeting, San Francisco, 2005, paper 05GM0313, pp.1-3.
[11] L. A. Barroso, B. Flach, R Kelman, B. Bezerra, J. M. Bressane, and M. Pereira. "Integrated Gas-Electricity Adequacy Planning in Brazil: Technical and Economical Aspects". Proceedings of IEEE 2005 PES General Meeting, San Francisco, 2005, paper 05GM0160, pp. 1-8.
[12] L.A. Barroso, H. Rudnick, S. Mocarquer, R. Kelman and B. Bezerra LNG in South America: the Markets, the Prices and the Security of Supply - IEEE PES General Meeting 2008, Pittsburgh, USA.
[13] H. Pistonesi, C. Chavez, F. Figueroa, H. Altomonte, "Energy and Sustainable Development in Latin America and the Caribbean: Guide for Energy Policymaking", Project CEPAL-GTZ-OLADE. Second Edition, Santiago, Chile, 2003.

13

Developments in Power Generation and Transmission Infrastructures in China

13.1 Introduction

The China electricity industry started in 1882. By 1949, the country had a small electricity system with 1.85GW installed capacity and 6,500km of transmission lines. The electricity system expanded rapidly over the last five decades or so. By the late 1990s, the expansion fundamentally changed the nationwide electricity shortage. The China electricity system now is the world's second largest with 338GW-installed capacity and generation was 1478TWh in 2001. Official statistics show power consumption growth in China averaging 7.8% annually throughout the 1990s. Starting from the second half of 2002, China electricity supply was far short of demand because of dry spells that decreased hydroelectric supply, a generator shortage, and unexpected demand from energy-intensive industries. During this period, twenty-one provinces, municipalities, and autonomous regions in China suffered large-scale electricity shortages. Some had to implement load shedding to limit electricity consumption to avoid blackouts. By the end of 2005, China accumulated a total installed capacity of 508 GW. China's electricity output reached 2474.7TWh. China Electricity Council (CEC) estimated that the electricity supply and demand would reach equilibrium in 2007. According to the International Energy Agency, to meet rapidly growing electricity demand, China will invest a total of nearly 2 trillion U.S. dollars in electricity generation, transmission, and distribution in the next 30 years. Half of the amount will be invested in power generation; the other half will go to transmission and distribution [1].

Developing fuel sources for electricity generation has been difficult due to the fact that energy resources are predominantly located in the west and north of the country, while large economic and load centers are in the east and south of China. Transportation of energy adds tremendous costs to electricity supply. This has been especially so in the case of already expensive hydropower development.

China's energy policy is shifting towards diversification of energy resources because heavy coal use has had an adverse impact on the environment. Developing hydroelectricity serves the government strategy to develop the poorer western region. Moreover, the government is also ready to develop natural gas as fuel for power generation. Close to 10GW natural gas-fired generation capacity was developed from 2001 to 2005, including 7.93GW in eastern China using piped gas from Xinjiang and 2GW in Guangdong Province using LNG shipped from Australia. Integrated gas combined cycle (IGCC) technology is a type of electricity generating technology with high efficiency and low pollution that can meet the need for environmental protection. Efficiency of electricity generation can reach more than 60%. Research on this key technology has been started in China. It includes the technologies of the IGCC process, coal gasification, coal gas cleaning, gas fuelling engines and residual heat systems.

In 2005, the per kWh electricity on average in the coal-fired plants consumed 374.00gce. The larger the generation unit, the smaller the amount of coal consumption per unit of electricity generated. For unit generating capacity of 300MW, the coal consumption rate is at 341.88g/kWh; for those units of 600MW capacity, the number is 326.34g/kWh. For super-critical units, the rate is at 320.58g/kWh, comparable to the OECD levels. In terms of power transmission losses, the average figure is about 7% for the national power grids.

In China, much of the renewable resources are in regions with low energy demand, such as Inner Mongolia and Xinjiang. Because the need for electricity could be hundreds or thousands of km away, there are serious questions about the ability of China's already shaky transmission system to handle the movement of these large amounts of electricity. Where transmission capacity is not sufficient, it will be impossible to invest in transmission lines. In fact, some laws limit the amount of renewable electricity that can be supplied to the local grid because of concerns about the additional burden on the transmission system.

Though use of hydro and nuclear power is growing, coal will still provide the majority of China's energy needs in 2030. Whatever the fuel mix, if economic growth in China stays on course, China is likely to account for 25% of the world's increase in energy generation in the next 30 years.

The China electricity policy is to achieve sustainable development of the power industry; to place equal emphasis on development and energy conservation; to attach great importance on environmental protection; and to deepen structural reform in the power sector. For the transmission grid, it plans to build West-to-East power transmission corridors with nationwide interconnection. The policy is to enhance regional and provincial grids interconnection and continue rural network construction and innovation. In addition, it strengthens construction of systems for protection, communication and automatic control. For power generation, China promotes energy conservation priority and the development of hydroelectric power. There are plans to optimize thermal power development and develop nuclear power and renewable energy steadily.

In October 2005, the "*Communist Party of China (CCP) Central Committee's Proposal on the Formulation of the 11th 5-year Plan for National Economic and Social Development*" was released. According to this, the China power industry should continue resource saving and environment friendly development, and realize sustainable development. The Proposal demonstrates that up to 2010, the China electric power industry will increase its installed capacity from 570 to 870GW. Investment of 125 billion US$ and 100 billion US$ will be needed in the power generation and power grid construction, respectively.

13.2 Main Transmission Projects

In China, the distribution of energy resources is quite uneven geographically. 82% of coal deposits are scattered in the north and southwest. 67% of hydropower is concentrated in the southwest. Therefore the north and west are called as the energy bases in China. But 70% of energy consumption is concentrated in the central and coastal areas of the country.

So transmitting electric power from the energy bases is one of the ways making up the deficits of energy in the central and coastal areas, and it is imperative to develop regional power systems interconnection. In addition, the comprehensive interconnection benefits, such as load leveling, emergency back up, peak load savings, improving operation performance can also be obtained. The construction of Three Gorges Hydropower Project has pushed the implementation of nationwide interconnection project. The nation's total installed capacity has reached around 510GW and 500kV AC lines or HVDC lines have interconnected all the regional electric power systems in the year of 2005. The main interconnection projects are shown in Table 13.1 [1].

In order to achieve a continual development in China, the policy, which is "*Developing hydropower actively, thermal power optimally, nuclear power appropriately, renewable energy suited to local conditions*", will be pursued.

05/2001	North- East China Power Grid (NECPG)	North China Power Grid (NCPG)	Transmitting power from NECPG to NCPG through 500kV AC
10/2001	East China Power Grid (ECPG)	Fujian Provincial Power Grid	Exchange power through 500kV AC
05/2002	Central China Power Grid (CCPG)	Sichuan & Chongqing Power Grid	Exchange power through 500kV AC
06/2003	Central China (Three Gorges) Power Grid	East China Power Grid (ECPG)	Transmitting power from Three Gorges to ECPG (3000MW) through ±500kV DC
09/2003	North China Power Grid	Central China Power Grid	Exchange power through 500kV AC
06/2004	Central China (Three Gorges) Power Grid	South China Power Grid (SCPG)	Transmitting power from Three Gorges to SCPG (3000MW) through ±500kV DC
07/2005	Central China Power Grid	West China Power Grid (WCPG)	Exchange power through Back-to-back DC (360MW)

Table 13.1 Main Power Transmission Projects

13.3 Power Grid Development

With the principle of unified planning for power grid development, China is making great efforts to implement coordinated growth of power grids at all levels including regional and provincial power grids, as well as those between power grids and power sources.

Now, China is planning to build a state bulk power grid with voltage level of 1000kV HVAC and ±800kV HVDC. From 2008 to 2020, the HVAC and HVDC hybrid grids will result in trans-regional, large capacity, long-distance and low loss transmission, as well as optimizing the resources allocation to a larger scope and relief the stress of power shortage [2].

Large amount of power will be transmitted from coal power base and hydropower base facilities in the north and southwest area to central and coastal areas of the country through HVAC and HVDC hybrid grids.

13.3.1 Trans-Regional Power Transmission

With efforts to strengthen Trans-regional power resource allocation, China is expanding the scale of Trans-regional power transmission to expand Trans-regional power transmission capacity. One of the measures is to speed up the upgrading of existing 500kV grids by advanced transmission technology.

13.3.2 Construction and Operation of HVDC Power System

There are six long-distance HVDC lines in operation in China. Through these HVDC lines, the power from the southwest area and Three Gorges is transmitted to South China and East China. The total transmission capacity of these HVDC lines is 15GW. In July 2005, the back-to-back DC project between Northwest power grid and Central power grid was put into operation. The exchange power is 360MW.

The HVDC projects under construction or in planning are as follows:

(1) The back-to-back ±500kV DC project between Northeast power grid and North power grid, with transmission capacity 1,500MW. It is be commissioned around 2008.

(3) The project of ±500kV HVDC from Ningxia in North-west China to Tianjing in North China. The transmission capacity is 3000MW. It will be available around 2008.

(4) The project of two ±500kV HVDC lines on one tower from Central China to East China. The transmission capacity is 6000MW. It will be available around 2009.

(5) ±500kV HVDC project from Hulunbeier Coal base in Hailongjiang province to Liaoning province in Northeast China. Transmission capacity is 3,000MW, and it will be available around 2009-2010.

(6) ±800kV HVDC project from Yunnan province to Guangdong province in South China. Transmission capacity is 5,000MW, and it will be available around 2009-2010.

(7) There will be three ±800kV HVDC projects for Xiluodu and Xiangjiaba hydropower station in south-west China, in which two will go to East China and one to Central-China. The transmission capacity of each project is 6,400MW. They will be commissioned from 2011 to 2016.

(8) ±800 kV HVDC projects from Jingping hydropower station in southwest China to East China. Transmission capacity is 6400MW, and it will be available around 2013.

(9) There will be ±800kV HVDC projects for Hulunbeier Coal base, in which one will go to Liaoning province in Northeast China and another will go to North China. The transmission capacity of each project is 6,400MW and it will become available from 2015 to 2020.

13.3.3 750kV and 1000kV AC Transmission and Substation Project

In September 2005, the first 750kV transmission project was commissioned in China. This project is regarded as a sample project. It is comprised of 146km transmission from Guanting of Qinghai province to East Lanzhou of Gansu province in North-west China. In 2007-2008, a 750kV power grid located in North-west China began to take shape. The 750kV transmission and substation project is of significance to acceleration of technology innovation on the power grid in China and the promotion of construction on the HVAC power grid, respectively.

In 2008, the first 1000kV AC transmission and substation project, as a testing and sample project, will be commissioned. It is the tie line between Central China power grid and North China power grid. The length of this line is about 650km.

13.3.4 Construction and Operation of Urban and Rural Power Grids

With urban and rural power grids construction and renovation, the grid structure is better reformed and the transmission line losses are decreased by a large amount. The reliability is improved greatly, with availability of urban and rural electrical power kept above 99.89% and 99.0%, respectively.

13.3.4.1 Enhancing international cooperation

Due to the policy of "*opening up*", China has built and continued strategic partnership with well-known enterprises in many countries and regions in the world. This has led to international cooperation in the field of power grid construction, mechanism reform, technical exchange, environmental protection, etc.

13.3.4.2 Improving environmental protection

China is giving close attention to harmonious development of power grid strengthening and environmental protection. High attention is being paid to protect the environment and landscape, water source and reduce waste. The government encourages development of Renewable energies and clean power, such as wind power in some islands of coastal areas, such as Xinjiang and Inner Mongolia, etc.

13.3.5 Opportunities and Challenges of National Grid

13.3.5.1 Strong growth in power demand

Despite rapid growth of the power industry as a result of the huge population, the per capita installed capacity and power consumption in China is only 0.3kW and 1,452kWh, respectively, which is less than half of the world average and 1/6 to 1/10 of that in industrialized countries. It infers that a huge development of power load will take place in the future. China is now building a well-off society in an all-round manner. The estimated annual GDP will reach 4000 billion USD in 2020. Sufficient power supply is necessary for fast and sustainable economic growth. It is expected that nationwide power consumption will reach 4,600TWh per annum, demanding total installed capacity of 1000GW by the year of 2020. It means that, in the following 15 years, the annual incremental capacity will be more than 33GW with annual growth of power consumption of 160TWh.

13.3.5.2 Economic performance of HV transmission

According to primary investigation, for same transferred power and transmission distance, the unit cost of 1000kV HVAC transmission is 73% of that of 500kV HVAC transmission. The unit cost of ±800kV HVDC transmission is 72% of that of ±500kV HVDC transmission. The advantage of HV in the transmission of huge quantities of electricity over long distances is apparent.

13.3.6 Construction of HV Transmission Grid

The State Grid is determined to construct HVAC pilot transmission lines and HVDC projects in coming years. Once the construction of the first series of HV projects is successful, the State Grid will promote extensive application of HV technology to construct a HV backbone network.

13.3.6.1 1000kV HVAC pilot project

HV transmission is an innovation of technology, and also a big challenge. All work on HV should be initiated from a practical pilot project. The purpose of building a pilot project is to test performance of the HV system and its equipment, accumulate experiences for HV research and operation, and improve the level of technology in HV equipment manufacture and power transmission.

The State Grid has completed the selection of pilot projects and the feasibility studies. The system scheme is provisionally determined. In this scheme, the total length of the HV transmission line is about 650km, including two HV substations and one switchgear station. The advantages of the pilot scheme include easy implementation of engineering construction, the wide testing of both the HVAC system and its devices, and extensive guidance for future application of HVAC in China.

13.3.6.2 Outgoing HVDC transmission line of Jinshajiang River

The advantage of HVDC is to transmit large quantities of power over very long distance. In China, ± 800kV DC will be mainly used to transmit large capacity over very long distance from huge hydropower bases and thermal power bases. Other application involves some long distance transmission projects with little support of power supply along the transmission line. There is abundant hydro resource along the Jinshajiang River. The exploitable hydropower is about 90GW, with annual power generation of 500TWh. At the first stage, there are two hydropower stations, viz. Xiluodu and Xiangjiaba. The total installed capacity of the two hydropower stations is 18.6GW, which is 0.4GW higher than that of the Three-Gorges project. In order to reduce transmission cost and power loss, and save transmission corridors for future development, the State Grid is determined to develop the scheme using 3 circuits at ±800kV UHVDC, one for the Central China Grid, the other two for East China. The total length of these UHVDC transmission lines is about 4,820km with transmission capability of 6.4GW over each circuit. At present, the feasibility study for the above project is on going. As a schedule, the construction of the first UHVDC transmission line will be started in 2008. It will be put into operation in 2011.

13.3.6.3 Prospect of national HV grid

In order to optimize allocation of energy resource, exploit the large-scale coal bases at Shanxi, Shannxi, Inner Mongolia and Ningxia, and harness remote hydropower in Southwest China, the State Grid will construct a nationwide HV transmission grid. At first, the State Grid will construct 1000kV HVAC transmission that links North Grid and Central China Grid. Then it will expand the 1000kV HV synchronous grid to East China; finally accomplishing a strong HVAC network that connects the North-Central--East China Grid. Combined with the HVAC and HVDC power grid for hydropower transmission in southwest China, it will form a strong HV grid that covers large energy bases and load centers. The total transmission capability of the HV and trans-regional grid will exceed 200GW.

13.3.7 South China HVAC/HVDC Hybrid Grid

13.3.7.1 The rapid growing-up of South China Power Grid

The South China Power Grid covers five provinces: Guangdong, Guangxi, Yunnan, Guizhou and Hainan. These provinces have an area of about one million square km and a total population of 220 millions. From 1980 to 2004, the total installed capacity of the South China Power Grid has increased by a factor of 10, up to 80.27GW (excluding that of Hong Kong and Macau). Annual generating capacity has increased by 14.3 times, up to 383.2TWh [2]. Total electricity consumption has increased by 15.6 times, up to 389.1TWh, accounting for 17.9% of the total in China.

In August of 1993, the power grids of Guangdong, Guangxi, Yunnan and Guizhou began interconnected operation. There was only one transmission passage from West to East; the transmission capacity was 600MW with sales of 2.1TWh/annum. Up to 2005, a solid framework of nine 500kV passages from west to east had been constructed, 6 AC lines and 3 DC lines with total capacity of 11.75GW. The annual trading amount reached 44.7TWh. These two figures have increased by 19.6 and 21.3 times accordingly. The west power ratio in Guangdong dispatchable total energy increased from 3% to 30%. The total length of transmission lines above 220kV is 39,283km and the total transformer capacity 140.05GVA, which is 3 times and 5.5 times, respectively, from the beginning of networking.

13.3.7.2 The unique features

The South China Power Grid is the one that has the most complicated structure and connections, the highest science and technology level, and at the same time, is the most difficult to operate in China. The grid can be summarized as follows:

1. *Long Transmission Distance and Large Capacity*
   The distribution of power resources and the load in the southern area are quite out of equilibrium. This characteristic requires implementing power transmission from west to east to optimize energy utilization. The distance of each of the nine long transmission passages from west to east is around 1000km and one pole capacity of the DC lines is 3000MW.

2. *Multi DC in feeds*
   DC channels and 6 AC channels of 500 kV from Tianshengqiao to Guangzhou Province and Guizhou Province to Guangzhou Province are operated in parallel. Three DC lines connect Three Gorges to Guangdong, TIanshenqiao to Guangdong and Guizhou to Guangdong, simultaneously supplying power to the Guangdong 500kV network. The electric distances between converter stations are very short.

3. *Various Types of Power Sources*
   There are various power sources within the grid, such as hydropower, coal fired power, nuclear power, pumped storage and storage by hydropower, oil-fired power, gas-fired power, wind power. The capacity of single units of nuclear and thermal power is quite large. Among those, the capacity of single unit of Lingao Nuclear Power Plant and Daya Bay Nuclear Power Plant is as large as 1,200MW.

4. *Wide application of new technologies*
   The South China Power Grid has centralized many advanced power transmission techniques in the world. The primary techniques include DC power transmission, electric trigger and light trigger of silicon controlled valve, thyristor controlled series capacitors, fixed series capacitors, high-altitude compact circuitry, and superconductor cable, etc. The secondary techniques include the largest and most advanced security stability control system in China, as well as the wide area measuring system which covers the whole grid and the online stability analysis and pre-decision system that has been primarily established

13.3.7.3 The challenges

For safe operation, the South China Power Grid is confronted with many risk issues.

1. *Outstanding problems of grid stability*

This problem consists of four aspects. The first aspect is power-angle stability. Once fault trips occur on AC transmission lines, it is possible to destroy power-angle stability and voltage stability because of large-scale power displacement. For a multi-feeding DC system, if the AC transmission problems cannot be isolated timely, it is possible that many DC lines are also disturbed that will destroy the system stability. The second aspect is dynamic stability. The west-to-east span of the South China Power Grid is nearly 2000km, the cross-area oscillation mode has relatively low damping, so it is a long-standing problem to control and eliminate the low-frequency oscillations. The third aspect is voltage stability. With rapid growth of load and inter-grid power transmission and receiving, as well as the formation of the multi infeed of several DC loops into Guangdong power grid, the problem of voltage stability has become more and more exacting, and this problem is quite unique. The fourth problem is thermal stability. During high peak load in summer, some circuits and equipment are nearly operated to the limit of thermal stability. In this case, it is possible that N-1 outage can threaten the security of equipment and the systems.

2. *The control systems security and stability*
Once the grid has serious faults, only if the security and stability systems takes measures of cutting-off transmission and load can the grid remain stable. It is technically very difficult to operate such a large-scale AC/DC stability control system, and also as the system is very sophisticated, it is very possible to make mistakes or for the protection fail to operate correctly.

3. *The dynamic voltage support*
Additional capacitors have been installed at many stations of the Guangxi Grid, and feasibility studies on installing SVC or SVG are still ongoing. It is estimated that after the Qinzhou power plant and the Fangcheng Bay power plants in Guangxi Province are connected to the grid of 500kV, the stability level for dynamic voltage stability may be considerably improved.

By means of various advanced technologies and management measures, stable and safe operation of the South China AC/DC hybrid power grid may be successfully enhanced.

13.3.8 Future of South China Power Grid

General planning of development of the South China Power Grid is to insist on scientific development to meet the need of power consumption for development of the economy and daily life; to meet the need of safe reliable and stable operation of the power system, to realize large-scale optimization of energy resources, to constantly improve technology and management level of the power system, to lower cost of the power system, to realize sustainable development, and to construct the South China Power Grid into a uniformed, open, reasonably structured, reliable modern power grid. To achieve this, the following priority areas are implemented.

1. *Speed up power grid development and technology upgrading*
   After two years deep investigation and research, HVDC application in South China has a good foundation. The reasons of HVDC application depend on power grid characteristic; depend on the need of increasing west to east transmission capacity as well as solving the problem of transmission passage space and land. It will effectively solve the problem of short circuits at load center-Guangdong Power Grid; it will effectively enhance the capability of multi-infeed of DC lines to promote safe and reliable operation of the power system. For the period 2010-2015, the first phase of ±800 kV DC lines from Yunnan to Guangdong are planned to be constructed of length 1600km and 5000MW capacity. The project will be put into operation before June of 2009.

2. *Optimize the allocation of power resources*
   From now on, the development of generation resources should be adjusted to satisfy requirements of load and environment protection. That is to optimize the coal-fire electric power, to develop hydroelectric power actively, to accelerate the speed of nuclear power

development, to develop natural gas and pump-storage for electric power reasonably, to develop new energy in accordance with regional features, and to construct peak load power plant in areas of intense load. It is planned that during the "*11th 5-years Plan*", newly installed capacity of the resources will be 67GW; while in 2010, the total capacity will reach 147.5GW. During the following 10 years, the new installed capacity will be 95GW. By the year 2020, total installed capacity will be 240.8GW. It is estimated that the average power per capita will be about 1kW.

3. *Increase power transmission*
During 2010-2015, according to *11th 5-year Plan,* west energy transmitted to east will increase 11.5GW to 13.5GW. By 2010, the transmission capacity from west to east will amount to 22.38GW-24.38GW. It is planned that during 2011-2030, 33GW-38GW energy will be added to the sum, that is, by the year of 2030, the total capacity of west to east transmission will reach 55GW-62GW.

Meanwhile, the South China Grid is actively promoting power cooperation with supply energy to Vietnam, Thailand and Burma. The South China Grid enhances cooperation on electric power with Hong Kong and Macau.

13.3.9 Wide Area Measurement System (WAMS)

With the development of GPS, computer and communication technology, the prototype of phasor measurement unit (PMU) was first developed in United States in early 1990s. It attracts great attention in China since its birth [4,5].

The installation of PMU in Chinese power grid can be dated back to 1995. The China Electric Power Research Institute (CEPRI) introduced a system that had the function of phasor measurement and was commissioned as PMU in the Chinese power grid. From 1995 to 2002, about 30-40 systems were installed and the main stations of WAMS were established in East China, South China, Northwest and Sichuan Power grid and State Power Dispatching Center (SPDC) successively. Such systems adopted modem as communication media and internal communication protocol where the data can be uploaded to the main station of WAMS every second. The installed system successfully recorded the dynamic process of low frequency oscillation that occurred in the Chinese power grid several times which revealed the significant value of synchronized phasor measurement technology in the area of power system dynamic monitoring and also pushed the development of prototype of PMU of Chinese manufacturer.

At the end of 2002, Chinese manufacturers have the commercial product of PMU that have been commissioned in the Chinese power grid since 2003. By the end of 2006, over 300 PMUs had been installed, which are mainly distributed at substations and power plants of the 500kV and 330kV voltage level. 7 regional WAMS are constructed in SPDC and North China, Northeast, Northwest, East China, Central China, South China power grids. 6 provincial WAMS are established in Jiangsu, Shandong, Guangdong, Guizhou, Yuannan and Shanxi power grids. Moreover, real time data exchange is realized among SPDC, North China and Northeast WAMS.

Since there are different approaches to achieve phasors and more than one manufacturer, a Chinese standard on PMU and WAMS was drafted by the State Grid Company and manufacturers in 2003, and finally issued in 2005. The standard supplements transmission protocol of historical data on the basis of IEEE Std 1344-1995 (R2001). The synchrophasor standard provides a technical specification for manufacturers and allows interchange of data between a wide variety of users of both real time and offline phasor measurements, which is of great importance for Chinese WAMS implementation. The blackouts that occurred worldwide in recent years also confirm the urgent needs of WAMS. Therefore, for the next five years, all 500kV substations and 300MW and above power plants in the Chinese power grid will install PMU according to *11th 5-year Plan*.

13.3.9.1 WAMS in China

In China the main station of WAMS is located at the regional or provincial dispatching center and composed of advanced application station, database server and data concentrator. It can be seen that the advanced application station retrieves data from data concentrator via LAN instead of Ethernet, which reduces time delay for the data. From this point of view, the data concentrator here is not the same as that in IEEE STD C37.118 although the names are same.

The data concentrator in the main station is one of key points in WAMS. Currently, some data concentrators already contain 5000 phasor measurements and the storage rate of the phasor is 100Hz. With the fast development of WAMS, how to construct the data concentrator with 100 PMUs and 10,000 phasor measurements and provide corresponding high-speed storage and enquiry technology is also a challenging work.

The functions of advanced application station include visualization of dynamic process and available transmission capacity, wide-area data recording and playback, and on-line low frequency oscillation analysis. Due to the long transmission distance and weak interconnection, low frequency oscillation is a quite severe problem in China. As an only tool catching the oscillation, WAMS played an important role in low frequency oscillation identification and control in China in 2005 and 2006. New identification and preventive control methods are undergoing development.

WAMS opens a new path for power system protections, especially for backup protections (the time delay of backup protections makes it possible to acquire and deal with the phasor data of power systems). Some research works has been launched into this area. To handle the cascading trip problem, the fundamental solution is to monitor the load and try to identify whether the overload is caused by flow transferring or internal fault. If flow transferring does occur in the system, then block the backup relay before the thermal limit of the line has been reached.

It should be pointed out that the philosophy of main protection should not be changed with the advent of WAMS. On one hand, acquisition of phasor measurements will inevitably increase the time delay of trip signal that has adverse impact to system component and stability. On the other hand, the introduction of WAMS information makes the main protection more complex that might correspond to lower reliability.

13.4 Diversity in Power Generation

Based upon the current situation of resource storage and the requirements of increasing power load in China, the basic principles of power supply is to develop hydropower vigorously, optimize the development of coal-power, promote the construction of nuclear electricity, develop natural gas generation steadily and speed up the exploitation of new energy resources generation. By 2020, it is estimated that the total capacity of conventional hydropower will reach 240GW, nuclear 40GW, gas 45GW, new energy sources 10GW, the other 600GW is still depended on the coal generation. Coal-fired power generation and hydropower still play a leading role in the future. Table 13.2 shows the generation mix in 2010 and 2020.

	2010	2020
Total installed Capacity [GW]	543-559(100%)	865-947(100%)
Coal	338-384(62-69%)	509-661(56-69%)
Oil	3-4(1%)	1-6(1%)
Natural gas	25-28(5%)	43-46(5%)
Hydro power	132-154(24-28%)	191-240(22-25%)
Nuclear power	9-15(2-3%)	31-40(3-4%)
Renewable sources	3-7(1%)	11-13(1-3%)

Table 13.2. Outlook of Generation Components from 2010 to 2020 [6]

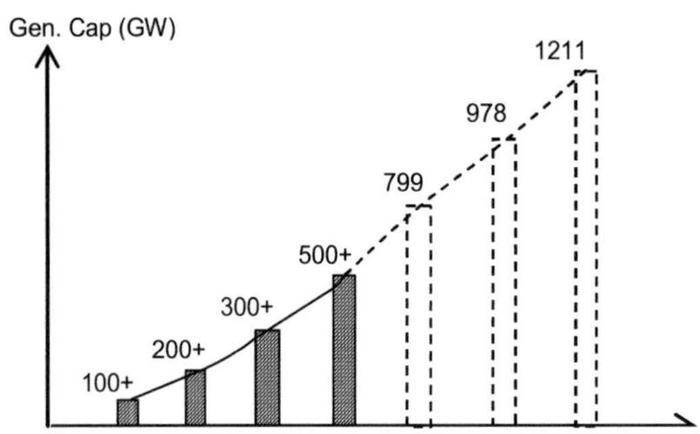

Fig. 13.1. Generation Capacity in China

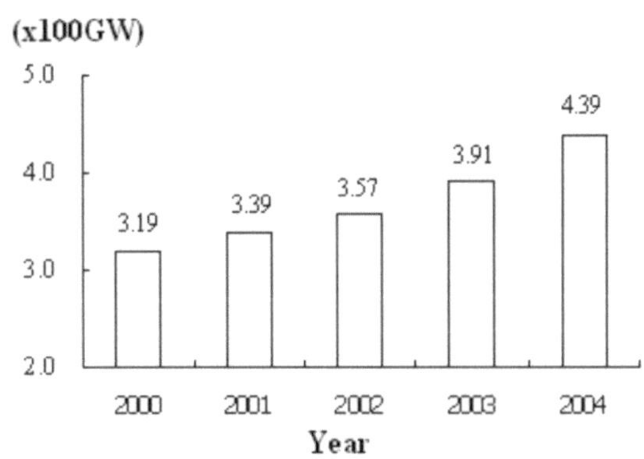

Fig. 13.2. Annual installed capacity in China since 2000

Figures 13.1, 13.2, 13.3 and 13.4 show generation capacity in China from 1990s to 2020, the annual installed capacity in China from 2000 to 2004 and utilization hours of coal-fired power from 2000 to 2004, generation capacity growth, demand growth and utilization hours of cola-fired plants respectively [7, 8].

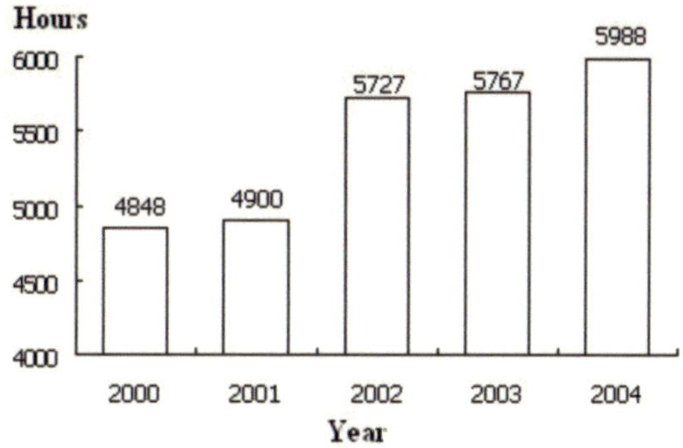

Fig. 13.3. Utilization hours of coal-fired power since 2000

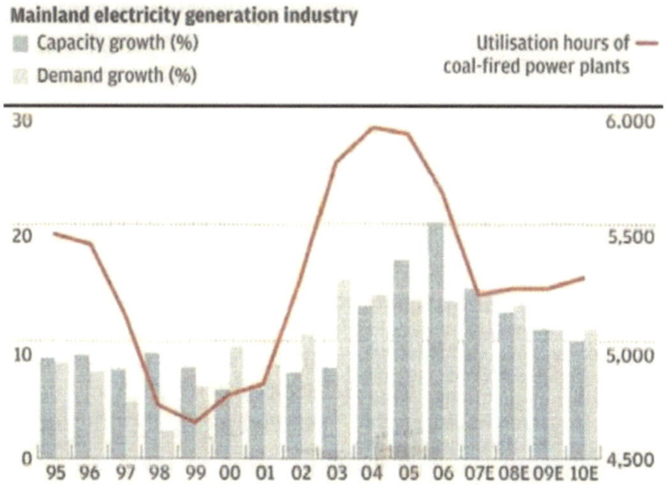

Source: J P Morgan [7]

Fig. 13.4. Generation capacity growth, demand growth and utilization hours

13.4.1 Higher Requirements on Resources Exploitation

To meet the future demand of increased power load, a sufficient supply of primary energy source should be guaranteed. There are three features of primary energy in China, i.e. a huge production but with a low quantity per capita and unbalanced in distribution. It results in higher requirements on resources exploitation. China is abundant in coal and hydro resources but lacks oil and gas. The gross potential of hydropower and the exploitable capacity rank number 1 in the world. And the coal reserves and its exploitable capacity rank number 2. But the gross oil and natural gas rank number 10 and 22 worldwide, respectively. It indicates the dominant roles of coal fired power generation and hydropower in the future power supply. However, the resource in per capita term is relevantly low and only about

40% of world average level. In order to provide a sustainable power supply, promote healthy development of the economy, we must increase the efficiency of power energy and promote application of energy-saving technologies. As a third feature, the primary energy bases are located conversely with those of productivity zones, i.e. areas with rich resource are far away from the developed economic zones. Two-thirds of hydropower bases are distributed in Sichuan, Tibet and Yunan, and two-thirds of coal resources are located in Shanxi, Shaanxi and Inner Mongolia. However, the economic developed zones are mostly located in the east area, consuming most of the electric power, but short of primary energy resource. It is about 500-2000km from the above resource bases in the west area to the load center in east zone. Therefore, it is indispensable to accomplish huge volume power transmission over long distance.

13.4.2 Biomass

China is active in generating electricity based on renewables. For example, by the end of 2004, the generation-installed capacity was 2GW. Naturally, there is tens of million tons of carbon dioxide reduction. In addition to biomass, there are many wind farm, hydro and solar projects in China, and they attract investments through carbon trading. In [7], it reported that the mainland power generation sector suffered a 93% drop in profit in the first two months of the year. The profit drop reflects spiraling coal costs, a prolonged tariff freeze and rising borrowing costs, all of which have squeezed profit margins.

Although the news increases pressure on Beijing to lift tariffs to restore profitability and offer investment incentives, analysts were not optimistic of an increase any time soon because of the time it will take for inflation to fall from the 12-year high it reached in February 2008. Huadian Energy, sister firm of Hong Kong and Shanghai-listed Huadian Power International, said it would report a loss as a result of falling plant utilization and higher coal cost. Of the economy's three worst performing sectors - power generation, power distribution and oil refining - the generation sector was the hardest hit, with profit falling 93% in January and February 2008, mainland newspaper The Economic Observer reported, quoting an unnamed source.

The National Statistics Bureau said recently that the power sector as a whole had recorded a 61% year-on-year profit decline in the first two months of 2008. While electricity distribution companies suffered revenue losses because of winter storm damage they were not affected by higher coal costs, the report said. Power companies generally lock most of their coal supply through one-year contract negotiated at the start of the year. However, some agreements are not fulfilled, exposing them to the risk that coal sellers will ask for higher prices and renege on existing contracts.

"Our concern is that as long as domestic spot coal prices remain at roughly the same level, there is a growing possibility that contract coal prices may need to be revised upwards throughout the year," JP Morgan head of Asia-Pacific utilities and infrastructure research wrote in a recent research report.

With domestic spot market coal prices 40% below international level, it was projected that coal costs of locally listed mainland power producers would rise between 20% and 30%

/year on the year in the third quarter. Analysts said unlisted power plants were believed to have fared worse than listed ones, because listed power plants were likely to be more efficient and receive priority in power sales. Figure 13.5 shows different weekly spot coal prices variations

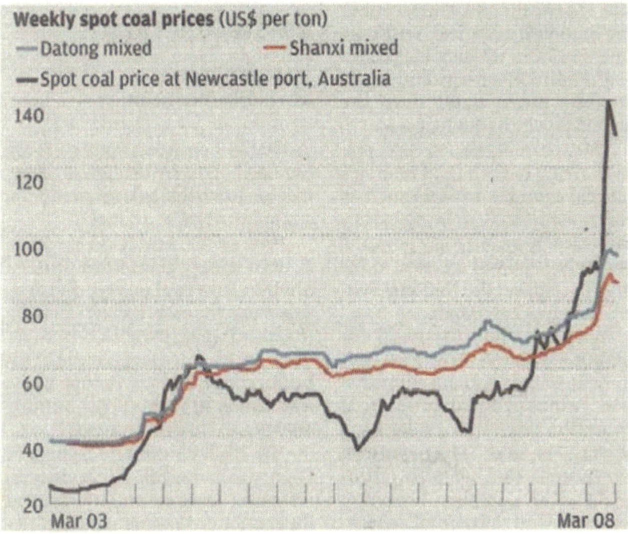

China's spot cola prices have corrected marginally by 5 to 10 RMB per ton since middle-February 2008 Australian prices have been corrected by 6.9% since mid-February 2008, but remain 30-40% above prices within China. [7, 9]

Fig. 13.5. Different weekly spot coal prices variations

This means lower utilization rates for less efficient plants, which tends to translate into lower profits as fixed costs such as depreciation and maintenance are spread over lower output. Listed power plants also tend to be in under-supplied or balanced markets. One exception is Huadian Energy, which operates 21% of the total generation capacity of over-supplied Heilongjiang province in northeast China. The province's overall power plants utilization is forecast to fall 9.27% in 2008 to 4890 hours from 5390hours in 2007, Huadian Energy said. The company's plant utilization eased 3.39% to 5270 hours in 2007.

Meanwhile, surging interest costs have bitten into the bottom lines of companies that sough to offset higher coal costs by building more plants and generating more power. Huadian Power International, whose debt-to-equity ratio is the highest among locally listed power firms, saw a 203.88% jump in interest costs in 2007 to 1.38 billion RMB, higher than its net profit of 1.19 billion RMB.

13.4.3 Natural Gas

Because small internal combustion engine power generation efficiency is 24-35%, the combined-cycle power generation efficiency may also reach 45-50%, this figure is higher than the conventional thermal electricity generation. This fact is different from the usual idea small unit is uneconomical to use. For cold-fired power plant, the bigger unit it is, the higher the economic efficiency will be. Presently, China develops the big centralized plant as the main direction, however developments in distributed generation will certainly expand rapidly too due to its higher efficiency, cleaner energy and short construction time, less transmission loss and reducing possibility of blackout to improve reliability.

13.4.4 Coal

Coal stands the dominant position in energy consumption. In 2003, coal accounted for 70-80% of China's primary energy consumption, a reality that is doubtful to change significantly in a short time. Also, a large amount of coal is directly consumed by end -users. Coal is responsible for providing 75% of energy used in industry and 80% of household energy.

The inefficient and obsolete way of coal usage in China is responsible for the country's severe environmental pollution. Carbon dioxide emissions are the second highest in the world, while sulphur pollution in the nation is the highest in Asia. The vast majority of the pollutants, smoke and dust, sulphur dioxide, carbon dioxide, carbon monoxide and nitrogen oxides, are the result of this by coal combustion. The pollution is not only responsible for several health hazards of the country; it has also created huge economic loss.

It is seen from the above that China, if it is to ensure energy supplies and to achieve sustainable development targets, has to improve energy efficiency in order to transform coal to a clean, high-efficiency, and convenient energy resource. Therefore, clean coal technology is one of the main focuses of China's energy strategy.

13.4.4.1 Low efficiency and large potential for energy saving

China as a developing country retains its dual economic structure of agriculture and industry. In the industry sector, modem industries co-exist with traditional industries; the main characteristic of the traditional small industries is that they are decades behind in terms of modernization. These small industries use obsolete equipment and as a result energy efficiency is low while energy consumption is high. It indicates that the estimated energy consumption of major industrial products is usually 30-40% higher than that of industrialized countries; while the efficiency of main equipment is 10-20% lower than that of industrialized countries. Therefore, because the levels of energy efficiency in China are very low, there is a great potential for energy saving.

13.4.5 Clean Coal Technology in China

Even though China is among the largest energy producing and consuming country in the world today, due to extremely large population it has a low consumption per capita. From 1949 to 2006, the quantity of China's raw coal production had increased from 32Mt to 1212.3Mt; crude oil, from 0.12Mt to 183.7Mt; natural gas, from 7Mm^3 to 50.6 billion m^3; electricity power, from 4.3TWh to 1,018.0TWh. In 2006, China's energy consumption per capita was accounting for less than the one-sixth of that of US. Due to its large population, the per capita is very unlikely to increase in short time. It is estimated that it will not reach the current average level of the OECD countries until 2050.

Coal preparation technology
Coal preparation is the most direct and practical approach to reduce pollution emission and is considered among the key interests in clean coal technology research. In 1995, the average designed wash capacity of key state-owned coal companies was 1.4Mt/y. The largest coking coal company was designed to wash 4.0Mt/y raw coal. The largest steam coal washery was with a capacity of 20Mt/y.

High-efficiency and low-pollution pulverized fuel combustion
The majority of China's power plants apply pulverized fuel combustion technology. The investment required is relatively small, and it is suitable for the majority of types of coal in China. The main boilers for power stations that are manufactured in China are 300-600MW sub-critical controlled cycle or natural-cycle coal combustion boilers with combustion efficiency of 93%.

The goals of China's development strategy by the year 2005 for the power plant boiler sector were to further incorporate 300MW, 600MW boilers combustors and combustion technology imported from foreign countries, and to develop combustors and combustion technology that have high-efficiency, good adaptability for different kinds of coal, low-load combustion stability, good performance of supply and low NOx emission. Also they had to be convenient to operate, economical and practical. Before 2010, all new power stations must adopt high-efficiency, stable combustion and low NOx emission pulverized fuel combusts as the boilers of old power stations will be gradually phased out. Also effective ways of reducing the emissions of the existing boilers will be applied.

Cycle fluidized-bed boilers
Cycle fluidized-bed coal combustion is a relatively new type of combustion technology developed in the past 20 years. It is characterized from its varied application, high combustion efficiency, and effective control of SO_2 and NOx emissions. China started R&D on cycle fluidized-bed boilers in the early 1980s. At present about 300 cycles fluidized-bed boilers less than 130 ton are manufactured based on domestic technology.

Coal combustion for combined cycle power generation
Coal combustion for combined cycle power generation technology includes IGCC, PFBC-CC and AFBC-CC. China started the research on pressurized fluidized-bed combustion combined cycle technology in the 1980s. Since then, a lot of research work has taken place for the design and operation of pressurized fluidized-bed combustion combined cycle technology. In 1991, the pressurized fluidized-bed combustion combined cycle power technology was listed among most important R&D projects of the *8th 5-year Plan (1991-1995).* Since then a pressurized fluidized-bed combustion combined cycle middle-stage experiment power station, was installed and tested in Xuzhou Jiawang Power Plant. As far as IGCC technology is concerned, China has fallen far behind that of developed countries. There are no IGCC plants currently in China even though IGCC equipment will be actually operating in China in the future.

13.4.6 China's Clean Coal Technology and Foreign Technology

When examining the gap between China and the developed countries in clean coal technology we should bear in mind two features. Firstly, China's technology is relatively immature and the application scale is rather small. For example, the largest production capacity of cycle fluidized-bed boiler in China is 220 t/h, while in developed countries, its production capacity can be as high as 700 t/h. PFBC technology stands in the 15MW experiment stage in China, while in Sweden, for example, is in the 350MW commercial demonstration stage. In the field of IGCC, 250MW demonstration equipment has been put into operation in the Netherlands and US, while in China, such technology is still in the stage of feasibility pre-research stage in demonstration plant.

Secondly, the application scope of clean coal technology is rather narrow. Even though some technology is mature, the application is very limited due to lack of effective policy instruments. As a result, the proportion of washed steam coal volume in China is less than 10%, in contrast, US is 45% and European countries 75%.

13.4.7 Reasons for Falling Behind in Clean Coal Technology

The main reasons for the existing gap in clean coal technology are the wide economic difference between China and developed countries, late environmental legislation, and the insufficient investment in clean coal technology research.

Overall deficiency in national strength

Clean coal technology is a highly capital-intensified and technology-intensified area. In US, in order to promote clean coal technology, the government provides funding to enterprises and other institutes for R&D activities of clean coal technology. In China on the other hand, because of the weak overall economic strength, the government is unable to invest considerable amount of capital on clean coal technology research. Also even if a technology is developed, the market for such capital-intensified clean coal technology is limited. For example, investment for IGCC is twice higher than that of the same scale power plant, which is a difficult sum for Chinese investors. Finally, even though China is the biggest coal consumer in the world, the R&D on clean coal technology began rather late.

Delays due to legislation

The market scale for clean coal technology is mainly determined by factors such as environmental legislation, enforcement of environmental laws and regulations, and public environmental awareness. As China's economy is still at a low level, and environmental legislation and enforcement are rather weak, it is very hard to stimulate innovative activities on environmental technology. Although public environmental awareness has somewhat increased, it is not strong enough to create pressures for enterprises to apply environment technology.

Disjunction between R&D and production

Because of the institutional system in the area of science and technology in China, R&D activities related to clean coal technology are primarily carried out in research institutes, which usually lack knowledge on actual needs of the industry. So, they find it difficult to develop products that meet the industrial actual needs. Also, for a technology to be commercialized, it has to go through a series of phases, such as medium scale experiment, pilot production and commercial demonstration. The lack of communication between institutions and manufacturers makes it difficult to link R&D results in the laboratory to the production practice.

Fossil Energy Protocol between the United States and China

The US-China Fossil Energy Protocol is intended to promote scientific and technological cooperation between the United States and China in the field of fossil energy, particularly activities related to research, development, demonstration, and deployment. A complementary objective is to create opportunities for US industry and power developers in China. The Protocol was originally signed in 2000 and was renewed in 2005 for an additional five years; signatories are US Department of Energy and China's Ministry of Science and Technology. A Permanent Coordinating Group governs the Protocol, chaired on the US side by DOE's Assistant Secretary for Fossil Energy and on the Chinese side by the Secretary General of the High Technology Bureau of MOST. Outreach and communications activities are underway to introduce Chinese corporations and research groups to the latest US technologies in the areas of coal gasification, coal liquefaction, natural gas technology, coal-bed methane recovery, acid rain control technologies, and carbon dioxide capture from power plants.

Analysis on China's acquisition of clean coal technology

According to their importance, the order for various means of clean coal technology acquisition in China is as follows: equipment import, license trade, and joint ventures. Some Chinese enterprises acquire technology by means of purchasing technology service and consultancy, or by cooperative production. However, the proportion is very small. Thus, they are treated together with license trade.

Importation of clean coal equipment

Investigating clean coal technology transfer in China thoroughly and obtaining original date and information are both difficult and time-consuming. This makes the identification of what kinds of aforementioned technology having been imported in China and what technology China imported being clean coal technology an extremely difficult task. Equipment imported from industrialized countries has advantages over Chinese equipment in terms of reliability, productivity and efficiency. Foreign power generators are also superior to domestic ones in term of efficiency and reliability. The thermal efficiency of foreign industrial boilers and central-heating boilers is at least 5% higher than that of average domestic equipment. As such, the above-mentioned equipment imported from industrialized countries is regarded as clean coal technology.

Power equipment

In 2006, the power industry consumed about 34% of all end-combustion coal in China, and there is a gap between China and industrialized countries in this sector. Therefore, it is very important for China to acquire clean coal technology from industrialized countries in order to use energy efficiently and to reduce emissions of greenhouse gases. The import of thermal power equipment occupied a relatively important position in China's thermal power industry. Desulphurisation technologies in China are mainly obtained directly from equipment imports. China has imported boilers for power stations that came mainly from US, Germany, Russia, Japan, Australia, UK and Switzerland. US is the most important supplier, providing about 70% of China's total imports.

Industrial boilers
Industrial boilers are the most important heat-generating power equipment in China. About 500,000 industrial boilers with 1,200,000 t/h have been used by the end of 1995. The production output of industrial boilers in China has increased since 1990; the average incremental percentage is 10.8% per year. China is the largest producer and user of coal combustion industrial boilers in the world. The coal fired by industrial boilers occupied 1/3 of the raw coal output. Because of lower operating efficiency, the actual thermal efficiency was only 60-70% during the period of the *8th 5-year Plan (1991-1995)*. It is 10-15% less than that of the international advanced level. This shows that energy waste in China is enormous and potential for energy saving is huge. Furthermore, more than 5 million tons of SO_2, 600 million tons of CO_2, large amounts of NOx and other harmful gases are emitted from coal combusted boilers every year in China. This not only causes heavy coal related air pollution in China, but also aggravates the global greenhouse effect. Therefore, it is significantly important to acquire advanced industrial boiler technology with high efficiency and low emission.

Factors influencing enterprises in purchasing clean coal equipment
There are many factors that influence enterprises decision in choosing which country, which company and what kind of clean coal technology or equipment to be imported. The major factors are as follows:

Reliability of equipment
Large utility boilers or industrial boilers are key equipment to supply power to importing enterprises. As a malfunction of the equipment causes the whole production to stop, one of the crucial concerns for the importers is the consistency of the equipment. So the history and experience of the supplier in providing reliable equipment is critical.

Consumer's assessment
In general, each industry sector has its own information communication system, by which users can evaluate the quality of certain equipment, usually in the forms of professional journals or conferences. For example, China's electricity power industry annually evaluates the reliability of electricity generation equipment, and the result is announced at a conference at the beginning of each year and in professional journals. Evaluations like the above have great effect on consumers purchasing new equipment.

Applicability and advancement
When acquiring clean coal technology, the government also considers if the technology is advanced or not. When it comes to the private sector though the prime factor is the applicability and the advancement of the technology secondary.

Equipment performance
Companies also focus on the level of automation and the easiness of operation and maintenance. As in all cases, capital cost is the most important constraint for importers, so they always balance the ratio of price and function of equipment or technology among different suppliers.

Environmental effects of equipment
As the government takes more measures to protect the environment and implements charges for disposing pollutants, the importers also have to consider the environmental aspects of the equipment.

Capital resources
As far as large-scale electricity generating equipment is concerned, the importer's decisions on which country's equipment will be utilized is closely related to the source of capital. If an international financial institution funds a project, the equipment supplier is decided by an international bidding process. If a foreign government or its commercial bank loans the capital, then the project is usually bounded by purchasing equipment from that country.

Traditional trade relationship
The longer the trade relationship in one country has with other countries, the easier it is to establish good trade and information channels with them. The good trade relations thus play a significant role to the importer's decision, as long as the equipment provided is advanced and reliable.

China's license trade in clean coal technology
Considering from perspectives of companies that acquired technology, the maturity of technology and the timing, China's license trade in clean coal technology has the following characteristics:

Most of entities that acquired technology are large companies. In China, most of medium or small enterprises are the major producers of thermal-energy equipment with high-energy consumption and high emissions. Although their technology level needs to be raised urgently, they are unable to participate decisively in the process of technology transfer. That is because, their limited capital and weak technology strength prevent them from assimilating the new technology effectively. Also, they lack the necessary information and technology transfer experience. Finally, the government's approval procedure and policies for technology transfer do not provide incentives for them to acquire clean technology.

The basic guiding principle for clean coal technology transfer in China is the improvement of outdated equipment and the development of new products and market through technology transfer. Under such guidance and with certain involvement of the government in technology transfer, the companies located in traditional industrial bases, as well as in the southeast areas, where the economic opening up and reform takes place, play a very important role in technology transfer in China.

Prior to 1992 technology transfer was entirely dominated by the government. As the Chinese economy reform progressed, the original mechanisms for technology transfer dominated by the government progressively ended and a new form of technology transfer depending on company's own choices was established in the beginning of 2000. As a result, the transaction activities of technology transfer were restrained to some degree.

The clean coal technologies that are obtained by China are mature technologies that have been fully commercialized for a long period, are more advanced than the domestic technology and can be quickly applied by the companies.

Clean coal technology transfer through joint ventures

Acquiring technology is one of the main purposes that China absorbs foreign capital and sets up joint ventures. In the area of clean coal technology, joint ventures have made a contribution to China's clean coal technology transfer, although the contribution is limited. Its main role is to promote equipment imports. Joint ventures involved in clean coal equipment are mainly in the power and heating sector and the manufacture sector.

In the power and heating sector, most joint ventures are cooperated with partners between Hong Kong and Mainland China, which stands 63% of the contracted-investment enterprises, enterprises run by US and Chinese partners account for 20%, by UK and Chinese partners 8%, with other countries 9%. Joint ventures invested by Hong Kong purchase equipment from both domestic and abroad markets, equipment of joint venture enterprises invested by industrialized countries are mainly purchased from abroad. In the power and heating sector, joint ventures have so far played a limited role in promoting clean coal technology transfer.

In the equipment manufacture sector for clean coal usage, based on the 1995 industry census, joint ventures involved in clean coal equipment manufacturing mainly concentrated in the coal-boiler manufacture industry. There are about 50 joint ventures in the coal-boiler (including power plant boilers) manufacture industry, 90% of them are small enterprises. The foreign investors for joint ventures of small-scale coal-boiler are mainly from Hong Kong and Taiwan. And what they invest is not technology, but mainly a little capital. The main purpose of the Chinese side is to share joint venture preferential policies. Technology transfer is only the secondary objective of their joint ventures.

13.4.8 Opportunities for China's Clean Coal Technology

The review and analysis of equipment imports license trade and joint ventures describe the present situation of China's acquisition of clean coal technology. But what opportunities and obstacles does China have in clean coal technology transfer?

Strengthening of air pollution control

The Chinese government has lately taken effective measures in order to mitigate the pollutants that are associated with the use of coal (acid rain, particulates, CO_2 and other greenhouse gasses). These efforts have produced results that have gained the rest of the world approval as the first positive step towards environmental protection. The strategies and measures adopted by the Chinese government can create new opportunities for China's clean technology acquisition.

In 1998, the Chinese government listed environmental protection as its top priority in infrastructure construction and greatly increased environmental investment. Also because it is difficult to alter the dominant percentage of coals in energy consumption, the 2010 Development Outline for Clean Coal Technology identifies priority fields and specific methods for the development of China's clean coal technology. Measures like the above create new prospects for clean coal technology acquisition.

Ongoing economical reform
From the early 1990s, China started reforming its system towards a market economy. By establishing market mechanisms under the law of the competition, companies with no competitive advantages other than a large amount of capital will gradually withdraw from the market. As a result, capital transfers to the more efficient companies. Also, the Chinese companies have to face for the first time pollution charges and penalties for environmental waste. As the protection of the environment and pollution control becomes one of the key factors of government policy, companies can be pushed towards the adaptation of clean coal technology through market forces.

Global environmental cooperation under Kyoto
With the Kyoto Protocol in 1997, it requires that developed countries have the responsibility to provide financial assistance and transfer environmental technology (including clean coal technology) to developing countries. Even though there has not yet been acceptable progress on the above field, for various political and financial reasons, the efforts that are currently taking place by international organizations lead to the assumption that developing countries will acquire in the end clean coal technology from developed countries in order to mitigate the effects of global CO_2 emissions.

Large foreign exchange reserves for Chinese companies
Before reform of the Chinese foreign trade system, there were severe restrictions into the amount of capital a company could use as foreign exchange for its imports, a critical factor that prevented companies from acquiring new technology from abroad. The reform of the foreign trade system removed these restrictions making the import of new technologies much easier. Also the Chinese government has reduced the import duties for clean coal technologies to very low levels in order to encourage its acquisition.

Increase of the environmental awareness of Chinese public
Environmental awareness of the Chinese public is increasing steadily after 2000 in economically developed city areas like Shanghai and Beijing. It is one of the indirect results of the increased income flow to Chinese citizens and improvement of the quality of life in these areas, when compared with the rural areas of China. The increase of that awareness can be pointed by the publication of environmental monitoring reports at these urban centers. The development of the public environmental awareness has two parallel positive results. It helps the individuals transform their life style particularly in terms of consumption. At the same time it stimulates the evolvement of social pressures, which forces the government to apply new laws and regulations for the protection of the environment. As a result pressure is applied to the companies to improve their environmental behavior with positive results for clean coal technology acquisition.

13.5 Obstacles in China's Clean Coal Technology Acquisition

Although there are some positive effects of the acquisition of clean coal technology in China, it should be noted that technology transfer is a two-way economic activity. The willingness, conditions of the supplier as well as policies, the availability of funds, resources, and the system for technology transfer in the recipient countries may all become obstacles for clean coal technology acquisition.

13.5.1 Fewer Joint Ventures in the Field of Clean Coal Technology

In the equipment manufacture industry for clean coal technology, there are fewer joint-venture enterprises, The main reasons are:

1. The market for clean coal equipment is different from the market for other commodities. Its scale is mainly influenced by non-economic factors such as environmental laws and enforcement, environment awareness, and not from technological potency of the investors. It is unlikely that investors will put a large sum of capital, until the uncertainty on China's environmental regulations seizes.

2. Because, China's production volume in the coal power equipment manufacture sector is surplus, with relatively low company profits and intense market competition, foreign investors are not easily attracted to the market. Under current environment standards and resource prices in China, the advantages in energy saving and emission reduction for foreign investors may hardly turn into advantages on the market.

3. Even though the technologies that have been transferred to Chinese companies from foreign companies are advanced for Chinese standards, they are not considered advanced for the rest of the world. Some of the technologies are even considered standard conventional amongst multinational companies. So because the costs of this kind of technologies are relative, foreign companies prefer to sell them through the market than through direct investment. With the improvement of environment law enforcement and intensification of environment sensation, many multinational companies are interested in China’s clean coal technology market and have expressed the intent of direct investment.

13.5.2 Restraints on Technology Transfer Strategies of Multinationals Corporations

The main obstacle that China faces in its efforts to acquire clean coal technology is the reluctance of foreign companies and organizations to transfer the most up to date technology. The main aim of these companies is their profitability and not the mitigation of environmental effects. As a result their policies focus on exporting equipment and not knowledge as these can lead to the development of local technological equipment. Technology will be exported to China only when it becomes outdated for their standards in order to maintain the dependence. It is a key element of the strategy of monopolizing the market through a monopoly of technology. For instance in the clean coal technology for electricity power boilers, transactional corporations firstly try to exclude the companies from developing countries from the worldwide bidding for power boilers. After they win the bidding they provide the machinery to native companies to manufacture through subcontracting with very competitive prices. As a result these corporations can obtain most of the profits through technological superiority while the companies who actually manufacture the products in the developing countries earn limited profits. Since the international market is structured in this way,

Chinese companies cannot obtain the latest technology in general and as a result in the field of clean coal as well.

Overall investment by foreign companies in the manufacture sector for clean coal equipment is relatively small. The reasons for less investment from abroad in the manufacture sector for clean coal equipment are non-economic factors such as Chinese environment laws, regulations and standards. They have not yet provided incentives to effectively stimulate the expansion of the market for environment equipment. With the improvement of environmental legislation and enforcement and the promotion of environment awareness in China, some multinational companies in the field of clean coal technology have expressed their interests in investing in China.

13.5.3 Low Electricity Price in China

Energy price is an important factor influencing energy consumption and the use of fuel. Before the economic reform took place, energy prices in China were under total government control as elements of the centralized economy. Today it is decided both by market and by government intervention. Even so, China's electricity price still remains relatively low. The low electricity price means that power plants with desulphurization technology have difficulties in recovering the investment costs. As clean coal technologies add more investment cost to the already required desulphurization, the power plants will not be able to recover this cost with the current electricity prices. Therefore, it is considered logical that there is no willingness from the part of the power plants to apply clean coal technologies.

Another factor is that the emissions charge in China is currently lower than the pollution control expenses. This means, that many companies prefer to pay the emission charge rather than applying pollution mitigation technologies. Because of that and of the low electricity prices that prevent the investment recovery from the power plants income, power plants that adopt pollution control equipment often have to stop operation due to high costs.

13.5.4 Weak Economic Strength of Chinese Companies

Chinese corporations have always been characterized by very low debt amortization rate. This results to a considerable lack of funds for the development of new technology. Therefore, they are unable to import clean coal technology due to low capital.

In addition to the above, the profits from the implementation of clean coal technologies are mainly long-term and indirect under the current condition in China. There are direct profits as the rise in energy efficiency lowers the production cost but the main part is the environment effects of clean coal technology. The profits of effects can only be realized over a long period of time. Currently, most enterprises attach great importance to the short-term effect while overlooking the positive environment effect achieved through the investment in clean coal technology equipment. Therefore they lack the motive to adopt clean coal technology.

13.5.5 Strategies for China to Transfer and Develop Clean Coal Technology

China's strategy for acquiring and developing clean coal technology must not only consider the need of environmental protection and economic growth, but also the long-term target of its own independent research, innovation and economic development. China faces the unique problem of environmental deterioration, shortage of resources, and undeveloped clean coal technology while reforming its political and economical system. If the country relies only in domestic supplies for clean coal technologies, the environmental conditions and the shortage of prime resources will be probably be accelerated. On the other hand, if China focuses solely on technology transfer, it will be a hostage to the willingness of transactional corporations and foreign governments, and be unable to develop its own technology. From the above Chinese strategy must try to combine these two aspects in order create its own. Thus the most advisable strategy is to integrate these two aspects so as to establish as soon as possible its own independent system of clean coal technology. To achieve this strategic goal, the following recommendations are proposed:

- *Strengthening of Chinese macro-control of technology transfer*
  As China is undergoing economic transition, the market mechanism is yet to be perfected. In order to establish its independent innovative system of clean coal technology, and achieve the coordinated development of economic growth and environmental protection, full play should be given to government's macro-control. The following steps should be adopted:

- *Define priority channels for clean technology transfer*
  The priorities of clean coal technology transfer should be changed. Instead of focusing primarily on equipment imports China should make license trading the primary medium for acquiring clean coal technologies. Among various ways of technology acquisition, the priority order should be license trade, foreign direct investment, and equipment imports.

- *Exchange the market for new technology*
  Equipment imports should be used as a medium of pressure for the transfer of environmental market and should also provide technological knowledge. In this way China can '*exchange*' its market for up to date technologies.

- *Support companies in technology acquisition*
  The government must support and encourage Chinese companies to implement clean coal technologies with economic and administrative measures. These measures can be in the form of subsidizing clean coal demonstration projects, or on sharing the costs for users of newly imported technology for a period of time.

- *Adopt effective measures to integrate technology transfer and domestic research and development*
  The importation of new technologies should not be done in random manner from private companies but under a general plan that must be co-ordinated by the government. In this way, the Chinese companies and research institutions, promoting domestic innovation, can assimilate the technology.

- *Take measures to improve the domestic clean coal technology research and development*
  Clean coal technology research and development projects are systematic engineering projects that involve various fields of technology. They require the cooperation of vari-

ous Chinese universities and technological institutions. Unfortunately, many institutions and universities do not have any knowledge transfers in order to protect their intellectual property and technological advantage. So the required cooperation between different organizations cannot be easily achieved. On top of that the universities and technology institutions tend to lack connection with the industrial sector, meaning that many of the developed projects do not consider the actual capabilities and needs of the industry and become inapplicable.

The most realistic approach to the above problem is if the industry is the one that provides the funding to research institutions in order to answer to specific technological problems. Such a change can promote cooperation and understanding between the different Chinese universities, institutions and companies and increase the applicability of research and development.

Promote the market for clean coal technology by means of law and policy
The Chinese environmental standards must be raised so that the clean coal technology market can be expanded. The intensification of environmental laws will force companies to adopt clean coal technologies. Also methods of increasing the public environmental awareness should be pursued.

At the same time the pollution charge must be raised and ways of increasing electricity prices must be pursued. The first is crucial to force companies to adopt clean coal technologies, the second is even more important as it will enable companies that apply the technology to survive economically. The prices must not only satisfy the current urge for development but also the needs of future generations.

Finally China should take measures to reduce the lack of legalization of rights for technology development. Legal protection of the intellectual property rights of domestic clean coal technology developers can encourage more research and development projects by institutions and companies.

China encounters the task sustaining its economic growth while adapting measures for environmental protection. This situation is truly unfair when compared with the developed countries that first proceed with their industrial and economic growth after the Second World War and then started establishing regulations for environmental protection. Under these conditions clean coal technology acquisition must not only obey the laws of the market but also be considered in the governmental planning and legislative control.

13.6 Development of Chinese Power Industry

In China, electric energy supply reaches 42% of all end users of energy supply as observed in 2002, which will grow up to about 70% in 2020 [6]. In order to meet the requirements of rapid development of the Chinese economy, the installed generator capacity has been increasing rapidly. In 1987, the national generation capacity was over 100GW. In April 2000 and May 2004, the total generation capacity reached 300GW and 400GW, respectively. And by the end of 2005, it reached 500GW. In the next decade, the electric load forecasted by the State Grid Co. (SGC) will be about 570GW, 752GW and 966GW in 2010, 2015 and 2020 respectively, and the corresponding generation capacity will reach 799GW, 978GW and 1211GW.

However because of the large population in China, the electric energy used per capita is still at a low level. In 2002 the kWh per capita of China was about 1,280, which is about the average level for the world in the 1970s. The current average level (2008) in China is 2,500kWh per capita. Therefore the power industry in China will continue to grow rapidly in the future [6]. Figure 13.6 shows a comparison of kWh per capital worldwide.

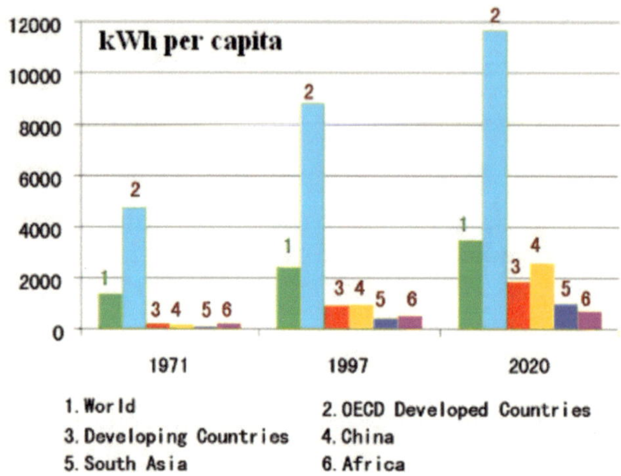

Fig. 13.6 Comparison of kWh per capita

The Chinese power systems have following features

- The load in China is mainly in the South and East of China, while coal and water resources are mainly located in North, Northwest and Southwest of China, which leads to the long-distance heavily-loaded transmission system from west to east and north to south of China.
- Since the resources of petroleum and natural gas are not rich in China and the coal production cannot meet the requirement of power generation, nuclear power plants constructed in the load center are under planning, but the total nuclear power capacity will be no more than 5% until 2020. In the meantime, the hydropower utilization will be accelerated especially along the Yangtze and Yellow Rivers. In general coal-fired and hydropower will be over 90% of total capacity in the next 20 years with a small amount of nuclear and other renewable energy.
- Based on the above, China will construct ±800kV HVDC and 1000kV UHVAC transmission systems to interconnect the 7 regional power grids of the Northeast, North China, the Northwest, Central China, East China, Chuan-Yu (i.e. the Grid composed of Si-Chuan Province and the City of Chong-qing) and South China in the near future.
- In the meantime, power industry deregulation is underway in China. In its first stage, the traditional vertical monopolistic operation has been converted. The former State Power Co. is now renamed the State Grid Co., SGC. Since the Ministry of Electric Power

does not exist anymore, the SGC will have the power to plan and construct inter-regional transmission systems, arrange their bilateral power exchanges and supervise the operation of regional grid companies, etc. (except that the S. China Grid Company is operating independently approved by the State Council for a certain degree of competition in the transmission sector). In the *11th 5-year Plan (2006-2010)*, the wholesale mode will be realized in regional power markets. The power utilities will form the power supply companies and join the competition in the power markets. The traditional transmission tasks will be implemented by the grid companies as transcos allowing open-access as for most power markets in the world.

- In the next decades, the key challenges in Chinese power systems are:
  - Optimal planning of nationwide power system interconnection with security consideration.
  - Key technology for ultra-high-voltage (UHV) transmission systems.
  - Optimal operation of the Chinese power system using advanced technologies (such as computer and communication engineering, applied mathematics and control theory applications, etc.).
  - Improvement of power system reliability and power transfer limits to avoid power system blackout and severe stability crises.
  - Strengthening power market operation at high-efficiency and encouraging energy saving of end-uses and environment protection to realize sustainable development of the power industry in China.
- To cope with the secure and optimal operation of the complex transmission networks in China, advanced technology will be widely explored, investigated and developed. International collaborations and academic exchanges should be greatly strengthened. It can be foreseen that along with the huge amount of investments entering the power industry, the R&D funds for power systems will increase rapidly.

Enhancing Demand Side Management

More attention is paid to enhance load forecasting, optimizing operation mode of the power grid, and concentrating on resources allocation optimization. Backed by the government, demand side management is actively implemented and gives priority to supply electric power needed by the people and key consumers, but also arranges the enterprises to practice peak load shifting so as to relieve the stress of power shortage.

Promoting mechanism reform and innovation of electric power industry

The mechanism reform and innovation of the electric power industry have been positively promoted in China. China enhances the construction of power markets. Meanwhile, it accelerates the establishment of modern enterprise mechanisms and perfects corporate governance structure.

13.6.1 The Clean Development Mechanism (CDM)

It was set up under the Kyoto protocol and requires rich countries to cut their emissions but allows them to do so by buying carbon credits from emission-cutting projects in poor countries. The market in carbon was worth £32bn in 2007. It is to be known as *cap-and-trade*.

The main provisions of the protocol expire in 2012, so carbon traders are watching closely the tense international negotiations on its successor. The negotiations are scheduled to finish by the end of 2009. It is likely that the US is willing to negotiate a new treaty and support the setting up of an emission market. The UN executive secretary is discussing the idea of '*climate bonds*' issued by governments to financial investors. This would be the best way to raise money to pay for efforts to reduce emissions in poor countries.

Such bonds would involve a developing country government setting a national target on emissions or energy, for example a certain amount of renewable energy by 2020. It could count on receiving carbon credits equivalent to this renewable energy under a successor to the Kyoto protocol, which would enable the financial ministry to issue 'climate bonds'. These would raise money from investors based on the promise that they would receive return from the sale of the resulting carbon credits.

This is not possible under the Kyoto protocol because the system for awarding carbon credits requires would-be investors to put money into emission cutting projects, often-small scale and carrying big investment risks. As a result, investment has been relatively small – the trading between the rich and poor nations was worth £7bn in 2007, compared to £32 trade in carbon in European Union.

The UN executive secretary believes that government-sponsored climate bonds would reduce investment risk, because government could act as a guarantor, and encourage much greater investment in carbon reduction in the development world. There is recognition that the instruments in place for carbon trading need to continue to be improved. At the moment the trading mechanism under the Kyoto protocol is about small investments in small projects that produce small reductions in emissions.

China has been by far the biggest winner from the Kyoto protocol, receiving tens of billions of dollars in investment to finance low carbon technology [10].

In 2007, 73% of carbon credits projects certified by the United Nations under the protocol were based in China. Under the Kyoto treaty, the 'Clean Development Mechanism' enables rich countries to meet part of the target to cut carbon emissions by funding the development of projects to reduce emission in developing nations. These include wind turbines, capture methane from sewage works and technology used in factory chimneys to destroy harmful industrial gases.

China has benefited most through projects that reduce industrial gases, particularly HFC, a by-product of making refrigerants. Most of the world's factories that produce HFC are in China. Installing equipment to destroy the gas is relatively inexpensive but generates a large volume of carbon credits because the gas is more than 11,000 times more harmful to the climate than carbon dioxide. Most emission-reducing projects now focus on energy efficiency.

13.6.1.1 Market economy development

In power generation, during the *11th 5-year Plan,* 300MW and above thermal plants will account for 50% of total generation capacity; only 600MW and above new thermal plants will be allowed to be built; the building of supercritical and ultra-supercritical power units will be encouraged. By 2020, nuclear power will be 40GW.

International ordering and bidding for supercritical, ultra-supercritical and large-scale gas turbine plants, and nuclear power plants will be expanded. 3GW new wind power plants will be installed; more than 70% of the wind power units will be domestically produced. Domestic production of wind power plants will be accelerated.

During the *11th 5-year Plan,* it aims to have: SO_2 emission: 12 million-13 million tons/year (10% reduction from 2005); soot emission: 3 million tons/year (more than 25% reduction from 2005); coal consumption/kWh: less than 360 g; transmission loss rate: less than 7 %; consolidated usage rate of particulates: 70%, etc.

The latent potential of energy conservation and environmental countermeasures in the power generation sector is great; consequently in the future, increased international (governmental and private sector) activity can be anticipated in the concerned sectors.

For transmission infrastructures, during the *11th 5-year Plan,* it aims to: construct 330kV and above power grids of length of 7500km; transformer capacity will reach 360GVA; and the investment will be around 1.5 trillion RMB. There are the State Grid Corporation of China (SGCC) and the China Southern Power Grid (CSPG)

Overseas orders for medium, high and ultra-high voltage transmission (1,000kV) and transformer facilities will be expanded; and technological collaborations (private sector) will be investigated with respect to network security and stable operations.

The State Electricity Regulatory Commission (SERC) was established in March 2003. Its major responsibilities are to develop laws and regulations; monitor electricity operations; propose tariff and adjustments to governmental pricing authority; investigate any possible violations of laws and regulations; and to supervise the implementation of universal service provisions, etc.

In the power market, the "*regulated power rate system*" and wholesale power market were introduced. For the retail sector, there was the introduction of the Renewable Energy Law and revision of the Electricity Rate System. Retail prices tend to continually increase. Implementation of the regional power market will be accelerated and it is anticipated that major generators will concentrate on the market due to fierce market competition.

In summary, the future power development plans from 2006 to 2010, may be summarized as follows:

- Proactively work on the development of hydropower
- Implement large-scale expansions of coal-fired thermal power, and upgrade technology

- Proactive development of nuclear power
- Appropriate introduction of natural gas power generation
- Introduce renewable energy
- Intensify energy conservation, and improve the efficiency of energy utilization
- Realize nationwide interconnection based on the concept of "*West-East Power Transmission*" and "*South-North Power Transmission*".

13.6.2 Electricity Tariffs

In [9], it reported that the National Development and Reform Commission (NDRC) issued three Measures establish a comprehensive tariff system for electricity in China, covering production, distribution and sale to end users.

The *On-Grid Tariff Measures* specify different rules for the determination of *On-Grid Tariffs* before and after the implementation of competitive pricing. Before the implementation of competitive pricing, the *On-Grid Tariffs* for power generating companies originally under the State Power Corporation were determined by the tariff regulating authority according to cost-compensation principles.

For power plants retained by the State grid enterprises that have an agreed *On-Grid Tariff* with the tariff regulating authority, their *On-Grid Tariffs* should remain unchanged. For those who have not agreed to an *On-Grid Tariff* with the tariff regulating authority, and wholly-owned power generating companies (State grid enterprises or non-State grid enterprises) and other power plants for which *On-Grid Tariffs* have not been agreed or confirmed by the tariff regulating authority, their *On-Grid Tariffs* shall be determined by the tariff regulating authority on the basis of the economic life span of the project, according to the principles of reasonable compensation, reasonable actual benefits and with consideration given to the tax payable under PRC law.

The On-Grid Tariff Measures provide that pricing systems such as peak and valley pricing, seasonal pricing etc. are to be introduced gradually. When the fluctuation in fuel prices is fairly large, the *On-Grid Tariffs* may be linked to the fuel price as long as the tariffs also reflect demand and supply.

After the implementation of competitive pricing and the establishment of competitive regional power markets, the *On-Grid Tariffs* of electricity generation companies which participate in the competitive market shall principally consist of two components: the capacity charge, which is to be determined by the tariff regulatory authority, and the energy charge, which is to be determined by market competition. The *On-Grid Tariffs* of electricity generation companies that do not participate in the competitive market will be determined by the tariff regulatory authority on basis of their economic life span.

The *On-Grid Tariff Measures* also deal with electricity generation companies with direct foreign investment. If such companies were constructed prior to 1994 and executed their power purchase contracts prior to 1994, or were established in or after 1994 with State Council approved undertakings regarding electricity tariff or the rate of investment return, new negotiations may be conducted in order to allow them to operate as much as possible in accordance with the new system.

The Transmission and Distribution Tariff Measures define "transmission and distribution tariffs" as the common term used for tariffs charged by State grid enterprises for the provision of access systems, inter-grid connection, electricity transmission and sales services.

The transmission and distribution tariffs are to be fixed by the government. The basis of the tariffs shall be considerations such as reasonable compensation of cost, reasonable determination of returns, lawful inclusion of taxes and fair sharing of the burden.

The *Transmission and Distribution Tariff Measures* distinguish between tariffs for electricity transmission and distribution services via public grids, tariffs for dedicated services and tariffs for ancillary services provided by an electric power enterprise. *The Transmission and Distribution Tariff Measures* detail how the various tariffs are to be determined.

The *End-User Tariff Measures* define the term "*Electricity Tariffs*" as the tariffs at which State grid enterprises sell electricity to end users.

The government fixes electricity tariffs. The tariffs shall be determined with consideration to the fair sharing of the burden, the efficient adjustment of the demand for electricity, and public policy objectives. The government will create a mechanism to link *Electricity Tariffs* to *On-Grid Tariffs*.

Electricity Tariffs are made up of four elements, the electricity purchase cost, electricity loss during transmission, electricity transmission price, and cost of government fund.

The *End-User Tariff Measures* aim to reform the various classes of tariff for end users into three categories: residential electricity; electricity used in agricultural production; and electricity used in industry, commerce or for other purposes. The tariff for each category is fixed per voltage class. For residential electricity and electricity used in agricultural production, the *Electricity Tariff* is calculated per kilowatt-hour. For industrial, commercial or other electricity users who have receiving transformers with a capacity of 100kVA or more or who have electrical equipment with an installed receiving capacity of 100kW or more, the *Electricity Tariff* comprises the kWh-hour tariff (calculated on the basis of actual use) and the basic electricity tariff (based on the user's consumption capacity).

It was reported in [11], because of the increasing cost of coal (the country's main source of energy for electricity production), the government in 2004 approved the mechanism linking coal and power prices. This allowed electricity prices to shift in line with coal price increases. Under the mechanism, if the price of coal rose by more than 5% in a six-month period, electricity prices could be adjusted. With the mechanism, 70% of coal price increases are transferred to end-users. Power generation firms bear the remaining 30%. However, electricity prices have not yet been fully linked to the market. The government will do more work on the electricity pricing system in order to link it more closely to movements in raw materials.

China is trying to adjust its power mix, hoping that clean power will make up at least 35% of the entire supply in 2010.

Under the *Renewable Energy Law*, which took effect in 2008, the Chinese Government announced rules for setting prices on electricity generated from wind, solar and biomass. As reported in China Daily in July 2006, power generators that use renewable fuels are allowed to charge higher rates to power grid operators than coal-fired plants, with the additional cost split among grid operators and partly passed on to retailers.

13.7 R&D on Power System Sponsored Funds by National Science Foundation China (NSFC)

In China R&D funds support power system projects through several ways. In the State Council, there are commissions that will support power system R&D projects relevant to their functions, including the State Development and Reform Commission, the State Economy and Trade Commission, and the State Electric Power Surveillance Commission.

The NSFC was founded in February 1986 with approval of the State Council, which is aiming at promoting and supporting basic research and key applied research in China. Now NSFC employs less than 200 staff members with over 90% as professionals at an average age of 42. It has established a solid system to make strategic plans on key and basic research, evaluate various proposals, and assess R&D outputs strictly, fairly and scientifically. Since its foundation, the budget of NSFC has increased dramatically from 80 million RMB in 1986 to 2 billion RMB in 2004, about 25 times higher. The projects sponsored by NSFC are now classified into the categories of major projects, key projects and general projects to be funded at different strengths and scales. Besides, special funds are launched for outstanding young investigators to promote their R&D excellence and for international exchange and collaboration.

In the traditional electrical engineering area (i.e. heavy-current area mainly on power engineering), 6 key projects were sponsored during 1996-2000 at 5.1 million RMB in total; while in 2000-2005, 6 major and key projects were sponsored at 11.3 million RMB in total. The NSFC sponsored general projects increased from 40 projects and 7.6 million RMB in 2001 to 98 projects and 26.4 million RMB in electrical engineering subjects in 2006.

The NSFC major and key projects in electrical engineering focus on the significant R&D areas in power engineering, such as FACTS devices (TCSC) design, development, operation and simulation, non-linear robust control theory and its applications in power systems, fundamentals of power markets, super-conductor technology and its applications in power systems, basic theory and key technology for power system security control, etc. A major project on key technology for HV AC/DC power systems is in the planning stage.

In a key project jointly sponsored by NSFC and SGC, EPRI-China has successfully designed and constructed a prototype of TCSC. The TCSC was installed on a 220kV AC transmission line in the Gansu Provincial Grid and commitioned in December 2004. The compensation level was 50% of line reactance with the capacity at 95.4 MVAr, which increased the transmission capacity by 33%.

Although the funds for power engineering is about 1/20 of all funds for engineering and material science, and 1/100 of overall NSFC funds, it plays an important role to give preliminary support on basic and fundamental research and lay a solid foundation for further supports from industry. Actually some key projects are sponsored by the power industry in parallel, such as the project on multi-infeed HVDC transmission system reliability and control, etc. In addition, other departments of NSFC (such as the information department, etc.) will also sponsor multi-discipline projects related to the applications of advanced technologies in power systems.

It should be mentioned that in parallel with NSFC, there are two National Key and Basic Research Projects (so-called '973' projects) on power engineering. One is "*Vital Research on Collapse Prevention and Optimal Operation of Large Power Systems in China*" (1999-2003) and the other is "*Fundamental Research on Improving the Reliability of Large-scale Interconnected Electric Power System*" (2004-2008). The project funds are 50 million RMB and 30 million RMB, respectively. The nationwide experts of power engineering are working together for the projects and have made great contributions to the Chinese power industry.

The facts mentioned above show clearly the impacts of fast development of Chinese power industry on power system R&D. The provincial governments, power companies and manufactures are all actively supporting R&D projects in power engineering so as to get benefits from the outputs.

13.8 Conclusions

An insight into the power generation and transmission infrastructure in China has been given in this chapter. It is clear that the future is full of uncertainty but there are many fronts that have been developed and advanced to minimize the risks occurred.

The coal-fired energy system can evolve in ways consistent with sustainable development objectives if public policies guide a high rate of innovation toward clean coal technologies. On the basis of present knowledge, it is possible to identify and describe advanced clean coal technologies that meet sustainable development objectives at reasonable cost.

Due to the enormous size of China's population, the country's coal reserves per capita are insufficient, and consuming a large quantity of coal causes tremendous environmental pressures. All these factors mean that, in order to ensure energy supplies and to achieve sustainable development objects, energy efficiency must be a priority.

It is believed that for many years to come, coal will still be the most important source for electricity generation. China's great potential market for clean coal technology is unquestionable. However, it must also be noted that there are many obstacles to clean coal technology acquisition in China. If no enforceable system and effective policy mechanisms are established, such potential will be unlikely to be turned into real opportunities.

13.9 Acknowledgements

Loi Lei LAI, Professor and Head of the Energy Systems Group, School of Engineering and Mathematical Sciences, City University London, UK has coordinated this Chapter. Thanks are due to Long Zhou and Ms Yingnan Ma of the Energy Systems Group at City University London for the detailed preparing this Chapter. Copyright release by the IEEE is very much appreciated.

13.10 References

[1] Yin Yonghua, "Electric power system in China: History of development, present status & future perspective", IEEE Power Engineering Society General Meeting, Tampa, June 2007.

[2] Yinbiao Shu, 'Current Status and Development of National Grid in China,' IEEE T&D Conference, 2005, Dalian, P R China, August 2005.

[3] Yuan Maozhen, 'South China HVAC and HYDC Hybrid Grid: Development and Challenges,' IEEE T&D Conference, Dalian, P R China, August 2005.

[4] Yusheng Xue, "Towards space-time cooperative defense framework against blackouts in China", IEEE Power Engineering Society General Meeting, Tampa, June 2007.

[5] Qixun Yang and Bi Tianshu, "WAMS implementation in China and the challenges for bulk power system protect", IEEE Power Engineering Society General Meeting, Tampa, June 2007.

[6] China's Energy Development Strategy and Policy Study, Development Research Center of State Council P.R. China, 2004, p.197.

[7] South China Morning Posts, April 2008.

[8] Shuyong Chen, Feili Huang, Baohui Zhang, Yixin Ni, Libao Shi and Zheng Xu, "Fast development of Chinese power industry and its impacts on power system R&D outputs", IEEE Power Engineering Society General Meeting, Tampa, June 2007.

[9] China: Electricity Tariff Reform, August 29, 2005 - Hong Kong http://www.hg.org/articles/article_805.html

[10] Fiona Harvey, Chinese reaps biggest Kyoto rewards, Financial Times, 27 May 2008.

[11] *China Daily* July 1, 2006, http://www.china.org.cn/english/2006/Jul/173343.htm http://www.fossil.energy.gov/international/International_Partners/China.html

14

Power Generation and Transmission Expansion Planning Procedures in Asia: Market Environment and Investment Problems

14.1 Introduction

This Chapter reviews the present status and prospective ideas of power generation and transmission planning procedures in a market environment. This problem is very important for different countries in Asia, as well as for other countries that have taken part in the liberalization of the electric power industry. Japan, Korea, China, India, Russia and others are now on the way of transition of their electric power industries to a new market structure and new economical conditions.

One of the main aims of electric power industry liberalization is the attraction of investment for new more effective generation and transmission development. The objective estimation of needed investment can be made on the basis of new advanced procedures of generation and transmission expansion planning in a market environment. This will be examined. There are many issues in the discussed problem. These include technology of power generation and transmission expansion planning procedures; methods of making decision on the generation and transmission network in a market environment; specific features of solving the discussed problem in different countries of the region; use of the correlation state regulation and market mechanisms for attraction of investment; and improvement of surplus generation for effective electricity market operation; etc.

14.2 Problems of Electric Power System Expansion Planning in a Market Environment and Procedures of Their Solution

In a liberalized environment different variants of the electric power industry structure are possible. These variants predetermine specific features of the problems of electric power system (EPS) expansion planning. In general this problem is solved by the rational combination of market mechanisms and state regulation, provided there are many subjects of relations (power supply companies, consumers, authorities, etc.) with a great number of commonly non-coincident criteria. And the uncertainty of future conditions for EPS expansion is responsible for a multi-variant character of possible decisions to be made and compared.

Complexity and multi-dimensionality of current extended EPSs, plurality of variants and criteria, availability of different preferences in decision choice all make it impossible to solve the EPS expansion problem as a general synthesis problem. In the centrally planned power industry this problem was solved by applying the hierarchical approach that was based primarily on the expert, but a posteriori technology of problem solution. In the liberalized power industry the problem is drastically complicated and the technology of its solution can be represented by different variants depending on specific features of the industry structure.

We will analyze different variants in structure of the power industry [1] that generate distinctions in the composition and character of EPS expansion planning problems. These variants comprise a regulated monopoly at all levels; interaction of vertically integrated EPS and open access to the main grid; a single buyer-seller of electricity (an electric network company) with competition of generating companies; competition of generating companies and a free choice of electricity supplier by selling companies or/and consumers when the main grid renders only transportation services; in addition to conditions of two previous cases (i) competition of selling companies in electricity supply to concrete consumers; (ii) intermediate and mixed variants based on the considered ones.

The general problem of EPS expansion planning can be divided into three groups of problems [2].

- The state strategies and programs for development of the power industry and EPS (the federal, interregional and regional levels);
- Strategic plans for development of power supply companies (vertically integrated, generating, network);
- Investment projects of electric power installations (power plants, substations, transmission lines).

In making decisions on EPS expansion different groups of subjects of relations have shown, largely non-coincident, interests that are expressed by the corresponding criteria. In particular:

1) Electricity producers or/and sellers (vertically integrated, generating or selling companies, an electric network company as the single buyer-seller of electricity) and also subjects of electric power industry rendering electric power services in the wholesale electricity market (maintenance of active and reactive power reserves, provision of system reliability, etc.) are interested in profit maximization as a result of their business.
2) Electricity consumers (selling companies of different levels, concrete consumers) are interested in minimization of the tariffs for electricity bought (in the wholesale or/and retail markets), provision of its quality and power supply reliability.
3) Interests of the authorities (federal and regional) are directed to maximization of payments into budgets of the corresponding levels, minimization of the environmental impact of electric power facilities, provision of the energy security of the country and regions, etc.
4) External investors (banks, juridical and natural persons) are interested in minimization of the period for return of investments in electric power installations, maximization of dividends, etc.

We will discuss the composition and specific features of EPS expansion planning problems from two points of view: technology and structure.

As to the technology an EPS is viewed as a technically single system consisting of power plants operating in parallel and connected with each other and consumers by an electric network. EPS can be modeled in different ways subject to the problem character and the level of consideration. For example, the structure and allocation of generating capacities of the

Unified Electric Power System (UPS) of Russia are chosen as a rule on the base of aggregated representation of large subsystems (e.g. interconnected EPSs--IPSs) and transfer capabilities of tie lines among them. If the same problem is solved for IPSs, their structure is described similarly in the form of aggregated subsystems and transfer capabilities of tie lines between them. To plan network expansion it is necessary to represent it in detail with generation capacities and their allocation that are determined at the previous stages. The UPS level usually deals with the UHV backbone network. At the IPS level the electric network is represented in greater detail considering transmission lines, substations of lower voltage classes. This set of problems on EPS expansion planning is a hierarchical sequence of problems, where decisions on system expansion are adjusted (or new decisions are made) at each stage by means of more detailed examination in the technological and territorial aspects.

As to the structure the technically single EPS in decision-making on its expansion is a set of structural units, i.e. companies, interacting with each other. If the expansion problems are solved depending on the structure, an EPS should be represented by vertically integrated generating and network companies that will expand based on their technological interaction within the system. When choosing decisions on generation and transmission network expansion the vertically integrated company, for example, has to take into consideration potential decisions of neighboring companies on their expansion. The generating company has to allow for prospects in expansion of competing similar companies and the network company as well. The network company, in turn, should have an idea on expansion of generating companies when analyzing trends in its expansion.

Each generating company in this case should consider both prospects for expansion of other companies and the state energy policy (at the federal, interregional and regional levels) and mechanisms of its implementation in the form of tax, credit, tariff and other policies. On working out the strategies and programs of power industry development the state, in turn, should implement its energy policy by taking into account incentives, possible behavior and interaction of generating companies in their expansion.

In general the problems of EPS expansion planning as applied to many subjects of relations that are guided by many non-coincident criteria are of a multi-criteria game character. Let us examine specific features of such statements for the mentioned three groups of problems.

The state strategy and programs of power industry development at the federal and regional levels are elaborated commonly on the base of the hierarchical game multi-criteria statements of problems [3]. Such problems appear when the state is at the upper level and the power supply companies are at the lower level. These problems are solved by the formal methods for creating the incentives for subjects' behavior at the lower level by the appropriate mechanisms foreseen at the upper level.

Here the hierarchical game multi-criteria problems may be cooperative or non-cooperative depending on conditions.

These problems can take place at interaction of the federal and regional levels, when the state strategies and programs are elaborated for the power industry development. Such problems are aimed to coordinate interests of the country and its regions. The state priorities in the industry development are formed at the federal level and then they are transformed into concrete trends in expansion of generation capacities and electric networks in the considered region. In general when the principles of authority sharing are adjusted and non-contradictory, the hierarchical multi-criteria game problems of a cooperative nature can be involved. The mechanisms of inducement or persuasion are applicable here, however, with somewhat different conceptual interpretation as against the previous case.

The indicated two problems can be studied jointly as one problem that reflects interactions among three groups of subjects: federal and regional levels of the country and power supply companies. Such problems are considered, in particular, as active systems with the distributed control and also reduce to hierarchical game models.

In individual cases the simpler statements of the hierarchical two-level problem as a two-stage sequence of multi-criteria problems of mathematical programming can be used. The strategy of national power industry development is considered at the first stage, the appropriate recommendations are adjusted at the level of strategies of regional power industry development.

An analogous two-stage sequence of problems can be analyzed in intersectoral terms, when the basic proportions in power industry development are determined at the first stage by the territorial-production model of the fuel and energy complex. Then these proportions are adjusted on more detailed models for decision making on power industry development.

Main attention in the considered problems is paid to mechanisms of interaction between the federal and regional or the energy and sectoral levels of elaborating the state strategies and programs of power industry development. Therefore, consideration of incentives for the behavior of power supply companies by one or another technique for representing uncertain factors becomes necessary. The key task for power supply companies in this case is to work out effective economic, legal and institutional mechanisms. They are to stimulate the companies to take into account priorities of the state policy in the electric power industry when elaborating strategic plans of their expansion and making decisions on investment projects. The optimal proportions of such mechanisms can be improved by solving the hierarchical game problems for the subjects "state - power supply companies" mentioned above.

Now we will analyze the next group of problems dealing with elaboration of strategic plans of power supply company expansion. At least three classes of such problems can be discussed here.

For the regulated monopoly without competition it may turn out necessary to solve multi-criteria problems of mathematical programming in terms of uncertainty and different preferences [4]. A rather simple way for considering uncertain factors is a scenario representation of combinations of their values. The game problems in the class of "games with the nature" may be analyzed on the base of ordinary and fuzzy payoff matrices in the other cases.

Elaboration of the strategic plan of the network company expansion, when there are vertically integrated or purely generating companies, refers to the second class of problems. Considering, in a certain sense, a subordinate role of the network company that reduces in the most general case to provision of competition for power producers and a free choice for power consumers, the problems of network company expansion can be studied in terms of "games with the nature". In this case the uncertainty in behavior of both power producers and consumers in the wholesale market is essential and taken into account by the appropriate payoff matrix of the game. For the network company as the single buyer-seller of electricity the conceptual meaning of uncertain factors is determined, as before, by the competition and at the power consumption level it depends only on demand uncertainty and elasticity. However, here the problem can also be examined in terms of "games with nature".

The coordination between generating companies especially under state regulation is possible. The problem takes the form of cooperative game [5].

And finally, the third class of problems is related to elaboration of the strategic expansion plans of competing vertically integrated or purely generating companies. Without the state regulation the problem reduces to a multi-criteria non-cooperative game. With state regulation the problem takes the form of a multi-criteria cooperative game, probably of a multi-stage character, i.e. it reduces to a positional game [6,7].

The problems of the third group dealing with decision making on investment projects of electric power installations (power plants, substations, transmission lines) work out a business plan for construction of the corresponding installation. Mathematically the problem statement depends on the investor position. If the power supply company (e.g. the network company) invests in the installation, the investment project may call for multi-criteria assessment. For an independent investor one should allow for an incentive for behavior of the other concerned subjects and the problem can be associated with the game statement. It can be either cooperative or non-cooperative depending on conditions.

14.3 Proposed Performance Criteria for Transmission System Planning based on Regulating Framework of TWBP in Korea

After the Korean Government unveiled "The Basic Plan for Restructuring of the Power Industry" in 1999, the Korean Electric Power Industry has been restructuring. Now, Cost Based Pool has been operated by KPX (Korea Power Exchange) from 2001, and market design of Two Way Bidding Pool is coming to the finish. As these circumstances change, power system planning is one of the most influenced parts by restructuring in the electric power industry in Korea. When the power system was operated by a vertically integrated utility; KEPCO (Korean Electric Power Company), planning was actually done by KEPCO on behalf of government, and the "long-term power development plan" has been made in accordance with the national electricity law. But, as Korean electric power industry has been restructured, various market participants already appeared and will appear in electricity market, so new regulating framework is established to guarantee the transparency of the electricity market, even though the Korean government decided to maintain government-leading resource planning even after restructuring of the electricity industry under the name of "electricity resource baseline plan" taking the demand/supply situation into consideration. Especially in the case of transmission system planning, it was discussed that objective and transparent criteria are required to be developed because a transmission company remains as a type of monopoly after restructuring in Korea, and the transmission network is strongly co-related to all market participants. Section 14.3 briefly reviews progress of restructuring in the Korean electric power industry, and examines the proposed regulating framework focusing on transmission network planning. In addition, Section 14.3 describes the proposed performance criteria for transmission system planning in Korea considering [8-12].

14.3.1 The Progress of Reconstructing in Korea

In "The Basic Plan for Restructuring of the Power Industry" unveiled by the Korean government in 1999, restructuring is scheduled as in the following steps (Figure 14.1).

Phase 1 (~ 2002) : Generation Competition
The generation sector of KEPCO was spilt up into six generation subsidiaries, five of which are to be privatized step by step. Gencos trade electricity by bidding through the Korea Power Exchange.

⇩

Phase 2 (2003~ 2008) : Wholesale Competition
The Distribution/retail sector is to be separated from KEPCO into separate companies, followed by the privatization of these companies as well. The transmission network will remain open to all market participants to ensure nondiscriminatory use of the national transmission network Introduction of consumer choice for large consumers; small, residential consumers will supplied by local distribution companies

⇩

Phase 3 (2009~) : Retail Competition
Every customer will be able to choose his or her own supplier of electricity

Figure 14.1 The plan for restructuring of power industry

Now, the market design of a Two Way Bidding Pool for introducing wholesale competition and retail competition into the electricity market is coming to an end in Korea though some issues like the separation of the Distribution/retail sector from KEPCO are still under discussion. The following Tables are summaries of what has been achieved to date.

14.3.2 Regulating Framework for Transmission System Planning

As competition and deregulation are introduced into the Korean electric power industry, it has been discussed that the following is necessary to make better transmission planning under the new competitive environment that is facing uncertainties in the electricity market (Table 14.1).

- Establishing rational and objective planning standard
- Developing transparent transmission planning process
- Sharing sufficient information on transmission planning between market participants and planner
- Providing sufficient information on transmission expansion plan to market participants are necessary
- Improving efficiency on transmission investment by transmission planning:

(1) Designing incentive mechanism for efficient transmission planning by transmission planner
(2) Designing regulatory mechanism for transmission business

Developments of the electricity network must be planned with sufficient lead time to allow any necessary statutory consents to be obtained and detailed engineering design/construction work to be completed.

Title	Contents
Grouping of Power plants for Gencos. (1999, 9)	› Thermal plants were grouped into five GenCos in consideration of balanced generating capacity ·revenue · asset value. › Nuclear and Hydro plants were combined into one group in the interest of ensuring safety and maintaining control of water resources
Market Simulation (2000. 4 ~ 2001. 3)	› One year of simulation prior to setting up the electricity trade market and establishing the Korea Power Exchange
Necessary Legislation Enacted.	› "ACT ON PROMOTION OF RESTRUCTURING OF THE ELECTRIC POWER INDUSTRY" enacted on December 23, 2000 enables the separation of generation sector into several companies. › "THE ELECTRICITY BUSINESS ACT" amended on Feb 24, 2002, mandates the establishment of an electricity trade market (Power Exchange) and a regulatory agency (Korea Electricity Commission)
Establishment of Korean Power Exchange (2001. 4)	› Korean Power Exchange was established as a non-profit independent organization, to facilitate transparent and fair management of the electricity trade market
Establishment of Korean Electricity Commission (2001.4)	› Korea Electricity Commission manages the privatization process and oversees market operation. It also takes necessary measures to protect consumers from unfair and deceptive business practices and to ensure fair competition among all participants
Establishment of Six Generation Companies (2001.4)	› The six independent generation subsidiaries of KEPCO are -Korea South-East Power Co. Ltd (KOSEPCO) -Korea Midland Power Co. Ltd (KOMIPO) -Korea Western Power Co. Ltd (KOWEPO) -Korea Southern Power Co. Ltd (KOSPO) -Korea East-West Power Co. Ltd (KEWESPO) -Korea Hydro & Nuclear Power Co. Ltd (KNHP)
Preparation for the Privatization of GenCos	› Five thermal power generation companies will be up for privatization. Hydro and nuclear generation company is excluded from the plan. › Plan consists of two stages of privatization) Stage 1 : Beginning in 2002, two generation companies will be consecutively privatized.) Stage 2 : After completion of stage 1, the remaining three will be privatized(Process will begin no later than 2005)

Table 14.1. The Progress of Restructuring in Korea

The electric resources development plan (ERDP) and the business plan for transmission network development (BPTND) are documents that describe the actual and predicted future changes and additions to the electricity network. These documents show the opportunities for future connections and indicate those parts of the transmission network most suited to new connections and to the transport of further quantities of electricity. This will assist in encouraging the promotion of competition and the development of the transmission network in a non-discriminatory manner.

The network planning committee (NPC) is a key element in the transmission network planning process ensuring that the requirements and proposals of the Korea Electric Power Exchange (KPX), other network service providers (NSPs) and users will be fully considered.

14.3.2.1 Network planning committee

In order to facilitate detailed input into the preparation of the transmission network development plan and to provide a forum for co-ordination of this plan with all interested parties the Transmission Asset Owner (TAO) will establish and chair a transmission network planning committee (NPC). The NPC will provide an informal forum for considering detailed developments to the transmission network. Membership of the NPC will include the KPX, other NSPs and users of the transmission network.

The NPC will not have any decision-making functions and its role will be only advisory. The responsibility for planning the transmission network rests completely with the transmission asset owner.

14.3.2.2 Business plan for transmission network development

Annually the TAO will prepare and issue the detailed BPTND for the next seven years. The BPTND will use the output of the ERDP and will be produced within three months of the publication of the ERDP. The BPTND will consider all the requirements outlined in the ERDP from the needs identified by the various parties represented in the ERPC and will describe how those needs are being or will be incorporated within the plan.

The TAO will be responsible for the preparation of the final version of the BPTND but the TAO will consult with the KPX and will take into consideration the needs of the other parties involved. The final version of the BPTND shall be submitted to Ministry of Commerce, Industry and Energy (MOCIE) for approval.

Ultimately the TAO will be held accountable for the quality of the planning activities and planning results included in the BPTND. In the event of the KPX being unable to operate the system to the required standards as a direct result of a TAO decision not to make an investment, then the KPX will take the necessary operational measures to secure the system and protect electricity supplies. The TAO will meet any additional constraint costs that occur due to this.

The parties may conduct independent assessments to evaluate alternatives. The KPX may also identify and suggest needs for investment based on its operational experience, engineering practice and professional estimation.

The flowchart for the transmission network planning process is shown in Figure 14.2.

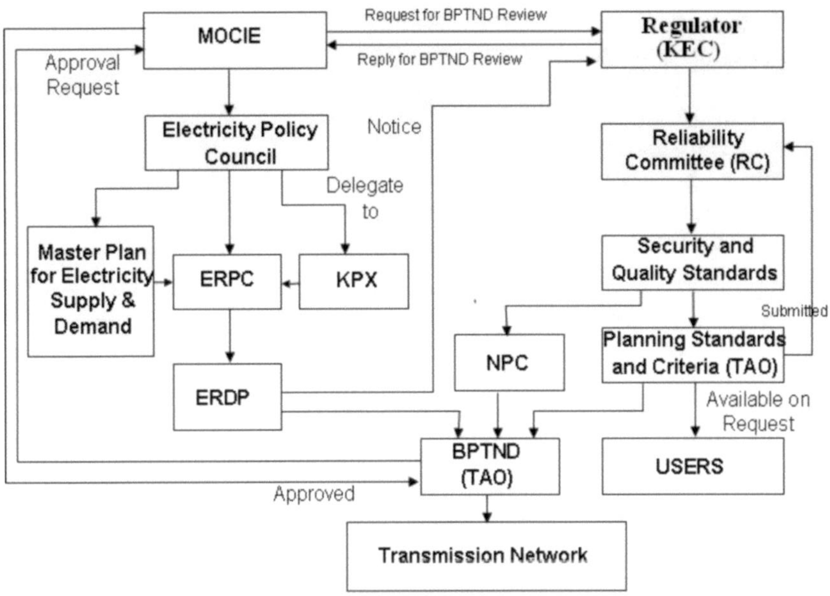

Figure 14.2. Transmission network planning flowchart

14.3.3 Background to performance criteria for transmission system planning

Based on the regulating framework for transmission system planning, this Section describes the proposed performance criteria as a part of planning standard and criteria.

14.3.3.1 Development of performance criteria

Performance criteria for transmission system planning is determined based on the extent that the transmission system can keep supplying electricity to loads when a disturbance occurs. However, load supplying may be interrupted by the adopted strategy of its system operator as well as by *deterioration* of electricity supplied. So, performance criteria should consider both aspects of these.

In Korea, the power system is operated at such a high reliability level that it never allows the loss of load on systems other than the one where a disturbance (including the failure of 1 route (2 circuits) 345 kV line) occurs. Thus, in this Section, performance criteria for transmission system planning is proposed to assure this principle.

14.3.3.2 Performance criteria for normal state

Normal state is a state where all system elements are in service after the power system is adjusted to supply load following specified operating procedures, and no faults or outages occur. For this normal state, performance criteria should meet the normal operating criteria that the system operator would apply to power system operation.

14.3.3.3 Performance criteria for abnormal state

1) **Classifying performance level**

Performance level is classified based on the allowable actions or conditions on systems other than the one where a disturbance occurs.

In this Section, considering the reliability criteria principle that is applied to Korean power system operation, any loss of load is not allowed in all performance levels. Table 14.2 shows each performance level classified by the allowable actions or conditions on system.

Level Allowance	PA-1	PPA-2	PPA-3	PPB-1	PPB-2	PPB-3
Tripping Generator	NO	NO	NO	YES	NO	YES
Generation Output Adjustment	NO	YES	NO	NO	YES	YES
Temporary Loss of Load*	NO	NO	YES	YES	YES	YES
Loss of Load	NO	NO	NO	NO	NO	NO

*Temporary loss of load: Case of the dropped load being restored within a short period of the time by switching actions at the station where the load is supplied

Table 14.2. Performance Level

2) Classifying contingencies into performance level

Selections of the considered contingencies are based on probability of that contingency happening. Moreover, special considerations are given to the characteristics of the Korean Power system.

In general, contingencies have different effects on the power system according to importance of the transmission system where contingencies are occurred. Thus, in this criteria, the transmission system is divided according to its main function and its voltage level.

- Generator connection system : transmission facilities connected to connection point of generating unit.
- Main system : transmission facilities connecting generator connection system and load supplying system.
- Load supplying system: transmission facilities connected to connection point of load.

Table 14.3 shows contingency classification in each sectioned transmission system.

Performance level	Load supplying system [kV]		M. Tr [kV]		
	154	345	154	345	765
PA-1	-	-	-	O	O
PA-2	O	O	-	-	-
PA-3	-	-	O	-	-
PB-1	OO	OO	-	-	-
PB-2	-	-	-	-	-
PB-3	-	-	-	-	-

Performance level	Generator connection system [kV]			Core system [kV]		
	154	345	765	154	345	765
PA-1	O	O	-	-	-	-
PA-2	-	-	-	O	O	-
PA-3	-	-	-	-	-	-
PB-1	-	-	-	OO	-	-
PB-2	-	-	-	-	OO	O
PB 3	OO	OO	O	-	-	-

O : Failure of 1 circuit or 1 Transformer bank
OO: Failure of 1 route (both of 2 circuits) line

Table 14.3. Contingency Classification

3) Transient voltage criteria

a) Overvoltage

It was reported that over-voltage criteria is not required as a performance criteria [8], and is not recommended since it is usually related to a local problem

b) Undervoltage

Table 14.4 shows voltage dip criteria applied by WSCC to avoid uncontrolled loss of load [8]. In this table, the values were based on the estimated response of electronic equipment such as computers to voltage dips. In this Section, it is assumed that Korean electronic equipment has a similar characteristic at least, so the values of this table can be applied as an under-voltage criteria. But, only A and B steps in this table would be applied as a criteria since the Korean power system does not apermit any loss of load.

Step	Instantaneous Voltage Drop	Maximum Duration of Voltage Dip Exceeding Minimum Drop	Loss of Load
A	25%	20[cycle]	No
B	30%	20[cycle]	No
C	30%	40[cycle]	Critical
D	30%	60[cycle]	Yes

Table 14.4.Voltage Dip Criteria Comparing Loss of Load

4) Transient Frequency Criteria

a) Over Frequency

Over-frequency problem is mostly associated to generators, but generators usually have local protection. So, it is reported that over-frequency criteria is not recommended [8].

b) Underfrequency

Under-frequency criteria is selected to coordinate with the operational strategy for UFLS (under frequency load shedding). UFLS is expected to arrest frequency decline and avoid the cascading as a result of a disturbance.

To do this, UFLS relay is set to be coordinated with under-frequency protection of generators and any other actions planned to occur when the frequency drops.

In the Korean power system, the UFLS relay is set at 58.8Hz. According to this strategy, automatic load shedding starts if system frequency drops below this value. In this Section, low-frequency criteria is proposed not to allow any loss of load considering this UFLS strategy.

5) Post transient voltage deviation

The criteria for post transient voltage deviation is set to provide some measure of the ability of the system to recover to acceptable operating conditions following a disturbance. It is also known that this criteria can provide some information about the incipient voltage collapse problem though it is not sufficient as a voltage stability criteria. 5-10% deviation is usually recommended for this [8].

14.3.4 Proposed Performance Criteria for Transmission System Planning

14.3.4.1 Performance criteria for a normal state

A normal state should meet the following performance criteria considering the operating criteria of the Korean power system.

- ◈ All transmission facilities should be kept within its thermal rating for normal state.
- ◈ System frequency should be usually at 60Hz, and adjusted within 60±0.2Hz otherwise any exception occurs.
- ◈ Voltage in the transmission system should be kept within the following guidelines indicated in Table 14.5.

Voltage level	Voltage Criteria	Remarks
154kV	156~164kV	Peak
	152~160kV	Off-Peak
345kV	336~360kV	-
765kV	746~785kV	-

Table 14.5. Voltage Criteria for a Normal State

14.3.4.2 Performance criteria for a disturbance

A disturbance means a fault or outage of system elements that is not expected. Response of the transmission system to this has to meet its performance criteria. Table 14.6 shows the proposed performance criteria for each disturbance.

Performance Level	Transient Voltage Dip (Measured in a Load Bus)	Transient Frequency (Measured in a Load Bus)
PA-1	-Maximum voltage dip : 25% -Max. Duration of V dip exceeding 20% : 20cycle	-Minimum : 59.6Hz -Max. duration of F below Min : 6cycle
PA-2	Same as above	Same as above
PA-3	Same as above	Same as above
PB-1	-Maximum voltage dip : 30% -Max. Duration of V dip exceeding 20% : 30cycle	-Minimum : 58.9Hz -Max. duration of F below Min : 6cycle
PB-2	Same as above	Same as above
PB-3	Same as above	Same as above

Performance Level	Damping	Post Transient V Deviation	Loading
PA-1	Positive	5%	Within nominated rating
PA-2	Positive	5%	Within emergency rating
PA-3	Positive	5%	Same as above
PB-1	Positive	10%	Same as above
PB-2	Positive	10%	Same as above
PB-3	Positive	10%	Same as above

Table 14.6. Performance Criteria for a Disturbance

Thus, as the electric power industry undergoes restructuring in Korea, it has been discussed that a reasonable regulation for guiding planning needs to be established, and the objective and rational criteria for transmission planning needs to be developed because a transmission company still remains as a type of monopoly after restructuring in Korea, and investment into the transmission system should be considered fairly and transparently. A transmission system determines a kind of infrastructure for trading electricity in the electric power market, so we have to promote transparent and rational circumstances to induce efficient investment. This Section reviews the progress of restructuring in the Korean electric power industry, and describes a regulating framework for transmission system planning. In addition, this Section examines the proposed criteria for transmission system planning. The proposed regulating frameworks provide various market participants with many chances to correlate them in planning the transmission system, and the proposed performance criteria became an objective standard by which transmission system is planned and maintained in the electricity market.

14.4 Power Generation and Transmission Planning in India – Methodology, Problems and Investments

India starting from an overall installed capacity of little above 1,300 MW at the time of its independence in the late nineteen forties, through its successive five-year planning periods could achieve a level of about 110,000 MW with energy generation of the order of 600 billion units in five and half decades. Though initial generation was concentrated to meet load in urban areas, at present vast countryside too has been covered under a massive electrification program.

Due to thrust on hydro development for the purpose of both irrigation and power, the nineteen fifties and sixties saw a good hydro-thermal mix in the generation front to meet the load along with the formation of Electricity Boards under the State Governments of the Republic of India and some river-valley project authorities for the implementation of projects. Due to the low level of load and the shorter distance of haulage of power, grids hitherto had been restricted to 132 kV and seldom going up to the 220 kV level. But with the setting up of quite a good number of mine-mouth large thermal power stations and few hydro power stations at the regional level by the Central Government-owned Companies starting from the 1980s, state grids were integrated to form the regional ones with simultaneous development of 400 kV networks for power evacuation from major stations as well as for strengthening of networks. Thus came into existence the five regional grids in the Northern, Western, Southern, Eastern and Northeastern part of the country. At present out of these further integration has made it possible to connect the Northeastern, Eastern and Western systems synchronously. The other two continue to operate through asynchronous mode of interconnection (HVDC) with the combined one, both in the form of back-to-back and bulk supply links.

In the mean time with restructuring vis-à-vis reform taking place in the power sector from the early nineties, some Independent Power Producers (IPPs) came into the arena and started feeding to the state grids coming under the concerned regional grid. At the same time in order to improve efficiency and performance in general in certain states vertically integrated state utilities were unbundled to form generation, transmission and distribution companies. Though the transmission sector was opened up for private sector participation in early 1998, it is yet to pick up. As traditionally in pockets, generation and distribution activities existed in the private sector, only these saw further expansion through setting up of new generation plants and acquisition of distribution companies. Transmission as a natural monopoly remains still under government-owned companies, both at central and state level, though at the beginning of 1998 it has been opened to private enterprises to build, own and operate from point to point. With the open access in inter-state transmission to any distribution company, trader, generating company, captive plant or any permitted consumer as per recent order of Central Electricity Regulatory Commission (CERC) certain changes are, however, expected in future.

With the Central Electricity Authority (CEA) of the Ministry of Power of the Government of India at the helm of affairs, the basic work of planning starts with load survey. State Electricity Boards (SEBs) at the grass-root level do the spadework with different agencies involved through collection of data concerning new demand in commercial, industrial, domestic, public service and irrigation areas and also growth for the existing systems in corresponding areas. CEA consolidates the projected figures on an all-India basis by working in close coordination with the SEBs and forecasts load and work out total requirement of electric energy and peak load to be met for the next few five-year plan periods based on a combination of partial end use technique and trend analysis, and computing long term projection by extrapolating the energy requirement at power station busbar. Various components, such as, Transmission and Distribution (T & D) losses (both technical and commercial), load factor, diversity factor, etc. are also taken into account state / system-wise along with growth rate. As the National Power Grid is under formation, long-term projection takes care of regional diversity factor considering the significant daylight time difference across the country from east to west. Recession in economy, restructuring of SEBs are the other pertinent factors that influence the overall scenario. These figures are, however, scrutinized by certain Departments of the Government of India including the Planning Commission keeping in mind the commensurate fund requirement vis-à-vis relative priority with respect to the other sectors of infrastructure of the country for investment under public sector.

Having made the blue print, CEA further works out the details of generation corresponding to various scenarios of load projected for a few five-year plans ahead. In the last two decades in order to boost the economy, for rapid addition to generation, setting up of thermal generation plants was given much preference. Though side-by-side hydroelectric power stations were added, the balance tilted very much in favor of thermal to push it to the level of about 72% of the total capacity installed. This is primarily on account of the long gestation period, as a result of rehabilitation and resettlement, and other environmental problems, the inter-state river-water dispute, etc. and consequent cost and time overrun. Of late more attention is being paid to this type of generation of electric power to strike a balance, which is needed for proper grid operation efficiently with standard parameters and in the most economic way.

With opening of the trade barrier, alternatives are also being looked at, like, whether to import fossil fuel or to produce indigenously. Exploration of new gas-wells as well as offshore drilling has lead to quick addition of gas-based plants, which of late are coming up and added to the grid with combined-cycle mode of operation. Thus under integrated resource planning, considering all possible sources to produce electricity in conventional ways including the nuclear one, most optimum solution is attempted for meeting the load requirement. In the process of planning for addition of generation, issues of system improvement to minimize T & D losses, raising of plant load factor, Renovation and Modernization (R & M) of old but still running power plants and also generation from renewable and non-conventional sources, etc. are considered to augment overall supply. Having known the load points, in the process it identifies the possible corridors of transmission of power vis-à-vis energy, though voltage level for it may be just an indicative one at this stage.

As applicable for the planning of any system, the basic philosophy of configuring transmission system is to achieve a level of operating performance with adequacy and security, which in turn requires a trade-off between cost and risk. It may be based on either a deterministic or a probabilistic approach. Though with an enormous amount of operational data available vis-à-vis past experience latter mode may be adopted for investing further, in the scenario of rapid development of system one may have to bank upon still on a deterministic approach very much or on a combination of both to expect ultimately acceptable system performance. Accordingly certain planning criteria have been evolved in India. Inputs in the form of possible generation sites with capacity available and so the loads in bulk, the process of transmission planning starts. It involves not only the corridors of transmission lines with voltage levels but also with the finding of locations of associated substations. Adequate transformation capacity in the substation with the possibility of future expansion, flexibility at the operation stage, etc. is the major guiding factors for such planning exercises. The systems so planned, hitherto on a regional basis are then presented in Standing Committee of the concerned region. Decision is taken to firm up the addition with the identification of agencies in the Central and State sector responsible for construction, owning and operation.

The problems encountered in having the feasibility of generation especially for hydro development is enormous. In the southern part of the country, while harnessing has been quite appreciable, most of the new generations from this type of source are expected further in the Himalayan region in the northern and northeastern part of the country. Besides rehabilitation and resettlement problem, issues concerning geological stability of the areas as well as consequent transportation of equipment and transmission of power to load centers pose great challenges.

Though in thermal generation expansion, the problem is somewhat less, indigenously available coal has high ash content resulting in somewhat unusually oversized designing of boiler and associated plants for large thermal sets. Of course with economy of scale not so an important factor, IPPs have come up with various sizes of units of different smaller sizes. But in the long run maintainability of equipment may require larger investment on account of spares. But due to high rate of growth in load, as inevitable for any developing country, in a shorter time frame, it is also highly desirable to add to the system larger units, particularly when solid fuel is used, to have quicker enhancement of installed capacity. However, this is being limited to some extent due to high-ash contained indigenous coal for use in thermal generation, as just mentioned.

On the other hand due to large population and environmental restrictions particularly due to forest coverage, Right-Of-Way (ROW) for the construction of Extra High Voltage (EHV) transmission lines is gradually becoming more and more difficult. In the early nineteen eighties due to the construction of single-circuit 400 kV lines quite a good amount of corridors have been lost with limited amount of flow of power. In fact subsequently with almost same amount of corridor width double circuit construction in hexagonal formation has paved the way for haulage of twice the amount of power. However, as a measure of enhancement of power flow, series compensation (both static and dynamic) is being implemented to increase loading capability of these lines. Also with the development of loads at intermediate locations, hitherto operating long lines (above 400 km) are being broken to form new substations in between, thereby improving structural stability and other operating parameters of the system in addition to enhancement of loading through these lines. Similarly on certain corridors through forest multi-voltage multi-circuit transmission is being attempted. But still some bottlenecks may be existing in the transmission system on account of inadequate compensation of reactive power at lower voltages, leading to burdening EHV system to run at lower voltage and consequently unable to deliver active power of the desired level. However, through various measures being attempted, the situation is improving.

On the investment front, if one looks right from independence it may be seen that the whole power sector has been primarily nurtured with funding by the state. Very little investment has come through private sources that concentrated in and around some of the large metropolitan cities. Only since the early nineties have the latter channels become active with opening up of the sector. Though transmission is yet to pick up, for generation and distribution, some progress has been made. The issue of cross-subsidy for the domestic and agricultural areas through revenue earning from commercial and industrial areas could not produce enough revenue on net basis for future investment, resulting in poor financial health of power utilities of the states. Due to this background, it failed to attract private investment even when high rate of return was permitted. Under this category whatever investment has been made so far in generation, it is restricted to thermal (mostly gas or liquid fuel based), barring one or two hydro projects.

In the power sector, like any other physical system, overall efficiency, is a point to be reckoned with. This naturally gives importance to the distribution system too. After giving due attention to generation in the initial stages, followed by the transmission system by making it sure to link with every generating system coming up, now the distribution system is being revamped under an accelerated development program with the public investment to a large extent through the states. Haphazard growth due to compelling requirement of delivering electricity to every place at the cost of deteriorating parameters resulted in a system with high losses, which when accompanied by low collection of revenue lead to almost bankruptcy of suppliers, mostly the public sector power utilities. To some extent it leads private investors too to shy away from the scene of running the distribution systems. However, with the concept of individual center of profit introduced to some extent, the situation is gradually improving with additional investment pouring in this vital part of the power sector.

14.5 Power System And Power Market Development In China Problems And Proposed Alliviation Measures

The economy of China developed at a quite fast rate with an average of 9.5% increase of GDP from 1978 to 2000. Most of the time, the power industry was under pressure of the capacity shortage. Only in recent years (especially 1997-1999) most power systems experienced capacity adequacy. Generally speaking, in years of 2000-2002 the demand and supply in China were almost balanced, except in some minor parts of power systems. However, starting from the summer of 2003, 21 provincial power systems experienced energy shortage, the situation has been getting worse during the winter. In 2003 the yearly installed generation capacity (including thermal, hydro and nuclear) was 30,000 MW, it is the fastest rate in the world; however, the rate of power consumption was still faster, it reached 15.3% in 2003. In 2004 37,000 MW of generation capacity was installed, still more was added in 2005 and 2006. The energy shortage problem was probably alleviated from 2005/2006.

The total installed generation capacity by the end of 2003 reached 384.5 GW, total yearly consumption reached 1,908 TWh, both numbered second place in the world.

The yearly total installed generation capacity in China from 1952 till 2020 is shown in Figure 14.3. Nationwide interconnection of regional power systems in China in 2005 is shown in Figure 14.4.

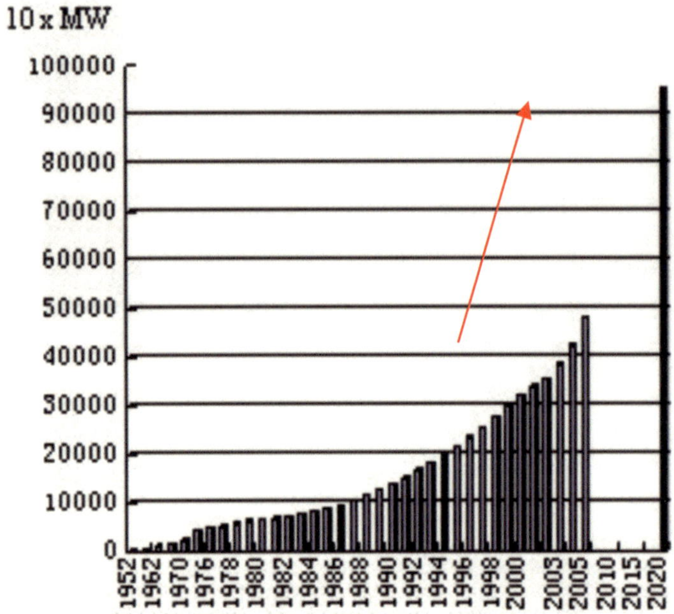

Figure 14.3. Yearly total installed capacity in China

The rate of installation of transmission lines has been very fast too. The total length of 220 kV and above transmission lines by the end of 2002 reached 188,700 km, in which 37,000 are 500 kV lines. At present there are six large regional power systems and 5 provincial power systems (usually they are not interconnected with the main grid). By the end of 2005 these power systems in China were interconnected into a huge nationwide power network linked by 500 kV and 750 kV HVAC/HVDC transmission lines except Xinjiang and Xizang autonomous region and Taiwan. This would further alleviate the energy shortage problem.

Considering the changing status for power shortage and power surplus, some key issues related to power systems security and generation capacity adequacy are discussed. One important issue is how to coordinate system generation and power grid construction under the environment of the power market with separated generation and transmission companies. Another issue is about how to setup power markets under the situation of continuously changing status of system generation capacity adequacy. In order to manage the generation adequacy problem, the basic conditions for the opening of a market and the requirement for market normal operation criterions are also discussed.

Along with the introduction of nationwide power systems interconnection and power market development, the most important issues are how to improve procedures for generation planning and transmission expansion planning, how to procure investment for building new power plants and new transmission lines, and how to alleviate energy shortage problems under a deregulation environment. These problems are discussed below [13,14].

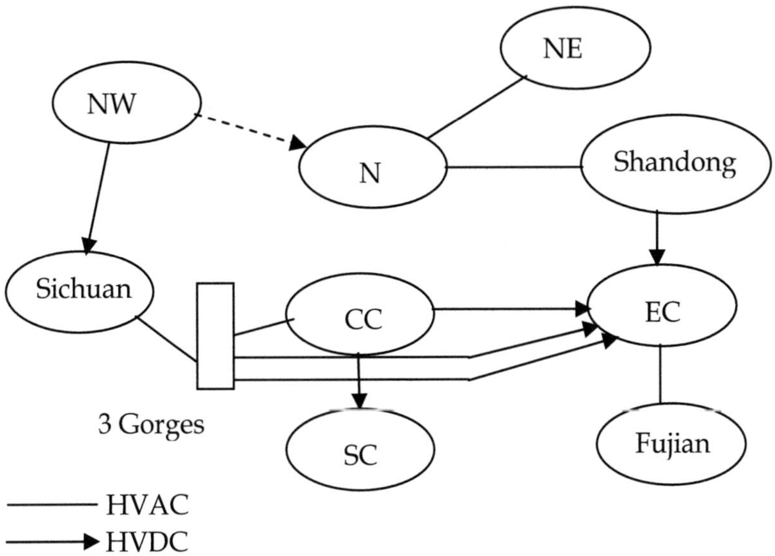

Figure 14.4. Nationwide interconnection of power systems in 2005 of China

14.5.1 Power Market Development

Since the first implementation of deregulation of the power sector in UK in 1990, more than 38 power systems in the world has restructured their power system and introduced competition in their electricity markets. There has been a lot of experience and lessons learned during their implementation, however, the restructuring of each later electricity market has benefited from the experience and failures of former electricity markets.

China has implemented her experimental electricity markets during 1999-2000. Six provincial/municipal power systems have been restructured to implement the electricity market, i.e. Shandong, Zhejiang, Shanghai, Liaoning, Heilongjiang and Jilin. Their experiences are:

- Ownership of power plants were not privatized, they were still owned by the state government or provincial government or foreign and domestic investors, but they were separated from the grid company and become independent power plants (IPP).
- About 80-90% of the demand are long-term bilateral contracts (usually one year) between IPPs and the grid company, the rest (10-20%) of the demand is under competition. This practice guarantees stable supply of energy to the user.
- They all established their power market support systems (including dedicated communication links, information system, bidding management system, schedule

application software, settlement management system), they were all developed by different Chinese vendors.

• The power systems have been in operation successfully and improved the efficiency of IPPs. Since the price of users cannot change, so the grid companies have gained some revenues through competition.

However, there are some limitations of deregulation at the provincial level, because deregulation at the provincial level can only optimize the utilization of resources in their own territory, they are not able to optimize the utilization of resource in the whole region, even when there are cheaper, cleaner energy in neighboring provinces. So what should we do next? Shall we continue to establish an electricity market in each province, or shall we establish an electricity market on a larger scale?

After evaluation of experiences of deregulation in China and abroad, the State Government of China made a decision in 2002 for restructuring of the power industry: The former State Power Company would be broken into five large generation companies (Huaneng Group, Huadian, Guodian, Datang, Power Investment), and two grid companies (State Grid Company-SG, South China Grid Company-SCG). There are five regional grid companies under the SG (NE China, North China, NW China, Central China, East China) and five provincial grid companies under the SCG in southern part of China (Guangdong, Guangxi, Yunnan, Guizhou and Hainan). The National Electric Power Regulation Commission (NEPRC) was established to monitor and regulate the forming and operation of six regional power markets. The forming of five generation companies and six regional grid companies including the SGC was completed by end of 2003.

According to the new policy, each regional grid company should establish one or several dispatch and trading centers (DTC) in their region. How can we establish regional power markets under the present energy shortage condition?

Fortunately not all regions and provinces has energy shortage problems, for example, the NE regional power system has no shortage problem, they are the first to establish their regional power market. After serious planning and design, their power market support system has been in trial operation since the middle of January 2004. Their special features are as follows:

• The regional DTC is responsible for all wholesale trading and dispatch of 500 kV transmission lines in their region. Provincial dispatch centers are responsible for the security of operation and retail trading in their provinces.

• The regional DTC begin to collect transmission service charges for 500 kV transmission lines.

• The regional DTC is responsible for the yearly generation market (consists of about 80% of the yearly demand forecast), monthly market (about 20% of the remaining demand forecast), and a day-ahead market for competition, a real time balancing market, and ancillary services.

• A two-part tariff system is implemented for IPPs, which include a capacity charge and an operation charge. The capacity charge would be paid to all available generators no matter whether it is dispatched or not, to compensate their equipment and installation cost.

The operation cost would be only paid to those generators whose bidding price is lower than or equal to the system marginal price.

- Other functions are similar to former provincial power markets.
- East China Regional Grid Company was the second regional power market; it was in trial operation in June 2004. Their special features are as follows:
- Provincial DTCs will coexist with regional DTC.
- The regional DTC will be responsible for day-ahead market and real-time market, provincial DTC will be responsible for long-term bilateral contracts (mostly one year contract between provincial DTC and IPPs in their provinces) and ancillary services.
- The regional DTC will be responsible for coordination between provincial DTCs.
- Other functions are similar to NE regional DTC.

The type of other regional markets will be similar to either NE China or East China. They are all actively planning and the design stage, and were in trial operation from 2005-2006.

There is a National Control Center in China; their future mission is coordination among six regional power markets, and settlement of 500 kV and above tie-line charges between regions.

14.5.2 Generation Planning, Transmission Expansion Planning and Investment in China

Generation planning. In the past, electric power design institutes usually undertook generation planning and transmission expansion planning studies. Take a regional power system for example. The usual way is, the regional grid company (RGC) provides the load forecast (5 years or 10 years), and entrust a design institute (usually a large design institute in the same region) to do the generation planning and transmission expansion planning. Based on the load forecast, the entrusted design institute would make a study on how much generation capacity will be needed each year, where these generation plants should be located, how many transmissions corridors would be needed to transmit the power to the load. The design institute would propose several schemes, make technical and economic studies and compare the results. The regional grid company and State Grid Company approves the final plan. Sometimes some very important planning projects (such as Three Gorges transmission project) can be entrusted to CEPRI to do detailed analysis, because CEPRI has advanced analytical tools and experienced experts.

If there are hydro resources available in the region, hydro generation planning would be entrusted to hydroelectric power survey and design institutes. They will make a survey on the water resources in the region, and make a proposal on the location and capacity of hydro stations to be built, and a yearly construction schedule. This proposal will be submitted to the regional grid company and Ministry of Water Resources as well as to the State Grid Company for approval. If it were a 5-year plan or 10-year plan, the hydro generation plan would be updated every year.
Transmission expansion planning. Usually regional grid companies are responsible for transmission planning between provinces (usually 500 kV or above); provincial power companies are responsible for transmission planning in their provinces (usually 220 kV). There are six large thermal power design institutes and six hydro-electric power design and survey institutes, one in each region, and small design institutes in each province, they do

the planning works for the regional grid companies. For very large transmission projects (such as Three Gorges transmission project or Nationwide Interconnection Transmission Project), many design institutes, together with the General Electric Power Planning and Design Institute (GEPPDI) and China Electric Power Research Institute (CEPRI) are working together to do the planning. Based on local load forecast and the national economic growth rate, and generation planning in the whole region, they would work out several transmission schemes, such as pure AC or DC, or hybrid AC/DC, what voltage level should be adopted, etc. for each scheme, technical and economical analysis will be made, such as load flow analysis, n-1 security analysis, transient stability analysis and dynamic stability analysis, sub-synchronous analysis etc. By comparison of technical and economical results of several feasible schemes, the best scheme will be determined. Again, the final transmission-planning proposal will be approved by the State Grid Company (former State Power Company).

After deregulation, the generation and transmission expansion planning procedure may not change much.

Investment. As for investment, in the past practically most of the power plants and transmission construction were funded by the former State Power or former Ministry of Power, except some large industry owned power plants. Since the national budget was limited, so the investment problem has been a limiting factor for the growth of generation and transmission expansion in China. To overcome the energy shortage problem, in the mid nineteen eighties the "provinces take the main role" policy prevails, some were invested by local banks, some by foreign loans, and some BOT plants were built. Before the separation of generation and transmission, the former SP manages state-owned electric power assets (almost half of the generating capacity) and ran the country's national transmission networks. After reform, the ownership of power plants is distributed to five large generation companies. They take responsibility to invest in new power plants. They have income from operating their IPPs, if necessary they may obtain loans from local or foreign banks. As for which generation company may obtain the right to build new plants for a certain regional company, there are two possible options: by bidding, or by contract.

The investment for new transmission lines came from various sources. Usually the investment for tie lines (330 kV and above) between provinces was shared by provincial power companies. Investment for transmission lines (500 kV AC/DC) between two large regions usually was shared by the two neighboring regional grid companies. Investment for outgoing lines connecting the power plant to the main grid was usually included in the budget of the power plant project. For very large power plant or transmission project such as Three Gorges project, the state government provides the investment; the National Congress approved it. We assume that the source of investment for new transmission line projects may continue to be so after the reform.

14.5.2.1 Energy shortage problems and proposed alleviation measures

In the introduction it is mentioned that China faced serious energy shortage problem in 2003. Twenty one provinces had to partly curtail their load which caused considerable damage to the industry and inconvenience to social life. The cause of energy shortage is due to the following reasons:

The rate of economic growth was faster than it was expected. The rate of economic growth for 2003 was planned at 7%, but the actual growth rate reached 9.1%, and the annual consumption of energy increased 15.3%. Of course energy shortage became inevitable.

There was drought in some of the provinces; consequently hydro-stations cannot produce enough power, which only make the energy shortage problem worse.

Although China produced 1.7 billion ton of coal in 2003, China still cannot provide enough coal for thermal power plants. This was another cause of the energy shortage.

14.5.2.2 The proposed counter measures

1) To increase annual installed generation capacity. China installed 37,000 MW of generation in 2004 and 48,000 MW in 2005 [1]. It seems that the investment was not a problem, because most provinces which suffered from energy shortage were very enthusiastic to invest and install new power plants in their provinces. The energy shortage problem became alleviated in 2005-2006.

2) To increase the reserve capacity requirement. According to the present design guideline, the requirement for reserve capacity is between 20-30% (including 8-15% for maintenance schedule); usually 25% reserve is used for planning purpose. Since the long-term load growth rate is difficult to predict accurately especially in a fast growing environment like China, it is recommended to use a higher figure for planning purposes to avoid future energy shortages.

3) More reserve capacity may result in a large number of power plants to become idle if the rate of load growth is slower than predicted. However, in China, a two-part tariff system in power market bidding was adopted. One part is capacity cost, which just cover the average cost of investment per kW of power plant no matter whether the generator is dispatched or not. The other part is bidding price, it depends on the efficiency of operation of the generator. A two-part pricing scheme is already being experimented with in the NE China Power Market. By application of this two-part tariff system the risk of investors for building new power plants can be considerably reduced.

4) To limit investment for new fast growing, energy intensive and low efficient industries. The state government has made a decision to limit new investment for steel, electrolytic aluminum, and the cement industries and construction of more luxurious houses. If the original schedule for installing steel plants was completed in end 2005, the yearly production capacity of steel in China would reach 330 million ton, it would be far more than necessary. The production capacity of electrolytic aluminum, and the cement industries are in a similar situation, if not worse.

5) To shave peak load, in East China and South China a two-stage energy tariff system has been implemented. It is likely more regions will adopt this system.

6) Demand Side Management (DSM). This technology has been implemented in Shanghai and other cities. A multi-stage energy tariff system is applied for some industries. It implies

that the higher the load, the higher the energy tariff, which automatically encourages the user to reduce their own load. This measure is very effective to help reduce the load; experience in Shanghai has led in a reduction of 2,000 MW in peak load.

7) Likewise, the average unit production cost per energy consumption is high in China, about 0.15 kWh/RMB, which is 3 times as high as in USA, and 5 times as high as Japan. In the industrialized part of China a special tariff could be added to each kWh of energy when the reserve capacity is lower than a certain threshold, say 5%. The smaller the reserve capacity, the higher the special tariff. If there is no spare and curtailment of load becomes inevitable, the tariff would be still higher. This measure would encourage all users to save their energy, and this tariff can be used for building new power plants.

8) Economic signaling in power market design. To encourage incentive of investment for building new power plants and transmission lines, Long Term Marginal Cost (LTMC) combined with MW-KM price system is proposed that would give an economic signal for investors to select the best site to build new power plants. And it is proposed to use a two-part transmission service charge system, including a fixed charge to compensate the investment charge, and an operation charge to cover the operation charge. This system would encourage the incentive of investors to build new transmission lines.

9) Implementation of more sustainable new energy source. For instance, in summer time the air conditioning load becomes very high. In Central China, the air conditional load reached 30% in summer of 2003. If solar energy generators are encouraged to be installed on top of all new buildings, it can be best utilized to feed air conditioning load, because the hotter the weather, the more power can be generated by solar generators. In the winter, solar energy can be also be utilized for heating purposes.

Thus, China's economic growth rate has been very fast. The average rate of growth from 1978-2000 was 9.5%. But the average rate of growth of total installed generation capacity was only 7.8%, which was the main cause for energy shortage. Lack of investment was the limiting factor for installing more generation plants and transmission lines.

China started her deragulation in late nineteen nineties. six provincial power markets were in trial operation in 1999-2000. Restructure of Chinese power sector was completed in 2003. five large generation companies and six regional grid companies were formed. NE China Grid Company was the first regional power market in China, it is characterized as a sole wholesale power market in the region. The second regional power market is East China, it is characterized as a hierarchical power market, where provincial power markets would coexist with the regional power market. Other regional power markets will fall into either NE mode or East China mode.

Past generation planning and transmission expansion plan procedures in China have been presented. Investment for expansion usually comes from various sources. Future procedures for generation and transmission planning and investment have been discussed. By installing more generation and transmission lines and limiting investment for energy intensive industries and by implementation of various demand elasticity measures, the energy shortage problems were alleviated by 2005-2006.

14.6 Generation Planning and Investment under Deregulated Environment: Comparison of USA and China

Since the 1980s, the electricity supply industry in the west and in South America has been undergoing rapid and irreversible change reshaping an industry that for a long time has been remarkably stable and had served the public well. A significant feature of these changes is to allow for competition among generators and to create market conditions in the industry, which are deemed necessary to reduce costs of energy production and distribution, eliminate certain inefficiencies, shed manpower and increase customer choice. While such restructuring also started in China several years ago, the situation is very much different from that in the west and other countries. As a result, a modified form of restructuring is occurring in China, driven by a need for rapid expansion of capacity in all three sectors, i.e. generation, transmission and distribution.

The power industry in China has rapidly developed in the past 25 years, and currently the generation installed capacity in China ranks the second in the world, only next to USA. Even so, the capacity is still not enough since the economics expand very fast. The annual electricity demand growth rate is around 15%-20% in the recent 3 years. In the summer of 2003, nineteen provinces suffered electricity supply shortage, accounting for two-thirds of the provinces in China.

The power industry in China will keep a rapid development in the next 20 years. The total annual power consumption will be 2,700 TWh by 2010, and the total installed capacity in China will be 600GW then. The annual increase of power consumption will be 4.5~5.5% from 2010 to 2020, the total power consumption will be 4,200~4,600 TWh in China by 2020, and the total installed capacity in China will be 900GW then. With the projects being completed, such as the "*Power transmission from the west to the east*", the "*Mutual transmission power between the north and the south*" and the "*Inter-connection of power grids over whole China*", there will be a nationwide power grid in China.

While the power industry restructuring represents a world-wide trend and the Chinese government has already decided to reform the power industry by separating the generation sector from the transmission/distribution sectors, there exist extensive debates concerning how competition and sustainable development of the power industry in China could be well balanced.

In fact, one of the key issues in restructuring of the power industry is to ensure that an adequate generating capacity will be available for reliable supply. Without sufficient supply, there will be no competition, and hence, the market will not work properly. This problem appears more serious in China since the load demand increases rapidly [15-17].

14.6.1 Reforming History of the Power Industry in China

The power industry in China used to be governed by the former Ministry of Electric Power with combined functions of regulation and enterprise practice before 1985. Due to rapid development of economy and hence increasing demand of electric power, electricity shortage became a very serious problem. However, at that time the government did not have much money to build power plants. To solve the problem, investment for power plants from different sources was encouraged by the government. To attract generation investment, the government guaranteed the recovery of generation investment and reasonable return. Specifically, there was a very stimulating mechanism for investment in generation plants at that time. For each IPP, the amount of electricity generated every year and the tariff associated were ensured by a long-term agreement signed by the governments. In general, the yearly return on investment was over 15%. With such a high and stable return and without investment risk, more and more capital was raised for the power generation sector, and as a result, a large number of independent power plants (IPP) have been built in China in the past 15 years.

In order to separate administrative functions from enterprises, the State Power Company of China (SPC) was established in 1997. It inherited from the former Ministry of Electric Power all the assets of power networks and around 50% of the generation installed capacity. By the end of 2001, SP had a total capital of 1,002.7 billion RMB, accounting for 72% of the total capital of China's power industry and ranking the sixtieth among the world 500 strongest enterprises.

In 2002, a significant reform with far-reaching influence was initiated for the power industry in China. The major objective of this round of reform is to make a break to the vertically integrated monopoly of the power industry, to introduce a market mechanism by organization restructuring, and to establish a regulatory system for the competitive electricity market.

For this purpose, it is necessary to split the assets of the SPC by separating power grids from power generation and the generation companies thus established. They must compete for supplying power in a competitive market. Five national generation group companies and two power grid companies were established at the end of 2003, by following fundamental principles: for each generation group company, the scale of assets and quality of installed capacity must be similar so that competition will be workable. With a rational geological distribution, they all have in principal a share of less than 20% in the regional electricity markets. In order to govern the operation of electricity markets, the State Electricity Regulatory Commission of China was established at the end of 2002. This committee is responsible for working out operating rules for the electricity market, regulating market operation and maintaining a fair competition among generation companies.

At the end of 2002, eleven power group companies were established in China, including two power grid companies - the State Power Grid Company of China and the South China Power Grid Company; five power generation groups companies - China Huaneng Power Company, China Power Investment Company, China Huadian Power Company, China Guodian Power Company, China Datang Power Company; four auxiliary companies - China Power Engineering Consultation Company, China Hydropower Engineering Consultation Company, China Water Resources and Hydropower Construction Company, China Gezhouba Company. The South China Power Grid Company is composed of five provincial power grids - Guangxi, Guizhou, Yunnan, Hainan and Guangdong. The State Power Grid Company of China is composed of the rest of regional power grids - North China (including Shandong province), Northeast China (including the east part of Inner

Mongolia), East China (including Fujian province), Central China (including Sichuan province and Chongqing City) and Northwest.

The present reform is only focused on the generation sector. It is expected that restructuring will be extended to the sectors of transmission, distribution and sales, but at this moment, the future picture is still not clear.

14.6.2 Generation Investment

14.6.2.1 Generation investment in the traditional power industry

Before deregulation, it was the responsibility of utility companies to assure that enough generation capacity was available and usually there was a centralized generation planning and investment associated. The traditional approach to this was to build planning capacities based on the forecasted load, loss of load probability (LOLP) calculation and estimates of the value of lost load (VOLL), and allocate the costs of the generation capacity implicitly among consumers. The investment and planning were regulated by the government or the regulator. As a regulated monopoly, investment recovery and reasonable profits for utility companies were guaranteed by the government or the regulator, and as a result, utility companies did not have real risks on generation investment.

14.6.2.2 Generation investment in the restructured power industry

In the restructured power industry, generally there is no central planning for new generation capacity additions and no guarantee is made anymore for recovery of generation investment and return. On the other hand, generation companies do not have any obligation for ensuring sufficient supply of electricity nowadays and in future. Each generation company makes its own independent assessments of the profitability of new generation projects, as for any other industrial investments. Since electricity markets are more akin to oligopoly rather than perfect competition and there exists strong entry barriers, the supply tends to be less than the socially optimal demand. Hence, it is a problem of extensive concern how adequate generation capacity can be secured in the long run under the electricity market environment. The electricity market failure in California has brought this subject to the forefront.

Investment on new generation capacity additions is a commercial and risky activity and is expected to become more prudent under the deregulated electricity market environment. This is because investors are more interested in short-term investment return, and are reluctant to invest generation capacity which requires large investment and long recovery period and has increasing uncertainties on load variation, restructuring policy and market management rules which influence their benefits. Investors are expected to spend a considerable amount of time and effort in analyzing the interaction between investment and the decentralized decisions by participants. In making a generation investment decision, expectations concerning future electricity demand, spot market prices, variations of regulatory policies, as well as the financial status are major considerations. The locations, capacities and timing of new power plants are basically at the generation companies' own discretion although an indicative generation planning may be provided by the regulator to guide the investment and planning, as is the case in several South America countries such as Chile, Peru, Bolivia and Argentina.

It is believed by many people especially regulators around the world that energy markets are not mature enough and cannot be entirely relied on for securing a desired system adequacy so that some supporting mechanism is needed, at least in the near future, in order to ensure adequate generation capacity. As a result, in many operating electricity markets there exist different forms of capacity payments for ensuring sufficient supply in short- and/or long-terms. Up to now, there are basically three approaches for capacity payments as detailed below.

1) Capacity Obligation Model: In this model, a capacity obligation is imposed to customers by forcing them, explicitly or implicitly, to sign long-term contracts with power suppliers. The regulators determine the amount of firm capacity that each one of the consumption entities has to buy, as well as the maximum amount that each generator is allowed to sell. In other words, a reserve is imposed on each load entity in proportional to its load. To meet this requirement, the load serving entity must enter into contracts with generation companies or procure its obligation through a 'capacity market' that is operated by an organization like ISO. This model is employed in Northeast of the USA including PJM, NYPP (New York Power Pool) and NEPOOL (New England Pool). In PJM and NEPOOL, capacity obligations are defined for "load serving entities" or participants. This is the recommended method in the standard market design (SMD) in USA.

A main advantage of this method is that system reliability can be assured according to well proved techniques and procedures that have shown to result in reliable system operation. A disadvantage is that some of the benefit of restructuring may be lost. One of the main problems of the traditional power industry is overinvestment since return-on-investment is guaranteed. By using capacity obligations based on forecasts, there is a risk of over-investment.

2) Administrative Payments for Capacity: In several countries such as Argentina, Chile, Colombia and Spain, administrative payments are employed with an explicit remuneration for the installed capacity as an economic signal intended to augment the volume of installed and available generation. Specifically, additional payment is made for available capacity during hours with high demand to motivate capacity investment.

Generation companies are offered a capacity payment based on their availability no matter if they get dispatched or not. The capacity payments are collected from customers as a prorated uplift similarly to other uplift charges such as the transmission charge.
This method has several major disadvantages:

It is not easy to determine an adequate level of capacity charge for inducing optimal capacity investments. Basically, the administrative payment for capacity is based on the expected cost of lost load, which is difficult to obtain. Overestimating this cost would create artificially inflated demand for capacity and result in high capacity prices, which in turn will lead to overinvestment on capacity.

A fixed capacity charge does not adapt to a varying balance between supply and demand for capacity.

3) Explicit Capacity Adder Payment Model: In this model the capacity is priced separately from energy and consumers are not required to procure capacity. At the initial operation stage of the England & Wales (E&W) electricity market, this approach was employed. In the E&W market, the capacity payment is set to the value of lost load (VOLL) multiplied by the loss of load probability (LOLP) on half-hourly basis and paid to all available capacity. However, determining an appropriate level of VOLL is very difficult, and in the E&W market it is administratively managed by the regulator through estimating the annual marginal cost of capacity required to meet expected demand at the required reliability standard. This is included in the uplift which is added to the spot market clearing price as the power purchase price. No matter if a generator is dispatched or not, it will receive the capacity payment. When the reserve is tight, the capacity payment will be very high. As a result, the price is very volatile. In this model, the capacity payment is basically determined administratively.

This approach has been criticized for distorting the market price and for being particularly easy to manipulate. Large generation companies have sufficient incentives to withhold capacity so as to magnify the capacity payment as well as to increase the market clearing price for energy, and such manifestations have been observed.

Since the capacity payment is dependent on the system operating conditions and hence is uncertain in the future, this method may not be able to induce sufficient investment on new capacity.

Another major problem of this method is the reliance of capacity payments and capacity requirement on VOLL. It has been criticized repeatedly that VOLL is administratively set and has no market base. A possible way to get a reasonable VOLL is through demand side bidding. Another problem is with the simplified calculation of LOLP.

Although it is difficult to state which approach is the best since different market models are employed in different places, determination of capacity payments by a market mechanism is generally believed a better way than by an administrative pricing method which can not adequately reveal the value of capacity and may result in overinvestment as is the case in the traditional power industry.

How to fairly allocate capacity costs among market participants is also a key issue to address. In the last two methods introduced above, capacity payments are shared by customers at a flat rate, which is usually determined based on energy usage and maximum capacity required. In the first method, it is determined by auction.

14.6.2.3 Generation investment in China

The California energy crisis makes the Chinese government more prudent in making the restructuring decision-making of the power industry, especially on generation investment issues. Although the generation sector has already been separated from the utilities in

China, investment and construction of new power plants are still under strict control of the government.

According to current energy situation, Chinese government has worked out a strategy for the development of the China power industry. That is, "to build up three channels in north China, central China and south China for power transmission from west to east, to further restructure the power generation to improve efficiency, to develop hydropower, to construct of mine-mouth coal-fired units, to shut down small capacity units, to properly develop nuclear power, to encourage heat-electricity cogeneration and power generation by waste material and renewable resources.

Recognizing the fact that a huge amount of investment will be required in the next 20 years so as to meet the need of rapid economic development and the disastrous event happened in California in 2000-2001, Chinese government will employ a so-called "two-part tariff" for generation companies. As the name denotes, the two-part tariff includes a capacity price and an electricity price. The capacity price will be determined mainly based on investment and construction costs of new power units, while the electricity price will be "discovered" through a pool-based electricity market or long-term contracts. In order to improve the investment efficiency, it is expected that a uniform capacity price will be applicable to the same kind of generation units, at least in a same region (in different regions, the investment and construction costs for the same kind of units may be different) and the capacity price will only cover part (such as 80%) of the AVERAGE investment and construction costs of the same kind of generation units. In this way, the investor will have some degree revenue stability and it will be easier for them to obtain financing for new generation capacity additions. The government hopes that by providing capacity prices for generation units, the recovery of investment and construction costs could be guaranteed to some extent, and as a result, some incentive could be provided for generation investment.

As expected, the "two-part tariff" has suffered strong objections from academics since this does not match with the market mechanism. However, the Chinese government insists that such a policy is very necessary at least at the stage with rapid economic development, since it will be disastrous if serious electricity supply shortage occurs in the future.

According to development plan of the Chinese government, it is estimated that the installation capacities of hydropower, natural gas combined cycle, nuclear power, wind power and coal-fired generation will respectively be 200 GW, 80 GW, 38 GW, 12 GW and 570 GW in 2020.

In China, it has already been determined that the generation planning will be conducted by grid companies although generation investment will be from different sources.

14.6.3 Risks in Generation Investment

As already mentioned before, there is actually no investment risk for the utility companies in the traditional power industry. The SMD of USA actually leaves the investment risk to the load service entities (LSEs) since LSEs have capacity obligations. While in China, the "two-part tariff' is actually used to reduce the investment risk of investors.

However, in any of the above mechanisms the ultimate risks are borne by the customers and/or investors.

Let us have a comparison of the effects of the above three mechanisms from several aspects such as risk distribution, efficiency incentives and economic signals, based on three scenarios:

a) normal level of power supply
b) over-supply
c) under-supply.

14.6.3.1 Risk distribution

The parts bearing higher risk under different supply scenarios with the above three mechanisms are listed in Table 14.7.

	Traditional method	The two-part tariff in China	Electricity market
Oversupply	Customers	Customers	Investor
Normal supply	No	No	No
Undersupply	Customers	Customers	Customers

Table 14.7. Risk Distribution

From the above comparisons, it can be concluded that the market based approach for generation investment could provide both high efficiency incentives and strong economic signals although investors have to bear higher risk. On the other hand, the approach employed in China through the "two-part tariff" is a mixture of regulation and competition, and a major difficulty associated with this approach is how to well balance regulation and competition and this is more a problem of art rather than science.

14.6.3.2 Efficiency incentives

The efficiency incentives of generation companies (utilities) under different supply scenarios with the above three mechanisms are listed in Table 14.8.

	Traditional Method	The Two-part Tariff in China	Electricity Market
Oversupply	Low	Medium	High
Normal supply	Low	Medium	High
Undersupply	Low	Medium	High

Table 14.8. Efficiency Incentives

14.6.3.3 Economic signals

The economic signals through electricity tariffs under different supply scenarios with the above three mechanisms are listed in Table 14.9.

	Traditional Method	The Two-part Tariff in China	Electricity Market
Oversupply	Weak	Medium	Strong
Normal supply	Weak	Medium	Strong
Undersupply	Weak	Medium	Strong

Table 14.9. Economic Signals

Thus, a reliable and reasonably priced supply of electricity is critical to the functioning of a modern economy and society. To achieve this goal, it is important to secure adequate generation capacity in the long run. Without sufficient supply, a market cannot lead to maximized social welfare, and sometimes even cannot work at all.

As already happened in California, capacity shortage will probably occur in those markets using the energy-only market model. An adequate capacity payment, in one way or another, appears necessary for ensuring the system adequacy. However, it seems that there does not exist a generally applicable so-called "best" method to solve the generation capacity adequacy problem, since a best solution to this problem is related to many factors such as the past, current and future scenarios of the power industry studied, social and economic development status of the country or region concerned. These factors must be well taken into account in designing a workable electricity market.

The balancing of competitive opportunities and regulatory measures is of great strategic importance in the power industry restructuring and should not be overlooked since the power industry is an important national infrastructure whose failure will have severe social and political implications.

14.7 Investment And Development Problems Of Russia's Power Industry

The federal law "*On electric power industry*" adopted in 2003 stipulates the basic principles of the state investment policy in the Russian electric power industry [18]. These principles need to be specified in the form of a system of measures and mechanisms of their implementation. The general structure of such a system is presented in Figure 14. 5. Let us comment on the scheme.

The most important objectives of the state investment policy in the sphere of electric power industry are to build a positive investment image of the country and develop a market investment infrastructure (a share market and bank system). It is necessary to essentially increase the level of state guarantees to investors which is possible only with the appropriate legal framework, stability and predictability of the state policy in electric power industry development, available economic mechanisms for implementation of state guarantees, etc. These are the problems of Federal Assembly and Government of the country and their solution will help essentially decrease the investment risk components.

Another important objective is to create an effective system of investment risk insurance including the so-called market insurance (options call, etc.). The role of the state in this context consists in formation of a legal framework for creation of the insurance system that includes economic mechanisms to support it and provide its stability.

The solution of the problems will also be fostered by an increase in the investment attractiveness of power companies, particularly of the generating companies being formed now, which can be achieved by providing financial transparence of their activity, economic stability, predictability and transparence of the management actions, other corporative measures. The role of the state here is also very important, as it should create the required legal framework to regulate the activity of the generating companies and their management. The framework should reflect the economic mechanisms that stimulate the management of the companies to act properly and employ economic sanctions in the case the regulating laws are violated.

Another aspect of the state regulation of market relations in electric power industry is the antimonopoly regulation that is carried out on the basis of the antimonopoly legislation by the Ministry of Antimonopoly Policy.

Elaboration of the state policy on electric power industry development and mechanisms of its implementation is a key problem and the role of the state in solution of this problem should be decisive. The government and, first of all, Ministry of Energy should be in charge for the state policy elaboration. This activity employs the elaboration of state strategies and programs of the electric power industry development that could determine the scale and proportions of the industry development, based on the need of stimulating some energy technologies, solving certain social problems, etc. The mechanisms for implementation of the state policy on the electric power industry development are tax, credit, export-import and other economic mechanisms.

Federal and Regional energy commissions should play an important part in the system of providing investment and development of Russia's electric power industry. Their main objectives are presented in Figure 14.5.

The considered components create the required conditions for investing into the electric power industry and for its development. The electric power industry development and the investments required should be substantiated in the frames of the management system of Russia's electric power industry and UPS development. This system includes:

- UPS and IPS expansion patterns that should be based on the system properties of the object;
- Strategic plans for expansion of electric power companies;
- Plans for commissioning of power facilities and investment programs;
- Programs for provision of shortage-free electric power industry development;
- Monitoring systems of investment programs.

All these lines require a methodology whose creation is an independent problem for further studies. Here let us focus on substantiating the elaboration of the programs to provide only the shortage-free development of the electric power industry.

The above components of the state policy on attraction of investments into generating facilities of the electric power industry call for a scrutiny. The complexity of their substantiation and implementation may result in inefficient and irrational decisions that will not lead to the desirable results in terms of external investment inflow. Besides under a free market the short-term objectives of the investors, including the generating companies themselves, dominate the long-term ones [19]. All this may lead to the inadmissible decrease in the reserves of generating capacities, generation shortage and, as a result, rise in electricity tariffs. Therefore there is a need for special measures to prevent such situations.

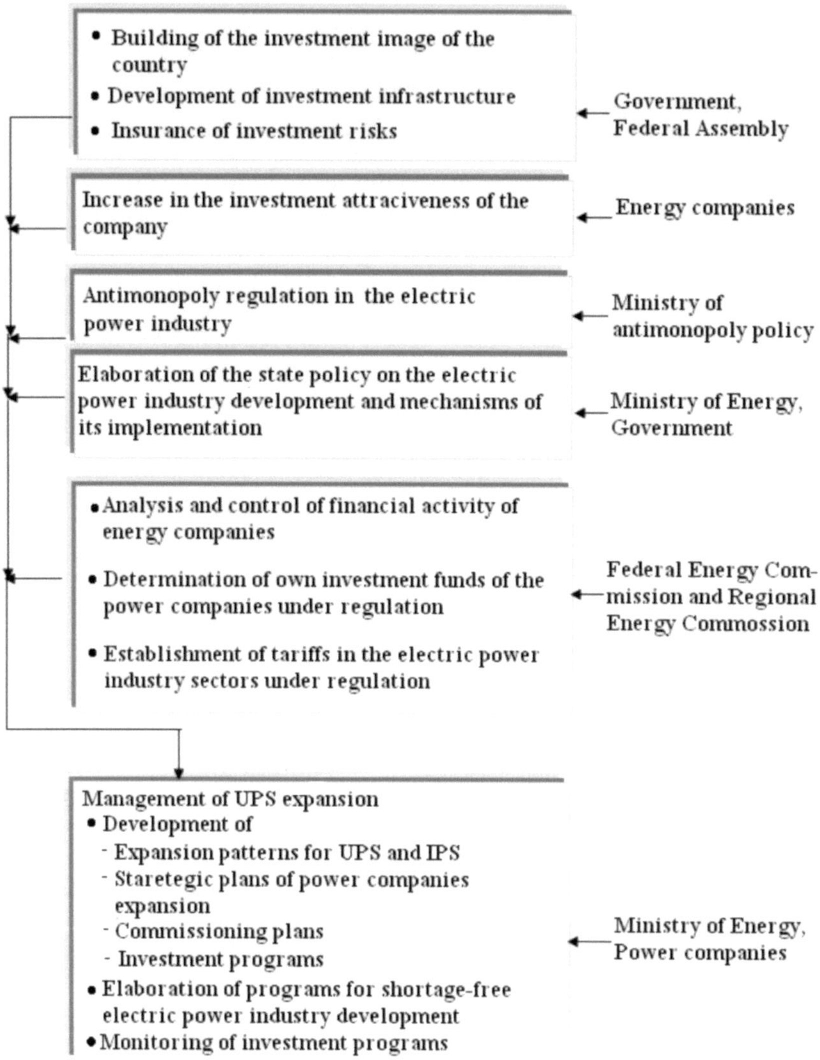

Figure 14.5. Structure of investment attraction and development of Russia's electric power industry

The federal law *"On Electric Power Industry"* envisages some short-term and long-term measures on prevention from power shortage and subsequent electricity tariff rise. The short-term measures include introduction of constraints on tariff rise (price caps) and introduction of regulation in the corresponding zone of the wholesale electricity market at generation shortage that may result in disappearance of conditions for competition.

The primary long-term measure in the considered context is creation of the state system of planning the additional generation expansion that could provide permanent availability of surplus generating capacities at the wholesale market and avoidance of their shortage. Nuclear energy that is entirely a property of the state can become a basis for development of such an *"insuring"* additional generation. Currently the nuclear energy has essential potentialities for commissioning additional generating units. Distributed generation, particularly small gas-turbine co-generation plants, may also play an important role, as there are great potentialities for their construction.

The considered state system of planning the additional generation expansion should form economic mechanisms to stimulate the generating companies and external investors to invest into construction of new power plants. However the main line is creation of the State Foundation for development of energy and energy conservation. Ministry of Energy, Federal Energy Commission and State Construction Committee forwarded the proposal on creation of the Foundation to Russian Government in July 2003.

Creation of the Foundation is particularly topical in 2003-2005 which is related to Russia's electric power restructuring. Further as the market relations grow stronger and the private investors appear the functions of the Foundation can be reduced to provision of system reliability and energy security of Russia.

14.8 North-East Asia Interconnection Scenario Map, and Power Reserve Strategy in South Korea

The purpose of this Section is to obtain the reserve power for the future power shortages faced by the metropolitan areas of the Korean Peninsula and by the southeastern area of the South Korea in North-East Asia. The assumed scenarios will be proposed in the cases of without or with connection to the Sinpo nuclear power plant, which is to be constructed with 2,000MW capacity in the future. In this connection, the types of a power transmission for interconnection consist of the 765kV HVAC and the HVDC. In Section 14.8, the various scenarios for providing the interconnection of the power systems among Far-East Russia, North-East China, North Korea, South Korea, and Japan are presented, and the resulting interconnected power systems are simulated by means of a power flow analysis performed with the PSS/E tool. The power flow map is drawn from the data simulated and the comparative study is done.

The interconnection of the power systems among North-East Asian countries (Russia, China, Mongolia, Japan, and Korea) has been proposed on numerous occasions, but little progress has been made due to the complicated political issues and economical problems involved. Now, the necessity for this power system interconnection is increasingly being felt due to the benefit of each country. Because of these reasons, Korea peninsula takes the role connect a bridge between different areas of Northeast Asia, such as Russia, Mongolia, China, and Japan. Therefore, the interconnection of the power systems within the Korean peninsula should proceed without delay in order to lay the foundation for a complete interconnection of the various power systems within North-Eastern Asia.

As the substitutes for relieving the problems of power imbalance and the shortage of power in the Seoul metropolitan areas in South Korea and the Pyongyang metropolitan areas in North Korea, five cases of scenarios as follows will be proposed according to the assumed scenario.

- The first case of the scenario involves 765 kV HVAC interconnection between Yangju bus of South Korea and Pyongsan bus of North Korea.
- The second case of the scenario concerns 765 kV HVAC interconnection between the Yangju bus of South Korea and the Pyongsan bus of North Korea to supply the Sinpo nuclear power plant which is to be constructed with 2,000MW capacity in the future.
- The third case of the scenario includes HVDC interconnection between the Busan area in South Korea and the Kyushu area in Japan to solve the power shortage in the Gyeongnam area of South Korea.
- The fourth case of the scenario consists of an HVDC interconnection between Sredne-Uchurskaya HPP in Far-East Russia and Wongi bus in North Korea without supplying Sinpo nuclear power plant in the future,
- In the fifth case of a scenario, the assumed scenario for an HVDC interconnection between Liaoning's power network in North-East China and Supung bus in North Korea will be proposed without supplying Sinpo nuclear power plant in the future.

14.8.1 Power System in South Korea

The South Korean electricity generation system can be divided into 7 geographical areas that take geographical boundaries into account. The transmission voltages used are 345kV for the major networks, and 154kV or 66kV for the local systems. Most 66kV lines are now either being removed or replaced by higher voltage lines. Power system on Jeju Island is now connected to the mainland via a 100km-long submarine transmission system, comprised of HVDC cables. Because the demand for power is increasing rapidly in the metropolitan area, 765kV facilities are in the process of being constructed and now come into operation in order to provide a stable large-scale power transmission between the large power generation plants and the areas where the consumers are located. The reasons for upgrading the highest system voltage to 765kV are to improve distribution between the large power plants and the load centers, to provides a stable supply of electric power, to meet the rapidly increasing demand in the metropolitan area, to resolve the difficulties involving transmission routes and substation sites, to interconnect power plants having a large capacity, and to create the necessary backbone for the transmission system in and around the ***Metropolitan area***. In South Korea, the potential increase in power demand is higher than that of any other country. The metropolitan area situated in the central parts consumed nearly 43% of the total electricity generated, and the southeast area consumed about 33%. However, most of the large-scale power plants have been constructed in the southern part of South Korea.

Consequently, the existing power grid includes multiple routes designed to supply the metropolitan area so that, by and large, the direction of power flow is toward the north. The future supply and demand for power is likely to become increasingly unbalanced, because the two regions in South Korea produce a much greater quantity of electricity than they consume. The 5th long-term power supply-demand study published in January 2001 recorded a current total demand of 41,007MW which will likely increase to 60,718MW in 2010. The total power generation capacity will have to be increased each year to keep up with the growing demand, and will in fact be raised to 1.5 times the current level for a total of 74,611MW in 2010. It is predicted that the electrical load will attain the level of the current developed countries, sometime between 2020 and 2030, representing a total power consumption of over 80,000MW. This represents doubling of the amount of electricity produced, compared with the present level.

14.8.2 Power System in North Korea

This study assumes that the power system in North Korea is divided into 5 areas. The power system in North Korea is smaller than that in South Korea. Most of the hydroelectric power plants are located in the hilly region of the northern areas in North Korea and most of the thermoelectric power plants are located in the metropolitan area. Moreover, power capacity in North Korea has been estimated to be approximately 7,000MW. Currently, it is known that transmission line voltage is composed of 110kV and 220kV.

14.8.3 Power System in Far East Russia

Installed capacity of Sredne-Uchurskaya HPP in Far-East Russia has been estimated to be approximately 3,300MW and average annual output to be about 15TWh. Its total half-wave length is approximately equal to 2,500km. Interconnection between Sredne-Uchurskaya HPP and South Korea will pass through territories of Far-East Russia, North Korea, and South Korea. The problem of bulk power transport over very long distances of 2,000-4,000km can be solved with the help of EHV & UHV transmission systems in both DC and AC. There are two types of AC transmission systems (TSs) over very long distances, namely line compensation TSs and Half-Wave TSs. The former includes line, terminal substations and reactive power equipment, which are placed at several intermediate substations of line. Power transfer capability of line compensation TS is primarily limited by stability, and the basic function of the reactive power equipment is to secure this. Half-wave line keeps its reactive power balance for all conditions and there is no need to have compensation devices. Transition from double-circuit 2-pole line to single-circuit 4-pole one improves ecological and economical parameters of DC TS at the expense of some increase of risk of forced outage only under extreme contingencies as mentioned above. For traditional AC TS a separate 3-phase line is an independent unit regardless of type of faults (line-to-ground or complete faults).

14.8.4 Liaoning Region Power System in North-East China

Liaoning's power network covering the 147,500 square kilometers of land is a modern power network with long history and full of vigor. From 1935 to 1941, the first 154kV and 220kV super-grid in China was formed in Liaoning province. This power network owns the first large 220kV-substation (Hushitai Substation) independently designed and constructed in China and the first extra-high substation (Dongjia Substation) in China that was independently designed and constructed, and adopted as domestic equipment. It installed the first 25MW fossil-fueled unit in China and the first set of domestic 200MW fossil-fueled units. The 800MW unit with largest single-machine capacity in China has been put into operation in Suizhong Power Plant. In the province, there are five overlarge thermal power plants with an installed capacity of 1,200MW including Suizhong, Qinghe, Tieling, Jinzhou, and Dalian. Huaneng has four sets of supercritical condensing generating units, and fourteen sets of thermal generating units with single-machine capacity at or above 300MW. Power network of 220kV has covered nearly all cities above county level and power networks of large cities in the province have formed two rings.

Hydraulic power units may adjust output automatically according to frequency. As a heavy industrial base of China, and with developed politics, economy, and culture, Liaoning province is the power load center in Northeast China. It has one 500kV line and six 220kV lines to connect with the power network in Jilin province. It also has two 500kV lines and one 220kV line to connect with eastern part of Inner Mongolia. By the end of 2000, the total installed capacity in Liaoning province was 15,185MW (hydropower: 1,156MW; thermal power: 12,559MW). The total installed capacity of the wholly-owned and holding power generation plants of Liaoning Electric Power Co., Ltd. is 2,854MW (hydropower:

456MW; thermal power: 2,398MW) and takes up 18.8% of the total installed capacity of the whole province. The independent power generation company has a total installed capacity of 10,861MW (hydropower: 488MW; thermal power: 10,373MW) and takes up 71.5%. The local self-supply power plants have a total installed capacity of 3,006MW, taking up 19.8%. The installed capacity of the plant at Sino-Korean boundary river is 545MW, taking up 3.6%. In Liaoning province, there are two hundred and one 220kV lines totaling 7,610km and five 500kV substations with a capacity of 6,000MVA. At present, there are two hundred and one 220kV lines totaling 7,610km, and ninety-four 220kV substations with a capacity of 18,426MVA.

14.8.5 Power System of Kyushu in Japan

Japan's power system is divided into nine regional companies serving the areas of Hokkaido, Tohoku, Tokyo, Chubu, Hokuriku, Kansai, Shikoku, Chugoku, and Kyushu, and transmission consists of 500kV, 220kV, 110kV, and DC250kV lines. The frequency used is 60Hz in the western part and 50Hz in the eastern part of the country. According to statistics published in 2001, the total generating capacity of the nine power companies is 33,765MW due to hydropower, 118,112MW due to thermal power, and 42,300MW due to nuclear power. The total capacity is therefore 194,177MW. The overhead transmission system consists of 46,692km of lines with a voltage of less than 110kV and 16,048km of lines with a voltage of more than 110kV and 19,324km of lines with a voltage of more than 187kV. The total length of these overhead lines is 82,065km. The underground system consists of 9,333km of cable with a voltage of less than 110kV, 909km of cable with a voltage of more than 110kV, and 624km of cable with a voltage of more than 187kV. The total cable length is a 10,865km. Among these regions, Kyushu has a total land area of 42,163 km^2 and is located in the southernmost part of Japan. The generating capacity of Kyushu's Electric Power Company is approximately 30,200MW. The backbone of its transmission system consists of

500kV, 220kV, and some 110kV lines. Kyushu's infrastructure is composed of nuclear, thermal, hydro, and geothermal power generating plants. The nuclear power plants are located both in the southwest coastal region and at the furthermost tip of Kyushu's northwest coast. The thermal power plants are located mainly on Kyushu's northeast and the northwest coasts. The hydro power plants are randomly distributed within the north and south central regions. The geothermal power plants are located in the north and south central regions.

14.8.6 Load Flow Calculations

Load flow calculations were made for different scenarios:

- Isolated load flow between South Korea and North Korea for North Korea - South Korea - Japan.
- Northward load flow - from South Korea to North Korea.
- Southward load flow - from North Korea to South Korea with Sinpo 2000 MW NPP power injection.
- Northward load flow (from South Korea to North Korea) for Far East Russia - Korea - Japan.
- Northward load flow (from South Korea to North Korea for North East China - Korea - Japan.
- Northward load flow (from South Korea to North Korea) after interconnection between South Korea and Japan.

Figure 14.6 shows an isolated flow map for the first scenario as an example of load flow calculations

14.9 Generation and Transmission Sector in Korean Power Systems

Throughout the world, restructuring and competition are being introduced into the electric power industry. In Korea, the act on the promotion of restructuring of the electric power industry was approved to allow the division of generation businesses on December 23, 2000. Furthermore, the Electricity Business Act (EBA) was amended in early 2002 so that it could mandate the establishment of an electricity trade market and the advent of a regulatory agency, Korea Electricity Commission. Amongst the radical changes since the restructuring, experiences associated with the investment and expansion planning in generation and transmission sectors is being examined. A long-term generation investment plan, taking into account scheduled and reasonably foreseen unscheduled outages as well as transmission installations is also being considered. Also, the construction, operation, and maintenance of the new facilities will be ensured by a closer examination of development plans for the network.

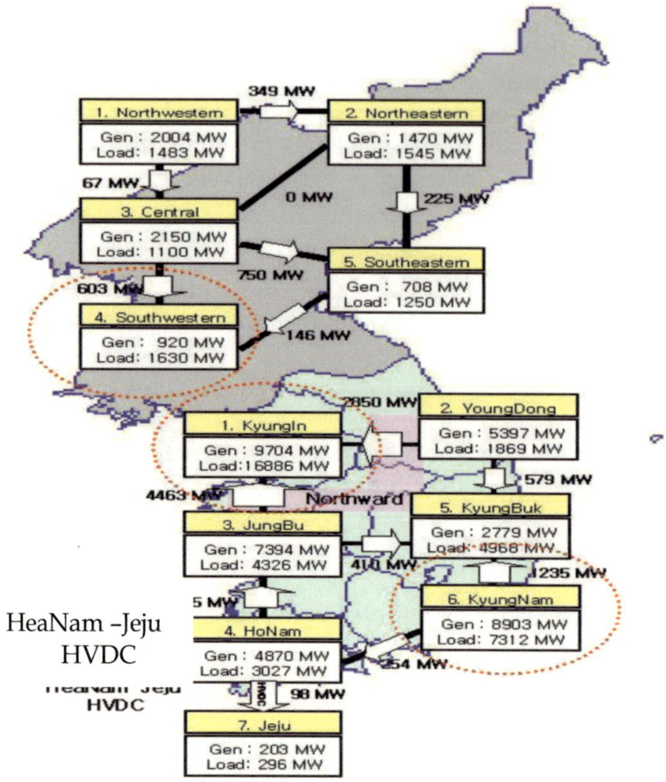

a) Flow in N.K and in S.K.

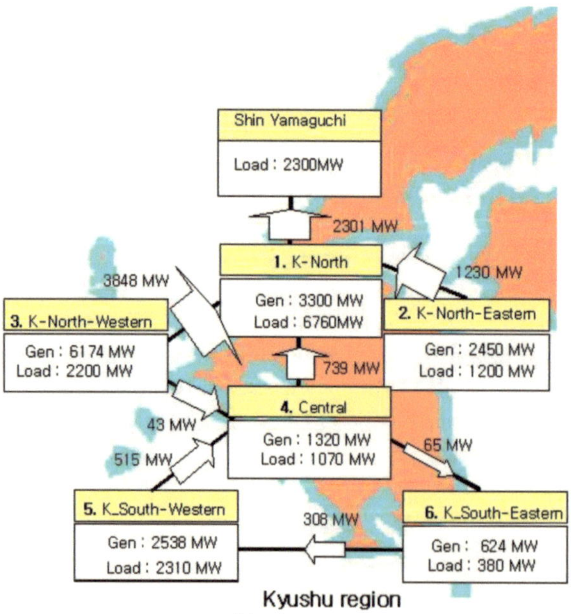

(b) Flow in Japan

Figure 14.6. Isolated flow map for N.K-S.K-Japan

14.9.1 Generation Sector in Korea

Before the electricity industry restructuring, the government had established a Long-term Power Development Plan (LPDP) and the Korea Electric Power Corporation (KEPCO), a vertically integrated utility, had implemented this LPDP in order to secure the affordable and reliable electricity supply. Under the domestic requirements and international pressure, the Korean government decided to gradually restructure the electricity supply industry and to promote consumers' rights. With electricity supply industry restructuring, the competitive market mechanism will be the crucial factor. In this context, the function of the former Long-term Power Development Plan inevitably started to make changes into non-binding guidelines or reference.

14.9.1.1 Restructuring plan for electricity industry

The Korean electricity supply industry restructuring procedures are summarized in the following [20]:

- Generation Competition (2001-2003): Separation and privatization of KEPCO's generation assets into six-generation subsidiaries and the introduction of competition in the generation sector
- Wholesale Competition (2004-2008): Separation and privatization of KEPCO's distribution assets and the introduction of competition in the wholesale electricity supply
- Retail competition (after 2009): Introduction of competition in retail electricity supply.

The Korean government, in consultation with the Korea Power exchange (KPX), biennially sets up the Basic Plan of the Electricity Supply and Demand (BPE) just as it had prepared the Long-term Power Development Plan. Here the KPX in Korea has three major functions [21].

- Facilitates operations of the electricity market by monitoring the process of bidding, metering and performing real-time dispatch to balance the supply and demand
- Ensures stable operations of the electric power system and security of electricity supply including maintaining the stability of transmission networks
- Supports the government in planning strategies to meet the demands for electricity by projecting market trends and forecasting demands for electricity.

However, the BPE will be established not as a binding force but as a tool providing market participants with appropriate information and market-based solutions.

14.9.1.2 Investment in generation capacity

Speaking of the recent electricity supply and demand trend in Korea, the electricity demand has steadily soared in accordance with the nation's high economic growth and the increase in the standard of living over the last 30 years. In 2001, a peak demand of 43,125MW and electricity sales of 258TWh ranked Korea the seventh among the Organization for Economic Cooperation and Development (OECD) members [22], see Tables 14.10 and 14.11.

Year	1999	2000	2001	2002
Peak Demand	37,293	41,007	43,125	45,773
Average Demand	27,320	30,328	32,552	34,974
Load Factor	73.3 %	74.0 %	75.5 %	76.4 %

Table 14.10. Trend of Demand and Load Factor in Korea (Unit:MW)

Year	1999	2000	2001	2002
Residential	34,581	37,102	39,211	42,278
Industrial	120,859	132,260	135,791	144,454
Commercial	58,775	70,173	82,729	91,719
Total	214,215	239,535	257,731	278,451

Table 14.11. Electricity Sales in Korea (Unit: GWh)

In spite of such a high demand growth, the continuous construction of generating plants and the promotion of demand side management have led to a stable electricity supply. Although there was no severe shortage of electricity supply capacity, the Korean electric power system has experienced periodical excesses and shortages of the generating capacity. To overcome this fluctuation, the government offers basic policy directions to the electricity industry including the reliability criteria, optimal fuel-mix, nuclear power plant construction policy, transmission and/or distribution expansion policy, the target of demand side management, and so on.

To sum up the directions of generating capacity construction, the government guides the investors to construct some generating capacity so that the reserve margin and energy resource mix can be satisfied and to adopt advanced generating technologies for additional capacity in order to cope with the strict environmental regulation. It is generally accepted that new generating plants are constructed at the existing sites as much as possible, and yet it is possible for them to be constructed in the vicinity of load centers, taking into the consideration the difficulties in the exploitation of land for transmission facilities.

In the event that the crisis of the electricity supply is predicted mainly due to the lack of generating capacity, the government will be ready to issue a contingency plan including generation capacity addition by the public company. In the near future, the government will reduce the regulatory intervention and organize the institutional foundations so that the electricity market can achieve the goal of the stable electricity supply and demand in the long run.

14.9.2 Transmission System in Korea

Improving transmission system operations will go a long way toward easing transmission bottlenecks by delaying or alleviating the need for construction of new transmission facilities. However, construction of new facilities cannot be avoided entirely. We must guarantee that needed facilities are identified in a timely fashion through open processes and that, once identified, they are constructed expeditiously.

14.9.2.1 Overview of current transmission network

Transmission voltages in Korea are 765kV and 345kV for trunk lines and 154kV or 66kV for local networks, while the 66kV lines are being phased out. Also, the power transmission network on Jeju Island is connected to the mainland system by submarine HVDC cables [23] (see Table 14.12).

Facilities	Line Length (C-km)		
	Overhead	Underground	Total
765kV	662	-	662
345kV	7,335	162	7,497
154kV	16,501	1,643	18,144
66kV	1,398	4	1,402
180kV(HVDC)	30	202	232
Total	25,926	2,011	27,937

Table 14.12. Transmission Facilities in Korea (as of 2002)

The expansion of the transmission system in Korea will now be focused on.. The voltage levels classify the objectives of the long-term transmission expansion planning:

- 765kV line: It will directly interconnect the generation plants to Seoul metropolitan area.

14.9.2.2 Transmission expansion planning

Several methodologies to improve the performance of transmission facilities are suggested as follows:

- To secure the stability of large-scaled transmission systems: timely expansion of transmission system and transmission voltage upgrades to 765kV
- To develop a countermeasure to the fault currents: strengthening of breaker standards and installation of series reactors
- To balance the supply and demand of reactive power: reinforcement of condensers for electric power, installation of shunt reactors and development of distributed generations.

Most of all, the sizes and date of transmission line construction are determined in such a way that the lines provide sufficient transportation capacity intimately in conjunction with the ultimate generating capacity at the plant sites. Enough has been said to find the authority that would answer the following question: What portion of the construction costs in building up the interconnection lines are charged to generators and/or transmission service providers? Further studies on this issue are needed and discussed in terms of various factors such as voltage level, economic assessment, and technology availability.

14.10 Acknowledgements

Nikolai I. Voropai, Professor, Corresponding Member of RAS, Director of Energy Systems Institute, Irkutsk, Russia has prepared and coordinated this Chapter. Contributors include colleagues at the Institute and Members of the IEEE PES W.G. on Asian and Australian Electricity Infrastructure.

14.11 References

[1] Hant, S., Shuttleworth, G., *Competition and Choice in Electricity,* Chichester, England, A March&McLennan Co., 1995.

[2] Voropai, N.I., Podkovalnikov, S.V., Trufanov, V.V., "Methodical Principles of Making Decisions on Electric Power System Expansion in Market Environment", *IEEE Porto Power Tech. Conf. Proc.,* Vol. 3, Porto, Portugal, Sept.10-13, 2001.

[3] Voropai, N.I., Ivanova, E.Yu.,"Hierarchical Game Theoretical Problem of Electric Power System Expansion Planning", *IEEE Bologna Power Tech Conf. Proc.,* Bologna, Italy, June 23-26, 2003.

[4] Voropai, N.I., Ivanova, E.Yu., " Multi-Criteria Decision Analysis Technique in Electric Power System Expansion Planning", *Electrical Power and Energy Systems,* 2001, Vol. 24, No. 1.

[5] Wu, F.F., Contreras, J., "Coalition Formation in Transmission Expansion Planning", *IEEE Trans., Power Systems,* 1999, Vol. 14, No. 3.

[6] Chuang, A.S., Wu, F.F., Varaiya, P., "A Game-Theoretic Model for Generation Expansion Planning: Problem Formulation and Numerical Comparison", *IEEE Trans. Power Systems,* 2001, Vol.16, No. 4.

[7] Voropai, N.I., Ivanova, E.Yu., "A Game Model for Electric Power System Expansion Planning in the Liberalized Environment", *MedPower'2002 Conf. Proc.,* Athens, Greece, Nov. 4-6, 2002.

[8] Kondragunta, J., "Supporting Document for Reliability Criteria for Transmission System Planning", SCE and WSCC Reliability Subcommittee,August 1994.

[9] "NERC/WSCC Planning Standards", NERC, 1997.

[10] "Planning Criteria for Long Term Power System Planning", KEPCO, 1999.

[11] "A Study on Criteria and Standards for Transmission System Planning", KEPCO/KERI, August 2001.

[12] Pandey, M. K., Billinton, R., "Electric Power System Reliability Criteria Determination in a Developing Country", *IEEE Trans. Energy Conversion,* 2000, Vol.15 No.3.

[13] China Electric Power News, April 13, 2004 (in Chinese).

[14] Wu, F.F., Wen, Fushuan, Duan, Gang: "Generation Planning and Investment under Deregulated Environment: Comparison of USA and China", *IEEE PES General Meeting,* Denver, USA, June 6-10, 2004.

[15] Yao, J.G., Tang, J., Li, X.Q., Zhang, B.H., "Two Bidding Models of Generation-Side Electricity Market for Application," *Journal of Automation of Electric Power Systems,* 2003, Vol.27, No.24.

[16] Li, G.H., "Separation of Power Plants From the Grids - A Theme of the Reform in China Power Industry," China Power Investment Group Corporation, Tech. Rep., March 25, 2003.

[17] Wen, F.S., Wu, F.F., Ni, Y.X.. "Generation Capacity Adequacy in the Competitive Electricity Market Environment," accepted for publication by *Int. Journal of Electrical Power and Energy Systems,* in press.

[18] Federal Law on Electricity of Russian Federation, Draft 202990, February 17, 2003.

[19] Voropai N.I., "Investment and Development of Electric Power Industry in Market Environment", *Power Con'2002 Proc., Vol. 1,* Kunming, China, October 13-16, 2002.

[20] Ministry of Commerce, Industry, and Energy of Republik of Korea, *Basic Plan for Restructuring of the Electricity Supply Industry,* January 1999.

[21] Korea Electric Commission, Restructuring of Electric Power Industry in Korea, July 2002.
[22] Korea Power Exchange, The 1st Basic Plan of Long Term Electricity Supply and Demand, August 2002.
[23] Korea Electric Power Corporation (KEPCO): www.kepco.co.kr.

15

Impacts of GHG Programs and Markets on the Power Industry

This Chapter provides an overview of the global responses to Climate Change (CC) and of the established and emerging GHG Markets and Programs arising from this. The impacts on the electrical power industry and how it is taking advantage of these programs and markets and adapting to CC is discussed. This includes the impacts on policy, strategy and decision-making in major players such as governments, manufacturers, utilities, contractors and consultants and how they are leading by example within their own operations..

15.1 Introduction

Global response to CC is well established and growing daily due to international initiatives such as the Kyoto Accord and the more recent Asia Pacific Partner Ship (APP) Climate Pact which the USA and Australia developed with a group of key Asian countries, notably China and India and which now includes Canada. International negotiations for the post Kyoto period got underway in Bali in December 2007 and a decision was reached to adopt the "Bali road map" for a future international agreement on CC. There are also many programs and initiatives at national, state and regional levels to monitor, record and reduce Greenhouse Gas (GHG) emissions. An effective tool or mechanism to accelerate the achievement of cost effective GHG targets is the concept of emissions trading or transfers among participants [1]. Essentially this involves treating GHG emission allowances and reduction/removal credit units like any other commodity in the marketplace. Arrangements are made for them to be traded on national and international exchanges. The marketplace sets the value of GHG emission credit units. These are bought and sold by countries and companies to facilitate meeting their GHG targets at lowest cost.

The main established markets for trading GHG credits and allowances are the European Emission Trading Scheme (ETS), the Chicago Exchange (CCX), the New South Wales Trading System in Australia and the International Emissions Trading (IET) scheme established as part of the flexibility mechanisms of the Kyoto Accord. There are many other markets and programs under development such as the Carbon Trust, the Regional GHG Initiative (RGGI) in North-east USA (discussed in section 4.); and the California Climate Action Registry CCAR).

15.2 International Response To Climate Change: An Overview

Access to modern forms of energy is fundamental to development and the eradication of poverty in the developing world, but energy is also responsible for much of the Greenhouse Gas (GHG) emissions that threaten stability of the climate system. Therefore the goal of GHG reduction efforts is not to deny people access to energy, but to reduce the carbon intensity of development and to moderate and stabilize the concentrations of GHGs in the atmosphere through a less carbon intensive energy supply, increased efficiency of energy use, improved carbon sinks and the capture and storage of carbon emissions at source. As different approaches to achieve this goal are carefully evaluated and international negotiations continue, most countries have implemented programs that promote research, tracking and reporting on carbon emissions, voluntary mitigation and adaptation measures, energy efficiency and renewable energy technologies. However, with the exception of a handful of European countries that have implemented some form of carbon/energy tax, governments have not banned or placed restrictions on fossil fuels or CO_2 emissions.

Efforts to reduce carbon-intensity of energy supply generally focus on increasing the efficiency of power plants and promoting low-carbon fuels and renewable energy (wind, solar, hydro, geothermal, etc.). Efforts to reduce energy use typically promote energy efficiency and conservation in the industrial, commercial, transport and residential sectors. Efforts to encourage improved management of agricultural and forest lands and the protection of forests tend to enhance the Earth's natural capacity to assimilate carbon and mitigate the impact of CO_2 emissions. Efforts to capture and store carbon emissions at source are focusing on coal fired power plants - so-called clean coal technology. An overview of international efforts to reduce the impact of GHG emissions through mitigation and climate adaptation programs is now presented [15.2.1 to 15.2.8].

15.2.1 Greenhouse Gases and Climate Change

The gases responsible for the strong atmospheric absorption of infrared radiation are called Greenhouse Gases (GHGs). The Greenhouse effect is illustrated in Figure 15.1. Water vapor and CO_2 are the most important GHGs and are responsible for the bulk of greenhouse warming. Both water vapor and CO_2 are naturally occurring as are other GHGs including methane, nitrous oxide, and ozone. Human activities, however, add to the levels of most of these naturally occurring gases, and are the sole source of other powerful classes of GHGs, including chlorofluorocarbons (CFCs), hydro fluorocarbons (HFCs) and per fluorocarbons (PFCs), among others.

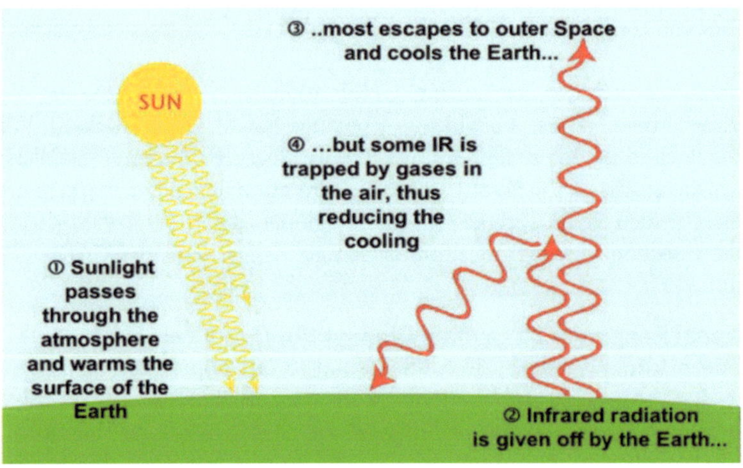

Source: Met Office Hadley Centre

Figure 15.1. The Greenhouse Effect

GHGs generally persist for long periods in the atmosphere. While many conventional air pollutants may persist in the atmosphere for only a matter of hours or days, many important GHGs persist for decades or even hundreds of years. For example, CO_2 has an estimated mean atmospheric persistence of 300 years and some CFCs may persist for as long as 400 years. As a result, these gases accumulate, become very well mixed in the atmosphere and have a global impact that is mostly independent of where they were emitted. GHG persistence has significant policy implications because the gases we emit today may impact the climate system for hundreds of years.

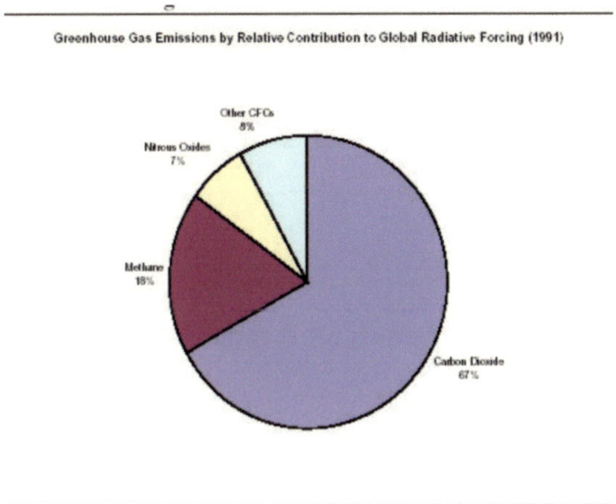

:: Greenhouse Gas Assessment Handbook, World Bank 1998

Figure 15.2. The contribution of selected Greenhouse Gases to Greenhouse Warming

GHGs differ in their ability to absorb infrared radiation. Among the most infrared radiation-absorbent are the CFCs, HFCs and PFCs. Other powerful GHGs include nitrous oxide and methane. For example, a molecule of CFC-12 is 15,800 times, CFC-11 is 12,400 times, nitrous oxide is 270 times and methane is 21 times as effective in absorbing infrared radiation as a molecule of CO_2. However, because atmospheric concentrations of these compounds are much less than concentrations of CO_2, they play a lesser role in greenhouse warming and CC. Figure 15.2 depicts the relative contribution to greenhouse warming of various GHGs when both their radiation absorbing characteristics and their relative concentration are considered. This excludes the effect of water vapour which is a major natural contributor to global warming. There is however a feedback effect of warming due to GHG emissions from human activity as warmer air can hold more water vapour. Increased water vapour also reflects more solar radiation and the net overall effect is not yet fully understood.

Carbon dioxide and other atmospheric GHGs absorb infrared radiation and create a natural greenhouse effect that warms the Earth. The natural greenhouse warming of the atmosphere keeps the Earth approximately 33°C warmer than it would be without an atmosphere. Humans have been emitting increasing quantities of these GHGs and now emit over 25 billion tons of CO_2 equivalent annually. Atmospheric CO_2 concentrations are now at their highest levels in more than 160,000 years. There is a scientific consensus that these steady additions of GHGs have begun to impact our climate and very likely may be the dominant force driving recent warming trends (see Figure 15.3). Solar cycles and oceanic currents are among the other factors that impact global climate.

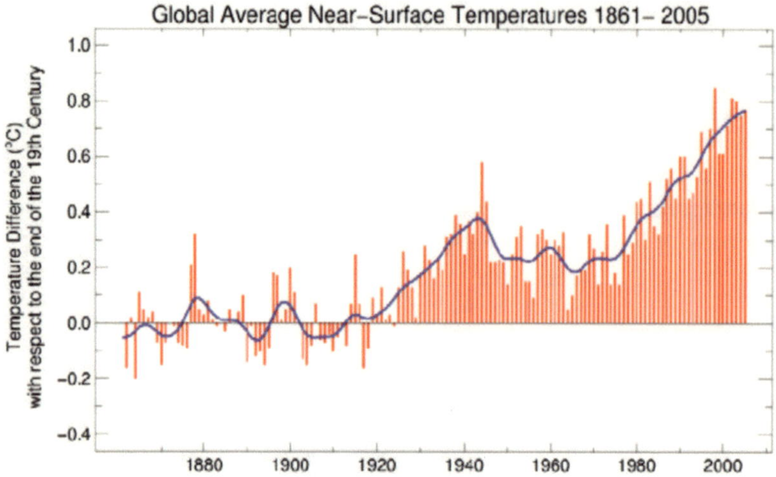

Figure 15.3. Global Average Near Surface Temperatures

As shown in Figure 15.4, global emissions are forecast to grow from all sources - transport and power generation growing fastest.

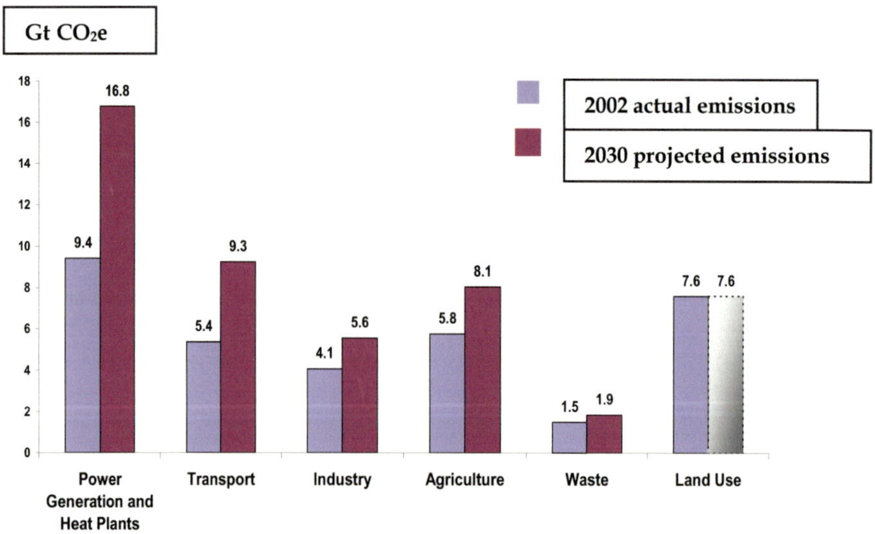

Source: Hadley Centre for Climate Prediction and Research Based on Folland et al (2000) and Jones and Moberg

Figure 15.4. Forecast growth in GHG emissions by Sector

Current emissions per capita are highest in the developed nations, the USA being highest at 20 tonnes of CO_2 per capita per year. However the larger fast-growing developing countries such as China and India account for much of the forecast growth in CO_2e emissions (see Figure 15.5). By 2025 China will be emitting GHGs at the same level as the USA. Thus the developed and developing nations must both be part of the solution and action must be taken urgently if we are to stabilize CO_2 emissions at 550 ppm or lower (see Figure 15.6). Stabilization at 550 ppm is projected to limit global temperature rise to 2^0C during the 21st Century. The Stern Review Report has estimated that this will require a 60% reduction in emissions from the energy sector by 2050 (see www.sternreview.org.uk).

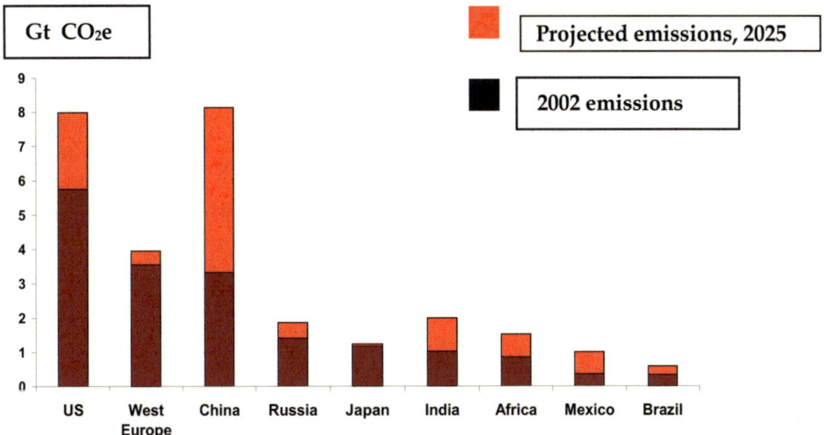

Source: World Resources Institute, CAIT Energy Information Administration Reference Scenario, Energy emissions only.

Figure 15.5. Forecast GHG emissions by major developing nations compared to US & Europe

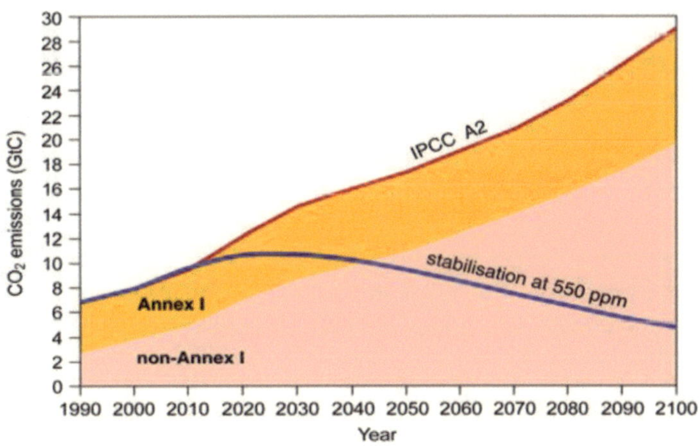

Source IPCC

Figure 15.6. Depiction of CO_2 emission reductions required to stabilize at 550ppm

15.2.2 Major Impacts on Power Systems

Some of the major impacts that CC will have on the power industry and systems include:

<u>Rising average and peak air, ground and water temperatures and variable river water flows</u>

- Impact on equipment/plant ratings and power system security
- Changes to seasonal demand patterns and peaks
- Impact on reserve margins and reliability of supply.

Extreme weather events (eg hurricanes)

- Increased risk to generation, delivery systems (Transmission and Disribution (T&D), telecommunications, and System Control Center reliability
- Emergency response and restoration needs and costs increased
- Need for improved extreme weather advance warning systems.

Forest Fires & Floods

- Increased risk to generation and delivery (T&D) infrastructure with impacts on reliability and costs.

Rising sea levels

- Risk to coastal generation and delivery systems (T&D) infrastructure and populations

There is a need to monitor and record these climate changes and impacts in order to establish sound databases on which to base the design and implementation of appropriate response and adaptation measures.

15.2.3 Major Global Programs

We will now take a look at some of the major programs and initiatives by the international community to mitigate and adapt to CC.

15.2.3.1 Kyoto protocol

The Kyoto Protocol developed by the UN Framework Convention on Climate Change (UNFCCC) was signed in December 1997 after two years of debate and negotiation about the inadequacies of the UNFCCC and its voluntary mechanisms and the need for more meaningful requirements. Much of the impetus for the Protocol came from the Intergovernmental Panel on Climate Change's (IPCC) Second Assessment Report which concluded that "the balance of evidence suggests a discernible human influence on global CC." The Kyoto Protocol commits developed countries which have signed the protocol to legally-binding emission reduction targets for six greenhouse gases - carbon dioxide, methane, nitrous oxide, hydro fluorocarbons, per fluorocarbons, and sulfur hexafluoride -- to be reached by the period 2008-2012. (CFCs are controlled under the Montreal Protocol.) These targets, which range by country from -8% to +10%, provide for a 5% emissions reduction from 1990 levels in aggregate.

With ratification of the protocol by Russia in the fall of 2004, the required level of "55% of developed country emissions" was reached and the protocol officially came into force on February 16, 2005. The Issue of the IPCC Fourth Assessment Report in 2007 strengthened the case for reducing GHG emissions.

15.2.3.2 Intergovernmental panel on climate change (IPCC)

IPCC was established by the World Meteorogical Organization (WMO) and United Nations Environmental Programme (UNEP) in 1988. It is open to all members of the UN and WMO. Its objective is:- "to assess scientific, technical and socio- economic information relevant for the understanding of CC, its potential impacts and options for adaptation and mitigation."

The IPCC produces Assessment Reports, Technical Papers; and Supporting Material. The Fourth assessment reports for Working Group I and Working Group II were issued in early 2007. They concluded that GHG forcing has *very likely* caused most of the observed global warming over the last 50 years. This strengthened the scientific evidence for anthropogenic global warming and the case for increasing adaptive capability to cope with the CC already occurring. The latter is particularly important for the poorest developing countries which will be hardest hit by CC and have the least capability to adapt.

15.2.3.3 Asia Pacific partnership on clean development and climate (APP)

The Asia-Pacific Partnership (APP) on Clean Development and Climate is an innovative new effort to accelerate the development and deployment of clean energy technologies. APP Partner Countries Australia, Canada, China, India, Japan, Republic of Korea, and the United States have agreed to work together and with private sector partners to meet goals for energy security, national air pollution reduction, and CC in ways that promote sustainable economic growth and poverty reduction.

APP was announced by President Bush on July 27, 2005. The initial six countries were Australia, China, India, Japan, Republic of Korea & USA which together are responsible for about 50% of world GDP and CO2 Emissions. Canada joined in 2007. The objective of APP is:- "To focus on practical measures to create new investment, build local capacity and remove barriers to the introduction of clean, more efficient technologies to improve national energy security, reduce pollution and address long term CC." The major power industry priorities are clean energy & high efficiency.

APP held their first meeting in January 2006 in Sydney, Australia. At this meeting a Work Plan was developed and eight Task Forces were setup with a focus on the power sector and energy intensive industries. This included:-

- Cleaner fossil energy
- Renewable energy technology and distributed generation
- Power generation and transmission efficiency (supply-side efficiency)
- Steel; Aluminum; Cement; and Coal mining
- Buildings and appliances (demand side efficiency).

The Task Forces will build on existing initiatives.
India hosted the second APP meeting in October 2007. Examples of APP successes include:

New Energy Efficiency labels used in China, similar to those in the U.S. ENERGY STAR program, are expected to encourage Chinese consumers to use more energy efficient appliances. This APP coordinated activity is projected to bring about an annual carbon emission reduction of 17.7 million tons of CO2, the equivalent of removing three million cars from the road for just one appliance, television set-top boxes.

- Solar Turbines, an APP private sector partner, has worked with Chinese partners to identify and setup units that provide 35 megawatts of clean energy technology to the coking industry in China. Initial projections indicate an annual savings of approximately 410,000 metric tons of CO_2 equivalent when all units are operational.

15.2.4 Other Programs and Initiatives

There are many other programs and initiatives at the regional, national, state/province and individual company/entity level. We consider the North American scene in the following and the UK Stern Review is noteworthy as it looks at the economics of CC both UK and global. Clinton's Large Cities Climate Leadership is also noteworthy - grass roots action in 22 cities.

15.2.4.1 Other programs and initiatives

Federal policies are driven by economy concerns, but the GHG lobby is pushing hard.

States are showing leadership in developing regulations and setting GHG reduction targets:

- NJ; MA; NY; NH; ME; CA have set reduction targets
- North-east US Initiative (RGGI and RGGR) (see Section 15.4. of this Chapter)
- Western Governors Alliance developing GHG policies
- The California Assembly passed the Global Warming Solutions Act (Assembly Bill 32) on August 30, 2006 and a companion bill for the electricity sector (Senate Bill 1368) which sets power plant emission performance standards
- Many states adopting Renewable Portfolio Standards (RPS) (see Section 15.4.2, and Energy Efficiency (EE) Programs).

There are several independent voluntary programs by Business, Individuals, and NGOs

15.2.4.2 Canada

The Conservative Government in Canada is developing a "Made in Canada" Plan. Canada has ratified the Kyoto Protocol but economic analysis shows that meeting Kyoto targets cannot be done without major impact on the economy (recession). Large industry emission reduction targets are expected with provision for "offsets". The focus is on technology solutions. For example The Early Actions Measures (TEAM) program has invested in leading edge projects. Also energy efficiency, renewable energy technologies, clean coal with carbon capture and storage, nuclear and hydrogen are priorities. Through Kyoto, Canadian entities have access to the Kyoto mechanisms of CDM & JI (see Section 15.2.6 for details).

15.2.4.3 Stern review report main conclusions

Doing nothing is not an option; action must be global, prompt and strong and we must mitigate and adapt. As already mentioned, the target **for the energy sector** is a 60% reduction in CO_2 emissions by 2050 to stabilize at 550 ppm (see www.sternreview.org.uk). The global economic impact is manageable "we can grow and be green". An important priority is to increase the adaptive capability of the poorest developing countries that will be hit earliest and hardest by CC and are least able to cope.

15.2.5 Renewable Energy

Renewable energy projects, particularly wind, small hydro and solar, offer compelling environmental advantages when compared to conventional fossil fuel-based power generation, including little or no conventional pollutant and GHG emissions. Renewable energy projects face serious challenges competing with conventional fossil fuel-fired power projects. They have achieved only limited success in the marketplace.

One of the most significant challenges facing renewable energy projects is the subsidy given by many governments to conventional forms of energy. Another challenge facing renewable energy development is the remote, decentralized nature of many renewable energy projects.

The wind industry now has a global installed capacity of over 50,000 MW and is growing at 35 to 40% per year. In 2006, for the first time, more new wind capacity was brought on line than nuclear power. The solar photovoltaics industry, which is now a $1 billion industry, is growing at 30% per year. The potential of renewables has not escaped the big conventional energy companies, including BP Amoco, ABB, GE, Enron and others, all of which have made considerable investments in the renewable sector. For example BP's alternative energy investments are valued at up to $7 billion. GE is investing heavily in its Ecomagination program launched in 2004. This is GE's commitment to imagine and build innovative solutions that solve today's environmental challenges such as climate change and benefit customers and society at large. The target investment in renewable energy is $6 billion by 2010. (See: http://ge.ecomagination.com/site/index.html)

15.2.6 Emissions Trading

An effective tool or mechanism to achieve cost effective GHG reduction targets is the concept of emissions trading or transfers among participants. Essentially this involves treating GHG emission allowances and reduction/removal credit units like any other commodity in the marketplace. Arrangements are made for them to be traded on national and international exchanges. The marketplace sets the value of GHG emission credit units. These are bought and sold by countries and companies to facilitate meeting their GHG targets at lowest cost. For this to work, just like any other commodity, there must be internationally accepted standards or a "common currency" for the measurement, monitoring, reporting, verification and certification of emission credit units [1]. The effectiveness of emissions trading schemes has been proven by the success of trading in acid rain gases (SOx and NOx) in curbing acid rain in North America. GHG trading schemes in the UK and Europe are already showing successful results for reducing CO_2 emissions (see: http://www.defra.gov.uk/environment/climatechange/trading/eu/operators/compliance.htm. and: http://ec.europa.eu/environment/climat/emission.htm

15.2.6.1 Emerging GHG markets

GHG markets can currently be split into two categories:

- The Kyoto compliant market
- The non-Kyoto compliant market.

The bulk of the current global activity in GHG trading is centered on the Kyoto compliant market. Developed countries, which have ratified the Kyoto Protocol and accepted their GHG emission reduction target, termed Annex 1 countries, may meet their commitments through domestic CC policy activity and the use of the Kyoto mechanisms. These "flexibility" mechanisms are Joint Implementation (JI); Clean Development Mechanism (CDM) and International Emissions Trading (IET).

Both JI and CDM are "project based mechanisms" and involve carrying out CC mitigation projects for the reduction or removal of GHG emissions. JI projects allow Annex I Parties to implement projects that reduce GHG emissions by sources, or enhance removals by "sinks", in the territories of other Annex I Parties, and to credit the resulting emission reduction units (ERU) against their own emission targets. CDM projects allow Annex I Parties to implement projects that reduce or remove GHG emissions in developing countries. Annex I Parties may use certified emission reductions (CER) generated by CDM projects in developing countries to contribute to compliance with their GHG emission commitments. The rules governing the CDM are available at: http://cdm.unfccc.int/ and those for JI projects are expected to be similar - see http://ji.unfccc.int/index.html. IET permits an Annex I Party to transfer (sell) part of its assigned GHG emission allowance (the amount of emissions the Party may emit during the commitment period) to another Annex I Party. It also permits trading of CERs and ERUs - see following web-site for background and rules: http://unfccc.int/kyoto_protocol/mechanisms/emissions_trading/items/2731.php.

Canada's Clean Development Mechanism and Joint Implementation (CDM & JI) Office was established within the Climate Change and Energy Division of the Department of Foreign Affairs and International Trade (DFAIT)) in 1998. The Office is the federal government's focal point for CDM and JI activities. It was created to enhance Canada's capacity to take advantage of the opportunities offered by the CDM and JI. Opportunities for Canadian industry can include: (i) generation of emission reduction credits; (ii) access to new markets and investment opportunities; (iii) an opportunity to demonstrate the viability of a voluntary approach; (iv) a showcase for environmental leadership. The services provided are aimed at reducing transaction costs for Canadian companies given the elaborate steps and procedures for these mechanisms.

The main non-Kyoto compliant markets are the UK Emission Trading Scheme (UK-ETS), the European Union-Emission Trading Scheme (EU-ETS), the Chicago Climate Exchange, and the New South Wales Trading System. The UK-ETS was launched in 2002 and was the world's first national economy wide GHG trading scheme. It is essentially a cap and trade scheme open to all entities in the UK, including 6,000 companies that already had CC Agreements. Full details of the scheme and results to date can be found on the web-site of the UK Department of Environment, Food and Rural Affairs (DEFRA) at:
http://www.defra.gov.uk/. See the following web-site for a full report on 2006 results:-
http://www.defra.gov.uk/environment/climatechange/trading/eu/operators/compliance.htm

In 2003, the New South Wales (NSW) Government in Australia introduced an emissions trading scheme building on an existing emissions benchmarking program in connection with electricity retailer licensing conditions. The benchmark system requires electricity retailers to reduce annual emissions from 8.65 to 7.27 tonnes $C0_2e$ per capita. All six GHGs expressed as units of one tonne of CO_2 are covered. They can achieve their targets by offsetting their liability with credits created from renewable energy and low emission generation, tree planting and energy efficiency. The system operates with a financial penalty of up to, but not higher than, AUS$15 (about US$8.5) per tonne of excess tonne CO_2e emitted.

The EU-ETS was launched in January 2005 and trades in EU Allowances (EUA) are already taking place. In this scheme each regulated entity in the scheme is assigned an "allowance" or amount of GHG it is permitted to emit. Entities may buy surplus allowances from other entities to meet their CO_2 commitments. The EU scheme may also be linked with the Kyoto CDM and JI project mechanisms. Details of the EU-ETS may be found at: http://ec.europa.eu/environment/climat/emission.htm. This includes reports on results to date and plans for the future of the scheme.

Although the former Presidential Administration in the U.S. did not seek ratification of the Kyoto Protocol, American companies are pursuing voluntary programs to reduce greenhouse gas emissions. Many are turning to emissions trading as a means of making reductions in their overall greenhouse gas emissions profile. Tradable units are Verified Emission Reductions (VERs) and have been trading since 1999. California and other West Coast states as well as Northeastern states are now entering the carbon constrained world through government mandates. Nine Midwestern states are also moving in this direction. In two years, it is highly likely the US Federal Government will mandate economy wide greenhouse gas emissions reductions that will focus on reducing the US carbon footprint of over 6 billion tons

Typical prices in voluntary GHG markets range from $1 to $10 per tCO2e and the EU market has ranged as high as $30 per tCO2e. Latest information on GHG market prices can be obtained by registering at the web-site of the Evolution Markets LLC: http://www.evomarkets.com

15.2.7 Mitigate and/or Adapt

While programs to reduce/remove GHGs will help mitigate the extent of change in global climate, there is still a need to adapt to the changes that have already occurred and may occur in the future. Thus adaptation programs are equally important to mitigation programs and there are many national and international initiatives for the assessment of CC variability and impacts and associated adaptation measures. An internet search for the term "adapting to CC" gives over 20,000 hits which is a measure of the global, extensive interest in this topic.

The Government of Canada Conference on Adapting to CC held in Montreal in May 2005 covered the following key topics which is indicative of the global scope of CC impacts: Coastal Zones; Forestry and Forest Ecosystems; Infrastructure; Communities; Industry; Engineering; The Arctic; Health and Vulnerable Populations; Tourism; Regional Water Impacts: Physical and Social Health Impacts; Agriculture; Water Resources Management; Fish and Aquatic Resources. There were also general sessions on Risk Management; Hazards and Extremes; Research Programs and Tools; Adaptive Capacity; Economics; Education and Awareness; and Taking Action on Adaptation.

The financial and insurance industries are particularly interested in the risks and impacts associated with CC. Reference [6] provides an overview of risks to the financial sector and stresses the need for international collaboration and research. Reference [7] provides the perspective of the insurance industry.

The IPCC Fourth Assessment Report Working Group II Report "Impacts, Adaptation and Vulnerability" has Chapter 18 discussing the inter-relationships between mitigation and adaptation measures and the trade-offs between the two. See:- http://www.ipcc.ch/ipccreports/ar4-wg2.htm

Striking the balance between mitigation and adaptation investments is an exercise in risk management. Focusing on technology measures for adapting to CC that has and may continue to occur is strategically important in managing those risks. Because of the complexities and considerable uncertainties in CC science and predictions, investment in adaptation measures to manage climate risks may prove to be of better value and have more certain, tangible benefits than CC mitigation (GHG reduction) measures. This is particularly important for the poorest developing countries which are least able to adapt and would be hardest hit. The risks of not developing the economies of these countries (that requires energy development as a critical driver) is far greater than the risks of CC. The human race has shown a great ability and propensity to adapt to climate circumstances beyond its control.

Figure 15.7 illustrates a classic cost/risk minimization approach to mitigation and adaptation. The mitigation curve is characterized by rapidly increasing costs and risks to the global economy the lower the target for CO_2e concentrations in the atmosphere. The adaptation curve is characterized by rapidly increasing costs and risks to the climate and the global economy the higher CO_2e concentrations are permitted to go. The sum of the two curves gives a range of CO_2e concentrations for minimizing cost and risk. This is estimated by some researchers to be in the range of 450 to 550 ppm of CO_2e.

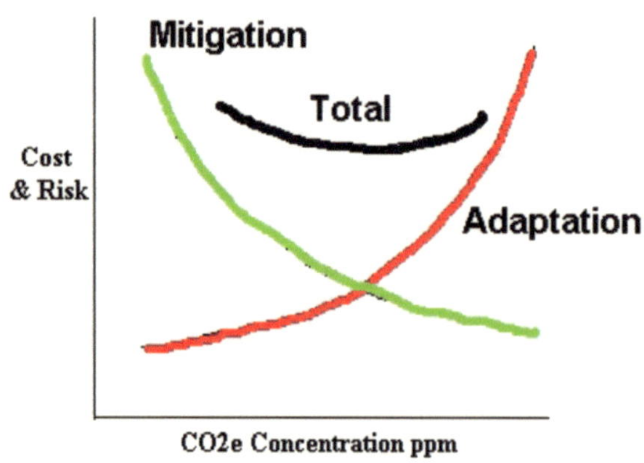

Figure 15.7. Cost/Risk Minimization Curves

15.2.7.1 Mitigation priorities for power industry

- No silver bullet: - Silver buckshot!!

The scale of the problem is so large that there is no single solution to reducing global GHG emissions. We will need all the options to achieve success, including:

- o Energy Efficiency and Conservation (End Use and Supply Side)
- o Low emission energy technologies (Renewable energy such as wind, solar, hydro, geothermal etc)
- o Clean Coal (Includes Carbon Capture & Storage -CCS)
- o Reducing dependence on fossil fuels
- o Development of LNG & Biofuels
- o Advanced Nuclear new build
- o Development of the Hydrogen economy.

15.2.7.2 Adaptation priorities for power industry

- Adaptation is essential to deal with CC that has already occurred
- Adaptive capacities need to be increased to deal with CC impacts, particularly in poor countries that will be hardest hit and least able to cope
- Power Sector Adaptation Measures: Examples
  - "Hardening" grid systems against extreme events
  - Coping with changed load patterns & plant ratings
  - Strengthening advance warning, emergency response & restoration plans
  - Improving back-up telecommunications and grid control
  - Extending climate monitoring and recording.

15.2.8 Section Conclusions

The global response to CC is diverse and major and covers both mitigation and adaptation technologies. Much more needs to be done and business and governments must work together on cost effective solutions to minimize risk. Major thrusts must be on clean, hi-efficiency technology for mitigation, and increasing adaptive capacity, particularly in the poorest countries that will be hit earliest and hardest by CC and are least able to cope. There may be funding challenges as will ensuring the skilled resources are available to implement the needed measures.

Climate science is hugely complex and still fraught with uncertainties and it is prudent to adopt a "no regrets" strategy at this time that makes good sense and minimizes costs and risks whatever the outcome on actual global climate change. We need a risk management approach that balances the costs and economic risks of overly severe CO_2 emission reduction targets against the costs and benefits of increasing our adaptive capability to cope with climate change This is particularly so in the developing countries which would be hardest hit by overly restrictive targets affecting their economic development and currently have the least adaptive capability.

15.3 Value of Non-Carbon Power and Emissions Avoidance

Estimates for the range of values to be ascribed to the avoidance and reduction of emissions using non-carbon or low emitting sources is now evaluated. This analysis utilizes published data to establish the values of the business and investor return, emissions avoidance, energy reduction, efficiency improvement, conservation and alternate technology deployment. It shows that there is no one unique, globally traded and valid value. The range of values ascribed to avoidance is coupled to the economic value of energy use. The range of costs of emissions reduction is highly dependent on the socio-politico-economic assumptions. Numerical results for both present and future energy scenarios are provided, explicitly including hydrogen and other non-carbon power sources in defining the economic value of a sustainable non-carbon future.

That carbon and emissions avoidance has value has been already understood and analyzed by the oil and gas industry, and carbon pricing has been assumed and undertaken in business planning [8]. In the UK, there is an ongoing formal review [9] that states: "*The economic challenges are complex. At its most basic level, CC is an externality: the emission of greenhouse gases damages others. But these costs will be felt over a long period and over the entire globe; their exact nature is uncertain; they interact with other market failures and imperfections; and those most affected – future generations – are not able to speak up for their interests. This points to a long-term international collaborative response. Effective collaboration will require a shared understanding of the incentives and institutions needed, and careful attention to the complex ethical issues involved.*"

In the UK there are future generational and moral issues to consider, with their own special emotive power and value. CC has already impacted commercial and industrial strategy. One leading oil and gas company has taken a position summarized as: "*We have worked for most of the last decade on the basis that one day carbon will be priced and that the application of technology which can reduce carbon will have a commercial value.*"[8].

To proceed with a transparent economic analysis, the existence and definition of two contributory values may be postulated and considered: an objective monetary value based on a market or trading of rights to emit GHGs and the associated emission avoidance costs; and a subjective social value based on the estimates of the probabilities of mitigation, of planet-wide changes to human lifestyle, and of species change, and their relevant costs. The true comparative "*value*" is therefore a composite estimate, including both tangible and intangible costs and risks, and depends on evaluation of the components contributing to these two types of values.

15.3.1 Nuclear Energy Example

To look at any alternate energy sources, it is necessary to define ones own costs and emissions, based on prevailing market and economic conditions. The potential impacts of GHG reduction and avoidance, and the opportunities and benefits from fuel switching that would be needed to stabilize the atmospheric GHGs to preserve economic growth and social progress, should be defined.

Illustrative estimates of the "*value*" to be assigned to avoidance and reduction using nuclear energy from the present zero value assigned to nuclear energy to the actual economic and social values derived from emissions avoidance that would still supply a sustainable energy future should be determined. These can then be directly compared to values derived from carbon credit trading, energy portfolio standards, and carbon sequestration, including the direct and indirect costs, risks and uncertainties.

15.3.2 Valuing Emissions Reduction

To value reduction and energy source substitution, it is necessary to value usage and emissions increase, which in present society are an acquired historical right. Then, the several different approaches to establishing a benchmark value for emissions avoidance by comparing it to the value of the original emissions themselves can be evaluated.

15.3.2.1 Economic value to a nation and the world

The value of carbon energy to the world is in providing economic growth. The purely *economic value* of the carbon emissions and power source is reflected in producing financial wealth for the country (such as the national GDP) using carbon energy. Energy is greatest in developed (rich) nations and a correlation between the growth in GDP to the growth in carbon energy use can be observed. This relationship also holds true at the global level. Hence, the global growth in GHG concentration in the atmosphere over the last 30 years (measured as ppmCO_2 at Mauna Loa, Hawaii where 1 ppmCO_2 ~ 9.10^{12} tCO_2) is directly and linearly correlated to the Gross World Product (GWP) (measured in teradollars, $\$10^{12}$ US).
GWP data (source:http://www.earth-policy.org/Indicators/Econ/Econ_data.htm) is compared with CO_2 concentrations from Mauna Loa in Figure 15.8. To reduce the effect of the year-to-year noise in the CO_2 concentrations, five-year averages for GWP are plotted against the change in CO_2 measured over those five years. Rather than plot ppm values of CO_2, the change is converted to Gt of CO_2 released based on 7.9 Gt of CO_2 required to cause a 1 ppm increase in the atmosphere accompanied by an equal release being absorbed in the oceans. 1 ppm was taken to be equivalent to a total of 15.8 Gt of CO_2 released. A linear fit of the data was calculated as:

$$CO_2\ (Gt) = 0.433\ GWP(t\$) + 8.70$$

The data can be interpreted as flattening over time, indicating diminishing energy intensity in the creation of value, but the average global economic value between 1950 and 2004 is 430 \$(US 2004)/t CO_2 (it is reasonable to use 1950 as the base year since the CO_2 build-up prior to about 1950 was relatively small).

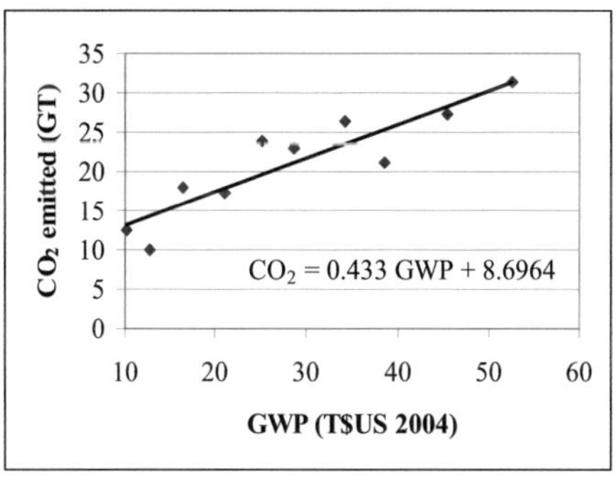

Figure 15.8. The global correlation.

15.3.2.2 Economic value to investors

In addition, the economics also involve the value to shareholders and investors in oil and gas companies: they have implicitly assigned a value by owning the company and taking a dividend on the profits.

To set the market value, it is noted that oil and gas already has an assigned market value, and hence so has the carbon content used for energy production, since 1 bbl oil contains ~115kgC (= 495 $kgCO_2$).

To set the order of magnitude, the value to stockholders and owners as profit from corporate sales is taken. At the 2006 BP Annual General Meeting (www.bp.com), and in the Financial and Operating Information for 2001-2005, it was reported that $19B was distributed to investors in 2005-2006 with a replacement cost /bbl in 2005 of ~$48/bbloe.
The profit per $/kgC = 41.8¢/kgC = 418$/tC translates to a present carbon emissions value to investors of 114 $/$tCO_2$, assuming no carbon is sequestered and all is used in combustion, oxidation and/or transportation.

The future potential or prospective distribution to shareholders is given as ~$65B over the three years 2006-2008. With a refining margin ~$850/bbl, the 2005 production was ~2.5Mbbl/d at a cost of ~$50/bbl ~$45B/a.

Returning about $65B over 3 years ~$22B/a, so the projected future profit/bbloe ~$22B/0.91Bbbl = $24/bbloe. Hence the investors' future Carbon value ~$48/$tCO_2$.

To attract investment or to be economically competitive without subsidy, any non-carbon alternate or carbon reduction scheme must have at least this substitute market investment value.

15.3.2.3 Assumed value of the right to emit

In a carbon-constrained system, the right to emit is governed by voluntary and/or regulated limits on total emissions. Thus emitting carbon can have a price or cost. The Kyoto Treaty targets are approximately a 5% global percentage reduction from prior years (1990 was taken as the baseline). To meet or encourage meeting this modest target, some nations invoked an *"emissions trading schemes"* either individually or collectively.

Many economic studies have attempted to determine or set limits on the assumed value, and establish the impact on the national, regional or local economy (e.g., Regional Greenhouse Gas Initiative (RGGI) report). Funds that are spent on carbon costs that raise energy prices cannot be spent on consumer goods. If promoting such a scheme a low value is assumed (typically $5-10/$tCO_2$), but the results are clearly sensitive to the assumed cost. In Canada, the impact on future national "scenarios" were examined under certain key assumptions.

These included the assumptions of a +2%/a base GDP growth, but also assumed a $10/$tCO_2$ cap guarantee with international permits from other countries who were below their agreed targets (e.g., Russia). The negative impact was about 3.5% over some 30 years, or ~0.1%/a lost economic growth. This has an estimated value of the fraction of the GDP, ~0.01x$1T/a ~$10B/a in GDP reduction.
Assuming a needed 200 $MtCO_2$ reduction to meet the target, this implies an allowed economic value ~$10B/200Mt = $50/t.

15.3.2.4 Actual trading value

A value can therefore be assigned from what emitters will actually pay to preserve or obtain the rights or credit of releasing GHGs. This value can be determined from a defined and hopefully market-driven *"emissions trading"* scheme, where the right to emit is established via some limit placed on the total allowed amount (a so-called cap-and-trade system). Within the pre-determined GHG emissions amount, which is distributed between emitters and energy market sectors, credits can be traded and exchanged for a price determined by credit supply and emissions demand.

Currently, it is estimated [10] that about 100 million tons of carbon credits are transacted in various markets worldwide. The World Bank report [11] stated: *"There are four active markets for GHG allowances as of May 2005; the EU-ETS, the UK Emissions Trading System, the New South Wales trading system and the Chicago Climate Exchange"*. Volumes exchanged on these allowance markets have increased dramatically compared with 2005, and is now comparable to the volumes exchanged through project-based transactions. Cumulative volume exchanged on these four markets from January 2004 to March 2005 is about 56MtCO_2e.

Of the four allowance markets, the EU-ETS is the largest, with an estimated 39MtCO_2e exchanged since January 2004, the bulk transacted since January 2005.

Unlike project-based assets, allowances are homogeneous assets, and purchase contracts for allowances are fairly homogenous. As a result, the spread of prices for one tonne of CO_2 of emissions (an EUA) at any given point in time is small.

The dominant trading is clearly in the EU, where an emissions trading schemed has been deployed which allowed trading of emissions credits (i.e., emissions rights) on the open market, within some overall limit or cap on the EU total. Presently, some 25 countries with some 6,000 participating companies constitute a trading volume of 2.1 billion allocated tons CO_2 per year.

In this European Trading Scheme (ETS) predictions have also been made of the effect of demand on the trading price [12].

The estimates ranged from \$20 to \$100\$/tCO_2 depending on actual US demand, which is presently zero. A useful conversion factor to bear in mind, since economic studies use different currencies, is that for 2006 currency conversion rates, \$100/tC = 20€/tCO_2. For 2005-2006 the ETS trading value range was between 10 and 30€/tCO_2 [13], and fluctuated widely.

This estimate is as close to an actual market value that is available. It is artificial as it refers solely in the EU, is not a global value, and is dependent on meeting arbitrary EU Kyoto targets.

15.3.2.5 Negative value of negawatts: conservation and efficiency relative socio-economic values

The *conservation cost* is obtained by adopting or encouraging restrictions in the energy demand (so-called demand-side management) and use, plus impact of efficiency and conservation measures versus adding energy sources.

There are more subtle social values also that can be determined from the so-called external impacts or from reduced use of carbon energy. The most popular are called conservation and efficiency improvements, and are presumed to value energy-use reduction, and hence emissions avoidance. Reduced energy usage is good if energy efficiency is also improved and there is also a net relative benefit. Reductions in energy use have been given the term "Negawatts" [14] to reflect the reduction attained.

There are two ways to improve economic efficiency: (i) in the production of energy, and (ii) in its use. By using a standard discounted cash flow model, as used for actual power plants and systems, the costs of saving electricity to their assumed new power plant generating costs, using consistent discount rates can be compared.

A test case (scientific data) for the claims of efficiency gains leading to energy and emissions reduction is taken from actual USA data. After extensive effort, the results of improvements in energy technology and efficiency are clear. The US Department of Energy (DOE) has had a large and important program of work on efficiency for many years. This shows the perverse market effect that as (carbon) energy is made cheaper, more is used, leading to actual increases in energy use and in emissions.

Consider the actual and projected energy intensities, energy use and emissions in the USA for 1990-2020. The data and projections are shown in *"Energy Outlook 2001"* [10].

The numbers and figures clearly show that energy use and emissions rise as energy technology improves and the price falls (similar trends appear in prior years), both in the past and into the future.

Improved efficiency (technology) was responsible for about 60% of the observed decline in energy intensity, is now declining and is more expensive to introduce. As a result of the continued improvements in the efficiency of end-use and electricity generation technologies, total energy intensity in the reference case is projected to decline at an average annual rate of 1.6 percent between 1999 and 2020.

The projected decline in energy intensity (1.6%) is considerably less than that experienced during the 1970s and early 1980s, when energy intensity declined, on average, by 2.3% per year. Approximately 40 percent of that decline can be attributed to structural shifts in the economy—shifts to service industries and other less energy-intensive industries; however, the rest resulted from the use of more energy-efficient equipment.

Although more advanced technologies may reduce energy consumption, in general they are more expensive when initially introduced. In order to penetrate into the market, advanced technologies must be purchased by the consumers; however, many potential purchasers may not be willing to buy more expensive equipment that has a long period for recovering the additional cost through energy savings, and many may value other attributes over energy efficiency. In order to encourage more rapid penetration of advanced technologies, to reduce energy consumption or carbon dioxide emissions, it is likely that either market policies, such as higher energy prices, or non-market policies, such as new standards, may be required.

Basically, the needed proven and projected efficiency improvements are more expensive, and cannot keep pace with increased carbon-based energy demand, so need policy incentives (tax and cost breaks) to be adopted. Therefore, only by adopting non-carbon energy sources can the trend of increased CO_2 emissions be changed, and therefore, a mix of non-carbon sources is needed, including nuclear, as is also assumed by the United Nations Intergovernmental Panel on Climate Change (IPCC).

In fact globally the situation is perversely made worse: the decreasing demand in one country attained by precious conservation measures causes some reduction in what would otherwise have been the cost of global energy production favoring increased demand by others, as these other economies grow. Thus, the developing economies of, say, India and China will use all the energy that others make available to the market place by conservation and efficiency measures. The most that can be claimed in world markets is a decrease in the rate of carbon energy growth, but not an actual decrease in the amount of carbon energy used. This is confirmed by the data and all authoritative projections.

15.3.2.6 The alternative or substitution value

This value can be estimated based on alternate energy technology options that reduce emissions but with added development, deployment and market costs that vary from technology to technology, and from sector to sector. In principle, it is possible to consider the value of emissions reduction versus emissions avoidance approaches (e.g., switching to hydrogen as an energy carrier).

It is not so simple to apply a value which is a composite based on relative health, emissions, land use, fuel supply, social and political aspects to arrive at relative rankings for differing substitute energy sources, emissions reduction technologies and GHG sinks in portfolio of options.

Consider the simplest case of power generation. Different sources and means produce differing amounts of emissions over their full *"life cycle"*, meaning from mining the raw materials, the construction and the operation, and finally the disposal and decommissioning. For any given source of power, there is a GHG emissions amount per kWh.

To evaluate the relative emissions value of any two options, a calculation can be made as follows:

Differential Value of Avoidance, \$ = $[\Delta gCO_2/kWh] \times [\Delta kWh] \times [\Delta\$/gCO_2]$

where,

$\Delta gCO_2/kWh$ is the difference between the emissions for any two sources
ΔkWh is the difference in the amount of power generated
$\Delta\$/gCO_2$ is the difference in the generating cost for any two sources.

Typical relative values are shown in Figure 15.9 for a variety of modern electric power units and a variety of studies, to illustrate the order of magnitudes.

For any given carbon value, for any given generation source, it is even more straightforward. For generation of 5TWh each year (by 600MW.a) avoiding approximately 3Mt/a @20\$/t, then the avoided emissions value may be assumed to be roughly \$60M/a.

These emissions differences may be translated into generating costs impacts, that is the price actually paid by a consumer (cf. gasoline). Avoiding 5Mt/yCO_2 @$30/t = 150M$/y. With a 1000MW(e) plant, approximately 7.8TWh/y will be generated, so the added cost of emissions, or conversely the benefit of avoidance is 1.9c/kWh, which is about a 30% increase in generating cost.

15.3.2.7 Avoidance, capture and sequestration value

The alternative is to eliminate, avoid or capture the emissions. Recently focus has been on establishing so-called Carbon Capture and Storage (CCS) as a viable option, which is essentially the immobilization of CO_2 in either: (a) a gas in natural or man-made geologic structures such as existing mines, deep saline aquifers, oil and gas wells, and salt domes; or (b) other stable chemical or physical forms. Also, pressurized re-injection into oil wells to recover additional oil (called Enhanced Oil Recovery (EOR)) is feasible at such sites, and CO_2 can also be collected elsewhere and piped to the injection location.

Electric Energy Technology	Switzerland PSI GaBe 2000 www.psi.ch	Canada Andseta & Gagnon HQ, 2000	IAEA Spadero et al. 2000 www.iaea.org	France (production only) Gouvernement de France, 2000
Natural Gas	605		696	500
Coal	1071	974	978	
Solar Panels	114 - 189		97	
Nuclear	16	3 – 15	21	0
Oil	855	778	811	701
Wind	36		36	
Hydro	4	15	16 – 23	

Figure 15.9. The relative life cycle emissions from differing sources.

Capture	Cost (£/teCO_2)**
Coal PF Retrofit	19
GTCC Retrofit	14
New IGCC	13-34***
New GTCC	21
Pipeline transport for EOR	7-8
Pipeline transport for storage in depleted gas fields	4-6
Injection for EOR	7
Injection for gas field storage	1

Figure 15.10. The comparative value of CCS. (Source: DTI, 2003.)

Since the amounts (volume and mass) of carbon are potentially very large, it is preferable to site CCS facilities near larger sources. The recent UK report [16] has costed many concepts, and derives a CCS cost range of some 10-30$/t$CO_2$. Perhaps unsurprisingly, this cost range is consistent with the trading value, implying that these are perhaps the two main competing options (i.e., CCS or buy emissions credits). The comparative value of CCS taken from the Department of Trade and Industry (DTI) report is indicated in Figure 15.10.

This range does perhaps underestimate the real cost since the figures do not usually include collateral CO_2 emission associated with the CCS operation. The use of combined EOR, CCS and gas recovery is presently being examined at full scale, combined with hydrogen production and power generation (see www.bp.com).

15.3.2.8 Value of alternate technologies

With continually rising emissions, there is a so-called "technology gap" to the desired goal of some reduced level. It would be of value if alternate technologies were some "magic bullet" that removed emissions, but a diversified portfolio of options is often recommended [17]. The costs to develop and deploy can be subsidized in the short term. But in a competitive marketplace, like the energy sector, the chance of success or market share for a new technology or product is heavily dependent on relative or comparative cost.

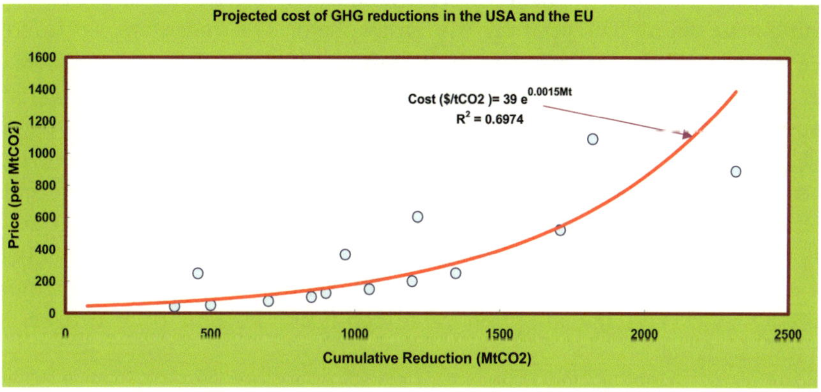

Figure 15.11. The Value of Technology

Recently, analyses of the emissions reduction potential of alternate technology pathways and scenarios have been published by the OECD's International Energy Agency at the specific request of the G8 countries [18]. This was to address the socio-political issues of environment, energy security, air pollution and poverty to determine a "clean, clever and competitive energy future". The study concluded that deployment of technologies that have an additional cost of *less* than a cost of \$25/t CO_2 could halve oil and electricity demand and stabilize emissions by 2050. Unfortunately sensitivity to the value was not studied, but it is clear that this value would exclude many of the technology options in Figure 15.11 and will not really impact transport emissions as it represents only some 1-2% of current fuelling costs.

15.3.2.9 Policy value: energy insecurity and carbon taxes

Government and national policy makers like to retain control over their own destiny and country. Since many of the major sources of carbon energy are focused in regions of relative geo-political instability, there is a value to be placed on having energy security and diversity of supply. The use of *"policy measures"* (a euphemism for taxes) is usual for governments, to raise revenue and/or provide fiscal incentives.

Thus, the recent Province of Quebec's *"Plan d'Action"* [19] is based on monetary incentives for GHG emissions avoidance.

For a cost of some \$200M in taxes plus \$328M in other measures, with a program total \$1.2b, the goal of the Plan is to avoid ~10Mt/aCO_2 in six years. The specific value assigned to carbon emissions avoidance is not stated, but can be estimated from the proposed program costs given above as within a range:

High ~\$1.2B/(10Mt/a x 6) = \$20/tCO_2
Low ~\$200M/(4.8Mt/a x 6) = \$7/tCO_2

If the cost or value is too high, the democratic election process usually solves this issue.

15.3.2.10 Global value of sustainable avoidance

As a final estimate of the value of emissions avoidance, some global limit or "target' for allowable emissions should be assumed. This is taken as a doubling of pre-industrial CO_2 concentrations to about 550 ppm in the atmosphere. The reduction achieved by any avoidance or technology means can be translated into an atmospheric concentration reduction. As a working example, the impact for a range of emissions reduction assumptions based on the UN's IPCC scenarios [20] for future energy use [21] has been evaluated, This was done using the MAGICC/SCENGEN [22] global model as an emissions scenario sensitivity tool. Any emissions avoidance could be assumed, but specifically we adopted the range covering high- and low-energy use by (the IPCC, the A1F1 and B2 base scenarios) [21]. These scenarios were modified by inclusion of significant added penetration of sources with low carbon dioxide emissions (including nuclear energy) for new power generation by 2030; and the adoption of a significant fraction of hydrogen in global transportation by 2040.

The results [21] show an emissions avoidance/reduction potential of 200 to 300 ppm CO_2 by 2100, using such a penetration of non-carbon power. This scale of emissions avoidance essentially allows for unconstrained economic growth, which is good for the developing nations pursuing this course course of action.

15.3.3 Results

Using existing data, estimates for the range of values to be ascribed to carbon emissions were evaluated and provided. This analysis utilizes published data to establish the values of the business return, emissions avoidance, energy reduction, efficiency improvement, conservation and alternate technology deployment. As a result, it is shown that there is no one unique, globally traded and valid value. The value ascribed to avoidance is coupled to the economic value of energy use; and hence the range of costs of emissions reduction is highly dependent on socio-politico-economic assumptions.

The use of alternate non-carbon energy is relatively of high value in typical schemes, including impact of conservation and efficiency measures. The results show a definite trend that confirms the considerable advantage of adding new-build advanced nuclear energy plants as potentially the lowest cost emissions reduction option with the highest value.

15.4 Impact of Regional Greenhouse Gas Initiative and Renewable Portfolio Standards on Power System Planning

Two developments in the Northeastern United States are having an impact on power system planning in that region. One is a cap on CO_2 emissions recently adopted by seven states. This is the result of a voluntary Regional Greenhouse Gas Initiative (RGGI) developed by nine states over the last two years. The second development is Renewable Portfolio Standards (RPS) that have been implemented in most states in the Northeastern US.

15.4.1 RGGI

The initial RGGI agreement[#] involved seven states (Maine, New Hampshire, Vermont, Connecticut, New York, New Jersey and Delaware) that signed a Memorandum of Understanding (MOU) in December 2005 to implement a cap and trading program for CO_2 emissions from power plants greater than 25 MW in those states. Massachusetts and Rhode Island joined in February 2007 and Maryland joined in April 2007. Pennsylvania, the Eastern Canadian Provinces, and New Brunswick are observers in the process. While participation in RGGI was voluntary, the MOU makes the cap mandatory.

The MOU establishes a CO_2 cap of 126.1 million tons for the initial seven states that would be implemented starting in 2009 and remaining at this level until 2014. In 2015, a gradual reduction in the cap would start and reach a 10% lower level by 2019. The cap would be implemented with a Model Rule as a framework for states to implement state regulations governing the details of the state cap and trading rules, compliance etc. The overall program would be administered through a Regional Organization, but would not have regulatory authority.

The CO_2 cap would be apportioned among the seven states and the states would apportion their caps to the individual generators in their state granting one CO_2 allowance for each ton of emissions. The trading of CO_2 allowances would be allowed across the seven states. To provide consumer benefits from this program the states would withhold 25% of the allowances from the generators. These could be sold and the funds used to support energy efficiency, renewable resources, carbon capture, or customer rebates.

A compliance flexibility feature of the RGGI program will be the ability of an affected generator to use offsets for up to 3.3% of its compliance obligation. Offsets are reductions in CO_2 or other greenhouse gases made outside of the electric sector that have been approved and certified by a regulatory process as to their legitimacy. These offsets can be created from a number of possible designated greenhouse gas reductions in the RGGI states on a one for one basis, or created in the U.S. outside of RGGI on a two for one basis. An additional flexibility aspect of the RGGI program is that it has two price triggers when CO_2 allowances reach price thresholds of $7/ton and $10/ton. With allowances at these price levels, more compliance flexibility is allowed in the use of offsets with an increase in the percentage use for compliance and a broader geographical area from which the offsets can be created and bought.

[#] www.rggi.org/agreement.htm

Massachusetts (MA) and Rhode Island (RI) also participated in the development of the RGGI program but did not sign the initial MOU. MA implemented its own CO_2 cap in 2006 affecting six fossil generating plants in that state. The MA cap is based on historical emissions (tons), and on a maximum emissions rate of 1800lb/MWH. It also established price caps so it has similarities to the RGGI program.

15.4.2 Renewable Portfolio Standards

RPS have been implemented by state legislation and regulation to encourage development of renewable resources. The RPS are percentage targets of the energy supplied that the load serving companies are required to meet on an annual basis. The percentage target generally increases each year and can be met with a range of renewable technologies. These typically include solar photovoltaic, wind, biomass, energy from wastes, and in some states fuel cells. The Northeast states with RPS include Maine, Vermont, Massachusetts, Rhode Island, Connecticut, New York, New Jersey, Pennsylvania and Maryland.

Compliance by the load serving entities generally is made from the energy from renewable projects across the region and is accomplished with the purchase of Renewable Energy Certificates (RECs[1]) associated with these projects. The value of a REC adds to the worth of the energy from a project, and provides greater incentives for investing in the development of renewable resources

15.4.3 Impacts on Power System Planning

Both RGGI and RPS have impacts on electric system planning in the region. The RGGI program would function similar to the SO_2 and NO_x cap and trade systems that have been functioning in the US and Canada. These systems provide regulatory certainty as to emission requirements for the generating plants affected. RGGI would be adding a third mandatory emissions cap for power plants in the seven participating states.

The RGGI Cap would function in the same manner like the SO_2 and NO_x caps, and cause dispatch or bidding adders that would increase the operating cost of fossil plants, especially coal and oil since these fuels have the highest CO_2 emission rates. These costs could change relative dispatch of the units and hence the system transmission flows.

In the modeling conducted during the development of the RGGI program, a wide range of natural gas price assumptions was examined for the electric system expansion to show feasibility of the cap. The results showed a very diverse set of generation additions to serve the energy and peak load growth out through 2024. For assumptions of more historical levels of natural gas prices the additions included a large amount of natural gas fueled combined cycle (NGCC) and onshore wind generation. For assumptions of higher natural gas prices, such as were experienced in 2005, clean coal plants were the major capacity addition with a lesser amount of NGCC and a similar amount of wind was selected in the model (to meet RPS) as with lower natural gas prices. The large amount of wind may not be feasible if the siting difficulties of current wind projects continue. These RGGI scenarios

---

1 A REC equals one MWH of renewable energy.

also assumed that the natural gas infrastructure would be expanded as needed. In the ISO/RTOs' regional planning processes, generation expansion scenarios will need to be examined with more detailed modeling to confirm that system reliability can be maintained and to determine the magnitude of the market costs of implementing the RGGI CO_2 cap.

RPS is providing some incentives for new renewable projects, especially wind and biomass. Based on the ISO/RTO system interconnection queues, wind and biomass appear to be the more attractive renewable projects being built. These renewable projects have to be sited where the energy source is located, which is usually not close to a major load centers, i.e. on remote ridgelines for onshore wind or where there are forested areas to provide wood harvesting with minimum transportation costs.

15.5 Conclusions

There is growing evidence of impacts of CC due to GHGs. Action is needed to reduce GHG emissions to mitigate risks of CC and to increase global capability to adapt. The power industry is a major part of the problem and must be part of the solution and show leadership. Much has been done through global and other programs, but there is urgency to do much more to reduce risks.

It is prudent to adopt a "no regrets" strategy at this time that makes good sense and minimizes costs and risks whatever the outcome on actual global climate change. The preferred risk management approach must balance the costs and economic risks of overly severe CO_2 emission reduction targets against the costs and benefits of increasing our adaptive capability to cope with climate change This is particularly so in the developing countries which would be hardest hit by overly restrictive targets affecting their economic development and currently have the least adaptive capability.

Major thrusts must be on clean, hi-efficiency technology for mitigation of emissions, and increasing adaptive capability, particularly of poorer developing countries. There are many opportunities for the power industry to show leadership in technology, processes and markets. There will be funding and skilled resources challenges, but there are many good investment opportunities.

Business and governments must work together on climate change mitigation and adaptation. GHG reductions can be realized through use of (i) market-based programs in which customers or manufacturers are provided technical support and/or incentives; (ii) mandatory energy-efficiency standards, applied at the point of manufacture or at the time of construction; (iii) voluntary energy-efficiency standards; and (iv) increased emphasis of private or public R&D programs to develop low emission energy technologies and more efficient products.

There is no one unique, globally traded and valid value for carbon. The value ascribed to avoidance is coupled to the economic value of energy use; and hence the range of costs of emissions reduction is highly dependent on socio-politico-economic assumptions. The use of alternate non-carbon energy is of relatively high value in typical schemes, including impact of conservation and efficiency measures. The results show a definite trend that confirms the considerable advantage of adding new-build advanced nuclear plants as potentially the lowest cost emissions reduction option with the highest value.

Adapting to climate changes will present challenges for all involved in infrastructure design and construction, health and medicine, water resources management, coastal zone management, agriculture, land use and forestry, and other areas. Increasing adaptive capability is a priority for the short term, particularly for the poorest developing countries which will be hit earliest and hardest by climate change and are least able to cope.

The RGGI CO_2 cap and the RPS requirements in Northeastern USA are adding new impacts and considerations for power system planning in that region. RGGI will most likely increase energy costs from fossil generators in the states where it will apply and possibly affect reliability. RPS will encourage smaller renewable resource projects, mostly onshore wind and biomass fuels that will interconnect at lower transmission or distribution voltage levels, and will not likely help serve large load centers. As larger amounts of wind projects are added, they could affect the need for increased operating reserve.

15.6 Acknowledgements

This Chapter has been compiled by James McConnach, Castle Hill Engineering Services, Bracebridge, Ontario, Canada; Chair of the IEEE PES W.G. on Climate Change. Contributing authors include; Romney B. Duffey (Principal Scientist with AECL, Pembroke, Ontario, Canada); Gilles Potvin (CDM & JI Office, Foreign Affairs Canada, Ottawa, ON, Canada); Alistair I. Miller (Researcher Emeritus, AECL, Pembroke, Ontario, Canada), and James E. Platts (ISO New Zealand, Inc., Holyoke, MA, USA). The Chapter is primarily based on an up-date of the papers presented at the Panel Session on "The Impacts of GHG Programs and Markets on the Power Industry" at the IEEE-PES 2006 General Meeting (GM2006) in Montreal ([23-26]).

15.7 References

[1] T J Hammons and J S McConnach. Proposed Standard for the Quantification of CO2 Emission Credits, Electric Power Components and Systems, Taylor & Francis, Vol. 33, (1), pp. 39-58, 2005.

[2] M. Jaccard, and W. D. Montgomery. 1996, "Costs of reducing greenhouse gas emissions in the USA and Canada," Energy Policy, vol.24, 10/11, 1996, pp.889-898.

[3] States' Guidance Document: Policy Planning to Reduce Greenhouse Gas Emissions, Second Edition (EPA, 1998).

[4] A. Chappell, and C. T. Agnew, "Modelling climate change in West African Sahel rainfall (1931-90) as an artifact of changing station locations," International Journal of Climatology, vol. 24, no 5, April 2004, pp. 547-554.

[5] Jim McConnach, Janet Ranganathan, Scott Rouse, Thomas Baumann and Namat Elkouche "Plans and Programs for Greenhouse Gas Reductions, Removals and Trading" Presented at PGI2004, Orlando, Dec. 2004.

[6] A F Zobaa, "Climate Change Risks and Financial Sector" IEEE-PES General Meeting, San Francisco, June 2005, Paper 05GM0044.

[7] Peter Hoeppe and Gerhard Berz, "Risks of Climate Change – Perspective of the Re-Insurance Industry, IEEE-PES General Meeting, San Francisco, June 2005, Paper 05GM0523.
[8] Lord Browne, BP Group Chief Executive, Speech to the World Petroleum Congress, Johannesburg, 29 September 2005. Available: http://www.bp.com
[9] Stern Review of the Economics of Climate Change, Discussion Paper 31, January 2006, p3. www.hm-treasury.gov.uk/Independent_Reviews/ stern_review_economics_climate_change/sternreview
[10] H. Hasselknippe, "Climate Change & Business", Auckland, PointCarbon, November 2004.
[11] International Emissions Trading Association (IETA), World Bank, "State and Trends of the Carbon Market", Washington, DC, May 2005.
[12] W.D. Nordhaus, "Life After Kyoto: Alternative Approaches to Global Warming Policies", Yale University, December 9, 2005.
[13] PointCarbon. Available: http://www.pointcarbon.com.
[14] A.B. Lovins, "Climate Change and Energy", presented at The Inaugural Lorne Trottier Public Science Symposium, McGill University, November 24, 2005. Available: http://www.mcgill.ca/gec3/symposium/
[15] US DOE, "Energy Outlook 2001", Energy Information Agency, Washington, 2001.
[16] DTI, "Review of the Feasibility of Carbon Dioxide Capture and Storage in the UK", UK Report, 2003.
[17] Battelle, "Global Energy Technology Strategy: Addressing Climate Change", 2004. Available: http://www.gtsp.battelle.org.
[18] IEA, "Energy Technology Perspectives: Scenarios and Strategies to 2050", International Energy Agency, OECD, Paris, 2006.
[19] Gouvernement de Québec, Plan d'Action 2006-2012, "Le Québec et les Changements Climatiques: Un Défi pour l'Avenir".
[20] IPCC, "Climate Change 2001: The Scientific Basis", Third Assessment Report, UN International Panel on Climate Change. Available: http://www.grida.no/climate/ipcc_tar/vol4/english/080.htm
[21] A.M. Miller, S. Suppiah, and R.B. Duffey, "Climate Change Gains More from Nuclear Substitution that from Conservation", Nuclear Engineering and Design, 236, pp. 1657-1667, 2006.
[22] T.M. Wigley, S.C.B. Raper, M. Hulme and S. Smith, 2004 "The MAGICC/SCENGEN Climate Scenario Generator, Version 4.1", Climate Research Unit. Available: http://www.cgd.ucar.edu/cas/wigley/magicc/index.html
[23] Dr. Ahmed Zobaa , Cairo University, Egypt and James McConnach, Ontario, Canada "International Response to Climate Change: An Overview" IEEE-PES GM2006, Montreal (Paper ID 06GM0027)
[24] Gilles Potvin, Senior Program Officer, CDM & JI Office, Ontario Canada "Canada's CDM and JI Office" IEEE-PES GM2006, Montreal (Paper ID 06GM0660)
[25] Dr. Romney Duffey, Principal Scientist, Atomic Energy of Canada Ltd. Ontario, Canada, "The Value of Non-Carbon Power and Emissions Avoidance"IEEE-PES GM2006, Montreal (PaperID 06GM0914)

[26] James Platts, ISO New England, USA, "Impact of Regional Greenhouse Gas Initiative and Renewable Portfolio Standards on Power System Planning" IEEE-PES GM2006, Montreal (Paper ID 06GM0920)
[27] The IPCC Fourth Assessment Report: Working Group II Report "Impacts, Adaptation and Vulnerability". See: http://www.ipcc.ch/ipccreports/ar4-wg2.htm

Reference websites include:
http://yosemite.epa.gov/globalwarming/ghg.nsf/actions/StateActionPlans/
http://www.nccp.ca/html/tables/pdf/options/Final_Options_Report_English.pdf
http://www.electricity-online.com/
http://europa.eu.int/comm/environment/climat/future_action.htm
http://www.grida.no/climate/ipcc_tar/wg2/461.htm
http://www.weforum.org/pdf/g8_climatechange.pdf
http://ge.ecomagination.com/site/index.html

UNFCCC	http://unfccc.int (includes CDM & JI)
IPCC	www.ipcc.ch
APP	http://www.asiapacificpartnership.org/
CCAR	www.climateregistry.org
RGGI	www.rggi.org
Canada CC	www.climatechange.gc.ca
TEAM	www.team.gc.ca/english/
Stern Review	www.sternreview.org.uk
CCX	www.chicagoclimatex.com
UK-ETS	www.defra.gov.uk/environment/climatechange/trading
EU-ETS	http://europa.eu.int/comm/environment/climat/emission.htm
UK Met Office	www.metoffice.gov.uk

16

Power Markets of Asian Countries in the International Markets Environment

This Chapter deals with the current state and problems of power markets in Asian countries in the international market environment. The process of restructuring the electric power industry and forming power markets in the world has almost a twenty-year history. Certain experience has been gained that reflects both the positive effects of market transformations in the electric power industry and some problems. Power markets in Asian countries are formed on the basis of world experience. However, in different countries this process progresses at different paces. Generalization of the experience in market transformations in the electric power industries of Asian countries, analysis of the benefits, and risks that may occur as a result of such transformations will help specialists solve the problems encountered in their countries.

16.1 Development Of Power Market In India

At the time of independence in 1947, the Indian power sector was merely concentrated in and around a few towns and urban areas to meet the need. In the following decade, it saw development of massive river-valley projects that led to some form of limited interconnected systems to provide power to the population along particular belts as a side-by-side benefit to the effort made for irrigation for agricultural need and flood control. However, the nineteen sixties gave proper status to development of the power sector both in terms of generating unit sizes and transmission voltage due to the requirement of rapid industrial development. This called for integration and evolution of the state grids. Attempt to join these grids to form the five regional grids became successful by the nineteen seventies and eighties with unit sizes going from 210 to 500 MW and transmission voltage from 220 to 400 kV as a consequence of transfer of a large amount of power from coal pit-head (mine-mouth) thermal power stations to urban conglomerations. Subsequent scenario of the power sector in the nineteen nineties and beyond has been quite bright from the point of view of development of HVDC systems, incorporated both for bulk power supply over a large distance up to about 1370 km, be it within a large state or region or for inter-regional transfer of power, and also for inter-regional back-to-back connection for limited transfer of power. Side by side to this, the sector was unbundled with the recognition of generation, transmission and distribution as separate and distinct activities so far as the power supply system is concerned. Both at state level and central level regulatory commissions were formed to decide tariff, grid code, etc. With opening up, the sector experienced participation of the private sector entities, mainly in generation and then to some extent in distribution. Transmission still remains a monopoly, with public holding terming it as State Transmission Utility (STU) or Central Transmission Utility (CTU) depending upon whether it belongs to any state or center. With Central Electricity Regulation Commission (CERC) permitting open access to the inter-state transmission facility from November 2003 [1], it opened vistas of power trading by state-owned companies or private traders or joint sector venture. It was an important step after promulgation of the Electricity Act 2003 [2]. Activities that followed and aimed at influencing scheduling and real time grid operation with pseudo Power Exchange [3] in place definitely pave the way for healthy trading in power. This is unlike other commodities in market. Electricity cannot be stored in its form and hence calls for supply-demand matching at every instant of time. Section 16.1 gives the status of such trading prevalent in India considering the market related to energy, generation capacity, transmission capacity and ancillary services one by one. Also, with the development in neighboring countries, the possibility of power and energy trading is examined.

16.1.1 Energy Market

For the Indian power sector, the bilateral energy market is on the basis of long term, short-term, day-ahead or intra-day commitments. With measurements logged at 15-minute intervals, a weekly cycle of settlement of energy is carried out. This is based on before the fact commitments at mutually agreed terms, but taking into account deviations settled at frequency actuated dynamic rate known as the Unscheduled Interchange (UI) rate [3]. However, the process has excessive reliance on the UI mechanism, though regulatory caps restrict the rate. The trend is encouraging with consensus being built for an organized market in this respect in the form of Power Exchange (PEX).

16.1.1.1 Generation Capacity Market

As one goes back in history, typically under Central Government regional power stations [termed as Inter-State Generating Stations (ISGS) (be it thermal – fossil fired or gas-based, hydro or nuclear)] established at different times have a common basis of sharing of power amongst the beneficiary states of the concerned region. An allocation is made of 85% of the installed capacity of the station by that procedure. Hence the capacity may be thought of as locked up in long-term bilateral contract between the producer and consuming states. The remaining 15% floating capacity is highly sought after during peak demand and it keeps changing hands subject to negotiating skill and political networking of the beneficiary causing a considerable amount of heartburning for the losers. What started as a flexibility margin to accommodate the seasonal demand pattern has degenerated into a discretionary instrument.

On the other hand, lackluster participation of private players in capacity addition (generation and transmission) could be attributed to lack of an organized capacity market. However, hope exists due to stray examples of capacity trade. One such case is with the Power Trading Corporation (PTC) brokering the sale of royalty share of the Himachal Pradesh State Electricity Board (HPSEB) in the Nathpa-Jhakri Hydro-Electric Project to Punjab State Electricity Board (PSEB) for the summer months.

16.1.1.2 Transmission Capacity Market

The Central Electricity Authority, an apex technical body of Government of India in the power sector, carries out perspective planning as a whole. With the data collected through load survey by its regional units in collaboration with the state electric utilities, long term-load forecasting is done. Based on the same matching, generation is formulated through the integrated resource planning approach identifying generation location and possible corridor for transmission of power from source to load. Thereafter, studies are performed to configure in detail the network for evacuation of power from the generating stations and consequent strengthening of the existing network, if required, with the level of voltage chosen with a view to have an adequate margin for future expansion. Transmission capacity expansion so planned is then deliberated in the Standing Committees region-wise through a consultative procedure to identify the utilities to build, own and operate the relevant expansions.

Accordingly, with the transmission system still being totally need-based and enjoying natural monopoly, the pricing is tightly aligned to long-term capacity allocations. Though open access is in vogue, in reality it has not been segregated yet as an independent facility under the fear of jeopardizing the existing setup. On the other hand, lack of addition may result in congestion sometime at some pockets during grid operation, Consequently, its management is totally based on the discretion of the concerned Regional Load Dispatching Center (RLDC). The long-term transactions have priority over short-term transactions. The RLDCs have discretionary powers over interstate dispatch and load regulation. Inter-regional (Pool to Pool) unscheduled interchange transactions are then used for easing congestion.

16.1.1.3 Ancillary Services Market

Ancillary Services are defined as those services that are necessary to maintain reliable operation of the interconnected / integrated transmission system. These services are required to effect a transaction. It includes reactive power and voltage control, loss compensation, scheduling, dispatch and settlement, load following, system protection, energy imbalance and black start facilities. In India a lot of work needs to be done in this area as described below.

Load Following-Primary Response

- Free Governing Mode Operation (FGMO) is mandatory as per grid code.
- Issue is diluted / scuttled under the garb of technical jargon / issues put forth by generators.
- Services are basically not priced and implicitly paid through capacity charges. Therefore, there is no incentive for Independent Power Producers (IPP).
- Frequency linked dispatch guidelines are for secondary response.

Voltage Control

- Reactive drawl and injection at interstate exchange points are priced.
- It is a simple mechanism. Issues in treatment are virtually of residual amount.
- Generators are not paid and very often they take refuge under a conservative machine capability curve.

Loss Apportionment

- Losses are shared by long-term customers in ratio of their subscriptions in ISGS.
- All energy transactions are discounted by estimated losses during scheduling.
- There is regulatory intent of moving towards the concept of incremental losses.

Scheduling and Dispatch

- RLDC coordinates as well as implements inter-utility contracts.
- Decentralized resource scheduling is in vogue with state load serving utilities having full operation autonomy of dispatching their generation resources.
- Though as per grid code there is a provision for 5% spinning reserve, due to perpetual shortage, in reality implementation has not been possible yet.
- Well-defined timeline exists for declaration of availability and requisitioning of energy up to capacity subscriptions of the shareholders.
- Expenses clubbed under the RLDC Operation and Maintenance (O&M) head is paid by long-term constituents only.
- At present a sum of Indian Rupees (INR) 3,000 / day/ transaction is charged for scheduling open access transactions.
- Inter utility settlement statement (Regional Energy Accounting taking care of UI and Reactive Accounting) is issued by the Central Pool Administrator. Capacity and energy charges are settled mutually while the unscheduled and reactive energy settlement is routed through a pool.

System Protection

Protection Coordination Committee (PCC) decides

- Equipment protection coordination at the regional level.
- System monitoring and supervision is carried out by RLDC.

Energy Imbalance

- It is addressed through unscheduled interchange mechanism.
- Weekly settlement cycle based on above is in vogue.
- It is the discretion of concerned RLDCs for arbitrage across asynchronous (HVDC) links.

Black Start

- It is purely voluntary.
- It is well documented under Regional Black Start Procedures.
- UI mechanism is suspended during period of disturbance and actual transaction is treated as schedule.

16.1.2 Possible Power and Energy Trading with Neighboring Countries

India is surrounded by the countries Nepal, Bhutan, Bangladesh, Srilanka. Pakistan creates major prospect in South Asia for trading in power and energy due to disposition of natural resources of different kinds for mutual benefits of all. Nepal and Bhutan are rich in Hydro resources, Bangladesh is rich in gas reserves and India is rich in coal resources, thus providing promising option for cooperation among countries. India can emerge as the main potential power / gas export market for the neighboring countries. Generation can be at source and trading through electrical interconnection. India can supply coal to the neighboring countries and can import gas from Bangladesh.

Issues to be addressed in the process of development are investment capabilities, lack of market information, viability of buyers, inadequacies in institutional mechanism, environment and social concerns. Cross border trading in electricity has technical considerations as well as political and economic ones. Pricing should be such that both sides benefit. For example, if one party has a lot of inexpensive hydropower, during monsoon seasons then it may benefit from selling it at lower price to a neighbor rather than having the water spill. There is necessity of a larger perspective while planning through an integrated approach for the entire SAARC (South Asian Association for Regional Cooperation) region. Both Generation capacity and Transmission interconnection capacity are to be enhanced. To be adopted is common principle / methodology for tariff determination, operational protocol, security / reliability and regulation. To be evolved also is the Contractual Agreement that addresses principal obligations that are equitable, risk sharing, issues related to financial and payment, commercial and legal, dispute resolution and arbitration.

Therefore, prerequisites for Regional Power Pool (RPP) may be summarized as –

- Technical solutions not difficult but Political will of the member countries important
- A cooperative mindset
  - -Willingness to reconnect the subcontinent
  - -Efforts to build trust / sensitize
  - -Greater sensitivities to issues
- A commitment from the member countries for
  - -Resources / manpower
  - -Reciprocal measures
- Success of Bilateral exchange will create the ground for multi-lateral exchange
- Regional economic prosperity should take precedence over political compulsion.

Thus, it is in the nascent stage, there is lot of promises in power trading in India with the participation of a number of players from public or private or joint holding companies. Permission for open access really has created opportunities for improving supply system through competition in terms of overall economy as well as ultimate efficiency. With the typical characteristic of the commodity (power) in the market that in its normal form cannot be stored and at every instant supply-demand matching is called for, inherent risk dictates necessity of well-laid principles of practices to be followed for short-term, mid-term and long-term contracts.

So far as power and energy trading with neighboring countries is concerned, presently power trading is based on bilateral agreements and although Energy Ring is high in SAARC agenda, the progress has remained slow. The strategies for promotion of trading can be through carrying out sector reforms, setting up suitable institutional arrangements, joint investment in project including Environment Impact Assessment (EIA), private sector participation, long term transmission planning and free exchange of information.

16.2 Russia's Power Industry Restructuring Current State And Problems

Russia's power industry is on the way to radical structural and managerial reformations. Reduction in electricity demand in the early 1990s provided good conditions for structural and managerial changes in the industry. However, the country is facing new industrial growth and the favorable period of excessive generation is coming to an end. The industry needs fast and effective reforms for economic efficiency improvements and investment attraction. This Section reviews the restructuring policy and describes the current state and development of a competitive environment in the Russian power industry.

Russia is an Eurasian country, and only about 30% of the power industry locates in its Asian part (in Siberia and Far East). But restructuring ideology is the same for each part of Russia. Therefore, the power industry state and restructuring will be considered in Russia as whole.

16.2.1 Technical Characteristics of the Russian Power Industry

The main distinctions of the industry are long-distant electricity transmission, unevenly allocated energy resources as well as electricity generation and consumption over vast territory.

Figure 16.1. Interconnected power systems in UES of Russia

The Unified Electric System (UES) in Russia is still operating as a single complex. There are 77 local utilities combined into 7 regional interconnected power systems [4-6] (Figure 16.1). Six of them (excluding the Far East) operate synchronously. The transmission network covers six time zones, which helps reduce the peak demand by 6%.

The Russian power industry had 219.2 GW of installed capacity up to the beginning of 2005, including 805.2 GW running synchronously within UES. 14.0 GW ran separately from UES in the Russian Far East and in the Arctic regions of the country. Taking into account dismantling units the industry had a slight growth of generating capacities in 1999 and a slight decrease in 2000.

All power plants in Russia generated a total of 953,100.0 GWh of electricity in 2005 including 340,700.0 GWh within UES. Thermal power plants on organic fuels generated 629,200.0 GWh or 66% of total production. Hydropower plants generated 174,400 GWh (18.3%) and nuclear power plants produced 149,500.0 GWh (15.7%).

The peak of electricity demand in 2005 in UES was registered on January 26 at 6 p.m. and reached 128.7 GW. The total excess of generating capacity in UES was 63.5 GW or 33% of installed amount. However, not all installed capacity could be used at peak hours. In winter

2005-2006 only 170 GW of generating capacity was available for operation. Taking into account a necessary capacity reserve of 21.4 GW (16.6% of maximum demand) the actual excess of capacity in Russia is 19.9 GW or 9.7% of available amount. UES of Russia supplies electricity to several countries in Europe and Asia (Figure 16.2).

12.5% of total generation was lost in the transmission and distribution networks. The annual electricity consumption grew by 4% in 2000 and 2.5% in 2001-2005.

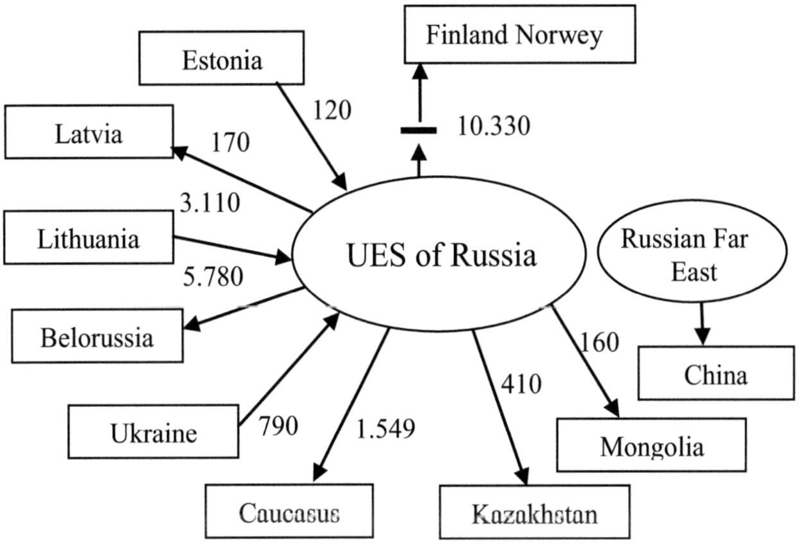

Figure 16.2. Export and import of electricity in 2005, GWh

16.2.2 First Steps of Restructuring

The restructuring process was launched in 1992 [4]. The ownership pattern was changed and the existing State-owned local utilities were rearranged into joint stock companies. Nuclear power plants with a total installed capacity of 21 GW were left under State control.

The joint stock company RAO *"UES of Russia"* was established in 1992 as a new managerial holding. 34 large power plants with a capacity of 57 GW (approximately 27% of total generating capacity) were withdrawn from local utilities to reduce the monopoly in power generation and became the property of RAO "UES of Russia". Also, the decision was made to concentrate the transmission network in RAO *"UES of Russia"* as well as Central and Regional Dispatching Offices. It was supposed that each new local utility had to delegate no less than 49% of its shares to the holding. The RAO was commissioned to set up and operate a wholesale market both technologically and commercially. Actually, not all the utilities delegated this percentage of their shares to the RAO. Nevertheless, RAO "UES of Russia" acquired too much property and authority.

Today RAO "UES of Russia" is the monopoly in the field of energy supply and is the financial and industrial holding, which consists of a number of regional power utilities, large federal power stations and intersystem electric grid. The State owns the controlling stake of the RAO. The Government appoints both the Board of Directors and the Chief Executive of the RAO. It regulates all activities of the energy holding and regulates the prices of its services and production.

Two kinds of electricity markets were established in the 1990s - the nationwide wholesale market and the local retail ones. Electricity prices for power producers on the wholesale and retail markets were fully regulated and set on the "*cost plus*" principle.

Large power plants, which were withdrawn from the utilities, nuclear power plants and local utilities with generation surplus could sell their electricity to utilities with deficient generation on the wholesale market. A few large industrial customers got the right to participate on the wholesale market. Each local utility had a choice either to generate power locally or to buy it on the wholesale market. The retail markets were established primarily within the territories under service of the local utilities and had to supply their end customers. The responsibility for the wholesale market operation was delegated to RAO "*UES of Russia*". The responsibility for retail market operation was imposed on the local utilities that provided distribution and customer services.

Wholesale electricity prices in Russia were based on generation and transmission costs. The Federal Energy Commission regulated the costs and profits of producers on the wholesale market under control of the Federal Government. The costs of utilities and prices on the retail markets were subject to regulation by Local Energy Commissions under control of local Governments. Electricity prices for the wholesale and retail customers were set as blanket (weighted average) prices calculated as a ratio of electricity cost to volume of electricity to be purchased. The prices on both the wholesale and retail markets were adjusted on a quarterly basis.

RAO "*UES of Russia*" was an operator on the wholesale market. It bought, as a single trade agent, electricity from producers using the pay-as-cost principle and sold electricity to the utilities at a blanket (average) price. The electricity price, additionally to the generation expenses, included transportation, operation and investment components. The same approach was used on local retail markets. The prices on the wholesale market were differentiated with respect to price zones according to the transportation and congestion costs. The prices on the retail markets were differentiated with respect to consumer groups, such as large industrial, small industrial, agricultural, residential and others.

Up to the middle of 2003 prices increased by 1.5-1.7 times [5]. Average prices greatly differ in the Siberian and Far-Eastern regions. There are cross subsidies between different groups of consumers. In most regions (except Urals and Siberia) industrial consumers pay above the average price, while rural residential consumers (except Siberia) pay less.

The funds for generation and transmission expansion were mostly filled up through the regulated investment component of wholesale prices. The funds for development of power plants and distribution networks that belong to local utilities were filled up through the regulated component of retail prices.

After the first steps of restructuring the situation in the power Industry remained very tense [4]. First of all, the UPS badly needed new investments. Financial injections into the industry were 5-6 times less than was required. This resulted in dangerous aging of generating and transmitting facilities. Nearly half the existing infrastructure exceeded its intended service life [7].

The efficiency of electricity and heat production decreased. Auxiliary electricity consumption in power plants together with electricity losses in networks increased to 23% in 1998 in comparison with 17% in 1991. The fuel component of electricity cost increased by 11% in the 1990s. The relative annual profit of the industry fell down from 25.5% in 1993 to 11.3% in 1999. This led to a hard and unstable financial condition for most of the local power utilities.

A huge debt of customers (more than $4.3 billion) was accumulated in the 1990s. The debt was comparable with the annual income of RAO *"UES of Russia"*. For years, the industry subsidized the Russian economy by supplying the electricity and heat to non-paying customers. A large portion of payments was accepted in barter and mutual debt write-offs. The situation with non-payment was improved considerably in 1998-99 as payments in money rose to 90%. But the accumulated debts pulled the industry down and did not allow local utilities to develop their generating and network facilities.

Wholesale and retail markets in Russia were highly regulated. However, the system of the State regulation was not effective. The pricing mechanism on the wholesale market was obsolete and did not encourage cost reduction. The regulation on local retail markets fell hostage to local governments and politicians. Electricity prices were too low (see Table 16.2) and did not often compensate even for the production costs. It led to generation reduction and consumer cut-offs. The legislation in the industry remained far behind the current needs.

Unreliable fuel supply was an urgent problem for the industry. Coal miners and natural gas suppliers did not provide power plants with enough fuel because the power plants and utilities often failed to pay for the fuel in time. This resulted in dangerous situations on some territories of Siberia and the Far East of Russia, particularly in the severe winter of 2001 when electricity and heat supply was interruptible. Unreliable fuel provision led to ineffective generating unit commitment.

Most experts considered the situation in the Russian power industry dangerous. In any case, the industry needed radical and wide-scale reforms based on the interests of different parties in the society.

16.2.3 A New Stage of Restructuring

A new stage of restructuring started in 2001 after Governmental resolution validation. The main goal of the stage is to provide stable work of the industry on the basis of competitive market development. The following measures are needed to succeed.

Development of a new legislative framework for the industry. The Parliament and the President enacted a new State Law *"About Power Industry"* in March 2003. The Law declared the market relationships in the industry as a main instrument for efficient and stable electricity supply.

Two existing Laws, namely "*State tariff regulation in the industry*" and "*State monopolies in Russian Federation*", were changed. Additionally the Government issued several new Resolutions on pricing of electric and heat energy, the wholesale market rules and boundaries of pricing zones.

The new Laws and Governmental Resolutions form a power policy that provides broad guidelines for coming reforms and enact the legislation necessary for implementing this policy. It helps to provide a transparent regulatory framework and to establish a market encouraging efficiency improvement.

Restructuring local power utilities. Local power utilities are under restructuring now. They separate "non-profile" businesses and create new entities for power generation, distribution and supply. They introduce new systems for bookkeeping and financial planning.

Small and combined electricity and heat production plants formed local generating companies (GenCos). Later, the local GenCos were combined into 14 territorial GenCos to improve their maintenance and to make cheaper financial credits for development and reconstruction. Local distribution companies (DistCos) are to provide an open access to the electric networks for end customers. Later, the local DistCos were combined into 5 inter-regional DistCos.

Several supply (energy-sales) companies are under creation in each region.

They will compete for the end consumers by lowering the price of its services and by offering more profitable and convenient conditions of energy supply. One of the supply companies is Guaranteeing Supplier (GS). GS is a company, which continues the obligation to provide supply service to all customers on the GS's service territory who request such a service.

Competition can be introduced in the generation and supply spheres, but is not feasible for transmission, distribution, and system control businesses because these are the functions of natural monopolies.

Restructuring the holding RAO "UES of Russia". It is evident that RAO "UES of Russia" dominates the industry and must be promptly and substantially restructured to improve efficiency and attract necessary investment. The following steps were made in 2002. Some of them are expected in the nearest future.

The Federal Transmission Company (FTC) was created to operate, maintain and expand the national transmission network. Lines and substations with voltage 220 kV and higher are combined under control of the new company. FTC will remain the sister company of JSC "*UES of Russia*" for two or three years. Later, 75% of FTC's shares will belong to the State.

New company called "*System Operator*" (SO) was established to operate UES. SO is targeted to provide reliable electricity supply and non-discriminatory access to the networks. SO combined the Central Dispatching Board in Moscow and seven regional Dispatching Centers. Later the dispatching divisions of the local power utilities will join SO. The ownership of the State in the capital of SO is more than 75% from mid-2005.

Administrator of the Trading System (ATS) was established as a non-profit entity for the wholesale market design and operation. ATS registers amounts of bilateral electricity trade, sets prices on the spot market for different buses of consumption, and supervises trading agreements and payments for electricity on the wholesale market.

Seven Wholesale Generating companies (GenCos) are designed and will be created with private property. The Wholesale GenCos have approximately equal installed capacities of 8.5-9 GW. Six of them will combine thermal power plants. One GenCo will consist of hydro power stations. Additionally, there is an independent GenCo consisting of 8 nuclear power plants. This company is supervised by the Ministry of Nuclear Energy and belongs to the State. Unbundled generating companies are established under dispersed ownership. Private investors are expected to bring financial resources to update running units and construct new ones.

Reforming the electricity markets. Significant developments are expected on the wholesale market. Two challenges will be available for electricity producers and customers. They may participate on the spot market or arrange bilateral forward contracts for electricity delivery. New companies such as the wholesale and local GenCos, local electricity suppliers, Federal Transmission Company and large end customers will participate on the wholesale market instead of vertically integrated utilities. System Operator will be responsible for safety and reliable operating conditions in UES.

A new pricing mechanism will be introduced in the wholesale market. Instead of blanket average prices a new market will introduce local marginal prices. The wholesale customers will pay at the margin what it costs to produce and supply electricity to them. On the one hand, it will encourage the power producers to improve the efficiency of production. On the other hand, it will result in price growth.

The spot market organization needs the development of sophisticated measurement system. Considering vast territory of the country this is a complicated technical problem.

Improvements in the State regulation. The State regulation of the industry should be concentrated on price regulation in the spheres of natural monopolies. On the wholesale market the focus of regulation is to prevent anticompetitive abuses and the market power exercises. On the retail markets regulation should be focused on balancing the interests of suppliers and customers. The prices in the natural monopoly spheres remain regulated including transmission tariffs, expenses of System Operator and ATS.

Prices in the competitive spheres are not regulated, but the Federal Energy Commission as the State Regulator can set price caps on the wholesale market. Also the State Regulator endorses the rules for customer's access to the networks and the procedures for tariff calculation. Local Energy Commissions regulate the tariffs for heat supply and electricity distribution as well.

Any cross subsidies between territories, groups of customers, different kinds of products (electricity and heat energy) and services should be eliminated step-by-step.

16.2.4 Investment Attraction into Russia's Power Industry

The volumes of new equipment commissioned in the last decade have been extremely small. The average annual input of generating capacities in the years 1991-2005 was 0.6-1.5 GW per year, while in 1976-1985 construction of new capacities was 6-7 GW. In China this rate was 17,000 MW annually during the last 15 years [8]. Financial injections into the industry were 5-6 times less than was required. This resulted in dangerous aging of generating and transmitting facilities [7,8].

Therefore attraction of investments into Russia's electric power industry is a key problem. The solution to this problem will somewhat differ for privately owned generating companies and for network companies that are regulated natural monopolies.

Attraction of external investors is important for generation expansion and calls for:

- Essential increase in the investment attractiveness of GenCos by providing financial transparency, economic stability, predictability of management actions and other corporate measures;
- Considerable increase in the reliability of State guarantees to the investor by clear legislature, predictability and stability of State policies in electric power industry, etc.
- Effective system of investment risk insurance.

The above measures can considerably decrease financial risks for investors, which will not require high electricity tariffs to compensate for these risks.

At the same time the world experience shows that in a free market environment the short-term purposes of power companies prevail over the long-term ones which, with time, may lead to an inadmissible drop in the reserves of generating capacities, formation of generation shortage and as a result growth of electricity tariffs [7,8].

The new Law on the electric power industry envisages prompt and long-term measures not to allow such situations. The prompt measures imply introduction of price caps and then regulation on the markets in the event that there are no conditions for competition due to the shortage of generating capacities. A basic long-term measure is development of a State Program for generation expansion that would provide permanent availability of surplus capacities on the wholesale markets. This can be based, for example, on nuclear power industry that belongs entirely to the State. Distributed generation can also play and an important role here. The Program should form economic incentives for GenCos and external investors to invest the capital in construction of new power plants.

As to the regulated network companies, their development can mainly be provided by the investment component included in the tariffs for power transmission service. At the same time, due to insufficient transfer capability of the electric network, and the need of its essential increase to expand the conditions for electricity market operation, it may turn out to be necessary to attract additional investments both governmental and private. The volumes, conditions and mechanisms of attracting such additional investments require additional thorough scrutiny.

16.2.5 A Transition Period in the Restructuring Process

A transition period has started since March 2003 after the new Federal laws were enacted. There are two sectors on the wholesale market during the transition period.

The first one is a competitive spot market with bids/offers submission on a per hour basis and with local marginal price setting. The other one is a fully regulated sector with "*cost plus*" principle of pricing for producers and average prices for customers within tariff zones.

Until 2006 the regulated sector worked with the single purchase agency. After January 2006 the sector was organized in the form of long-term bilateral contracts. The State Regulator will consider the prices of producers and define the set of suppliers for each wholesale customer. Prices within the bilateral contracts will be set for three or five years. Every customer will be able to dissolve the whole package of bilateral contracts once a year and join the competitive spot market.

Firstly each producer could sell up to 15% of its generation in the competitive sector. Year by year the share of sales in the competitive sector grows. In the end of transition period all electricity trade will be competitive. The Government defines duration of the transition period.

16.2.6 Regional Problems of the Market Development

The wholesale market of electricity in Russia has been established as the nationwide market. Nevertheless, the unity of the market exists only «dejure» [9]. The wholesale market is divided in fact into several large regions due to the high cost of electricity transmission and limitations in the transmission capabilities.

The European part of Russia including Ural, Siberia and the Russian Far East are three large regions. The structure of the economy, electricity consumption and climatic conditions in these regions are quite different. These circumstances bring additional troubles for using a nationwide model of the wholesale market. The Federal wholesale market in Russia should have distinctions at least in three large regions.

The European region of Russia has 72% of total installed capacity and 74% of total electricity consumption. Fossil fuel power plants in the region burn mainly natural gas. There are nuclear and peak (semi-peak) hydro power plants among producers. Generation is relatively equally spread over the territory. Transmission network is well developed and provides alternative ways for power supply. The European section of UES has the surplus generation and transmission reserves.

It seems reasonable to have the following model of the wholesale market:

a) the independent Administrator of the Trading System (ATS) in the region organizes and operates the day-ahead wholesale spot market. ATS works as a «flexible» entity arranging transactions as a brokerage system in the power exchange. The bilateral trade for a year and a quarter is permitted in addition to the centralized spot market;
b) large fossil-fired power plants and hydropower plants are combined into private generating companies. Nuclear power plants are combined into the State concern "*Rosenergoatom*",

which plays as an independent producer on the wholesale market. Customers on the market are local supply companies and large power consumers;
c) ATS calculates the local marginal prices for electricity based on free bids/offers on the spot market. Contract partners define prices within the bilateral contract. The State Regulator sets tariffs for electricity transmission and the System Operator's services for a year. It sets the price caps for the spot market as well.

The Siberian region has 22% of the Russian generating capacity and 21% of total consumption. More than 50% of electricity is produced by large hydropower plants. These plants generate very cheap energy and have a strong influence on the economy and ecology of different territories. Almost 30% of total amount of electricity is generated in cities at cogeneration plants. Electricity generation at these plants greatly depends on heat consumption. Almost all fossil-fired power plants burn Siberian coals.

The transmission network has weak ties with the European sections of UES and is isolated from the Far-Eastern one. Power grid in the region is spread on a vast territory and has limitations on power flows in some cut sets. That is why it is difficult to provide an open access to the transmission network for all market participants.

Electricity generation and consumption in Siberia are subject to unstable natural impacts due to variations in river inflows and long droughty periods. The climate is severe with annual and seasonal temperature variations.

Considering specific conditions of the Siberian region the following market organization can be proposed as rational:

a) establishment of the Siberian branch of ATS with the same main principles of work as the ATS in the European part. Proportions between bilateral trade and spot market sales may differ from other sections of UES. Regional Dispatching Center, which is the sister-company of System Operator, supervises reliable operation of the power grid in Siberia;
b) electricity transactions should be provided among the market participants located in Siberia;
c) introduction of a technique for long-term generation scheduling and pricing for hydropower plants. The technique will take into account the availability of water in the reservoirs, ecological impacts and interests of water users;
d) introduction of a procedure for price setting on the spot market based on competitive prices of electricity and regulated prices of generating capacities. The procedure should reduce the "*jump up*" of the wholesale prices due transition from the average to the marginal principle of pricing;

The Far Eastern region has less than 10% of electricity generation and consumption. There are a few relatively large power plants combined with a weak and extended network. Electricity supply of many territories is separated from UES. There are no sufficient conditions yet for the effective competition among producers and suppliers. It seems reasonable to save the State regulation for electricity producers for the nearest future. Considering the importance of the Far East for the interests of Russia the power industry in this region needs special governmental support.

16.2.7 Expected Effects of the Market Reform

New markets may result in:
1. Introduction of non-discriminatory relationships among market participants. It follows from unbundling the generation, transmission and supply businesses and from the new legislation enacting;
2. Removal of the cross subsidies among businesses, territories and customers. It will provide proper economic signals for electric power systems operation and development;
3. Encouragement of the electricity producers to reduce their production cost due to marginal pricing introduction;
4. Improvement in financial conditions for generation, transmission and distribution companies due to electricity price growth;
5. Attraction of domestic and external investments.

At the same time there can be some negative impacts including:
a. Increase in the wholesale and retail electricity prices due to introduction of marginal principle of price setting instead of average regulated approach;
b. Decrease in electricity supply reliability for the nearest future due to lack of experience in operation and control of new unbundled companies.

Thus, the following can be concluded:
1. The Russian power industry does need an improvement in efficiency of operation and new investment inflows. The industry and the Government see the main way out from current problems in restructuring of the holding "*UES of Russia*" and local power utilities. The market rules and market infrastructure should be improved for competition introduction.
2. The new State Law opens good perspectives for further progress in the restructuring process. The industry has entered a practical stage of restructuring. Power utilities are unbundling for separation of generation, transmission, and supply services. New market rules have been developed to improve the wholesale and retail markets.
3. Two challenges will be available for electricity producers and customers. They may participate in the spot market and arrange bilateral forward contracts for electricity delivery. There will be competitive and regulated sectors in the wholesale spot market during the transition period. Up to 15% of generated electricity is selling in the competitive sector now. Year by year the share of the competitive sector will increase.
4. The restructuring of the industry is accompanied by apprehension of significant increase in electricity prices and possible decrease in supply reliability. The restructuring process should be well prepared and staged over time. Any new reforming decisions have to be made after detail analysis of consequences and based on available potentialities. Different regions of the country may choose their own restructuring models and determine the terms and rates of restructuring.
5. Attraction of investments into Russia's power industry is a key problem. The new legislation in the power industry envisages prompt and long-term measures to develop generating facilities and electric networks. The long-term measure stipulates creation of the State Program of expansion planning that would provide permanent availability of surplus capacities in the wholesale markets. The Program should form economic incentives for GenCos and investors to invest the capital in power plants. Development of the market infrastructure can mainly be provided by the investment component in the tariffs for power transmission service.

Russia has unique conditions for the competitive environment development. Its restructuring policy may be useful for large countries where the power industry is on the way of reforms.

16.3 Power Industry Restructuring In China

Restructuring and reform of the traditional integrated power industry has been started all over the world for more than 10 years. In various deregulated power systems, the market structures, the degree of privatization and the sequence of reform stages are different. China started the process of power industry restructuring since the late 1990s. In this section, the current situation of China power industry reform is introduced and the following issues will be focused on: market structure, electricity price policy, investment and planning, renewable energy in the new market.

16.3.1 China Power Industry Structure and its Restructuring

China has been experiencing significant economic growth since the 1980s. Electricity consumption increased more than three times due to the booming market and manufacture industry. At the end of 2004, the installed capacity in China has reached 440GW, and the annual generation has reached 2180TWh. Both installed capacity and generation are ranked the second highest in the world. Within the 440GW, the proportions of thermal units, hydro station, nuclear and wind power are 73.7%, 24.5%, 1.6% and 0.17 %, respectively [10].

16.3.1.1 Power Industry Restructuring

In February 2002, the State Department issued the Power Industry Structure Reformation Program. The program includes three main points: 1) Restructure state-owned generation assets, and establish several independent generation companies; 2) Restructure transmission assets, and establish two grid companies; 3) Establish State Electricity Regulatory Commission (SERC) to monitor and regulate the electricity markets.

On the 29th of December 2002, two grid companies, State Grid Company and South China Grid Company were established. The State Grid Company has five subsidiary regional grid companies. They are North China Regional Grid, North East China Regional Grid, Central China Regional Grid, East China Regional Grid and North West China Regional Grid. The geographical areas of the regional grids are shown in Figure 16.3.

On the same day, five independent generation companies were established. They are:

- China Huaneng Group,
- China Datang Corporation,
- China Huadian Corporation,
- China Guodian Corporation, and
- China Power Investment Corporation.

Figure 16.3. China regional grids

The installed capacities of the five-generation companies are given in Table 16.1. Each of the five-generation groups has a certain amount of share of any local market [11]. After the restructuring of generation assets, the State Grid and South China Grid own 24.5GW generation capacity, of which, 15.9GW capacity is used for frequency regulation and load following to ensure the reliable system operation [10].

The installed capacities owned directly by the two grid companies and all regional grid companies in 2004 are given in Table 16.2. The total generations of all grids in February 2005 are given in Table 16.3 [12].

Companies	Capacity (GW)
China Huaneng Group	40.98
China Datang Co.	4.95
China Huadian Co.	8.64
China Guodian Co.	29.30
China Power Investment Co.	27.96

Table 16.1. Installed capacities of the five generation companies in 2002

	Power Grids	Thermal Capacity (GW)	Hydro Capacity (GW)	Total Installed Capacity (GW)
State Grid	Own by State Grid	9.8	2.1	7.7
	North China Grid	77	74.5	2.5
	North East Grid	41.5	35.9	5.6
	East China Grid	78	64.4	11.6
	Central China Grid	65.4	41.9	23.5
	North West Grid	27.1	18.8	8.3
South China Grid		--	--	80.3

Table 16.2. Installed capacity in each grid company at the end of 2004

Power Grids	Generation (TWh)	Increase Rate (%)
North China Regional Grid	32.6	5.67
North East Regional Grid	15.4	2.21
East China Regional Grid	30.3	1.62
Central China Regional Grid	22.9	6.45
North West Regional Grid	10.2	13.7
South China Grid	17.5	1.16

Table 16.3. Generations of all grid companies in February of 2005

16.3.1.2 Regional Electricity Markets

The structure of China electricity market is a regional electricity market based on regional grids. The first regional market, North East electricity market, started generation auctions in January 2004. Four months later, East China regional market started its market operation [13]. The market participants of a regional electricity market are power system operation and exchange center, Grid Company, the five-generation companies and local provincial generation companies. The regional regulator commissions monitor and regulate the operation of regional markets. At the current stage, the system operation and exchange center is the single buyer of the market. The market settlement mode is Pay-as-bid. The Single buyer model has been used in many countries as a transition stage toward competitive market.

In the energy markets, generators provide long-term offers and short-term offers in the auction market. Long-term offers include annual transaction offers and monthly transaction offers. A Short-term day-ahead spot market will be established in the next stage of the regional markets.

Some hydro plants owned by the grid companies provide ancillary services, such as reserve, frequency regulation and voltage support. Some compensation mechanisms have been applied for the ancillary services provided by the units owned by the generation companies.

The mid-term goals of China regional electricity markets are: establish ancillary service markets; run day-ahead spot markets; consider unit emission criteria in the auction markets; start bilateral contract markets and generation right markets. The long-term goals are: open customer-side markets, establish financial markets, such as, futures, forward and option markets.

16.3.2 Electricity Price Policy

The electricity price policy plays an important role in China power industry reformation. In May 2005, the National Development and Reformation Commission (NDRC) issued three regulations about electricity prices, Generation Price Regulation, Transmission and Distribution Price Regulation, and Customer Electricity Price Regulation [14].

16.3.2.1 Generation Price

Generation price is composed by capacity price and energy price. Energy price represents the variable costs of a power plant. The price is obtained by competition through regional auction markets. It is the price for a certain amount of power during a time period. Generators can submit 3-segment offers or 5-segment offers depending on the market regulations. In some regional markets, the regional grid company provides a reference energy price for each provincial generation company.

Capacity price represents the fixed costs of a power plant. It can be calculated as following:

$$Capacity\ price = capacity\ payment\ /\ installed\ capacity.$$

Capacity price is determined by the installed capacity of the plant and the capacity payment, which is decided by the NDRC from time to time according to the power supply and demand of the time. The purpose of utilizing capacity price is to guarantee the basic benefits for power plants. The ratio of capacity price and energy price can reflect the degree of competition in a market. The higher percentage the energy price in a generation price, the higher competitive level the market is. In one of the regional markets, the capacity price vs. energy price can roughly be 1: [2~3].

Electricity price reformation in one of the core parts of power system deregulation. Price reformation will be accomplished gradually in a longer term in China. The current two-part generation price is a transition pricing mechanism towards completely competitive electricity market.

16.3.2.2 Transmission and Distribution Price

In the Transmission and Distribution Price Regulation issued by NDRC, the T&D price is composed by transmission price, ancillary service price, and special service price. Transmission price covers the network expansion and maintenance costs. Ancillary service fee is paid for the regional grid companies for their provision of ancillary services, such as, reserve, frequency control, voltage control, black start, etc. Grid companies use grid-owned units to provide such ancillary services. The special service prices are charged to market participants that require special services from the grid company. Such as, connect to the network; build a transmission line for a special project, etc. The government and NDRC decide the transmission and distribution prices.

16.3.2.3 Customer Price

Electricity price for customer is regulated and decided by NDRC. Customer prices can be classified into three categories: residential customer price, industry customer price and agriculture customer price.

16.3.2.4 Electricity Price Interlink with Coal Price

Most areas of China have experienced a high increase of electricity consumption in 2003, and one third of the provinces are suffering power shortages [11]. In some areas, electricity demands are higher than power supplies. System operators sometimes have to shed load to balance the power demands. On the other hand, the fossil-fuel price goes up significantly at the time. Around 74% of installed capacities in China are fossil-fuel thermal plants. The high coal price limits the profits of power plants by a maximum extent.

In April 2005, NDRC announced a scheme to interlink electricity prices with coal prices. According to NDRC, customer side electricity prices can be adjusted with the change of coal prices. A few months later, the customer electricity prices in all provinces increased from 0.95 to 3.99 cents (Chinese Cent) per kWh. The average electricity price of the whole country has increased 2.52 cents per kWh [15]. This price increase is around 5%.

On the other hand, generation prices all over the country are adjusted with the increased coal price. The generation prices of all provincial power companies increased from 0.33 to 3.1 cents per kWh. Other Independent Power Producers (IPPs) followed provincial power companies and increased their generation prices to keep up with the coal price increase [16].

At the current stage of power system restructuring, government and NDRC are in a leading position regulating electricity prices. The government and NDRC determine both transmission and distribution price and customer side electricity price. Only generation prices can be partly decided by auction market.

16.3.3 Renewable Energy Policy

In China power system, coal-fired thermal power plants dominate the electric power generations (around 74%). The generation costs and the electricity prices rely on coal prices to a certain extent. With the coal prices continuously going up in 2004 and 2005, the profits of most generation companies drop down. On the other hand, hydro generations are around 24% of the total generations. Renewable energy generation sources are not more than 1% of the total generations. The current renewable energy sources are mostly small-scale wind turbines. The slow development of renewable energy was due to the lack of market regulations and pricing mechanisms for high-cost renewable energy entering electricity markets.

In February 2005, the Renewable Energy Act was issued by NDRC. The Act indicates that the Chinese government encouraged renewable energy generation. The detailed regulations of the Act were enforced in January 2006. A proper renewable energy policy and a reasonable pricing mechanism would facilitate the development of renewable energy sources in China power system.

The renewable energy generations currently used in China are mainly solar energy and wind energy. The solar energy generation production reached 230MW in 2005. Some of the solar energy generations are installed at remote areas. In Shenzhen, a photovoltaic power station of 860kW has been interconnected to the transmission grid. It is estimated that the total solar generation production could reach 500MW in 2010.

The total installed capacity of wind power in 2005 is 500MW. Besides this, the wind turbine capacity is going up at an annual increase rate of 40%. The capacity is expected to reach 100GW in 2020, which will be around 10% of the total installed capacity of that year. Wind power and solar energy generations are the potential new generation sources in the future China power industry.

One of the factors that limit development of renewable energy is the high production cost. The cost of renewable energy is much higher than that of the regular coal-fired generation. In China, the cost of small hydro generation is around 1.2 times of the cost of thermal generation; the cost of biomass generation is about 1.5 times, the cost of wind power is about 1.7 times, and the cost of photovoltaic generation is about 11-18 times [17]. The average generation price for traditional thermal sources is around ¥0.5/kWh, the cost of wind power is usually around ¥0.8/kWh and the cost of solar energy could me more than ¥5/kWh. Besides the high cost of renewable energy, the uncertainty of the renewable energy pricing mechanisms and the unclear renewable energy policy also raised the investment risk. Investors are hesitated to invest in renewable energy generation unless they can see the benefits.

The Renewable Energy Act and its regulations provide detailed rules about interconnecting renewable energy units to the grids and the detailed generation prices for various types of renewable energy.

In the new regulations, two parts compose renewable energy generation price: government regulating price plus market auction price. The wind power generation price decided by the government is expected to be the fossil-fuel generation price plus ¥0.25 per kWh. The generation prices for photovoltaic might be classified into two categories: ¥3.2/kWh for those located at open areas and ¥3.4/kWh for those located in buildings. The prices for other renewable energies, such as biomass energy, are also stated in the regulations [10]. The government-authorized renewable energy generation prices will insure the companies recover their generation costs and the return rate of investment.

To facilitate renewable energy generations, the government authorized some regulations as well as the regulating prices. In the Act, it has been explicitly stated that the future renewable energy policy will be a quota system. The grid companies must sign purchase contracts with renewable generation companies and buy all the contracted renewable energy generations in their grids as well as providing transmission services. On the customer side, the large customers are compulsory to pay for the shares of the renewable energy allocated to them. The future regulations might fix a required proportion of the renewable energy in a grid. On the other hand, some preferential policies will be issued to encourage generation companies developing renewable energy sources. For example, some perquisites might be provided for renewable energy, some funds are founded especially for renewable energy, providing tax reduction or low interest loan, etc.

In Shanghai, the government has started to enforce the regulations on special generation prices and purchase prices for Green Power. The residential customers are required to buy Green Power at least 120 kWh per year.

16.3.4 Investment and Planning

The electricity consumption in China has been increasing in recent years. The total consumption in 2004 is 2.17PWh, which is a 15% increase of 2003. In 2005, consumption increased by another 10% and reached 2.4PWh. On the generation supply side, 50GW new generation capacity was installed in 2004, and 70GW new generation capacity was installed in 2005. However, it is still not easy for all investors to enter the market of generation investment. The five independent generation companies and local government owned generation companies play the dominant roles in generation investment, although the individual investors and overseas investors hope that they can hold a share in the investment market.

16.3.4.1 Investment

After the power industry restructuring, government regulates transmission and distribution. The generation side is deregulated, and independent power companies are able to participate in generation side markets. Each generation investment project is examined and discussed by the government before it is approved.

Currently, state-owned generation assets are around 90%, and private investors and overseas investors own the other 10%. Of the 90% state-owned assents, 35% is owned by the five-generation companies (including the 40GW approved in 2004), and the other 65% is owned by the local government generation companies [18].

The generation investors in China are mainly in five categories:

- The former state-owned five generation companies.
- Some other state-owned companies.
- Local government owned generation companies. They own around 55GW capacities in total.
- Overseas investors. They own 37GW in total.
- Other private investors.

In the new generation investment projects, the state-owned generation companies still occupy a large share. For example, the five-generation companies mostly invest in the 40GW generation capacities approved in 2004.

16.3.4.2 Overseas Investment

At the end of the 1980s, electrical consumption demands increased dramatically. To attract oversea capital investments to the China generation market, the China government has provided a very high rate of return for overseas generation investments. The annual rate of return was as high as 15% - 20%. In some provinces, the local governments sign long-term contracts with investors.

The contracts guarantee that the generators can get fixed generation prices for a number of years. More than 30 overseas companies invested in power plants in China. In 1997, their total capacity reached 14.5% of the total installed capacity of the country. In 1999, the government started to abolish the fixed high rate of return. Some of the investors do not see the high profits any more and withdrawed their investment from the market. The overseas generation investment reduced from 14.5% of 1997 to 7.5% of 2002.

16.3.4.3 Power System Planning

In the traditional integrated power system, planning is performed centrally. For some years, power system planning has been under a situation that generation planning always leads transmission planning.

After restructuring of the power industry, the integrated power system has been separated into some generation companies, and some grid companies. Who will perform power system planning is still not clear. The State Grid Company has the most possibility of being authorized by the government to perform transmission network planning for the whole country grid. Each regional grid company is responsible for its' own regional grid expansion and planning. Generation companies are responsible for generation expansions. Where and what size to invest new generations are decided by the government and the State Grid Company. Generation companies bid for building new generation capacities. The government and NDRC examine and approve all the expansion projects.

16.3.5 Challenges

There are some issues that need to be emphasized in development of the China power industry after system restructuring [19]:

- The structure of generation sources is not yet fully optimized. The proportion of fossil-fired generation sources is much higher compare to other clean renewable energy generation sources.
- The high proportion of generation relying on coal makes the electricity prices affected by coal prices significant.
- The transmission network expansion and planning are lagging the generation expansion. How to perform generation planning and transmission planning after unbundling generation and transmission is an emerging issue.

The challenges to the industry and government in the restructured power system come from several sides. From the investment side, efficient policies would facilitate the investment in renewable energy sources and in optimizing generation source structure. From the planning side, generation planning and transmission planning coordination is a big challenge for the unbundled system. An authorized body for power system planning might be a good option to solve the problem. From the development side, the current single-buyer regional electricity market is still in a transient stage towards the competitive electricity market. Electricity market models in all countries and areas are various. Finding the best electricity market model for the China power system will be one of the goals for the future China power industry.

16.4 Restructuring of the Electric Power Industry and the Current State of the Power Market in Japan

The restructuring plan for the Japanese electric utility industry has first focused on the major objective of lowering electricity rates, which are rather expensive by international standards, to levels comparable to those prevailing in foreign countries. However, the plan has called for achieving this objective through competition while also addressing public-interest issues such as environmental protection, with due consideration of the conditions specific to Japan, such as the nation's current heavy dependence on imported resources (Figure 16.4). Institutional reforms have therefore been pushed forward in accordance with the Basic Law on Energy Policy Making. A Japanese-style electricity deregulation model designed to secure fair competition while retaining the integrated power generation/transmission/distribution regime has been created with the existing electric power companies playing a central role in working to promote the public interest.

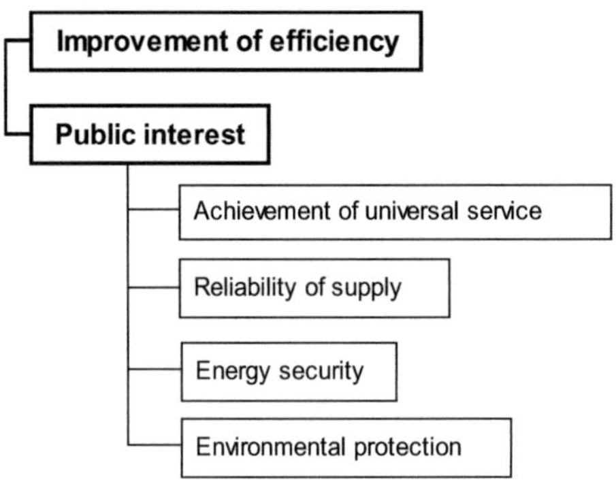

Figure 16.4. Basic principle of the restructuring

The Japanese electric power industry is being restructured in stages as discussed later. In the stage begun in April 2005, just over 60 percent of the retail electricity market was opened up, while basic mechanisms for the Japanese-style deregulation model, such as a neutral body and an electric power exchange, were put in place. Although it is still too early to judge the results of the restructuring, some appreciable benefits such as lower electricity rates have already been achieved. In fiscal 2007, discussion commenced on progressing toward complete retail deregulation. In order to provide inputs for this discussion, the efforts made so far to restructure the electric power industry were assessed.

This Section reviews the progress of restructuring and discusses the characteristics of the adopted Japanese-style electricity deregulation model. It then assesses and verifies the ongoing power industry-restructuring plan and reports on the results of the industry restructuring efforts [20,21].

16.4.1 Progress in the Restructuring of the Japanese Electric Power Industry

The restructuring plan for the Japanese electric power industry has made slow but steady progress through three broadly divided steps. The major milestones in these steps are as follows (Figure 16.5):

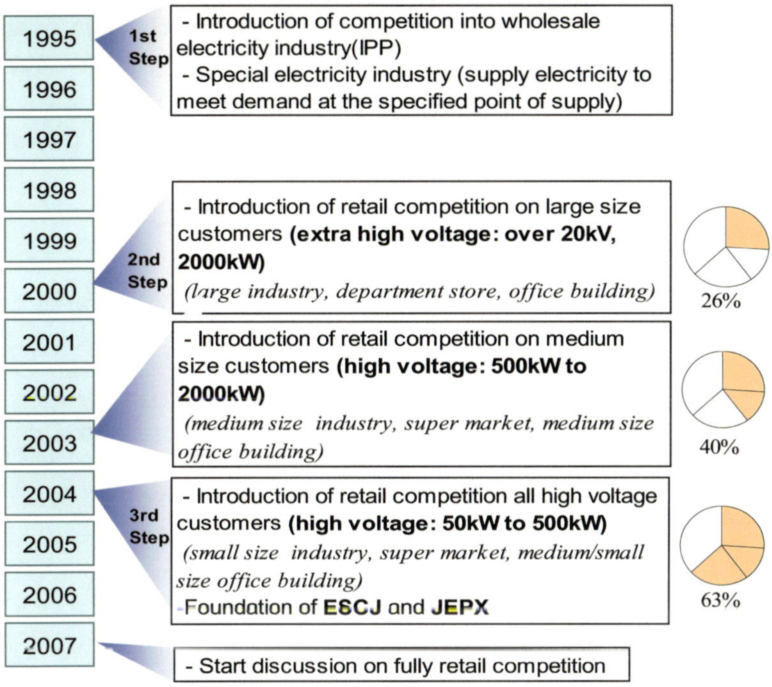

Figure 16.5. Steps of the restructuring

1st Step: 1995
- Liberalization of access to wholesale electricity industry (entry by IPPs)
- Introduction of a wholesale power supply bidding system
- Institutionalization of a retail electricity business at specified points of supply (Special electricity industry).

2nd Step: 2000
- Liberalization of retail electricity sales to extra high voltage customers (customers receiving electricity at 20 kV and with contract power of 2,000 kW or more). Newcomers with power generating facilities are referred to as Power Producers and Suppliers (PPSs).

3rd Step: 2005
- Liberalization of retail electricity sales to high voltage customers (customers with contract power of 50 kW or more)
- Establishment of a neutral body, the Electric Power System Council of Japan (ESCJ)
- Establishment of the Japan Electric Power Exchange (JEPX)
- Elimination of pancake rates
- Revision of the balancing rule (the imbalance system).

In Japan, in order to fully consider the characteristics of electricity and perform public welfare tasks, an electricity deregulation model designed to achieve structural reforms within the framework of electric power companies' integrated power generation and transmission system has been adopted. This is instead of the vertically unbundled electric utility model prevailing in the United States and European countries. The changes made in the third step are described in more detail as these represent the salient features of Japan's structural reforms.

16.4.2 Outline of the Institutional Revisions Effective in 2005

Figure 16.6 outlines the institutional revisions made since April 2005. The main measures implemented include:

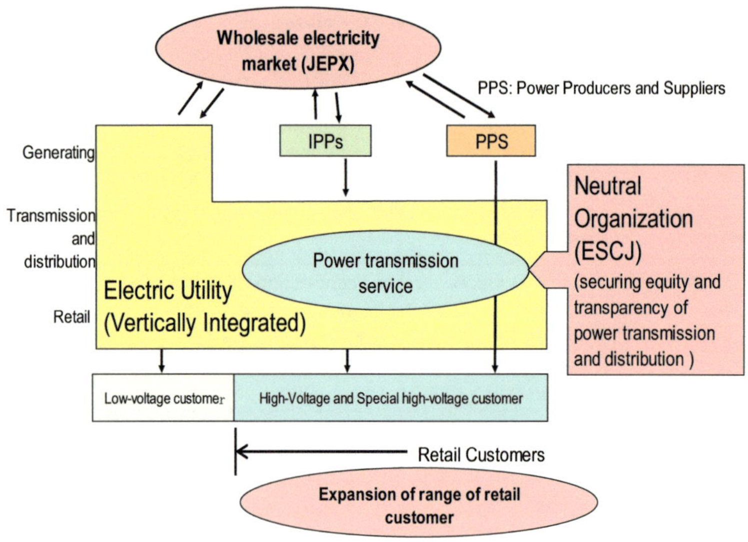

Figure 16.6. Institutional revisions made since April 2005.

1) Establishment of a neutral organization and the enforcement of activity regulations in order to secure fairness and transparency in the network sector within the integrated power generation/transmission framework
2) Creation of the Japan Electric Power Exchange
3) Expansion of retail electricity deregulation to cover all high voltage customers
4) Elimination of the pancake rate system in order to facilitate the use of power grids across the country.

16.4.2.1 Neutral Agency

The neutral agency is called the Electric Power System Council of Japan (ESCJ), and its main functions are:

1) Rule making for using power grids
2) Rule monitoring (settlement of disputes)

3) Support for the operation of interconnected transmission lines (management of ATC, maintenance work coordination, and congestion management)
4) Provision of information about transmission systems.

As shown in Figure 16.7, the ESCJ is composed of a governing board and various expert committees, such as a rule-making committee and a rule monitoring committee under the control of a general meeting of members who are the general power utilities, PPSs, wholesale electricity suppliers and neutral parties, each having an equal voting right. A consultative board conveys the views of non-members to the Council.

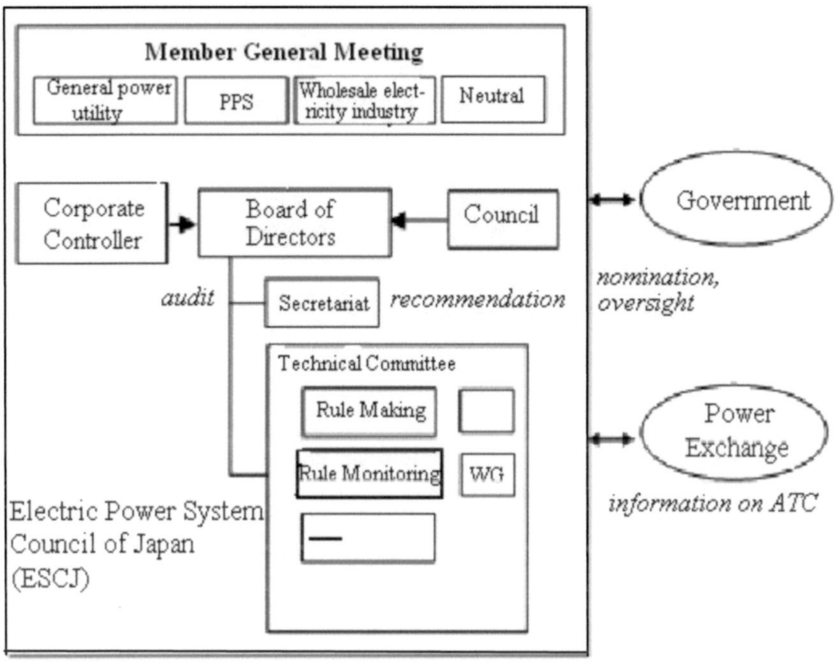

Figure 16.7. Structure of the ESCJ

16.4.2.2 The Japan Electric Power Exchange

The Japan Electric Power Exchange (JEPX) is a privately managed voluntary institution that was founded as Japan's first nationwide power exchange. The JEPX performs risk management functions, including the formation of a benchmark price, and the provision of selling and procurement means in the event of a mismatch of supply and demand.

Anyone who is capable of handling the generation and has assets worth 10 million yen or more can participate in the trading market. The JEPX deals in the following types of market.

1) Day-ahead market (spot deals):

Spot deals are deals for selling and buying electricity in units of 30 minutes to be used the next day and for 48 different commodities of electricity. Contracts are closed on the single-price auction basis.

2) Forward market:

Forward market deals sell and buy electricity for one year ahead from the following month in one-month units. Contracts for 24-hour delivery and for daytime delivery from 8 a.m. to 10 p.m. on weekdays are traded. Transactions are performed on a continuous session (price-oriented, time-oriented) basis.

16.4.3 Assessment of Institutional Reforms

A study on deregulating retail electricity for all customers started in fiscal 2007. The institutional reforms implemented so far are now being assessed and verified. The assessment looks at:

1) How macroeconomic policy objectives, such as enhanced efficiency, stable supply and environmental protection, have been achieved; and

2) What kind of results have the implications of individual institutional reforms, such as the elimination of pancake rates and the establishment of a neutral agency and an electric power exchange brought?

The results of the assessment were made available by the summer of 2006. This section summarizes the areas of assessment and the items assessed.

Regarding the macroeconomic policy objectives in 1) above, as shown in Table 16.4, the effects of the current institutional reforms on enhancement of the efficiency of electric power supply services, supply stability, and environmental protection is verified. Essential points in the assessment include the robustness of the electricity rate and trading market, capital spending, congestion of interconnected transmission lines, investment in the development of technologies, and investment without consideration of the environment.

On the other hand, the effects of individual institutional reforms are assessed in terms of the objectives mentioned in Table 16.5.

Objectives	Issues	Areas of Assessment	Items of Assessment
Enhanced efficien-cy of electric power supply service	Retail electricity market	- Changes in electricity rates - Factors contributing to cost reductions and rate cuts - Initiatives other than electricity rates	1) Electricity rates - Changes in the levels of electricity rates - Quantitative analyses of the effects of institutional reforms on electricity rates 2) Factors other than electricity rates - Efforts in terms of services - Initiatives designed to strengthen business bases, including technological innovation and development of overseas business operations 3) Options for customers - Number of suppliers and shares - Customer satisfaction
	Wholesale electricity market	- Robustness of the power generation and wholesale electricity markets - Performance of JEPX's initial purposes	1) Wholesale electricity market - Trading volume by electric power companies and PPSs, and prices - Trade channels (cross trading with JEPX, self-supply and purchase from other companies, etc.) - Correlation among wholesale electricity prices, generation costs and retail prices 2) JEPX - Position of JEPX in the wholesale market - Trading volume and prices in spot market and forward market, and liquidity compared with bilateral trading
Supply Reliability	Construction, maintenance and renewal of facilities	- Facility investment - Maintenance and renewal of facilities - Attainment of suitable generating mix	- Supply reliability - Changes in the amount of facility investment - Facility maintenance levels and measures for extension of life of facilities - A future power generation mix, transmission line construction plans, and electric power development plans by PPSs, etc.
	Interconnected transmission line	- Limitation of line capacity	- Construction of interconnected lines - Maintenance and operation of interconnected lines

Table 16.4. Assessment of the Macroscopic Policy Objectives (continued on next page)

Objectives	Issues	Areas of Assessment	Items of Assessment
Supply Reliability	Load dispatching control and system management	- Burden on load dispatchers - Appropriateness of ancillary services	- Changes in electric power companies' load dispatching control and system management activities following institutional reforms - Harmony between protection systems such as special protection schemes and short-circuit relays and newcomers' facilities - Provision and operation of ancillary services
	Safety and disaster restoration	- Changes in the functions of safety and disaster prevention systems	- Response to safety and disaster restoration needs
	Technology development and succession of skills	- Necessity of development of new technologies in the power transmission and distribution sector, etc.	- Implementation of technology development projects in response to institutional reforms
Environmental protection	Choice of power sources	- Orientation toward a generation mix attaching importance to economic efficiency only	- Changes in the generation mix and CO2 emissions in the power sector
	Development of environment-related technologies	- Efforts toward environmental development that have no direct contribution to profits	- Investment in the development of environment-related technologies
	Interconnected transmission line	- Limitation of line capacity	- Construction of interconnected lines - Maintenance and operation of interconnected lines

Table 16.4. Assessment of the Macroscopic Policy Objectives (continuation)

16.4.4 Evaluation of Institutional Reforms in the Early Stage

Institutional reforms are now being systematically evaluated from the above mentioned perspectives. This section evaluates the institutional reforms in their early stage based on currently available data. Note that some factors, such as electricity rates, can be accurately evaluated as the result of institutional reforms as sufficient time has passed since the early stage of the reforms, while it may be too early to evaluate other factors such as the robustness of transactions on the electric power exchange that were commenced in the third step of the reforms.

	Areas of Assessment	Items of Assessment
Wheeling service system	- Elimination of pancake rates and revitalization of wide-area power transaction - Functions of the new imbalance charging system	- Effects on wide-area transactions and on JEPX trade - Recovery of transmission line costs, an appropriate cost reimbursement mechanism, and restrictions on the siting of power sources in remote places - The incidence of imbalances and the shouldering of charges
Activity regulations	- Securing of fairness and transparency in the transmission/distribution segment through activity regulations	(a) Information firewall - Management of information related to wheeling service, and formulation and announcement of internal rules (b) Prohibition of discriminatory treatment - A survey of cases of discriminatory treatment in access to wheeling service for specific power suppliers (c) Prohibition of cross-subsidization (keeping of different account books) - Preparation and disclosure of documents needed to verify an income and expenditure account statement and the process of preparing the statement
Neutral agency	-Securing of fairness and transparency in access to power grids, construction of facilities and disclosure of information	- Establishment of an organization and a mechanism designed to achieve fairness, transparency and neutrality - Rule-making procedures securing fairness, transparency and neutrality - Consistency of load dispatching communication and system information disclosure systems (OASIS) with the neutral agency's rules - Fairness and transparency in the scheme for rule monitoring - Current situation of rule monitoring - Implementation of load dispatching communications - Appropriateness and security of the information disclosure systems (OASIS)

Table 16.5. Assessment of Individual Institutional Reforms

16.4.4.1 Electricity Rates

The wide gap between electricity rates in Japan and in foreign countries, which triggered the structural reforms in the Japanese electric power industry, has narrowed compared with before restructuring of this industry was started, as shown in Figure 16.8.

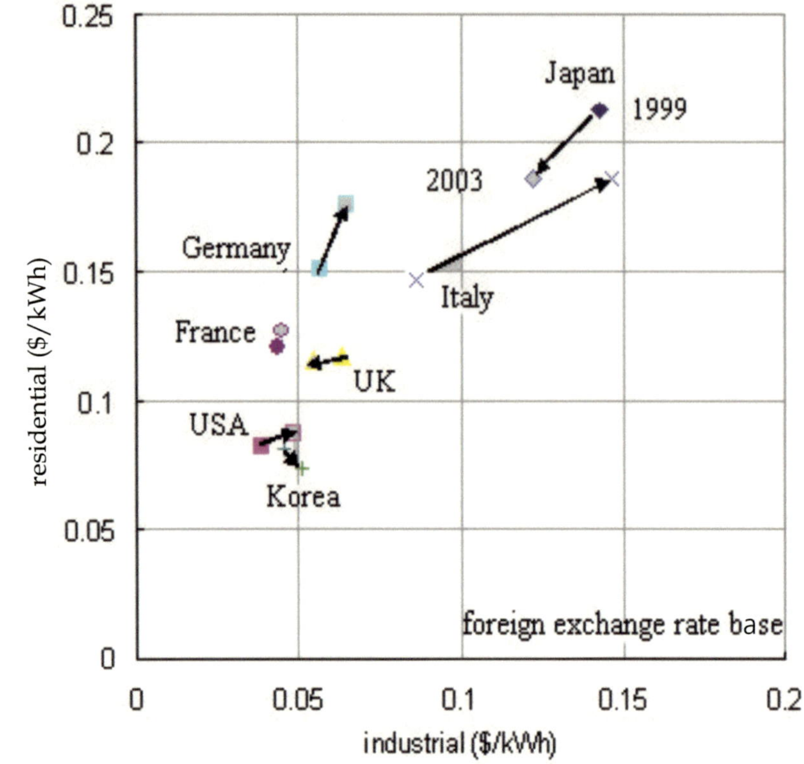

Figure 16.8. Comparison of electricity charge before and after the institutional reform

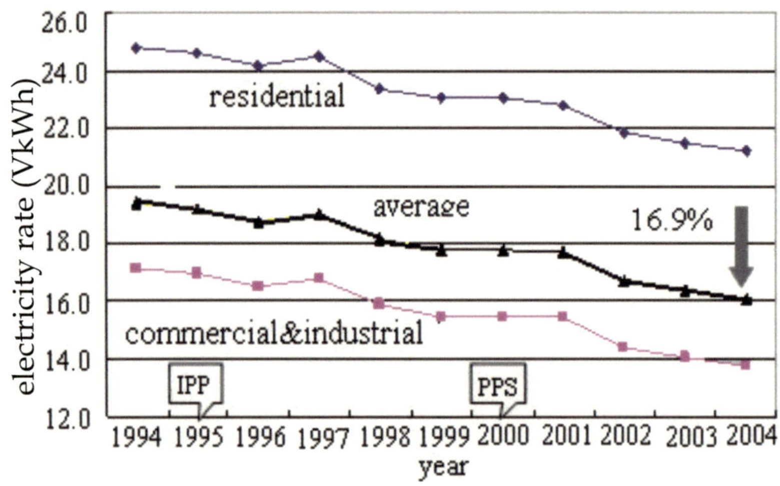

Figure 16.9. Changes in electricity rates over the past decade

Changes in electricity rates over the past decade are shown in Figure 16.9: electricity rates dropped by 1.8% annually on average from fiscal 1994 to fiscal 2004 and a simple comparison between electricity rates in fiscal 1994 and in fiscal 2004 shows a decline of approximately 17%. As the graph shows, residential charges, which are not liberalized, have declined to sufficiently lower levels. In the liberalized segment, a simple comparison between electricity rates in 2000 and in 2004 revealed a significant drop of nearly 26% in electricity rates for commercial customers.

16.4.4.2 Situation of Newcomers

A look at newcomers (PPSs) in the liberalized sector shows that although their share is still small at around 2%, PPSs have been steadily increasing their power sales (Figure 16.10). PPSs' shares vary according to electric power companies. In Tokyo Electric Power Company's service area, where the utility meets an immense demand for power, newcomers hold a share of more than 5%, whereas PPSs account for no more than 1% in the service areas of some local electric power companies.

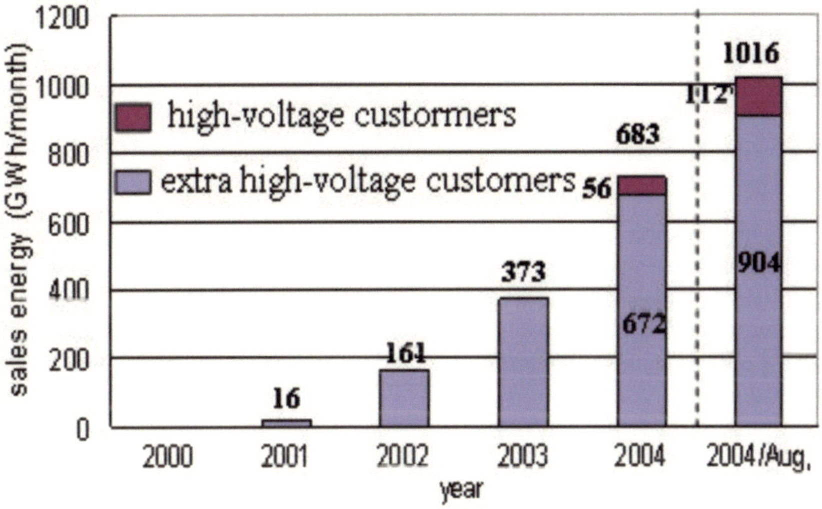

Figure 16.10. Share of power producers and suppliers (PPSs)

16.4.4.3 The Japan Electric Power Exchange

The Japan Electric Power Exchange is still immature as it was opened only in April 2005. As shown in Figure 16.11, over the five months following its opening, the exchange has had thin trading and prices are relatively high in summer, but subsequently transactions have increased gradually and prices have become steady. Reasons for this tendency include the emergence of relatively large sellers, a rise in oil prices, and the suspension of large power sources.

At the time of the exchange's opening, an outlook for trading volume was provided as shown in Figure 16.12. The actual volume of transactions already surpassed the estimate for the first year in mid November.

Challenges that the exchange must tackle in the years ahead include:
Measures to increase the trading volume: Increase in the number of exchange members, such as in-house power generation facility owners.

Introduction of new commodities meeting needs: Introduction of short-term forward delivery contracts.

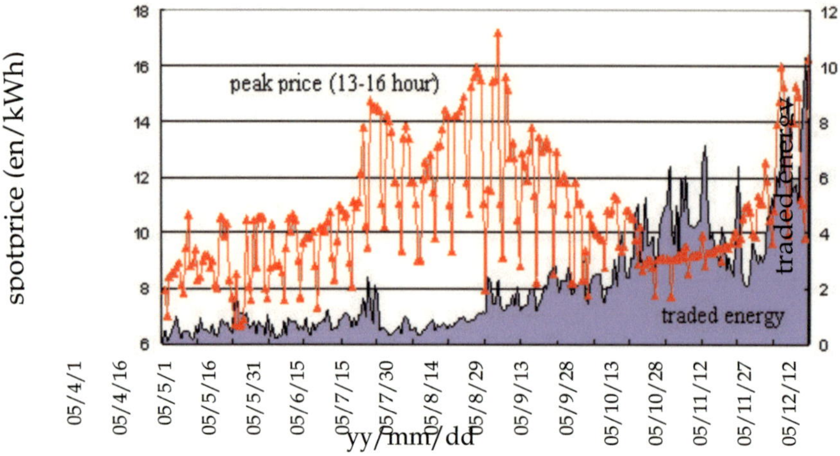

Figure 16.11. Price and traded energy in day-ahead market of JEPX

16.4.4.4 ATC and Transmission Line Congestion

Available transmission capacity (ATC) for the next ten years in Japan is shown in Figure 16.13 [22]. In Japan, due to the geographical conditions and historical background, interconnections between electric power companies are loosely connected and transmission congestion poses a problem in domestic interconnected lines. ATC is therefore presented for each interconnected line.

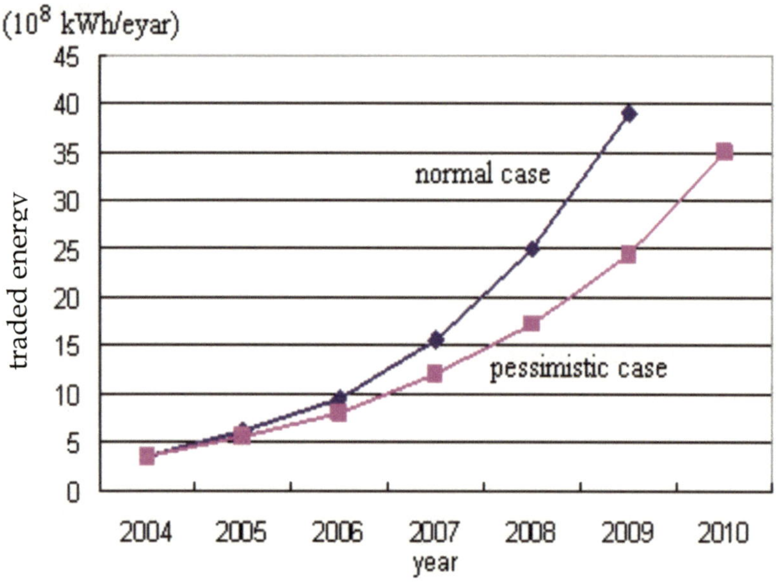

Figure 16.12. Prospect of trading volume (at the opening time of JEPX)

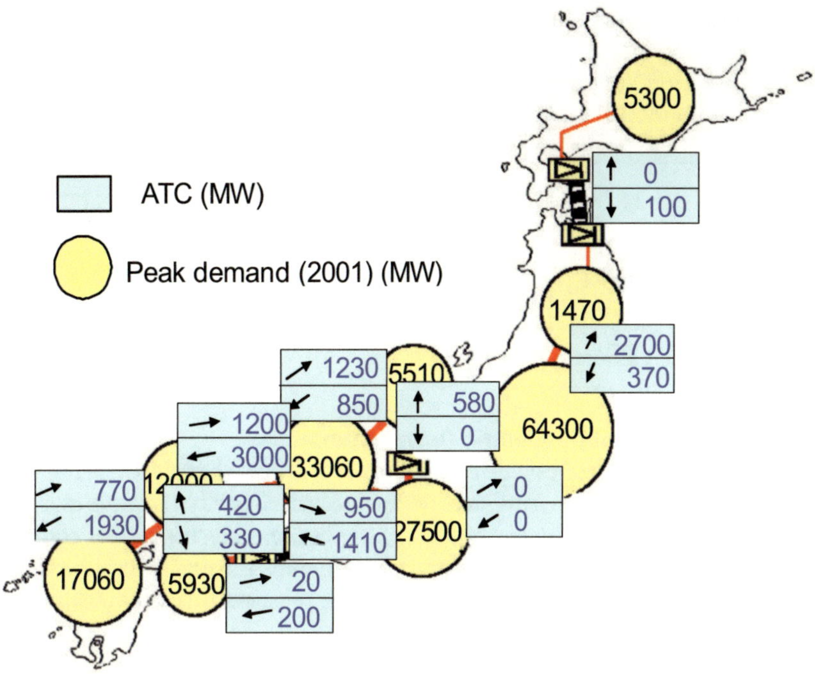

Figure 16.13. ATC of the interconnected line (from 2005 to 2014)

The frequency and duration of transmission line congestion are shown in Table 16.6. Congestion occurs several times a month. In September and October 2005, transmission congestion occurred due to a decrease in the counter power flow (a power flow in the opposite direction) resulting from the extended period of repairs on large power sources. The ESCJ has undertaken a technical study on ways of managing interconnected transmission lines, including reinforcements.

Month	Frequency		Duration	
	Planned	Emergency	Planned	Emergency
April	0	1	0	10.82 h
May	1	0	130 h	0
June	4	2	90 h	26.5 h
July	3	3	123.5 h	56.97 h
August	1	1	80 h	744 h
September	20	2	660 h	257 h
October	31	0	824 h	0
November	3	1	123.5h	39.6h

Table 16.6. Congestion of Interconnected Lines (2005)

16.4.4.5 Adequacy of Generation Capacity

One of the ESCJ's functions is to assess reliability of the power system. An assessment of the adequacy of generation capacity is given in Table 16.7. The adequacy of power sources is expected to be maintained at an appropriate level for the next ten years or so.

	Demand/Supply Balance	
	2009 (August)	2014 (August)
Peak demand	18,200	19,246
Planned capacity	19,838	21,268
Reserve ratio	1.09	1.11

Table 16.7. Prospect of Generation Adequacy

16.4.4.6 Capital Investment by Electric Power Companies

Capital investment by electric power companies has fallen sharply over the past several years as shown in Figure 16.14. This decline has occurred partly because demand growth has slowed down as shown in the Figure and partly because electric power companies have virtually completed their bulk power transmission systems through investments made over the years. In fact, reductions in capital spending have released financial resources that have allowed the electric utilities to lower electricity rates.

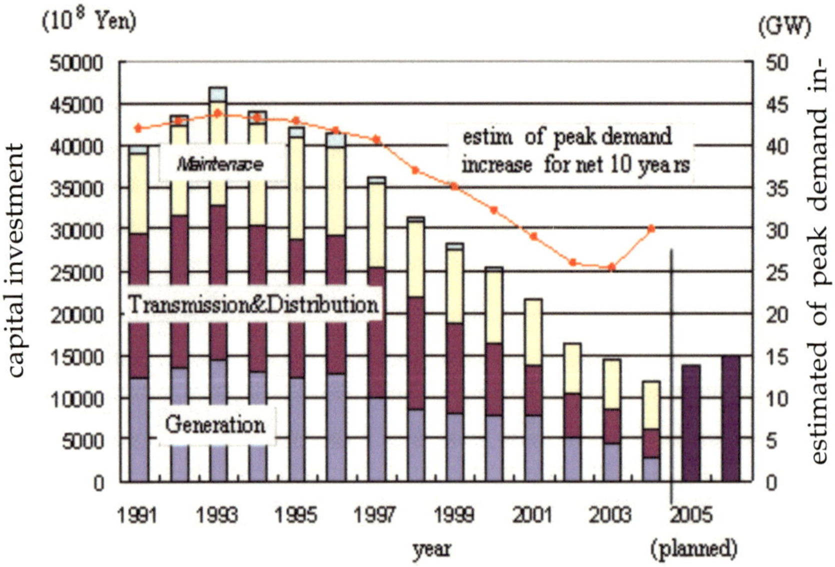

Figure 16.14. Trend of capital investment in utilities

The population of Japan began to decline in 2005, much earlier than expected, and so electric power demand cannot be expected to increase significantly in future. Still, the facilities that were constructed during the growth period will have to be renewed by around 2010. How the facilities should be managed within the context of electricity deregulation is an important question that remains to be answered.

Thus, this Section has outlined the institutional reforms in the Japanese electric power industry and an assessment of the reforms already made. The structural reform of the Japanese electric power industry is characterized by a step-by-step approach as well as by reforms being implemented within the framework of electric power companies' vertically integrated power generation and transmission structure. Although it is still premature to draw conclusions about the results of the structural reforms, electricity rates have dropped and the differential between electricity rates at home and abroad has narrowed considerably. A national-level assessment of the institutional reforms was made and the results were produced in the summer of 2006. Based on these results, discussion has commenced in order to establish a Japanese-style electricity liberalization system.

16.5 Pricing for Transmission Services in Korean Electricity Market

In Korea, the trend of heavier real power flows into densely populated load centers from several vast power plants in remote locations will continue or become profound, leading to their national-interest in transmission bottlenecks during some periods of the year. The first step toward increasing the role of market forces in managing transmission system operations is to develop the role of location price signals to direct the actions of market participants toward outcomes that improve operations when congestion occurs on the bulk power grid. In this overall perspective, we need to thoroughly investigate how best it would be to send an adequately accurate location price signal with the congestion costs incorporated into the transmission pricing rule when the electricity market is not unduly maintained. This Section attempts to make a supportive and self-explanatory proposal that it could fit the Korean Cost-Based Pool (CBP) that satisfactorily sharpens the location price signal.

16.5.1 General Characteristics

16.5.1.1 The Evolution of Competition in Korean Electricity Industry

It used to be assumed that electricity generation, transmission, distribution and supply enjoyed significant vertical economies that would be lost if the functions were placed under the control of different companies. Such long-held belief made it possible that the Korea Electric Power Corporation (KEPCO) had monopoly power – supported by legal protection. Since the 1997 financial crisis, economic policy in Korea has aimed to remove barriers to trade and competition. Network industries like electricity and natural gas, which were historically sheltered from competition and operated within national or regional boundaries, have experienced radical change as a consequence. National pressure to liberalize electricity markets reflected the perceived benefits of introducing market forces into the electricity industry previously viewed as a natural monopoly with substantial vertical economies. In the meantime, the generation sector was split into six subsidiaries that will be privatized. Still, the KEPCO is being engaged in monopolistic business activities of the transmission and distribution systems alike. In an attempt to help mitigate potential negative prospects about which the hasty reform drive could bring, a new transitional electricity market, dubbed 'Cost-Based Pool (CBP)', was set up in 2001.

16.5.1.2 The Distinctive Feature of CBP

The characteristics of the Korean CBP market can be briefly summarized as follows:

- The Korea Electric Power Corporation (KEPCO) is the single purchaser. Exceptionally, large consumers (above 50MVA) and district electricity businesses can either buy electricity directly from the Pool or KEPCO.
- The generators added at peak load are given the short-term marginal price (SMP) and the capacity payment, 7.17 won/kW as the fixed cost, while the base-load power plants receive both the base-load marginal price (BLMP) and the capacity payment, 21.49 won/kW as the fixed cost. When the CBP was first designed, greater portions of the windfall profits from the high SMP for each generation company were envisaged because of the shortage of base generation capacity. Thus, the so-called BLMP was adopted to evade the excessive revenues of the base-load generators.
- As such, there is no location price signal. If the generators inevitably change their output due to the congestion, they have been paid the uplift to make up for their generation cost.
- The generation cost is a priori known by the committee's actual test. Individual generation companies offer their available capacities alone, not prices. Then the Korea Power Exchange (KPX) performs economic dispatch based on the generation fuel costs.
- Indeed, the transmission price seems to be nominal in that it is published every year but not really applied to the market participants. In the CBP, the KEPCO collects the electricity price from the customers on a regulated tariff and provides the variable costs (SMP and BLMP) plus the capacity payment for the generation companies.

Originally, the futuristic model, namely *Two Way Bidding Pool* (TWBP) in which the market-clearing price would be determined from the bids of customers and the offers of generation companies in the unconstrained dispatch, was supposed to commence in 2004 in order to overcome the shortcomings of the current CBP. However, the government halted it and accordingly the CBP is expected to continue for the time being.

16.5.2 Pricing for Transmission Services under the Cost-Based Pool

Now, the main drawback in the CBP is an essential absence of the location price signal. Though either nodal pricing that recognizes different prices at every location or zonal pricing that creates administrative aggregations to reallocate costs is a nearly dominant answer to the prospective price signal in energy markets [23,24], it would be really time-consuming to rectify the current uniform pricing regime and, at the same time, redesign the market in terms of future network infrastructure costs. The access charge such as license plate or postage-stamp method is widely accepted to meet revenue expectations. In some European electricity markets, the power tracing method has been introduced for full cost recovery and location price signal [25].

16.5.2.1 The Basic Structure of Pricing for Transmission Services

In general, the overall equipments of transmission system are grouped into connection assets, common and location components, respectively. The pricing for transmission services in the CBP mainly consists of the access charge and the transmission usage charge as seen in Figure 16.15.

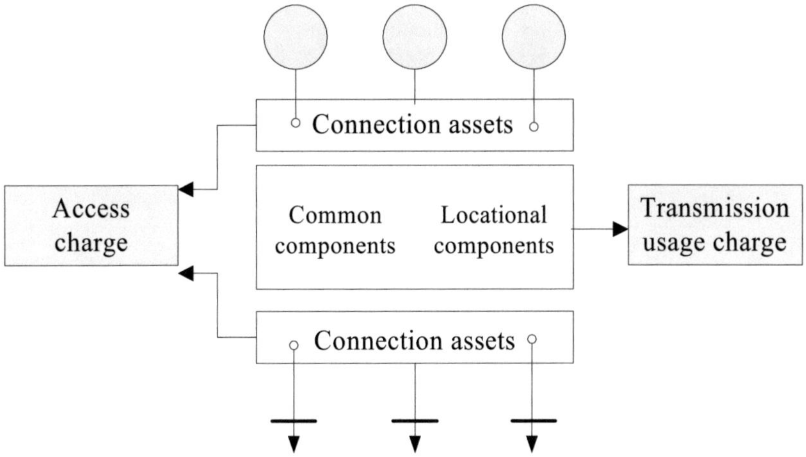

Figure 16.15. The basic structure of Korean transmission pricing

• *Access charge*: All the market participants who would likely use the transmission system should pay the charge for the connection assets that directly interconnect them with the substation. As is well known, a 'deep' basis is better rather than a 'shallow' basis in case of the access charge. From a practical standpoint, the CBP comes to have a clear preference for a "shallow" basis since it is a relatively good tool for elaborating the stringent criteria.

• *Transmission usage charge*: The electricity passing the connection assets is transmitted from generators to loads through the core of transmission system. Two components, i.e. the location and the common service components, are included hereof. In principle, the common component is made up of facilities related to the reactive power, non-operational land holdings, communication equipment, whereas the location component encompasses transmission lines and circuit breakers, etc. In the CBP of Korea, postage-stamp rate method is traditionally used for the common component, while the power tracing method associated with the location component is poised to give a location price signal to the participants. In Figure 16.16, the transmission charges in the CBP are represented for better understanding.

16.5.2.2 Power Tracing Method

The gist of the tracing method is to evaluate the contribution of transmission users to transmission usage of location components. This method may be employed to determine which generators are supplying corresponding loads, how much use each load is making of specific transmission lines and what are the annual costs of individual network elements to be recovered.

Especially, the cost of substation with which several branches are connected is actually divided into each line. In this method, more detailed load flow analysis of the system and its operation at the peak time are usually required to allocate the estimated annual costs of network elements to all the participants who use them. In an early stage, the fault current based power tracing method in the Victoria pool was scrutinized, but at this time, the methodology set forth by Felix Wu is in popular use [26].

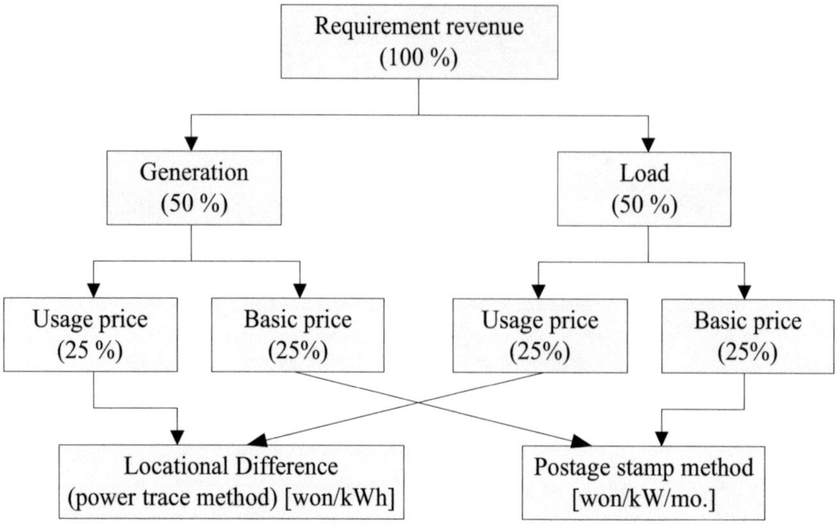

Figure 16.16. Illustration of transmission charges in the CBP

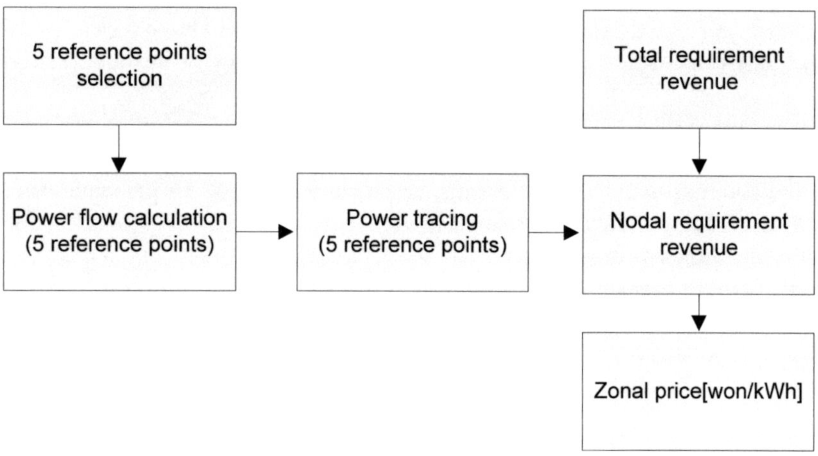

Figure 16.17. Procedure of usage price calculation

A natural implementation of the proposed algorithm is described in the following procedures (see Figure 16.17):
1) Choose 5 reference points when each load reaches 100, 90, 80, 70 and 60 % of peak load.
2) The power tracing method is applied for 5-averaged load-demand and generation level at buses, allowing for a whole set of 5 reference points.
3) The cost of each transmission line is calculated using equipment replacement cost which deals with what it will cost to replace the piece of equipment in the future and then converts the future cost into today's pricing.
4) All 5-transmission prices as to each bus are created from a pair of the averaged demand and supply at the respective reference points. A weighted average of 5 transmission prices against the elapsed time is equal to the final transmission price at each bus.

5) The zonal price, or a weighted average of the node price against the corresponding load-demand within the same zone, is derived.

16.5.3 Case Study

Data from Korea electric power system is used to calculate transmission prices. Figure 16.18 is a schematic showing locations of major generation and transmission facilities in the Korea electric power system. From the geographical point of view, the system can be characterized in the following [27]

- Island system
- Concentration of load demand in the metropolitan region
- Location of major generation plants in non-metropolitan regions
- Environmental concerns and restrictions on regional transmission siting.

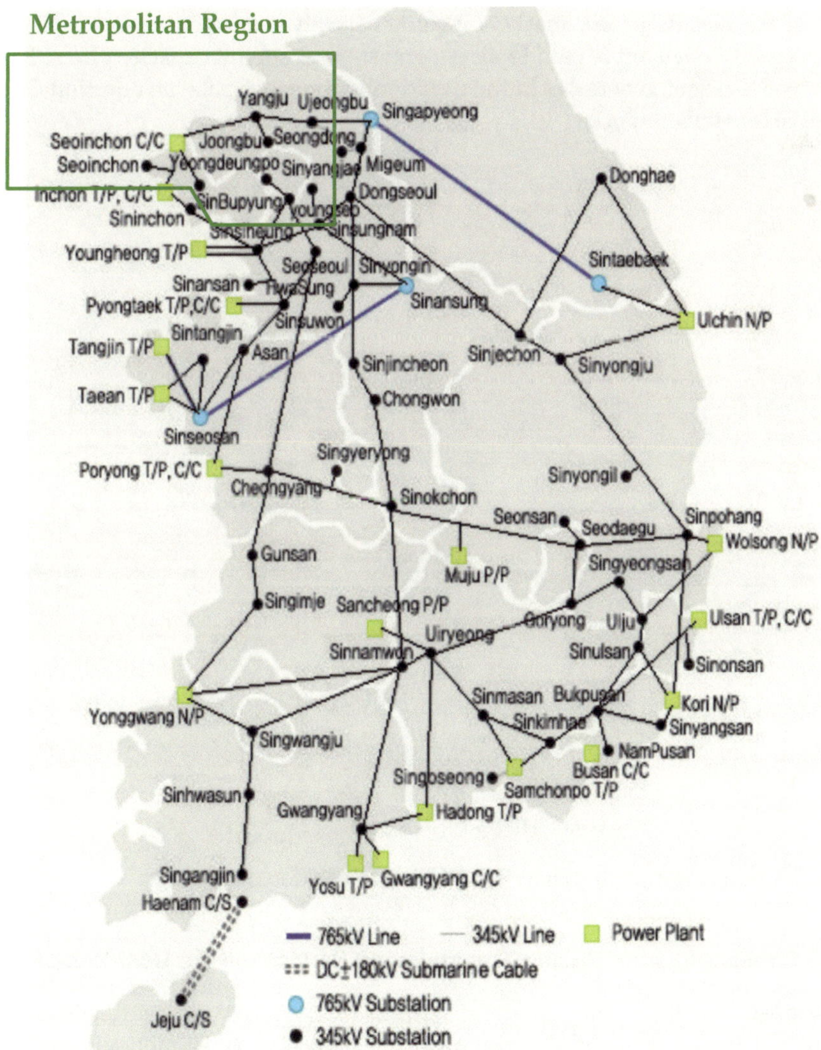

Figure 16.18. Schematic of major transmission facilities in Korean electric power system

The transmission price calculation algorithm used in NETA and Ireland is applied to the Korea electric power system for comparison with CBP method. In the results of simulations, it shows that the methods of NETA and Ireland provide stronger location signal than the CBP's. In Figure 16.19, the left side of the horizontal axis indicates the nodes of metropolitan region and the right side indicates the nodes of non-metropolitan region. While there are negative prices in the metropolitan region for the methods of NETA and Ireland, there are all positive prices in the CBP's method.

Even though providing location price signals is an important role of pricing transmission services, it is undesirable that these methods should be applied to the Korean pool model. The reasons are as follows:

- The energy markets of NETA and Ireland is run by contracts without location price signals.
- Most of the electric power markets include capacity market or capacity payment. In CBP, capacity payment is paid to all generators who submit capacity offers. For example, lower efficient generators in the metropolitan region make an additional profit on negative transmission price.

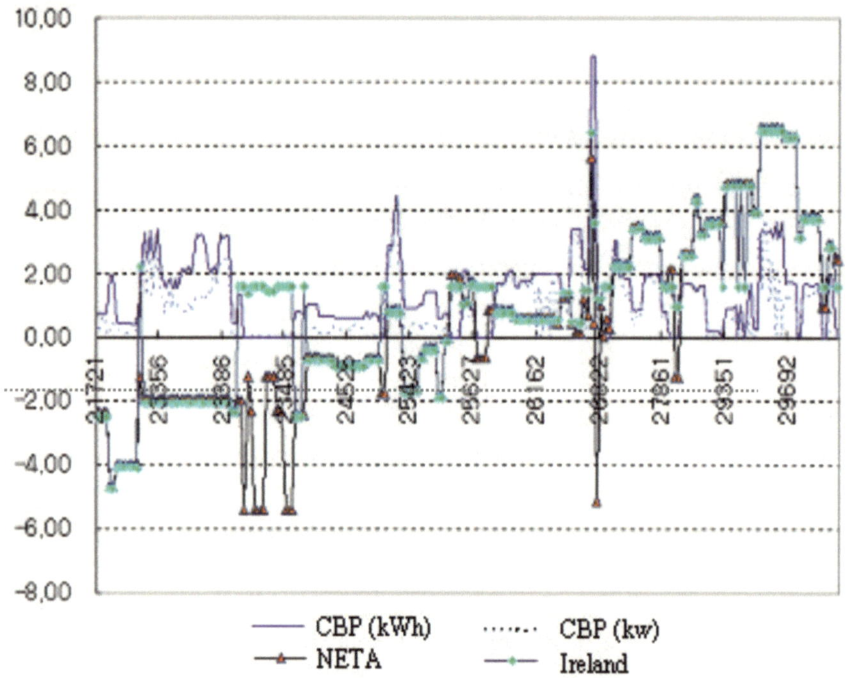

Figure 16.19. Comparison with transmission price calculation algorithms

16.5.4 The Challenges from Pricing Transmission Services in the Cost-Based Pool

16.5.4.1 Access Charge

- Though it is clearly stated that the already established generation companies in the CBP should take the responsibility for the access charges, they do not pay for these charges as it is. Conversely, a group of independent power producers with the power plants in course of construction is installing the connection assets as long as their budget allows. That is, new entrants are being outright discriminated from the old ones in the CBP model.

- It is even difficult for the existing generation companies to be imposed on the access charge in the CBP environment. That is why the market should reward those companies with the SMP, CP and infinitesimal reserve-related charge which are earmarked according to the generation fuel costs and installed generation capacity, and hence they can afford no extra payments for access charges, let alone the transmission prices. This critical defect is fairly associated with not only the access charges but also the transmission pricing in itself.
- For some generation companies, they are linked with the common components via the transmission line whose voltage might be as high as 765 kV. Those companies are facing a heavy burden with the enforcement of the capacity investments of the past.
- Today and increasingly in the future, some parts of connection lines may be constructed across quite long-distance routes so as to be utilized for the common components.

16.5.4.2 Transmission Usage Charge

In relation to the transmission usage charge, the following questions can arise:

- Further work will be needed since the allocation rules, with rating the shares of the generation companies and load entities or common and location components at fifty-fifty, are totally unconvincing forceful standards.
- In fact, the extent or systematic methodology of yielding location price signal should be discussed in a meticulous way and it should ultimately have a bearing on the energy market. But neither the CBP energy market nor capacity price provides location price signals between two areas. Therefore the pricing for transmission services to provide different location price signals between two areas is necessary.
- The appropriateness or usefulness of the power tracing method will be analyzed in detail.

The states of power systems are subject to change, relying solely on the planned maintenance overhauls for generating units, new entry of power plants and the configuration of the infrastructure. Nonetheless, the ongoing method does not capture the changes of conditions in power systems, with the transmission price through the year fixed at a single value. It needs to be identified how much impact each of the factors have on the transmission price so that the transmission pricing should be able to enhance fairness and efficiency.

Thus, in the face of much complication, the Korean government has been committed to putting the final touches on the public services in the electricity industry, reiterating its willingness to go ahead with the policy. The fruit of these incessant efforts has proved bitter as yet. In these staggering surroundings, it is a brilliant future task to see the inherent limitations of the current *Cost-Based Pool* and find fundamental solutions without delay. To be sure, it will be allowed with the nation's consent if there are attractive alternatives to the aforementioned tantalizing problems. Apparently, the transmission pricing should be a reasonable economic indicator used by the market to make decisions on resource allocation, system expansion and reinforcement [28].

16.6 Economic Convergence Points of Russian, CIS and Asian Power Markets

The convergence of Asian power markets crosses political and geographical boundaries. This Section provides an analysis of the technical and economic commonalities involved. The two primary goals of power sector reforms have been liberalization and the establishment of a competitive power market.

This Section discusses the fundamentals of cooperation, and the primary goals of power sector reforms. The main reform priorities and primary strategic areas are discussed in the context of the developing power market in Russia. Also the Section makes the point that despite the large geographical area, and economic disparity of Asian countries, there are commonalities to be considered in power sector reform. Each country should create a reform methodology in a national and regional conceptualization.

It is explained that there are five common steps on which this can be based. Additionally, when determining the correct model to apply to an economy to bring about the desired efficiencies, it is important to consider 13 transitional issues.

16.6.1 Economic Analysis of Russian and CIS Power Sector Interconnections and Markets

The Commonwealth of Independent States* (CIS) power sector has been characterized by cooperation, liberalization, privatization, investment, and cross border integration of electricity markets [29]. The fundamentals of cooperation have been stated in the CIS agreement entitled "On coordination of interstate relations in the field of electricity of the Commonwealth of Independent States" signed in February 1992. Arising from this agreement, cooperation has been implemented in the areas of structural reforms, power system integration, legislation and price liberalization. This is in response to serious power sector problems experienced by CIS countries in financial, technical, operational, and investment areas. As a group, the CIS countries faced additional crises because of the 12 members, only four; Azerbaijan, Kazakhstan, Russia, and Turkmenistan have enough fuel and energy resources to cover internal demand.

The primary goals of power sector reforms in the CIS have been liberalization and the establishment of a competitive power market. This has been conceived, since 1996, as a complex process, based upon relevant technological, structural and legislative foundations [29]. Many of the CIS countries have phased-in the process of power sector reforms. These reforms have followed previous global experiences by de-integrating vertically integrated monopolies into distinct distribution, transmission, and generation companies. This has established the foundational single-buyer model of market reform, in preparation for more liberalization. The single-buyer model was introduced in developing countries in the 1990s.

---

* The CIS is composed of: Armenia, Azerbaijan, Belarus, Georgia, Kazakhstan, Kyrgyzstan, Moldova, Russia, Tajikistan, Turkmenistan, Ukraine and Uzbekistan. In October, 2000, the heads of five countries (Belarus, Kazakhstan, Kyrgyzstan, Russia, and Tajikistan) signed an agreement on the creation of the Eurasian Economic Community. Armenia, Moldova and Ukraine have observer status under EAEC. In October, 2005, Uzbekistan agreed to join this organization. In September 2003, four countries (Belarus, Kazakhstan, Russia and Ukraine) signed an agreement on the Formation of CES (Common Economic Space).

It provides exclusive rights for a single transmission and dispatch company to purchase all electricity from generators and then sell it to distributors [30].

The single-buyer model became popular in the CIS countries and elsewhere because of technical, economic, and institutional reasons. These include:

- Balancing of electricity input and output is facilitated by a single dispatch method in real-time.
- The "*contract path*" problem is avoided.
- The single-buyer model is usually responsive to the sector Ministry in the areas of generation capacity investment and state-owned company's financial affairs. The most influential stakeholders favor this.
- Price regulation is simplified, maintaining a unified wholesale price.
- The single-buyer model is favored by politicians who do not agree with a complete withdrawal of the state from wholesale electricity trading.

There is evidence, however, that it is more efficacious to adopt a market model that has multiple buyers formed immediately after unbundling, in areas such as the CIS. The single-buyer model is criticized because it tends to place generation capacity expansion decisions in government officials' hands, which do not assume the financial consequences. This model also makes the state ultimately responsible for poor power purchase agreements that cannot be honored. This is regularly part of the contract agreement. This model also does not respond well to reductions in electricity demand. Wholesale electricity prices will rise in response to demand reduction because of fixed capacity charges, which must be applied to a reduced volume of electricity purchases. Another drawback is the under development of cross-border power trade. This is because the state-owned single-buyer has very little profit motive. This can cause long-term problems when a neighboring country or region develops a more liberalized power market model. Incentives for distributors to collect payments from customers are also reduced. Again, politically unpopular decisions are difficult for a state-owned entity. This is important because in the single-buyer model, delinquent payments from distributors are assumed by the aggregated cash proceeds of the single-buyer. Paying and non-paying distributors are treated basically the same, weakening the resolve of distributors to enhance the collection of payments. Another political intervention in this model has been experienced in the Ukraine and Poland. In these countries, groups have lobbied on behalf of coal miners to provide special treatment for coal-fired power plants. Interestingly, because of the political advantages of the single-buyer model, some governments have significantly delayed the next "*phase*" of fully liberalized markets [30].

The CIS has experienced advances in the privatization of the power sector. For example, Armenia, Azerbaijan, Georgia, Kazakhstan, Moldova and the Ukraine have either contracted the management of electric companies, or sold generation companies and distribution electric grids. Modernization through the construction of energy facilities has been facilitated by the creation of a legal basis. This has attracted direct investments in improvement projects from internal and external sources. For example, Armenia, Azerbaijan, Georgia, Kyrgyzstan, Tajikistan and Uzbekistan have received investments in projects ranging from hydro power plants, steam gas turbine units to electricity grid restoration [29].

There has been ongoing integration, and re-integration of the CIS power systems. For example, in June 2000, the power system connections of Russia and Kazakhstan were restored. Since the autumn of 2001, 11 of the 12 CIS countries have re-synchronized their power systems as an interconnected power system of the Commonwealth of Independent States (IPS-CIS) [29]. Additionally, the CIS is involved with power exchanges and trade with the neighboring countries of Afghanistan, China, Finland, Hungary, Iran, Norway, Poland, and Slovakia. The CIS power authorities believe that cross-border synchronization is beneficial because it places better use of existing generation capacity, provides emergency assistance options, and harmonizes standards in the areas of safety, ecology, and technology.

As part of an economic analysis of Eastern Europe and CIS transitional development from centralized planning to free market forces, a framework is used to evaluate the impact of infrastructure upgrades [31]. Research has indicated that "*The potential for developing competitive markets in transition economies had been inhibited by the inadequacy of both the institutional and physical infrastructure inherited from the socialist era.*" Although the following are general recommendations for transition economies, the following economic contributions of market-enabling infrastructure can be applied to the power sector:

- *Market Selection*: By increasing market competition, infrastructure investments that reduce transaction costs reduce the market share of higher-cost firms. This raises the average production efficiency of the economy.
- *Infrastructure Investments*: If these investments increase product market competition, incentives for firms should change to help reduce their costs by engaging in a restructuring plan.
- *Increasing Market Share*: If lower cost firms enter the market, they have probably been attracted by an infrastructure that lowers transaction costs (European Bank, 2000).

Therefore, each investment project in the CIS power sector can be evaluated in terms of the contributions of direct market selection, restructuring and entry.

The primary problem with the power sector in Russia, in the context of the CIS, has been inadequate investment [32]. Reasons for this are found historically, primarily being a lack of investment in generation capacity, and increased industrial activity and subsequent higher demand. A main goal has been to attract internal and external large-scale investments to the potentially competitive elements of the power sector. Although Russia has formed a plan, crafted in 2001, to eventually liberalize wholesale and retail electricity tariffs and to privatize segments of the sector, other contiguous CIS members have taken a lead in reformation. Kazakhstan was a reform leader by unbundling transmission, distribution and generation. Most generation and distribution assets have been privatized, but transmission networks are still owned by the government. In April 1999, the government of Kazakhstan approved a program to develop the electric power sector, with a view to 2030 [33]. The main priorities are:

- To have economic and population self-reliance in electric power, and to have energy independence as part of national security.
- To create competitive resources for electric power export to supply energy markets of contiguous and third countries.
- To develop a competitive electric power market on the basis of electric power transport and distribution networks accessible for generators and a system to control power flow.

The primary strategic areas are:

- To create a Kazakhstan unified energy system (UES).
- To restore the synchronous network with Russia, and other energy networks of central Asia.
- To further develop an open and competitive market for electric power.
- To commission new generation capacity to offset power imports.

Assuming these priorities, it is useful to view Kazakhstan as a precursor to significant, similar changes in Russia. In 1990, the electricity shortage in Kazakhstan reached a level of 17.3 billion kWh [33]. This shortage was made up by power imported from Kyrgyzstan, Russia and Uzbekistan. In 1996, the state generation and power grid monopoly was the object of reform, with the aim to create a competitive power market. That year, the government initiated a program to privatize and restructure the power sector formalized by Decision No. 663 of May 1996. Because of this decision, large generators were sold to investors, and a grid company was created to own and maintain the high voltage grid assets, including voltage levels of 1150, 500 and 220 kV. Additional decisions (1188 of September, 1996 and 1193 of July, 1997) provided a model for Kazakhstan to follow for the creation of an electricity market. Designed for competition, the single electricity market would have two levels, wholesale and retail. The trade of electricity would be based upon fixed date bilateral purchases and transactions, and agreements on grid capacity. The competitive model would be brought about by pro-active changes in the electricity sector:

Restructuring to create an effective, competitive market based on forwards contracts under the supervisory control of a centralized dispatch.

A testing period for competitive market principles and enhanced quality indicators, especially regarding current frequency.

The creation of a pool of reserves – comprising a market of electricity reserves.

As of 2003, the wholesale power market is functioning on the basis of bilateral contracts. The electricity exchange for day in advance power purchases was established in February 2002. The total volume of sales at the exchange is considered to be 7 % to 10 % of the total supply of electricity for Kazakhstan [34].

Hirschhausen and Waelde [35] have posited that an economic transition period from socialistic to capitalistic markets, especially in energy, no longer exists. Interestingly, the authors make the critical point that experience with emulating institutional models has shown that they often work very differently for the emulator than the emulated. Utilizing an institutional interpretation of energy sector reform in the CIS, the authors contend that the transition from socialistic to market based economies has led to diverse outcomes. These differing outcomes are based primarily upon the pre-existing formal and informal institutions that dominate the particular country. The CIS countries have looked to Western market economies as models for the restructuring of their power sectors. Two diametric cases that have been studied are the UK and French approach. The authors qualify this statement by suggesting that there is no theoretically or empirically discernable best practice to structure and regulate the global energy sector. The CIS, including Russia, however, have overwhelmingly chosen the UK (British) approach which has been reproduced in Australia, Canada, New Zealand, Scandinavia, and several U.S. states. This approach has been marked by very direct privatization, corporatization, and competition. The French system, in contrast, has essentially maintained integrated monopolies protected from competition. Despite the similarities found in the French and formerly Soviet approaches, linked by a dominating public service obligation and a close relationship between management and politics, the French system has not been the model of choice for the CIS. The initial elements of the approaches found in most CIS countries have been to:

- Identify appropriate reform models from international experiences,
- Attempt to re-produce those models in an effective way domestically,
- Allow the domestic model to emerge and evolve with conceptualized principles [35].

Interestingly, after surveying international examples, some countries such as Russia and the Ukraine have crafted early proposals that have included additional competitive elements. This enthusiasm for the UK approach can be tied to what is perceived as superior wealth generation, technological innovation and swiftness of economic response to the needs of the market. Possibly most importantly to consider, Hirschhausen and Waelde [35] indicate that the AC connection of Poland and other east European countries to the West European grid was an external impetus to modernization. This is an important recognition of the relation between the physical connections of electricity grids of distinct systems, and the subsequent need to reform such areas as security, technical requirements, quality and communication. Indeed, this is demonstrated by the relationship of technical and economic needs of the Japanese power grid interconnection (PGI) considered with Eurasia. Arakawa [36] suggests that a major issue concerning Japan's PGI is the "*eventual interconnection of the Japanese power system with the Eurasian mainland [which] will be achieved with restructuring of Japan's electric power market to be freely competitive. In addition, the relationship between Japan and Russia must be improved to a point that the neighboring nations will be able to cooperate, for example, in mutual development of Siberian natural resources. Apart from political and economic concerns, no major technical difficulties in PGI are anticipated*".

Projects such as the potential Japanese-Eurasian interconnection can be financed through three main approaches, (1) public ownership, (2) public-private partnership, and (3) private ownership. Though not the primary focus of this Section, more rigorous analyses of potential Asian interconnections should be carried out before assuming the existence of financial, capacity, or reserve capacity incentives. For instance, if the potential interconnections are between two monopolies, then analyses can be carried out using bilateral monopoly trade theory. This would be particularly useful with a monophony - monopoly power market scenario. However, in this scenario, the Nash equilibrium theorem cannot be used because the asymmetry found between distinct systems discounts the determination of a unique price.

16.6.2 Convergence of Asian Power Markets

Asia-Pacific Economic Cooperation*, (APEC), created in 1989 to continue to stimulate regional economic growth, has remained a consensus-based entity, without binding agreements [37]. APEC, although a non-treaty based organization, is having a significant impact on the reform of the Russian economy, including the Russian power sector. The primary purposes of APEC have been to encourage economic growth, trade, investment and cooperation in the Asia-Pacific region. APEC is a 21-member organization, each called a Member Economy, representing 47% of global trade. The APEC Member Economies include the Russian Federation as an integral participant. Particular themes can be found in APEC interests, including:

- Reduction of tariffs.
- Efficient domestic economies.
- Increased exports.

Other APEC activities include policy creation and economic cooperation to facilitate the exchange of products across regional borders [37].

Established, and ongoing studies of the APEC economic area electricity supply industry have indicated that reform in the developed economy members has been designed to improve efficiency in the sector. In contrast, the developing economy members have sought reform because of partial access to electricity supply, low infrastructure investment, uneconomic pricing of electricity, and inability to manage high demand growth [38]. The progress of microeconomic reform of the APEC electricity sector has been steady across both developed and developing member economies. Despite the large geographical area, and economic disparity of member economies, an APEC energy-working group has concluded that there are commonalities to be considered in electric sector reform:

- Governments remain responsible for the outcome of reform, even after industry restructuring takes place.
- If governments cannot demonstrate the success of reform programs, there may be political consequences.
- Reform needs to be conceptualized to each member economy based on needs, circumstances, and resources.
- Despite advances in a competitive electric market, electric security and stability of the entire system must always take precedence.
- Restructuring has been defined as a "break-up" of generation and supply, which are contestable, and mostly non-competitive elements of transmission and distribution.
- The introduction of a competitive electricity market will create high risk for consumers if supply is unreliable, or becomes unreliable. High levels of competition are not appropriate for developing economies in the early stages of electric sector reform.
- If the impetus is to keep the electricity market at a marginally competitive level, other aspects of the system can be made more competitive. The separation of the transmission system and the creation of a competitive system to acquire primary and secondary energy are examples.

---

* APEC's 21 Member Economies are: Australia, Brunei Darussalam, Canada, Chile, People's Republic of China, (Hong Kong, China), Indonesia, Japan, Republic of Korea, Malaysia, Mexico, New Zealand, Papua New Guinea, Peru, The Republic of the Philippines, The Russian Federation, Singapore, Chinese Taipei, Thailand, United States of America and Viet Nam.

- The historical emergence of competitive electricity markets has arisen "*naturally*" from trade between distinct power systems via sub-national and national power system interconnections.
- Under pricing of electricity removes much of the incentive to invest in infrastructure.
- There is no "*reform standard*". This being the case, it is important to create an objective setting, monitoring and measuring system to manage reform.
- Highly valuable empirical reform experience can be lost. It is important to retain talented experience for ongoing institution building [38].

Importantly, it can be maintained that each member economy should create a reform methodology in a regional and national conceptualization. However, there are five common steps that this can be based upon:

- Create national policy objectives.
- Identify reform risks.
- Assume and implement reform strategies that are realistic.
- Project-manage the reforms.
- Create an institutional unit to consistently monitor the progress of reforms.

Interestingly, the research methodology used by the APEC energy working group was composed of a literature review, interviews of reform participants in APEC economies, cross-sectional assessments of the current status of reform, and applications of practical experience in case studies of developed and developing economies in APEC. Overall reform of the electricity sector has been divided into eight primary sections:

- Policy objectives
- Management
- Structure
- Framework for law and regulation
- Mechanisms for wholesale market
- Transmission and distribution
- Retail tariffs
- Privatization [39].

By utilizing mixed research methodologies to provide analyses, each of these eight sections has strategic principles designed to provide common points of reference for reform, and to also provide the basis for a conceptualized approach. Interestingly, Principle 35, under Wholesale Market Mechanisms, states: "*The establishment of a competitive wholesale market should only be considered in the context of overall market structure and design*" [39]. Wholesale market mechanisms should be able to interface with regulatory frameworks, transmission and distribution planning, and access. This should account for the need to provide cost effective supply from generation to the distribution connecting points. Competitive wholesale market mechanisms are integral to overall power system reform planning and implementation. The primary aim of power system reform should be to improve economic efficiency. This can be divided into three areas:

Productive efficiency, which is the relationship between production input and output, and the implementation of best practice concepts.

Allocated efficiency, which are the way resources are utilized, in the light of appropriate signals for investment and consumption.

Dynamic efficiency, which is the extent that innovation and productivity increases are encouraged over a specific duration [39].

Critically, when determining the correct model to apply to an economy to bring about the desired efficiencies, it is important to consider transitional issues. These transitional issues include:

Ownership
If the current model of the power sector is a state-owned monopoly, full privatization should be delayed until it can be determined that the new companies will be viable, and after new markets have been tested.

Incumbents
If competition is introduced, incumbent utilities are under an obligation to reduce operating costs, and to develop new strategies and new markets to supply power to. The restructuring of the industry creates new risks for incumbent utilities. This includes the areas of finance, regulation and politics. As an example, generators can have competition amongst local power suppliers. Additionally, as the electricity grid becomes more interconnected, with more room for capacity, interstate and international generators then add to the competitive marketplace.

Stranded Costs
Stranded costs are the historical costs, mainly through generation plant construction, that may not be recoverable in the new prices charged in a competitive market. Generally, there are four types of stranded costs:
1) New generating plants could out-compete old plants, necessitating the decommissioning of older plants.
2) Competition can provide lower cost, long-term fuel or power purchasing contracts. Utilities bound to uneconomic contracts have higher input costs, and consequently a loss in earnings, making recovery difficult.
3) Utilities that have been obligated to invest in "*regulatory assets*" can find that the regulatory regime can change under a competitive environment, reducing their protective "*extended payment plans*" by deregulated, lower prices.
4) Other public policy programs such as Demand Side Management (DSM) programs can be stranded, as their cost recovery will become impossible in a deregulated system [39].

Although full recovery of stranded costs are rare, most experts in the field agree that governments should pay utilities for stranded costs, on a case by case basis, based upon specific attributes of the utilities in question.

Security of Supply
Two primary aspects of security of supply must be considered when planning and implementing a transition to competitive markets. They are the securing of long duration sources of generation fuels and the reliability of the power system.

Investment
It has been found that competition in generation can reduce incentives to invest, and lower operating costs. In building the reform of the power system, it is important to create a system for signaling the need for generation capacity and power grid improvement.

Reliability
Reliability, by definition, is moving from strictly technical reliability to economic reliability. Economic reliability is composed of a contractual arrangement between generators, distributors, and consumers, reflecting a certain level of reliability.

Transparency
Pricing mechanisms in a monopolistic system often are subsidized to appease social policy requirements. However, in a deregulated market, consumers should eventually be able to choose their own retail supplier. In this arrangement, consumers need accurate and transparent information regarding price and service.

Social Policy
Although government is responsible for social policy, it is recommended that energy policy not be an element of social policy. A newly designed regulatory system can be crafted to synchronize public and private interests.

Equity
The government as a consequence of deregulation may no longer maintain universal service obligations. Low-income customers may need to be assisted by direct government funding.

Consumer Protection
It is possible that industrial customers will benefit more from deregulation than residential ones. Regulation will need to be applied to enforce appropriate competition laws.

Pricing Issues
It is important to manage prices as a country transitions from one power market model to another. Two areas are commonly found: price increases and reduction of employment levels. Price increases may be made more gradual, especially by utilizing subsidies, efficiency improvements, enhanced competition and disbursement of consumer information.

Reform and the Environment
As a wholesale power market develops to be more competitive, there is more pressure to run plants that have a lower cost of production. It is possible that these lower costs of production plants emit the most harmful contaminants into the environment. Though not intrinsic to the reform process, it is recommended that environmental policy evolve at the same time as economic and social policy.

Regulation
Market power is a central issue in deregulation, and requires a regulatory regime to constrain non-competitive manipulation of the market.

Reflecting the theme that suggests that each country, including the APEC members, should conceptualize a reform program designed on one or many models, the APEC energy working group states, "*The circumstances of each APEC economy present unique issues and problems in designing regulatory aspects of market reforms. What has succeeded in other economies may not be appropriate in a particular economy in Asia.*" This is substantiated by the prominence that Russia holds as a case study when looking at APEC area electricity sector deregulation [40].

Indeed, Belyaev, et. al. [41], foreshadowing Arakawa [36], indicates that the interconnection of electric power systems, to broaden power markets, is expanding internationally. The North-East Asia region, composed of China, Japan, North and South Korea, and the Far East area of Russia hold the highest potential to design and implement interstate interconnections. The east Siberian area, including the Krasnoyarsk and Irkutsk electric power systems, has a surplus of capacity. This extra capacity is because of reduced exports to the European and Ural part of Russia as the industrial economy has been in decline. Consequently, and for example, energy experts at the Irkutsk, Russia electric utility have concluded that it is possible for an economically efficient interconnection between Irkutsk and China, exporting, respectively, up to 3 GW [41]. Plans such as this are supported by published policies of APEC [42]. Cross-border transmission networks, as planned in APEC, are perceived to strengthen the security, quality and flexibility of energy supply.

Thus, it is important to adopt a market model that has multiple buyers formed immediately after unbundling, in areas such as the CIS. CIS cross-border synchronization is beneficial because it places better use of existing generation capacity, provides emergency assistance options, and harmonizes standards in the areas of safety, ecology, and technology.

It is thought that the economic transition period from socialistic to capitalistic markets, especially in energy, no longer exists. Experience with utilizing institutional models has shown that they often work very differently for the emulator than the emulated.

Wholesale market mechanisms should be able to interface with regulatory frameworks, transmission and distribution planning, and access. This should account for the need to provide cost effective supply from generation to the distribution connecting points.

The primary aim of power system reform should be to improve economic efficiency. The economic milieu of each economy presents unique issues and problems when designing regulatory aspects of market reforms. What has succeeded in other economies may not be appropriate in a particular economy in Asia.

16.7 The Globalization of Energy Markets in Asia

Energy market globalization is deepening and broadening, not only through international trade but also through cross-investments, deregulation of domestic markets, and industrial restructuring that links the older energy industries to the new global political economy. This transformation of energy industries and markets is apparent around the world, and it offers great promise in terms of economic efficiency, technology development, and consumer choice.

The process of energy globalization is uneven, however, and some of its impacts will present new challenges for strategic planners. What new relationships are developing between producers and consumers, and between buyers and sellers? Who are the winners and losers? In a context of opening energy markets, why is there renewed concern about energy security around the world today? What types of security challenges will energy globalization present during the next two decades?

There are varying approaches to energy security in a context of market globalization. The United States supports market-oriented energy policies at home and abroad that open traditionally closed markets to new forms of competition and restructuring. Asia, a region where the United States has vast security stakes and where the most rapid increases in oil and gas imports are projected in the next two decades, deserves special attention. Policymakers in Asia and other countries worry that the market alone will not ensure energy security. The United States has generally pursued energy security on a different track, making Persian Gulf security a high priority.

To promote the cooperation and mutual interdependence that open energy markets require, it will be necessary to explore different approaches to energy security, analyze some of the unintended security risks that globalization of energy markets entails, and draw conclusions about the implications for U.S. security. Although the United States has already made large investments in Asian security, new multilateral approaches will be needed to pre-empt and mitigate the energy-related disruptions that may lie ahead. Defending the sea-lanes, to take an example, will be more important than ever in the future, but ensuring freedom of transit will require new multilateral efforts that cannot be simply subsumed under traditional alliances. Although the United States will have adequate access to energy supplies, it may be drawn into energy-related disputes, as weak states fragment, and producers and others seek to exert political leverage via energy supplies and infrastructure. Despite the uncertainties and difficulties of multilateral initiatives, it will be necessary to use them to address myriad energy-related security problems that are likely to arise as unintended consequences of energy market globalization. This Section presents the globalization of energy markets in Asia [43-51].

16.7.1 Energy Challenges in Northeast Asia

16.7.1.1 Recent Progress in Energy Integration in Northeast Asia

In the IEA [WJL1] programs, Northeast Asia, Figure 16.20, includes Northeast China (Heilongjiang, Liaoning and Jilin Provinces), East Siberia and the Far East of Russia (mainly Sakhalin Islands and Saha Republic), Mongolia, and Democratic People's Republic of Korea (DPRK). These countries and areas form a unique region in terms of energy supply lines together with Japan and Korea.

Recently, there has been increasing interest to integrate this area with various cross-border energy development projects and, in fact, several regional and international organizations have begun to put in serious effort to tackle this issue. Since June 2000, the United Nations Development Plan (UNDP) has held the international program called TRADP (Tumen River Area Development Program) among five Northeast Asian countries (China, Russia, Mongolia, Korea and DPRK) has organized annual meetings among the National Coordinators to discuss the cooperative work in energy, trade and investment, transportation, telecommunication and tourism. The UNDP is providing administrative and consultancy service for this series of meetings, and the IEA was invited to become a collaborating organization in the energy part of this program.

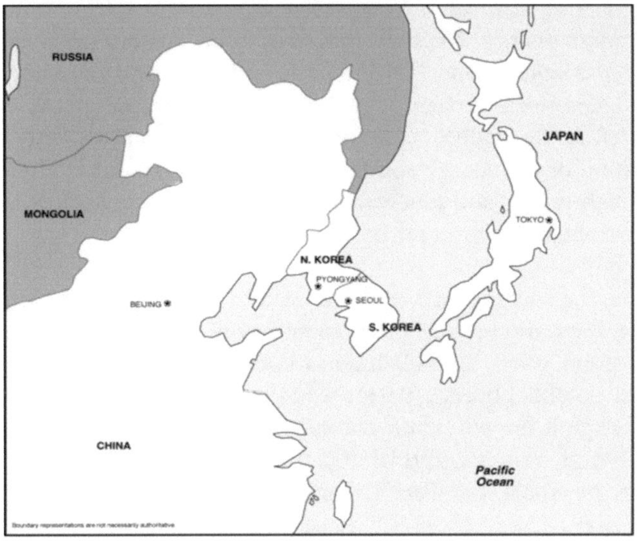

Figure 16.20. The potential Northeast Asia power market

In June 2001, an international symposium on Northeast Asian energy cooperation was held in Seoul jointly by UN/ESCAP (United Nations Economic and Social Commission for Asia and the Pacific), KEEI (Korea Energy Economics Institute), and IEEJ (The Institute of Energy Economics, Japan). In this conference, a necessity to build an inter-governmental communication channel was suggested. Based on this discussion, a senior government officials meeting among Russia, Mongolia, Korea and DPRK was held in Vladivostok on 8-10 April 2005. The IEA was invited as a guest speaker to share the Member countries' experience in energy trade with Russia. A second session of this senior officials meeting was scheduled in November 2005.

In addition to the internal energy cooperation within the Northeast Asian region, the cooperation between ASEAN and Northeast Asian countries is getting stronger. Since the ten member states of ASEAN have limited resources to nourish their own economic development, the partnerships with Japan, China and Korea have great significance. In this case, a collaborative relationship with ASEAN+3 can even benefit DPRK, Mongolia and Far-East Russia as well. There have been continuous talks and meetings since 1999 among these countries not only on energy cooperation but also all possible economic partnerships. These efforts were accelerated by ASEAN+3 Summits and will have substantial impacts on energy security of Northeast Asian region.

Considering that the Northeast Asian region accounts for more than one fifth of the world's energy consumption and is expected to account for one-third of the world's total energy demand increase over the next 20 years, it is certain that the energy integration of this region will have a significant impact on world energy security. This is good ground for IEA Secretariat attention to the energy development in this area.

16.7.1.2 The On-going Energy Integration Projects in Northeast Asia

One of the distinguishing features of Northeast Asia is that, in spite of both immense energy demand and abundant energy resources that exist in the region, the region remains segregated. On the demand side, the most critical issue of the three major energy importing countries- Japan, China, and Korea- is their high dependency on imported oil, especially from the Middle East. Therefore, the common denominator for these large energy consuming countries is diversification of the energy sources, i.e., the need to shift from oil to other energy resources such as natural gas and renewables, and to move from Middle East towards other energy suppliers for their energy imports.

On the supply side, the energy reserves in the eastern region of Russia are crucial to Northeast Asia. To date, most discussions of Russian energy have focused on its exports to Europe, but in the last five years, Russia's interest in developing energy relations with its eastern neighbors and potential energy partners has grown. The main reason for this "*Russia East*" policy is to exploit the abundant natural gas and oil resources in the east of Russia such as Sakhalin Island, Yakutsk, and Irkutsk near Lake Baikal, which could supply Northeast Asia and even the whole Asia-Pacific market.

Under these circumstances and in order to ensure energy security in the region, there have been many private sector efforts to connect these abundant energy resources with the demanding countries. The following are the major ongoing projects and plans in this region. Some of these initiatives are progressing more actively than others and bear a close review.

16.7.2 Developing International Power Markets in East Asia

The Greater Mekong sub region (Figure 16.21), Cambodia, Lao People's Democratic Republic, Myanmar, Thailand, Vietnam, and the Yunnan Province of southern China—has significant potential for cross-border power trade. The sub region is well endowed with low-cost hydro resources—the Mekong River Basin is the world's twelfth largest river system—and China, Lao PDR, Thailand and Vietnam have large coal and natural gas reserves. The potential for trade stems from imbalances in costs and in supply and demand between countries in close proximity: the low cost hydro potential is in Lao PDR, Myanmar, and Yunnan Province, but the main markets are Thailand and the more distant Malaysia-Singapore grid about 1,000 kilometers away.

Recent studies comparing scenarios of electricity self-sufficiency in each country with a full trade scenario show that full trade could yield cost savings of at least US$10.4 billion in 2001–20 and a reduction of airborne pollutants valued at US$160 million a year (these estimates assume a significant slowing in power demand over the next few years in Thailand as a result of the current financial crisis.) The savings would arise from:

- Lower operating costs due to economic power exchange, postponed and lower investments in generation due to least-cost development of regional energy resources, and reduced spinning reserve costs.

- Lower coincident peak load (compared with the sum of individual peak loads), mutual access to generation reserves for interconnected systems, a more robust power supply to meet such unexpected events as load growth above forecast or delayed commissioning of generation and transmission projects, and increased system reliability.
- Lower greenhouse gas emissions and other pollutants, largely due to a shift from thermal to hydro generation in the long-term.

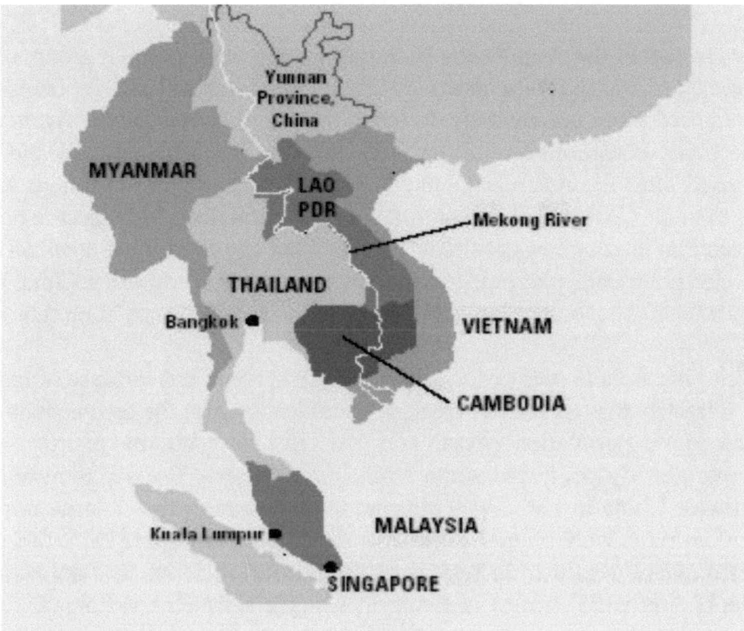

Figure 16.21. The potential East Asian power market

There is growing interest in cross-border bilateral power trade in the sub region, spearheaded by private developers in Lao PDR selling power to Thailand. The government of Thailand has agreed to buy 3,000 MW from these private power developers from 2006, and several independent power producer (IPP) projects are moving ahead. China's Ministry of Electric Power is encouraging studies of the export potential of Yunnan's planned Jing Hong hydropower plant and associated transmission lines to Thailand, through Lao PDR, with the support of the Lao and Thai governments. The Vietnam and Lao governments have signed a memorandum of understanding on purchases of about 2,000 MW of power by 2010.

16.7.3 Energy Market Globalization

Does it matter that globalization is unfolding unevenly and that policy priorities for enhancing energy security are defined differently in the United States, Europe, and Asia? Traditionally, analysts have focused on the potential for military conflict over energy resources as the primary threat. Extrapolating 20 years ahead, based on consensus supply-and-demand projections that show sharp increases in Asia's energy requirements; a number of energy-related issues are likely to generate new types of problems and unintended consequences associated with deepening globalization. To the extent that globalized energy markets more deeply integrate economies in the region, investment resources, entrepreneurial skills, and experience in governance will be available to mitigate the downsides. At the same time, U.S. officials responsible for security as well as for economic policy need to anticipate problems—many of them unintended consequences of globalization—that they may be required to address. Focusing on Asia, where there is no overarching, institutionalized security framework and where energy market globalization offers perhaps the biggest un certainties as well as great promise, brings potential problems into sharper view.

Among the countries of the Asia-Pacific Economic Cooperation (APEC) group, electricity demand is projected to increase 60 percent by 2010, with China's electricity demand likely increasing by almost 6.4 percent annually. In India, the International Energy Agency (IEA) forecasts that electricity consumption will be more than double between 1995 and 2010. These forecasts (revised after the Asian economic downturn) imply major additions to generating capacity and to grids. Coal will likely continue to play the major role in electric power generation, but substantial increases in gas-fired generation are expected. Asia now has only limited intercountry electricity trade and pipeline systems. A number of countries, China in particular, have substantial energy resources located far from industrial and population centers.

Most of developing Asia is part of the global energy system, but because of inadequate investment in infrastructure as well as weak political leadership, the connections are in some cases tenuous. Rapid population growth and pressures for economic restructuring and deregulation have already produced some wrenching changes. Twenty thousand miners rioted in Northeast China in early 2000 after an announcement that a large mine had gone bankrupt, and workers were offered a one-time severance package equal to $68 per working year. The army was brought in to restore order, but the incident was not reported in the press for weeks. Industrial unrest is rising in China's resources sector, where inefficient plants must be closed in line with government restructuring plans and ambitions to enter the World Trade Organization.

Russia exemplifies another type of political complication associated with market integration. In Russia, the country with the world's largest natural gas reserves, a good portion of which is located in the Far East, there are frequent blackouts. Gazprom cut gas supplies to RAO Unified Energy Systems (UES) recently in response to nonpayment. Gazprom is not investing enough to keep its gas flowing, and UES has warned that its old network of power stations and lines needs $75 billion in investment if Russia is to avoid blackouts. Europeans and Asians hoping to import more Russian gas are rightly concerned about supply security in light of Russia's status as a non-signatory of the energy charter, which includes transit provisions.

These examples illustrate the potential political fallout when energy market globalization occurs in developing and transitional economies that lack experience with market competition. As markets and infrastructure are connected across national borders, fuel substitution and economic benefits accrue. At the same time, new vulnerabilities are created. Energy infrastructure such as power grids can be the target of terrorists and opposition groups. These concerns are not unique to developing countries, of course.

The President's Critical Infrastructure Commission has outlined serious threats to the U.S. energy system from a number of sources—including hostile governments, terrorist groups, and disgruntled employees—as well as accidents.

For some groups in developing economies, the sharp changes in fortune that accompany restructuring and global energy market integration can create a political backlash that threatens the security of neighbors who buy energy from them or import it through their territories. Intense discussions are now under way in Northeast Asia about cooperation in pipelines and high-voltage transmission lines extending from Russia into China. According to some estimates, Eastern Russia could supply half of Northeast Asia's natural gas needs by 2020. These projects offer great promise in meeting energy demand and in hands-on cooperation among countries that have been historical competitors and enemies. The United States and countries in the region need to discuss the security implications of growing and asymmetrical interdependence, however, at an early stage. Joint planning and scenario analysis involving government as well as private sector organizations will be needed to anticipate and mitigate risks. The United States could lend support for discussions involving public officials and private sector representatives from Japan and South Korea, but Russia and China also need to be involved. In addition to high-level discussions on rules of the road for cooperative energy development, there is a need for joint efforts among environmental experts to assess potential effects, among regulatory authorities to discuss harmonization of equipment and industrial standards, and among legal experts to clarify issues such as transit rights and reciprocal tax treatment.

In developing Asia, where energy market integration is uneven, energy demand will grow sharply; because the infrastructure is inadequate and vulnerable, security-related problems are likely to grow. Attacks on energy infrastructure in friendly nations could lead to requests for U.S. assistance—both official and private. U.S. cooperation in the APEC and other regional initiatives to promote common standards and shared infrastructure are, in this light, a good investment. Although U.S. support for APEC energy market liberalization initiatives has been strong, energy security concerns have been treated with less urgency. U.S. industry and government could make this a higher priority and share expertise for assessing and mitigating risks.

A second dimension of uneven globalization—Asia's growing dependence on Middle East oil—also will present new challenges. The United States has made great investments in Persian Gulf security and has gone to war to ensure the stability of the region and its oil production. In the future, the narrow, shallow Straits of Malacca and the sea-lanes between the Middle East and Asia will be more congested with tankers and other ships carrying fuel and commodities. Today, 90 percent of Japan's oil imports and most of South Korea and Taiwan's oil imports flow through these waters. More than 200 vessels pass through the Malacca, Sundra, and Lombok Straits and the South China Sea daily. In 1994, more than $1 trillion in international trade passed through these waters, which have seen an increase in serious accidents since the early 1990s. Piracy, kidnapping, and other acts of violence by non-state actors, such as leftwing rebels in the Philippines, are also on the rise. China has fortified small islets in the South China Sea with fort-like structures, and the number of incidents involving fishing and naval vessels from Southeast Asian countries has increased.

Although some argue that territorial chokepoints such as these narrow water passageways are no longer security concerns in an age of globally integrated electronic markets that permit re-routing of cargo and fuel switching, securing freedom of the sea-lanes may well be more of a security challenge in the future. Competing claims among six claimants to the Spratly Islands, differing interpretations of the United Nations Law of the Sea, and the inability of the International Maritime Organization to establish safety and environmental standards of sufficiently high quality all contribute to a sort of maritime anarchy. At the urging of the Philippines and other Southeast Asian states, the Association of Southeast Asian Nations (ASEAN) Regional Forum has agreed to take up the question of a code of conduct for the South China Sea; however, China opposes legally binding agreements and prefers to deal separately with each country. Other countries favor demilitarization and joint development, with the geographically closest claimant country taking stewardship over disputed areas. In this context, the potential for military conflict remains significant. By supporting efforts of regional states to address these issues, the United States can add momentum and expertise.

In the future, accidents and acts of terrorism and piracy will be even more likely throughout the region. Some have called for a change in the transit passage law enshrined by the Law of the Sea separating commercial and military traffic. The objective would be increased regulation of commercial vessels in the Straits of Malacca to ensure navigation safety without affecting military or government vessels. Such a regime would involve not only the key states but also shipping concerns and user states such as Japan, China, and the United States. A working group has led another approach on maritime security cooperation of the Council for Security Cooperation in the Asia Pacific, a nonofficial organization that provides input to the ASEAN Regional Forum. The working group has developed guidelines for maritime cooperation and plans to examine the Law of the Sea to identify areas that need clarification in order to ensure maritime security in South Asia. These efforts suggest that addressing maritime security problems in Asia will be a challenging task, but arguably a good investment in preventive diplomacy. Cleaning up after a major oil spill and relief efforts to deal with terrorism or piracy could be much more costly after the fact.

Another way to address vulnerabilities in energy transportation through the sea-lanes is to develop regional emergency response mechanisms. Japan, Australia, and New Zealand [WJL1] are the only Asian members of the IEA, although South Korea is following IEA activities closely, and programs for nonmember states such as China have recently expanded. Asia lacks a viable region-wide program of emergency response or oil stockpiles. Although the impulse is strong for many of the Asian countries to pursue old-style resource diplomacy to secure supplies of Middle East oil, a more effective approach would be to build cooperative emergency response measures.

Market-oriented approaches can also contribute to solutions. Asian countries could permit cross-investment in downstream facilities so that refinery operations could be streamlined and efficiencies improved, encouraging Middle East countries to consider establishing storage facilities in the region. In addition, government involvement in emergency response and stockpile development is needed. American political support, technical expertise, and approvals to use international development assistance funding would help significantly in addressing energy security concerns in Asia and in bolstering the confidence and mutual trust required to sustain energy market liberalization policies over the long haul.

International corporate linkages in Asian energy markets are most extensive in the upstream resource exploration and development areas. Japanese firms have for years been mining coal in Australia, developing natural gas resources in Indonesia, and purchasing oil from China. With greater openness come new possibilities. Tokyo Electric Power has stakes in new power-generating ventures in Malaysia and Vietnam. Enron has teamed up with ORIX Leasing to compete in Japan's energy services and electric power markets. Marubeni, a Japanese trading company, and Sithe Energies, an independent U.S. power producer, plan to buy power plants and market electricity in Japan. Gas and electric power are the focus of networks of growing international joint ventures that include firms from many Asian countries, as well as from the United States.

These corporate linkages today extend further and deeper into the domestic economies and, in some cases, can stimulate market-oriented corporate restructuring and advanced technology development. They can also lead to new security challenges. In 1996, Japan imported almost one-fifth of its natural gas from Indonesia, a country where violent independence movements have threatened central authority in some regions. Electric power, gas, and steel companies have long-term contracts for liquefied natural gas (LNG) imports from Indonesia that stretch more than a decade ahead in some cases. Two-fifths of Indonesia's LNG exports come from Aceh, at the western end of Sumatra. Aceh is overwhelmingly Islamic; its rural people resent the wealth of the Javanese who run the industrial enclave. Disputes and violence have erupted. The potential fragmentation of energy- and resource-rich regions poses problems not only for central government but also for the importers whose investments become vulnerabilities. The United States, Japan, and others have an interest in developing multilateral approaches toward assistance that leverage the resources of the international community and address the basic grievances that have led to strife and tension.

Advanced technology is diffusing through energy development, presenting another double-edged sword from a security perspective. Japan, South Korea, Russia, China, Taiwan, India, and Pakistan have commercial nuclear power programs, and four of these states have tested and/or developed nuclear weapons. For Japan, nuclear power has been the central pillar of its energy policy—seen as Japan's only hope for gaining a degree of autonomous control (through technology indigenization) and for meeting environmental commitments. However, the serious criticality accident that took place recently at a fuel fabrication plant shook Japan's energy policy leadership enough for the government to announce a comprehensive review. Japan's ambitious plan to develop the complete fuel cycle has proved to be expensive and technically difficult. Such problems aside, Asia has become the new center of gravity for the global nuclear industry, as additions to capacity in this region are projected to make up three-quarters or more of the world's total over the next two decades.

For safety, environmental, and nonproliferation reasons, advanced technology cooperation in energy among Asian nations is essential. Working with other nations around the world, the industrial operators and research institutions of Asia need to develop a stronger safety culture. In addition, governments will need to work to strengthen nonproliferation norms (a very difficult task in South Asia) and to build cooperation in material protection, accounting, and export controls. Weapons of mass destruction proliferation are clearly a major threat to the stability of a region where the security framework is weak. Two of the benefits of addressing the North Korea problem have been an expansion of security cooperation between Japan and South Korea and a broadening of dialogue involving China.

Other forms of cooperation are also needed to make the most of new technologies that are coming on stream. They include micro turbines and fuel-efficient vehicles that offer promise not only for industrialized countries but also for many developing nations. Regulatory barriers, as well as established business practices, may present obstacles to the application of new equipment and systems. Government leadership in eliminating regulatory obstacles and in supporting international partnerships could speed up penetration and assimilation of technologies—with environmental gains for all concerned.

Thus, as energy market globalization proceeds in Asia, the likelihood that the United States will be forced to deal with threats that stem from unintended consequences will increase. Multiple actors will be involved, and solutions will in most cases need to be constructed—at least in the near term—in the absence of established frameworks and institutions.

The potential security risks stem in large part from the unintended consequences of uneven globalization in a context of partial market liberalization. In the current transitional phase, critical choices are being made about financial investments, partnerships, technology development, and fuels that will affect evolving and multidimensional interdependent relations among actors. Addressing energy security concerns, rather than dismissing them, is a requirement for promoting market oriented policies.

In this fluid context, the United States should take pre-emptive action, investing resources in preventive diplomacy and building security communities on specific issues in order to avoid the need for military force deployment down the road. Despite the uncertainties and inadequacies of multilateral approaches, there is really no alternative. The investments will be costly (not so much in terms of hardware, but in terms of time) and will challenge the skills of strategists trained to deal with more traditional security threats. Security specialists will need to work more closely with economic policymakers and the private sector, bridging the traditional separation between security and economic policy domains.

Asia offers the most striking example of both the potential risks of neglecting these issues and the tremendous gains that can come from devising new ways to address the concrete problem of energy security. China and India, the emerging new energy giants, will need assistance in meeting energy requirements and addressing concerns about energy security—if they are to contribute to, rather than detract from, Asian security. The United States will need to work proactively with them and with other countries in the region, forming new communities to deal with specific energy security concerns. In many cases, doing so will require focused dialogue not only with close allies and friends but also with other countries. Issues that require attention include disputes over energy-rich areas such as the South China Sea, the absence of an emergency response program in Asia to deal with oil supply interruptions, and the need for cooperation in resource development and efficient and environmentally sound energy use in the Russian Far East and China, as well as the potential for expanded energy cooperation involving South and North Korea, if progress continues in building trust and reducing threats on the Korean Peninsula.

16.8 Acknowledgements

Nikolai I. Voropai, Professor, Corresponding Member of RAS, Director of Energy Systems Institute, Irkutsk, Russia has prepared and coordinated this Chapter. Contributors include colleagues at the Institute and Members of the IEEE PES W.G. on Asian and Australian Electricity Infrastructure.

16.9 References

[1] "Open Access in Inter-State Transmission", Central Electricity Regulatory Commission, New Delhi, India, Nov. 2003.

[2] "Electricity Act 2003", Ministry of Power, Government of India, New Delhi, India, June 2003.

[3] Mukhopadhyay, S., K. Dube, S., "Status of Power Exchange in India: Trading, Scheduling, and Real Time Operation of Regional Grids", in *Proc. 2005 IEEE PES General Meeting,* San Francisco.

[4] Palamarchuk, S., Podkovalnikov, S., Voropai, N., "Getting the Electricity Sector on Track in Russia", *The Electricity Journal*, USA, October, 2001

[5] Lamourex, M.A., Palamarchuk, S.I., Voropai, N.I. "Current Developments in Russian Power Industry Restructuring", *Proc. 2003 Intern. Workshop "Liberalization and Modernization of Power Systems: Congestion Management Problems"*, Irkutsk, Russia, Aug. 11-14, 2003.

[6] Reshetov, V.I., Semenov, V.A., Lisitsin, N.V., *Unified Electric System in Russia at the Edge of Centuries. Current State and Perspectives of Development.* Moscow: ENAS Publishers, 2002 (in Russian).

[7] Voropai, N.I." Investment and Development of Electric Power Industry in Market Environment", *Proc. of the POWERCON'2002 Conf.*, Vol. 1, Kunming, China, Oct. 13-17, 2002.

[8] http://old.rao-ees.ru/en/news/pr/show.cgi?020404udal.htm

[9] Voropai, N.I., Palamarchuk, S.I., Podkovalnikov, S.V., "Power Industry in Russia: State-of-the-Art and Reformation Process", *Proc. of 4-th Intern. World Energy System Conf. (WESC-2002),* Toryo, Japan, March 4-6, 2002.

[10] State Power website, http://www.sp.com.cn

[11] Wu, F. F., Fu, Shuiti, "China's Future in Electric Energy", IEEE Power *& Energy Magazine*, 2005, Vol. 3, No. 4, pp. 32-38.

[12] State Power Information website, http://www.sgcc.com.cn/

[13] China Electric Power News, October 19, 2005.

[14] China Power News website, http://www.cpnn.com.cn

[15] "Report on 2005 China Electricity Markets and Electricity Price"

[16] China Electric Power News, August 19, 2005.

[17] China Energy website, http://www.china5e.com.cn

[18] China Power System Information website, http://www.chinarein. com

[19] Xizheng, Zhao, "Optimize the Power Source Structure and Upgrade the Power Industry" *China Industry News*, 2005, No. 10.

[20] "Detailed design of the Japanese desirable restructuring of electric power industry", METI, June 2004 (in Japanese).

[21] Materials submitted to the "Subcommittee on the evaluation of the Institutional Reforms" METI, 2005 (in Japanese)

[22] "ATC of the interconnected transmission lines (Prospect of future 10 years)" , ESCJ , June 2005 (in Japanese)

[23] Schweppe, F.C., Caramanis, M., Tabors, R., Bohn, R., *Spot Pricing of Electricity*, Kluwer Academic Publishers, Boston, 1988.
[24] Hogan, W.W., Contract Networks for Electricity Power Transmission. *Journal of Regulatory Economics*, Vol. 4, 1992.
[25] 25Leveque, F., *Transport Pricing of Electricity Networks*, Kluwer Academic Publishers, Boston, 2003.
[26] Wu, F.F., Ni, Y., Wei, P., Power Transfer Allocation for Open Access Using Graph Theory-Fundamentals and Applications in System without Loopflow. *IEEE Transactions on Power Systems*, Vol. 15, No. 3, 2000.
[27] Lee, B., Song, H., Kwon, S. H., Jang, G., Kim, J. H., Ajjarapu, V., "A study on determination of interface flow limits in the KEPCO system using modified continuation power flow (MCPF)," *IEEE Trans. Power Syst.*, 2002, Vol. 17, No.2.
[28] Jeong, H.S., Moon, Y.H., Oh, T.K., Hur, D., Park, J.K., "Status and Perspective of Transmission Pricing Scheme in Korean Electricity Market," *CIGRE/IEEE PES International Symposium*, Session 2a, Seoul, Korea, 2005.
[29] CIS Electric Power Council. "Current Status and Development of CIS Countries' Power Sectors," presented for the Energy Charter Seminar Liberalizing Trade and Investment in the Eurasian Power Sector, Brussels, 2002.
[30] Lovei, L., "The Single-Buyer Model," Public Policy for the Private Sector. Washington, D.C., The World Bank Group, 2000.
[31] European Bank, Power Sector Regulatory Reform in Transition Economies: Progress and Lessons Learned, London, European Bank for Reconstruction and Development, 2003.
[32] International Monetary Fund (IMF), Cross-Border Issues in Energy Trade in the CIS Countries, Washington, D.C., IMF European II Department, 2002.
[33] Energy Charter Secretariat (ECS), Kazakhstan Investment Climate and Market Structure in the Energy Sector, Brussels, Belgium, ECT, 2002.
[34] Eurelectric, Electricity Sector Reform: The Pan-European, CIS and Mediterranean Dimension, Brussels, Belgium, Union of the Electricity Industry - Eurelectric, 2003.
[35] Hirschhausen, C. van., Waelde, T., "The End of Transition - An Institutional Interpretation of Energy Sector Reform in Eastern Europe and the CIS," *MOCT Economic Policy in Transitional Economies*, 2001.
[36] Arakawa, F., "Restructuring Needed for PGI," presented at the Workshop on *Power Grid Interconnection (PGI) in Northeast Asia*, Shenzhen, China, 2002.
[37] Asia-Pacific Economic Cooperation (APEC). (2004) *About APEC.* [online]. Available: http://www.apecsec.org.sg/apec/about_apec. html
[38] Asia-Pacific Economic Cooperation (APEC), *Electricity Reform in APEC Economies - The Way Ahead*, Eter Smiles & Associates, Resource Law Inter-national, APEC Energy Working Group, 2003.
[39] Asia-Pacific Economic Cooperation (APEC), *Strengthening Operational* Aspects of APEC Energy Micro-Economic Reform, Manual of Strategic Principles, APEC Energy Working Group, Singapore: APEC Secretariat, 2001.
[40] Asia-Pacific Economic Cooperation (APEC), *Electricity Sector Deregulation in the APEC Region*, Asia Pacific Energy Research Centre, Institute of Energy Economics, Japan, 2000.

[41] Belyaev, L. S., Voropai, N.I., Podkovalnikov, S.V., Shutov, G.V, "*Prerequisites and Directions in Formation of North-East Asian Power System*," *Perspectives in Energy*, 1997-1998, Vol. 4.
[42] Asia-Pacific Economic Cooperation (APEC), *Cross-Border Power*, APEC Energy Working Group, APEC Secretariat, Singapore, 2002.
[43] IEA '01, 2001 World Development Indicators.
[44] Neff, T. L., "*Improving Energy Security in Pacific Asia: Diversification and Risk Reduction for Fossil and Nuclear Fuels*," Centre for International Studies, Massachusetts Institute of Technology, 1997.
[45] Ofgem '99, *The new electricity trading arrangement*, Volume 2, July 1999.
[46] OECD '98, Energy Balances of OECD Countries, IEA/OECD Paris, 1998.
[47] UN Energy Statistics, several volumes between 1970 and 1997.
[48] International Energy Outlook, 1999.
[49] Enron Energy Outlook 1999–2020, 9.
[50] Asia Pacific Energy Research Center, APEC Energy Demand and Supply Outlook (Tokyo, Updated September 1998).
[51] Harris, M. C.Chapter 12, "The Globalization of Energy Markets," *Global Century: Globalization and National Security*, 2001, Vol. I & II.

17

Market Mechanisms and Supply Adequacy in the Power Sector in Latin America

Latin America has emerged in the recent years as one of the most dynamic regions for electricity markets. The region is characterized by high demand growth rates and strong hydro share. In the 1990s the region has been one of the world leaders in private investments in the electricity sector. The process of reform of the electricity sectors so far has not reached all countries in the region and occurred in "waves" in the countries with deregulated markets. The example of Chile was firstly followed by Argentina in the early 1990s and shortly later by Bolivia and Peru. By the mid-1990s it had spread to Brazil, Colombia, and several Central American countries, concluding the "first wave" of the power sector reform. From 2004-2008 some countries in the region made adjustments in their regulatory frameworks aiming at keeping the positive aspects of the first stage of their reforms, but correcting the issues that have not worked as expected. A convergence to a common market model was observed and ingredients of this "second generation" include: (i) incentives to forward contracts to induce the entrance of new capacity (the so-called competition "for the market", instead of "in the market"), (ii) in some countries these contracts need to have physical coverage and (iii) the use of procurement auctions to incentive efficient contracting and setting up pass-through price, (iv) the possible complementation of state and market in the expansion of generation and transport.

This Chapter reviews the design and outcomes of this second generation of power sector reforms in Latin American countries, aiming at identifying the common challenges and the schemes adopted to tackle them such as the use of contract auctions, strategies to deal with demand uncertainty, capacity expansion and degree of state intervention, among others. With these diverging approaches, their primary challenge is to ensure sufficient capacity and investment to reliably serve their growing economies.

17.1 Latin America: Economic Growth and Energy Supply

All developing countries require high investments to respond to a continuous increase in electricity demand, directly linked to economic growth. Only South America requires about 70 thousand million dollars of investment in the power sector in the next ten years. Because electricity consumption per capita is relatively small, it is not surprising that while industrialized countries have had an average annual growth of electricity consumption between 1 to 2%, the Latin American subcontinent has experienced an average growth of over 5% during the last decade. There are 270,000 MW installed in Latin America (2007), with Brazil dominating with 34%, followed by Mexico with 19.5%, Argentina with 12% and Venezuela with 8%. Electric systems in the region are often of a radial nature, with weakly meshed networks and only few incipient international interconnections. The region is rich in resources (hydro potential, oil and gas). Hydro generation is the dominant supply source in the region with a share of 52% of the total installed capacity (68% as far as energy generation) and often with plants within complex cascades over several river basins with diverse hydrological patterns. The strong share of hydro generation results in a "clean" energy matrix and "leverages" other renewables which have seasonal (e.g., cogeneration from biomass) and intermittent (wind) production pattern: hydro reservoirs are used to compensate for the variability of wind power production and the seasonality of biomass energy production with no need for expensive and polluting thermal plants as backups to the renewable resources. Natural gas has an increasing role in the region, and cross border electricity and gas links are important resources for the region.

A profound transformation took place in the Latin America electricity supply industry organization in the 1990s. Essentially, three electricity markets were developed in the region: the Central American market, the Andean market and the Common Market of the Southern Cone (Mercosur). The one with the largest per capita energy consumption is Mercosur, an integrated market of 230 million people over an area of 12 million square kilometers, comprising the countries of Argentina, Chile and Brazil plus Paraguay, Uruguay and Bolivia (see Figure 17.1).

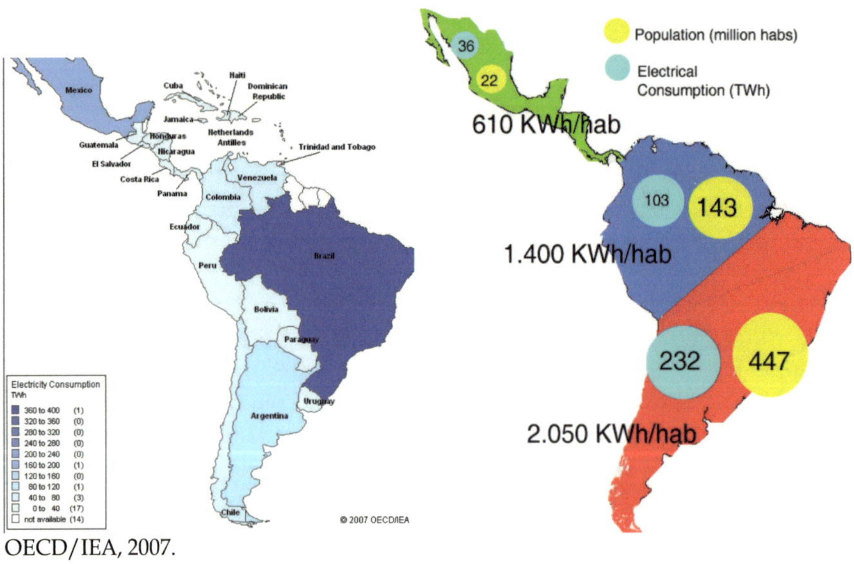

Source: OECD/IEA, 2007.

Figure 17.1. Overview of Latin America

17.2 The First Generation of Market Reforms in Latin America

While the motivations of the power sector reform in developed countries was introduced to facilitate competition as means to reach the ultimate goal of greater efficiency, the reforms in Latin America had the additional motivation of relieving the government to fund investments in new generation capacity that were required to match predicted load growth. Although differing in the degrees and details of implementation, the power sector reform in the region shared a common two-stage process [1].

The first stage of the power sector reform in the region was purely based on market mechanisms to achieve these goals, inspired by similar reforms in more developed countries. Pure market mechanisms formed the backbone of the regulatory frameworks to achieve the objectives of reliable and efficient guarantee of energy supply and adequate tariffs. In particular, the key driver for decisions was the spot prices in the short-term market, which would then be used to provide the correct economic signal for the entrance of new generation: if there is an imbalance between supply and demand, then these prices should increase and thus create incentives for the construction of new plants. The market risk resulting from the spot price volatility would be managed through pure risk-management instruments, such as forward contracts, options, etc. The only "non financial" instrument would be the capacity charge, whose main objective would be to ensure the remuneration of peaking units and reserve generation. With those market mechanisms in place, State-owned utilities were privatized and consumer choice was introduced in different degrees.

Most Latin American countries reformed their power sectors based on these principles, obviously with differences among the implementation details. The accumulated experience so far has shown many positive aspects, such as a greater efficiency of the private utilities, the positive effect of the eligible consumers as market benchmark and the transparency brought by the regulatory agencies, which provide confidence for investors.

On the other hand, some important difficulties (power crisis, rationings) have appeared, in particular with respect to the security of supply. A recent evaluation carried out by the World Bank [2] shows that about 20 countries around the world had energy supply difficulties (power crisis, rationings) in the late 1990s and early 2000s. Argentina, Colombia, Brazil and Chile are among these countries. A first reason for these supply difficulties is that the economical signal provided by the spot market is too volatile to correctly indicate and stimulate the entrance of new capacity. This is especially true for the countries with a strong hydro share, where the occurrence of conjuncture favorable hydro conditions can drive downwards the spot prices even if there are structural problems with supply. It has also been observed that in hydro systems the energy spot price increases substantially only when the system is "too close" from a power crisis, when there is not time anymore to make investments. A second reason is the combination of a strong demand growth but with a large volatility in the growth rates ("stop and go" economies that can be heavily affected by international crisis). This makes the generation activity very risky and makes difficult the closing of "project finance" by the financing institutions for new projects, which ends up constraining the entrance of new capacity (see [19]).

Figures 17.2 and 17.3 next show the historical record of energy spot prices in Brazil as well as the historical electricity load and GDP growth. It is possible to see the volatility of prices and load growth, which discourages the fully functioning of an energy-only market and challenges the implementation of merchant projects (load growth volatility may disrupt market growth expectations).

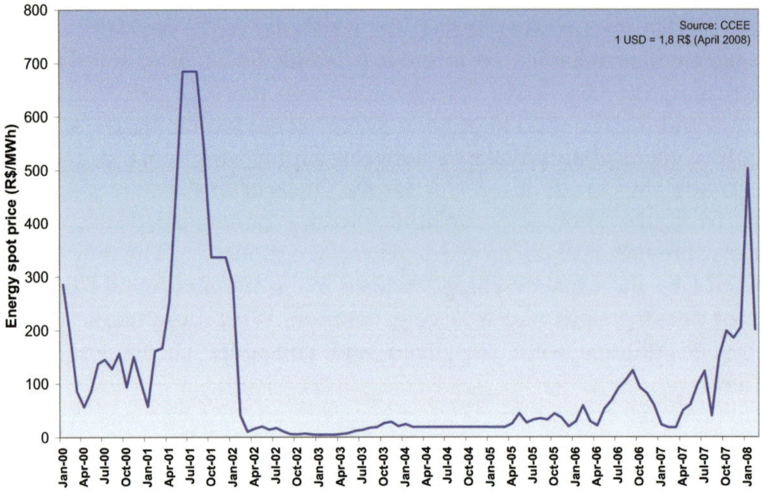

Figure 17.2. Brazil energy spot prices (R$/MWh)

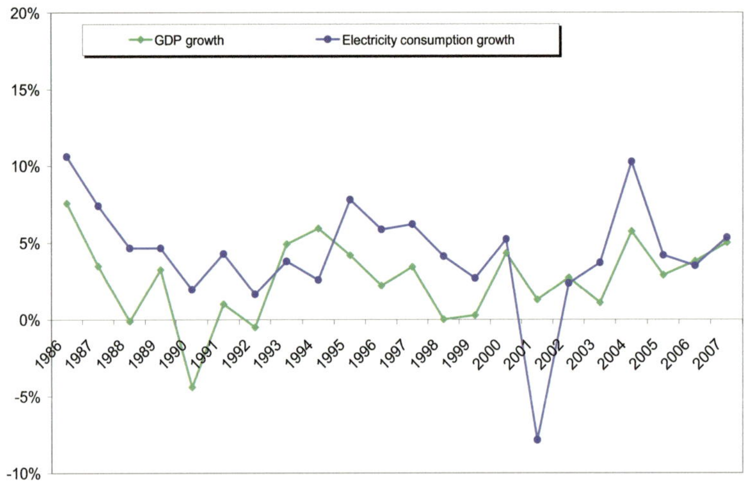

Figure 17.3. Brazil: Load growth and GDP growth

In summary, the fast (and uncertain) electricity load growth and energy spot price volatility introduce challenges for the supply adequacy of a region where projects are financed under a "project finance" basis and long-term stable cash flows are needed to allow the structure of the financial loans (for Greenfield generation).

17.3 The Second Generation of Market Reforms in Latin America

Due to the aforementioned difficulties and challenges, many countries in the region have made adjustments in their regulatory frameworks, adopting special mechanisms to stimulate the entrance of adequate capacity and ensure resource adequacy with a reliable supply. These adjustments aim at keeping the positive aspects of the first stage of their reforms, but correcting the issues that have not worked as expected.

A convergence to a common market model was observed. The characteristics of this "second stage" include: (i) the key point of the competition is not in the spot market but in the contract market with the demands that result in the entrance of new capacity (the so-called competition "for the market", instead of "in the market") and (ii) requirement of coverage of these contracts by physical generation capacity. The core of these proposals lies in two main rules:

- The first rule is that all consumers, both regulated and free, should be 100% contracted;
- The second rule states that all contracts, which are financial instruments, should be covered by "firm energy" or "firm capacity" certificates (FEC and FCC).

For the regulated users, the procurement of new capacity is carried out through public auctions. In one scheme, such as the one adopted in Brazil, the distribution companies (Discos) are required to inform their load forecasts and a contract auction is jointly carried out to meet the total load increase. In an alternative scheme, such as the one adopted in Chile and Peru, each Disco manages its own auction. Aggregating the demand of various interested parties may be possible. In a third scheme, adopted in Colombia, a firm energy option is purchased in an auction for all consumers.

Overall, the advantage of the auction of contracts is to recognize that a Power Purchase Agreement (PPA) provides a degree of certainty in the generators' cash flow, allowing access to long-term financing. The process can be cost-competitive if procured through international competitive bidding (competition for the market). The next sections described how these challenges in electricity supply are being faced by Brazil, Argentina, and Chile.

17.4 Brazil: Auctions of Options and Forward Contracts

The Brazilian power system is the largest in Latin America, with an installed capacity of 105 GW. Almost 90% of the energy produced comes from hydroelectric plants; the remaining generation mix includes natural gas, coal, nuclear and oil. Bioelectricity (co-generation from ethanol production, which uses the sugarcane bagasse as a fuel) is emerging as a competitive new source.

The main hydro system is composed of 140 plants, with capacities ranging from 30 MW to 14,000 MW (Itaipu binational plant at the border with Paraguay), and located in a dozen river basins throughout the country. Because of Brazil's large area (equal to the continental USA plus half of Alaska), the basins have a wide variety of weather and stream flow patterns. For example, when the well-known climate phenomenon "El Niño" occurs, the Northeast region faces droughts, whereas the Southern region has an increased rainfall. In order to take advantage of this diversity, the Independent National System Operator (ONS) dispatches the whole hydro system as a "portfolio", with transfers of huge energy blocks from the "wetter" regions to the "drier" ones. Hydro plants are dispatched with basis on their expected opportunity costs ("water values"), which are computed by a multi-stage stochastic optimization model that takes into account a detailed representation of hydro plant operation and inflow uncertainties (see [4]). As a result, the overall supply reliability is increased and the use of fossil fuels in the thermal plants is minimized. This type of countrywide dispatch optimization requires an integrated and robust transmission network, able to accommodate different export/import patterns among regions. The main grid has 90 thousand km, with voltages from 230 kV to 765 kV AC. There are also two 600 kV DC links: a 900-km line, which is part of Itaipu's transmission system and a 2200 MW back-to-back interconnection with Argentina (for more details, see [3]).

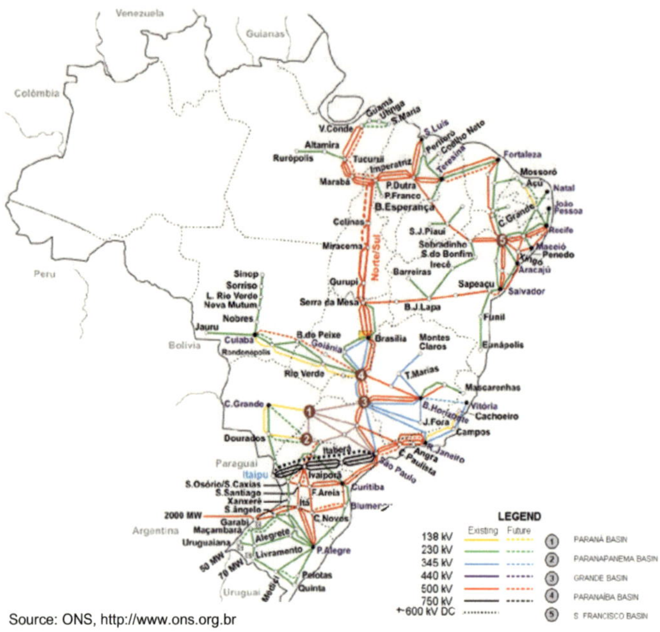

Source: ONS, Brazil, 2007.

Figure 17.4. Brazil - physical system

17.4.1 The Market Design

The first rule in the Brazilian regulation is that all consumers, both regulated and free, should be 100% contracted. The contract coverage is verified ex-post, comparing the cumulative energy (MWh) consumed in the previous year with the cumulative energy contracted. If the contracted energy is smaller than the consumed energy, the user pays a penalty related to the cost of building new capacity. For regulated consumers, it is allowed to be over contracted in up to 3%. The second rule states that all contracts, which are financial instruments, should be covered by 'firm energy certificates' (FEC)[1]. For example, in order to sign a contract for 1000 average MW[2], the generator or trader must show that it

---

1 This is because the Brazilian system is 85% hydro and is energy-constrained, not peak-constrained.

2 Average MW = GWh/#hours

possesses firm energy certificates that add to the same amount. The FECs are tradable and can, along the duration of the contract, be replaced by other certificates; the only requirement is that the total firm energy of the certificates adds up to the contracted energy.

FECs are issued by the regulator for each generator in the system, and reflect their sustainable energy production capacity. For hydro plants, for example, the FEC corresponds to the yearly (firm) energy production capacity in dry years. For thermal plants, the FEC is given by the available capacity (discounting average maintenance and forced outage rates), adjusted by a 'derating' factor that depends on the variable operating cost[3].

The joint requirement of 100% coverage of loads by contracts and 100% coverage of contracts by firm energy certificates creates a link between load growth and construction of new capacity. For example, if an industrial user forecasts that its load will increase by 200 average MW three years from now, it will start to procure contracts, in order to avoid the penalties. If supply is tight, the generators and traders will not have enough FEC to cover the contract; the only option will be to offer a contract for an investor to construct a new plant. In other words, the contract coverage requirement makes load growth the driver for the construction of new capacity, and the FEC coverage requirement ensures that the constructed capacity is adequate to cover this load growth. If the system is 100% contracted and contracts have physical coverage, then supply reliability is assured.

In this scheme, deregulated consumers can negotiate bilaterally their energy needs and are free to contract their "new energy" as they wish. On the other hand, in order to promote the most efficient purchase mechanism for regulated (captive) consumers, the contract obligation scheme for distribution companies was coupled with the use of regulated procurement auctions as the main mechanism for contracting energy. Regulated energy contract auctions are therefore the backbone for the induction of efficient purchases on behalf of captive consumers by distribution companies. The auctions for energy purchase by distributors are separated into regulated auctions for contract renewal (existing energy auctions) and auctions for new ("Greenfield") energy (new energy auctions). Two public auctions carried out every year for new energy delivery 3 and 5 years ahead Auction for renewable energy (wind, solar, small hydro and biomass) can also be called. Note that the Brazilian approach bundles two products: the FEC and the energy contract. This section discusses the new energy auctions.

17.4.2 Regular Auctions for Existing and New Energy

17.4.2.1 Auctions for new energy

The new energy auction is intended to exclusively promote the construction of new capacity to cover distribution companies' load increase. In these auctions, standard long-term energy

---

3 The idea is that an expensive plant, for example, diesel-fired, is only dispatched late in a drought situation, whereas a cheaper plant, for example combined cycle natural gas, is dispatched earlier. As a consequence, the cheaper plant's contribution to the overall 'firm supply' is more significant than that of a more expensive plant. Taken at the extreme, a thermal plant whose variable operating cost was equal to the rationing cost would have a firm energy certificate of zero average MW.

contracts (15 – 30 years) are offered to potentially new generators. Every year, two types of new energy auctions are carried out:

• *Main auction* (A-5) – this auction offers bilateral contracts for new capacity with duration between 15 and 30 years, which will enter operation in five years' time (hence the name), such as hydro plants. The idea is that this contract will allow the investor (auction winner) to obtain the project finance, and have enough time to build the plant. Besides ensuring the expansion of supply, the decrease in the investor's risk should lead to a reduction in new capacity costs. One of the main aspects of this auction is that distributors will have to estimate energy demand five years in advance in order to better estimate its energy needs on A-5 auction.

• *Complementary auction* (A-3) – this auction also offers bilateral contracts with duration between 15 and 30 years for new capacity. In this case, however, the plants should enter operation in three years' time (such as a thermal plant), not five. The idea is that this auction complements the A-5 auction carried out two years earlier, because "now" there is less uncertainty about future load growth.

In order to avoid a contracting strategy where "everything" is contracted in this auction, the amount contracted by a distributor on an A-3 auction is limited to 2% of its total load observed 2 years before. This "two-stage" auction process is a strategy for dealing with load growth uncertainty, which is an important issue in Brazil, and also to account for the needed construction time for the new plants. One important consequence of this scheme is that new capacity additions in the system will be a consequence of the willingness of the distributors to buy energy in these auctions. Since the regulated market (captive consumers) accounts for about 75% of the total market, the results of the new energy auctions will drive the system supply expansion for the regulated market. Finally, the new energy auctions are not exclusive for distribution companies. Large consumers and self-producers can have access to new hydro concessions through these auctions by making bids for the available projects.

17.4.2.2 Auctions for existing energy

Existing energy contracts complement the new energy contracts so as to cover 100% of the load. They are also auctioned. The objective of these auctions is to recontract yearly the existing energy, i.e., to supply the current market. The contract duration is between 5 and 15 years and they start on January of the following year. For this reason, the Existing Energy auctions are known as "A-1". Besides the duration, the existing energy contracts have other special characteristics that make them different from the new energy contracts: the existing energy contracted amount can be reduced, at any time, to match the distribution utilities' load reduction in case a qualified captive consumer becomes a deregulated consumer. In addition, the existing energy contracted amount can be reduced, at the utilities' discretion, up to 4% in each year to make up for demand uncertainty. In other words, the existing energy contracts are more similar to financial "put" options.

17.4.2.3 Adjustment auctions

These auctions are intended to "fine tune" the match between contracted supply and load. The adjustment contract duration is up to two years and auctions are carried out three or four times a year, with the contract starting within the same year. For this reason, they are also known as "A-0" contracts. The amount contracted by a distributor on an adjustment auction is limited to 1% of its total load contracted.

17.4.2.4 Summary of energy auctions

Figure 17.5 presents the overall scheme of the regular existing and new energy auctions, which are yearly offered:

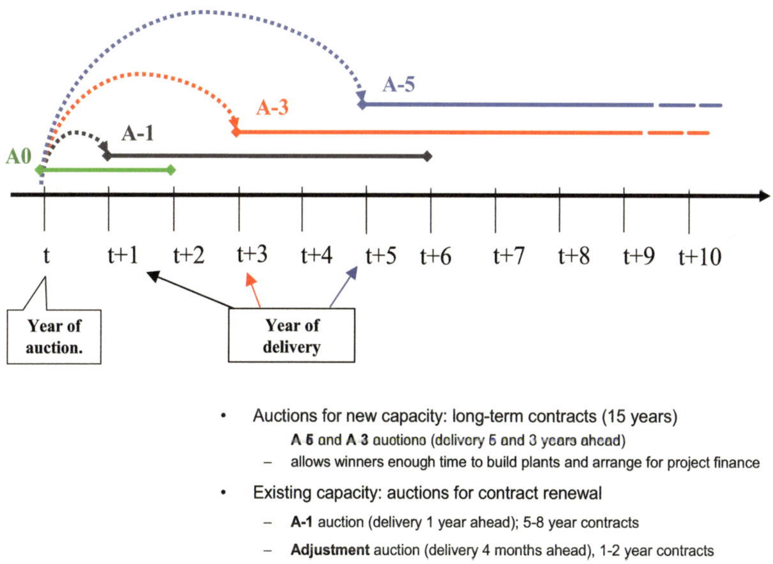

Figure 17.5. Existing and new energy auctions products and delivery dates

The portfolio of energy supply contracts of a distribution company is then formed by a mix of existing and new energy contracts.

17.4.3 Auction Procedure

Each auction (new energy, existing energy and renewables) is jointly carried out by all distribution utilities. Sixty days before each auction, each utility declares the energy demand (MW average) that it wants to contract. The auction announcement then calls for offers that will cover the total demand (sum of all declarations). The joint contracting scheme is a mechanism for creating economies of scale in the contracting process of new energy for small Discos as well as for equalizing tariffs among consumers aiming at assuring tariff adequacy.

Also, each separate winner of an auction will sign individual bilateral contracts with each distributor participating in the auction, being the energy amount of each contract proportional to the Disco's declared demand, and the total contracted quantity for each generation company (GenCo) matches its offered quantity. This allows the benefit of cheaper energy to be shared by all consumers. Although a "central procurement" is made, Discos are responsible for deciding how much energy they want to contract (i.e., responsible for load projections), thus avoiding the 'optimistic' government bias that in many countries has led to over-capacity and expensive energy contracts. Contract costs can be passed through to customers up to a benchmark price (overall resulting weighted price of the auction), and winners of an auction will sign individual bilateral contracts with each Disco. In other words, this is not a single buyer model, the Government does not interfere in the contracts nor provides payments guarantees.

In both existing and new energy auction, the objective is to contract energy at the lowest possible cost to consumers. Therefore, the auction design is chosen accordingly. Auctions carried out so far have used a two-phase hybrid auction, where in the first phase an iterative descending-price clock auction design is applied and the auction ends with a final round of bids using a pay-as-bid scheme (second phase). Figures 17.6, .7 and .8 show the main steps of the auction mechanism. In the case of auctions for new capacity, the country has used two contract types: standard financial forward contracts and energy call options.

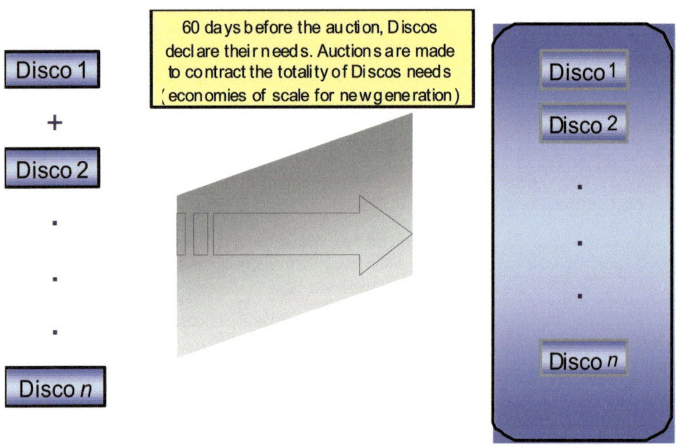

Figure 17.6. Auction scheme: before the auction

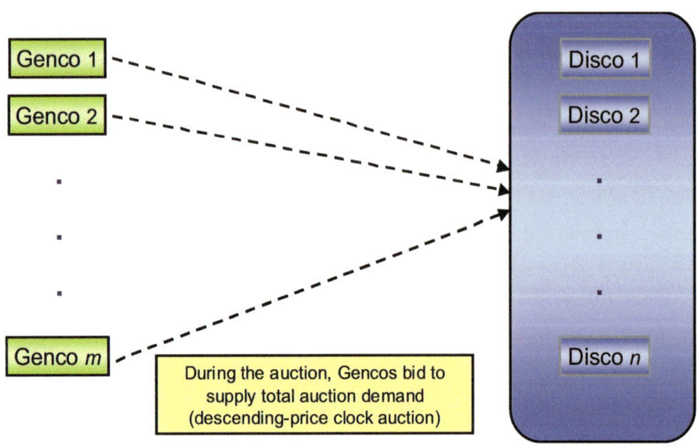

Figure 17.7. Auction scheme: during the auction

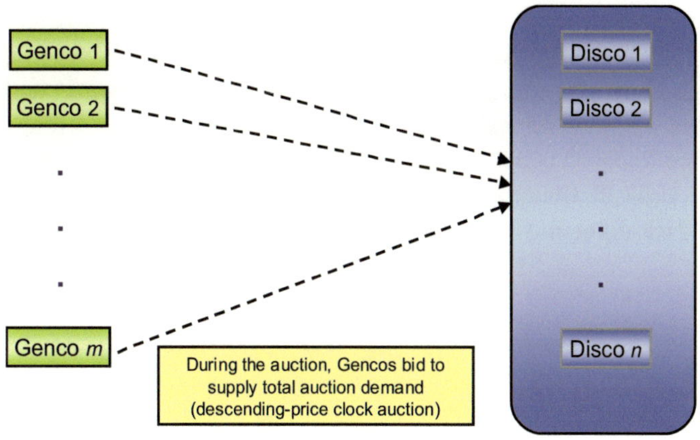

Figure 17.8. Auction scheme: after the auction

The Ministry of Mines and Energy sets up a committee to organize these auctions. This committee is in charge of proposing all the relevant documents, including auction design, design of energy contracts and price caps for each auction.

17.4.4 Types of Contracts

The contracts auctioned in the new energy and existing energy auctions are financial instruments and can be of two types:

(i) *standard financial forward energy contracts*, also known as contracts "by quantity". These are standards "take or pay" energy contracts in which the buyer pays a fixed $/MWh for the energy contracted and the seller has the delivery risk, clearing the difference between energy produced and energy contracted at the spot market;

(ii) *energy call options*, also known as contracts "by availability". These are contracts where the consumer "rents" the plant from the investor, paying a fixed amount ($/kW.month), and reimburses the plant for the variable operating costs ($/MWh) whenever its flexible part is dispatched or the consumer bears the spot market transactions costs otherwise. For details, see [5].

Contract prices are adjusted every year for inflation and have fuel price indexation. The government has the right to decide which type of contract will be offered in each auction. The objective is always to provide to distribution companies the best portfolio of contracts to minimize the consumer costs. Overall, MME has been applying the contract type (i) for the existing energy auctions. As for the new energy auctions, type (i) has been applied for hydro plants and type (ii) for thermal plants.

17.4.5 Auction Results

The implementation of the regulated auctions started in 2004, when the first existing energy auction was carried out. This represented the largest electricity auction in world history. Since then, several other auctions for existing and new energy were carried out, involving a total energy volume of almost 31,000 average MW (firm energy, not peak capacity) and involving about 85 billion USD in financial transactions. A summary of the blocks contracted and weighted average resulting auction prices are depicted in Figures 17.9 and 17.10.

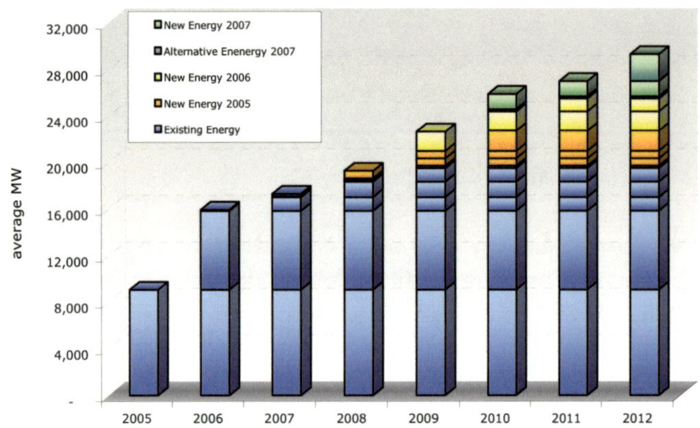

Figure 17.9. Energy blocks contracted (energy average MW, not peak)

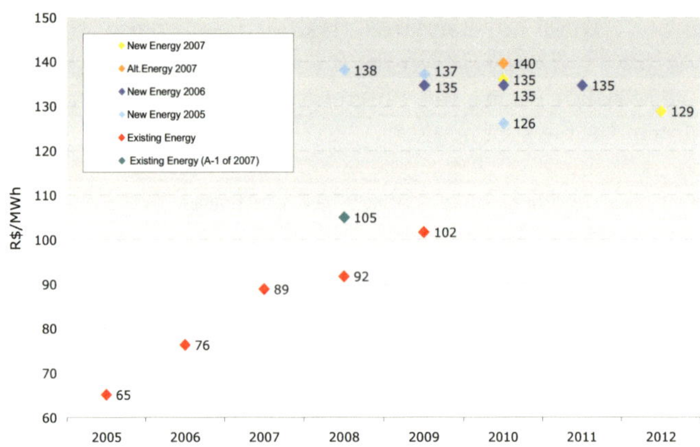

Figure 17.10. Average contract prices (1 USD = 1,85 R$)

Overall, the auctions for new capacity in Brazil have been of great interest to local and international investors looking to South America's energy market: the candidate suppliers list has been high and contracted generation has included a mix of a wide variety of technologies, comprising new hydro projects, gas, coal and oil-fired plants, sugarcane biomass and international interconnections.

17.5 Argentina: Successful Reform Clogged By Government Intervention

Argentina has 24,000 MW of installed power capacity for a peak load near 18,000 MW, with additional 2,200 MW of firm exportations committed to Brazil should be added to. Domestic natural gas demand averaged 90 Mm^3/day in 2004, while exportations represented near 25 Mm^3/day extra. Roughly 50% of total energy requirements are covered by natural gas. Although the country was completely energy self-supplied up to 2004, the hydrocarbons reserves horizon was significantly reduced in the last years mainly due to small investment on exploration. Natural gas reserves have now a horizon near 13 years versus 20 years in 1999. Oil reserves present a similar trend but smoother: current horizon is near 12 years. Alternative energy resources to natural gas for power generation include potential hydro developments mainly concentrated in plain rivers, which imply high investment requirements. Use of other energy resources is limited. Historically, coal represented a small proportion of energy balance, while 1,000 MW in two nuclear power plants were developed in the 1970s.

At the beginning of the 1990s Argentina reformed its energy sector as part of a wider economic reform whose main basis was the implementation of a fixed currency exchange rate regime that tied local currency 'Peso' to the US Dollar at a 'one to one' ratio combined with a free regime of importation and exportation of capitals. Inefficient performance of vertically integrated state-owned utilities during the previous decades led to an integral transformation of the energy sector. This process included the implementation of a completely new regulatory framework established by both the Electricity and the Natural Gas Acts, passed in 1991 and 1992 respectively. State-owned utilities were vertically and horizontally unbundled and further privatized or given in concessions. Wholesale markets for natural gas and electricity based on private participation were implemented. Transportation and distribution mainly remained as regulated monopolies within their concession areas, with the only exception of electricity transmission expansions for which an innovative scheme based on market participants' decisions was adopted. Production was completely deregulated, allowing entry of private companies in oil and gas exploration and production, as well as in electricity generation. Oil sector reform based on a new Hydrocarbons Act also included the privatization of public company YPF during 1992, which had the monopoly on upstream activities, and the deregulation of retail fuel prices.

Performance of the Argentinean energy sector after reform was largely reported as successful, and was often cited as a model of deregulation. In power generation, Argentina developed one of the most competitive markets worldwide. Wholesale electricity prices decreased from near 5 cts/kWh in 1992 to 3 cts/kWh in 1994 and less than 2.5 cts/kWh in 1997, while domestic consumption grew at an average annual rate of 5.7% between 1992 and 2000. Also, two private interconnectors of 1,000 MW each were built to export electricity to Brazil. Increase of energy exports also included oil and gas. The country stopped importations of gas from Bolivia in 1994 while started exportations to Chile and Brazil. Thus, Argentina became the benchmark for successful deregulation processes worldwide. This rosy situation worsened and decayed under a severe economical crisis that affected all the country's economy sectors at the end of 2001.

17.5.1 The Crisis

As described in [14], after nearly 10 years of a fixed currency exchange rate regime, Argentina faced a severe political and economical crisis at the end of 2001. President De la Rua resigned on 20 December 2001. Within the next 10 days it defaulted on its international debts. On 6 January 2002 the Congress passed a special law that gave the "emergency" status to the economy and abolished the fixed currency exchange regime. Since most of public and private contracts signed during the last decade were at prices and/or tariffs nominated in US Dollars, this law established the legal basis for unilateral government's intervention on such prices, what included tariffs of regulated activities. These actions further motivated foreign investors to litigate against the Argentine government on international institutions such as CIADI.

To meet the economic crisis, the Peso was allowed to float. Within first six months of 2002 it had fallen from parity with the US dollar to 3.6 pesos/dollar, although several months later it stabilized around 3 pesos/dollar with some intervention of the government in order to avoid a higher appreciation of the Peso.

17.5.2 Energy Policy after the Crisis

Under the umbrella set by the "Emergency Act" passed in early 2002, which is still in force, the Government took several decisions regarding the energy sector aiming:

- to minimize devaluation effects on end user's prices, that in practice meant frozen tariffs in case of gas and electricity, and the implementation of withholding taxes on exports, that reduced the market reference price for oil and gas exporters in order to avoid increasing domestic prices and, at the same time, increase government income.
- to guarantee end users' supply, ensuring covering of operational cost to existing producers but not fixed costs recovery, and promoting new expansions, most of them still in project status.

Frozen tariffs of regulated activities were implemented subject to future renegotiation of concession contracts, which in practice has not happened yet. Consequently, the devaluation augmented relative competitiveness of the Argentine economy with respect to the rest of the world. Local industry was benefited from frozen tariffs of gas and electricity and distorted oil-derivatives prices.

An agreement between natural gas producers and the government was signed in 2004. The latter committed to increase regulated tariffs to industrial customers in order to allow a gradual recovery of natural gas prices (wellhead prices in the Neuquina basin had decreased from 1.40 US$/MBTU in 2001 to 0.40 US$/MBTU in 2002). The energy sector, with frozen or distorted prices, would undoubtedly contribute to finance the local industry's higher competitiveness in the post-crisis years, in what seems to have been a political decision.

17.5.3 Consequences of the Post-Crisis Policy and Later Developments

The energy sector faced, and still faces, an economic long-run mismatch between what the economy needs from the energy industry and what this industry can offer to the economy under the current "relative prices scenario". In practice, this has meant lack of investments in all energy sub-sectors since end of 2001. Consequently, domestic demand grow was gradually absorbing installed capacity, including those investments originally committed to exportations, as the horizon of hydrocarbons reserves was significantly reduced, particularly on natural gas. The next figures illustrate these effects.

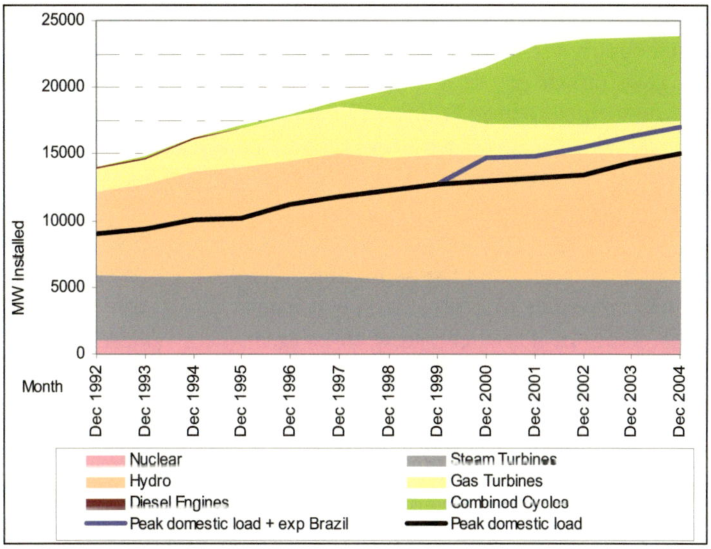

Source: CAMMESA and Mercados Energeticos

Figure 17.11. Argentina - installed power capacity vs peak load

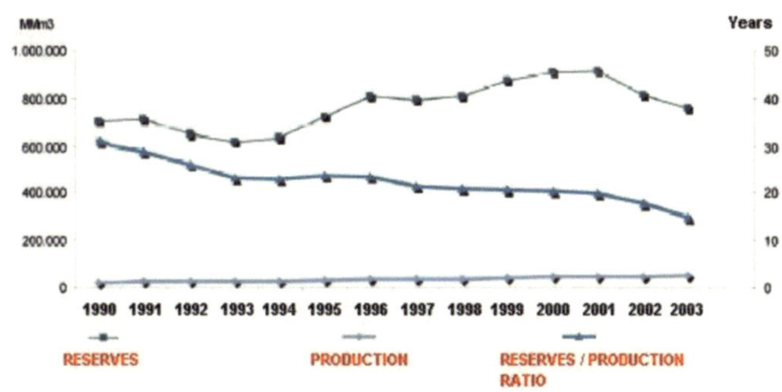

Source: Annual Report on Hydrocarbons Reserves 2003 – Energy Secretariat

Figure 17.12. Argentina - Performance on natural gas E&P

These facts were evidenced in April 2004, when the government announced reductions on natural gas exports to Chile in order to avoid curtailments on domestic demand. Consequences on electricity exportations to Brazil are yet unknown, since exportations contracts roughly have the characteristic of an option for the Brazilian demand: while electricity prices in Brazil are lower than the price of Argentine energy, which works as a strike price, interconnectors are not dispatched. Given the fact that the Brazilian power market had lower prices since 2002, Argentine options actually have not been significantly exercised. In case they do in the future, similar restrictions to those applied to gas exports to Chile should not be discarded.

But restrictions to exports were not enough to supply the domestic energy demand. In view of this situation, the government restarted permanent importations of natural gas from Bolivia in 2004, as well as occasionally imported electricity from Brazil. In addition, significant quantities of fuel oil and diesel were imported from Venezuela during 2004 in order to ensure full fuel supply for thermal power plants in case natural gas was not available. In mid 2003, the government together with private Argentine companies announced the construction of a new pipeline from Bolivia to Buenos Aires, which would allow an increase of natural gas offer by 20 mm^3/day. This project was recently discarded in light of the severe institutional and political crisis in which Bolivia is currently involved. Frozen tariffs and distorted prices blocked most of investment recovery for those existing companies at the time that crisis started. In the particular case of the power market, measures adopted lead to a significant imbalance between what the demand paid and what generators had to receive that resulted in a significant credit requested from generators. The government proposed to swap such credits with shares of a new company to be created for building and operating a new power plant. It should be noted that all the described actions, most included in the denominated "Energy Plan 2004-2008" published by the government, were oriented to ensure full supply of future energy demand reducing the expected average total cost by allowing special tariffs for them and, simultaneously, avoiding recovering of 'old' investment costs by private investors. More than 4,000 MW of new combined-cycle thermal plants were installed in Argentina between 1997 and 2001. Investors questioned that these plants were considered as 'old' investments, less than five years after they were installed.

17.5.4 The Government as a Leader in Energy Development

A new state-owned company promoted by the government, ENARSA, was created in October 2004. Main initial assets of ENARSA were full exploration and exploitation rights of most of oil offshore areas, but its business scope covers all energy-related activities. It is argued in Argentina that withdrawal of the government from the energy sector during the 1990s was excessive, and consequently more significant presence is now required. However, the question arises if the optimal way to achieve such presence is through a company that, in theory, is able to develop any energy business, and consequently compete with the private sector under unknown rules that, in addition, can be changed by the government itself.

The government said that ENARSA will allow them to follow what happens in the energy sector 'from inside', and consequently evaluate whether private energy companies' behavior is adequate or not. On the other side, many private companies see ENARSA as a tool by which the government may press them to agree conditions that, otherwise, would not be accepted. An agreement signed between ENARSA and PDVSA for acquiring retail network of gas stations currently owned by Dutch-British company Shell increased this perception in the private sector, since this is part of a wider strategic agreement between Argentinean and Venezuelan governments on energy matters that gives other dimension to the ENARSA's threat.

17.5.5 Domestic Problems Dominate the Energy Agenda

The energy plan presented by the government just seems to be a palliative for the expected consequences of about four years of lack of investments, rather than a strategic positioning of the country towards the possible international scenarios. Recent history seems to show a country that, worried by its self-created problems, perhaps has not given adequate importance in the last years to the development of its own energy resources as a strategic positioning of the country towards the complex possible international scenarios. This could represent a high cost for the country in the next years, but nothing indicates that this situation can be reverted in the near future.

17.6 Chile: The Difficulties of Modernizing the Reform Process

The Chilean power sector, that started a deregulation process back in 1982, has been another example in the region of sound sector reforms that have kept private power investment flowing, while reducing prices of electricity. The main difficulty in Chile has been to modernize its original outdated reform. The power sector has experienced several crises over its developments that have surfaced the weaknesses of its market model. The most recent crisis started when, as indicated in Section 17.5.3, the Argentinean government started facing problems with its gas supply and in April 2004 decided to reduce gas exports to Chile.

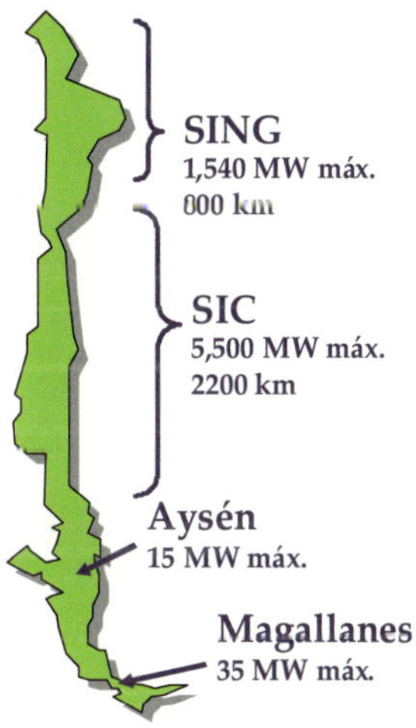

Figure 17.13. Chile

Chile, with 12,000 MW installed capacity in its two main interconnected systems (SIC and SING), is a country with limited energy resources except for its hydro reserves in the Andes Mountains. Its own oil only provides less than 10% of the country's needs, while its coal is of poor quality, so that imported coal has to be used for electric generation. Hydroelectric generation has been developed by using most of the low cost resources in the central part of the country. More expensive remaining significant reserves are over two thousand kilometers south from the main load center (Santiago). Argentinean natural gas arose as an attractive abundant cheap alternative and so an energy integration protocol was signed in 1995 with this neighboring country. Under that protocol, both governments agreed to establish the necessary regulations to allow free trading, export, import and transportation of natural gas. Private investors were strongly behind this process, and heavily invested in several pipelines that crossed the Andes and defined an energy supply path that would significantly rely on the efficient combined cycle generation plant technologies. The protocol worked very well and Chile fully relied on Argentina to provide the necessary energy required to sustain its important economic growth. Gas exports steadily grew through several pipelines. The petrochemical industry and the thermoelectric generation became the main users of natural gas. The arrival of this cheap fuel and the efficient generation technologies meant a significant reduction in the electricity prices in the main interconnected systems as shown in Figure 17.14. As explained before, these good days were finished since the rise of partial gas curtailments in 2004. The crisis has growing effects, as partial curtailments have become total curtailments from 2007. This situation has led to a sharp electricity price increase as shown in Figure 17.14.

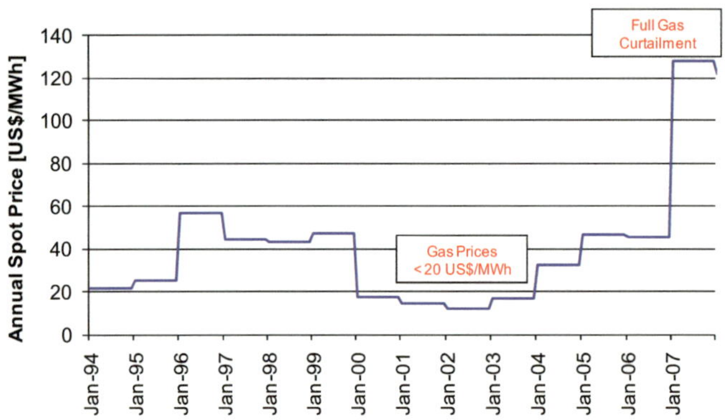

Figure 17.14. Spot price in the Chilean Central Interconected System (CIS)

17.6.1 Looking for Market Alternatives to Face the Crisis

The crisis brought by the reduction of Argentinean gas left Chile with no alternatives. Although, next-door Bolivia has significant natural gas resources and it has increasing exports to Brazil and Argentina, it denies the fuel to Chile due to its long-term border disputes with Chile (Bolivia lost its access to the Pacific Ocean in a 19th century war with Chile). In addition, Peruvian gas is not yet an alternative, given the distance from the Camisea gas fields to the main consumption centers in Chile. Chile was not prepared for the surfacing conditions. As a demonstration, the National Energy Commission, in its Indicative Plan of April 2004, projected the construction of seven combined cycle natural gas plants in the next ten years, all fed by pipelines from Argentina. Based on this fact, large expansions of existing electric transmission corridors were included in that plan. Also, major new hydro plants and interconnections with other systems were postponed until 2010 or later and therefore gas continued to be the major driver of expansion in a market with demand growing around 7% year. With the rise of the crisis, the Indicative Plan of October 2004 introduced radical changes to the energy supply government's point of view and so only one combined cycle plant based on Argentinean gas was considered. The government decided to bet on liquefied natural gas (LNG) as the alternative and defined a project to build the necessary installations to import it from abroad (Indonesia, Australia and Algeria being supply alternatives). But in the deregulated privatized Chilean power market, where private capital is the one making investment decisions, there is little space for the government to act, unless changes of laws were introduced. This is what happened in 2005. But changes were towards market mechanisms.

17.6.2 Chile: The New Market Model

In Chile, according to the 1982 regulatory model, the energy price for the regulated consumer was calculated by the government every six months as a unique value that represented the expected marginal cost of generation and losses in the transmission system. It was computed for each node of the interconnected system by means of penalty factors. This centralized calculation of prices, the volatility of the spot market due to the high hydro participation and the curtailments of natural gas imported from Argentina since 2004 (22% of the capacity of the main Chilean interconnected system corresponds to natural gas turbines) created a very risky environment for generation investment when new capacity was strongly needed. Therefore, the government looked for solutions by exploring long-term contracts at a price fixed by a free bidding process in order to ensure profitable cash flows for investors and thus stimulate the entrance of new capacity. Thus, as described in [16], a new regulatory model was implemented in the country by incorporating a real market signal in consumer prices through auction mechanisms. The old energy price calculation will fade out, as auctions replace existing contracts. The aim is to reflect cost expectations of generators and investors and the existence of an attractive market with high, but competitive, yields. Although, each distributor must auction its own demand, the new law allows them the accomplishment of a large auction, in which generators and new agents can bid for the added demand of several distributors. As in the Brazilian case, the Chilean auction process also obeys the rules described in section 17.3. Figures 17.15 and 17.16 describe the functioning of the Chilean system before and after the new scheme.

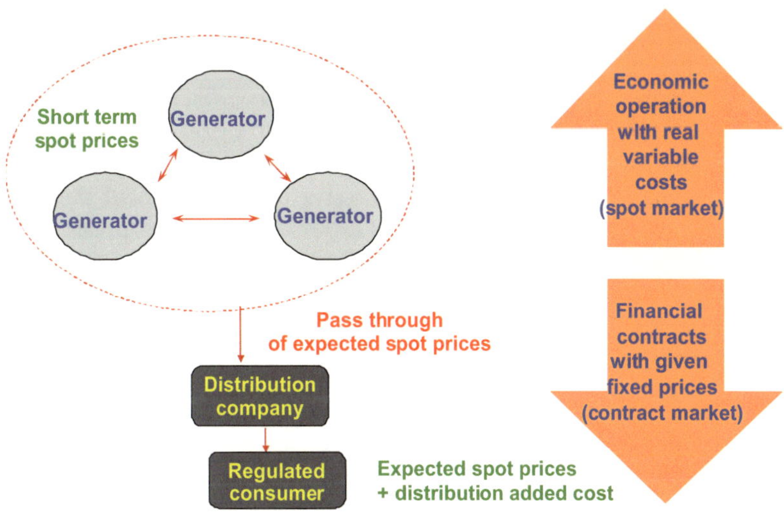

Figure 17.15. Previous Chilean model: spot prices all through the chain

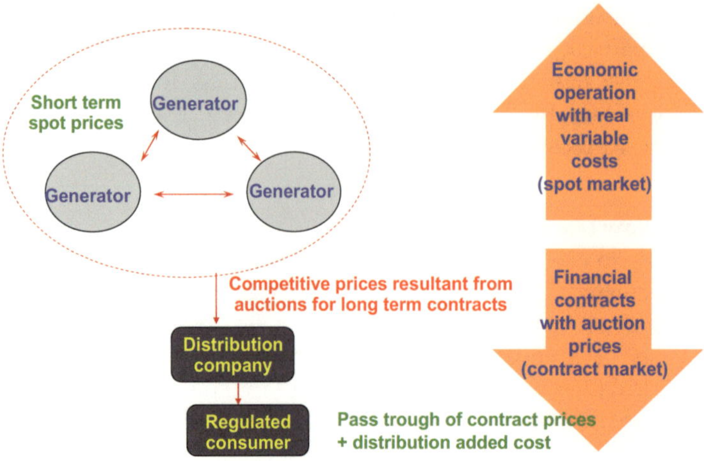

Figure 17.16. Current Chilean model: Contracts through market to stabilize revenues

Specific characteristics of the Chilean energy auctions are:

- Distributors must be 100% contracted all the time, at least for the next 3 years
- Distributors must contract their energy through auctions. Auctions must be public, open, transparent and without discrimination
- Each distributor auctions its consumption requirements according to its own needs
- Each distributor must design and manage its own auction. However, several distributors can organize a process to auction their added demand
- Distributors can offer contracts for 15 years at a fixed price (indexed according to changes in main variables)
- The government set a price cap for the auction
- A capacity price is fixed by the government (indexed according to CPI)
- Generators offer a price and an amount of energy (the amount of capacity is computed by means of a load factor)
- Auction winners will be the agents who bid the cheapest energy price alternative.

One of the most important aspects of the Chilean framework is that distributors design and manage their own auctions. This fact has opened a discussion about the incentives for distributors to design a mechanism that obtains lower end-consumer prices. It is important to consider that contract prices are passed directly to the consumers by using a pass-thought mechanism. Thus, distributors have a constant yield for their assets, irrespective of the auction results. Distributors auction their demand at any time, depending on their needs. Although distributors design their own auctions, the regulator must approve the final designed mechanism.

Generators must give a yearly justification to the National Energy Commission (NEC) of their firm capacity to supply all the regulated contracted demand (unlike the firm energy used in Brazil, firm capacity is required in Chile). Generators can use a mix of existing plants and new ones to justify their capacity. Thus, the general auction process is not divided into existing capacity and new capacity auctions as in the Brazilian case.

The new regulatory model has a complex methodology to determinate the adequacy capacity (or firm capacity) of a plant:

1) Firm capacity of hydroelectric plants is computed by using the two driest historical hydrology profiles and their regulation capacities among others. So, run of river plants and reservoir plants could present very different firm capacities for the same amount of nominal capacity.
2) Firm capacity of thermal plants is computed by using the available capacity (discounting average maintenance and considering force outage rates). Gas plants consider gas supply curtailments.

Finally, the new model considers contracts with energy delivery, at least, 3 years ahead. It allows investors to obtain project finance and have sufficient time to build new plants. Hence, the new mechanism represents a business opportunity for new investors in the generation business. The generators that are participating in the auctions compete by offering energy prices, which are indexed during the contract period. NEC administratively defines capacity price previous to the auction, and it is indexed according to changes in CPI during the contract period.

In the Chilean mechanism, each bidder together with its supply offer proposes indexing formulas. The mentioned formula must be built according to the power source of the bidder. However, it is important to highlight that, according to the designers (Discos), due to the unpredictability of fuel prices, these formulas are not taken into account by the auctioneer during the auction process. This fact has caused several discussions in the Chilean electricity market because contract allocation can change dramatically if price projections are incorporated into the mechanism. Consequently, generators that present expected fuel prices dropping in time need to bid high prices at the beginning of the period in order to get enough revenues. On the other hand, generators with high-expected prices can bid a low price at the beginning of the period. Thus, when indexing formulas are not taken into account for the allocation mechanism, bidders with high-expected fuel prices are favored, and vice versa.

Although generators bid only quantities and prices of energy, the final contracts include volumes and prices of both capacity and energy. Thus, every block of energy auctioned contains the capacity needed by each Disco that is computed before the auction by means of a load factor. The existence of a capacity payment included into the contract motivates Discos to manage their loads in order to present a higher load factor and, consequently, a better use of the system capacity.

17.6.3 The Auction Design

The Chilean bidding process allows distributors to auction their demand in one single simultaneous process, in which every generator bids for a specific set of products (a Chilean product corresponds to a specific block of demand from a distributor). Generators can bid for a net amount of demand higher than their capacities. Nevertheless, each of them must specify its maximum capacity and the process could assign at most this amount. All blocks of demand are assigned to every generator at the same time by means of a combinatorial sealed bid mechanism as shown in Figure 17.17.

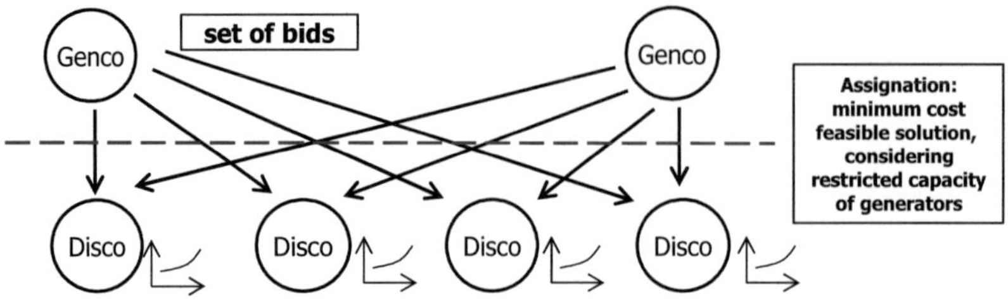

Figure 17.17. Auction mechanism

This mechanism has led to a large price differential among different products and distributors due to generators can choose a diverse set of bidding strategy for each auctioned contract.

As explained in [9], the auction design is crucial to get a good performance of the market, differences in price (and allocation) can be observed when applying various bidding rules.

17.6.4 Results, Difficulties and Next Steps

Overall, the first auction was carried out on October 2006, where about 1,300 average MW (30% of energy sales of the main interconnected Chilean system expected for 2010, 90% of the auctioned demand) was covered by the main Discos at an energy price of 53 US$/MWh in average, involving about 7 billion US$ and supply contracts until 2024. No new agents made bids in the first auction process due to the short time given to prepare the offers. After this, two more auctions have been carried out for contracting energy by 2010. Whilst new generation projects have been activated due to the new framework, difficulties have arisen such as:

- High prices driven by the indexing formulas which are not considered in the allocation process
- Lack of new investors participating in the auction and high presence of the current agents
- Large price differential and level of competition among contracts (distributors).

In despite of these facts, the new framework has been well evaluated by the market in order to ensure adequacy in generation by including real market signals in regulated contract prices. Future auction improvements must increase competition and the entrance of new agents, as explained in [10], along with a better policy to spread prices among distributors.

17.7 Colombia: Auctions for Long-Term Reliability Options

Colombia is located in the North West corner of South America. It is interconnected with Ecuador to the south and with Venezuela to the east and to the northeast. An interconnection line is in the last study stages with Panama and Central America, to the North West. The installed capacity in 2007 was about 14,000 MW, of which 66% was hydro, 27% gas, 5% coal. The remaining 1% corresponds to cogeneration and wind. Total demand is about 50 TWh, growing at a 5% annual rate. The electric energy in Colombia comes mainly from hydro-generation plants (77%) and a minor proportion from thermal-generation plants (18%). The dependency of the Colombian electricity market on hydro resources makes it necessary for the electric sector to have enough generation plants with firm energy to replace hydro-generated energy in dry periods that occur during climate phenomena such as El Niño. Without these alternative resources, demand would have to be rationed, implying high costs on the national economy.

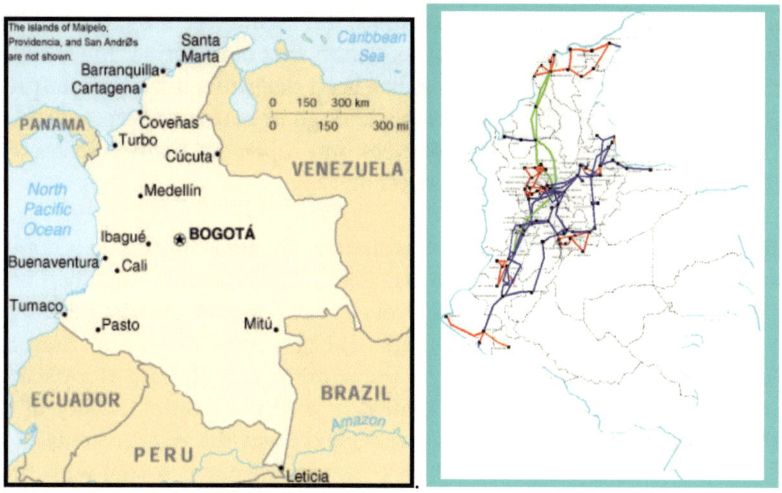

Source: EIA and XM, 2007.

Figure 17.18. Colombia

Following the pattern of the electricity markets in the region, the price volatility in the energy Spot Market (see Figure 17.19), largely explained by the huge hydraulic component of supply and the periodic occurrence of El Niño phenomenon in Colombia, poses a considerable risk for generation companies that need financing for their projects. This situation calls for the implementation of hedging mechanisms to mitigate the risks for generation companies and new investors.

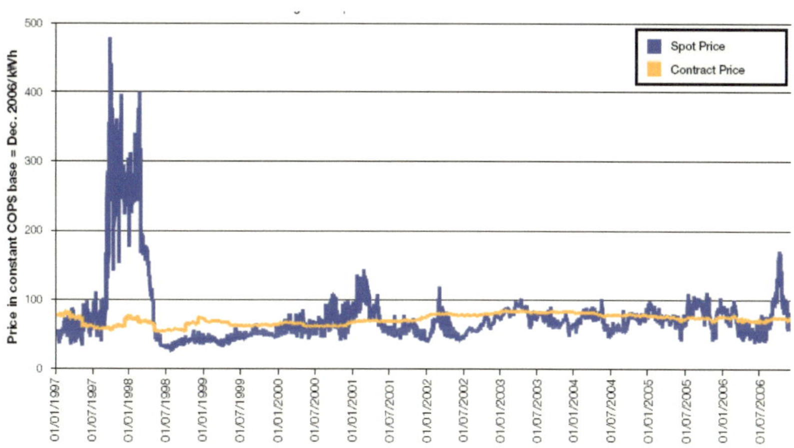

Source: XM and CREG, 2007.

Figure 17.19. Energy spot and contract prices in Colombia

17.7.1 The Previous Scheme: Capacity Charge

During 1992, Colombia experienced the most serious electrical rationing that the country has known. Direct costs were estimated of the order of three billion US dollars that the Colombian society paid in various ways. Rationing was mainly due to shortages of water resources brought by an El Niño event. This event precipitated the formation of the electric market (July 1995) and therefore, from its origins, the regulation of the Colombian electrical market is mainly focus on the potential consequences that may derive from a new rationing. Consequently, the regulation of the market has been determined by the interpretation that was made of the main cause of the rationing: shortage of hydro resources. Then, the efforts have been centered in preserving the resources and replacing them with more expensive resources that are complementary and more reliable.

Implementing a remuneration scheme that promotes income stabilization is considered as a fundamental issue by the regulatory body (CREG). Therefore, making investment in generation resources viable to efficiently cover the demand requirements, particularly during critical periods of low hydraulic supply [11,12], arises as an important task for the regulator. The first mechanism adopted was the administratively settled capacity charge: in general terms, it is a regulated income oriented to guarantee the reliability of the system, based on the remuneration of the plants established from the requirements of generation during the summer season estimated by an economic dispatch model with transmission, having as reference a critical hydrologic scenario and a demand projected for the year in reference. Initially the hydrologic scenario was associated to the critical biennium 91-92, later, this scenario was changed to an artificial "hyper dry" hydrologic event.

The capacity charge scheme in Colombia has always faced several challenges and implementation difficulties such as the administratively setting of the payment and the calculation of the firm energy, among others. With the intention of correcting these distortions and centrally replacing established procedures by market mechanisms, changes were introduced in 2006. They will be described next.

17.7.2. The New Scheme: Reliability Charge

Following ten years of uninterrupted application of the Capacity Charge scheme, CREG considered it beneficial to replace it with a market scheme, which conveys a long term signal that promotes new investments in generation resources in Colombia, to guarantee the availability of electric energy at efficient prices in periods of scarcity. A new method was designed, based on a market mechanism denominated Reliability Charge, which has been in place since December 2006. This new mechanism preserves the essential factors of settlement, billing and collection that guaranteed the successful payment to generation companies in the previous scheme. It is fully described in [12,13].

One of the essential features of this new scheme is the existence of the Firm Energy Obligation (OEF), which is a commitment on the part of generation companies backed by a physical resource capable of producing firm energy during scarcity periods. This new scheme aims to ensure the reliability in the supply of energy in the long run at efficient prices. To achieve this purpose, the OEFs needed to cover the demand auctioned among generation companies and investors. The generator who wins the OEF allocation receives a stable compensation during a specific time period, and in exchange commits to deliver a determined quantity of energy when the energy spot price is higher than the pre-determined level, the Scarcity Price. Such compensation is settled and collected by the system and is paid by all the end-users of the interconnected system, through the fees charged by commercialization companies.

17.7.2.1 Firm Energy Obligation, commitment and scarcity price

The Firm Energy Obligation is an option product designed to guarantee the reliability in the supply of energy in the long run at efficient prices. When the spot price surpasses in at least one hour during the day the value previously established by CREG, which is known as the Scarcity Price, it reflects a critical electric energy supply situation. When this occurs, it serves as a trigger factor for generation companies with OEF allocations to produce, as required in the ideal dispatch, a determined daily quantity of energy. The OEF can be acquired through centralized transactions in the wholesale energy market. The OEFs are auctioned and allocated uniquely among generators or investors that have or are planning to own generation resources. Only those generators with their corresponding firm energy at a determined time can participate in the OEF auction.

The firm Energy for the Reliability Charge (or ENFICC) refers to the maximum electric energy that a generation plant is able to deliver on a continual basis during a year, in extreme conditions of hydro inflows. The Scarcity Price, which is established by the CREG and updated monthly based on the variation of the Fuel Price Index, has a double purpose. On the one hand, it indicates the time when the different generation units or plants will be required to fulfill their OEFs, which happens when the spot price exceeds the scarcity price, and on the other hand, it is the price at which this energy will be paid. The commitment period of the OEF is decided by the owner or the commercial representative of the generation resource that backs up the OEF. If the generation plant is new, meaning at the time of the auction its construction has not started, the obligation to generate energy can be between a minimum of one year and a maximum of twenty years. If it is a special resource or at the time of the auction the generation plant or unit is in the process of construction, the obligation to generate energy is between one and ten years. Finally, if it is an existing resource, which implies that it is ready to operate (or it is already operating) in the wholesale energy market at the time of the auction, the commitment period of the OEF is one year.

During the commitment period of the OEF, the generator receives the Reliability Charge remuneration, a value that is determined in the auction where the generating company participated to obtain its OEF. The owner of the OEF commits to generate daily, as required in the ideal dispatch, a certain quantity of energy up to the amount specified in the OEF. When the Spot Price exceeds the Scarcity Price, in order to verify that each generator has fulfilled its commitment, all the energy generated from all its plants at each hour of the ideal dispatch are added up.

The generator who acquires an OEF will receive a fixed remuneration during the commitment period of the OEF, whether the fulfillment of this obligation is required or not. The price for each kilowatt-hour of the OEF corresponds to the clearing price in the auction in which the generator sold its firm energy. This price is denominated as Reliability Charge.

When this firm energy is required, which happens when the Spot Price surpasses the Scarcity Price, aside from the Reliability Charge the generator also receives the Scarcity Price for each kilowatt-hour generated associated with its OEF. In case the energy generated is more than the obligation specified in the OEF, this additional energy will be paid or rewarded at the Spot Price.

In summary, the "Reliability Charge" acts like an option with an exercise price equal to the "Scarcity Price": a generator with a given firm energy allocation, should make this energy available to the spot market at the scarcity price, whenever the value of the spot market is equal or above the scarcity price. Plants can generate above their firm energy commitment, selling this spare energy at the prevailing spot market price.

17.7.2.2 Allocation of the firm energy options: auctions

The allocation of the OEF among different generators and investors is done through a dynamic auction. In this transaction in the wholesale energy market, generators and investors participate actively, while the electricity demand of end-users connected to the system is represented by a price-quantity function previously established by the CREG. For this purpose, an auction to allocate the OEFs is undertaken three years before the firm energy obligation can be called. The Auction for the allocation of OEF is a one sided process. This means that the generators and potential investors, who have complied with all the requirements necessary to participate in the auction, can actively bid. The demand of end-users connected to the system is represented by an aggregate demand curve that is previously established by the CREG and made known to the public before the auction is conducted.

The mechanism employed is a descending clock auction and is carried out as follows:

- The auctioneer opens the auction at a price equivalent to two times the Cost of Entrant; a value calculated by the CREG and made known to the bidders (generators and investors) before the auction. Likewise, the auctioneer announces the floor price at which this first round will close.
- Between these two prices the bidders build their firm energy supply curve and this information is sent to the auction administrator. Figure 17.20 describes the auction methodology.
-

Taking into account that the purpose of conducting auctions is to acquire firm energy, these auctions only take place when it is estimated that the demand for energy three years from now cannot be covered during scarcity periods of power supply by the firm energy production of existing generation resources and new resources that will enter into operation during the next three years. Annually, the CREG evaluates the balance between the supply and demand of firm energy and if the CREG deems it necessary to convene an auction, it communicates this decision through a Resolution, and indicates the timetable of the activities required before and after the auction to enable bidders to participate in the process and to formalize the allocation of the OEFs.

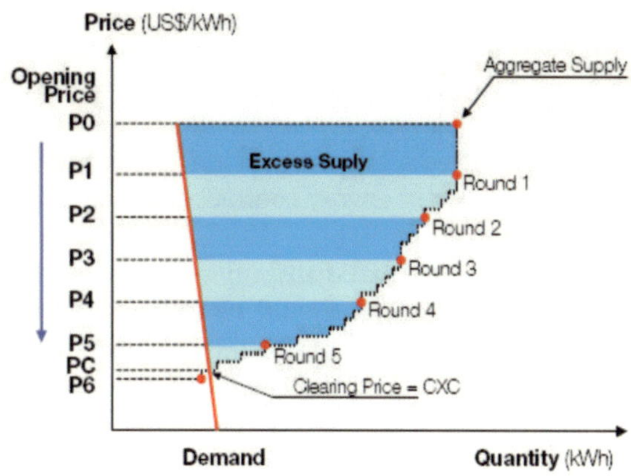

Source: CREG, 2007.

Figure 17.20. Descending price-clock auction with intra-round bids in Colombia

In summary, the unit price ($/kWh) paid for each kWh of firm energy allocated, as well as the firm energy allocated to each generator, are the result of a "descending clock auction" with an elastic demand curve (Figure 17.20), that takes place three years before the regulator estimates that the firm energy will be required, or when the Regulator so decides. The price obtained as a result of this auction is guaranteed to new investors for a period of up to 20 years, to help them in firming up their cash flow and thus to facilitate project finance. For existing plants, the price is valid only for the following year.

17.7.2.3 Results

The first auction under the reliability charge scheme was carried out in May 2008 with a successful result, guaranteeing a capacity coverage to Colombia until 2018. In parallel with the new "Reliability Charge", the regulator is replacing bilateral contracts by short term (up to three years) energy contracts in which all the demand will be auctioned in concurrent auctions for regulated and unregulated clients. In order to reduce risks, these auctions will be rolling, periodic with a certain percentage of the demand being auctioned each time.

17.8 Peru and Central America: Towards Energy Auctions

17.8.1 Peru

In 2007, Peru had 7.0 GW of installed generating capacity. In the same year, the country generated 25.0 TWh of electricity, while consuming 22.6 TWh. Even though installed capacity is evenly divided between hydroelectricity and conventional thermal, 80 percent of Peru's total electricity generation comes from hydroelectric facilities: conventional thermal plants generally operate only during peak load periods or when weather factors dampen hydroelectric output. The power sector underwent vertical and, to a lesser degree, horizontal restructuring initiated in 1994, following enactment of a new Electricity Concessions Law in 1992. The country first market design followed the principles adopted in Colombia and Chile: capacity payments assigned by the regulator and the energy spot market as the marketplace for energy trading and provider of signals for new investment.

As in the Brazilian, Chilean and Colombian cases, Peru has undergone a drought and several difficulties with the current scheme came up. Therefore, as described in [17], a proposal of reform has been elaborated in 2006 to ensure generation adequacy and to reduce the exposition of the Peruvian electric system to the risks of excessive prices and a prolonged deficit of energy by introducing competition "for the market". These reforms are mainly based on the implementation of energy contract auctions mechanisms.

An energy Law passed in 2006 defines that distribution companies must be 100% contracted for the next three years and that auctions should be called to ensure the entrance of new generation. The contracts to supply electricity for the medium or long arranged under the terms of the tender process, i.e. employing regulated electricity rates fixed as a result of the best bid received, will reduce the levels of risk as much for the consumers as for the producers and will make more feasible the new investments, possibly increasing with new agents the generation supply and as a consequence the competition in and for the market in Peru.. Figure 17.21 next describes the transition process:

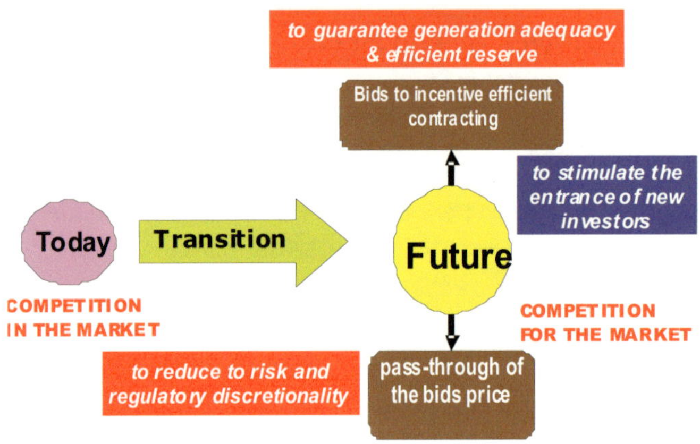

Figure 17.21. Peru: Competition for the market.

By the time of this writing (October 2008), the regulations that specify the guidelines of the Law are still being prepared and the first auctions to contract new energy are expected to be called in 2008. However, the principles that will guide the Peruvian auctions are similar to the ones contained in the Brazilian and Chilean frameworks.

17.8.2 Central America: Towards Regional Energy Auctions

Central American nations have a population of 41 million inhabitants, a GDP of some US$ 90 billion and installed generation capacity of about 9000 MW.

During the early 1990s the Power Sector in Central America (CA) was basically managed by vertically integrated state-owned utilities that concentrated the production and supply of electric power. As in the vast majority of Latin-American countries, the main characteristics of the power industry before the restructuring were electric power shortages, vertically integrated state-owned utilities, lack of fresh funds, poorly maintained power plants and unavailability of public financing resources. As a consequence, the reform of the power sector took place in many Central American countries, even though with distinctive features in each case, following the basis of the model applied in other South American countries such as Chile, Argentina and Brazil. The aim of the Reform was to provide a new institutional and regulatory framework based on private investments, competition in the generation segment, and adequate regulation of monopoly services (transmission and distribution) that could ensure a reliable and economic supply of electricity.

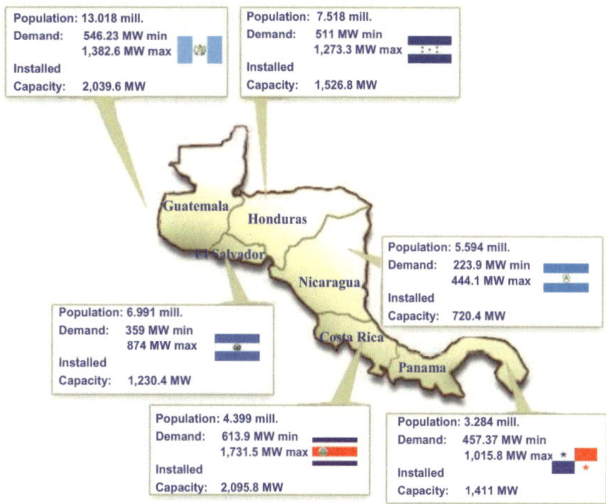

Source: EOR, CEPAL

Figure 17.22. Central America.

17.8.2.1 Supply adequacy schemes

The region is a net importer of liquid fuels. Therefore, international fuel prices directly affect the electricity market. This fuel price increase has introduced tariff problems and has put pressure to modify the rules of competitive markets. Electricity tariffs vary across the region (with variations up to 70 %), showing different energy policies between countries.

There are several models of organization of the electricity sectors in CA. Countries in the early stages of liberalization in CA (Honduras, Costa Rica) have primarily used PPAs to support generation expansion. El Salvador, initially with an energy-only market, did not include in its original market design any capacity support mechanisms. The other three countries (Guatemala, Nicaragua and Panama) have taken a market approach for energy trading, but with obligations to distributors and large users to buy in advance their expected demand through forward contracts. The electricity markets in these countries has been designed based on (i) energy prices related to variable (regulated) generation costs (cost-based pool) with a relatively low (capped) value during shortage conditions, (ii) a generation capacity obligation for loads to cover in advance through contracts their participation in the system peak load, (iii) auctions to cover reserves for the next year (only in Panama), (iv) a daily capacity market to settle imbalances (deficiencies and excesses) in generation capacity due to differences between expected peak load and actual peak load and, finally, (v) differences between committed generation capacity and actual availability.

In Guatemala, El Salvador, Nicaragua and Panama, at the time of market opening, distribution companies were allocated existing PPAs as a form of "vesting" contracts, however, additional new capacity is secured through supply contracts with shorter periods and without recourse to sovereign guarantees.

In summary, the electricity market reform panorama is:

• Guatemala, Nicaragua and Panama have organized competitive markets with a high level of regulatory intervention to ensure adequacy, through mandatory requirements to distribution companies and large users (final users authorized to find their own source of supply in the market) for contracting forward their expected peak demand plus some defined security reserve margin.
• El Salvador organized initially an energy-only market, with little regulatory intervention to ensure adequacy, although in 2002, as a result of decreasing reserve margins during several years, it introduced amendments to the law to incorporate some level of intervention on this and other topics.
• Costa Rica and Honduras created single buyer competitive markets, maintaining a centrally planned system, and allowing private participation in generation through Power Purchase Agreements (PPA) with Independent Power Producers (IPPs). In Honduras a new plant is basically built through PPAs which in December 2004 represented 64% of the total system installed capacity. On the other hand, in Costa Rica private participation is limited to 30% of the country's total installed capacity.

17.8.2.2 A full regional marketplace

The most interesting part of Central America is the high degree of integration among the different countries. The Governments of Costa Rica, El Salvador, Guatemala, Honduras, Nicaragua and Panama, in the framework of the Central American Integration System (SICA), initiated in 1996 a gradual process of electrical integration by developing a competitive regional electricity market through transmission lines which interconnect their national grids and by promoting regional generation projects. They implemented the project known as the Central American Electrical Interconnection System (SIEPAC) and so the Framework Treaty for the Central American Electricity Market was agreed in 1996. Two regional agencies were created to better fulfill the purposes of the Treaty: the Regional Electrical Interconnection Commission (CRIE) and the Regional Operating Agency (EOR). This regional market allows spot transactions and will allow regional firm contractual arrangements once the new Interconnection in 200 KV will be in service (expected for 2010).

One of the main objectives of the new Regional Electricity Market (MER), once implemented, is to enable the construction of regional generation projects, which will take advantage of economies of scale and provide cheaper electricity to consumers in the region. To support this type of projects -and in general the fulfillment of national capacity and energy obligations with sources originating in other countries, the MER market design provides for firm regional supply contracts that will be required to acquire firm transmission rights in order to be accepted by local regulators as a comparable source of supply to generation located within the country's borders. The MER will effectively allow an integrated approach to adequacy, through the concept of regional firm contracts. A second objective of the MER is to increase effective competition. The possibility of distributors of CA to procure in the MER energy to fulfill their obligations, rather than in their national markets, increases substantially the level of competition. The main characteristics of the MER are:

- It constitutes a "seventh" market "superposed" on top of the national markets
- A regional regulatory agency (CRIE) and an independent system and market operator (EOR) are created
- Countries preserve national regulations and interact with the MER through "interfaces" (a feedback mechanism between the national markets and the MER)
- A short-term market is established, with ex-ante (day ahead) and ex-post (real time balance) hourly nodal prices (reflecting energy, congestion and losses prices), for each node of the regional transmission network (RTR - Red de Transmisión Regional)
- A contract market is established, with firm and non-firm contracts
- Transmission rights (financial and physical) are auctioned by EOR.

Since November 2002 the MER has been operating using a "transitory" code - an hourly day-ahead energy and transmission dispatch with hourly nodal prices at the tie-lines substations [18]. The "new" regulations have been recently approved by CRIE and a SCADA/EMS system and models (pre-dispatch, transmission rights auctions, settlement, etc.) are being developed.

Figure 17.23. SIEPAC line.

At the regional level, the augmentation of transmission capacity between countries, the existence of firm contracts and the associated transmission rights, opens up an opportunity for the coordination of the distribution companies procurement auctions (individually the distributors are very small) to incentivize efficient contracting (i.e. through the entrance of regional generators, which are too big for a single country, and even more so for a single distribution company).

MER firm contracts have priority of supply at the buyer's node; they must be approved by the national regulators involved (the seller's and the buyer's countries), and must hold the corresponding firm transmission rights. New transmission rights are created/acquired by either increasing (building new lines) the transmission capacity between the seller and buyer's locations, or through long and short-term auctions organized by the EOR.

17.8.2.3 Regional contracts and firm transmission rights

A firm regional contract means "iron on the ground" for both generation (capacity and energy) and transmission. In a firm regional contract the seller agrees to deliver firm energy at the RTR node declared by the buyer.

A firm regional contract offers the buyer security of delivery for the contracted energy, limiting the risk of energy provision, price and the associated variable transmission costs - except when, due to conditions on the RTR, it is technically impossible to deliver the energy. The objectives of firm regional contracts are:

- Give both, buyer and seller, greater security and obligations of fulfillment of the commitment;
- Assure the buyer the delivery of the contracted energy;
- Promote the development of regional generation plants;
- Promote interchanges of greater term and volume; and
- Promote the development of the RTR.

Due to their characteristics, firm contracts will be, in general, long term commitments. Nevertheless, their terms and duration are decisions of the parts and not subject to regional regulation. A firm regional contract establishes a priority of supply different to that which would "naturally" arise from the physical location of the generation committed. A firm regional contract "locates commercially" the contracted energy in the country where the retirement is committed (Figure 17.23). The contracted energy has priority for the supplying of the demand of the buyer at the RTR node declared for the energy retirement, instead of having priority of supplying for the demand of the country in which the seller (generator) is located.

The seller, or the agent whom the parts decide, must hold firm transmission rights (between the node of injection and the node of retirement) for the transmission capacity required by the contract. Firm transmission rights give not only financial protection against the variability in the difference of nodal prices between the agents' locations, but also guarantee that firm regional contracts can be physically accommodated by the RTR.

The energy committed in a firm regional contract cannot be offered (to sell) in a national contract to guarantee the supplying of the demand of the country in which the seller (generator) is physically located, i.e. the same energy cannot be committed simultaneously in a national and a firm regional contract. To avoid the risk of supplying or undue national dependency, the amount of energy that an agent qualified in the MER will be able to buy or to sell in this type of contracts will depend on:

- the energy authorized according to the regulation of the respective country, considering CRIE regional criteria for the firm energy estimation (that take into account generation

capacity, availability of energy resources, peak demand of each national system, exiting regional and national contracts and reserve requirements); and
• the associated firm transmission rights held.

The selling agent in a firm regional contract will be able to optimize the delivery of the energy to the buyer, from purchases in the regional opportunity market. Additionally, and as a reflection of the firmness, the selling agent must have injection offers to the regional opportunity market for, as a minimum, the totality of the firm commitments acquired in the MER. If the delivery of energy to the buyer is not possible due to the unavailability of the seller's energy, the seller will assume the penalties that are derived from the breach of the contract. If firm energy cannot be delivered due to RTR constraints (security, quality) firm contracts will be reduced proportionally.

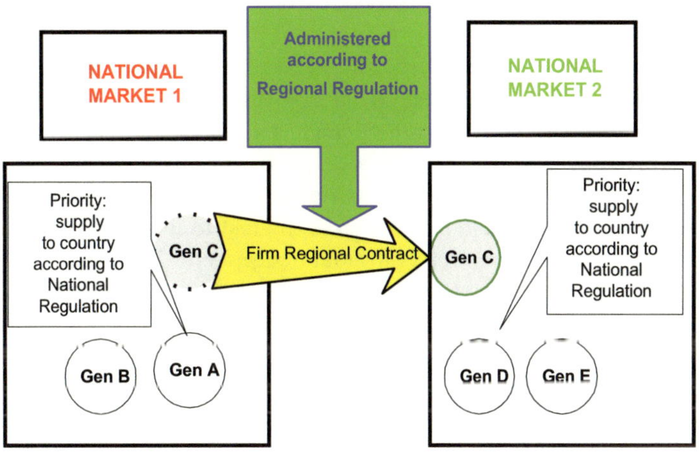

Figure 17.24. Priority of supply of firm regional contracts in MER.

17.8.2.4 Perspectives

In summary, several electricity markets in Central America have implemented capacity obligations mechanism to support the prices of the energy markets, which by itself have shown limited success, at least in markets with a small participation of the demand, in maintaining an adequate level of supply adequacy through timely investments. In light of the experience in the region, it can be inferred that in small electricity markets, as is the case of each national market in the Central America region, competition is very difficult to achieve (individual competitors would need to be very small) and energy-only schemes do not seem to be able to guarantee supply adequacy. Even markets with "complementary" controls for supply adequacy (obligation to contract and capacity payments) have been somehow intervened (e.g. price caps). The Regional Electricity Market -being implemented- has been designed with features such as firm transmission rights that are expected to support the region-wide compatibilization and optimization of the supply adequacy objectives.

17.9 Further Reading

Further reading on Latin America market mechanisms and supply adequacy together with electricity resource adequacy planning is given in References [23-24].

17.10 Conclusions

The primary challenge for Latin American countries is to ensure sufficient capacity and investment to serve reliably their growing economies. Although converging in the need of a "second stage" of measures to ensure generation adequacy in the region, some countries (Brazil, Chile, Colombia, Central America and Peru) retained the market scheme and improved the rules to ensure the entrance of new capacity. Other important countries in the region, however, went back to the state-controlled scheme. This is the case of Argentina (already analyzed in this chapter), Bolivia, Ecuador, Paraguay and Venezuela. Therefore, an "ideological split" is observed in the region.

The reform processes, including these recent auction mechanisms, have aimed at creating conditions to respond to growing demand with economic investment and operation, but with key decisions made by private actors, with a limited role being played by central governments. The priorities of the private actors are essentially business oriented, responding to their strategies and their risk assumptions. Overall, the new capacity auctions in Brazil, Colombia and Chile have been of great interest to international investors looking to South America's energy market: candidate suppliers include a wide variety of technologies, comprising new hydro projects, gas, coal and oil-fired plants, sugarcane biomass and international interconnections. Peru seems to be following the same path and Central America presents an innovative regional market with cross-border supply adequacy schemes. With these diverging approaches, this is how these countries are facing the challenges of electricity supply.

Among the issues that still need to be reviewed are the social and environmental constraints, which are an inherent part of electric markets and cannot be swept under the rug. As discussed in [20,21], the concern with the environment is absolute legitimate but in some cases has resulted in the construction of more expensive equipments this disrupting an efficient system expansion. The most fundamental challenge is to allow the society to know, through lively participation on the studies and licensing process of hydro and thermal plants, that there is no competitive energy without environmental impact. A policy of zero environmental impact has obviously a very high economic cost and the society must be aware about this tradeoff, so that the best choice to conciliate environment, economic growth and social justice can be chosen. The rapid and hardly predictable changes in the sector, including national and international interconnections of the power and gas networks, strategic considerations by firms, availability of fuels and increasing public participation, make this a complex task.

For further and updated details on the Latin American deregulation, we refer the reader to [22].

17.11 Acknowledgements

This Chapter has been compiled by Dr. Luiz A. Barroso, PSR, Rio de Janeiro, Brazil; Chair of the IEEE PES W.G. on Latin America Infrastructure; Rodrigo Moreno from Systep, Chile &

Imperial College London, UK and Dr. Hugh Rudnick, Pontificia Universidad Catolica de Chile. Contributing authors include B. Bezerra (PSR, Brazil), M. V. Pereira (PSR, Brazil), J.Rosenblatt (PSR, Brazil), S.Mocárquer (Systep, Chile), , J. Karacsonyi (Mercados EMI, Spain), M. Tinoco (SNC-Lavalin, Canada), R. Ríos (consultant, Mexico), F. Montoya (SNC Lavalin, Canada), D. Cámac (Osinergmin, Peru), V. Ormeño (Osinergmin, Peru), L. Espinoza (Osinergmin, Peru), L. V. Sbértoli (SIGLA, Argentina), R.Varela (SIGLA, Argentina) and M.Madrigal (World Bank, USA). The Chapter is primarily based on an update of the papers presented at the Panel Session on "Market Mechanisms and Supply Adequacy in the Second Wave of Power Sector Reforms in Latin America" at the IEEE-PES 2006 General Meeting (GM2006) in Montreal ([14-19]).

17.12 References

[1] H. Rudnick, L.A. Barroso, C. Skerk, and A. Blanco. "South American Reform Lessons - Twenty Years of Restructuring and Reform in Argentina, Brazil and Chile". IEEE Power and Energy Magazine, Vol. 3, (4) July/August 2005, pp. 49-59.

[2] Implementing Power Rationing in a Sensible Way: Lessons Learned and International Best Practices; L.Maurer, J.Rosenblatt, M.Pereira - ESMAP - World Bank, 2005

[3] L.A. Barroso, L.M. Thomé, M.V.Pereira and F.Porrua, "Planning Large Scale Transmission Networks in Competitive Hydrothermal Systems: Technical and Regulatory Challenges", IEEE Power & Energy Magazine, Vol. 5, Issue 2, Page(s): 54-63, Issue March-April 2007.

[4] M.V. Pereira, , N. Campodónico, and R. Kelman, "Long-term hydro scheduling based on stochastic models", Proceedings of EPSOM Conference, Zurich, 1998 - Available at http://www.psr-inc.com

[5] B.Bezerra, L.A.Barroso, M.V.Pereira, S.Granville, A.Guimarães, A.Street, "Energy Call Options Auctions for Generation Adequacy in Brazil and Assessment of Gencos Bidding Strategies", Proceedings of IEEE General Meeting 2006, Montreal.

[6] C. Vásquez, M. Rivier, I.J. Pérez-Arriaga, "A market approach to long-term security of supply" IEEE Transactions on Power Systems, Vol.12, N°2, May 2001.

[7] S. Oren, "Generation Adequacy via Call Options Obligations: Safe Passage to the Promised Land", 16, UCEI Energy Policy and Economics, September 2005.

[8] Oren, Shmuel. S. "Capacity Payments and Supply Adequacy in Competitive Electricity Markets". VII Symposium of Specialists in Electric Operational and Expansion Planning. Curitiba, Brazil. May, 2000.

[9] Moreno, R., Rudnick, H. and Barroso, L., "First Price and Second Price Auction Modelling for Energy Contracts in Latin American Electricity Markets", 16th Power Systems Computation Conference, July, 2008, Glasgow, Scotland

[10] Barroso, L, Rudnick, H., Moreno, R., Bezerra, B., "Ensuring Resource Adequacy with Auctions of Options and Forward Contracts", paper 07GM1368, IEEE Power Engineering Society 2007 General Meeting, Tampa, Florida, June 2007

[11] J. V. Bermúdez. R.C. Pinzón, "The Colombian electricity market and its impact in hydrothermal expansion", 2008 PES General Meeting, Pittsburg, USA.

[12] Reliability Charge - regulatory scheme to guarantee the reliability in the supply of electric energy in Colombia, CREG official report, available at http://www.creg.gov.co/cxc/index_e.htm

[13] P.Cramtom, S.Stoft, "Colombia Firm Energy Market", Proceedings of the Hawaii International Conference on System Sciences, January 2007
[14] L. V. Sbértoli, "The State and The Market in The Power Sector Planning", 2006 IEEE PES General Meeting, Montreal, Canada (paper 06GM0209)
[15] L.A.Barroso, J.Rosenblatt, B. Bezerra, A. Resende, M.Pereira, "Auctions of Contracts and Energy Call Options to Ensure Supply Adequacy in the Second Stage of the Brazilian Power Sector Reform", 2006 IEEE PES General Meeting, Montreal, Canada (paper 06GM0205)
[16] H. Rudnick, S. Mocarquer "Contract auctions to assure supply adequacy in an uncertain energy environment", 2006 IEEE PES General Meeting, Montreal, Canada (paper 06GM1192)
[17] D. Camac, V. Ormeño, L. Espinoza, "Assuring the efficient development of electricity generation in Peru", 2006 IEEE PES General Meeting, Montreal, Canada (paper 06GM0208)
[18] J. Karacsonyi, M. Tinoco, R. Rios, F. Montoya, "Supply Adequacy Mechanisms and Cross-Border Contracts in the Central American Regional Electricity Market", 2006 IEEE PES General Meeting, Montreal, Canada (paper 06GM0207)
[19] M. Madrigal, J. Millan, R. Robboy, J. Molina, "Financing Challenges for Generation Investments in New Supply Adequacy Mechanisms in Latin America", 2006 IEEE PES General Meeting, Montreal, Canada (paper 06GM0206)
[20] L.A. Barroso, H.Rudnick, S.Mocarquer and B.Bezerra, "A Delicate balance in South America: the Challenges of Hydroelectricity in South America", IEEE Power & Energy Magazine, Vol. 6, Issue 4, July-August 2008, Page(s): 22 - 35.
[21] L.A. Barroso, H.Rudnick, S.Mocarquer and T. Castro, "Creating harmony in South America: the challenge of conciliating energy development & environmental constraints in South America" IEEE Power & Energy Magazine, Vol. 4, Issue 4, July-August 2006, Page(s): 32 - 46.
[22] Information on Latin-American deregulation in www.ing.puc.cl/power
[23] T.J. Hammons, L.A. Barroso and H.Rudnick, "Latin America: Market Mechanisms and Supply Adequacy in Power Sector Reforms", Energy Systems, Vol.2, February 2011, pp. 83-111, DOI 10.1007/s12667-011-0027-8. (Available-on-line).
[24] T.J. Hammons, L.A. Barroso and H.Rudnick, "Integrated Natural Gas—Electricity Resource Adequacy Planning in Latin America", International Journal of Power and Energy Systems, Vol. 30, (1), 2010, pp. 1-14.

18

Energy Issues under Deregulated Environment

18.1 Introduction

Until two decades or so, electricity was generally provided by vertically integrated utilities with a single utility providing electricity generation, transmission and distribution. Introduction of competition in electricity markets promoted interconnectivity of transmission systems across utilities, and "wheeling," which occurs when one utility provides transmission services across its lines for another utility. The benefit of increased transmission interconnectivity is that when one or two-generation facilities or transmission connections fail, other connections provide back up through alternative sources of generation and transmission capacity. Such breakdowns do occur on a regular basis, and usually are barely noticeable by consumers. Interconnections of generation and transmission systems across utilities mean that more back-up capacity is available for the local utility when it experiences an individual breakdown. The drawback to interconnectivity, however, is that when failures are more widespread, larger system failures are able to migrate across the entire region rather than be confined to the local utility [1].

It is believed that deregulation of electricity and lack of investment in the transmission network, particularly in transmission interconnections, are the main causes of major blackouts. While greater investment in the existing energy system could prevent blackouts, a better solution to the problems would be to introduce a cleaner, more efficient and more decentralized energy systems. Switching to energy efficient and renewable energy technologies, and the development of distributed generation systems could increase the reliability of the electrical delivery systems, making it less vulnerable to blackouts.

Power blackouts cause widespread havoc as well as losses to business. This highlights the dependency on electricity, as well as the general lack of strategic business contingency plans to deal with blackouts. Losses experienced as a result of a power failure include the loss of critical data, loss of productivity, lower efficiencies, damage to a company's reputation, as well as inability to deliver products and services.

Information and communication technology will provide the tools to monitor, measure, and assess grid performance in real time, route power flows, reduce loads, and take measures needed to maintain grid stability. Including secure networks of sensors, communication links, information processors and dynamic algorithms would make the grids largely self-healing. Developments that are under way include semiconductors capable of handling high power flows: improved solid-state ac switches, controllers, inverters, and converters. Superconductivity cables and better sensor and communication networks to detect and control disturbances on the grid are also being developed. Modern information and communication technologies could help power utilities to achieve highly secure energy systems.

Current research and development budgets for transmission and distribution (T&D) are relatively small and under pressure as a result of industry deregulation. As a consequence, it remains unclear how soon the information and communication technology developments supported under the Flexible AC Transmission System (FACTS) and other programs will become affordable to T&D operators and widely deployed.

Internal combustion engines, micro-turbines, fuel cells, and photoelectric arrays are in development and are becoming more cost competitive with technologies for centralized power plants. Locating distributed generation close to the load can mitigate or avoid grid congestion, reduce T&D line losses, and produce heat that may be recoverable for cogeneration. Distributed generation and distributed storage can be combined with onsite power conditioning to deliver good power quality and reliability for digital loads. Greater concern about energy and infrastructure security increases the value of distributed generation as a source of emergency or standby power.

In this new world of open competition, prices will be unbundled with lot of variability in the pricing options available to customers. These developments have created a requirement for substantial additional resources in the area of new product development, pricing and competitive intelligence. Many utilities must balance the need to exist in a competitive environment with the remaining obligation to serve some customers. They are exposed to greater price risks as long-term contracts that guarantee price and quantity are replaced by shorter-term transactions, including a thriving spot market in some fuels. Consumers, who are accustomed to stable electricity rates, now see prices that vary with supply and demand conditions. Due to the open competition, many utilities have recognized the need to develop substantially stronger marketing and risk management skills so as to effectively compete with the new power marketers and brokers in the marketplace.

Risk management is important in an open competitive environment because most firms which compete in an open competitive environment have shareholders and they would like to be sure that the earnings from their company is steady and reliable. Companies who do not manage the risks they face and consequently do not have a reliable earnings outlook may be viewed less favorably in the capital markets. Equally important is the fact that many firms entering the open market competitively are trading and selling electricity in new and different ways from anything they have previously experienced. With new operating methods and environments, it is important to know what new risks these firms now face, and how to deal with those risks.

Pricing models allow power providers to decide when to enter into mid term, long term or spot pricing deals. Without a model, these suppliers may become passive price takers and may be unable to compete with sophisticated trading/dealing market makers. Modeling of consumer behavior helps the generators and marketers of electricity to more accurately forecast demand in the various market segments and thus manage the risks of over/under production and buying/selling into unfavorable market conditions.

18.2 Harmonics Generated from Railway Operation

Nowadays, there are many critical infrastructures, including those industries, institutions, and distribution networks and systems that provide a continual flow of the goods and services essential to a country's defenses and economic security and to the health, welfare, and safety of its citizens. These infrastructures are experiencing an important evolution, increasing their performances by the introduction of a series of new technologies. As a result, the interdependence between different kinds of infrastructures is increased and in many cases their vulnerability may also be increased.

Rail transportation can be an example of such evolution. Rail transportation is considered to be a critical infrastructure in many countries, since that much of their economy relies on it to supply the necessary components for its production. Railway infrastructure will certainly have to be upgraded to support the corresponding traffic increase. This upgrading may be realized by the introduction of new technologies on the existing infrastructure, avoiding therefore the construction of new infrastructure. In particular power electronics compensators are proposed as an interesting alternative to the construction of new lines and substations [2].

The electric power supplies all over the world are becoming under pollution with harmonic currents caused by modern electronic equipment, such as many kinds of electronics compensators. These harmonics can cause interference with communication systems, generate extra losses in the wiring and transformers or even overload electrical systems. This problem is especially emerging in the networks of railways. Representing a non-linear load, trains generate harmonic currents, which therefore lead to a high level of reactive power. The national grids are in face of more charges on the bills for electricity due to this pollution. In the UK, the railway system plays a very important role. In some railway systems, to provide the required power to the trains, there are three 400kV connections to the National Grid Company (NGC); one of them is a dual connection. Due to the nature of the traction load, i.e. single phase load with high content of harmonic currents, NGC have placed strict restrictions on the quality of supply at the intersections. In addition, there are restrictions on the voltage profile along the centenary system. Due to these requirements, the need for load balancing and voltage regulating equipment are essential.

18.2.1 Static VAR Compensator

In most of the industrial applications, thyristor-based and shunt connected systems have been proposed for railway VAR compensation. These devices are known as SVCs. They are composed of a capacitor, which is the VAR generator, and a Thyristor Controlled Reactor (TCR), which behaves as a variable VAR absorbing load (depending on the firing angle of the thyristor valve). The branch current is controlled by phase angle controller by the firing pulse to the thyristors, which is the voltage across the reactor and is the full system voltage at 90° firing angle and zero at 180°. The current through the reactor is the integral of the voltage, as thus it is fully controllable with the thyristor valve between the natural value given by the reactor impedance and zero. Thus, the SVC can inject or absorb a variable amount of reactive power to the railway network, adapting the compensation to the load conditions at each instant.

In the SVC case, the amount of harmonics injected into the line depends on the firing angle. The harmonics flowing on the railway system can provoke some problems not only on the railway system but also in other systems related to it, for example, the electrical public utility.

Harmonic filter performance studies have to in addition to include transformer and grid impedances, since these make significant difference in the model. Unfortunately, when using SVCs as filter banks, SVCs themselves generate significant levels of harmonic currents that additionally burden the filters and increase harmonic voltages on the network.

In order to account for contribution from utility network background distortion to harmonic filter ratings, the utility grid background harmonic voltage has to be obtained from the grid. These harmonic voltage sources are considered as ideal voltage source, feeding the harmonic filter banks through the grid transformer.

For calculation of harmonic distortion it is essential to know the impedance characteristics of the system. For a power system, which can have a number of different configurations, it is impossible to specify the impedance as a complex number for each harmonic frequency. Such impedance would be valid only for one specific configuration during one specific load condition. Therefore, standard practice is to specify the impedance as area in the R/X - plane. This area, which covers every system configuration and load condition circumscribed by its perimeter, is often given as a circle.

As a practical Example, two SVCs are connected to the trackside 25kV busbar, one to the catenary and the other to the feeder. The SVCs are rated 3.5MVAr inductive to 41.5MVAr capacitive at 27kV. The SVCs consists of three filter banks and one TCR. The filter banks are tuned to the 3rd, 5th and 7th harmonic sized 26.5MVAr, 7.5MVAr, and 7.5MVAr respectively, i.e., 41.5MVAr in total. The TCR is rated 45MVAr-giving 3.5MVAr on the inductive side. By the aid of phase angle control of the TCRs, a continuous variable output ranging from 3.5MVAr inductive to 41.5MVAr capacitive is obtained.

18.2.2 Data Analysis

Measurements on performance were carried out on a typical railway operation. Figure 18.1 shows the variation of traction transformer catenary current that lasted for about 22 minutes and 30 seconds.

Fig. 18.1. Waveform for transformer catenary current

By using FFT and digital signal processing techniques, harmonics are determined and the second to fifth harmonics are included in Figures 18.2 to 18.5 respectively below:

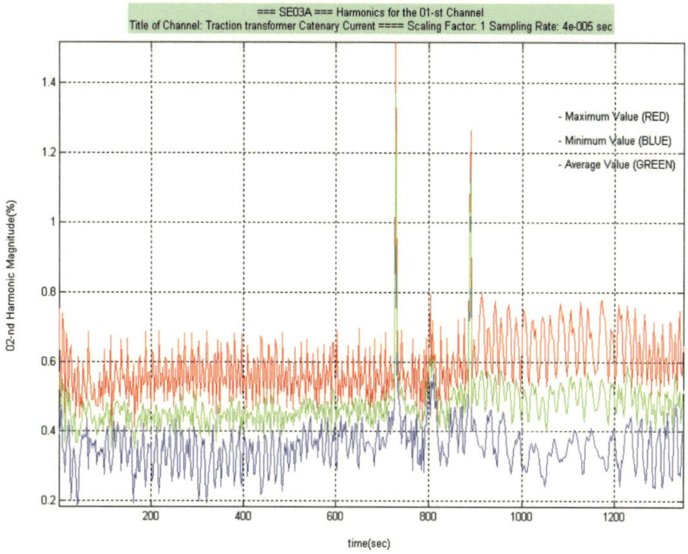

Fig. 18.2. Second harmonic

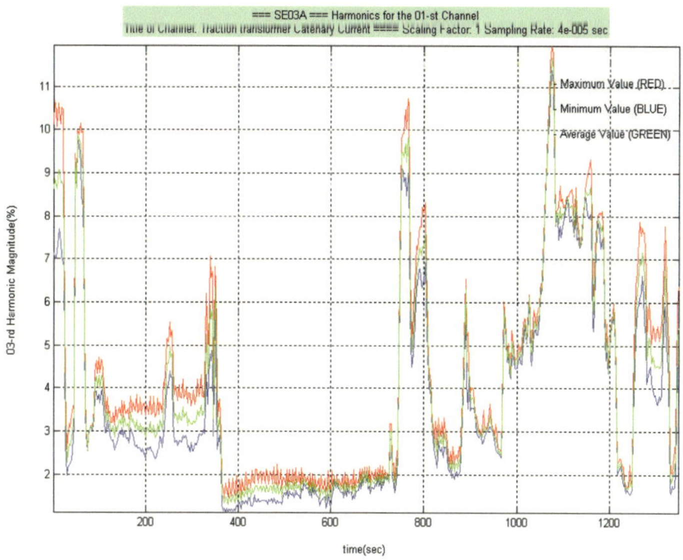

Fig. 18.3. Third harmonic

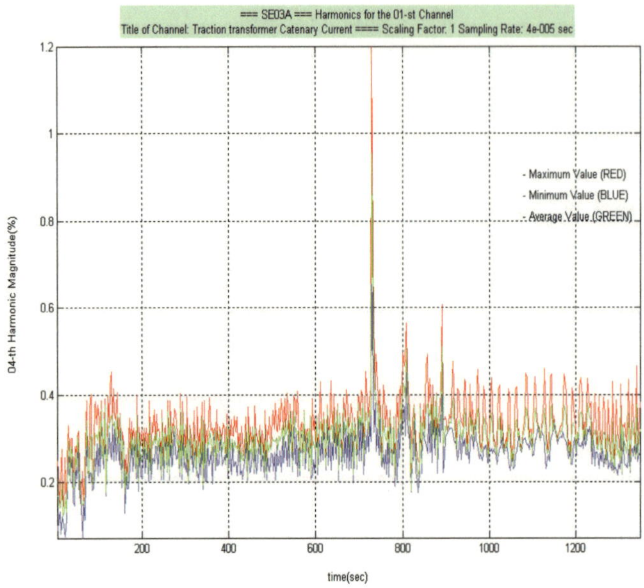

Fig. 18.4. Fourth harmonic

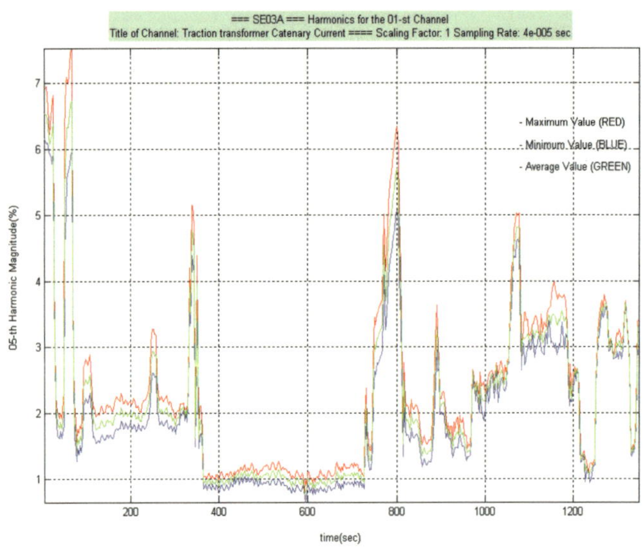

Fig. 18.5. Fifth harmonic

18.3 Practical Application of Wavelet to Power Quality Analysis

Power quality has become a major concern for utility, facility and consulting engineers in recent years. International as well as local standards have been put in place to address the power quality issues [3].

To the facility managers and end users, frequent complaints by tenants/customers on occasional power failures of computer and communication equipment, and the energy inefficiency of the low voltage (LV) electrical distribution system are on the management's agenda. Harmonic voltage and current produced by nonlinear loads would cause extra copper loss in the distribution network, which on one hand will increase the energy cost and on the other hand would increase the electricity tariff charge. The benefits of using power electronic devices in the LV distribution system in buildings, such as switch mode power supplies, variable speed drive units, etc. to save energy are sometimes offset by the increased energy loss in the distribution cables by current harmonics and the cost of remedial measures required. Voltage harmonics caused by harmonic voltage drops in the distribution cables are affecting the normal operation of voltage sensitive equipment as well.

In order to improve electric power quality and energy efficiency, the sources and causes of such disturbance must be known on demand sides before appropriate corrective or mitigating actions can be taken. In the past harmonic distortion is predominantly due to integer harmonics. Nowadays the levels of sub-harmonics and inter-harmonics are rising significantly, which make the harmonics problem even worse.

Since harmonics are steady state phenomenon, corrective measures available are basically by filtering and/or isolation. Yet, before deciding what corrective measures are to be adopted, the nature of the harmonics problems needs to be identified. A traditional approach is to use Fast Fourier Transform (FFT) to analyze harmonics contents contained in the power signal. The FFT has many attractive features. That theory of FFT has been fully developed and is well known; scientists and engineers are familiar with the computation procedures and find it convenient to use as many standard computation tools readily available. It is however easily forgotten that Fourier Transform is basically a steady state analysis approach. Transient signal variations are regarded by FFT as a global phenomenon. One example is that FFT transforms an electrical impulse into frequencies ranging from zero to infinity in the frequency spectrum.

As power quality issues such as sub-harmonics, integer harmonics, inter-harmonics, transients, voltage sag and swell, waveform distortion, power frequency variations, etc. are commonly experienced by electricity users, this Section attempts to develop an algorithm based on wavelet transform to identify power frequency variations, sub-harmonics, integer harmonics and inter-harmonics.

18.3.1 Wavelet Transform and Analyzing Wavelet

Wavelet Transform (WT) has been drawing many attentions from scientists and engineers over the years due to its ability to extract signal time and frequency information simultaneously. WT can be continuous or discrete. Continuous Wavelet Transform (CWT) is adopted for harmonic analysis because of its ability to preserve phase information.

The wavelet transform of a continuous signal, f(t), is defined as:

$$Wf(u,s)=\langle f,\psi_{u,s}\rangle=\int_{-\infty}^{+\infty} f(t)\frac{1}{\sqrt{s}}\Psi^{*}(\frac{t-u}{s})dt \quad , \tag{1}$$

where $\psi^*(t)$ is the complex conjugate of the wavelet function $\psi(t)$;
s is the dilation parameter of the wavelet; and
u is the location parameter of the wavelet.

The wavelet function must satisfy certain mathematical criteria. These are

- a wavelet function must have finite energy; and
- a wavelet function must have a zero mean, i.e., has no zero frequency components.

The simplified Complex Morlet Wavelet (CMW) is adopted in the algorithm for harmonic analysis, which is defined as

$$\Psi(t)=\frac{1}{\sqrt{\pi f_b}}e^{\frac{-t^2}{f_b}}e^{j2\pi f_c t} \quad , \tag{2}$$

where f_b is the bandwidth parameter and;
f_c is the center frequency of the wavelet.

The CMW is essentially a modulated Gaussian function. It is particularly useful for harmonic analysis due to its smoothness and harmonic-like waveform. Furthermore CMW is an analytic wavelet therefore it is able to separate amplitude and phase information.

Strictly speaking, the mean of the simplified CMW in (2) is not equal to zero as shown in (3) below.

$$\int_{-\infty}^{+\infty}\Psi(t)dt=\frac{1}{\sqrt{\pi f_b}}\int_{-\infty}^{+\infty}e^{j2\pi f_c t}e^{\frac{-t^2}{f_b}}dt=e^{\frac{-f_b}{4}(2\pi f_c)^2} \tag{3}$$

However the mean of the CMW can be made arbitrarily small by picking the f_b and f_c parameters large enough. For example, the mean of the CMW in (3) with f_b=2 and f_c=1 is $2.6753\text{x}10^{-9}$ which is practically equal to zero. The frequency support of the CMW in (2) is the value of the entire frequency axis.

The time range support of the CMW in (2) is from -8 to 8. The value of f_b should not be larger than 9, otherwise the CMW cannot decline fast enough to zero within the time range support.

18.3.2 Harmonics Frequency Detection Algorithm

Given a signal f(t) represented as

$$f(t)=a(t)\cos\phi(t) \tag{4}$$

The wavelet function in (2) can be represented as

$$\Psi(t) = g(t)e^{j\eta t}. \tag{5}$$

The dilated and translated wavelet families are represented as

$$\Psi_{u,s}(t) = \frac{1}{\sqrt{s}}\Psi\left(\frac{t-u}{s}\right) = e^{-j\xi u} g_{s,u,\xi}(t) \tag{6}$$

Where $g_{s,u,\xi}(t) = \sqrt{s}\, g\left(\frac{t-u}{s}\right)e^{j\xi t}$; and $\xi = \frac{\eta}{s}$.

The wavelet transform of the signal function f(t) in (4) is given as

$$Wf(u,s) = \frac{\sqrt{s}}{2} a(u)e^{j\phi(u)}\left(\hat{g}(s[\xi - \phi'(u)]) + \varepsilon(u,\xi)\right). \tag{7}$$

where $\hat{g}(\omega)$ represents the Fourier Transform of the function g(t).

The corrective term $\varepsilon(u,\xi)$ in (7) is negligible if a(t) and $\phi'(t)$ in (4) have small variations over the range of $\psi_{u,s}$ in (6) and if $\phi'(u) \geq \frac{\Delta\omega}{s}$. If a power signal contains only a single frequency waveform, the corrective term can be neglected safely. However for a power signal containing harmonic frequencies from low frequency to high frequency, the corrective term will contribute to the wavelet coefficients, making the frequency detection not so straightforward.

The instantaneous frequency is measured from wavelet ridges defined over the wavelet transform. The normalized scalogram is defined by [4] as:

$$\frac{\xi}{\eta} P_W f(u,\xi) = \frac{|Wf(u,s)|^2}{s} \tag{8}$$

It is calculated as:

$$\frac{\xi}{\eta} P_W f(u,\xi) = \frac{1}{4} a^2(u) \left| \hat{g}\left(\eta\left[1 - \frac{\phi'(u)}{\xi}\right]\right) + \varepsilon(u,\xi) \right|^2. \tag{9}$$

Since $|\hat{g}(\omega)|$ in (9) is maximum at $\omega = 0$, if one neglects $\varepsilon(u,\xi)$, (9) shows that the scalogram is maximum at

$$\frac{\eta}{s(u)} = \xi(u) = \phi'(u). \tag{10}$$

The corresponding points $(u, \xi(u))$ calculated by (10) are called wavelet ridges. The analytic amplitude is given by

$$a(u)=\frac{2\sqrt{\frac{\xi}{\eta}P_W f(u,\xi)}}{|\hat{g}(0)|}=\frac{2\sqrt{\frac{|Wf(u,s)|^2}{s}}}{1}=\frac{2|Wf(u,s)|}{\sqrt{s}} \quad (11)$$

18.3.3 Discrimination of Adjacent Frequencies

The Fourier Transform of a dilated CMW in (6) is represented as

$$\Psi(sf)=\sqrt{s}e^{-\pi^2 f_b (sf-f_c)^2} \quad (12)$$

The function $\Psi(sf)$ can be regarded as a band pass filter centered at the frequency f_c. The CWT of a signal is the convolution of the signal with a group of band pass filters that are produced by the dilation of the CMW.

Suppose that (12) is represented as

$$\Psi(sf)=x, \quad (13)$$

where x represents an arbitrary magnitude to be defined later.

Combining (12) and (13) gives

$$f=\frac{f_c}{s}\pm\frac{1}{s\pi\sqrt{f_b}}\sqrt{\left|ln\left(\frac{x}{\sqrt{s}}\right)\right|}, \quad (14)$$

where $\frac{f_c}{s}$ is the center frequency of the dilated band pass filter; and the bandwidth is $\frac{2}{s\pi\sqrt{f_b}}\sqrt{\left|ln\left(\frac{x}{\sqrt{s}}\right)\right|}$.

Figure 18.6 shows the plot of the frequency support of two dilated CMW at scales s_1 and s_2 respectively.

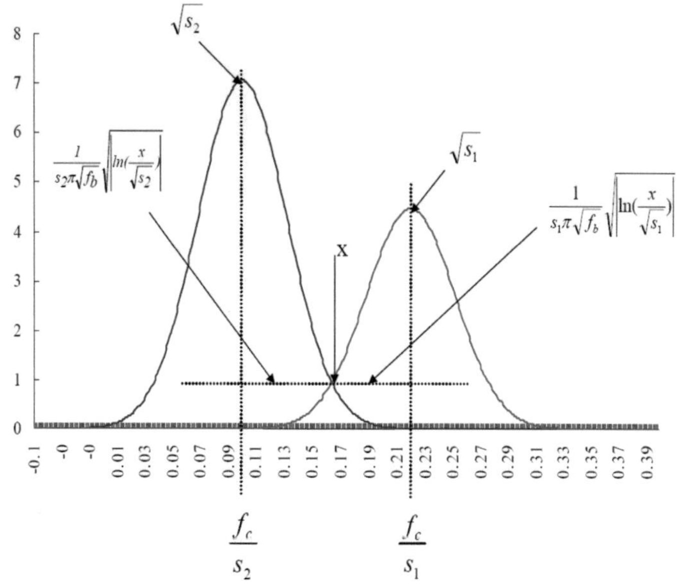

Fig. 18.6. Frequency plot of (14) for two CMWs at scales s_1 and s_2

If the two CMWs are used to detect two adjacent harmonic frequencies in a signal, with their frequencies represented as

$$f_1 = \frac{f_s f_c}{S_1} \ \& \ f_2 = \frac{f_s f_c}{S_2}, \tag{15}$$

where f_s represents the sampling frequency, then

$$\frac{f_c}{S_2} - \frac{f_c}{S_1} \geq \frac{1}{S_2 \pi \sqrt{f_b}}\left(\sqrt{\left|ln\left(\frac{x}{\sqrt{S_1}}\right)\right|} + \frac{1}{S_1 \pi}\sqrt{\left|ln\left(\frac{x}{\sqrt{S_2}}\right)\right|}\right) \tag{16}$$

Assume that $s_2 > s_1$, (16) is simplified to

$$f_c \sqrt{f_b} \geq \frac{1}{\pi}\sqrt{\left|ln\left(\frac{x}{S_2}\right)\right|} x \frac{f_2 + f_1}{f_2 - f_1} \tag{17}$$

For $s_2 \leq 300$ and $x \leq 0.01$, (17) becomes

$$\frac{1}{\pi}\sqrt{\left|ln\left(\frac{x}{S_2}\right)\right|} \leq 0.87 \tag{18}$$

Substituting (18) into (16) gives

$$f_c\sqrt{f_b} \geq 0.87\,x\,\frac{f_2+f_1}{f_2-f_1}\,. \qquad (19)$$

It is estimated that the magnitude of x should not be larger than 0.01. Equation (19) is used to determine the values of f_b and f_c in (2) for the continuous wavelet transform with complex morlet wavelet that is a necessary condition to discriminate adjacent harmonic frequencies in the power signal.

18.3.4 Harmonics Amplitude Detection Algorithm

Theoretically, once the algorithms developed before identify the harmonic frequencies presented in the power signal, the corresponding harmonics amplitudes would be determined readily by (11).

The values of $2\sqrt{\frac{|Wf(u,s)|^2}{s}}$ in (11) are produced in the process of generating the scalogram.

Due to the imperfection of the filters produced by the dilated CMWs and aliasing, the amplitudes detected are corrupted by noise. Simulation results show that the amplitudes for harmonic frequencies ranging from 50Hz to 1000 Hz have errors of the order of ±5%. Figure 18.7 below shows a plot of the absolute coefficients generated by CWT for the harmonic frequency at 991.5Hz.

In Figure 18.7, the vertical axis represents the magnitude of the absolute coefficients and the horizontal axis represents the data points. The small fluctuations as shown in the absolute coefficients plot are due to filter imperfection and aliasing.

Discrete Stationary Wavelet Transform (DSWT) is adopted to remove the fluctuations appeared as noise superimposed on the absolute coefficients plot in Figure 18.7.

The Symlet2 developed by Daubechies is used for the DSWT of the absolute coefficients. It is found that a decomposition level of 5 is sufficient for harmonic frequencies up to 1000Hz.

Figure 18. 8 shows the DSWT output of the absolute coefficients shown in Figure 18.7 which clearly shows that the superimposing fluctuations are removed resulting in an accurate detection of the harmonic amplitudes.

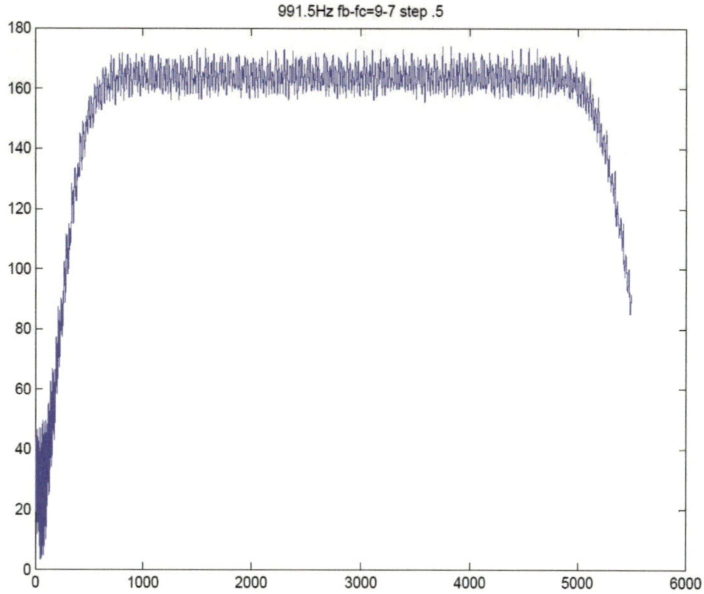

Fig. 18.7. Absolute coefficients plot generated by CWT (using Complex Morlet Wavelet, f_b=9, f_c=7) for harmonic frequency at 991.5 Hz

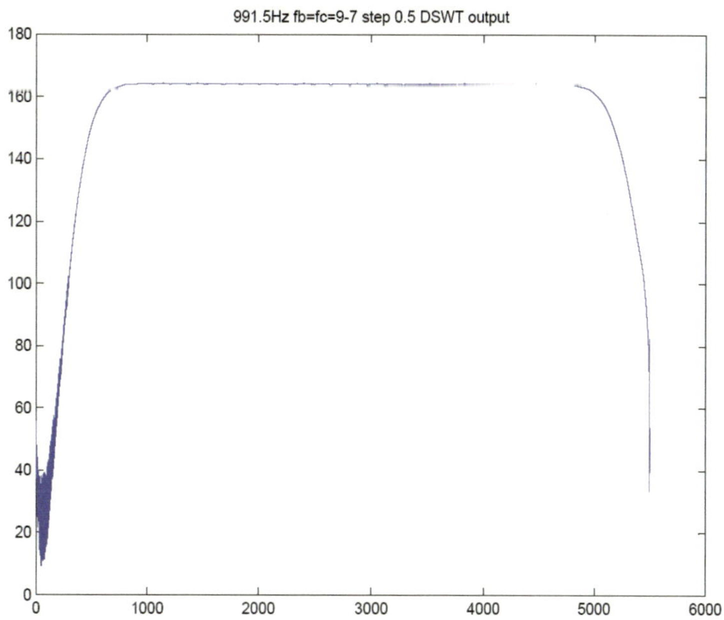

Fig. 18.8. Coefficients generated by discrete stationary wavelet transform (using Symlet2 wavelet, level 5 decomposition) from the absolute coefficients generated by CWT (using Complex Morlet Wavelet, f_b=9, f_c=7) for harmonic frequency at 991.5 Hz

18.3.5 Simulation Setting

A simulated signal is used to test the validity and accuracy of the harmonics detection algorithm. The simulated signal contains a combination of the harmonic frequencies as shown in Table 18.1.

Harmonic Frequency (Hz)	Amplitude	Phase Angle (Degree)
50.1	311	0
102	280	5
149.5	248.8	7
249	217.7	10
371	186.6	15
412	155.5	20
550	155.5	25
620	124.4	-30
770	93.3	42
891	62.2	-61
991.5	31.1	82

Table 18.1. Harmonic Frequency contained in the simulated signal

The simulated signal is sampled at 20kHz. Since the highest harmonic frequency in the simulated signal is 991.5Hz, the number of data per cycle for 991.5Hz is approximately 20.

This is the minimum data size required for accurate amplitude representation. A higher sampling frequency would give a better representation of the harmonic amplitudes, but more data points are produced subsequently resulting in slow computation. For faster CWT computation, the simulated signal will be down-sampled for the detection of lower harmonic frequencies. The down-sampling settings are as shown in Table 18.2. In any case a minimum of 20 data per cycle is maintained. The data size for CWT computation is set at 5000.

Frequency (Hz)	Sampling Frequency (Hz)	Data Size
50.1	1000	5000
102	2500	5000
149.5	5000	5000
249	5000	5000
371	10000	5000
412	10000	5000
550	16000	5000
620	16000	5000
770	20000	5000
891	20000	5000
991.5	20000	5000

Table 18.2. Sampling frequencies and sample data size for harmonic frequencies of the simulated signal

The necessary condition discussed before for discrimination of adjacent frequencies requires that the complex morlet wavelet should be set at $f_b = 6$ and $f_c = 7$.

18.3.6 Simulation Results

The simulation results for harmonics detection is shown in Table 18.3. It can be seen that the frequency detection by the proposed algorithm is very promising, especially at high harmonic frequencies. At low harmonic frequency detection, the scalogram plot is corrupted by high frequency components which exhibited as noise. It is proved that the necessary condition established before is successful in distinguishing adjacent frequencies.

Harmonic Frequency (Hz)	Detected Frequency (Hz)	% Error
50.1	50.14	0.08%
102	102.04	0.04%
149.5	149.51	0.01%
249	249.10	0.04%
371	370.96	0.01%
412	412.00	0%
550	549.83	0.03%
620	620.16	0.03%
770	770.07	0.01%
891	891.15	0.02%
991.5	991.50	0%

Table 18.3. Harmonic frequencies detection results

The accuracy in the detection of harmonic amplitudes depends on the accuracy in harmonic frequencies detection. As seen from the results shown in Table 18.4, the harmonic amplitude detection results are very satisfactory. Except for 50.1 Hz, the amplitude detection errors for all the other harmonic frequencies are smaller than 0.5%.

Harmonic Frequency (Hz)	Harmonics Amplitude	Detected Amplitude	% Error
50.1	311	309.07	0.62%
102	280	279.17	0.29%
149.5	248.8	248.57	0.09%
249	217.7	216.72	0.45%
371	186.6	186.32	0.15%
412	155.5	155.35	0.10%
550	155.5	155.47	0.02%
620	124.4	124.32	0.06%
770	93.3	93.17	0.14%
891	62.2	62.21	0.02%
991.5	31.1	31.23	0.42%

Table 18.4. Harmonic frequencies amplitude detection results

The larger amplitude detection errors are found to have happened at 50.1 Hz and 249 Hz respectively. Table 18.5 shows a comparison of errors in harmonic frequencies detection and the corresponding amplitudes detection. It is observed that the frequency detection errors for these two frequencies are also comparatively higher. Therefore it is concluded that the accuracy in amplitude detection is affected by the accuracy in frequency detection.

Harmonic Frequency (Hz)	% Harmonic Frequencies Detection Error	% Amplitudes Detection Error
50.1	0.08%	0.62%
102	0.04%	0.29%
149.5	0.01%	0.09%
249	0.04%	0.45%
371	0.01%	0.15%
412	0%	0.10%
550	0.03%	0.02%
620	0.03%	0.06%
770	0.01%	0.14%
891	0.02%	0.02%
991.5	0%	0.42%

Table 18.5. Comparison of detection errors in harmonic frequencies and harmonic amplitudes

Further refinements on both frequency and amplitude detection would be achieved by a careful choice of f_b and f_c of the complex morlet wavelet.

18.4 Energy Risk and the Management

Risk may be defined as the potential harm that may arise from some present process or decision or from some future event, the hazard to which we are exposed is uncertainty. Harm can take different forms in the electricity industry, whether one is an investor in generation, a sale organization, or an end-user.

Uncertainties could be due to some of the following items:

- Electricity price (volatility)
- Fuel price
- Fuel availability (uncertain availability)
- Economic conditions (inflation, floating exchange rates, interest rates)
- Volume (uncertain ability to balance supply and demand of electricity)
- Financial risk (credit and settlement)
- Weather load dependence
- Environmental constraints
- Transmission restrictions
- Technology changes
- Regulatory conditions (consistency and stability of regulations, transparency)
- Political decisions (political tariff setting).

There is a need to adequately assess and manage risk. Some risk analysis methods for energy include [5]:

- Position reporting - monitoring of portfolio positions
- Deterministic scenario analysis
- Sensitivity analysis
- Value-at-risk models
- Maximum loss model.

Risk management is the process of assessing risk and developing strategies to manage the risk.

There is a need for investment in new capacity due to electricity demand continues to grow and reserve margins have declined as markets have been liberalized.

For the UK, when extending System Operation to Scotland there were a number of differences. The two Scottish systems have been essentially vertically integrated since Nationalization - England and Wales has always split Supply and Distribution from Generation, Transmission and System Operation.

The Scottish demand is only 10% of GB demand but they have 10GW generation. Transmission is at 400/275 to the major stations and load centers but there is a lot of low capacity 132kV transmission supporting remote areas with a lot of small generating stations - hydro and new wind [6].

We need to observe Generation down to 5MW (30MW in South Scotland) to ensure secure operation of the Transmission system. This means modeling a number distributed stations (up to 100MW), exempted from having a generation license. This highlights the issues that arise with increased penetration of distributed generation to the system.

A change in any output or load will immediately affect all the others. The frequency controls the magnitude of each demand by direct relation to nominal (50Hz) and it will change immediately if a large demand or generation source changes magnitude.

Inertia on rotating synchronous generation will release extra output on falling frequency and vice versa, while frequency sensitive units will increase or decrease steam flow to turbine of a frequency change.

The balance has to be maintained at a level such that so that frequency does not go outside statutory limits (± 0.5Hz, 1%) or operational limits (± 0.2Hz 0.4%). To do this it is essential to know or predict the export and import of all elements of generation and demand.

At any time, the level of ordered plant may turn out to be insufficient and expensive but fast reacting generation must be used.

Observability and predictability of demand and Generation output is crucial to maintaining a viable balance, keeping spare and reserve levels and thus fuel burn and emissions down.

Conventional Plant is slow to start to compensate for market errors. Much expensive part loading is required to provide response and plant has to be warmed to be available as spare. Better and more frequent forecasting is being developed and use of frequent countrywide observation seems appropriate for better accuracy in market timescales. Also try and forecast when such errors are likely. The difference between the predicted and the actual output has a marked effect on the level of demand to be met by other plant.

The use of other flexible plant to cover renewable shortfall risk is inefficient; that capacity should be scheduled to help meet demand and reduce main generation. Large demand variations need to be tackled to improve efficiency of delivery and reduce capacity requirement. The energy efficiency argument for electricity is not just to reduce demand it is also to control when optional demand is used.

Modern ICT systems can improve information flow to and from large numbers of distributed locations. This can be used to improve observation and predictability. Work on demand side management (DSM) shows that such information systems can have major benefits in improving load factor and thus reducing emissions and unnecessary fuel burn.

The 21st century vision for information systems should have all the information (including 'real-time') to make sound asset management and investment decisions; a fully populated asset register linked to network management, financial, work management, SCADA, and other key information systems; comprehensive modeling, reporting, and decision support tools, ability to predict future performance accurately across short, medium and long timeframes; uncertainty (over asset condition and time-to-failure) reduced such that performance can be predicted within acceptable levels of confidence and last but not least a comprehensive network and asset knowledge to optimize risk and maximize business performance.

There are the risks and challenges, such as re-investment is not like green-field investment; managing outage risk is an enormous challenge; successfully integrating new technologies with old during the transition stage is another enormous challenge. But the greatest challenge is in deciding what exactly should be done, for example, to do something different or to take the 'safe' option and rebuild an infrastructure that was designed in the 19th and 20th centuries. It would be nonsense to apply the same approach for the same problem but to expect to have a different result.

Turning to the economic consideration, replacing like with like is always the easiest - and cheapest - option in the short-run, but the assets installed today will still be there in 50 to 70 years time. There is a big question that they meet future requirements for performance. Basically redesigning networks will require significant up-front costs and some existing assets may need to be retired early. Therefore incentives to invest for long-term sustainability rather than short-term gains are essential.

However, presently an inadequate core-skill base is a big issue [7]. Since 1996, the number of electrical engineering students on degree courses has fallen by 29%. Only 13% of these opt for power-related modules. Overall, less than 150 students graduate with a degree in power engineering in the UK each year - and most of these are overseas students (about 80% at the post-graduate level). Presently, more effort has been spent to encourage young (school-age) people such that they perceive engineering to be an exciting career/profession.

Liberalization also leads to uncoordinated cross-border trades; increased transmission distances and transmission systems run closer to their limits due to commercial pressure.

The future will have more decentralized but coordinated operation. Coordination requires effective IT tools to exchange of real-time information. Modern control room facilities should be able to provide accurate and sufficient system information for the operators' safety operating; all available remedies for different security issues and suggestion for operators for the best remedies. Monitoring system can be designed to display machine angle separations and damping factors for transient stability and dynamic stability, respectively. Secured network of sensors, communication links, information processors and dynamic algorithms are developed for intelligent grid.

Pricing Models are developed for mid-term, long-term or spot pricing deals. They must manage the risks of over/under production and buying/selling into unfavorable market condition.

In summary, factors that need to be taken into consideration for the electricity industry include asset age profiles and condition trends indicate the need for a sustained ramp-up in asset replacement - perhaps over 15 years. Electricity demand continues to grow and this has implications for plant ratings. Customer expectations regarding security, reliability, and quality are increasing. Environmental considerations are placing new constraints on network design and operating criteria. Networks must support the low greenhouse gas emission which means being accessible to distributed (renewables) generation. To achieve this, it is essential to have long-term national commitment to infrastructure investment and management, an effective incentive-driven regulation, a creation of appropriate skills and experience through the wider education system and the development of appropriate technologies by suppliers for demand-side management and smart metering.

18.5 Insurance Issues for Energy Risk

The initiation of deregulation in the energy sector, which introduced competition to the electric power industry, has triggered the creation of Investor-Owned Utilities (IOUs). For IOUs to be profitable, they require good risk management, good strategy and reduction of overheads. Because of this, energy companies have been subject to significant restructuring programmes, acquisitions and mergers, asset divestitures and other forms of corporate restructuring.

Energy prices are volatile and following the introduction of the European Emission Trading Scheme (EU ETS) in January 2005 energy generators were attempting to diversify into low emission energy resources such as wind, hydro and nuclear. Most of the low emission energy resources are based on proven technology but are difficult to insure since not much historical data is available. The insurance industry, on the other hand, has been somewhat slow with the amendment of their insurance products to follow the trends in the energy sector. As a result, the insurance sector is losing money with out-of-date insurance products in terms of underwriting profitability and new products that are required for new technology in the energy sector e.g. fuel cell power plants.

It can be costly for insurers to calculate premiums bases on underwriting procedures that have been used for similar equipment but where significant differences exist. For example, new wind power generation companies have no claims history and little experience with new turbines, which have similar design as proven products but differ internally. For precise underwriting it is critical to monitor technological changes and ensure experienced staff is available.

Furthermore, energy companies vary in their business strategies and are therefore vertically and horizontally integrated businesses. Vertically integrated energy companies e.g. transmission can be served by the insurance industry with a single line of business written or risk group. Companies that are horizontally integrated and possess their own generation, transmission and distribution companies require a mix of business lines from a single or multiple insurers.

With this, the risk manager of an energy company needs to understand what types of insurances are available on the market place and which part of their risk portfolio can be covered by risk transfer to insurers. The shortfall of suitable insurance cover in the energy sector has forced the formation of industry specific mutual insurance companies that offer coverage not available in the insurance sector or with favorable conditions. Such lack of insurance cover opens the doors for Alternative Risk Transfer (ART) products that are often not as stringently regulated as the insurance sector.

18.5.1 Risks to Energy Complaints

Energy companies are exposed to a multitude of risks; therefore good risk management is required. The first steps required for risk management are risk identification, assessment, avoidance, reduction, retention and transfer. Not all risks are transferable to insurance companies since an insurer can only accept "pure risks". In order to transfer risks, risk managers need to identify insurable risks within their operation. To choose from a portfolio of available insurance policies, possible events and their likelihood need to be identified. With this in mind, policies that guard events that give rise to an insurable primary cause need to be bought into consideration. There are several risk categories to consider [8] and they are summarized as follows:

A. Property Risk

Energy companies may own pipelines, refineries and buildings or invest into real estate or property developers.

B. Legal Risk
Energy companies are exposed to corporate governance, joint venture disputes, contract risk negotiation, directors' and officers' liability, pollution liability, and exposure to third party contracts.

C. Political and Regulatory Risk
There are sovereign or political risk factors to deal with such as confiscation, nationalization, expropriation, kidnap and ransom, antitrust/collusion, transmission confiscation, currency inconvertibility, war and civil disturbance.

D. Operational Risk
Maintain security, guards, fences, CCTV, fire, well blowout, explosion, business interruption, reputation, supplier failure, industrial espionage, and shutdown risks.

E. Environmental Risk
Exposure to climatic changes, population explosion and pandemics, changes in the legal system, nuclear proliferation and geological affects such as earthquakes and tsunamis need to be evaluated.

F. Intellectual Property Risk & Keyman Insurance
Losing a key employee and their specialized knowledge, failure to comply with best employment practices, and incidents involving gross misconduct at the workplace can have a profound effect on your company's performance.

G. Financial Risk
An IOUs operation is also affected by a range of financial and economic risks such as fluctuations in the stock market, commodity prices and GDP, interest and exchange rates. Economists as well as evaluation of the fiscal regime that the host country is offering evaluate market risks.

18.5.2 Option for Cover
Energy companies have several options on how to manage their risk portfolio. This section illustrates the four basic options that risk managers may have to transfer insurable risks.

A. Do Nothing
Keep uninsured risk with investors and shareholders. Cover losses from current revenue stream account as operating costs. This is useful for large companies that have multiple $billion turnover. As a result, cash flow can be interrupted and expensive short-term loans may be required. If the risk is retained, issues relating to risk avoidance and risk reduction should be addressed.

B. Buy Insurance
Pay insurance premium to a direct insurer, mutual insurer or via broker to insurance company. Claim on insurance cover if loss occurs. Investors like to protect their capital and many banks require insurance cover for their investment. Note that cover is cheaper if it has high excess but investors may disapprove and request lower excess.

C. Capital Build up

Energy companies can build up capital reserves as part of a risk management contingency plan. This works well for smaller amounts and where the maximum amount of the risk is known and can be calculated.

D. Self Retention

Absorb the first part of the loss via capital build up and cash flow and pass on losses exceeding a limit to insurance companies. This is referred to as self-insurance or captive insurance and can be considered for large risks. Captives can purchase insurance and have tax benefits on cash build up.

18.5.3 Insurance Cover for Energy Companies

The purposes of an insurance cover is to reduce risk exposure to investors, keep the company focused on their core business, avoid diversification, and increase the cash flow by reducing loss reserves and to cover unexpected events that lie outside core activities. As well as providing cover, insurance companies give expert advice for identifying and managing risks with insurance experts that can suggest changes to business operation for BI risk reduction. They may suggest keeping spares on site to reduce delivery times to remote sites e.g. remote transformer sites to keep spare fuses, cables and connectors.

Multiple-line insurance companies or energy brokers have energy insurance divisions that are measuring exposure to potential loss of energy companies. They are able to make suggestions on how to minimize liability exposure through transfer of risk to an energy underwriter. Once the business is placed, periodic reviews of the insurance portfolio will be carried out since legal and technological changes may require contract alterations. Insurance professionals will also audit policies for *accuracy to ensure adequate coverage.*

A. Insurable Risks

Traditionally, "pure risks" are beyond the core competencies of the subject-matter possessor. They should not offer an opportunity of gain and, in case of a loss; the possessor should be indemnified to the value of the subject matter at a point in time. However, some insurance companies now cover speculative risks that were traditionally not insurable, such as the exchange of currency, weather insurance, environmental impact and damage assessment.

B. Uninsurable Risks

Speculative, non-financial and fundamental risks are uninsurable risks because the insured would gain from a claim. This is because speculative and non-financial risks are uninsurable as a matter of principle while fundamental risks are uninsurable because of lack of willingness or capacity. Thus, fines are uninsurable because it is against public interest.

In some countries failing in deliver power or congestion mismanagement is fined by the regulator. Imposed fines caused by power blackouts or improper congestion management will have to be paid by the energy company. As far as liability goes, energy companies are currently not liable for the widespread losses to business and individuals. Therefore businesses that rely heavily on energy supply should consider business interruption cover.

8.5.4 Insurance and Deregulation

Prior to deregulation, contracts between generation and transmission companies did not exist since both entities belonged to the same government-owned institution. Since more and more energy companies are now in private and located in different countries, some contracts require insurance cover for business interruption and political risks.

Manipulative traders using criminal tactics are another risk that energy companies should protect themselves from them by means of risk transfer. Rogue energy traders have the potential to take part in fraudulent reporting of sales transactions, megawatt "laundering", fake power delivery scheduling, conspiracy and price fixing.

By purchasing D&O cover, energy trading companies can protect themselves from claims made by victims of market manipulations. With the recent upsurge of corporate scandals and insider trading, e.g. Enron and directors being sued more frequently, premium rates of D&O cover had increased. But the recent introduction of International Accounting Standards (IAS) and the increased transparency of financial reporting have seen D&O premiums fall back to competitive levels. Increased energy trading in a deregulated market causes transmission line congestion and thus frequency and voltage unreliability that in extreme cases can cause damage or loss of distribution infrastructure. Such damage, combined with rogue trading, may not be recoverable from an insurance contract.

State owned energy companies had little requirement for private insurance to secure their business. This has saved costs since the government has the financial capabilities to supply support in case of an accident. Privatized energy companies are no longer under the umbrella protection of the government and therefore require insurance cover from the commercial or mutual sector. Governmental energy companies were largely self-insured via means of a captive.

The total cost of insurance coverage for an energy company that operates its own generation, transmission and Utility Company is usually smaller compared to individual companies. One of the reasons for this lays in the inherent contract uncertainty between separate companies which cannot avert business interruption to any of the partners caused by a supplier. Additionally, basic, non-specialist cover that is required by all companies, such as employer's liability, needs to be purchased for each entity, thus raising the overall cost caused by duplication compared to a single entity. The separation of large state-owned companies will introduce duplication and therefore increase overheads. Businesses are now reducing their cost base via mergers and acquisitions to lessen overheads and duplication. Such enterprises use this as a means for diversification and growth opportunity. Nevertheless, services and functions have been separated into entities by generation, transmission and distribution that resemble part of the enterprise. With this, each entity has now unique needs in terms of their insurance portfolio.

Energy companies that retain their risk are referred to as captives. Captives are wholly owned insurance subsidiaries of non-insurance parents that are permitted to write admitted cover of the parent in many or all EU countries. Their advantages are that the insured has no need to expose sensitive information to an external third party, profits during a soft market will remain in the enterprise, storing funds in pools without paying for risk transfer, investment income from funds, direct access to reinsurance and tailored XL programme and potential tax advantages if a loss occurs. Disadvantages are the up front and running costs of an insurance company subsidiary, funds for initial capitalization, fees, taxes, reinsurance and wages.

There are covers a captive cannot provide e.g. workers compensation, automobile liability and general liability. Such risks can be insured via a fronting arrangement. In fronting, an external licensed insurance company provides the cover and the unlicensed captive will provide reinsurance to the fronting company, gaining large policy discounts.

18.5.5 Relevant Cover Types

Insurance covers are very complex and contain many clauses on the primary cause and the insured subject matter. This section summarizes a few cover types that are common in the energy sector. Each cover type listed outlines its basics and is subject to variations in their policy wordings.

A. Property and Casualty (PC, PI)

Classes that are generally covered range from utilities and chemical operations to alternative energy sources, oil and gas, and pipeline and refinery risks. Additionally risks such as construction, property damage (PD), transportation, equipment breakdown and communications can be included.

Casualty insurance is generally segmented by the class of business clients operate since there are distinct differences in their insurance needs, for example in mining, oil and gas and power utilities all have different cover requirements.

B. Statutory Liability

Companies require Employer's Liability (EL) insurance in many countries. It protects the insured against liability arising from bodily injury or disease sustained by their employees out of and in the course of their employment in the business. Many companies have Public and Product liability and Professional Indemnity insurance to protect against claims arising from third parties.

C. Business Interruption (BI)

The purpose of BI policies is to protect an operation from loss of revenue. This cover is beneficial as part of a risk management portfolio because lost income in a monetary form during a predetermined period of time after the loss occurrence can be recovered. Many insurance policies are limited by the sum insured or the policy's limit of liability. Therefore BI will stop paying when normal operation resumes or the limit has been reached.

D. Boiler & Machine (B&M)

Industrial boilers are generally excluded from all cover types and must be insured separately.

The electrical machinery insurance contract covers losses caused by the breakdown of electrical machines. It is primarily to indemnify loss resulting from property damage to the insured and others for which the insured may be liable.

E. Advance Loss of Profits (ALOP)
ALOP cover is usually used to protect anticipated revenue from projects when their completion has been delayed. This cover is beneficial when construction work or machinery or equipment is delayed e.g. for a new power plant or transmission lines. Anticipated revenue from the electricity generation can be recovered from the insurance company. The claim amount is difficult to calculate since no past income figures are available. This policy terminates when construction has been completed. E.g. to protect an operating plant from loss of anticipated income, a BI policy is required. ALOP can be part of a Construction All Risks (CAR) or Erection All Risks (EAR) policy.

F. Sabotage and Terrorism
This cover is excluded in many insurance covers, especially in property policies. Only a few insurers are offering this cover which can protect against Malicious Damage (terrorism), Mutiny, Revolution, Strikes and War and protects Property, BI, CAR and PD. UK insurers will reinsure terrorism cover with the Pool Re insurance scheme. Pool Re will ensure that terrorism insurance availability for commercial property would continue after the withdrawal of reinsurers from the market. The HM Treasury is the reinsurer of last resort for Pool Re in the event that all funds are exhausted. Similar State Compensation Funds have been set up in the US.

G. Directors & Officers (D&O)
Directors and Officers policies will protect director's personal assets from claims to the organization. In UK law, companies are allowed to indemnify director's legal costs if they have been found not guilty. Such costs can be recovered from a D&O policy.

H. Nuclear Cover
Operators of nuclear power plants in the EU are liable for any damage caused by them, regardless of fault. Their liability is limited by both international conventions and by national legislation, so that beyond their financial limit the 1998 Paris/Brussels Convention dictates how claims responsibility is handled.

The 1998 Paris/Brussels Convention operates in three tiers of compensation payable to claimants. The 1st tier corresponds to the operator's liability amount of 700 million euro. This is followed by a payment by the state in which the liable operator's installation is located for up to 500 million euros. Followed by the 3rd tier where the contributions from all of the contracting parties must pay up to 300 million euros. With this, the Paris/Brussels regime will provide for up to 1.5 billion euros of compensation.

I. Forced Outage Cover
All players in a deregulated wholesale power prices environment that buy and sell electric power are exposed to an outage risk. If an outage occurs when spot market power prices for replacement power are high, the financial loss can be covered by electricity outage insurance.

J. Weather Risk Programs
Amount of Rainfall: Protect hydroelectric companies from draught. Money is paid for every inch or cm below expected rainfall average up to a certain point.

***Demand Management*:** Protect utilities by paying fixed amount for every degree below average thresholds to offset lost revenue caused by low demand.
***Generator Start-up*:** Fixed amount is paid if temperature change causes power demand.
***Wind Generation*:** Pay if wind speeds fall below threshold levels so that power can be bought from the spot market.

18.5.6 Obtaining Cover

Energy companies have several options on how to obtain cover for insurable risks.

A. Private or Governmental Cover

The following factors may be used to support a decision for a national or multilateral insurance policy. Consider private insurance cover from an insurer or via a broker if there is a nationality requirement for governmental cover. If a project does not represent a new investment, private insurers are most likely to offer coverage since governments generally insure only new investments. Additionally, private cover is more flexible when it comes to negotiating contract wording but governmental contracts often have a higher contract certainty.

The government insurance is usually cheaper and solvency is assured but it may take longer to process an insurance policy from negotiation to inception and settlement of claims. Claims from a governmental insurance company can be easier to recover given that private companies are more aggressive when it comes to claims payment. And because both governments, host and foreign, will be aware of the insurance contract, claims non-payment or intent to instigate damage may lead to conflict between governments.

The government insurance companies will usually write a policy for a longer period, fifteen or twenty-year term, while private ones will write policies as short as annual. This is particularly important for long-term projects to avoid escalation of insurance costs.

It is important to be aware of the fact that a private insurance contract will be invalidated if disclosed to the foreign government since it may lead to a claim caused by the foreign government (de facto principle).

B. Mutual Insurer

Mutual insurance companies have been formed for risks that are difficult to insure or cannot be placed elsewhere. Such insurers are referred to as industry mutual such as Aegis (casualty, management liability and property), Energy Insurance Mutual (excess casualty and management liability), Nuclear Electric Insurance Limited (NEIL) (nuclear property) and Oil Insurance Limited (OIL) (energy property). They were formed to fill needs not met by commercial insurers and require clients to be members of their organization. Problem is that increase in members does not guarantee long term stability or success and that some companies do not wish to pass on their earnings to some of their rivals if they claim.

C. Partnerships

Insurance companies are working in partnership with energy companies, their clients. Such partnerships are a pragmatic way to confront challenges that are too big and risks that are too complex for any one energy company to go it alone.

Partnership is in effect an insurance contract that can be created by strategic partnering or planning. If a claim occurs and damage needs to be repaired, a partner may offer favorable terms or fast response times to minimize their own affects from the claim. If the partners are located in the foreign country where the project is based, the political risk is reduced since they are more aware of the local law and political developments.

18.5.7 New Technologies

Energy underwriters had massive losses because they did not understand the impact of new technology to their business. With the introduction of new gas turbines, business interruption and machinery breakdown cover was not correctly adjusted and cover started after a very short period of interruption e.g. 7 days. With increased complexity of machinery, specialist materials and long delivery distances, a relatively simple fault on a turbine took up to 6 weeks to repair. This has caused large losses on business interruption and machinery breakdown policies.

As a consequence, insurers now have a better understanding of the new technology they insure and store hard-to-get items locally and ensure that staff is appropriately trained.

With the onset of new technology in the alternative energy line of business, new hi-tech products require insurance to protect against losses arising from mechanical breakdown, fire, damage and theft.

With generating technology being more proven with fewer defects, rate reductions can be negotiated. But price increases or defect clauses in contacts are expected for new unproven generation technology e.g. renewable.

18.5.8 Recent Disasters

The number of major disasters may have dropped but their severity has increased. With the hurricane disasters, 2005 has been the most expensive year for the insurance industry.

After the losses in energy lines in 2005, the market can still accommodate demand but the combination of increasing volatility and exposure has resulted in a hard market and there may be no viable alternative to self-insurance or going captive for many. Without changes to pricing and contracts, the direct and mutual insurance market may lose some of its bigger and better clients for good.

This development has been seen as a start-up opportunity for new reinsurance companies with a clean balance sheet to provide reinsurance contracts, as they do not have the loss experience from the past years. Such reinsures offer short tail coverage (1 year) with high deductibles to take advantage of possible high earning from increased rates. Established reinsures that have been in business for many years and lost money are now increasing their capital base to benefit from the hard market to recoup previous losses. This competition between old and new is good news for cedants.

18.5.9 Claims Payments

With insurance, the quality of service cannot be evaluated until a claim is made, therefore claims processing in terms of speed and accuracy is paramount. Insurance companies can settle a claim via cash, repair or replacement.

The amount of the loss is generally the net book value of the insured investment. The book value is an important factor in determining how much will be recovered in the event of a loss e.g. the book value to be utilized can be from a foreign entity or a local parent company.

18.5.10 Impact of Energy Price

Commercial insurance has a significant impact on energy companies risk management strategy and cost base. Many energy companies have cited the availability and cost of insurance as negatively impacting their business profitability. Increase in running costs of an energy company is reflected in the price of energy, thus driving energy costs up.

Energy companies with captives that buy reinsurance should be aware that reinsurance pricing is not regulated and therefore can increase by several factors for high-risk energy lines. Such increases should be part of their risk management during the annual renewal season.

Energy companies can avoid this annual insurance renewal cycle with its unpredictable pricing by joining mutual insurance organizations and gain more stable pricing as a long-term alternative risk funding strategy.

The sheer size of the 2004/2005 Hurricane season and the World Trade Center (WTC) in 2001 has meant that many reinsurers have had to reconsider the acceptance and pricing of single large risks. E.g. Munich Re increased its premium rates for oil platforms in the Gulf of Mexico by 400% in November 2005.

Premium increase in BI and other turnover related covers could be expected for electricity companies since electricity is related to the overall energy price. Energy companies should try to base BI on transmission volume, not pricing.

Insurers that had losses on the upstream market will try to retrieve the losses in the downstream energy market, creating a competitive environment that can drive current renewal prices for electricity utilities down. In the past years, there is rate reductions of 40-50% as compared to 2002 .

The deregulated energy sector can manage some of its risks by means of risk transfer via insurance. In an environment of global climate change, hurricanes, floods, false accounting and volatile energy prices they must compose innovative risk management portfolios at competitive terms. Before deregulation, state owned utilities had the financial support from governments to cover losses, now with smaller IOUs that large capital base has become unavailable and energy suppliers are forced to pursue their own risk management solutions. Professional risk management, preparation and presentation of risks, can pay dividends on insurance contract renewal.

18.6 Wind Energy Generation System

18.6.1 Introduction

This section presents a unique Axial-Flux Permanent Magnet Synchronous Generator (AFPMSG), which is suitable for both vertical-axis and horizontal-axis wind turbine generation systems. An outer-rotor design facilitates direct coupling of the generator to the wind turbine, while a coreless armature eliminates the magnetic pull between the stationary and moving parts. The design and construction features of the AFPMSG are reviewed. The flux-density distribution is studied, with the aid of a finite element software package in order to predict the generated e.m.f. waveform. The performance equations of the AFPMSG are derived, and the condition for maximum efficiency is deduced for both constant-speed and variable-speed operations. The experimental results, in general, confirm the theory developed [9].

The past few decades have witnessed rapid development in the use of alternative energy resources for electrical power generation, which plays a key role in rural electrification and industrialization programs. Power generation utilizing wind energy, in particular, has received great attention in countries all over the world. In remote areas where a central grid connection is not feasible, small-scale autonomous wind-energy power-generation systems may be developed for supplying to local consumers, reducing the connection cost, and avoiding the transmission and distribution losses. The market potential of wind-energy generators is considerable in view of the surging power demands in China and Southeast Asia. Self-Excited Induction Generators (SEIGs) have been widely used for wind energy power generation. Although induction machines are robust and inexpensive, they need capacitors to provide excitation, and their satisfactory operation requires an excitation controller. Over-voltage and over-current are operational problems that need to be resolved under variable speed operation. The space-consuming capacitors are bulky and expensive.

Greater availability and decreasing cost of high-energy permanent-magnet (PM) materials, neodymium-iron-boron (NdFeB), in particular, has resulted in rapid permanent magnet generator development, especially for wind energy conversion applications. PM machine advantages include lightweight, small size, simple mechanical construction, easy maintenance, good reliability, high efficiency, and absence of moving contacts. More importantly, PM generators can readily deliver power without undergoing the process of voltage build-up and there is no danger of loss of excitation.

Many small wind-turbine manufacturers use direct-coupled PM generators. Compared with a conventional, gearbox-coupled wind turbine generator, a direct-coupled generator system eliminates mechanical reduction gear, reduces size of the overall system, lowers installation and maintenance costs, lessens component's rapid wear and tear, lowers noise, and quickens response to the wind fluctuations and load variations.

However, a direct-coupled generator has to operate at very low speeds (typically from 200 r/min to 600 r/min) in order to match the wind-turbine speed, and, at the same time, to produce electricity within a reasonable frequency range (25–70 Hz). The generator is physically bigger in size and must be designed with a large number of poles. Various PM machine topologies have been proposed for direct-coupled wind generator applications, namely outer-rotor design, modular design, axial-field machine, the TORUS generator and coreless generator.

These machines have been developed mainly for use with horizontal-axis wind turbines. In this section, a unique axial flux permanent-magnet synchronous generator (AFPMSG) that can be used in a horizontal-axis wind turbine (HAWT) or a vertical-axis wind turbine (VAWT) system will be investigated.

Application potentials include a power source for rural farms, villages, and home energy for remote-area weather monitoring equipment, and a portable power supply for nomadic people.

This section is organized as follows. Two direct-coupled wind turbine systems that may employ the proposed AFPMSG are introduced in sub-section 18.6.2. The design features and construction of the prototype generator are presented in sub-section 18.6.3. The analysis of the flux density distribution of an experimental AFPMSG using a two-dimensional finite element package is presented in sub-section 18.6.4. Steady-state performance analysis is discussed in sub-section 18.6.5. Experimental results are presented and discussed in sub-section 18.6.6.

18.6.2 Wind-Turbine Generator Systems

Two small-scale wind-turbine generator systems are proposed here. Figure 18.9 shows a horizontal-axis wind turbine system (HAWT) that employs the proposed direct-coupled AFPMSG. To facilitate direct coupling of the generator to the turbine blades, an outer-rotor machine configuration is used. The rotor rotates about a stationary shaft, which is supported on a tower by means of a yaw mechanism. The turbine blades are attached on the flange surface of the rotor. For simplicity in construction, a single-sided AFPMSG configuration is adopted. As shown in Figure 18.10, the disk armature winding is attached to the shaft via a metal coupler and is sandwiched between the two rotor frames, one of which carries surface-mounted magnets.

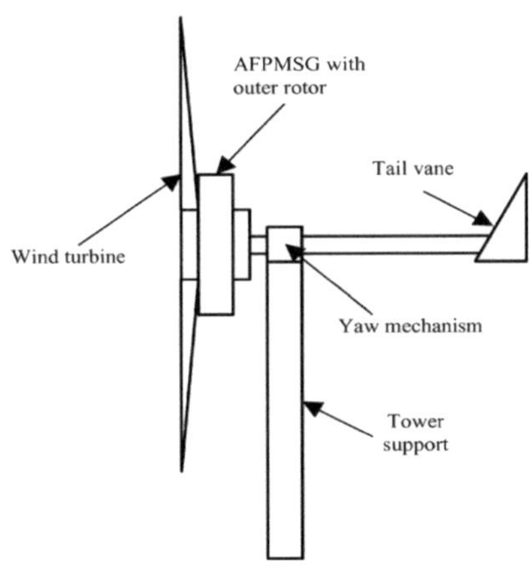

Fig. 18.9. Proposed arrangement of a micro-horizontal-axis wind turbine (HAWT) system using an outer-rotor AFPMSG

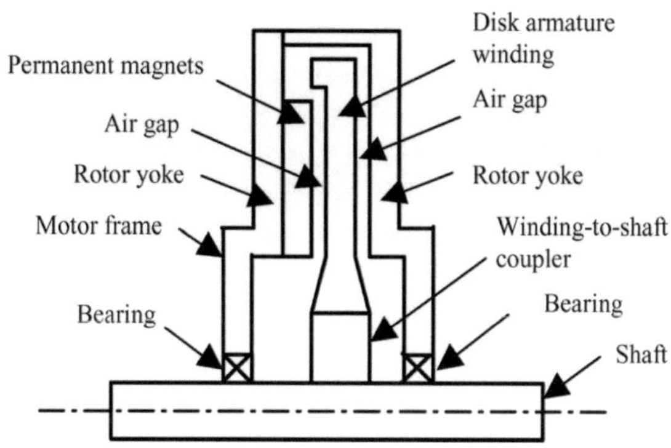

Fig. 18.10. Cross-sectional view of the proposed outer-rotor AFPMSG

The totally enclosed design will keep off rain, dirt, and foreign matter, therefore, a nacelle is not required, and system cost and weight is minimized. The rotor frames also serve as the yokes by completing the magnetic circuit. The proposed outer-rotor AFPMSG design may also be applied to a form of vertical-axis wind turbine (VAWT) system, as shown in Figure 18.11. This turbine has recently received some attention for possible deployment in a rooftop wind generation system. The turbine consists of a circular disk that spins on a stationary shaft. The rotatable shutters, when driven into the wind, will cause the circular disk to spin about the hollow shaft (shown shaded), thereby turning the rotor of the AFPMSG, which is attached to the disk.

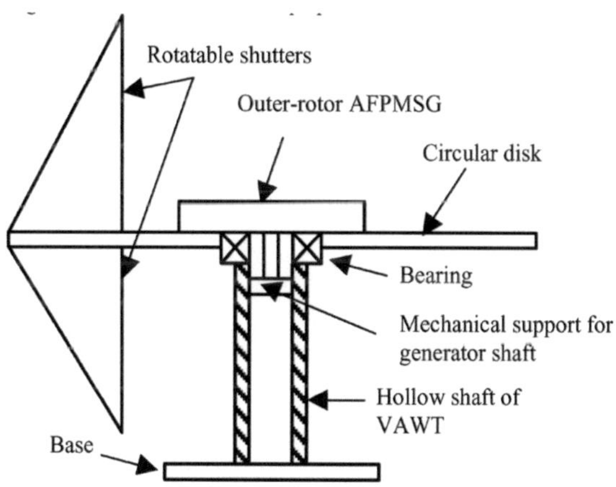

Fig. 18.11. Vertical-axis wind turbine (VAWT) using the proposed outer-rotor AFPMSG

18.6.3 Design and Construction of AFPMSG

A. General Design Considerations

The AFPMSG's weight is reduced by using a large number of poles and high-energy neodymium-iron-boron (NdFeB) magnets for the rotor field. When driven by a low-speed wind turbine, the poles enable generation at a reasonable frequency range. This also reduces yoke thickness and the length of armature coil overhang. The low-speed generator design poses a less stringent demand on the mechanical strength of the rotor magnets. Since high-energy

NdFeB magnets are used, an air gap disk winding design is feasible. The coreless armature design results in zero magnetic pull between the stator and rotor, eliminates iron loss, and improves generator efficiency. There is no cogging torque so smooth running is assured. The number of poles of the AFPMSG is determined by the intended operating speed of the wind turbine. Most small-scale wind turbines have nominal speeds in the range of 400–800 r/min. Hence, for an output voltage at a reasonable frequency, the number of poles will probably be in the range of 10–18. The NdFeB magnets, which are approximately trapezoidal-shaped and have a short length in the direction of magnetization, can be easily manufactured, and are readily available in the market.

B. Principal Machine Dimensions

For a given output power and operating speed, the AFPMSG principal dimensions may be determined using an approach similar to conventional machine design approaches, based on specific magnetic and electric loadings. For the special geometry of the AFPMSG, the output power *P* is given by:

$$P = 1.11\xi\sigma_f K_{w1} n \overline{B} ac \frac{\pi^2}{4} D_1 (D_2^2 - D_1^2) \tag{20}$$

Where

ac specific electric loading at the inner circumference of the armature;

*D*1 inner diameter of rotor magnet;

*D*2 outer diameter of rotor magnet;

*Kw*1 winding factor of armature;

B specific magnetic loading;

n rotor speed;

ξ ratio of output voltage *V* to open-circuit voltage *EF* ;

Of correction factor to account for flux fringing in the radial direction at the inner and outer peripheral regions.

For small machines supplying a pure resistive load, the ratio *ξ* may be chosen to be 0.7–0.8.

In order to maximize the output power for given values of specific loadings, the ratio of *D*2 to *D*1 should be chosen to be √3. From (20), the optimal output power of the AFPMSG may be expressed as:

$$P_{\text{opt}} = 1.11\xi\sigma_f K_{w1} n \overline{B} ac \frac{\pi^2}{2} D_1^3 \tag{21}$$

By equating *P*opt in (21) to the desired power output, *D*1 (and hence *D*2) can be determined.

The total axial length of the AFPMSG is given by:

$$l_{\text{axial}} = l_m + l_g + l_{y1} + l_{y2} \tag{22}$$

lm thickness of magnet along direction of magnetization;
lg effective air gap;
*ly*1 thickness of rotor yoke with magnets;
*ly*2 thickness of rotor yoke without magnets.

The proposed AFPMSG has a coreless armature configuration. For magnetic circuit computations, the effective air gap *lg* should include the axial thickness of the disk winding, i.e.

$$l_g = l_{\text{wdg}} + 2g \tag{23}$$

where *l*wdg is the thickness of disk armature winding and, *g* is the physical clearance between disk armature winding and rotor surface (assumed to be equal on both sides of the winding).

For a given voltage and output power, the number of turns and cross-sectional area of armature conductors may be determined, subject to the limits of current density. The thickness of the armature winding may be determined from:

$$l_{\text{wdg}} = \frac{Z_c A_c}{\zeta \pi D_1} \tag{24}$$

where *Zc* is the total number of armature conductors, *Ac* is the cross-sectional area of each conductor, and *ξ* is the space utilization factor.

The factor *ξ* should allow for the space occupied by the epoxy resin to form a disk armature of sufficient mechanical strength.

A sufficiently large physical clearance *g* between the armature winding and the rotor yoke should be chosen in order to avoid physical contact between the winding and the rotor during normal operation. For small machines *g* is in the range of 0.5–0.8 mm.

To minimize the weight of the magnets, one should aim for an operating point that gives the maximum energy product. This is achieved when the magnet flux density *Bm* is equal to one-half of the remnant flux density *Br*. From a consideration of the magnetic circuit and assuming no fringing, the magnet length *lm* and the effective air gap *lg* are related by:

$$l_m = \frac{A_m l_g}{\alpha A_g} \tag{25}$$

where *Am* is the area of magnetic pole, *Ag* is the area of air gap, and *a* is the magnetic flux leakage factor (i.e., ratio of the flux in magnet to the flux in air gap).

If the maximum allowable yoke flux density is *B*max, the thickness of yokes *ly*1 and *ly*2 may be determined as follows:

$$l_{y1} = \frac{\Phi_{y1}}{(D_2 - D_1)B_{\max}} \tag{26}$$

$$l_{y2} = \frac{\Phi_{y2}}{(D_2 - D_1)B_{\max}} \tag{27}$$

where Φ*y*1 and Φ*y*2 are the total flux entering each rotor yoke.

Equations (20)–(27) enable the principal dimensions of the AFPMSG to be determined in a machine design program.

C. Prototype AFPMSG

A 16-pole design was adopted for the prototype AFPMSG built in accordance with Figure 18.10. An output frequency of 60 Hz is obtained when the machine operates at a nominal speed of 450 r/min. The pertinent technical details are given in the Appendix.

Figure 18.12 shows the shape of NdFeB magnets and their positions on the rotor back-plate (which is part of the motor frame). To facilitate assembly of the magnet poles, two circular arrays of nonmagnetic spacers were fitted onto the back-plate at the interpolar axes. The magnets were then inserted into the regions between adjacent radial rows of spacers. A bonding adhesive was next applied to the edges between each magnet and the rotor back-plate for better mechanical strength.

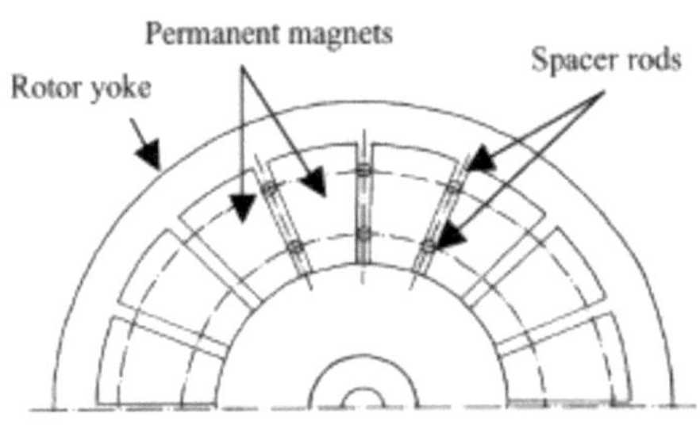

Fig. 18.12. Schematic diagram showing the layout of rotor magnetic poles

A star-connected, double-layer, full-pitch armature winding with 48 coils was used. Construction of the disk armature winding required a special technique. A total of 48 pegs were arranged, equally spaced, as a circular array on a winding workbench as shown in Figure 18.13. The armature coils were then assembled, the pegs providing proper positioning. The wire used has a special coating, which softens and becomes an adhesive when treated with a solvent. As the coils were laid, the solvent was applied and the coils were pressed together.

The coil ends were then connected to produce the phase windings, after which the whole winding was put into a circular mold and impregnated with epoxy resin.

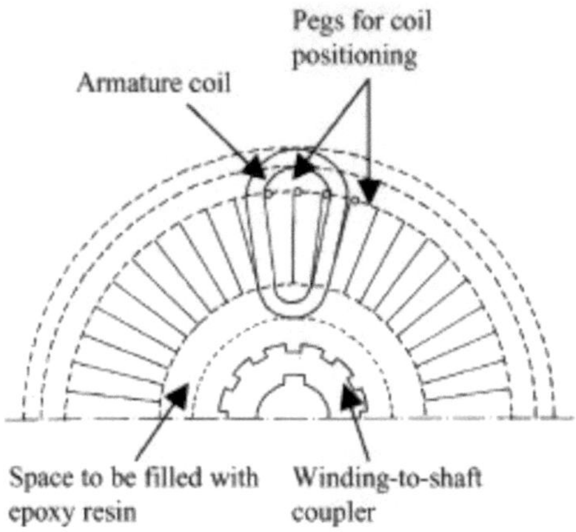

Fig. 18.13. Schematic diagram showing construction of the disk armature winding

During the impregnation stage, the winding-to-shaft coupler was also placed in the mold with its axis coincident with that of the armature winding. The coupler was held in this position throughout the thermo-setting period so that the coupler and the disk winding became an integral unit. Finally, the inner bore of the shaft coupler was trimmed to ensure that the plane of the disk winding was normal to the shaft.

Assembly of the generator involved sandwiching the stator disk winding between the two rotor frames (one with surface mounted magnets and one without). Jacking bolts were used to control the separation between the rotor frames by using the screwed holes provided on each motor frame (which are visible in Figure 18.20 later). This prevented the two rotor frames from accidentally snapping into each other during the assembly process due to the strong magnetic pull.

18.6.4 Flux Density Distribution

The flux density distribution in the AFPMSG affects the voltage waveform and the losses, and hence the efficiency. Strictly speaking, magnetic field analysis of the AFPMSG is a three dimensional (3-D) problem and requires a 3-D finite element method (FEM) software. In order to save modeling time and computation time, a 2-D FEM package is used in this study instead. Since the prototype machine being investigated has a large number of poles, there is only a slight loss in accuracy if the 2-D analysis is performed on a cylindrical surface at the mean diameter of the AFPMSG. Figure 18.14 shows the 2-D model constructed for the analysis of the experimental machine's flux density distribution.

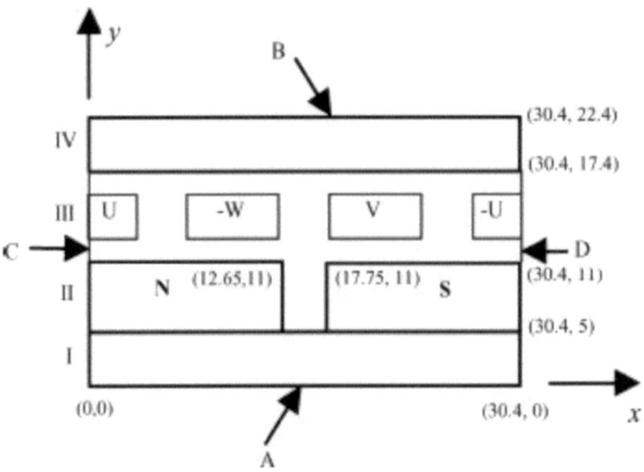

Fig. 18.14. Model for 2-D magnetic field computation of the experimental AFPMSG, all coordinates being expressed in millimeters (I, IV—steel yokes; II—NdFeB magnets, III—armature winding)

The variable *x* (Figure 18.14) denotes the circumferential distance measured from the centerline of a north pole, and the variable *y* denotes the axial distance from the bottom surface A of the lower rotor yoke with magnets. The magnetization of the rotor NdFeB magnets is in the axial direction. All the flux is assumed to be confined within the motor frame, hence tangential boundary conditions are assigned to surfaces A and B of the rotor yokes. In other words, the normal components of flux density are forced to zero at these surfaces. Periodic conditions are assigned to the surfaces C and D at the centerlines of the rotor magnets. The armature winding (region III) is modeled as rectangular conductor areas in the air space between the magnets and the upper rotor yoke. The machine phases are denoted by U, V and W, while the positive and negative signs indicate respectively a "go" and a "return" coil-side. For the time instant being modeled, the centerline of phase U coincides with the centerline of a rotor north pole. If the generator is on load, each conductor area is excited by the instantaneous value of phase current that corresponds to the specific rotor position shown in Figure 18.9. This model was solved using the 2-D static solver of the finite element software MagNet, Version 6.

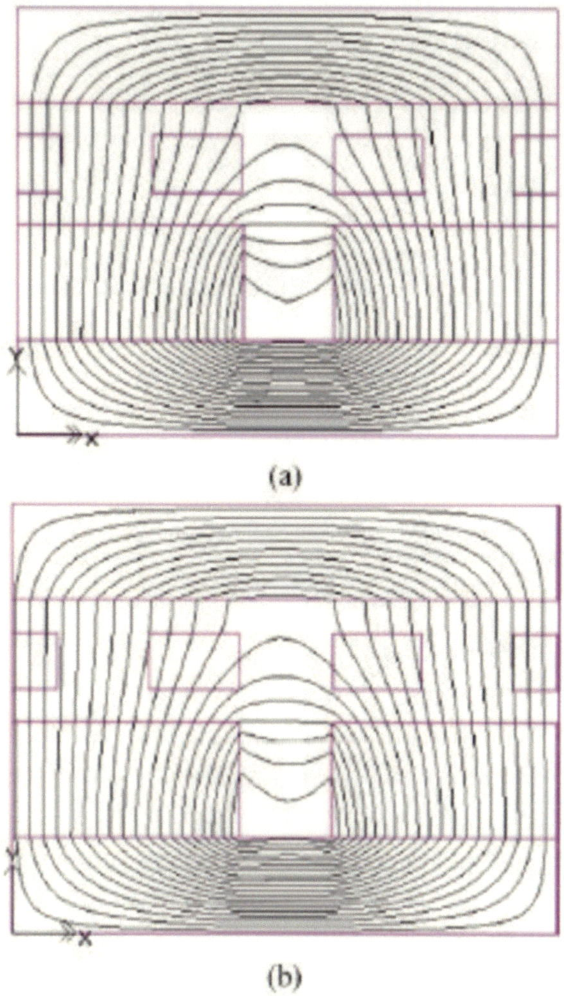

Fig. 18.15. Flux plots of AFPMSG. (a) No load. (b) Full load at unity power factor

Figure 18.15(a) and 18.15(b) show the computed flux plot of the prototype AFPMSG on no load and full load at unity power factor, respectively. It is observed that the air gap flux density in general has an axial component *By* as well as a circumferential component *Bx*. The flux plot also reveals that there is considerable leakage flux between adjacent magnetic poles. Besides, the flux lines are most dense in the bottom rotor yoke on which the magnets are mounted.

As shown in Figure 18.16, the flux density in the bottom yoke (at y = 4.8 mm) reaches 1.8 T. In the prototype machine, however, the actual flux density is slightly low since the radial length of the rotor yoke is larger than that of the magnets (Figure 18.12). Due to the relatively long effective air gap, armature reaction effect is suppressed and the flux density distribution of the AFPMSG at full load differs only slightly from that at no load, as observed from Figure 18.15(a) and 18.15(b). The air gap flux density, leakage flux, and saturation level of the generator are thus primarily determined by the rotor magnetization, and these do not vary significantly with normal load current.

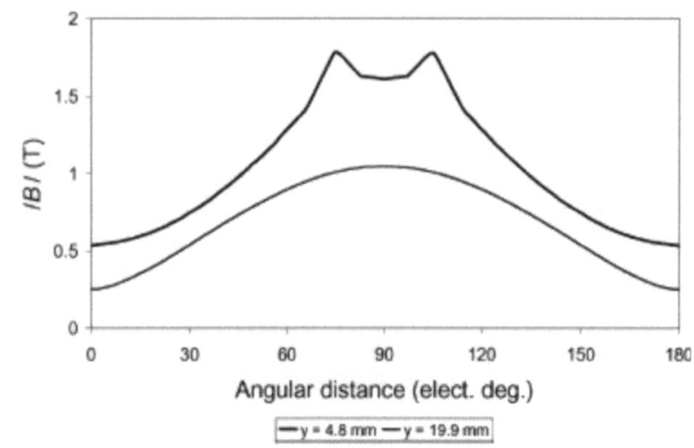

Fig. 18.16. Variation of absolute value of flux density with angular velocity with annual distance lower and upper rotor yokes

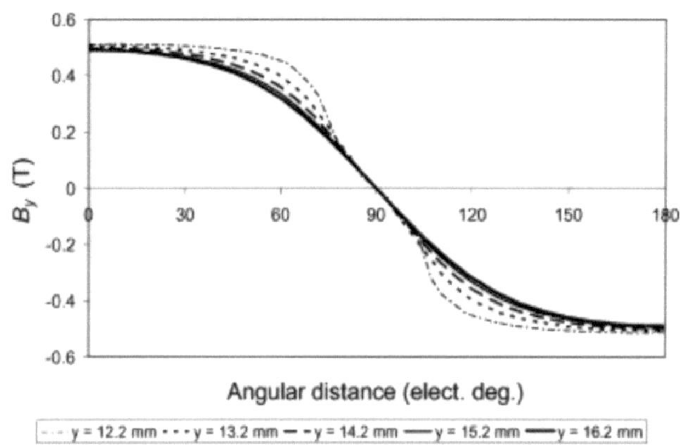

Fig. 18.17. Computed axial component (By of no-load air-gap flux density

Figures 18.17 and 18.18 show respectively the computed variation of *By* and *Bx* with angular distance along the circumferential direction. Both waveforms vary considerably with the axial distance *y*, measured from the lower surface A of the rotor yoke with magnets. It should be noted that *Bx* will not contribute to any rotation electromotive force (e.m.f.), but, together with *By*, will cause eddy currents to flow in the armature conductors and hence will result in eddy current losses. The eddy current loss *Pe* may be computed using the method discussed in [10].

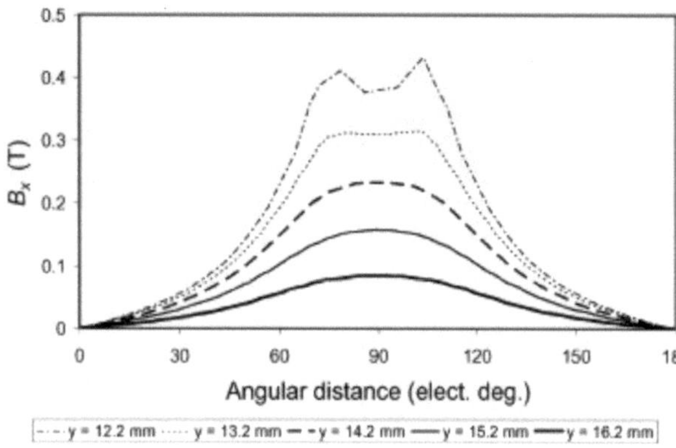

Fig. 18.18. Computed axial component (By of no-load air-gap flux density

An examination of the *By* waveform reveals that slot harmonics are absent. For smaller values of *y* (i.e., at axial positions closer to the rotor magnets), the waveform becomes approximately trapezoidal and there is considerable harmonic distortion.

From Table 18.6, it is observed that the fundamental of *By* decreases only slightly with the axial position *y*, but in the conductor region it is approximately equal to 0.56 T. The amplitudes of lower-order harmonics (up to the 11th) also decrease with *y*. At y = 14.2 mm (i.e., the center plane of the disk winding), *By* contains a 3rd harmonic component of 12.2%, a 5th harmonic of 1.6%, and a 7th harmonic of 0.5%, while higher harmonics are negligible. With triplen harmonics excluded, the total harmonic distortion (THD) in *By* is about 1.7%. In the experimental machine, the armature conductors are located in regions around the mean air gap plane. Hence, it is expected that the THD in the output line voltage will also be 1.7%, and the voltage waveform should be quite sinusoidal.

The results in Table 18.7 show that while *Bx* is comparatively small compared with *By*, the percentage harmonic contents are much larger. At the mean air gap plane (y = 14.2 mm), the 3rd, 5th and 7th harmonics in *Bx* are, respectively, 29.1%, 3.9%, and 1.3% of the fundamental.

Harmonic	Axial position *y* (mm)				
Order *n*	13.2	13.7	14.2	14.7	15.2
1	0.5822	0.5704	0.5603	0.5520	0.5448
3	0.0089	0.0077	0.0682	0.0609	0.0552
5	0.0143	0.0110	0.0091	0.0072	0.0063
7	0.0053	0.0040	0.0030	0.0023	0.0022
9	0.0061	0.0032	0.0019	0.0006	0.0002
11	0.0069	0.0037	0.0025	0.0019	0.0014
13	0.0018	0.0004	0.0003	0.0006	0.0007

Table 18.6. Principal harmonics in B_y at no-load

Harmonic Order n	Axial position y (mm)				
	13.2	13.7	14.2	14.7	15.2
1	0.2377	0.2080	0.1789	0.1502	0.1222
3	0.0765	0.0626	0.0521	0.0420	0.0330
5	0.0122	0.0093	0.0069	0.0051	0.0039
7	0.0048	0.0034	0.0024	0.0018	0.0012
9	0.0075	0.0047	0.0031	0.0020	0.0015
11	0.0056	0.0032	0.0020	0.0014	0.0008
13	0.0035	0.0019	0.0009	0.0008	0.0004

Table 18.7. Principal Harmonics in B_x at No-load

18.6.5 Steady-State Performance

A. Prediction of Terminal Voltage

Since the surface-mounted NdFeB magnets have recoil permeability close to that of air, the AFPMSG may be regarded as a cylindrical-rotor synchronous machine with a constant field excitation. Figure 18.19 shows the per-phase equivalent circuit of the generator when supplying an isolated resistive load, where *EF* is the no-load terminal voltage, *R* is the armature resistance, *Xs* is the synchronous reactance, *RL* is the load resistance, and *V* is the terminal voltage; all being per-phase quantities.

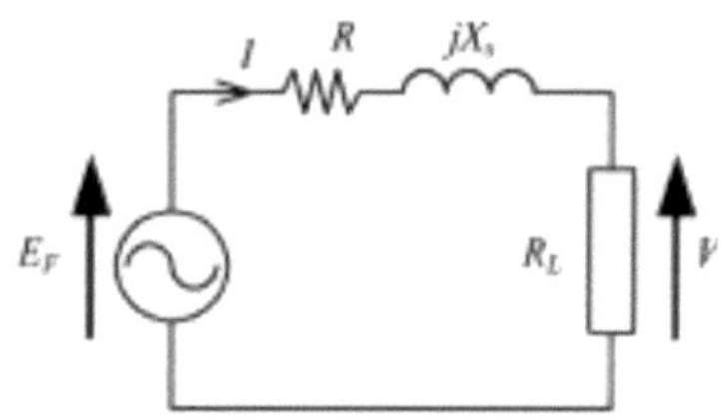

Fig. 18.19. Per-phase equivalent circuit of AFPSGM

From the circuit, the current, terminal voltage, and output power of the generator may be determined as follows:

$$I = \frac{E_F}{\sqrt{(R+R_L)^2 + X_s^2}} \tag{28}$$

$$V = E_F \frac{R_L}{\sqrt{(R+R_L)^2 + X_s^2}} \tag{29}$$

$$P_{\text{out}} = \frac{3E_F^2}{(R+R_L)^2 + X_s^2} R_L \tag{30}$$

By using (28) and (29), the load characteristics of the AFPMSG can be computed provided the synchronous impedance is known.

B. Determination of the Synchronous Reactance

The armature resistance *R* may be determined from a dc resistance test, while the e.m.f. *EF* due to PM excitation may be determined from an open-circuit test. Performance of a short-circuit test, however, may not be feasible due to thermal limitations and the possibility of irreversible demagnetization of the rotor magnets. A more convenient method to determine the synchronous reactance *Xs* is by carrying out an inductive load test. This test gives a fairly accurate prediction of the synchronous reactance without the need for load angle measurements. The AFPMSG is driven at constant speed and a variable three-phase inductive load is connected across the armature terminals. Readings of armature current *I* and terminal voltage *V* are taken. From the voltage phasor diagram, *Xs* may be determined as follows:

$$X_s = \frac{\sqrt{E_F^2 - (IR)^2} - V}{I}. \tag{31}$$

C. Losses and Efficiency

Since the AFPMSG has no armature core, there is no armature iron loss and the losses mainly consist of the copper loss $Pcu(= 3I^2R)$, the friction and windage loss *Pfw*, and the eddy current loss *Pe* in the armature conductors. The sum of *Pfw* and *Pe* is equal to the mechanical power input to the generator shaft under no-load conditions. The efficiency of the AFPMSG is thus given by:

$$\eta = \frac{P_{\text{out}}}{P_{\text{out}} + \text{Losses}} = \frac{3I^2R_L}{3I^2R_L + 3I^2R + P_{fw} + P_e}. \tag{32}$$

The efficiency of the AFPMSG may be increased by designing it with a small armature resistance *R*, i.e., by using conductors of a large cross-sectional area. Multiple-circuit coils may be used in order to minimize the eddy-current loss *Pe* [10]. For operation at a given speed, *Pfw* and *Pe* may be assumed to be constant. Under this condition, the maximum efficiency of the AFPMSG will occur at a load resistance *RL* given by:

$$R_L = \sqrt{\frac{3E_F^2R}{P_{fw} + P_e} + Z_s^2} \tag{33}$$

For a given load resistance *RL*, the condition for maximum efficiency under variable speed operation may be estimated by assuming that the friction and windage losses *Pfw* vary linearly with speed over the speed range being considered. The eddy current losses *Pe*, on the other hand, vary as the rotor speed squared [10]. We can therefore write:

$$E_F = k\omega \tag{34}$$

and

$$P_{fw} + P_e = k_1\omega + k_2\omega^2 \tag{35}$$

where k, $k1$ and $k2$ are constants and ω is the angular frequency of output voltage. Equation (32) can thus be written as follows:

$$\eta = \frac{3k^2\omega R_L}{3k^2(R+R_L) + (k_1 + k_2\omega)\{(R+R_L)^2 + (\omega L_s)^2\}} \tag{36}$$

where Ls is the synchronous inductance per phase.

Maximum efficiency will occur when the derivative $d\eta/d\omega$ is equal to zero, from which the following equation may be derived:

$$2k_2\omega^3 + k_1\omega^2 - k_1\left(\frac{R+R_L}{L_s}\right)^2 = 0. \tag{37}$$

Equation (37) may be solved analytically or numerically to give the angular frequency at which maximum efficiency occurs, and, hence, the value of the maximum efficiency.

If, however, the armature winding is made of conductors with a small cross-sectional area, the eddy current losses may be ignored and maximum efficiency will occur at an angular output frequency given by:

$$\omega = \frac{R+R_L}{L_s}. \tag{38}$$

The value of maximum efficiency is

$$\eta_{\max} = \frac{R_L}{R+R_L}\left(\frac{3k^2}{3k^2 + 2k_1L_s}\right). \tag{39}$$

18.6.6 Experimental Results and Discussion

Figure 18.20 shows the setup for experimental investigations on the AFPMSG. The shaft of the machine was mounted on a special test rig so that the rotor frame can turn freely. Provisions were made on the motor frame for coupling the generator rotor to a dynamometer motor drive by means of a belt transmission. The turbine is therefore emulated by varying the speed of the dynamometer motor.

Fig. 18.20. Test rig for experimental investigations on the AFPMSG. The outer rotor being driven by a dynamometer motor via a belt transmission

The armature resistance of the AFPMSG was measured to be 0.58Ω from a dc resistance test. Table 18.8 gives the no-load test and inductive load test data obtained at a rotor speed of 450 r/min. From (31), the synchronous reactance *Xs* of the AFPMSG at 450 r/min (or 60 Hz) was determined to be 0.25Ω per phase. Due to the large effective air gap, armature reaction is relatively weak in the AFPMSG and hence *Xs* is small compared with the armature resistance *R*. The machine performance, such as the voltage drop, therefore depends primarily on *R* in this type of machine.

Test	Line voltage (V)	Line current (A)	Power Input (W)
No-load test	25.0	0	17.0
Inductive load test	21.7	5.9	–

Table 18.8. No-load and inductive load test data for AFPMSG

A check of the 2-D finite element computation is in order. Using the measured no-load line voltage, the fundamental component of *By* was found to be 0.564 T. This value agrees very well with the average value of *By* in the conductor region, computed at the mean radius of the machine (Table 18.6). This shows that the 2-D model is sufficiently accurate for obtaining a good engineering solution.

Figures 18.21(a)–(c) shows the variations of terminal voltage, output power, and efficiency of the experimental AFPMSG with line current under constant-speed operation. The voltage-current characteristics are practically linear from no load to full load. At a speed of 600 r/min, the voltage drop between no load and full load is 25% and an output power of 340Wcan be delivered at rated current. When the speed is reduced to 300 r/min, the voltage drop between no load and full load increases to 50%, and at rated current the output power is only 110 W. Due to the relatively large armature resistance, maximum efficiency of the machine occurs at low values of load current. A maximum efficiency of 79.0% can be achieved at 600 r/min.

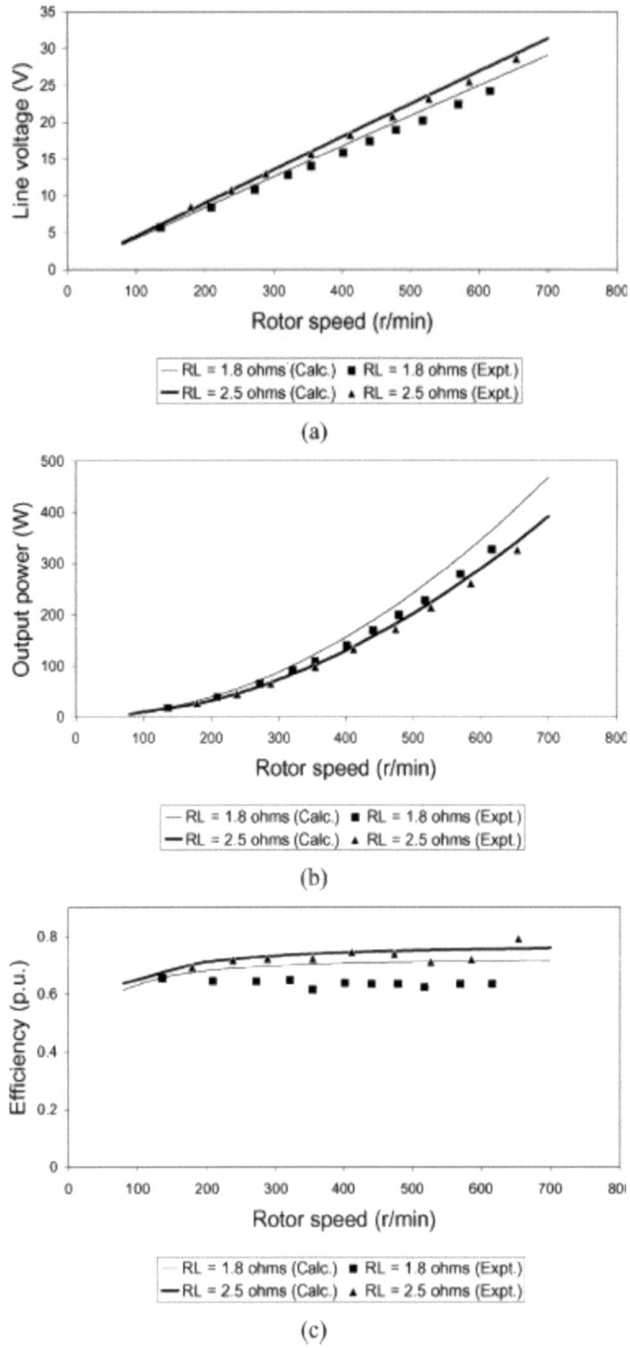

Fig. 18.21. Performance of AFPMSG under constant-speed operation (a) Variation of voltage with current. (b) Variation of output power with current. (c) Variation of efficiency with current

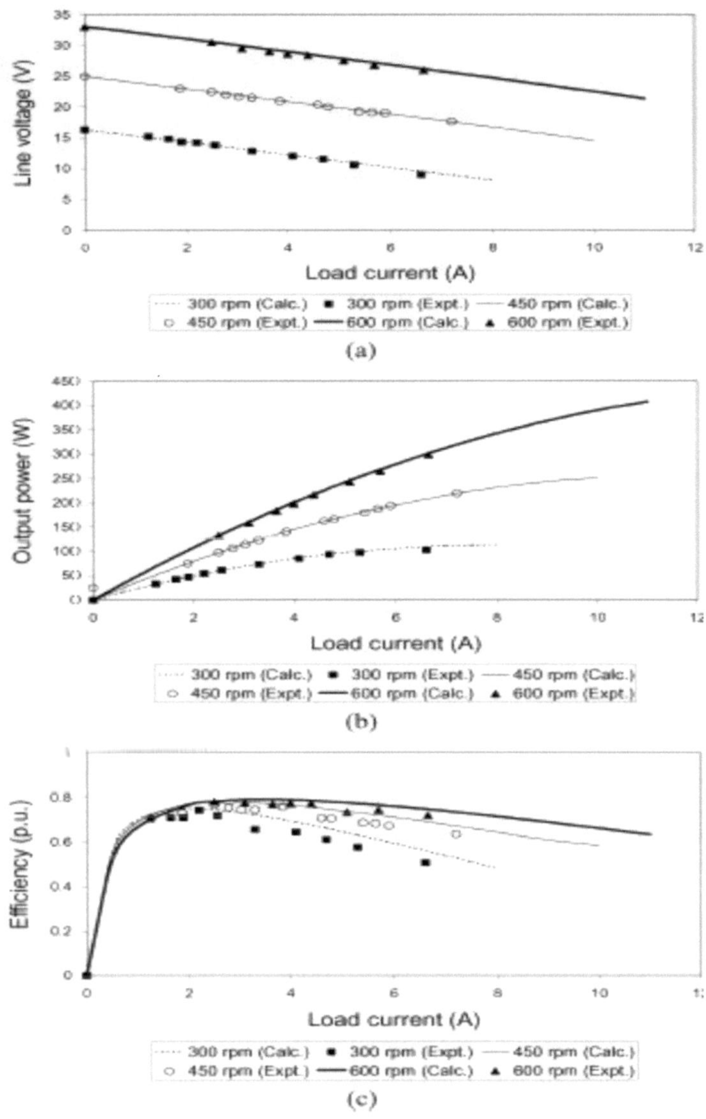

Fig. 18.22. Performance of AFPMSG: variable-speed operation (a) Variation of line voltage with rotor speed. (b) Variation of output power with rotor speed; (c) Variation of efficiency with rotor speed

Figures 18.22(a)–(c) shows the variations of terminal voltage, output power, and efficiency of the experimental AFPMSG with speed when the load resistance is constant. The output voltage varies almost linearly with the rotor speed, while the output power varies approximately with the square of the rotor speed. The efficiency, however, is only slightly affected by the speed. With RL = 2.5Ω, the computed efficiency varies from 74% to 76% when the speed increases from 350 to 700 r/min.

The close agreement between the voltage-current and power-current characteristics confirms the theory developed in sub-section 18.5 of this chapter. Experimental values of efficiency, however, correlate less well with the computed values due to the difficulty in accurately determining the losses in the belt transmission.

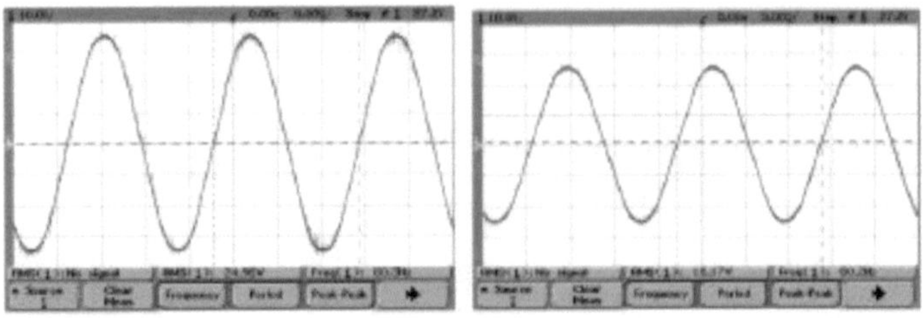

(a) (b)

Fig. 18.23. Line voltage waveforms of AFPMSG (a) At no load. (b) When delivering a current of 6.1 A to a resistive load (voltage scale: 10 V/div; time scale: 5 ms/div)

Figure 18.23(a) and 18.23(b) show, respectively, the line voltage waveforms of the AFPMSG at no load and when delivering a current of 6.1 A to a resistive load. The waveforms are practically sinusoidal.

From a measurement using a harmonic analyzer, it was found that in each case there was mainly a 1.4% 5th harmonic and a 0.2% 7th harmonic, with a total harmonic distortion of 1.6%. This experimental result is consistent with the harmonic analysis of the By waveform in Table 18.6, the slight reduction in THD being due to the spread of the conductors for each coil side. The voltage waveforms have confirmed that the proposed AFPMSG is an excellent source of sinusoidal power. The measurements also revealed that the total harmonic distortion was not sensitive to the variation of load. This is due to the fact that the armature reaction m.m.f. has only a slight effect on the resultant flux density distribution in the air gap, as observed from the flux plots in Figure 18.15.

18.7 Systematic Losses and Smart Electricity Use

The economic efficiency of today's Electricity Supply System (ESS) is driven by two mutually supporting key factors: the load factor; and fuel conversion efficiency. The higher the load factor, the more efficient the running of the plant and the more investment in generation efficiency is worthwhile. The higher the generation efficiency the more the plant will be called on and so the higher load factor.

This sub-section presents two ways by which smart appliances enhance both these factors, one quite rapidly but the other needing deeper change in the way we market electricity.

The role of refrigerators is in providing frequency response. In a population of refrigerators, there will at any instant be many that are on and many that are off. On average, in the UK, the average load from domestic refrigeration is over 1GW, although, if they were all replaced by the most efficient of modern appliances, it would be less than this. One technology, when fitted into new refrigerators, influences the numbers that are on or off depending on system frequency so as to balance the continuous small and occasional large fluctuations in supply and demand. In this, they displace the frequency response service largely supplied from thermal power stations.

The frequency service from power stations is particularly carbon intensive, creating significant losses or inefficiencies in the plant. To provide low frequency response, generation has to be operated at part load, but with capability to increase or reduce load very rapidly – typically within 5-10 seconds. While there is little published on how great are the resulting losses, they are significant, with resulting emissions perhaps in the order of over a million tons of CO_2 p.a. in the UK.

A population of refrigerators, on the other hand, can react to a change of frequency within about 1/10 of a second. While refrigerators shift their consumption by seconds and minutes, and this marginally reduces the need for peaking capacity, much bigger rewards are available if load can be shifted by hours or even days from the current peaks. In the UK (and in many other countries), the peak load (which is what dictates the needed generation capacity) arises at around 6.00 pm in winter, and is followed by a decline as the nation settles down for the evening and the night.

In an idealized example case, if the overall daily Great Britain load could be spread evenly across the day in the winter, then the peak generation need would be reduced by some 15GW. While this is, in practice, unrealizable, even a proportion of this would save many billions in new investment as existing plant reaches the end of its life and electricity consumption continues to grow. In addition, some of the plant that was run would be able to run at a constant output and so more efficiently. These are big rewards.

Nobody really knows how much peak load can be shifted to other times. What can be said is that many appliances, such as dishwashers, laundry machines and the like, have enough microprocessor intelligence to manage a shift of their consumption to when it would be cheaper, so long as the user's deadline for clean dishes or laundry is met. All they need is guidance and reward for doing so.

Smart tariff appliances do this by price, although one proposal using frequency where the frequency is deliberately run below its nominal level over peak periods is being investigated. It was proposed to use broadcasting, by suppliers, of a continuously varying expected price over the next day, days and even weeks. Appliances see this expected price, and so can optimize their consumption to meet the users' deadline at the minimum and cost, known and displayed when the users set the deadline. Smart meters also track the consumption and its cost at each moment, and add up the bill for presentation to the user and to the utility from time to time.

Of course, as more wind generation penetrates the system, the price will depend, in part, upon the weather, so as wind forecasts change, so can prices and therefore appliance plans and their consumption. You end up doing your laundry when the wind is blowing.

It does not know quite how big will be the rewards from this. It is assessed that a variant of the refrigerator technology, applied to water heating in South Africa, can lead to peak capacity reductions of around 3GW, and do so within about 3 years. Some work in the US (seemingly based on big appliances and big consumptions) suggests a saving of the order of $200 p.a. per household. Is this $200 p.a. per household an amount that could be afforded to waste?

18.8 Combined Heat and Power and District Heating with Thermal Storage

The combined production of heat and power (CHP) or put in another way, the utilization of the waste heat from power production for heating homes and buildings and meeting process heat demands are experiencing growing interest in the UK. CHP and District Heating (DH) or Community Heating (CH) as it is often referred to as in the UK, provide the following advantages: high flexibility of fuel usage; high efficiency; high environmental quality, less effluent/waste disposal problems; reduced air pollution and increased cost-efficiency.

Denmark decided to explore the opportunities and invest in CHP and DH almost 30 years ago. Today 60% of the Danish housing stock, whether privately owned or council dwellings is connected to a DH scheme and more than 95% of this heat comes from waste heat, i.e. from CHP. Over 50% of electricity production in Denmark is CHP. In the UK less than 3% of dwellings are connected to a DH scheme and only around 7% of electricity production is CHP.

Denmark has broadly seen three scales of CHP that were mainly implemented in the following chronological order:

- Large scale CHP in cities (>50 MWe).
- Small (5 kWe – 5 MWe) and medium scale (5 – 50 MWe).
- Industrial and small scale CHP.

Denmark's ten major cities (the smallest has a population of around 50k the largest around 1.1M) have citywide DH where most of the heat (95-98%) is produced by large CHP plants. The five largest are gas-fired combined cycle plants; the others are using natural gas, biomass, waste or biogas. Several hundred towns and communities are supplied with DH created by local initiatives back in the 1960s, which now have been modernized and are mostly supplied by CHP.

In a DH system, it includes the production, the distribution and the customers. A DH network can be split into three levels:

- Branches and connections to consumers
- Distribution heat network, e.g. 100 °C/40 °C
- Transmission heat network, e.g. 120°C/70°C.

A DH scheme can consist of a distribution network only or a combination of distribution and transmission. One reason for the transmission/distribution concept is that the transmis-

sion system can be run with a higher pressure level, the transmission network is often a 25 bar system, while the distribution system can be a 6, 10 or 16 bar.
The cost of installing the heating network depends on four factors:

- The design operating temperature and pressure
- The complexity of existing services
- The length of the network
- The peak heat demand.

Thermal storage has been used in DH systems for decades, the main aim being to separate time-dependent demand and occurrence of heat and electricity from one another.

Practically all CHP plants of the backpressure type, as well as small-scale plants only producing heat and electricity in fixed ratios are equipped with a thermal store. CHP plants of the extraction type have earlier only to a limited extent been operating with a thermal store.

Operating a CHP plant in a liberalized electricity market increases the need for more flexibility of the plant in order to operate in the most economical way, serving both the heat consumers as well as the electricity market.

The thermal store is used for short-term storage of water-based energy. Basically there are two main purposes for having a thermal store:

- To save operational cost in the form of heat production cost.
- To save investments (in the form of investments in peak load capacity and network capacity). The investment in a thermal store should be carefully compared to that of establishing a peak load unit in the network.

In Denmark the thermal stores are mainly installed in order to save heat production cost as most of the DH systems are supplied from CHP plants. This means that the heat production cost is not only related to the fuel cost but also to the selling price of electricity. For many years the selling price of electricity from decentralized CHP plants has been based on a triple tariff. Now the selling price of electricity may have many values during the day and may change hour by hour.

As the selling price of electricity reflects on the heat production cost, the heat storage tanks in Denmark are mainly utilized in order to optimize the power production and are mainly related to the power production in two ways:

- Back pressure production: The proportion between electricity production and heat production is fixed; an increase in electricity production will result in an increase in heat production. Typical production equipment is backpressure steam turbines or piston engine installations
- Extraction production: An increase in the heat production will decrease the power production. Typical production equipment is extraction steam turbines.

How the heat storage tank is utilized depends on the types of production units in the DH network.

Thermal stores in Denmark are centralized and connected to the DH system between the CHP plant and the network. Decentralized thermal stores have been implemented in the Netherlands. The purpose of these thermal stores can be to save heat production cost but the decentralized placement of the tanks indicates that the purpose of the tanks also has been to reduce the pipe diameters of the network and thereby the network investments. Furthermore, the decentralized placement of thermal stores may be a possibility if the space at the central production unit does not allow for a large storage tank.

Thermal storage does not generally reduce fuel consumption and therefore has no impact on CO_2 emissions. As the purpose of the tank is mainly of an economic nature, the specification of the size of thermal store must therefore be based on an economic analysis.

The following information is required in order to accurately size a thermal store:

- Daily load variations during the year over a number of years.
- The possible savings related to the production units due to the installation of the thermal store. The possible savings will typically be related to an optimized power production in relation to large changes in the selling price of electricity.
- Prices for the heat storage tank installation including pumps, pipes, valves etc.

Due to heat loss it is important that the surface area of the water in the thermal store is as little as possible in relation to its volume. The optimal would be a ball but due to considerations for the construction a cylinder where the diameter is equal to the height is ideal. Normally, a height/diameter ratio above 1.5 is used. It should be taken into account that there will be an unusable separation layer of typically one meter. In order to secure a good separation between hot and cold, diffusers should be placed at the top and bottom.

The connection of the thermal store to the DH system will depend on the type of DH system.

The main issue here is the maximum temperature of the DH system where there is a distinction between systems with supply temperature of up to 97 ^{o}C and systems with higher supply temperatures.

If the maximum temperature is lower than the boiling point of the water (in Denmark which is near the surface of the sea, this in practice means 95-97°C) a non-pressurized tank can be used. A minor gauge pressure is maintained at the top of the tank as nitrogen or steam cushion prevents the oxygen from the air to penetrate into the DH water.

Sometimes a non-pressurized tank is also used for maintaining the static pressure in the DH system. The non-pressurized tank is often used in the smaller or medium sized systems and they will typically have a height of between 15 and 25 meters. Depending on the topography of the network and on whether the tank is placed either high or low compared to the other parts of the DH system, the water column in the tank may be sufficient to provide the required static pressure in the DH system.

In large DH transmission systems the maximum temperature is often above 100°C (typically up to 120°C). If water with that temperature should be stored in a non-pressurized tank, permanent boiling would take place at the top of the tank that is highly undesirable. In order to avoid this, temperature lowering installations or pressurized tanks may be used.

18.9 Conclusions

It is clear there are many issues in the energy sector. The energy field is fuelled with excitement. Public interest and awareness is heightened in topics like energy efficiency, global warming and climate change, the carbon footprint, smart metering, intelligent grids, renewable energy sources, distributed generation, power quality, risk management and asset management. Observation, predictability, and reliability are the key to efficiency. Predictability in Market timescales is key in efficient scheduling. Cheap advanced ICT systems to monitor distributed systems and communicate dynamic tariffs. Volatile energy prices are a fact of life. Regulatory stability and a clear energy policy are essential for future security of supply. Investment in new generating capacity is essential - not only in renewable. Some topics have been introduced with examples.

This chapter has also presented a form of axial-flux permanent magnet synchronous generator (AFPMSG) that is suitable for a direct-coupled wind turbine system. The application of this machine to horizontal-axis and vertical-axis wind turbine generator systems is discussed. An analysis of the magnetic flux density distribution in the AFPMSG has been made with the aid of commercial, finite element software. The computed results show that the output line voltage is practically sinusoidal for the proposed machine configuration. Experiments performed on a prototype generator have confirmed the feasibility of the AFPMSG design and the validity of the equivalent circuit model.

18.10 Acknowledgements

Loi Lei LAI, Professor and Head of the Energy Systems Group, School of Engineering and Mathematical Sciences, City University London, UK has coordinated this Chapter. Thanks are due to Mr. Long Zhou and Ms Yingnan Ma of the Energy Systems Group at City University London for the detailed preparation of this Chapter. Copyright release by the IEEE is most appreciated

Appendix

The prototype AFPMSG has the following design specifications:

1) Output at nominal speed
230 W, 16.7 V, 8.0 A, 60 Hz
2) Rotor field
Number of poles 16
Outer diameter 200 mm
Inner diameter 110 mm
Thickness of magnets 6 mm

Pole arc at mean radius 150°e
Remanent flux density 1.128 T
Recoil permeability of magnets 1.03
Thickness of each rotor yoke 5 mm
3) Armature winding and air gap
Winding type three phase star, double layer, full-pitch
Number of coils 48
Turns per coil 7
Armature resistance at 75°C 0.58 Ω
Thickness of winding 5 mm
Physical air gap length 2 × 0.7 mm
4) Mass of active materials
Copper 0.8 kg
Iron 2.0 kg
Magnets 0.56 kg

18.11 References

[1]. Loi Lei Lai and T. J. Hammons, "Energy issues under deregulated electricity energy markets," IEEE PES 2006 General Meeting, Montreal, Canada, 18-22 June 2006.
[2]. Davor Vujatovic and Qingping Zhang, "Harmonics generated from railway operation," IEEE PES 2006 General Meeting, Montreal, Canada, Paper No. 06GM0852, 18-22 June 2006.
[3]. Norman Tse, "Practical application of wavelet to power quality analysis," IEEE PES 2006 General Meeting, Montreal, Canada, Paper No. 06GM0790, 18-22 June 2006.
[4]. Rene A. Carmona, Wen L. Hwang and Bruno Torresani, "Multiridge detection and time-frequency reconstruction," *IEEE Trans. on Signal Processing*, Vol. 47, No. 2, Feb., 1999.
[5]. Hugh Rudnick, "Energy risk in Latin America: the growing challenges," The IEE International Conference on Energy Trading and Risk Management, City University London, ISBN: 9 78086 341591 3, 21-22 Nov 2005.
[6]. S R Browning, "BETTA and distributed generation - invasion and uncertainty," The IEE International Conference on Energy Trading and Risk Management, City University London, ISBN: 9 78086 341591 3, 21-22 Nov 2005.
[7]. Dave Openshaw, "Balancing risk & opportunity - An infrastructure manager's perspective," The IEE International Conference on Energy Trading and Risk Management, City University London, ISBN: 9 78086 341591 3, 21-22 Nov 2005.
[8]. Harald Braun and Loi Lei Lai, "Insurance issues for energy risk," IEEE PES 2006 General Meeting, Montreal, Canada, Paper No. 06GM0285, 18-22 June 2006.
[9]. T F Chan and L L Lai, "An axial-flux permanent-magnet synchronous generator for a direct-coupled wind turbine system," IEEE Transactions on Energy Conversion, March 2007.
[10]. R.-J. Wang, M. J. Kamper, K. V. Westhuizen, and J. F. Gieras, "Optimal design of a coreless stator axial flux permanent-magnet generator," *IEEE Trans. Mag.*, vol. 41, no. 1, pp. 55–64, Jan. 2005.

19

Africa: The African Union and New Partnership for Africa's Development (NEPAD)-The Power Footprint

19.1 Introduction

The last century has demonstrated that every facet of human development is woven around a sound and stable energy supply system with the electricity grid in its various forms being the most optimized motive source for maintaining a sustainable standard of living and quality of life. The last century has also created a vast knowledge base of man's delicate relationship to the environment and its relationship to his own survival. The historical continuance of resource wars represents the clearest indication of this imbalance and which has spurred environmental protection in all its facets. Global economic growth will drive electrification, both in developed and developing countries. Almost without exception, the major technology trends depend upon an advanced electricity infrastructure. In Africa, widespread access to electricity will be a prerequisite for sustaining economic growth. Developing countries need clean, affordable electricity to grow their economies and meet the aspirations of their people. Thus, electrification will be a key factor in global stability.

Particularly important are the four linked goals of protecting earth's life support systems, improving human welfare, eliminating poverty, and stabilizing population. Only when the world's citizens have achieved a minimal quality of life will they have the will and resources to participate in the global economy. Without this headroom, economic development will be sub marginal.

In last 20 years there has been a concerted effort by African countries in formulating strategies to address the continent's social economic crisis and integration into 21st century development challenges. Africa the world's richest continent in terms of readily accessible natural resources and 13% of the world population accounts for about 2% of world economic output. Real gross domestic product (GDP) in Africa as a share of the world total remained constant at about 2% and earlier projections suggested it would remain at 2% through 2020. In the past five years however, there has been a dramatic increase in economic activity largely driven by a shifting geopolitical landscape brought on by an intense demand for energy and other natural resources to feed the Asian '*Tiger*' economies of China and India. This comes as African statesmen have sought to consolidate various development initiatives such as the OMEGA, New African Initiative (NAI), and Lagos Action Plan (LAP) into an efficient body. The African Union (AU) and New Partnership for African Development (NEPAD) have emerged as the umbrella organizations positioned to steer the continent for effective change from its recent historical past of post-colonial development into an African Renaissance. As part of this vision for development there are a number of firsts being introduced in the planning and implementation of some of these projects over a large resource mix of natural gas, hydro, coal, petroleum based fossil fuels and various alternative energy resources to support the vast rural development needs.

New high voltage (HV) transmission lines are at the top of the agenda for many energy planners in the various regions of Africa. These lines are identified within the power pools of Southern Africa (SAPP) and West Africa (WAPP) and the Mediterranean ring of North Africa with links to the Middle East Gulf States. The Central Africa Power Pool (CAPP) is also in the process of identifying its new HV lines and the East Africa region (EAPP) has been formulated with projects identified. The proceeding chapter looks at the centrality of the huge hydropower potential of the Grand Inga project and the sensitivity of pricing electricity exports as they relate to transmitting power across Africa's proposed new HV lines. Capacity-planning in Africa's power pools will be significantly affected by the proposed new HV lines and as such the costs of interconnection across Africa draws upon comparisons with HV line networks in other locations.

The Southern African Power Pool (SAPP), West African Power Pool (WAPP), Egypt, and the East African Power Pool (EAPP) have each expressed their interest in continental interconnection, a major step forward in development. The hydropower potential from the River Congo is a big attraction to regional planners and the key development project of Grand Inga (39GW potential, located 150km from Kinshasa) necessitates the planning of very long HV lines. The Central African Power Pool (CAPP) and the Democratic Republic of Congo (DRC) have much to gain from exporting the potential hydropower at the right price. Despite the challenges, it is also clear that there has never been throughout human history a period of more potential to harness 21st technology for rapid and sustained infrastructure development across all sectors for a modern sustainable economy. The new benchmarks set by petroleum prices reaching above USD$120 per barrel significantly influence the economics of comparative energy solutions such as natural gas combined cycle plants and other fossil fuels in favor of the capital intensive hydroelectric energy models. These changing dynamics are the challenges faced by governments and the private sector in arriving at practical solutions.

The chapter contributions as such summarize many of the efforts to date in addressing strategic approaches to Africa's development organized through the IEEE PES International Practices Sub Committee forums. The various viewpoints while sharing a fair amount of redundancy in some areas remains it's most welcome aspect as it satisfies the intent to produce an optimization of the various ideas to produce functional, balanced and sustainable economic models. The emphasis throughout the various viewpoints borrow from *'lessons learned'* in the global power industry such as deregulation, the 2001 California energy crisis, environmental activism to curtail Green House Gas (GHG) emissions and the technical challenges of Power Systems operations in a dynamic market environment. A great emphasis is also placed on the human development factor and methods for its integration into new knowledge based industrial systems brought on by the 21st century ICT revolution currently driving the global economic system.

The opening paragraphs outline the vision of the AU and NEPAD organizations emphasizing the various regional power pools at the core of their development roadmap.

Commentaries on these programs from economic, technical and environmental perspectives follow the program vision and status after which commentary from selected academia through comparative studies outline various roadmaps for developing the all important human capacity factor to develop, manage and sustain these complex systems in a complex environment.

19.2 The Need for Infrastructure - New Partnership for Africa's Development

19.2.1 Why Regional Infrastructure

The need for Infrastructure--New Partnerships for Africa's Development is summarized in Figure 19.1 [1].

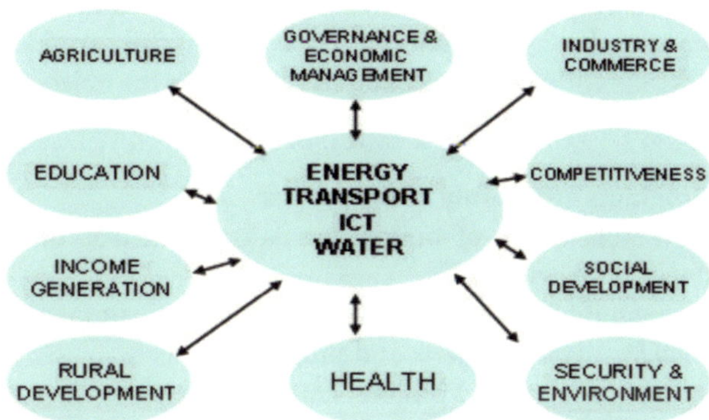

Figure 19.1. Need for Infrastructure in New Partnership for Africa's Development

The continuing demise of the human social condition is what has spurred the leadership through the formation of the African Union and NEPAD. At its core infrastructure deficiency found throughout the continent is the clear focus for development. There is also the realization from experiences in their individual national economic programs and other global economies that greater efficiencies can be derived through regional and ultimately continental cooperation articulated as follows:

- Regional and International trade are central to economic growth and development
- Efficient infrastructure network has the effect of generating new investments in other sectors
- African countries, individually, are too small to generate economies of scale found in larger markets
- Weak infrastructure linkages condemn the region to low competitiveness in the global market
- Regional infrastructure leads to larger project sizes capable of attracting more private sector investments.

19.2.2 Approach Adopted

The approach adopted by NEPAD in Infrastructure is two-pronged:

i) A short-term action plan (STAP) based on a survey of countries and RECs
ii) A Medium-Long Term Action Plan/Strategic Framework, which is linked to and complements the short-term action plan. It will take up projects and initiatives that require more time for preparation and development as well as institute an enabling framework for future development of infrastructure.

19.2.3 The Role of NEPAD and the Typology of STAP

NEPAD tasks to ensure the successful implementation of Short Term Action Plan:

- Mobilizing political will:
  a) Facilitate the mobilization of resources.
  b) Facilitate knowledge sharing, networking and dissemination of best practices among countries, Regional Economic Cooperatives (RECs) and technical agencies
- Underpinning all NEPAD infrastructure programs is the objective of strengthening *sector governance.*

STAP Projects and Programs are mainly of four types:

- *Facilitation* – establishment of policy, regulatory and institutional framework to create a suitable environment
- *Capacity Building* initiatives to empower particularly the implementing institutions
- *Physical/Capital Investment* projects and programs
- *Studies* to prepare new priority projects and Programs.

STAP Projects/Program Selection Criteria

The STAP project selection process was guided by the following criteria:

- Projects that are at an advanced stage of preparation and that can be fast-tracked
- Projects that support both a regional approach to infrastructure provision and regional integration
- Projects that have stalled for various reasons and where NEPAD's intervention could be expected to make a difference
- Initiatives that offer solutions to regional policy, regulatory or institutional constraints.

19.2.4 NEPAD Energy Flagship Projects

- Regional Pools and Inter-connections
- Greater Inga Integrator Study
- Establishing Regional Linkages for African Energy Commission
- West African Gas Pipeline.

Projects under Implementation

- Electricity Master Plans for sub-regional interconnections in West, East (EAPP) and Central Africa (CAPP)
- West African Power Pool (WAPP)
- Southern Africa Power Pool (SAPP)

- WESTCOR Power inter-connect (DRC - Angola - Namibia - South Africa)
- Nigerian Benin Electricity Interconnection
- Mozambique-South Africa Gas Pipeline
- Electricity Networks Interconnection in Central Africa Study
- Kenya-Burundi-Rwanda-DRC Electricity Interconnection Study
- Morocco-Spain Electricity Interconnection
- Eastern Nile Power Trade Program Study
- Zambia-Tanzania-Kenya Electricity Interconnection
- Ethiopia-Djibouti Electricity Interconnection
- Benin-Togo-Ghana Electricity Interconnection
- Mozambique-Malawi Electricity Interconnection (Kenya-Uganda Oil Pipeline and West African Gas Pipeline).

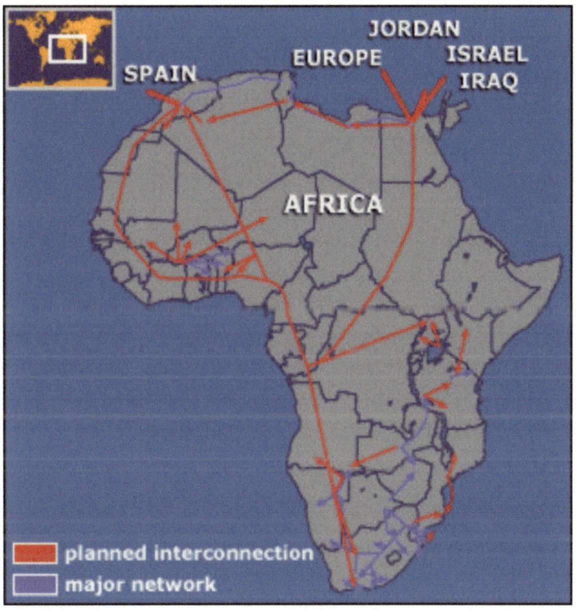

Figure 19.2. Planned Interconnections and Major Networks in Africa

19.3 The Future of SAPP, WAPP, CAPP, and EAPP with INGA

19.3.1 African Power Pools and the Centrality of INGA

Over the past decade there have been major initiatives taken by African governments to improve reliability and reduce costs by promoting the development of regional power pools.

The Southern African Development Community (SADC) created the SAPP in 1995 and the Economic Community of West African States (ECOWAS) created the WAPP in 2001 (see Chapter 10). Each of these power pools covers a very extensive area including 12 countries in the first instance and 14 in the latter (transmission lines being built (Figure 19.3).

Most recently the CAPP was created in early 2005 and there is currently discussion for developing an EAPP. These regional initiatives for improving trade among states all depend on new international HV lines being built.

Figure 19.3. Africa Regional Power Pools, CAPP, EAPP, SAPP, and WAPP

Power Pool	Total Existing Generation (MW)	Sub-Sahara Generation (Percentage)
CAPP	4,561	8%
EAPP	3,092	5%
SAPP	42,324	72%
WAPP	8,579	15%
Total	58,556	100%

Table 19.1. Sub-Sahara Regional MW Totals [2]

Africa's largest regional power pool is the SAPP with over 42GW of generation capacity (Table 19.1). Total electricity generating capacity of Sub-Sahara Africa is about 59GW (7% of U.S. total of 983GW). With Africa's much larger area and smaller generating capacity there is the question of whether such a large spread-out continental grid, involving expensive long transmission lines with large line losses, can be economically justified? There is an ever-growing interest, in spite of the economic challenges, to transmit the enormous hydropower potential of the River Congo to the north, south, east and west of the continent. The Purdue modeling team has built models for SAPP and WAPP. A preliminary CAPP model has now been built and a proposal prepared for modeling the East Africa region. These modeling initiatives will provide top level planners with quantitative economic assessments of the new regional interconnections, demonstrating the magnitude of the gains from joint construction and trade.

What are the most critical new lines required in each of these four regions of Africa and how can the experiences of the United States and other large interconnected networks assist in the planning of a network across Africa?

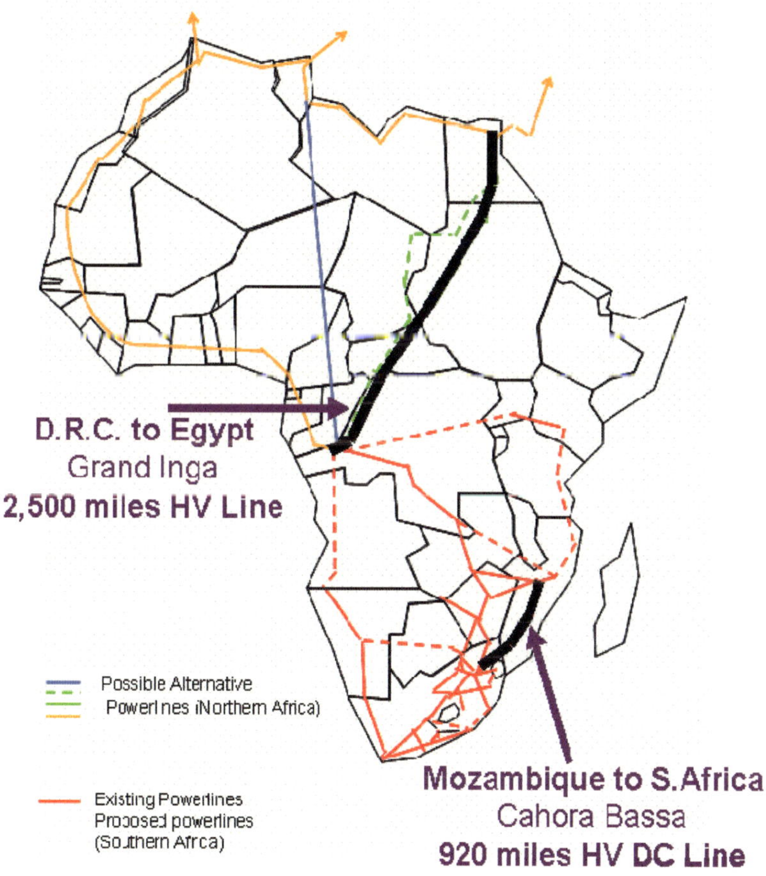

Figure 19.4. Long-Term Transmission Planning in Africa

The great hydropower potential of the River Congo, especially at Inga, can certainly play an important role in providing power regionally. Located at the heart of Africa (150km from Kinshasa) it is at the center of a future continent-wide power network (Figure 19.4). DRC-Inga currently exports and wheels power to SAPP countries including Zambia, Zimbabwe,

Botswana and South Africa. Power from Inga is transmitted to the Zambian grid along a 500-kV direct current (DC) line from Inga to Kolwezi in southern DRC, and a 220-kV line from Kolwezi to Kitwe in northern Zambia [3]. Viability of a second southern interconnection, from DRC to SAPP via Angola and Namibia, rests solely on expanding the generating capability of the Inga facility. Expansion of Inga 3 (3,500MW) coupled with the rehabilitation of Inga 1 and 2 can provide enough excess generating capacity that will justify the creation of an expanded regional electricity export scheme. The Western Energy Highway will connect DRC-Inga to Nigeria and WAPP, providing 1,000 MW of electricity. The fully implemented Grand Inga scheme will be the largest generating facility in Africa with 39,000 MW and feasibility studies indicate that it's interconnector to Egypt would be viable with the construction of the Northern Energy Highway, passing through Congo, the Central African Republic, and Sudan to Egypt, a distance of about 2,500 miles.

There are striking differences in the amounts of HV transmission lines in Africa and North America. Sub Sahara Africa is about 2.5 times the size of USA. The SADC has an almost equal area to that of USA. Its 5,710km of international HV lines together with South Africa's 25,180km of HV lines amounts to 12% of the HV lines in USA (Table 19.2). The high demand centers in Africa are mostly concentrated in the capital urban areas and are very widely dispersed making a marked difference with the much higher number of high demand centers in USA.

Region/Country	Surface Area (1000 km^2)	HV Transmission Line Length Above 110kV (km)
Sub-Sahara Africa	24,267	N/a
SADC	9,275	5,710
Rep. South Africa	1,221	25,181
Nigeria	924	11,000
USA.	9,629	248,648
Canada	9,971	N/a
Mexico	1,958	23,500

Table 19.2. HV Transmission Lines in Africa and America [3,4]

19.3.2 Electricity Trading Between the North American Interconnects

Long distance electricity shipments in USA were originally reserved for unexpected outages in generation. An interesting exception to this comes from Canada's net flows of hydropower exports to New England and the west coast states. Net power flows between the three U.S. interconnections tends to be very limited. Canada's exports account for 5% to 10% of its total generation. In the case of CAPP these numbers will become reversed, with domestic consumption taking 5% to 10% of Inga's total production, assuming new continent wide interconnections will be constructed.

In the 1990s, the wholesale trade of electricity in USA was promoted and the Federal Energy Regulatory Commission (FERC) established procedures to ensure the availability of non-discriminatory transmission access. It had been the formation of the North American Electric Reliability Council (NERC) in 1965 that ensured compliance with guidelines for providing overall reliability and system security. In Africa there is going to be need for a similar organization, as several countries will be involved with the proposed long HV lines.

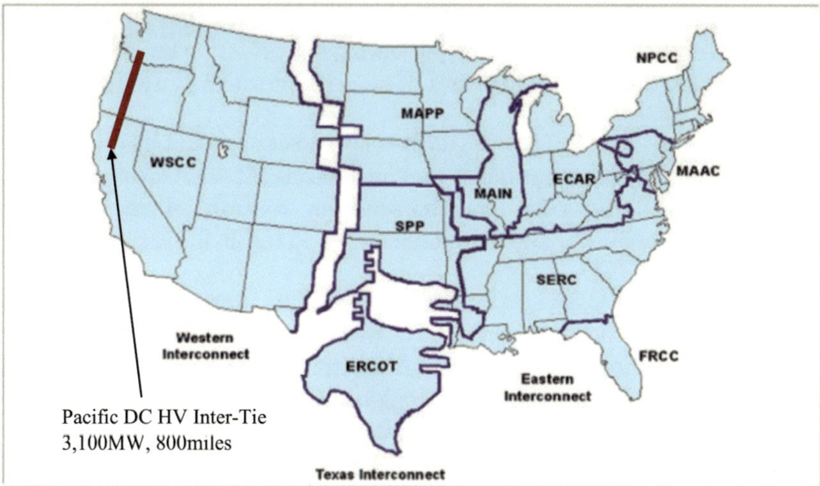

Figure 19.5. The Main Interconnections of the U.S. Electric Power Grid and the 10 North American Electric Reliability Council Regions

North America's three interconnected networks (Figure 19.5) are the Eastern Interconnect (the largest), Western Interconnect (second largest, west of the Rocky Mountain ranges) and the Texas Interconnect. There is very little load carrying capability between these three regions. Is it a technical problem, economics, or simply no demand exists at present? Each regional grid operates as a single large utility with a common set of operating rules. The Texas System is not interconnected with the other two networks (except by certain direct current lines). The other two networks have limited interconnections to each other. Both the Western and the Texas Interconnect are linked with different parts of Mexico. The Eastern and Western Interconnects are completely integrated with most of Canada or have links to the Quebec Province power grid. Virtually all U.S. utilities are interconnected with at least one other utility by these three major grids. Mexico has a national interconnected grid with four regional divisions and about 23,500 miles of HV lines. It connects with the U.S. at several points over the border and in 2003 imported about 72GWh and exported 953GWh.

Voltage	1990	1999	Change
230kV	70,511	76,762	6,251
345kV	47,948	49,250	1,302
500kV	23,958	26,038	2,080
765kV	2,428	2,453	25
Total	**144,845**	**154,5033**	**9,658**

Source: EIA, Electricity Transmission Fact Sheet [5]

Table 19.3. US High Voltage AC Transmission Mileage - Selected Years

In planning Africa's new HV lines, control of the lines is to be an important issue. The FERC expects new regional transmission organizations (RTO) to improve power grid reliability while reducing discriminatory transmission practices, and increasing investments in the transmission infrastructure. The issue of exactly who will control transmission of electricity under a nationwide system of RTOs needs resolving [6]. During this debate, in the 1990s, over 9,500 miles of new HV transmission lines were built in the U.S. giving approximately a 7% increase (Table 19.3).

The early 21st century has seen less new HV lines being constructed and this is becoming of great national concern especially for the summer peaking seasons. At what level of administration is Africa to debate the construction and transmission controls of the inter-power pool interconnections? To date the HV lines have been limited to within the regional power pools.

Most electricity trade in the U.S. takes place not between the three interconnected systems but among the power pools in each interconnect. The main exception to this, as already noted, is in the case of Canada exporting its hydropower. Major transfers of more than 3,900MW of peak demand takes place between the two NERC regions ECAR and MAAC, for example. Typical tie line transactions between U.S. power pools can vary between about 30 MW and over 3,000 MW (Table 19.4) but the lines are shorter than those being proposed for Africa.

In the U.S. a 765kV line might carry 3.8 GW but it will only be 100 miles long (Table 19.5). Extra long lines as being considered in Africa will need further technical study and will be much more expensive. In the U.S. the transfer of 3,000MW over several hundred mile and more will normally involve several lines. Transmission lines that are 1,000 miles long or more, similar to the Mozambique to South Africa DC line, are special designs for which the capital costing and operating costs require extra evaluation.

Exporting electricity from Mozambique's Hydro Cahora Bassa (HCB) to South Africa, and Canada's hydropower generation to the U.S. provides significant revenues to the exporting countries. In the case of DRC the export revenues could become substantial from building Grand Inga (Stages 1 and 2) with initial exports of 8,000MW (56,000 GWH/year). This could raise annual export revenues of $1.5 Billion or more once the full demand is being supplied.

Interface	Peak Demand (MW)
NEPOOL to NYPP	27
NYPP to NEPOOL	888
Net, NYPP to NEPOOL	861
NYPP to MAAC	1,261
MAAC to NYPP	1,684
Net, MAAC to NYPP	422
MAAC to ECAR	969
ECAR to MAAC	3,908
Net, ECAR to MAAC	2,939
Total Gross Transactions (Four NERC Regions)	**8,737**

Source: Office of Integrated Analysis and Forecasting, PowerWorld® model runs [7]

Table 19.4. US. Electric Transmission Network - A Multi-Region Analysis Interregional Gross and Net Tie Line Transactions

Voltage (kV)	Length (miles)	Maximum Capacity (GW)
765	100	3.8
	400	2.0
500	100	1.3
	400	0.6
230	100	0.2
	400	0.1

Table 19.5. Capacity Limits for Electrical Transmission Lines [8]

In the case of Canada it is one of the world's largest producers of hydroelectricity, generating over 315,500 GWh (2002). Very similar to DRC it is estimated that Canada has 180 GW of hydroelectricity potential remaining, although only 34 GW is currently deemed economically feasible. The economic analogy of building more hydropower in Canada with the DRC's Inga might help planners in Africa. Export potential for sending power to the U.S. from Canada has the attraction of further massive energy revenues but the capital-intensive nature of new hydro capacity could overwhelm benefits from trading. This is an issue that confronts the Inga project. Correctly pricing Inga's electricity exports is going to be essential for the successful launching of the project as it looks towards providing mutual benefits to consumers in Africa's power pools as well as to DRC.

19.3.3 The Preliminary CAPP Model

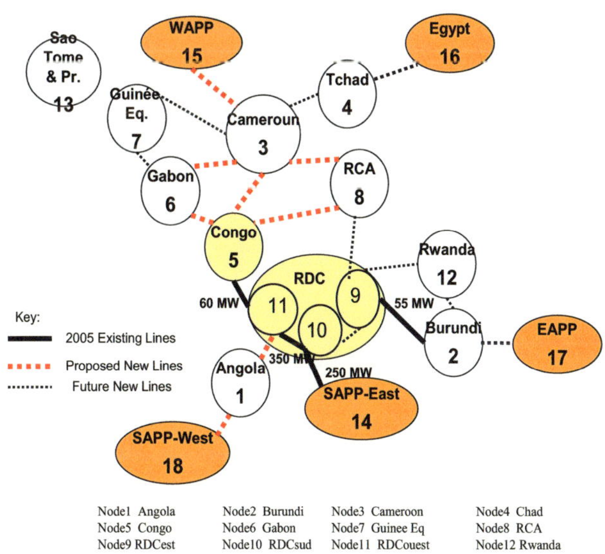

Note: RDC is French for DRC, similarly Tchad for Chad

Figure 19.6. The Preliminary CAPP Model - With 18 Nodes Including 5 Export Nodes [9]

Recently, Purdue's Power Pool Development Group's (PPDG) long-term planning software has been utilized to explore the economic gains that could be expected from the future development of the CAPP with its' 10 connected countries as indicated in Figure 19.6.

As Figures 19.3 and 19.6 indicate, the central location of CAPP allows it to consider exports to each of the two major Power Pools already in existence, SAPP and WAPP, as well as possible sales to Egypt and EAPP. These export opportunities, along with the well documented advantages of common operation and expansion of the grid within the 10 country regions, should make the establishment of CAPP a top priority for any Pan-African electricity generation planning project.

The model simultaneously cost minimizes expansions in both the generation and transmission sectors [10,11]. The water cost was set at $0.5/MWh which was the value stipulated by the SAPP some years earlier [12]. For demonstration purposes initial export demands were set at 1,000 MW each for SAPP west, WAPP, and EAPP, 250 MW for SAPP east, and 4,000 MW for Egypt. A general growth rate was assumed of 5% for CAPP as well as at the export nodes of SAPP, WAPP, Egypt and EAPP.

The 18-node model provides an optimal planning strategy for new lines emanating from Inga (node 11, DRC west, Figure 19.6). It is a 20-year long-term capacity expansion and electricity trade model as developed over the past several years for the SAPP and WAPP [13]. Unserved energy costs are set at $140/MWh and unmet MW at $3M/MW. The unserved energy and unmet MW costs could be argued for being raised but these values have been used in SAPP and WAPP and were therefore employed in the preliminary CAPP model.

While the CAPP modeling report [9] is still a work in progress, the model predicts the need for major transmission construction projects to serve the need for power flows within CAPP, and even larger investments in HV lines to allow power flows from the Inga sites to the five export markets shown in Figure 19.6. As the demand from Egypt, SAPP, WAPP, and EAPP increase, as well as demand within CAPP, then a portion of the larger expansion capacity envisioned at Grand Inga appears to be justified.

However, the CAPP data still needs careful compilation and validation, a task planned for the next phase of the project.

19.3.4 Investment and Electricity Pricing Issues

The determination of the electricity demand growth rates, demand forecast figures, and electricity prices are critically important in the planning process for new capacity. Improved forecast training in many countries of Africa, with more detailed data collection, will improve the determination of such critical numbers. The less industrialized nations frequently have problems with inadequate power supplies. These are reflected in the growth rates data as "*hoped for rates*" and do not provide satisfactory input data for planners.

The problem with all the plans to utilize the enormous hydro power potential of the Congo lies in, unlike distributed generation projects having short construction times and small construction costs, centralized hydro projects require very large initial investments in dams and the transmission lines long before any project revenues are generated. The demand growth numbers for projects like Inga have significant affects.

A realistic model of constrained growth will improve the forecasting technique (Figure 19.7). The demand numbers significantly affect the attraction of suitable investments for the two Inga projects (Inga 3 and Grand Inga Stage 1) being modeled. The growth rates of 5% and more are often considered as reasonable but looking at the historic numbers for the instances of Egypt, Nigeria and South Africa this is higher than what has been happening.

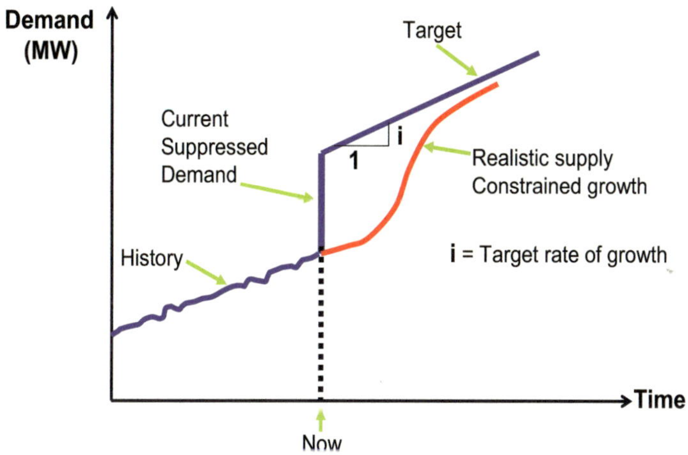

Figure 19.7. Electricity Growth Rate and Suppressed Demand

The average historic electricity demand growth rates for the largest national utilities in Africa over the past 10 years or more has been in the order of about 2%. This rate has been considered as a "*low case*" expansion scenario in the SAPP and WAPP. The numbers in Table 19.6 show the historic and average growth rates for these three countries.

Consider an illustration of the magnitude of the problem with having low demand growths. The two Inga hydro projects, the 3500MW Inga 3, and the 4000MW Grand Inga Phase 1 project - which are the driving forces behind much of the power pool activity in Africa - have estimated capital costs of roughly $4 Billion each. To this must be added the estimated transmission costs of $8.7 Billion to hook up the Inga sites to the export markets within SAPP ($1 Billion estimation at $1M/MW), WAPP ($1 Billion estimation), EAPP ($1 billion), and Egypt ($5.7 Billion). Thus the total upfront investment costs of the two Inga projects are in excess of $16.7 Billion.

Assuming a capital cost 10% and a project lifetime of 40 years, a range of $2.12 to $1.67 Billion dollars a year in returns to the investors must be assured for the projects to be financially viable. Further, all these export markets, each a functioning or planned power pool in itself, have local base load combined cycle generation construction options whose capital and operating costs are in the range of $30 to $40 per MWh (gas price range of $2.00 to $3.00 per MBtu), depending on the price of natural gas in these regions. These gas prices are reasonable estimates of current gas prices in Africa. If opportunities for LNG exports develop then these prices could increase. These domestic regional options will determine the maximum price these markets would be willing to pay for hydro electricity imported from the Inga projects. A further complication is that many of these regions already have capacity expansion projects on-going to satisfy near term needs for new capacity.

If we make the optimistic (for Inga) assumption that all projected growth in demand beyond 2005 in the four regions would be met by Inga power, as long as the price does not exceed the $/MWh range indicated above, we have the basic structure of a procedure to determine if the Inga projects make economic sense.

Billion kWh	1992	1993	1994	1995	1996	1997	1998
Egypt	40.45	44.41	46.56	48.44	48.13	51.65	55.6
Nigeria	13.15	12.84	13.74	12.92	13.36	13.47	13.5
S.Africa	144.6	149.37	156.2	160.89	168.3	175.5	175

Growth Rates	1993	1994	1995	1996	1997	1998
Egypt	9.80%	4.80%	4.00%	-0.60%	7.30%	7.70%
Nigeria	-2.40%	7.00%	-6.00%	3.40%	2.30%	-1.10%
S.Africa	3.30%	4.60%	3.00%	4.60%	4.30%	0.10%

Billion kWh	1999	2000	2001	2002
Egypt	60.59	66.86	72.93	75.58
Nigeria	13.63	13.9	16.13	18.43
S.Africa	178.14	183.76	185.9	189.36

Growth Rates	1999	2000	2001	2002	Total	Average
Egypt	8.90%	10.30%	9.10%	3.60%	25.90%	2.6
Nigeria	2.40%	-5.20%	23.00%	14.30%	18.40%	1.8
S.Africa	3.2%	1.2%	1.0%	9.0%	14.4%	1.4

Table 19.6. World Total Net Electricity Consumption and Demand Growth Rates for 1993-2002 [14]

Figure 19.8 shows the yearly net revenue stream available to the investors in the Inga projects assuming a range of demand growth rates from 2% to 4% in the four markets, using the base electricity consumption in 2005. The revenue stream, obtained by extrapolating the kWh figures in Table 19.5, is what remains as a return for investors, after having subtracted from the revenue estimates hydro operating costs of $2/MWh, and assuming no line loss. Also shown in Figure 19.8 are the annual required returns to the investors, assuming two alternative lifetimes for the Inga projects of 20 years, and 40 years, and capital cost of 10%.

Figure 19.8 also shows the most optimistic assumption with a 4% growth rate in demand being well in excess of historical rates as shown in Table 19.5. This results in the project yearly cash flows not covering the yearly-required returns until year 19, while the pessimistic assumption with a 4% growth rate results in the annual revenue stream equaling the required annual return only after 25 years have passed. Note that if the growth rate is 2%, the revenue stream never generates the required annual revenue stream during the lifetime of the Inga projects. Does this mean that the Inga projects should be abandoned? Not at all, but it simply means that much more analysis must be undertaken before any investor group will look seriously at Inga as a viable investment option with these export assumptions.

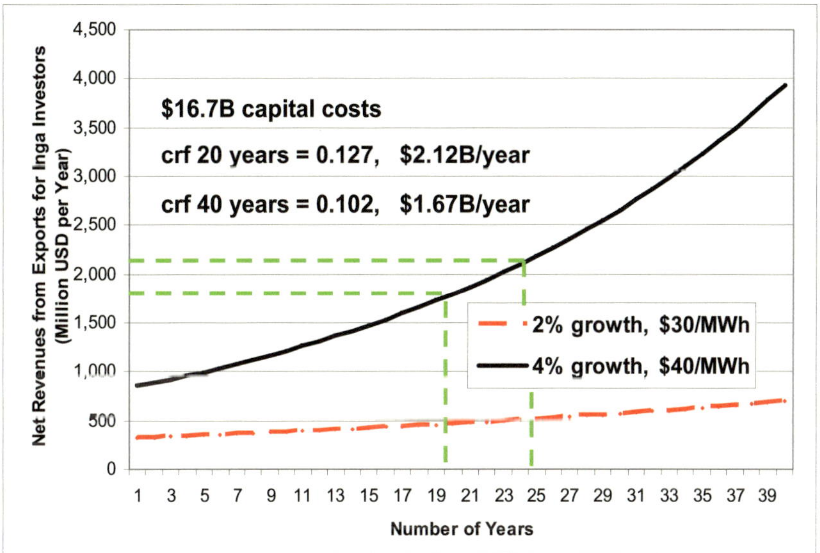

At $30/MWh & 2% demand growth the annual cost of capital does not break even with revenue from exports within a 20 year or 40 year planning horizon. At $40/MWh and 4% demand growth the annual cost of capital breaks even at 18 years (40 year horizon) and 24 years (20 year horizon)

Figure 19.8. Net Revenues from Exports for Inga Investors With 2% and 4% Demand Growth (USD)

Comparative assessment to similar sized projects can always help if it were possible to obtain the growth and cost data involved. Certainly Mozambique's exports to South Africa are more appropriate for the Inga project than say Canadian hydropower to USA. The level of risk in North America is less and the cost of borrowing capital therefore reduced. High electricity growth rates elsewhere in the world make a major difference and China comes to mind. The huge Three Gorges project can be justified with the 8% to 10% historic growth rate but can the much smaller African growth rates justify the construction of such large projects?

Perhaps it is Egypt and the Mediterranean region with its large and growing demand for electricity that is the additional market for an enlarged Inga. If this is the case then the expansion costs of the DRC to Egypt line together with Inga, and the electricity export prices appear to be the first two most important issues for consideration. Secondly, firm power contracts as well as wheeling rates will need to be agreed upon among all the players and stakeholders to secure adequate investments.

Without the Egyptian export gateway it is hard to justify the capacity expansions as growth rates as high as 4% or higher for many African countries are not taking place. The suppressed demand has to be remembered but still massive rural and urban electrification programs are required to take place to see the needed growth levels. These are some of the opportunities and challenges facing those energy planners promoting the substantial expansions for Inga and the inter-regional power grid of Africa.

In summary, the vision of a continent wide HV power grid across Africa with Inga at the heart of the network has inspired African electricity planners for many years. The concepts and documented benefits of integrated African power pools, as demonstrated by the studies done by Purdue's PPDG for the SAPP and WAPP, support the impetus towards implementing the Pan-African HV network plan. Central to a strong future continental network is the creation of an efficient CAPP because of its location and the potential of Inga.

While the results of current work by PPDG support the general economic feasibility of the vision, this section (Section 19.3) questions the approach taken by some supporters in their promotion of several very large projects, rather than a series of smaller ones. Both economic theory and industrial practice tell electricity planners that in situations, as in Africa, where capital costs are high and demand growth rates are low, it is best to forgo the scale economies present in constructing a few large projects, and choose instead to expand capacity slowly to allow the expansion in capacity to better match demand growth.

There might be enormous revenues and benefits from building Grand Inga and major new HV lines across Africa. The time has arrived for a combined in-depth analysis of the three broad development scenarios referred to: (a) building Grand Inga for power exports to the Mediterranean, (b) building Grand Inga as a power source for all Africa, and (c) planning for massive urban and rural electrification. Each scenario holds great potential but each one needs to be considered within the complementary inclusiveness of all three scenarios combined, if sustainable development is Africa's goal.

19.4 Security Considerations of Modern Power Pool System Interconnections [15]

The lessons brought on by the advent of deregulation and the resulting weaknesses that unfolded during the 2001 California Energy Crisis have introduced caution to planners in how best to adapt market forces to the requirements of power system operations. In evaluating the planned synchronous interconnection of the African regional power pools interconnection, security issues should be taken care of before any economic analysis is undertaken. Power system security studies have traditionally been used in industries only during planning. This Section will address future real-time security systems for operation of '*deregulated*' power pools with an emphasis on technical aspects, namely: voltage stability, transient stability, and dynamic stability that are three approaches to power system security studies.

	Transient Stability	Dynamic Stability
Synchronous Stability	Observe transient response (1~2 swings after disturbance) of generators	Observe dynamic response (1~6 seconds after disturbance) of generators
Voltage Stability	Observe transient response (1 or 2 swings after disturbance) of bus voltages	Observe dynamic response (1~6 seconds after disturbance) of bus voltages

Table 19.7. Categories of Power System Stability

Stability is categorized as synchronous stability, angular stability, and voltage stability depending on the quantities in which one is interested. Synchronous stability is focused on rotor angles or frequencies of generators and voltage stability is focused on bus voltages. Depending on the time-period one is observing, stability can be categorized as transient stability and dynamic stability. The observed time-period for transient stability is about a couple of swings after the disturbance. That of dynamic stability is normally 1 to 6 seconds after the disturbance. However, if necessary, the simulation time can be longer than 10 seconds. Table 19.7 summarizes different categories of stability.

19.4.1 Transient Stability

For transient stability, one is interested in the system ability to survive a large disturbance, such as a fault, or a sudden change in generation, load, or system configuration, without prolonged loss of synchronism. In this part, rotor angle behavior of generators after the disturbance is looked at. The system is said to be stable in transient state if rotor angle separation between any two machines tends to keep in a certain range after the disturbance. On the other hand, the system is transient unstable if the rotor angles keep running away.

19.4.2 Dynamic Stability

Dynamic stability is the ability of a power system to return to its initial state or reach another steady state after a small disturbance. Dynamic stability of a system can be understood by looking at damping of the system oscillation, which can be triggered by either an unexpected disturbance, or a regular operation of the power system. The oscillation should die out in several seconds after the disturbance for a strong system. If the oscillation is decaying slowly, then the system has bad dynamic characteristics. If the system has a lasting oscillation or even diverging oscillation, the system is dynamically unstable. The poor damping of the oscillation signals a narrow stability margin of the power system. With this small margin, the system operators must be very careful when they are executing system operations because the operation criteria might not apply. The same operation might drive the system to an unstable area though it was good previously.

A power system may be stable during the transient period but unstable during the dynamic period.

One of the phenomena of dynamic stability problems is the oscillation. The frequency of the oscillation is between 0.2 and 1.0 Hz. Thus, this oscillation is called low frequency oscillation since its frequency is low compared to system frequency.

System oscillation is often found in an interconnected system. Power oscillation begins when the power flow on the tie line linking two areas in the system increases to some level. The oscillation disappears if the power flow is reduced. Therefore, the transmission capacities of tie lines are subject to a stability limit and, unfortunately, the stability limit is normally lower than thermal limit.

System dynamic characteristics are changing not only with system expansion, but also with system operation. The power system might have spontaneous low frequency oscillation without any known specific disturbance if the system is operating near the dynamic stability margin. Actually, there must be some but not special disturbance. Because the system is near the margin, this disturbance happens to drive the system to an unstable area and triggers spontaneous oscillation. Once the disturbance disappears, the oscillation will be decaying slowly.

Figure 19.9 is an example of a spontaneous low frequency oscillation observed on April 1991 in a power system. The oscillation frequency is about 0.8 Hz. Figure 19.9(a) is the system frequency oscillation and Figure 19.9(b) is the real power oscillation of a nuclear unit.

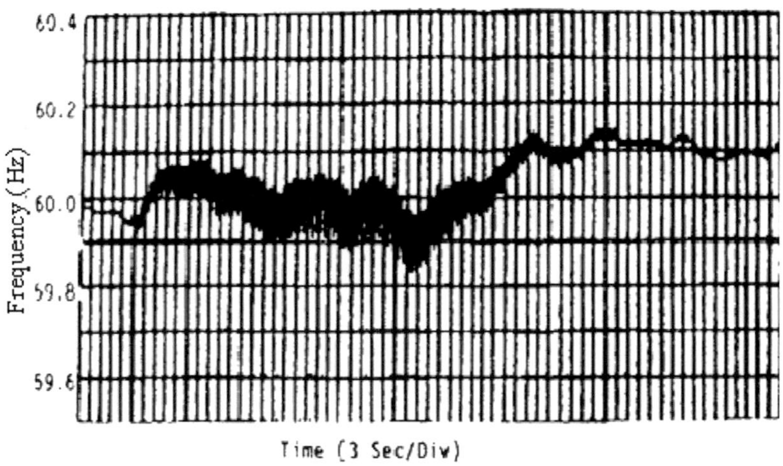

Figure 19.9(a). System Frequency Oscillation

Since the power system is always subject to small disturbances, dynamic stability is essential for system operation, particularly if the system is operated at a tight dynamic stability margin. This is important for the countries of the Sahel region now contemplating Nuclear power as an option to address the looming water crisis through desalination systems. The above example is an important consideration and one to be analyzed as these countries will also be interconnected to the emerging West African Power Pool (WAPP) and the North African Power Pool.

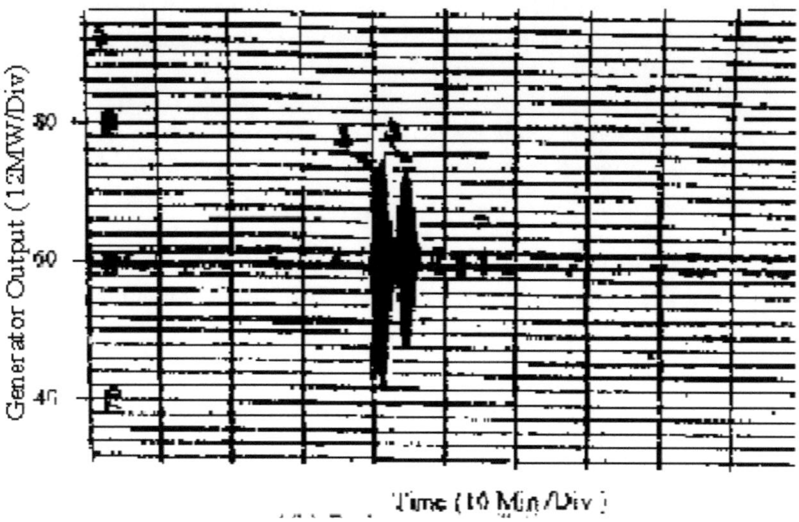

Figure 19.9(b). Real Power Oscillation

19.4.3 Voltage Stability

The power system should be operated securely not only in usual conditions, but also when there exist disturbances. The power system is stable when the system is able to restore to its initial condition or reach another steady state, which is acceptable in terms of operational standards after experiencing a disturbance. For voltage stability, the system should be able to maintain the magnitudes of bus voltages when experiencing the disturbance.

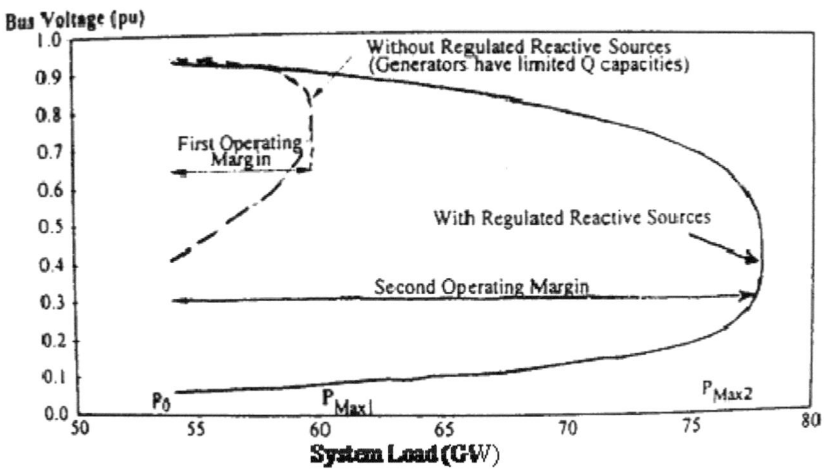

Figure 19.10. PV Curve of a Power System

Power failures caused by voltage instability or so called voltage collapse that lead to power system blackouts have been reported in many countries, such as Sweden, France, Japan and USA (including the August-2003 blackout). The affected area, because of voltage instability, can be metropolitan or nationwide. The effects are not only in power interruption to customers, but also in mass transportation and industrial manufacturing. The losses to the power utility are in both finance and customer service.

Increasing load and deficient reactive power supply generally causes voltage collapse. It is a combinatorial problem affected by both system and load characteristics. The power-voltage characteristics of a power system are represented by a P-V curve (Figure 19.10).

It is clear that the degree of stability cannot be judged based on how close the bus voltage is to the normal level. The power industry is basically just using the magnitude of bus voltage as a measurement of voltage stability.

19.4.4 Today's Congestion Management

- Only check the thermal limit
- Real-time limit is not there
- No voltage stability limit
- No transient stability limit
- No dynamic stability limit
- The responsibility of system congestion due to single contingency is not clear.

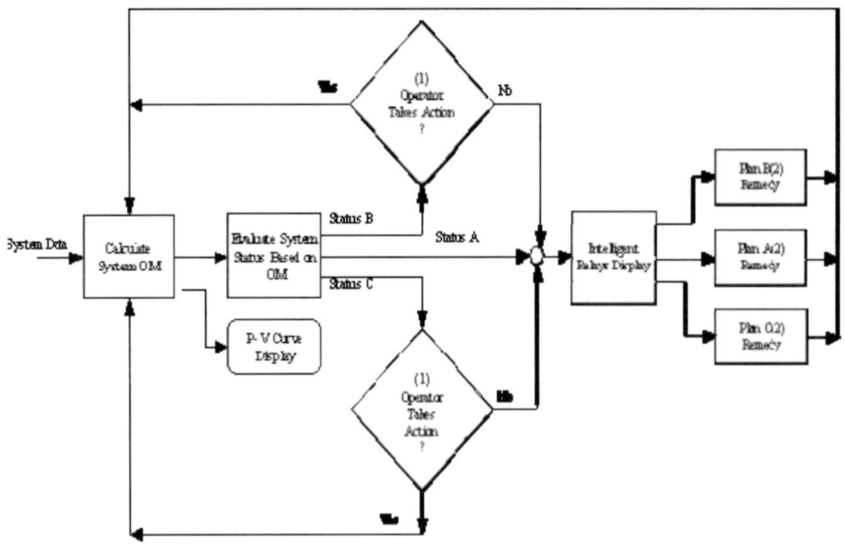

Note: (1) Preventive actions include Re-dispatch, Shedding interruptible loads, Shedding weak bus loads, etc.
(2) Plans A, B, C remedies

Figure 19.11 Conceptual Design of an Intelligent Control-Room Operating System

If the detailed story of the August 2003 US. blackout is examined; the security issues of power system must be handled using a different approach. As a result of market driven dynamics introduced under deregulation in today's control room, the power system operator may not know the system margin. All utility operation control systems should provide the following:

- Real-time data of system margin
- Tools to ensure the Operator is informed and involved
- System identifies weak bus(es) of system
- Dynamic load-shedding (if necessary)
- An intelligent system should be installed immediately at all ISO facilities.

A conceptual design of an intelligent control room operating system is illustrated in Figure 19.11.

19.5 Area Control Considerations for the Emerging WAPP [16]

The full implementation of the WAPP project will be staggered over four phases over a period of more than 20 years. Each phase comprises an institutional development component and an infrastructure component.

The ECOWAS countries have been divided into two main zones: Zone A includes Benin, Burkina Faso, Côte d'Ivoire, Ghana, Niger, Nigeria and Togo, and Zone B includes Cape Verde, the Gambia, Guinea, Guinea Bissau, Liberia, Mali, Senegal and Sierra Leone (Figure 19.12).

Phase 1 covers most of the Zone A countries (except Niger, Nigeria and Togo) and Mali from Zone B

Benin, Burkina Faso, Mali, and Togo have been identified as the main prospective importers of electricity due to high generation costs, while Côte d'Ivoire and Ghana have been identified as the main prospective exporters of electricity in the region during phase 1 implementation.

Note: WAPP Demonstration Results only

Figure 19.12. WAPP Zone A and B countries: WAPP MW reserve exports, period 2

19.5.1 Financial Issues

In general, synchronous interconnection must be accomplished through multiple large capacity transmission paths placed in service simultaneously. A thorough analysis of the optimal number of lines necessary to accomplish reliable interconnection depends upon the anticipated transfers over the lines and requires engineering and economic analyses. For those originally isolated systems, construction of new transmission facilities and improvement of existing transmission facilities would be necessary to provide the infrastructure to facilitate desired power transfers. Investments in transmission facilities have historically been funded by utilities. The facilities for interstate connection and the required infrastructure improvements may fall outside the traditional paradigm of transmission funding by utilities. The investment for the construction of the required facilities must have a reasonable expectation of recovering the associated costs from their customers or users of the facilities. The issue of providing the necessary economic incentives for construction of new transmission facilities in an environment where transmission owners must provide open access is common to synchronous interconnection investments. However, incentive for cost recovery and profit for investment may defeat the purpose of interconnection to provide cheap and clean energy in Africa.

Synchronous interconnection could impose additional operating cost on utilities and other owners of electric generating facilities. In order to maintain reliability, generators may have to adjust operations to accommodate those of utilities elsewhere on the interstate grid. The magnitude of these additional costs is difficult to quantify due to uncertainties over the operating characteristics of the interconnected grid. Any additional operating costs caused by synchronous interconnection raise two issues. First, the additional operating costs must be offset against estimates of gains from trade considered as benefits from synchronous interconnection. Second, there must be some mechanism for beneficiaries of power flows to compensate those entities that are forced to bear additional costs to accommodate those flows. Though initial evaluations suggest that any additional operating costs are probably not very large, there is considerable uncertainty and controversy over the significance of these costs and it would probably not be prudent to ignore them.

19.5.2 Technical Issues

Interconnection enhances the ability to import power when there is a shortage due to extreme weather or generator outages is a reliability benefit. However, interconnect AC network will increase the complexity of the system that is subject to various reliability, security, and stability problems due to the interactions among the increasingly prevalent automatic generator voltage and speed controls, system frequency, tie line flow, and critical bus voltages. The analysis of system dynamic performance and the assessment of power security margin have correspondingly become more complex. This may threaten reliability and lead to wide area power outages. The social and economic cost of power outages, especially extended outages over a wide geographic area can be significant, as was learned in the North America Northeast blackout in August 14, 2003. It took only nine (9) seconds for the blackout to spread across Canada and several states in the US, effecting more than 50 million people. Some went without power for more than three days. Understanding the behavior and fundamental characteristics of the system are critical for secure operation.

19.5.3 Cascade Failure and Protection Coordination

If the failure of equipment may trigger other events and cause other devices to trip out of service, the system is threatened by the possibility of cascading outages. As an example, the condition might be precipitated by a transmission line failure caused by a falling tree branch. In response to the outage, all remaining transmission line flows adjust to carry more loads. This may result in tripping another overload line and worsen the system situation that lead to system blackout. The interconnected system is more susceptible for this type of situation since the under-frequency protection may not function properly. These cascading overloads are a threat to secure system operation, and were the main reason for the spread of the Great Northeast Blackout in the 2003. Regular evaluate and update the protection scheme is necessary when expanding the interconnection networks.

Interconnecting these planned AC network and or HVDC networks will increase the complexity of the system that in turn will increase system reliability, security, and stability problems due to the interactions of equipment and control actions. Therefore the primary reliability threats in a transmission system of Voltage stability, Dynamic/Transient stability, and Cascading failure and protection coordination as discussed in 19.4 should remain predominant during the planning phases particularly for emerging economies with marginal system parameters.

19.6 Hydropower and African Grid Development: Rights Based Perspective

19.6.1 Hydropower and African Grids

African energy needs are indeed vast. Africa is home to 13% of the global population, but has the lowest energy consumption per capita of any continent. Most grid energy generation is in three countries: South Africa, Egypt, and Nigeria. Even then, a disproportionate amount of those using grid based energy live in urban areas. Vast disparities exist between grid energy available for commercial and non-commercial use. Grid based energy continues to be available overwhelmingly in urban areas, benefiting commercial use and those able to afford it. Concerns exist that financing grid-based development displaces resources available for energy development that could better promote poverty alleviation.

Hydropower is a significant source of existing and planned grid-based energy in Africa. Statistics are often used declaring that Africa's hydropower potential has gone virtually unexploited. While hydropower can generate significant electricity for grid systems and provide effective peak load power, hydropower projects are often proposed with overstated benefits and understated costs. Hydropower projects also have a history of poor implementation that has resulted in inequitable sharing of project costs and benefits. Beneficiaries of hydropower projects tend to live away from the hydropower site, and receive the grid based electricity, generally in urban areas or large towns. Those bearing the costs of hydropower projects may be directly displaced, have negative impacts to their livelihoods (such as fishing or agriculture), have increased health risks from water-borne disease, and face disruptions to social systems by temporary migration into the area during project construction. Those bearing the costs often do not benefit directly from the projects, or receive adequate compensation that recognizes all the social costs endured. Without genuine participation in the decision making process, communities often do not receive project benefits that outweigh their share of the costs.

Reservoirs of hydropower dams often displace thousands of people. The Kariba Dam shared by Zimbabwe and Zambia displaced 57,000 in the 1950s. These are people for whom adequate compensation has never been granted, and whose lives and livelihoods were expensed for this addition to grid development. Currently, the Merowe Dam is displacing 20,000 villagers in Sudan without receiving proper compensation. They have been denied participation and genuine access to the justice system. In the past 50 years, some 40-80 million people have been forcibly resettled for large dams, and millions more face such a fate as we speak.

There are many current proposed hydropower projects across Africa. The largest is the NEPAD-backed Grand Inga scheme, which would be the core of a continental grid system. Over-simplified statements are made that if only Inga could be developed, the whole continent would be lit up. There is little discussion occurring, however, about how to develop the demand in rural areas for this type of project. With fifty-two generating units, it would be the largest hydropower project worldwide. Including transmission, it would cost an estimated cost of $10 billion. Grid development like Grand Inga contradicts the goals of small-scale sustainable energy projects that were discussed at the World Summit on Sustainable Development in 2002.

In the SADC region, there are many other projects proposed or underway. Mphanda Nkuwa in Mozambique is another NEPAD backed project that would fulfill the country's effort to attract energy intensive business. Significant hydropower development, such as Tekeze and Gojeb, is occurring in Ethiopia with expectations to export power. Other significant projects include the 520 MW Capanda Dam in Angola, the Kafue Gorge Lower Dam in Zambia, and the 400MW Bui Dam in Ghana.

Current plans to develop the African grid system include the promotion of regional transmission lines in order to develop power pools and numerous large-scale energy projects that will feed specifically into grid systems. The grid system, as currently planned, primarily benefits industry and wealthy communities in urban areas. There is virtually no benefit to rural areas, or the urban poor. Local industry and small business generally do not benefit from grid development to the extent that major commercial and industrial customers do. These large businesses, often foreign-owned, benefit from increased power generation, but often wield enough power to receive electricity at rates providing little profit margin for the government, if at all. In some cases, major end users pay rates subsidized by residential customers.

Power grids are not designed to reach the hundreds of millions of Africa's rural poor. Grid systems can create a greater divide between those with and without access, generally increasing the disparity between rural and urban areas. Mass grid development may even encourage greater urbanization, causing cities to develop at increased rates, leading to other negative economic impacts that cities must then address (such as increased water, sanitation and other infrastructure needs, increased crime, and increased spread of HIV and other diseases). Many Africans live outside of the formal economy, living on subsistence and small enterprises that are often overlooked by development planners and policy makers. Designers of grid systems must be acutely aware of consumer demand and affordability. Whether in urban areas or in more distant communities, grid connection does not alleviate poverty for those unable to afford the electricity.

The New Partnership for African Development (NEPAD) has been described as a blueprint for Africa's self-determined economic propulsion out of poverty and toward sustainable development. NEPAD recognizes that half the Africa population lives on less than $1 per day, and that infrastructure is desperately needed to improve people's lives. However, NEPAD continues to be a top-down entity made up primarily of African elites with only token input from civil society. In a rush to promote foreign investment, economic growth, and NEPAD's political success, the body is virtually blind to the fact that its activities are in direct contradiction to its mission of African sustainable development.

Regional economic development planning and power pools are both gaining ground in Africa. More and more countries are receiving World Bank advisement to privatize their energy systems and promote competitive markets. Significant manipulation of market circumstances happens by those with the greatest market power: suppliers and major end-users. Civil society is rarely, if ever, in a position to benefit from the liberalized market. Where there is supposed to be greater consumer power, industrial consumers wield the most power, often to the detriment of residential and small business users.

World Bank and IMF loans are regularly conditioned to include privatization of government enterprises and promotion of free market systems, including liberalizing capital markets, promoting market-based pricing and free trade. Unfortunately, these measures only move political economic powers from government bodies and politicians to private, often foreign, companies. None of these changes provides increased political economic power to civil society.

19.6.2 Using a Rights-Based Approach

Utilizing a rights-based approach can bring more effective, more sustainable, more rational and more genuine development decisions. The inclusion of civil society in decision-making promotes transparency, which will likely decrease corruption. It will ensure that poverty alleviation happens, rather than poverty displacement, or even poverty generation. It will ensure appropriate solutions are found that fit the problems at hand because project analysis will be more complete. Most importantly, local participation and ownership of decisions helps safeguard against harm done by development projects, and will promote the sustainability of solutions found.

A rights-based approach allows for a positive transformation of power relations between various stakeholders involved in decision-making. There are four primary criteria to a rights based approach. First, it must include a linkage to human rights and accountability. Second, it includes equity of benefits and costs allocation. Third, it also includes empowerment and public participation, with attention to marginalized groups. And finally, it includes a transparent process. A primary concern over development projects in Africa is the external control of projects that affect internal peoples. Those within Africa need to be given decision-making control in their own development.

19.7 Targets and Technologies for African Electrification

19.7.1 Global Energy System Vision

Over the next 50 years, universal access to at least a minimum level of electricity and related services can contribute to dramatic improvements in the quality of life (education, economic justice, public health and safety, and environmental sustainability for the world's under-served populations). In 2000 the United Nations General Assembly adopted a comprehensive set of "*Millennium Development Goals*" to help create a more coherent worldwide focus on the truly pressing tasks for the coming fifteen years [18]. Global electrification can greatly assist the effort to achieve those UN goals, such as halving the incidence of extreme poverty or reducing the waste of material resources.

The World Summit on Sustainable Development held in Johannesburg reaffirmed those goals and gave particular attention to the need for assuring a greater supply of modern energy services, notably electricity, electricity, to the entire world's population [19]. This report affirms and adopts that goal. For the benefits we envision, electricity will have to meet reasonable standards of quality and reliability be available for commercial, industrial and residential uses, be affordable, and cause minimal environmental impact. A diverse portfolio of generation options will be required, including advanced clean fossil, renewable, hydroelectric, and nuclear power sources, plus high-efficiency end-use technologies and applications to support both environmental and economic sustainability. Our vision for the 2050 global energy system is therefore one of worldwide new capabilities and opportunities for quality of life, dignity, and environmental sustainability, enabled by universally available electricity.

What is needed is a global vision for realizing electricity's essential value to 21st century society, a plan to set strategic technological priorities, and an outline of the associated research, development, and delivery requirements needed to achieve this vision. In this context, EPRI's Electricity Technology Roadmap outlines a vision for the future based on broad stakeholder input to spur debate, consensus, leadership, and investment that will enable electricity to continue to fulfill its potential for improving quality of life on a global scale. The initial version of the Roadmap, released in 1999, describes a series of destinations for the power system of the 21st century [20]. A companion volume that supplements the initial report is now available [21]. This report expands the original by identifying three comprehensive high-priority goals that are most essential to assuring global economic and environmental health. They are:

- **Smart power** – the design, development, and deployment of the smart power system of the future
- **Clean power** – the accelerated development of a portfolio of clean energy technologies to address climate change
- **Power for all** – the development of policies and tools to ensure universal global electrification by 2050.

These characteristics reinforce the Roadmap's original destinations and provide a basis for a new planned initiative to include a series of detailed recommendations for technology development.

19.7.2 Improving Efficiency of the Energy Supply Chain

As societies strive to improve access to modern energy services, they must also find ways to make the energy system more efficient. The efficiency of the full energy supply chain (extraction, conversion, delivery, and consumption) has only reached about 5%; therefore, large opportunities for improving efficiency remain at every stage in this chain. For example, using today's energy sources and technology, achieving universal supply of at least 210 Mega Joules per day per capita by 2050 would approximately triple the current global rate of energy consumption. Fortunately, realizing technological advancements that are now visible throughout the energy supply chain could reduce the 210 Mega Joules per day threshold by 2050 to as little as 125 Mega Joules per day with no loss in economic productivity or quality of life potential. The efficiency of electricity generation, for example, now typically in the 30% range, could easily reach, on average, 50–60% by 2050, based on modest technology improvements over current practice. Even greater performance is possible if step function technology advances occur, as seems likely. For example, the emergence of low wattage lighting and appliances aimed at the developing world suggests rapid technological progress in household energy efficiency. Even the automobile is on the threshold of tran formative change.

19.7.3 Electrifying the World

As a practical matter, electricity must form the backbone for the transition to a globally sustainable energy system and the modernization process it enables. Electricity's ability to transform the broad array of raw energy and other natural resources efficiently and precisely into useful goods and services, irrespective of scale, distinguishes it from all other energy forms. Electricity also serves as the unique energy prime mover enabling technical innovation and productivity growth—the lifeblood of a modern society. One need look no further than rural North America in the 1920s and 1930s – regions that were transformed from economic backwaters through active rural electrification programs – to see the importance of electrification as the precursor to economic opportunity and well-being. Further, as electricity's share of "*final energy*" in USA. increased from 7% in 1950 to nearly 20% today, the energy required per unit of GDP dropped by one third. Such important achievements, which occurred throughout the industrialized world, remain elusive in the least developed world regions. Over the last 25 years, about 1.3 billion people have been connected to electric service, but even this achievement has not kept pace with global population growth. Today, the International Energy Agency estimates that 1.6 billion people lack access to electricity. To keep pace with the world's growing population, electrification must reach at least an additional 100 million people per year for at least the next 50 years. This is about twice the current rate of global electrification.

A roadmap for destinations is indicated in Table 19.8.

Destination	Summary
Strengthening the Power Delivery Infrastructure	An advanced electricity delivery system that provides additional transmission and distribution capacity and *"smarter"* controls that support dynamic market activity and the rapid recovery from cascading outages, natural disasters, and potential terrorist attacks
Enabling the Digital Society	A next-generation power system that delivers the power quality and reliability necessary for sophisticated digital devices and seamlessly integrates electricity systems with communications systems to produce the *"energy web"* of the 21st century
Enhancing Productivity and Prosperity	New and far-reaching applications of the energy web that increase productivity growth rates across all sectors of the economy
Resolving the Energy/ Environment Conflict	Clean, cost-effective power generation technologies combined with workable CO_2 capture, transport, and storage options
Managing the Global Sustainability Challenge	Universal access to affordable electricity combined with environmentally sound power generation, transmission, and delivery options

Table 19.8. Roadmap Destinations

19.7.4 Setting Electrification Goals

Equally important as universal access to electricity is assuring adequate levels of electric service for those who have access. Our work suggests 1,000 kWh per person per year as a benchmark goal for minimum electric services—an essential milestone in the pathway out of poverty. This target is similar to the electric consumption in emerging modern societies that use a mix of fuels (some directly, others via electricity carrier) to satisfy their needs. It lies between very low levels of electrification (100 kWh per person per year) insufficient for measurable economic benefits and the 10,000+ kWh per person per year of the current US economy. Achieving this target can help meet personal needs for basic lighting, communication, entertainment, water, and refrigeration, as well as provide electricity for the efficient local production of agriculture and goods and services.

In choosing the 1,000 kWh per capita per year goal, we are mindful that improved energy efficiency and complementary innovations would allow delivery of basic energy services using less electricity. Nonetheless, the benchmark reveals that, under current trends, perhaps 90% of the world's population in the next 50 years will be born into conditions that fall short of the 1,000 kWh goal. Based on country averages, about 3.7 billion people today live in countries where the average per capita consumption of electric power is below the 1,000 kWh threshold. Over the next 50 years, it is likely that another 3 billion people will be added in these electricity-deficient areas.

Table 19.9 below presents anticipated trends in energy and economic statistics over the next 50 years for Africa and other parts of the globe. Actual data for the year 2000 are presented along with two projections, one representing a "*business as usual*" scenario and the other a world driven by sustained efforts to use electricity as the engine of economic growth in Africa and around the world. These data are derived from the US DOE Energy Information Agency International Energy Outlook for 2004[22], from a World Energy Council study of energy futures [23], and from other sources. Africa trails all other regions in economic growth, in energy and electricity growth, and in carbon emissions. Moreover, Africa attains the target of 1,000 kWh per person only in the electrified case. The extreme poverty of much of Africa is a key factor in limiting the pace of electrification, but the failure of reforms and other political issues also play a role.

Providing power to a global population in 2050 of 9 billion—including minimum levels of 1,000 kWh per person per year to the very poorest people—will require roughly 10,000 GW of aggregate global generating capacity, or three times the current level, based on today's technology. That corresponds with at least a 3% annual rate of increase in global electricity supply. Even with major efficiency gains in the generation and use of electricity, the aggregate global requirements for electricity generation will still be prodigious. Therefore, a critical priority is the development and deployment of an advanced portfolio of clean, affordable, generating technology options—fossil, nuclear, and renewables—that reflects the diverse resource, environmental, and economic realities of the world, while enhancing efficiency and productivity throughout the energy supply chain.

19.7.5 Crucial Issues in Global Electrification

Global Electrification Prospects in Africa are summarized in Table 19.9. To build the necessary momentum toward global electrification, research initiatives must address the whole electricity supply chain—from market policies through generation, transmission and distribution. In some cases, technology development will be required, but first some improvements in basic understanding are essential to meeting global electrification goals. Studies are urgently needed to quantify the value proposition of electrification under a variety of policy and technology scenarios. This information will play an important role in helping policymakers develop incentives as well as regulatory and market frameworks that will encourage private sector investment in electricity infrastructure for underserved areas. Also necessary are analytic tools that can improve this understanding and lead to development strategies specific to individual regions, to accommodate the differences in resources, human needs and cultural norms. The availability of these and other analytical tools will help avoid the mistakes that have occurred in recent African electrification initiatives. This body of work is beyond the scope of this chapter, but significant problems in African electrification have arisen due to poor management practices, political corruption, counter-productive cross subsidies, ineffectual reform programs, among others [24,25].

These issues must be resolved to assure the success of electrification programs.

	GDP per capita (10^3 US $PPP per year)	Primary Energy per capita (10^6 J per day)	Electricity Consumption per capita (kWh per year)	Electricity (% of Final Energy)	Carbon Emissions (MTC/yr)
2000					
Sub-Saharan Africa	1.7	70	840	7	140
3rd World	2.4	70	1,550	7	900
Industrialized World	28.0	650	7,300	18	3,200
2050 Reference Case					
Sub-Saharan Africa	2.0	90	900	10	400
3rd World	3.5	110	1,900	11	2,700
Industrialized World	39.0	690	11,000	3	2,950
2050 Electrified Case					
Sub-Saharan Africa	4.0	120	1,460	31	350
3rd World	5.3	130	2,930	31	2,300
Industrialized World	39.0	460	16,100	48	1,420

Table 19.9. Global Electrification Prospects in Africa

19.7.6 Highest Priority Actions

The highest priority should be assigned to activities in two areas. First, additional research is needed on the *"value equation"*—the costs and benefits associated with universal electrification. This section proposes some global goals and strategies, but work is needed to understand the implications of those global goals for particular localities and regions and to outline specific strategies for achieving the goals. For example, the goal of 1000 kWh per person per year will vary with local conditions (e.g., heating requirements) as well as the potential for increasing efficiency and the competition between electricity and other energy carriers.

These questions require local and regional attention. Such analytical work must be done in a way that reflects appropriate local policies and the emerging new reality that electrification is increasingly funded with private capital and operated as a partnership between private firms and public institutions. In that emerging market, assessing the value equation requires attention to public values and policies as well as private incentives.

Second, work is needed on specific technologies that will be essential to meeting the goal of universal electrification. Improvements across a broad portfolio of generation and delivery systems will be needed. Especially for service in remote rural areas there is a need to create or adapt relatively clean, low-cost, and readily deployable off-grid distributed generation options. For service in most other areas improvement of grid-based systems will be needed, with special emphasis on improving the reliability of distribution infrastructure.

Work on these topics will require attention to the interplay between technological capabilities, the goals that particular regions and localities may set for electrification, and demographic change. Low-power distributed generation may be adequate for achieving universal access to electricity. But if the goal is extended to include large consumption of high quality electricity then today's rural distributed generation systems may be unable to supply the level and quality of power demanded. New higher power systems with intelligent metering that complement distributed and grid-based power may be required.

19.7.7 Outlook for Generation Technologies in Africa

The electrification of Africa offers the opportunity for a fresh look at designing a 21st century power system. For example, systems for the developing world are expected to rely on distributed generation for many applications, rather than the focus on central generation that is typical of countries that electrified during the 20th Century. Distributed designs may be the least costly and quickest way to get power to rural areas in developing countries using readily available indigenous resources. Distributed energy resources will also have a role in supplying the electricity needs of urban areas in developing countries. Note, however, that the markets for power in urban areas of the developing world dwarf the demand in rural areas. This suggests that there will be a continued role for central station generation in many developing countries that must necessarily rely on indigenous resources to control costs.

The distributed generation portfolio for developing countries is essentially the same as for the developed world. Moreover, petroleum-based liquid fuels may have an advantage in rural settings, because of the high volumetric energy density and the potential for upgrading existing refineries and building new ones to refine coal and crude oil into clean fuels. Liquid fuels are also valuable because they can be used both for stationary power requirements and for motor fuels (e.g., synthetic diesel oil).

Renewables will have an especially important role in developing countries. In general, technologies addressing the needs of the developed world can be adapted for use in developing countries. Examples include solar photovoltaic, wind generation, and biomass. To use these technologies effectively in the developing world, technology advances are needed in several areas, such as reducing the capital and operating costs of the equipment, reducing maintenance requirements, and improving the efficiency of end-use technologies. End-use efficiency improvements can lead to substantial reductions in the power requirements and capital cost of the generation equipment. Work is also needed to develop low-cost storage options—batteries, flywheels, and ultra capacitors for example—to deal with the intermittency problems of wind and solar power.

In many circumstances, power systems in developing countries will be designed to fill the needs of single users. However, village systems will probably require some version of a multiply connected mini-distribution grid, because simple radial distribution schemes will be unable to handle more than one generator on a system.

End-use technologies can also be designed to meet the needs of rural settings. Direct current end-use equipment—lights and power supplies for electronic applications—can be connected directly to DC generators, such as PV systems and fuel cells, without the need for

AC inversion of the generator output, and conversion back to DC at the point of use. Other considerations include the need for standardization of voltage levels, interconnection standards, and safety measures such as current limiters. Finally, guidelines for the initial electrification of developing countries can speed the process by summarizing the case histories of other organizations and countries, recognizing that no single solution will suffice for all applications.

19.7.8 Technology Portfolio

African power producers, transmission companies, and distribution companies have several options for introducing electricity and expanding its reach. There are two principal options. The first is to implement current technologies. The advantages of this approach are low initial cost, a reliable, proven technology, and technicians skilled in operation and maintenance requirements. However, these advantages are mitigated to a degree by the relatively low efficiency and high emissions of some designs. In addition, purchasing today's technology may lock the purchaser into yesterday's solutions, and in the future it may be difficult to retrofit a more modern solution. A second class of power systems incorporates new technologies with higher efficiencies, better environmental performance, and lower life-cycle cost. Frequently, the superior performance and low life-cycle cost may be offset by a higher initial capital cost.

One key attribute of new technologies is the potential to address climate change concerns through the implementation of a portfolio of zero- or low- carbon emitting generation systems. In the African context, this suggests a growing reliance on distributed generation, fueled by natural gas or renewable primary energy sources, in addition to clean coal technologies and nuclear generation.

The portfolio strategy offers the greatest flexibility and resiliency in meeting the uncertainties of the future, as well as the opportunity for different regions of the world to adjust the portfolio balance to suit their circumstances. A number of factors can shift the balance of the portfolio, including the availability and price of fuels, the pace of technological advancement, capital requirements, regulation, and policy. One critical factor will be the growing pressure to internalize the environmental costs of fossil energy, which will increase the relative importance and attractiveness of renewable and nuclear energy.

There is general agreement that we will have to continue to use coal as a fuel resource in South Africa. The issue here is the design of the next generation of coal plants. There is a significant opportunity to improve the environmental performance of coal by "*refining*" it into clean gaseous fuel or chemical feedstock. The gasification process can provide both high-efficiency power generation and hydrogen. This process is also amenable to carbon capture and sequestration.

Natural gas is also an option for African electrification. The reserves in Algeria and Nigeria can be tapped to provide fuel for gas turbines, and ultimately for fuel cells. Gas imports can supplement the indigenous reserves. Key technological issues include the need for liquefied natural gas (LNG) infrastructure for shipping and handling.

Distributed Energy Resources (DER), which includes generation, storage, and intelligent control, will become an integral asset in the African electricity supply system. As DER grows, it could fundamentally change the relationship between power supplier and consumer, and over time, the network architecture of the distribution system.

The portfolio of DER generation technologies includes reciprocating internal combustion (IC) engines (500 kW-5 MW), small combustion turbines (5-50 MW) and even-smaller micro turbines (kW-scale), and various types of fuel cells. Photovoltaic, small wind turbines, and other renewables are often considered DG technologies. Commercial DER storage technologies include batteries and capacitor banks. These technologies should find ready application in the African context. Advanced and novel DER concepts under development include Stirling engines, various generating technology hybrids, flywheels, "*ultra capacitors*," and super conducting magnetic energy storage systems. Related R&D is addressing DER-specific power conditioning equipment. Implementation of these technologies in Africa will require substantial site-specific evaluations. "*Ruggedized*" equipment that resists breakage and has minimal maintenance and repair requirements is likely to capture much of the market for rural areas.

19.7.9 Mitigating Greenhouse Gas Emissions

Addressing potential global climate impacts is becoming an urgent priority for the energy industry and policymakers alike. This reflects the fact that atmospheric CO_2 concentrations have increased 33% over the last 200 years, and are continuing to increase.

Changing from a global system where more than 85% of the energy used releases CO_2 to a system where less than 25% is released requires fundamental improvements in technology and major capital investments. A robust portfolio of advanced power generation options—fossil, renewable, and nuclear—will be essential to meet the economic aspirations of a rapidly growing global population.

There is no single solution to the climate change conundrum. Activities on all nodes of the electricity value chain—from fuel extraction to power generation to end use—are contributing to the buildup of CO_2 and other greenhouse gases (GHGs) in the atmosphere, with a potential impact on precipitation and other important climactic factors.

Addressing today's and tomorrow's complex climate issues will require a multidisciplinary carbon management strategy on three broad fronts:

1. Decarbonization, defined as reducing the carbon content of the fuel. Renewable generation, biomass, and nuclear power are the principal means for decarbonization. However, some petrochemical processes are available that produce liquid fuels with a high hydrogen content that could be used in gas turbine generators.
2. Sequestration, which consists of removing CO_2 from the product stream at the point of production, is a commercially available technology, but reducing the high costs of the technology would probably be required to make sequestration a viable alternative in developing countries.
3. Efficiency improvements reduce the energy required to produce a dollar of economic output. Efficiency improvements can be found throughout the energy supply chain, from mining and transporting fuel, converting the fuel to electricity or other energy carrier, power delivery, and end-use efficiencies.

Developing countries, including African countries, pose a particularly difficult challenge in addressing climate issues. As discussed earlier, the economic development of these countries depends on expanding electricity consumption, and most low-cost generation technologies emit greenhouse gases. However, as technologies are deployed in coming decades, solutions that meet the needs of the developing world will almost inevitably become viable.

19.7.10 Outlook for the Intelligent Power Delivery System

Although this Section focuses on the supply side of the electricity equation, the ultimate force pulling the electricity sector into the 21st century may turn out to be the technologies of electricity demand—specifically, intelligent systems enabling ever-broader consumer involvement in defining and controlling their electricity-based service needs. This will be true in developed and developing countries alike. It is important to remember that supply and demand in the electricity industry still rely on the same system design and much of the same technology in use since the dawn of electrification. This is a remarkable record of performance, but one that can no longer be sustained through merely evolutionary changes in the status quo.

Historically, the power delivery issues of security, quality, reliability, and availability (SQRA) have been measured and dealt with in a fragmented manner. In the future, they will almost certainly become a highly integrated set of design criteria to meet the evolving power requirements of consumers. Fortunately, the suite of advanced technologies that can be used to improve the security of the power delivery system can also be used to improve power quality and reliability, and transform the power system to meet the needs of the 21st century. These technology developments will first be manifested in the industrialized world, but developing countries will be able to leapfrog many of the intermediate steps in the development process. Consequently, their cost and time requirements to offer commercial solutions will compare favorably with the developed world.

The result will be dynamic technologies that empower the electricity consumer, stimulating new, innovative service combinations emphasizing speed, convenience, and comfort, with different quality levels and types of electric power. A vigorous, price-sensitive demand response from an increasing class of consumers whose energy choices reflect both electricity prices and power quality will become an integral part of the electricity marketplace.

The shorthand for this new system is the intelligent power grid, or "*Intelligrid*", conceived of as an electricity/information infrastructure that will enable the next wave of technological advances to flourish. This means an electricity grid that is always on and "*alive*", interconnected and interactive, and merged with communications in a complex network of real-time information and power exchange. It would be "*self-healing*" in the sense that it will constantly monitor its condition and self-correct at the speed of light to keep high quality, reliable power flowing. It could sense disturbances and counteract them, or recons the flow of power to cordon off any damage before it can propagate. It would also be smart enough to seamlessly integrate traditional central power generation with an array of locally installed, distributed energy resources (such as fuel cells and renewables) into a regional network.

The smart, self-correcting power delivery system will become the conduit for greater use of productivity-enhancing digital technology by all sectors of the economy, leading to accelerated productivity growth rates. The power system will enable new energy/information products and services across the board, and reduce or eliminate the parasitic costs of power disturbances characteristic of, for example, the US economy today.

To complete the picture, digital technology will also be able to open the industrial, commercial, and residential gateways now constrained by the meter, allowing price signals, decisions, communications, and network intelligence to flow back and forth through the two-way "*energy/information portal*". The portal will provide both the physical and logical links that allow the communication of electronic messages from the external network to consumer networks and intelligent equipment. For consumers and service providers alike, this offers a tool for moving beyond the commodity paradigm of 20th century electricity service. It will complete the transformation of the electricity system functionality, and enable a set of new energy information services more diverse and valuable than those available from today's telecommunications industry.

The Intelligrid may appear to be a distant dream when compared with the near-term needs of African electrification. However, the ability of the developing world to leapfrog intermediate technologies may allow implementation of elements of the Intelligrid system as they become available. In particular, a wireless information network will be able to provide much of the communications support needed for a power system based on distributed energy resources. The hardware and software needed for distributed energy systems in Africa are already available. Implementation of a distributed Intelligrid will be limited by financial considerations rather than technology considerations.

19.8 Providing Electricity Services to Rural Africa [26]

19.8.1 Understanding the Challenge

The communiqué from the G8 meeting in Gleneagles, Scotland in the summer of 2005 called for major action to support economic development in Africa. Even with the World Bank instituting a Clean Energy Investment Framework, the task is still daunting. The Action Plan for meeting Africa's energy service needs to include:

(a) Access to clean cooking, heating and lighting fuels, coupled with sustainable forest management
(b) Scaled up programs of electrification
(c) Additional generation capacity to serve newly connected households and enterprises, including through regional projects
(d) Provision of energy services for key public facilities such as schools and clinics, and
(e) Provision of stand-alone lighting packages for households without access to the electricity grid.

While ambitions to meet the Millennium Development Goals by 2015 are laudable, in terms of energy infrastructure design, finance and implementation, and developing the local capacity to operate and maintain those systems, 2015 is very close.

Several understated challenges that technology and finance companies, government agencies, and local communities face is how to design and implement new electricity services in time and space. For example, not all businesses and households in a village or town will receive electricity at the same time. Initially small village scale systems may only electrify community buildings, and then for only several hours per day from a diesel or biogas genset, or micro-hydropower system. However, we know that as communities develop, demand for modern energy services may begin to grow rapidly.

Several other daunting challenges have to do with a) how quickly can electric service be provided - at any level, b) what requirements are there from the viewpoint of grid extension, or the development of parallel fuel supply infrastructures to support generators, and how to maximize economic benefits/ and reduce cost and availability risks as local economies become more dependent on electric service. New tools for optimizing the configuration of village scale power systems, especially those that incorporate renewables are now readily available. However, energy demand is far from static and may vary significantly by time of year (climate, agricultural energy demand), as well as time of day showed how alternative configurations of wind-diesel systems could meet different levels of village electricity demand (with different economic values).

If local communities wish to tap multiple local energy resources, then their dynamics must also be taken into account. In the case of sun and wind, these may be more predicable than other resources, such as hydropower and biomass, especially if areas are drought prone–or worse–if vegetation is poorly managed.

So, if electricity is to be delivered to rural areas in the near-term, with factors of demand growth, resource dynamics, and system expansion taken into account, a new design approach is needed.

19.8.2 Designing Robust Solutions

Integrated energy technology demonstration projects must become models both for dissemination and training, recognizing full well that local communities must adapt these technologies and systems to suit their own needs and resources, including the ability to keep them running, and pay for operating costs. There are now good templates from small villages in India and elsewhere on how to collect costs from users, and task residents for routine maintenance tasks.

One challenge is to identify a range of "*basic systems*", based upon demand levels and pattern, and renewable resource dynamics that can then be adapted to actual village conditions.

19.8.3 Growing with Time

As hinted above, once electric service becomes available, demand for electric service is likely to grow rapidly. As illustrated in Figure 19.13, over time neighboring villages will install their systems, grow from intermittent to 24 hour service, and with enough planning and coordination, link up their villages to one another and where applicable linked to a centralized grid.

This poses a challenge to standards makers, so that *"plug and play"* village systems can easily be linked together and operated in a coordinated fashion without facing service quality impacts. The day of resistive loads has passed, and the power quality requirements of *"electrified"* villages must be respected, and planned for.

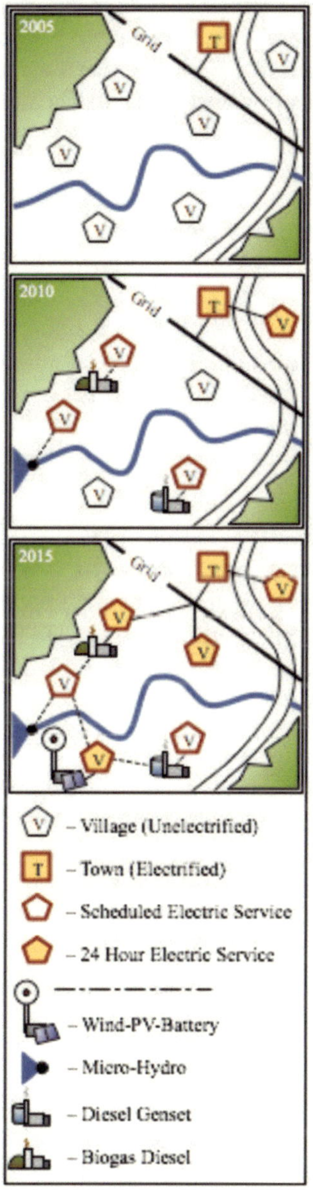

Figure 19.13. Diversity and Growth of Village Scale Power Networks over Time.

Concurrent with the development of design tools, on illustrative demonstration projects, is the need to collect quality information on changes in and drivers of electricity demand, as well as the patterns and variability in numerous renewable resources (wind, solar, hydro/precipitation, crop yields and forest productivity).

19.8.4 Building the Context and the Capacity

Taking the into consideration the various aspects of the challenge outlined above, it is clear that goals put forth by the UN and OECD can only be pursued by developing numerous "*capacities*" ranging from international finance and access to "*best practice*" technologies, to the development of operation, maintenance and small business skills down at the local level.

From a strategic planning viewpoint, "*context building*" is needed such that the initial provision of electric service cannot only be maintained, but expanded through time in a manner that maximizes both the use of local resources, and puts these new energy services to best economic use.

Building this integrated capacity to electrify 1.6 billion people, whether through grid extension to growing urban areas, to far from grid small population centers, is a very large task. It will take a huge commitment in time, people and money. However, with modern communications and information tools, "*best practices*" from design to operations should rapidly penetrate the industry and propagate from one local to another.

19.8.5 Modified Micro Grids Alternate Models [27]

Systems suitable for future grid connection should also be contemplated particularly in urban areas are in many cases just as deficient as the rural sector. This section proposes a concept of "*Olympic Ring*" type Micro grids to illustrate the bottom-up development. This Micro Grid power network architecture has incorporated the following power system and power electronics technologies:

- Advanced power network control techniques that allow for deployment of a wide range of DG and power network solutions into real-world applications.
- An advanced power converter system to enhance the capabilities of DG and storage systems. The distributed energy resources are able to interact directly over the power network to provide power sharing, power flow, and control.
- An open-platform energy management system that can provide remote monitoring, data collection, and aggregation of distributed power systems into "*dispatch able*" blocks of capacity.
- A static isolation switch system that manages the interface of Micro Grid power systems to the utility, and allows for seamless transitions between stand-alone and grid-connected operation.

Depending upon the loading condition and available resources, a single-ring (Figure 19.14) or double-ring (Figure 19.15) local node could be formulated. Each ring will be equipped with a local node controller for local optimization and control.

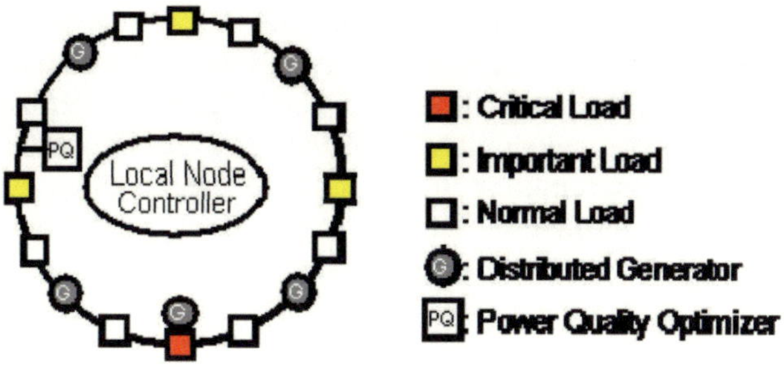

Figure 19.14. Formation of Single-Ring Local Node

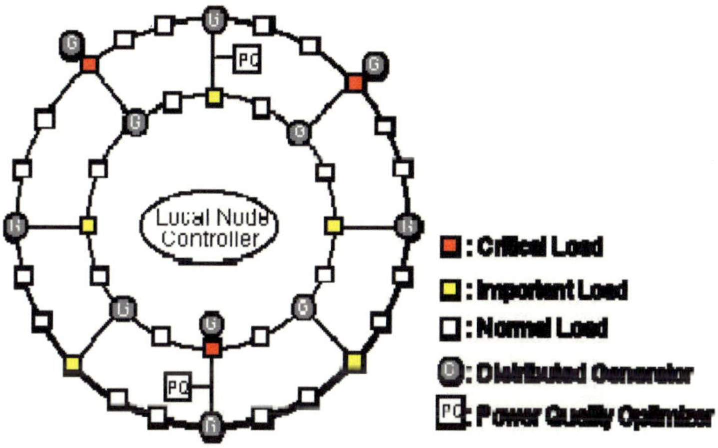

Figure 19.15. Formation of Double-Ring Local Node

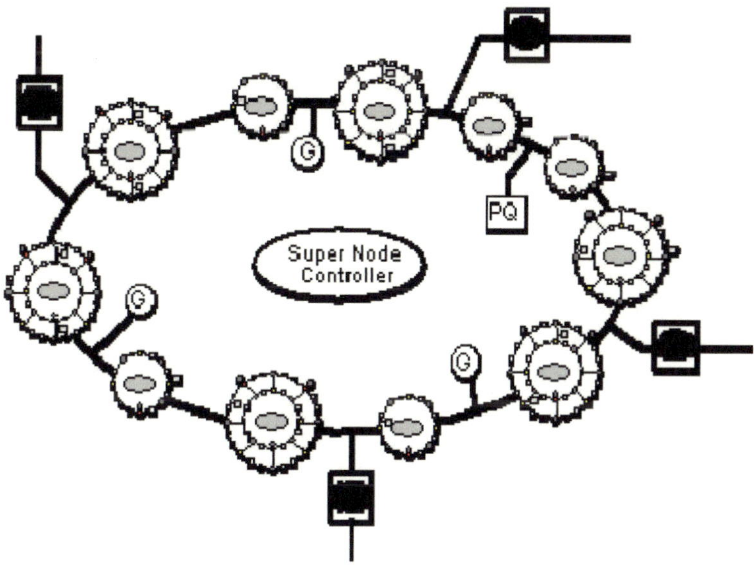

Figure 19.16. Formation of a Super Node and Grid Connection

Clusters of local nodes will then form a super node. As shown in Figure 19.16, a ring configuration is preferable since it offers higher system reliability. Super nodes may equip larger scale distributed generation facilities and power quality controllers. The protection schemes between local nodes are similar to the branch protection within the local nodes. Depends upon the operation condition, the super node can be either autonomous or non-autonomous. Super node controller will communicate with SCADA/EMD of the local utility or ISO, local node controllers, distributed generators, and PQ optimizers. Super node will serve as basic building block of the future "*Smart Grid*".

In summary, an uncontrolled unbundling of utilities and the resulting wave of growth in small distributed generation Independent Power Producers (IPPs), would lead to the creation of un-standardized meshed networks should the technical and economic planning associated with larger integration not be contemplated with today's technical standards. Building this integrated capacity whether through grid extension to growing urban areas, to far from grid small population centers, is a very large task. It will take a huge commitment in time, people and money. However, with modern communications and information tools, "*best practices*" from design to operations should rapidly penetrate the industry and propagate from one local to another [26].

19.9 The Kenyan Electric Power Sector –A Case Study [28]

In many African countries there is a tension between grid and off-grid electric service provision and it is unclear whether centralized or decentralized power system architecture will emerge. This Section explores some of the dynamics of system development in Kenya, where poor grid infrastructure has resulted in a thriving private market for photovoltaic panels and a growing number of industries are investigating shifting to on-site generation. The research is based on ethnographic interviews and observations in Kenya and uses System Dynamics modeling tools to analyze qualitative and quantitative feedback in the system.

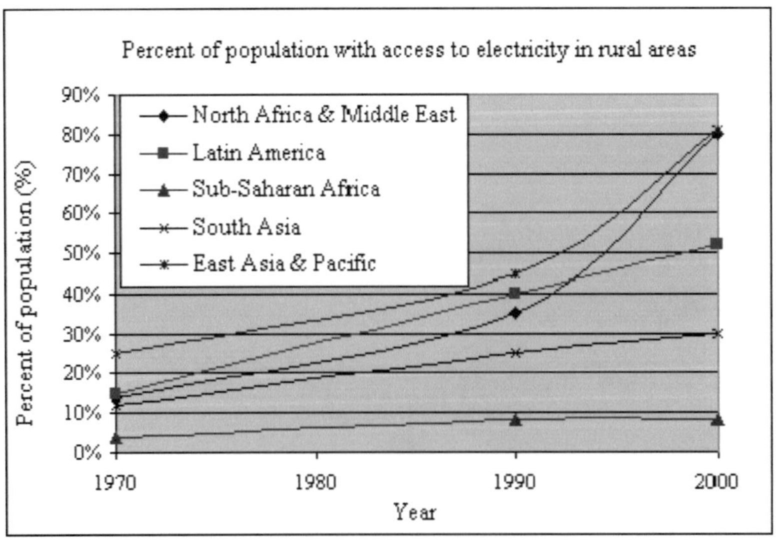

Figure 19.17. Percent of Rural Population with Access to Electricity by Regions

In the developing world Sub-Saharan Africa (SSA) and India are the least electrified regions of the world and they continue to fall further and further behind (see Figure 19.17). Although the lack of modern energy services in these regions is well documented, the underlying reasons are not well understood. The growth of the electric power system in the SSA country of Kenya is now focused on where both the dynamics that have led to the low availability of power as well as the drivers that could enable greater access in the future are explored.

Seventy percent of people in SSA live in rural areas and rural electrification rates are extremely low. This presents a challenge for electrification because it is expensive to connect a diffuse population. Both the line losses and cabling costs due to long transmission distances make installing the infrastructure very costly. This technical limitation, added to the fact that the majority of the rural population has little ability to pay for electric service, makes it economically impossible to extend the grid to all areas. The only justification for rural electrification has been the social necessity.

Lack of electricity and modern fuels can be linked to an increase in disease and environmental degradation, and economic stagnation. Homes without electricity continue to use biomass and kerosene for cooking and lighting, which leads to respiratory and eye infections. These households also deplete biomass resources, which can increase desertification and cause land erosion. Lack of modern energy sources can inhibit education due to poor lighting conditions and inhibit economic growth due to the time used gathering traditional fuels and the inability to expand businesses using more efficient energy sources.

Even in urban and industrial areas electricity access is low. While most industries are located near the central grid, many must invest in back-up power supplies and power smoothing equipment to manage the frequent outages and inconsistent voltage supply in the network.

19.9.1 Electricity in Africa as a Complex System

Discussions of complex systems usually focus on computer networks, transportation systems, or manufacturing logistics. However, African electric power systems are complex infrastructure where the architecture is not already determined. While most complex systems research focuses on existing complexity, in Africa there is an opportunity to study the system as it develops. So far there has been little research to understanding system development in this area. Karekezi and Kimani [29] and Pandey [30] have noted the lack of research in African power systems and the insufficient use of modeling in developing countries, respectively. Hammons has reviewed Recommendations for Power Pools in Africa [31] and Hammons et al. [32] also cite this need with reference to the World Bank, saying that *"[it] has not yet found a reliable model for dealing with the special needs of sub-Saharan Africa electricity infrastructure"*.

Africa faces a choice between following the traditional model of centralized generation, and developing a decentralized model. There are benefits and detriments associated with both options. While a decentralized model may make it easier to provide service to remote populations, it may limit system growth in the future. Were an inexpensive bulk power supplier to come online, such as the Grand Inga hydropower station in the Democratic Republic of Congo, a country with a decentralized system might have difficulty benefiting from this source.

If the choice were simply a technical one, the system could be analyzed and optimized according to the least cost or most technically efficient model. However, there are several non-technical issues that add complexity such as most governments now seeing electricity as a social right. If an optimization model were to show it is uneconomical to provide any access to certain areas, this would not meet the desired goal of the system. There is also an issue of complexity due to corruption. Any planning which ignores the presence of corruption does not reflect the true cost of implementation. This problem of non-technical complexity highlights the need for new approaches to system analysis in Africa.

19.9.2 Selection of Case Study

Kenya typifies the difficulties of energy development in eastern Africa, with its low population density and an installed capacity of only 1147 MW. Kenya is also a regional economic and political anchor and ideally development in Kenya will positively impact Uganda and Tanzania [33][1], as well as other countries in the region.

The scope of this Section covers the range of electric power consumers and generators in Kenya, as well as the organizations that sell and regulate power. Kenya has privatized power generation with roughly 70% of generation by the Kenya Generating Company (KenGen). The remainder is provided by independent power producers (IPPs). Electricity is sold to Kenya Power and Lighting Company (KPLC), who sells to consumers, and the Electricity Regulatory Board (ERB) regulates the sale on both sides. Consumers who are not connected to the national power grid have the option to buy off-grid generating equipment from dealers. Figure 19.18 shows the scope of the case.

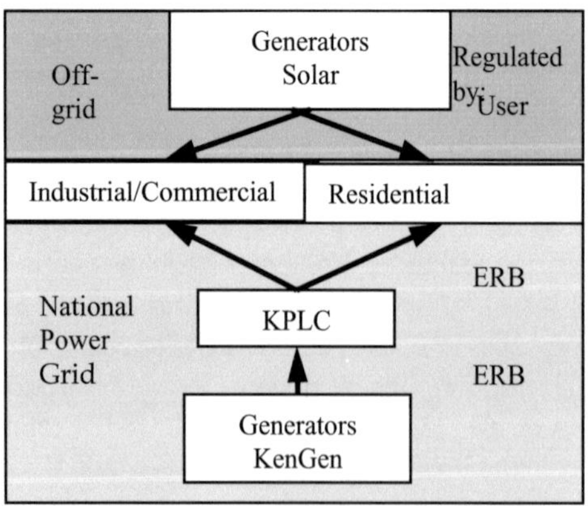

Figure 19.18. Scope of Analysis of Kenyan Electric Power System

[1] Kenya, Uganda, and Tanzania have existing cooperation agreements under the East African Community alliance. Kenya and Zambia are also working together to create a link that will bring power from the Southern African Power Pool into East Africa.

The case study concentrates on the interaction of the actors in the system and how their decisions feed back into the system and affect its development. The method used was selected because it seeks to understand qualitative, as well as quantitative, aspects of the system.

19.9.3 Method

The goal of the method is to understand why electricity access is stalled in SSA, and what can be done to enable growth. System dynamics modeling is appropriate method in this case because it can represent the range of technical and non-technical feedback in the system. For the model to be useful, however, it must be grounded in reality. Sterman [34] found participant interaction and interaction with clients essential to formulating the non-linear functions of a model, which points to the use of interviewing and observation as methods. The fieldwork for this study followed the standard method for system dynamics modeling [34]. This includes attention to stakeholder interaction, causal loop diagramming, calibration, and sensitivity analysis.

The interviews conducted in Kenya included residential and industrial consumers, representatives from KPLC, KenGen, and the ERB, and off-grid service providers. A final source of information was quantitative data collected in Kenya, both concerning the operation of the power system and the socio-economic status of the population. The data gathered from these sources are being used in the creation of the system dynamics model.

As stated earlier, this model is not intended to be predicted. Rather it could be used to identify points were policy could have an impact. Saeed and Prankprakma used system dynamics to study the link between technological development and economic growth [35]. They found that technological development has the potential to be a policy lever for economic growth in developing economies but only if a feasible path can be determined. Similarly, this study is attempting to find policies for inducing development in the energy system.

19.9.4 Preliminary Findings

The interviews have given some preliminary insight into the dynamics of the electric power system growth in Kenya. One of the key findings may be that Kenya, and Africa in general, is not so different from the rest of the world in terms of electrification. Instead of focusing on what makes Africa different, perhaps policy-makers should be focusing on how it is the same.

Grid infrastructure in Kenya is characterized by high fees and long waits for connections, large voltage fluctuations, and relatively common outages. Standby power supply in Kenya has become so common that commercial and residential customers accept frequent interruptions in the power supply. Even in very modern commercial centers or tourist hotels, power interruptions are not met with surprise, rather the customers simply pause while the generators automatically come online and then go about their regular activities. This is not the case with industrial consumers. Manufacturing and production processes frequently cannot simply restart if there is a power interruption. A food processing plant outside of Nairobi estimated that for every power interruption they lost four hours of productivity due to spoilage of the product and the need to reset and clean all processing equipment. In this case the feedback is that as power interruptions become more of a burden to the customer, the more likely they are to seek other sources of electricity.

Most commercial and industrial consumers that have been interviewed have said that if there were a standby power supply that could compete on cost with the grid, they would consider producing their own power. Already several large consumers, such as sugar, tea, and paper manufacturing companies, generate a portion of their own power. The Kenya Tea Development Authority (KTDA) has assessed the feasibility of on-site generation at 20 more of its tea factories and Mumias Sugar recently signed an agreement to expand its boiler capacity to generate 35 MW on site.

If a significant portion of industry disconnects from the grid, or generates the majority of their own power, it will reduce the revenue to the Kenya Power and Lighting Company (KPLC). If this happens, it could hinder KPLC's ability to invest in infrastructure, which would in turn encourage more consumers to move off-grid. This dynamic has already been seen in the telecom sector in Kenya. The national provider, Telkom, was ill equipped to manage the introduction of competition from mobile phones.

Residential consumers are similarly choosing to go off-grid. Estimates vary as to the total number of PV panels sold, but consensus says it is well over 100,000 units. In most rural cities the electrical appliance shops sell PV panels and systems and several large retailers and wholesalers operate across the country. According to interviews with dealers, customers choose PV in most cases because they are not close enough to the grid to be connected. However, some are buying PV even after having paid for a connection to KPLC, because after waiting several years they have still not been connected. Others decide to keep using PV even when the grid comes to their village since they have already made the investment.

Kenya, like many African countries, is at a critical point in its electric power infrastructure development. Depending on where investment is focused, the system could grow as an interconnected grid with generation flowing out of power stations, or it could become a decentralized system where industrial and residential power consumers generate power on-site. Even if the on-site generators remain connected to the grid, the technical and financial structure of the system will shift. In the US city of Chicago in the late 1800s, there was a surprisingly similar tension between dedicated power suppliers and industry and businesses generating their own power on site. In that case, the system shifted to a centralized utility when Samuel Insull was able to cut costs for power producers through increasing load factor and diversifying customer demand. In Kenya, it is important to understand how these types of policies and investments could impact the system development. Finally, this study will also question whether Africa is really all that different from other regions in its power system development. If the dynamics are similar to other regions which have already gone through this process, then that may lend insight into how to spur development.

19.10 The African Power Development Footprint: Accelerating the Technical Skills Factor

DEFINITIONS:
AAU - The Association of African Universities,
ICT - Information, Communications, Technology,
AVU - African Virtual University,
EPRI - Electric Power Research Institute,
SST - Strategic Science & Technology

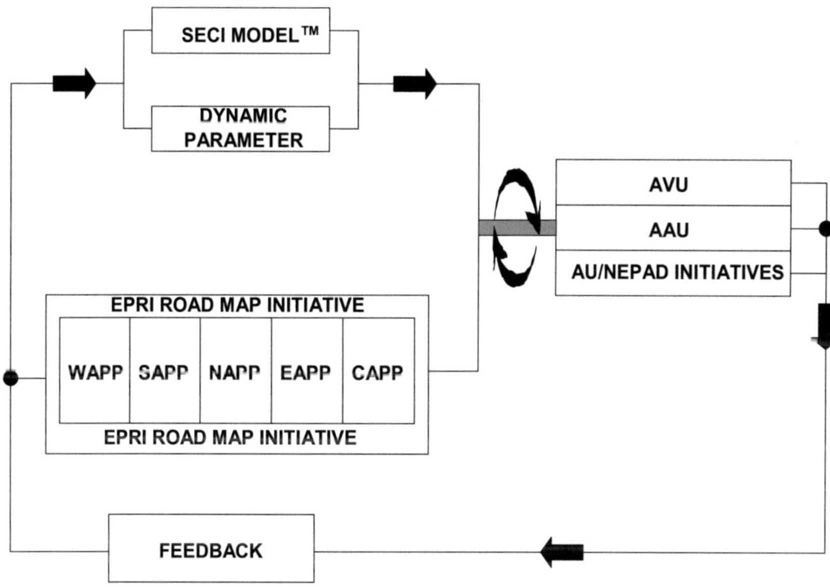

Figure 19.19. The African Power Development Knowledge Engine

The multidisciplinary nature of Power Generation and Transmission projects provide an interesting synthesis of knowledge generation and potential for its capture. This also includes the converging and diverging nature of geopolitical issues, humanitarian crises, infrastructure and human capital deficiencies. '*Think-outside-the-box-solutions*' (TOTB) is therefore necessary to effectively capture and apply this knowledge. The proposed African Power Development Knowledge Engine model Figure 19.19 is created in broad terms from an examination of various programs and studies from around the world and is then configured to synthesize elements from these various INPUTS to address an African context. The sheer ambition of attempting to converge such divergent disciplines into something practical leaves one open to skeptics were it not for the exigencies of disciplines such as System Dynamics and systems engineering broadly. To quote J.W. Forrester, the founder of system dynamics; "*Interest in System Dynamics is spreading as people appreciate its unique ability to represent the real world. It can accept the complexity, no linearity, and feedback loop structures that are inherent in social and physical systems*". In educating the individual, the objectives of a systems dynamics education might be grouped under three headings: 1) developing personal skills, 2) shaping an outlook and personality to fit the 21st century, and 3) understanding the nature of systems in which we work and live [36]. The subsequent sections and proceeding paragraphs will attempt to indicate the possibilities even though concrete models (simulation) that are at the core of the studies have not yet been tested specifically for this model.

19.10.1 Model Overview

The African Sectors of focus are the South African Power Pool (SAPP), West African Power Pool (WAPP) and the initiatives in North Africa with interconnections to the Middle East and Europe (NAPP). Studies such as the Purdue long-term economic model and R&D programs from EPRI's Road Map Initiative and SST are the proposed foundation candidates, from which interdisciplinary synthesis over a wide range of applications can be generated. The information density contained within these sector initiatives provides sufficient '*Synthesizing Capability*' for creating knowledge enabling infrastructures [37]. The Purdue University Power Pool development group commissioned by the Economic community of West African States (ECOWAS) and the Southern Africa Development Community (SADC) addresses regional and country specific power generation and transmission opportunities over a variety of economic scenarios and generation resources such as hydro, fossil fuel, thermal, natural gas etc. Other members of the African Union in North Africa also have large development foot prints as illustrated earlier in Figure.19.2 and are candidates for focusing on regional specific analyses and interconnections such as Egypt's ties with Jordan and the wider Mediterranean countries [38,39] (Figure 19.20). This Northern Grid will eventually interconnect with the Gulf Coordination Council states that have already implemented a successful multi grid integration plan with many firsts in the application of new technologies (Figure.19.21).

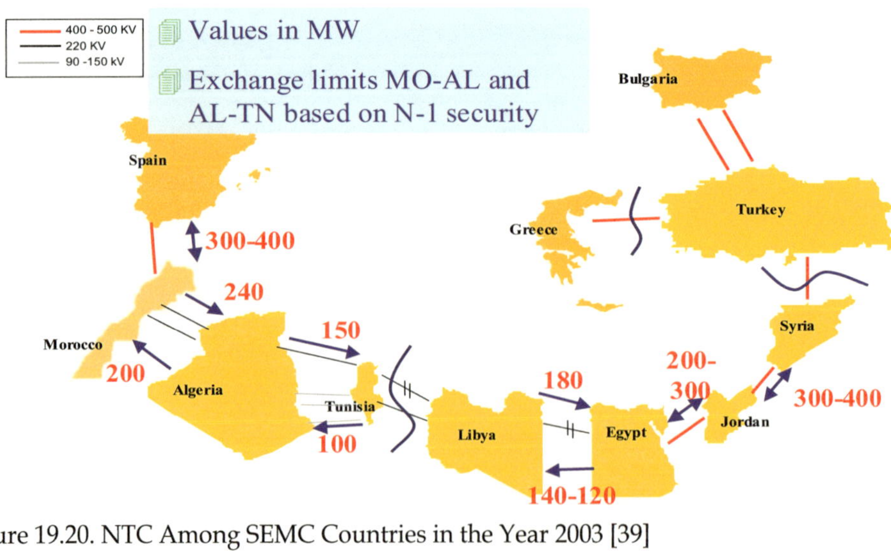

Figure 19.20. NTC Among SEMC Countries in the Year 2003 [39]

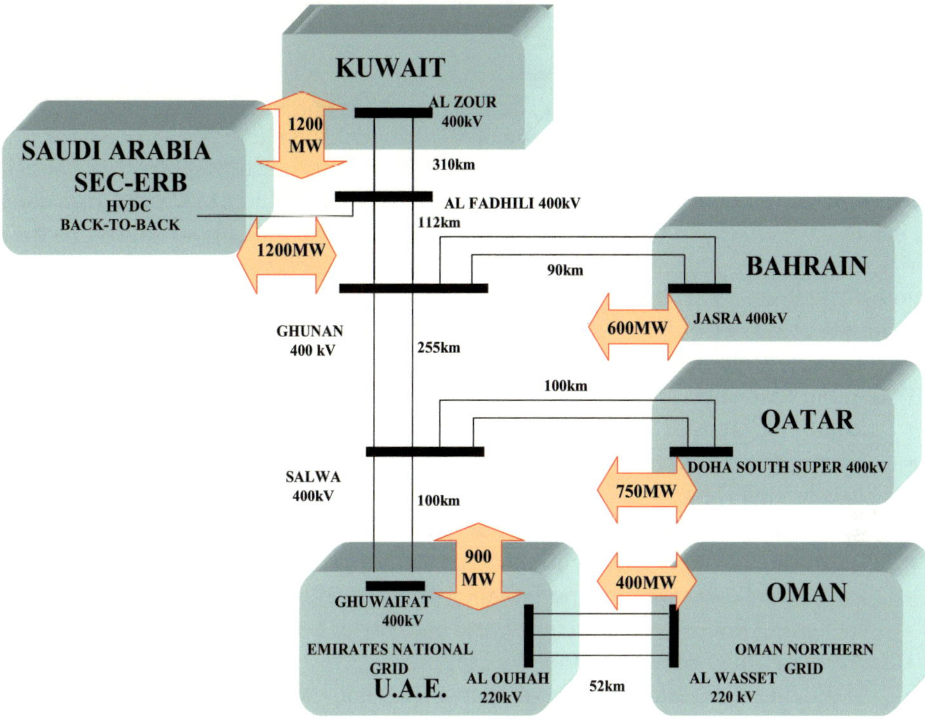

Figure 19.21. Conceptual Diagram of the Interconnection System of the Gulf States [39]

Within this context of knowledge capture, the EPRI Electricity Technology Road Map Initiative, The Alliance for Global Sustainability (AGC) and institutions such the Moscow Power Engineering Institute (MPEI) with an international focus on 21st century energy issues, represent a major convergence or synthesis of global Industry Experience and R&D. US-African organized programs hosted for example by the College of Engineering, Architecture and Computer Sciences (CEACS) and the Center for Energy System and Controls (CES&C) at Howard University have received prior research funding for supporting international workshops on power system operation and planning in Africa. Over the past ten years, the National Science Foundation (NSF) has supported the hosting of the International Conference on Power System Operation and Planning (ICPSOP) in various African countries namely, Nigeria, Ghana and Ivory Coast, Durban, South Africa and in Cape Verde [40]. Towards this end the importance of these advanced research institutions to enhance programs such as these cannot be overstated.

EPRI has over 150 participating electricity stakeholder organizations participating in the EPRI Road Map program [21]. The Roadmap Initiative seeks to develop a comprehensive vision of opportunities for electricity-related innovation to benefit society and business. The Roadmap also translates that vision into asset of technology development destinations and ultimately the needed R&D pathways. The Creation of the Roadmap began with the exploration of opportunities in five distinct topical areas:

- Sustainable global development
- Electricity and economic growth

- Power delivery Infrastructure
- Power Production
- Environmental Knowledge Base.

EPRI has also adopted a strategy whereby SS&T provides the strategic resources for EPRI's integrated R&D planning process, helping connect the specific technical objectives of EPRI's sector programs with the broad societal goals defined by the Road Map. SS&T concentrates on a set of 15 *limiting challenges* representing critical issues and opportunities facing the electricity enterprise and society along with the associated gaps in knowledge and technological capability. The limiting challenges link the destinations identified by the Road Map with the objectives of EPRI's sector programs.

The 15 limiting challenges are the following:

- Improved Transmission Capacity, Grid Control, and Stability
- Maintain and Strengthen Portfolio of Generation Options
- Accelerated Development of Carbon Capture and storage Technologies
- Creation of the Infrastructure for a Digital Society
- Improved Methods for Communicating and Applying Scientific Knowledge
- Improved Power Quality and Reliability for Precision Electricity Users
- Increasing Robustness, Resilience, and Security of Energy Infrastructure
- Advances in Enabling Technology Platforms
- Exploiting the Strategic Value of Storage Technologies
- Transformation of Electricity Markets
- Ecological Assessment Management
- High Efficiency End Uses of Energy
- Maintaining and Improving Water Availability and Quality
- Global Electrification
- Development of Electricity-Based Transportation Systems.

19.10.2 Standardizing Curricula as a Strategic Factor

For the ambitions set forward by the AU and NEPAD manpower development in the technological, health, agricultural and management field remain core to any kind of institutional integration. Previous IEEE PES International Practices for Energy Development and Power Generation presentations [41] have recommended creating Communities of Practices (COP) utilizing *knowledge creation* models such as the SECI™ Model from Japan (see Figure 19.22) which states that in order to "*make these things happen*", there must be "**Ba**" for knowledge creation [42]. **Ba** means internal communities of groups of technologists or knowledge workers who share the same interest or purpose. More specifically, they are cross-functional human networks or groups including virtual relationships on intranets, extranets or the Internet. These communities are called COP (communities of practice). Participants understand the contexts of others and oneself, and through interaction, change/create the contexts. Hence, it is constantly moving.

The key to understand context is interaction. Knowledge does not just reside in one's mind. Knowledge emerges through shared contexts that are created through interaction [42]. This author feels that further efficiency can be achieved from the proposed Knowledge Engine by incorporating systems dynamics to mine it's inherent complexity. In so doing a key objective would be the formulation of a "Renaissance Man" or woman as defined by Forrester [36].

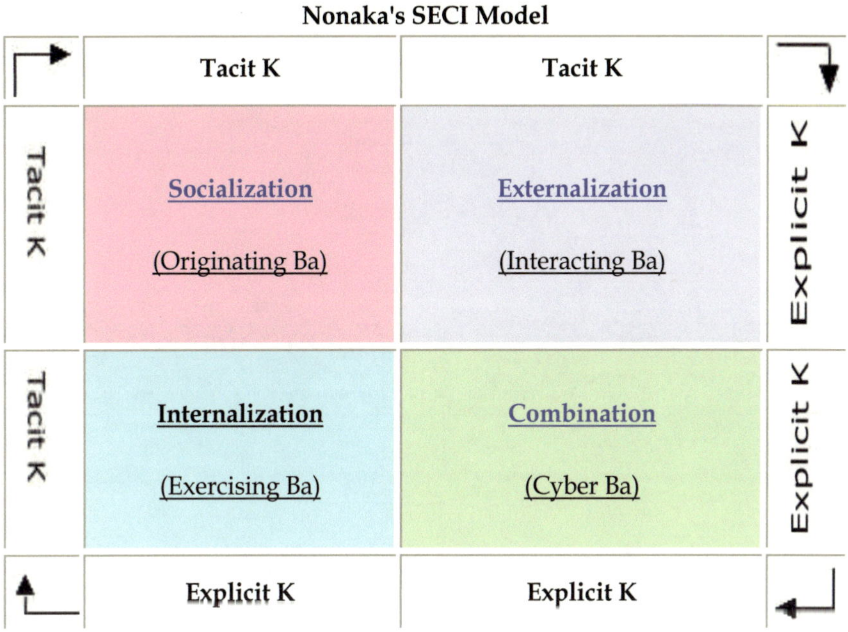

© Nonaka, Takeuchi, Konno, Toyama

Figure 19.22. SECI™ MODEL [41]

We have previously stated our proposed knowledge engine is introduced to create knowledge from local African and International continental conditions as a paradigm to improve efficiency. This we strongly believe can be achieved by strengthening institutional integration around specific and standardized curricula. With today's ICT advantages the challenge then becomes creating curricula that create Virtual Worlds used for improving the learning process Sterman [43]. Figure 19.23 is a basic representation of how a Systems Dynamics approach would initiate a basic level of evaluation of a system.

"System dynamics trains an individual to see the interrelationships in systems as being far more interesting and important than separate details. The interrelationships reveal how the feed back loops are organized that produce behavior. Students with a background in systems modeling should be sensitized to the importance of how the world is organized. They should want to search for interconnecting structure that gives meaning to the parts. System Dynamics provides a foundation underlying all subjects" [36]

Figure.19.23 represents an idealized learning process whereas effective learning involves continuous experimentation in both the virtual world and real world. Feedback from both informs the development of mental models, formal models and the design of experiments for the next iteration.

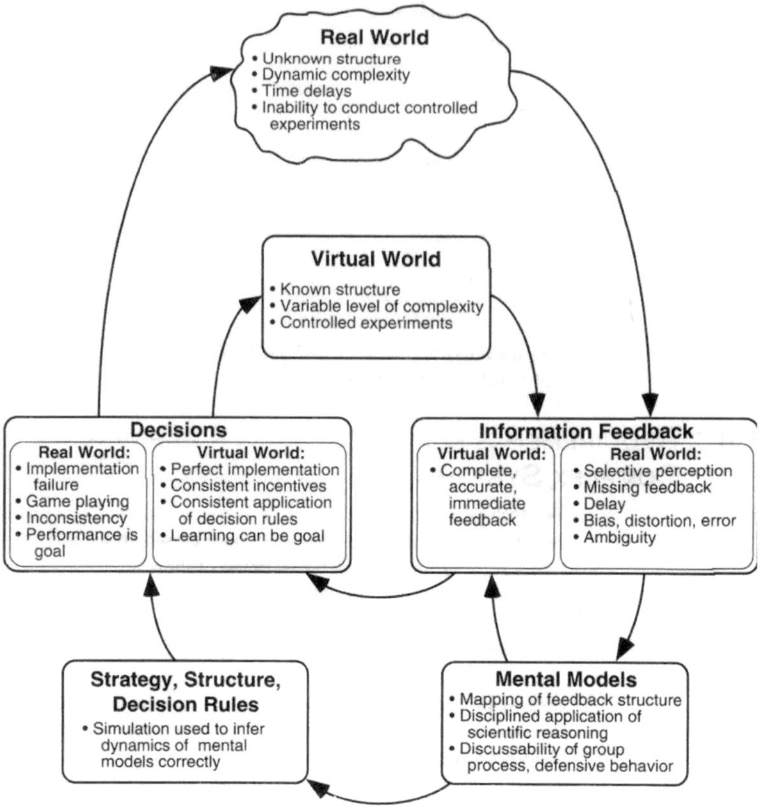

Figure 19.23. Systems Dynamics Representation []

19.10.3 The Dynamic Parameter Model [44]

In the Dynamic Parameter study a *"Dynamic Parameter"* is an evolved entity abstracted from any selected object (subject) that has significant information density or that lends itself to information convergence in a given sub domain or context as represented by Figures 19.24 and 19.25. In this study *"core"* **C** as shown represents the recommended curriculum to be derived from mapping the EPRI road map and its integration with models from other global experiences. In Figure 19.24. The different backgrounds represent different location contexts.

In this instance the object or context can also be projects within one of the power pools being looked at namely the WAPP, SAPP & NAPP. Figure 19.25 illustrates a given *'Domain'* where the continuous dynamics of relationships are created from knowledge exchange leading to constant flows of feedback and action. The resulting ICT network infrastructure or COP will produce behavior mimicking that of a neural network constantly under training/development/refinement. And while there is no precise agreed-upon definition among researchers as to what a *neural network* is, but most would agree that it involves a network of simple processing elements (*neurons*), which can exhibit complex global behavior, determined by the connections between the processing elements and element parameters [45].

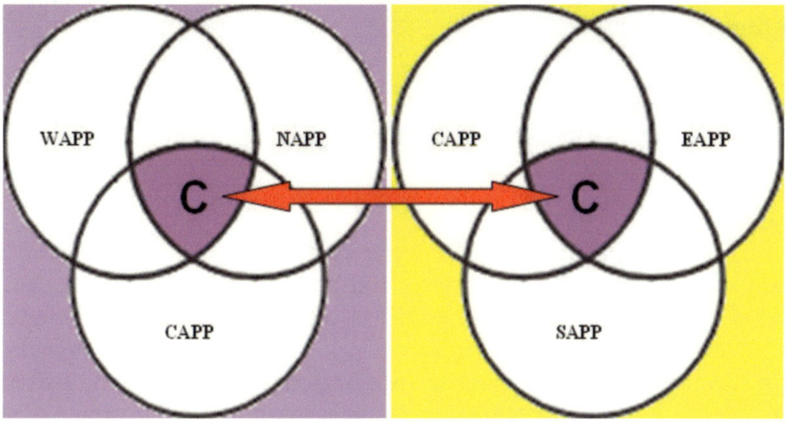

Figure 19.24. Dynamic Parameter Model

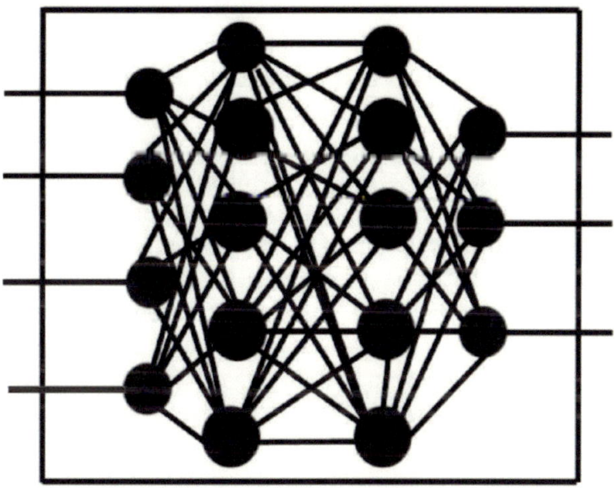

Figure 19.25. Dynamic Parameter-Sub Domain Relationships

As mentioned earlier specific programs such as the National Science Foundation (NSF) sponsored US-Africa Workshops produced a variety of dynamics from such collaborations as noted below [40]:

- US utility companies who attended have won opportunities to do business with Africa power industry as consultants, vendors, etc.
- Recruitment of outstanding doctoral students into United States universities with power programs, for example, Howard University, Tennessee Tech, Texas A&M, and Washington State University (WSU)
- Joint projects between United States power industry and African universities and industries. For example, joint projects between Howard University and Kwame Nkrumah University of Science & Technology (KNUST), Monsura University, etc.

- Exchange program between the United States and Africa
- US industry and power consultants have become interested in business in Africa. This includes K&M Engineering and Consulting Corporation and the US Education Institute, Inc. (AEI).

19.11 Further Reading

Further Reading on African Electricity Infrastructure is available in References [46-49].

19.12 Conclusion

The focus on power Generation as the driving knowledge base is because of the societal impact information generated *before*, *during*, and *after* its development. The simulation of societal impact to address societal problems is itself a contribution to the student body's social consciousness and source of ensuing feedback. Sharing that consciousness/knowledge among the body of coordinated participants in a local regional or international context creates a dynamic of new ideas and the potential for multiple solutions from a variety of sources and disciplines. The WESTCOR project in South West Africa is one such '*opportunity project*' whose large footprint covering a large resource laden geographical area can influence the training of a large student body in Systems Thinking throughout the continent and elsewhere. The same holds true for the North African and Middle Eastern systems. Within the same context of these larger projects, the development of less capital intensive Microgrids integrated with strategic new technologies, e.g. low heat input advanced refrigeration and heating such as the Thermosorber ™ , biomass plasma gasification, integrated solar thermal systems and wind power also become important economic development strategies. Micro grids in particular are an essential development '*growth incubator*' and can be setup as "*energy malls*" strategically placed for optimized development solutions such as Millennium goal villages and towns.

Further, the value derived from these small Distributed Generation micro grid systems for rural and urban development are no less complex as they too introduce logistics problems for systems installations, operations and maintenance (O&M) requirements, fuel supply dynamics in instances where renewable energy sources may not be easily accessible or available. Their often small-sizes being less capital intensive can be seen as being '*charity sustainable*' where applicable. The development of these small distributed generating systems can introduce a sustainable-strategic dynamic between charities, manufacturers, government agencies, NGOs and universities to penetrate society with greater efficiency and robustness. These small systems can also further the active role of universities and technical schools to provide '*energy peace corps*' in support of the development and maintenance of these systems.

19.13 Acknowledgements

This chapter has been complied by Bai K Blyden, CEO BBRM Group LLC, California USA; Co Chair IEEE PES International Practices WG on African Electricity Infrastructure, Interconnections and Electricity Exchanges. Contributing Authors include; R.T Mochebelele (Advisor Infrastructure, NEPAD Secretariat, South Africa); F.T. Sparrow (Director, Power Pool Development Group, Purdue University, West Lafayette, IN, USA); Brian H. Bowen (Potter Engineering Center, Purdue University, West Lafayette, IN, USA); Moshin Chen

(Professor Emeritus Energy Systems Research Center, University of Texas Arlington, USA); Bruno Kapandji Kalala, (Permanent Secretary, PEAC CAPP Central African Republic); Stephen Gehl (Director EPRI Road Map Initiative and Strategic Sciences and Technologies, Palo Alto California); Wei Jen Lee (Director Energy Systems Research Center, University of Texas Arlington, USA); Bai K Blyden (BBRM Group LLC, California USA); Stephen Connors (Director MIT Energy Lab, Cambridge, USA); Kate Steel (Systems Engineering Division, MIT, Cambridge, USA); Terri Hathaway (Director International Rivers Network, Berkeley, CA, USA); Lori Pottinger (Editor World Rivers Review, Director Africa Program--International Rivers Network, Berkeley); Adnan Al-Mohaisen (General Manager, GCC Interconnection Authority, Saudi Arabia); Luc Chaussé (Project Director, Transmission & Distribution Division, SNC–Lavalin Inc.) and Satish Sud (Vice President, Power Systems Energy Division, SNC–Lavalin Inc., Montreal, Quebec, Canada); and James A. Momoh (Program Director, Electrical & Communication Systems (ECS), National Science Foundation, Arlington, USA).

19.14 References

[1] R.T. Mochebelele, "New Partnership for Africa's Development" , NEPAD Secretariat, South Africa. www.ieee.org/ipsc International Practices Sub Committee Panel session Presentation (2004-2007)

[2] SAPP & WAPP Reports, Purdue University, https://engineering.purdue.edu/IE/Research/PEMRG/PPDG// and http://www.eia.doe.gov/cneaf/electricity/page/fact_sheets/transmission.html

[3] "Electricity in Southern Africa", Financial Times Energy Publishing, 1996

[4] Federal Ministry of Power and Steel, 2003 http://www.nopa.net/Power_and_Steel/messages/19.shtml, and [2]

[5] EIA, Electricity Transmission Fact Sheet http://www.eia.doe.gov/cneaf/electricity/page/fact_sheets/transmission.html

[6] "Proposed Pricing Policy for Efficient Operation and Expansion of Transmission Grid", FERC, January 2003, Washington DC http://www.google.com/search?q=Eric+Hirst,+Brandon+Kirby,+%E2%80%9CTransmission+Planning+for+a+Restructuring+U.S.+Electricity+Industry&hl=en&lr=&filter=0

[7] "The Electric Transmission Network: A Multi-Region Analysis", Office of Integrated Analysis and Forecasting, PowerWorld® model runs http://www.eia.doe.gov/oiaf/analysispaper/tbl_5.html

[8] Eric J. Lerner, "What's Wrong with The Electric Grid?" http://www.aip.org/tip/INPHFA/vol-9/iss-5/p8.html

[9] F.T. Sparrow, Brian H. Bowen, "The CAPP Demonstration Long-Term Planning Model", Purdue University, PA Consulting, April 2005.

[10] Brian H. Bowen, F.T. Sparrow, Zuwei Yu, "Modeling Electricity Trade Policy for the Twelve Nations of the Southern African Power Pool, SAPP", Utilities Policy, Vol. 8, pp. 183-197, December 1999

[11] Zuwei Yu, F.T. Sparrow and Brian H. Bowen, "Pooling Resources in Southern Africa", Transmission & Distribution, February 2001

[12] Brian H. Bowen and F.T. Sparrow, "Pool Planning", Water Power, December 2001

[13] F.T. Sparrow, Brian H. Bowen, "User Manual for the Long-Term Model", 7th Edition, Purdue University, February 2001, https://engineering.purdue.edu/IE/Research/PEMRG/PPDG/SAPP/user_manual
[14] EIA, World Total Net Electricity Consumption 1980-2002 http://www.eia.doe.gov/pub/international/iealf/table62.xls
[15] Mo-Shing Chen, Professor Emeritus, University of Texas at Arlington, TX, USA. Security Issues of Power System Interconnection, www.ieee.org/ipsc International Practices Sub Committee Panel session Presentation (2004-2007)
[16] Wei-Jen Lee, Director, Energy Systems Research Center, The University of Texas at Arlington, Arlington, TX, USA. Technical Issues: Area Control Considerations etc in the WAPP and SAPP Areas, www.ieee.org/ipsc IEEE International Practices for Energy Development and Power Generation Sub Committee Panel Session Presentations (2004-2007)
[17] Terri Hathaway, Hydropower and African Grid Development: A Rights Based Perspective, International Rivers Network, Berkeley, CA, USA. and Lori Pottinger, Editor World Rivers Review, Director Africa Program--International Rivers Network, Berkeley, CA, USA. Paper 05GM0884
[18] United Nations General Assembly, 2000, "United Nations Millennium Declaration" Resolution A/Res/55/2, 18 September.
[19] United Nations, 2002, "Report of the World Summit on Sustainable Development," Johannesburg, 26 August – 4 September, A/Conf. 199/20
[20] EPRI, "Electricity Technology Roadmap, 1999 Summary and Synthesis Report," EPRI Report No. CI-112677-V1, July 1999.
[21] EPRI, "Electricity Technology Roadmap, 2003 Summary and Synthesis Report," EPRI Report No. 1010929, 2003.
[22] U.S. Department of Energy, Energy Information Agency, "International Energy Outlook 2004," www.eia.doe.gov/oiaf/ieo/index.html.
[23] N. Nakicenovic, A Grubler, and A. Macdonald, editors, "Global Energy Perspectives" International Institute for Applied Systems Analysis and World Energy Council, Cambridge University Press, 1998.
[24] S. Karakezi. and J Kimani, "Have Power Sector Reforms Increased Access to Electricity among the Poor in East Africa?" Energy for Sustainable Development, Vol. 8, No. 4, December 2004, pp10 – 25.
[25] O. Davidson. and S. Mwakasonda, "Electricity Access for the Poor:
[26] Stephen R. Connors, Providing Electricity Services to Rural Africa, Director, MIT Laboratory for Energy and the Environment, MIT, Cambridge, MA, USA. www.ieee.org/ipsc International Practices Sub Committee Panel session Presentation (2007).
[27] B.K. Blyden and Wei Jen Lee. Modified Microgrid Concept for Rural Electrification in Africa (paper 06GM1159), www.ieee.org/ipsc International Practices Sub Committee Panel Session for Energy Development and Power Generation Presentation (2006)
[28] IEA. World Energy Outlook. Chapter 13: Energy and Poverty. 2002..
[29] S. Karekezi and J. Kimani, "Status of Power Sector Reform in Africa: Impact on the Poor", Energy Policy, Vol. 30, 2002, pp. 923-945.

[30] R. Pandey, "Energy Policy Modeling: Agenda for Developing Countries", Energy Policy, Vol. 30, 2002, pp. 97-106.
[31] T.J Hammons. Status, Plans of Action, Further Developments, and Recommendations for Power Pools in Africa, Paper TPWRS 0028, IEEE Transactions on Power Systems, Vol. 21, (2), 2006, pp. 673-682.
[32] T.J. Hammons, B.K. Blyden, A.C. Calitz, A.B. Gulstone, E. Isekemanga, R. Johnstone, K. Palukuka, N.G. Simang and F. Taher. "African Electricity Infrastructure, Interconnections and Electricity Exchanges" IEEE Transactions on Energy Conversion, Vol. 15, (4), 2000, pp. 470-80.
[33] World Energy Council and FAO-UN. The challenge of rural energy poverty in developingcountries. 1999.
[34] J.D. Sterman, Business Dynamics: Systems Thinking and Modeling for a Complex World, Irwin-McGraw Hill, Chicago, 2000.
[35] K. Saeed and P. Prankprakma, Technological Development in a Dual Economy: Alternative Policy Levers for Economic Development. Paper presented at the 1994 International System Dynamics Conference, Sterling, Scotland, 1994.
[36] Jay W. Forrester. Learning through System Dynamics as Preparation for the 21st Century, Keynote Address Systems Thinking and Dynamic Modeling Conference for K-12 Education June 27-29, 1994
[37] Ikujiro Nonaka Synthesizing Capability: 'A Key to Create a New Reality'. The Graduate School of International Corporate Strategy, Hitotsubashi University, Sept 11, 2001
[38] T.J. Hammons. International Interconnections and Electricity Deregulation in Africa and the Gulf States. International Journal on Power and Energy Systems, Paper 203-3728, Vol. 28, (4), 2008, pp. 372-383.
[39] B. Cova. Progress of the Mediterranean Ring and the Interconnection with Europe, Power Engineering Society General Meeting,2004.IEEE Volume 2, Issue 10, 10 June 2004, pp. 1229-1234
[40] A.M.H.A. Karim, N.H. Al Maskati and S. Sud, "Status of Gulf Co-Operation Council (GCC) Electricity Grid System Interconnection", IEEE, Power Engineering Society Energy Development and Power Generation Committee Panel Session: Status of International Interconnections and Electricity Deregulation in Africa, Denver, June 6-10, 2004.
[41] James A. Momoh, US African Power Research and Education Activities: Challenges, Experiences and Opportunities, www.ieee.org/ipsc International Practices Sub Committee Panel Session Presentation (2004)
[42] Nonaka Ikujiro, Konno Noboru , The concept of "Ba': Building foundation for Knowledge Creation. World Resources Institute ,California Management Review Vol 40, No.3 Spring 1998.
[43] John D. Sterman. Business Dynamics, Systems Thinking and Modeling for a Complex World. McGraw-Hill/Irwin (February 23, 2000)
[44] B. Blyden. The Integration of Plant Expert Systems and the concept of a Dynamic Parameter, IEEE International Conference on Systems, Man, and Cybernetics. Vol.1, pp 928-930, 1992
[45] Artificial Neural Network, From Wikipedia, The Free Encyclopedia

[46] T.J. Hammons and P.Naidoo. Africa—Integrated Gas and Electricity Transmission Planning in Power Generation and HVDC Engineering in Harnessing Large-Scale Hydroelectric Sites for International Regional Power Systems, Energy Systems (ENSYS), Vol.1, (9), 2009, pp. 79-112, Available on Springer Link: http://www.springerlink.com/openurl.asp?genre=article&id=doi:10.1007/s12667-009-0002-9

[47] T.J. Hammons. Status of African and Middle East Renewable Energy Projects—Infrastructure, Developments and Cross Border. International Journal of Power and Energy Systems, Vol. 30, (2), 2010, pp. 73-85.

[48] T.J. Hammons. Status of Implementation of Generation, Transmission and Interconnection Projects in Southern Africa Power Pool, Int. J. Agile Systems and Management (IJASM), Vol. 4, (3), pp. 238-260.

[49] T.J. Hammons, P. Naidoo and L. Musaba. Run of River Bulk Hydroelectric Generation from the Congo River without a Conventional Dam. Natural Resources, Vol. 2, (1), 2011, pp. 18-21, Available Online (http://www.scirp.org/journal/nr).